IET TELECOMMUNICATIONS SERIES 99

Metrology for 5G and Emerging Wireless Technologies

Other volumes in this series:

Volume 9	**Phase Noise in Signal Sources** W.P. Robins
Volume 12	**Spread Spectrum in Communications** R. Skaug and J.F. Hjelmstad
Volume 13	**Advanced Signal Processing** D.J. Creasey (Editor)
Volume 19	**Telecommunications Traffic, Tariffs and Costs** R.E. Farr
Volume 20	**An Introduction to Satellite Communications** D.I. Dalgleish
Volume 26	**Common-Channel Signalling** R.J. Manterfield
Volume 28	**Very Small Aperture Terminals (VSATs)** J.L. Everett (Editor)
Volume 29	**ATM: The broadband telecommunications solution** L.G. Cuthbert and J.C. Sapanel
Volume 31	**Data Communications and Networks, 3rd Edition** R.L. Brewster (Editor)
Volume 32	**Analogue Optical Fibre Communications** B. Wilson, Z. Ghassemlooy and I.Z. Darwazeh (Editors)
Volume 33	**Modern Personal Radio Systems** R.C.V. Macario (Editor)
Volume 34	**Digital Broadcasting** P. Dambacher
Volume 35	**Principles of Performance Engineering for Telecommunication and Information Systems** M. Ghanbari, C.J. Hughes, M.C. Sinclair and J.P. Eade
Volume 36	**Telecommunication Networks, 2nd Edition** J.E. Flood (Editor)
Volume 37	**Optical Communication Receiver Design** S.B. Alexander
Volume 38	**Satellite Communication Systems, 3rd Edition** B.G. Evans (Editor)
Volume 40	**Spread Spectrum in Mobile Communication** O. Berg, T. Berg, J.F. Hjelmstad, S. Haavik and R. Skaug
Volume 41	**World Telecommunications Economics** J.J. Wheatley
Volume 43	**Telecommunications Signalling** R.J. Manterfield
Volume 44	**Digital Signal Filtering, Analysis and Restoration** J. Jan
Volume 45	**Radio Spectrum Management, 2nd Edition** D.J. Withers
Volume 46	**Intelligent Networks: Principles and applications** J.R. Anderson
Volume 47	**Local Access Network Technologies** P. France
Volume 48	**Telecommunications Quality of Service Management** A.P. Oodan (Editor)
Volume 49	**Standard Codecs: Image compression to advanced video coding** M. Ghanbari
Volume 50	**Telecommunications Regulation** J. Buckley
Volume 51	**Security for Mobility** C. Mitchell (Editor)
Volume 52	**Understanding Telecommunications Networks** A. Valdar
Volume 53	**Video Compression Systems: From first principles to concatenated codecs** A. Bock
Volume 54	**Standard Codecs: Image compression to advanced video coding, 3rd Edition** M. Ghanbari
Volume 59	**Dynamic Ad Hoc Networks** H. Rashvand and H. Chao (Editors)
Volume 60	**Understanding Telecommunications Business** A Valdar and I Morfett
Volume 65	**Advances in Body-Centric Wireless Communication: Applications and state-of-the- art** Q.H. Abbasi, M.U. Rehman, K. Qaraqe and A. Alomainy (Editors)
Volume 67	**Managing the Internet of Things: Architectures, theories and applications** J. Huang and K. Hua (Editors)
Volume 68	**Advanced Relay Technologies in Next Generation Wireless Communications** I. Krikidis and G. Zheng
Volume 69	**5G Wireless Technologies** A. Alexiou (Editor)
Volume 70	**Cloud and Fog Computing in 5G Mobile Networks** E. Markakis, G. Mastorakis, C.X. Mavromoustakis and E. Pallis (Editors)
Volume 71	**Understanding Telecommunications Networks, 2nd Edition** A. Valdar
Volume 72	**Introduction to Digital Wireless Communications** Hong-Chuan Yang
Volume 73	**Network as a Service for Next Generation Internet** Q. Duan and S. Wang (Editors)
Volume 74	**Access, Fronthaul and Backhaul Networks for 5G & Beyond** M.A. Imran, S.A.R. Zaidi and M.Z. Shakir (Editors)
Volume 76	**Trusted Communications with Physical Layer Security for 5G and Beyond** T.Q. Duong, X. Zhou and H.V. Poor (Editors)

Volume 77 **Network Design, Modelling and Performance Evaluation** Q. Vien
Volume 78 **Principles and Applications of Free Space Optical Communications** A.K. Majumdar, Z. Ghassemlooy, A.A.B. Raj (Editors)
Volume 79 **Satellite Communications in the 5G Era** S.K. Sharma, S. Chatzinotas and D. Arapoglou
Volume 80 **Transceiver and System Design for Digital Communications, 5th Edition** Scott R. Bullock
Volume 81 **Applications of Machine Learning in Wireless Communications** R. He and Z. Ding (Editors)
Volume 83 **Microstrip and Printed Antenna Design, 3rd Edition** R. Bancroft
Volume 84 **Low Electromagnetic Emission Wireless Network Technologies: 5G and beyond** M.A. Imran, F. Héliot and Y.A. Sambo (Editors)
Volume 86 **Advances in Communications Satellite Systems Proceedings of the 36th International Communications Satellite Systems Conference (ICSSC-2018)** I. Otung, T. Butash and P. Garland (Editors)
Volume 87 **Real Time Convex Optimisation for 5G Networks and Beyond** T.Q. Duong, L.D. Nguyen and H.D. Tuan
Volume 89 **Information and Communication Technologies for Humanitarian Services** M.N. Islam (Editor)
Volume 90 **Communication Technologies for Networked Smart Cities** S.K. Sharma, N. Jayakody, S. Chatzinotas and A. Anpalagan (Editors)
Volume 91 **Green Communications for Energy-Efficient Wireless Systems and Networks** Himal Asanga Suraweera, Jing Yang, Alessio Zappone and John S. Thompson (Editors)
Volume 92 **Flexible and Cognitive Radio Access Technologies for 5G and Beyond** H. Arslan and E. Başar (Editors)
Volume 93 **Antennas and Propagation for 5G and Beyond** Q. Abbasi, S.F. Jilani, A. Alomainy and M.A. Imran (Editors)
Volume 94 **Intelligent Wireless Communications** G. Mastorakis, C.X. Mavromoustakis, J.M. Batalla and E. Pallis (Editors)
Volume 95 **ISDN Applications in Education and Training** R. Mason and P.D. Bacsich
Volume 96 **Edge Caching for Mobile Networks** H. Vincent Poor and Wei Chen (Editors)
Volume 98 **Artificial Intelligence Applied to Satellite-Based Remote Sensing Data for Earth Observation** M.P. Del Rosso, A. Sebastianelli and S.L. Ullo (Editors)

Metrology for 5G and Emerging Wireless Technologies

Edited by
Tian Hong Loh

The Institution of Engineering and Technology

Published by The Institution of Engineering and Technology, London, United Kingdom

The Institution of Engineering and Technology is registered as a Charity in England & Wales (no. 211014) and Scotland (no. SC038698).

First published 2021

The Institution of Engineering and Technology
Michael Faraday House
Six Hills Way, Stevenage
Herts, SG1 2AY, United Kingdom

www.theiet.org

British Library Cataloguing in Publication Data
A catalogue record for this product is available from the British Library

ISBN 978-1-83953-278-8 (hardback)
ISBN 978-1-83953-279-5 (PDF)

Typeset in India by MPS Limited

Contents

About the editor

Tian Hong Loh is a principal research scientist at the National Physical Laboratory (NPL), UK. He leads work at NPL on a wide range of electromagnetic and wireless communication metrology research areas in support of the telecommunications industry. He has authored and co-authored over 160 publications and hold 6 patents. He is also a visiting professor at University of Surrey, and a visiting industrial fellow at University of Cambridge, UK. He is a UK representative of URSI Commission A (Electromagnetic Metrology), a topic editor of Electronics, an associate editor of IEEE Journal of Electromagnetics, RF and Microwaves in Medicine and Biology (J-ERM), IET Microwaves, Antennas & Propagation (MAP) Journal, IET Communications (COMMUN) Journal and International Union of Radio Science (URSI) radio science bulletin, a member of the IET and a senior member of the IEEE. He was the project coordinator of an European Association of National Metrology Institutes (EURAMET) European Metrology Programme for Innovation and Research (EMPIR) project on 'Metrology for 5G Communications', a guest editor of IET MAP special issue on 'Metrology for 5G Technologies', and was the TPC chair of the 2017 IEEE International Workshop on Electromagnetics (iWEM 2017). His research interests include metamaterials, computational electromagnetics, small antennas, smart antennas, multiple-input–multiple-output (MIMO), electromagnetic compatibility, wireless sensor networks, body-centric communication and 5G and beyond communications.

Preface

Focusing on user experience, the emerging wireless communications for fifth generation (5G) and beyond will not just deliver a much faster connection and higher data capacity but will also promise to deliver low energy consumption and seamless connectivity between billions of people and trillions of devices, even in a crowded scenario, together with ubiquitous communication for the Internet of things (IoT). The rollout of the 5G and the definition of emerging wireless technologies such as the sixth generation (6G) is in progress. At present, with strong global momentum, multiple worldwide communities are working on various research and development activities for 5G and beyond.

As new technologies increasingly emerge in the arena of modern devices, antenna and system design for 5G and beyond, the standards bodies, industry and research communities are facing the challenge of diverse 5G technological requirements, and on verifying products that meet desired performance parameters. Metrological supports are essential for underpinning all aspects of 5G and beyond in response to high demand for universal, dynamic and data-rich wireless applications. Measurements of the signals, devices, antennas, complete system and radio frequency exposure assessments will be needed for development, manufacture and deployment of these new wireless communication technologies.

This book presents the state-of-the-art measurement capabilities, testbeds and the relevant research activities for 5G and emerging wireless technologies at a wide range of frequencies. Some topics to highlight are

- device nonlinearity and multiphysics characterisations;
- fully connected millimetre-wave hybrid beamforming testbed;
- investigation of plane wave generation using different facilities;
- International comparison campaigns;
- 5G wearable antenna methodologies;
- 5G NR OTA test metrology using different techniques;
- 3GPP/CTIA conformance test of 5G/6G base station and mobile terminals;
- OTA and channel sounding for V2X autonomous vehicles communications;
- channel sounding – empirical measurement and contribution towards ITU-R;
- world's first real-time mMIMO trials;
- use cases on railway, train, drone and IoT;
- THz channel characterisations;
- RF exposure of 5G NR base station.

This book is first of its kind to focus on metrology for current and future wireless communication technologies. It provides timely support to industry, academia, standard bodies and national measurement institutes during the development of 5G and emerging wireless technologies. The book is divided into six inter-link parts: (1) waveform metrology; (2) device metrology; (3) antenna metrology; (4) system OTA metrology; (5) propagation channel metrology; and (6) RF exposure metrology.

Acknowledgements

The editor would like to express his gratitude to the following individuals who reviewed the various chapters in this book, many drawn from the pool of authors for other chapters of this book.

- Akram Alomainy – Queen Mary University of London, UK
- Andrew Smith – National Physical Laboratory, UK
- Elijah I. Adegoke – University of Warwick, UK
- Erik Kampert – University of Warwick, UK
- Fabien Heliot – University of Surrey, UK
- Haiming Wang, Southeastern University, China
- Jeffrey R. Guerrieri – NIST, USA
- Martin Salter – National Physical Laboratory, UK
- Sohail Payami – VIAVI Solutions Inc., UK
- Tim Brown – University of Surrey, UK
- Wael Boukley Hasan – University of Bristol, UK
- Yunsong Gui – National Physical Laboratory, UK

The editor would also like to thank all colleagues who have contributed to this book. I have enjoyed working with you all and appreciate your support. Finally, the editor would like to thank the IET for their help in finalizing this book.

Acknowledgements

Abbreviations and acronyms

2D	two-dimensional
2G	second-generation wireless network
3D	three-dimensional
3GPP	Third Generation Partnership Project
4G	fourth-generation wireless network
5G	fifth-generation wireless network
5G NSA	5G non-standalone
5G SA	5G standalone
5G-PPP	5G Infrastructure Public Private Partnership
5GIA	5G Infrastructure Association
5GIC	5G Innovation Centre
5GMF	The Fifth-Generation Mobile Communications Promotion Forum
6G	sixth-generation wireless network
6GIC	6G Innovation Centre
AAS	active antenna system
AC	anechoic chamber
ACE	Antenna Centre of Excellence
ACF	autocorrelation function
ACS	adjacent channel selectivity
ADAS	advanced driver-assistance system
ADC	analog-to-digital converter
AiP	antenna-in-package
ANFR	The National Frequency Agency
AoA	angle of arrival
AoD	angle of departure
AP	access point
ARIB	Association of Radio Industries and Businesses
AS	angular spread
ASA	angular spread of arrival

ASD	angular spread of departure
ATF	antenna test function
ATIS	Alliance for Telecommunications Industry Solutions
AUT	antenna under test
AWG	arbitrary waveform generator
AWGN	additive white Gaussian noise
BBA	broadband amplifier
BBU	base band unit
BEIS	UK government's Department for Business, Energy and Industrial Strategy
BEL	building entry loss
BER	bit-error-rate
BFN	beamforming network
BIPM	International Bureau of Weights and Measures
BIRMM	Beijing Institute of Radio Metrology and Measurement
BLER	block error rate
BPF	band-pass filters
BPSK	binary phase-shift keying
BS	base station
BSE	base station emulator
BSM	basic safety message
BSS	basic service set
BTS	base transceiver station
BW	bandwidth
C-ITS	cooperative intelligent transportation system
C-V2X	cellular vehicle-to-everything
CA	carrier aggregation
CAD	computer-aided design
CAM	connected and automated mobility
CAP	continuous aperture phased
CATR	compact antenna test range
CBW	channel bandwidth
CCAV	Centre for Connected and Autonomous Vehicles
CCD	charge-coupled device
CCDF	complementary cumulative distribution function
CCSA	China Communications Standards Association

CDF	cumulative distribution function
CDL-A	clustered delay line A
CDM	code division multiplexing
CE	channel emulator
CEM	computational electromagnetic
CENELEC	European Committee for Electrotechnical Standardization
CEPT	European Conference of Postal and Telecommunications Administrations
CI	close-in
CIR	channel impulse response
CMI	Czech Metrology Institute
CNT	carbon nanotube
COST	European Cooperation in Science and Technology
COTS	commercial-off-the-shelf
CP-OFDM	cyclic-prefix orthogonal frequency division multiplexing
CQI	channel quality indicator
CSI	channel state information
CST	computer simulation technology
CT	computed tomography
CTF	channel transfer function
CTIA	Cellular Telecommunications and Internet Association
CW	continuous wave
CWE	cylindrical wave expansion
CXR	cross polar reduction feed
D2D	device-to-device
DAC	digital-to-analog converter
DC	direct current
DCM	dual carrier modulation
DDCMS	Department for Digital, Culture, Media and Sport
DENM	Decentralised Environmental Notification Message
DFF	direct far-field method
DGBE	diethylene glycol butyl ether
DIO	digital input/output
DL	downlink
DLA	discrete lens array
DMA	direct memory access

DMP	Dimatix Material Printer
DoA	direction-of-arrival
DoD	direction-of-departure
DoFs	degrees of freedom
DPA	Doherty power amplifier
DS	Doppler spread
DSP	digital signal processing
DSRC	dedicated short range communications
DSS	directional scanning sounding
DTU	Technical University of Denmark
DUT	device under test
E-field	electric field
E-UTRA	Evolved Universal Terrestrial Radio Access
E2E	end-to-end
ECC	envelope correlation coefficient
ECU	electronic control unit
EIRP	effective isotropic radiated power
EIS	effective isotropic sensitivity
EM	electromagnetic
EMC	electromagnetic compatibility
EMF	electromagnetic field
eMMB	enhanced mobile broadband
ENL	equivalent noise level
EO	electro-optic
EOS	electro-optic sampling
EQC	equivalent currents
ERP	effective radiated power
ESA	European Space Agency
ETSI	European Telecommunication Standards Institute
EuCAP	European Conference on Antennas and Propagation
EurAAP	European Association on Antennas and Propagation
EURAMET	European Association of National Metrology Institutes
eV2X	enhanced direct vehicle-to-everything
EVM	error vector magnitude
F/B ratio	front-to-back ratio
FCC	Federal Communications Commission

FD	frequency domain
FDCS	frequency-domain channel sounding
FDM	frequency division multiplexing
FDTD	finite difference time domain method
FEM	finite element method
FF	far-field
FFT	fast Fourier transform
FI	floating intercept
FIB	frequency invariant beamformer
FIFO	first-in-first-out
FoM	figure of merit
FPGA	field-programmable gate array
FR1	frequency range 1
FR2	frequency range 2
FSPL	free-space path loss
FWA	fixed wireless access
FVT	full-vehicle antenna testing
GaN	gallium nitride
GaN-on-SiC	gallium nitride on silicon carbide
Gbps	gigabits per second
GBSMs	geometry-based stochastic models
GCF	Global Certification Forum
GHz	gigahertz
GO	geometrical optics
GPS	Global Positioning System
GSCM	geometry-based stochastic channel model
GTD	geometrical theory of diffraction
HAP	high-altitude platform
HD	high definition
HEMT	high electron mobility transistor
HPA	high-power amplifier
HPBW	half-power beamwidth
HPC	high-performance computing
HR	highly reflective
ICI	inter-carrier interference

ICNIRP	International Commission on Non-Ionizing Radiation Protection
ICTFT	inverse continuous-time Fourier transform
ICTs	Information and Communication Technologies
IDFT	inverse discrete Fourier transform
IEC	International Electrotechnical Commission
IEEE	Institute of Electrical and Electronics Engineers
IF	intermediate frequency
IFF	indirect far-field
IHS	Information Handling Services
IIC	Industrial Internet Consortium
IID	independent and identically distributed
IIoT	industrial Internet of things
IMD3	third-order intermodulation distortion
IMT-2020	International Mobile Telecommunications-2020
InH	indoor hotspot
IoE	Internet of everything
IoT	Internet of things
IR	infrared
ISI	inter-symbol interference
ISM	Industrial Scientific Medical
ITIS	Swiss Foundation for Research on Information Technologies in Society
ITS	Intelligent Transportation System
ITU	International Telecommunication Union
ITU-R	International Telecommunications Union Radiocommunication sector
ITU-T	International Telecommunications Union Telecommunication Standardisation sector
IUI	inter-user interference
KPI	key performance indicator
KRISS	Korea Research Institute of Standards and Science
LDPC	low-density parity-check coding
LED	light-emitting diode
Li-Fi	light-fidelity
LMS	least mean squared
LNA	low noise amplifier
LO	local oscillator

LoS	line-of-sight
LS	least square
LSNA	large signal network analysers
LSOP	large signal operating point
LSP	large-scale parameter
LTE	long-term evolution
LWA	leaky-wave antennas
M2M	machine-to-machine
MAE	mean absolute error
MBPGMs	measurement-based pseudo-geometric models
MCS	modulation and coding schemes
METAS	Federal Institute of Metrology
METIS	Mobile and wireless communications Enablers for the Twenty-twenty Information Society
MF	middle-field
MIMO	multiple-input–multiple-output
ML	maximum likelihood
MLE	maximum likelihood estimation
MLFMM	multilevel fast multiple
mm-wave	millimetre wave
MMCX	micro-miniature coaxial
MMIC	monolithic microwave integrated circuit
mMIMO	massive multiple-input–multiple-output
MMSE	minimum mean square error
mMTC	massive machine-type communication
MoM	method of moments
MOS	metal oxide-semiconductor
MoU	memorandum of understanding
MPAC	multi-probe anechoic chamber
MPC	multipath components
MRI	magnetic resonance imaging
MSU	measurement supply unit
MT	mobile terminal
MUF	Microwave Uncertainty Framework
MU-MIMO	multi-user multiple-input–multiple-output
MVG	Microwave Vision Group

MVI	Microwave Vision Italy
NAIC	National Automotive Innovation Centre
NF-FF	near-field to far-field transformation techniques
NGMN	next-generation mobile networks
NGV	next-generation vehicular
NIM	National Institute of Metrology
NIST	National Institute of Standards and Technology
NLoS	non-line-of-sight
NMI	National Metrology Institute
NMS	National Measurement System
NPL	National Physical Laboratory
NR	new radio
NRB	number of allocated resource block
NVNA	nonlinear vector network analyser
NYU	New York University
O2I	outdoor-to-indoor
OAM	orbital angular momentum
OATS	open area test site
OCXO	oven-controlled crystal oscillator
OEW	open-ended waveguide
Ofcom	office of communications
OFDM	orthogonal frequency division multiplexing
OFDMA	orthogonal frequency division multiple access
OTA	over-the-air
PA	power amplifier
PADP	power angle delay profile
PAE	power-added efficiency
PAP	power angle profile
PAPR	peak-to-average power ratio
PAS	power angular spectra
PC	power class
PCB	printed circuit board
PCIe	Peripheral Component Interconnect express
PD	power density
PDCCH	physical downlink control channel
PDF	probability density function

PDP	power delay profile
PDSCH	physical downlink shared channel
PEC	perfect electric conductor
PET	polyethylene terephthalate
PHY	physical
PIMMS	primary impedance microwave measurement system
PL	path loss
PLE	path loss exponent
PLL	phase-locked loop
PM	polarisation-maintaining
PN	pseudo-noise
PO	physical optics
PPS	pulse-per-second
ProSe	proximity-based service
ps	picosecond
P2P	peer-to-peer
PSS	primary synchronisation signal
PTB	Physikalisch-Technische Bundesanstalt
PTF	power transfer function
PWE	plane wave expansion
PWG	plane wave generator
PWS	plane wave synthesis
QAM	quadrature amplitude modulation
QoE	quality of experience
QoS	quality of signal
QPSK	quadrature phase-shift keying
QZ	quiet zone
R&D	research and development
Rb	rubidium
RBs	resource blocks
RC	reverberation chamber
RCS	radar cross section
RE	resource element
RF	radio frequency
RH	relative humidity
RIMAX	maximum likelihood parameter estimation framework

RIS	reconfigurable intelligent surface
RMa	rural macrocell
RMS	root mean square
RMS DS	root mean square delay spread
RMSE	root mean square error
RPSD	received power spectral density
RRH	remote radio head
RRM	radio resource management
RSRP	received signal received power
RSU	road side unit
RT	ray-tracing
RTDO	real-time digital oscilloscope
RTS	radiated two-stage
RTSO	real-time sampling oscilloscope
RV	random variable
Rx	receiver end
S-parameter	scattering parameter
SA	spectrum analyser
SAE	Society of Automotive Engineers
SAGE	space-alternating generalized expectation–maximization
SAM	specific anthropomorphic mannequin
SAPR	shadowed antenna power ratio
SAR	specific absorption rate
SC	short-circuit
SCDF	Shadowed Cumulative Distribution Function
SCM	spatial channel model
SCME	spatial channel model extended
SCS	subcarrier spacing
SDR	software defined radio
SE	spurious emission
SF	shadow fading
SFP+	small form-factor pluggable plus
SGH	Standard Gain Horn
SI	International System of Units
SIMO	single-input–multiple-output
SINR	signal-to-interference-plus-noise ratio

SISO	single-input–single-output
SLL	side-lobe-level
SMa	suburban macrocell
SMART	small antenna radiated test range
SNR	signal-to-noise ratio
SOLT	short-open-load-through
SP4T	single-pole four-throw
SS/PBCH	synchronization signal of the physical broadcast channel
SSB	synchronization signal block
SSID	service set identification
SSPC	share sync pulse clock
SSS	secondary synchronisation signal
STRC	share trigger reference clock
SU-MIMO	single-user multiple-input–multiple-output
SWCNT	single-walled carbon nanotubes
SWC	spherical wave coefficient
SWE	spherical wave expansion
TD	time-domain
TDCS	time-domain channel sounding
TDD	time division duplex
TDM	time division multiplexing
TEM	transverse electromagnetic
TFVF	test fixture validation factor
THz	terahertz
TIS	total isotropic sensitivity
ToA	time of arrival
TR	thermoreflectance
TRs	technical reports
TRL	thru, reflect, line
TRP	total radiated power
TS	technical specification
TSDSI	Telecommunications Standards Development Society India
TSG	Technical Specification Group
TTA	Telecommunications Technology Association
TTC	Telecommunication Technology Committee
Tx	transmitter end

UAV	unmanned aerial vehicle
UDP	user datagram protocol
UE	user equipment
UGW	unit gate widths
UHF	ultra-high frequency
UK5G	5G national innovation network
UL	uplink
UMa	urban macro-cell
UMi	urban micro-cell
URLLC	ultra-reliable low-latency communication
USIR	user shadowed intensity ratio
USRP	Universal Software Radio Peripheral
UT	user terminal
UV	ultra-violet
UWB	ultra-wideband
V2I	vehicle-to-infrastructure
V2N	vehicle-to-network
V2P	vehicle-to-pedestrian
V2V	vehicle-to-vehicle
V2X	vehicle-to-everything
VANET	vehicular ad-hoc network
VNA	vector network analyser
VNA-DUO	Vector Network Analyser Dynamic Uncertainty Option
VSA	vector signal analyser
VSG	vector signal generator
VSWR	voltage standing wave ratio
WAVE	wireless access in vehicular environments
WBAN	wearable body area network
WCDMA	Wideband Code-Division Multiple Access
WG	working group
WHO	World Health Organization
WLAN	wireless local area network
WPAN	wireless personal area network
WSSUS	wide stationary and uncorrelated scattering
XPD	cross-polarization discrimination
XPR	cross-polarisation ratio
ZF	zero-forcing

Chapter 1
Introduction to metrology for 5G and emerging wireless technologies

Tian Hong Loh[1]

This chapter introduces (1) the scope of the book, including the definition of metrology and its relationship to measurement traceability, standardisation and national measurement system (NMS); (2) the global initiatives, standardisation activities and technology trends in fifth generation (5G) and emerging wireless technologies, as well as an overview of global metrological research and development (R&D) activities; and (3) the structure of the book, including brief highlights of each chapter.

1.1 Scope

The digital economy is essential for wealth creation and is increasingly underpinning all aspects of social and business activities. As 'Connectivity' becomes more and more ubiquitous, more reliable and offers higher capacity, it will become the new 'lifeblood' of the digital economy and connected society [1] with over 8.9 billion mobile subscriptions by 2025 [2]. Information handling services economics have estimated that the 5G networks and beyond will enable USD12.3 trillion of global economic output in 2035 [3]. This new wireless network promises seamless connectivity between trillions of devices serving billions of people, millisecond latency, multi-Gbps (Gigabits per second) data capacity and low energy consumption, which will be part of the Internet of things (IoT). The rollout of 5G networks and large-scale deployments of cellular IoT will lead to fundamental changes in our society, impacting not only consumer services but also industries embarking on digital transformation. Also, these new technologies are expected to transform how we interact with the world – providing opportunities for wide-ranging and, as yet, unforeseen new applications, business models, enhanced lifestyles and increased productivity [4]. The global telecommunications industry is playing a crucial role in the development of emerging wireless technologies for IoT, 5G and sixth-generation (6G) mobile networks.

[1]5G and Future Communications Technology Group, Electromagnetic and Electrochemical Technologies Department, National Physical Laboratory (NPL), Teddington, United Kingdom

The growing demand for high-speed reliable mobile and fixed connectivity communication for a wide range of diverse new applications such as virtual and augmented reality, autonomous driving, remote surgery, unmanned aerial vehicles, vehicle-to-everything (V2X), high-altitude platforms and satellites, has driven the global digital agenda to better exploit information and communication technologies to foster innovation in emerging wireless technologies. Such demand is expected to grow by a factor of 1,000 over a decade with an attendant increase in energy consumption and the number of connected users to multiply 10- to 100-fold [5]. The immediate challenge for communication networks will be to scale up to these traffic requirements and serve new kinds of elaborate machine-to-machine systems and applications (e.g. smart manufacturing, smart cities, tracking/locating and autonomous vehicles) using emerging wireless technologies, many of which are still to be developed.

Metrology is the scientific study of measurement and is defined by the International Bureau of Weights and Measures [6] as

> "the science of measurement, embracing both experimental and theoretical determinations at any level of uncertainty in any field of science and technology"

whereas metrological traceability, also known as measurement traceability, is defined as [7]

> "the property of a measurement result whereby the result can be related to a reference through a documented unbroken chain of calibrations, each contributing to the measurement uncertainty".

Measurement traceability is important because it gives assurance that the measurement results agree with national or international standards within the statement of uncertainty in measurement. The establishment of measurement traceability and improvement of measurement science and accuracy enable manufacturers to provide confidence in their specifications. This plays a key role in customer/supplier relationships, for which products need to be demonstrated as 'meeting specification', regardless of who is carrying out the test or when/where the test is being performed. Figure 1.1 shows the measurement traceability pyramid demonstrating the levels of metrology along with the way that traceability and calibrations run from the realisation of the International System of Units (SI) down to the measurements performed in the industrial environment. The traceability chain works upwards from the bottom of the pyramid, where measurements done by industry and testing laboratories can directly be related to the unit definition at the top through the traceability chain created by calibration [8].

Figure 1.2 shows the structure of an NMS, which provides an infrastructure between National Metrology Institute (NMI) laboratories, calibration facilities, accreditation bodies and industry, whereby measurements made in industry using test instrumentation can be made traceable to the SI. In many countries, national standards are maintained by an NMI which provides the highest level of standards for the calibration/measurement traceability infrastructure in that country [9]. Some examples of such government agencies include the National Physical Laboratory (NPL) in the United

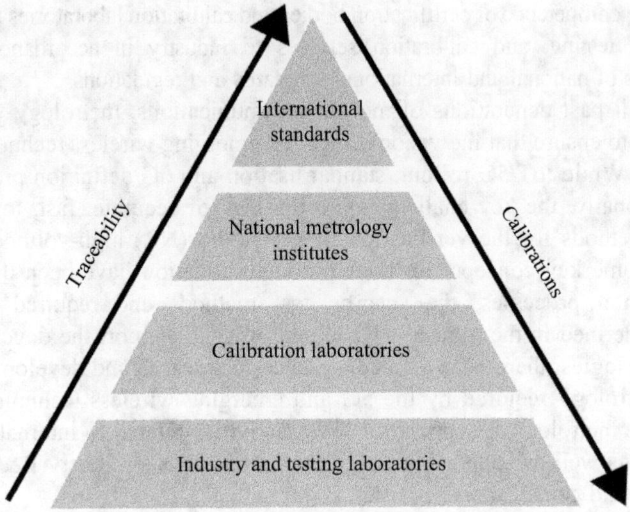

Figure 1.1 Measurement traceability pyramid (from [10])

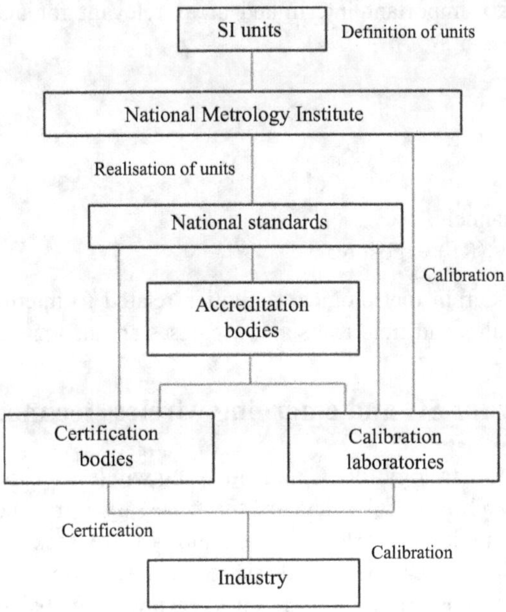

Figure 1.2 An overview of a national measurement system (from [11])

Kingdom, the National Institute of Standards and Technology (NIST) in the United States, the Federal Institute of Metrology (METAS) in Switzerland, the Czech Metrology Institute (CMI) in the Czech Republic and the Physikalisch-Technische Bundesanstalt (PTB) in Germany. Accreditation bodies are generally responsible for assessing the competence of certification bodies and calibration laboratories that provide certification, testing, and calibration services to industry in accordance with the requirements of national and international standards and regulations.

As in all past generations of mobile communications, metrology will have a pivotal role to ensure that the vision of 5G and emerging wireless technologies will be realised. While IoT/5G rollout, standardisation and 6G definition processes are ongoing globally, the key challenges are the lack of accurate, fast, low-cost and traceable methods for the verification of new radio (NR) high-volume products. Although some key concepts for their product validation have been discussed in standardisation processes, the specific test methods and required equipment remains undefined in the industry. Hence, in order to support the development of these technologies, there is an immediate need to research and develop the underpinning metrology required by the 5G and emerging wireless technologies. This will provide metrological support for NMS activities related to international standardisation as well as supporting the telecommunications industry R&D, product verification and certification activities.

This book presents the state-of-the-art measurement capabilities, testbeds and the relevant research activities for 5G and emerging wireless technologies across a wide range of frequencies. The focus is on measurement techniques and metrology for the following six important interlinked areas relevant for 5G and emerging wireless technologies:

- waveforms
- devices
- antennas
- systems
- propagation channel
- radio frequency (RF) exposure

These areas underpin metrological activities related to international standardisation. Several real-world field trials and use cases are also presented.

1.2 Metrology for 5G and emerging wireless technologies

The development of metrology for 5G and emerging wireless technologies requires a global-scale research effort. To develop the necessary infrastructure and international standards, multiple worldwide communities are seeking a global consensus for the vision, applications, standards, and identification of suitable spectra [12]. The following subsections provide an overview of global initiatives, standardisation activities, technology trends, metrological R&D activities and state-of-the-art measurement capabilities.

1.2.1 Global initiatives

At present, with strong global momentum, multiple worldwide IoT/5G/6G research initiatives and projects are developing and defining the emerging wireless technology

Table 1.1 Global initiatives

Regions	Examples
Europe	• The 5G Infrastructure Public Private Partnership (5G PPP) [21] • Mobile and wireless communications Enablers for the Twenty-twenty Information Society (METIS) [22] • 5G Infrastructure Association (5GIA) [14] • 6G Flagship [23] • Hexa-X [24] • Industry 4.0 Working Group [25] • 5GIC [26] (now 6GIC [27]) • UK DCMS 5G testbeds and trials programme [28] • UK5G [29]
Asia	• International Mobile Telecommunications-2020 (IMT-2020) 5G promotion group [16] • 5G Forum [17] • Association of Radio Industries and Businesses (ARIB) [30] • The Fifth Generation Mobile Communications Promotion Forum (5GMF) [18] • Telecommunications Standards Development Society India (TSDSI) [20]
North America	• 5G America [15] • New York University Wireless Research Center [31] • Next G Alliance [32] • Industrial Internet Consortium (IIC) [33]
South America	• 5G Brazil [19]

required to meet existing and future demands for higher data rates and greater capacity. Some examples are shown in Table 1.1. The global 5G multilateral memorandum of understanding (MoU) was officially signed at the 2020 Institute of Electrical and Electronics Engineers (IEEE) 5G World Forum [13] to formalise a worldwide alliance to support multilateral collaboration on 5G systems. The partners include 5GIA (Europe) [14], 5G Americas (North America) [15], International Mobile Telecommunications-2020 (IMT-2020) 5G promotion group (China) [16], 5G Forum (Korea) [17], The Fifth Generation Mobile Communications Promotion Forum (5GMF) (Japan) [18], 5G Brazil (South America) [19] and Telecommunications Standards Development Society India (TSDSI) (India) [20]. The MoU will stimulate collaboration between governments and industries in technology projects and encourages trials and deployment across the world.

1.2.2 Standardisation activities

The establishment of international standards for industrial implementation and deployment of wireless communication systems depend heavily on global and regional regulation. For example, the spectrum use is crucial for their ensuring interoperability. On a global level, the International Telecommunications Union

Radiocommunication Sector (ITU-R) [34] defines the spectrum for different services in the RF spectrum. They have issued the requirements for 5G radio interface technologies in international mobile telecommunications (IMT)-2020 standard [35]. In parallel with activities carried out in ITU-R, the Third Generation Partnership Project (3GPP) [36] defines and has written the technical specifications for mobile communication radio access technologies since the second-generation (2G) mobile network. The 3GPP is formed by the following regional and national telecommunication standard organisations:

- European Telecommunication Standards Institute (ETSI) [37] in Europe
- Alliance for Telecommunications Industry Solutions (ATIS) [38] in the United States
- China Communications Standards Association (CCSA) [39] in China
- Telecommunications Technology Association (TTA) [40] in Korea
- Association of Radio Industries and Businesses (ARIB) [30] in Japan
- Telecommunication Technology Committee (TTC) [41] in Japan
- TSDSI [20] in India

and a variety of other organisations, for example, Cellular Telecommunications and Internet Association (CTIA) [42], Next Generation Mobile Networks (NGMN) [43] (see [36] for full list).

The IEEE, along with other organisations such as the European Committee for Electrotechnical Standardization (CENELEC) and the International Electrotechnical Commission (IEC) also contribute to the 5G standardisation process in various aspects. The 3GPP has defined the technical specification for both 5G Non-standalone (5G NSA) and 5G Standalone (5G SA) NR deployments. The main difference between them is that in NSA, the 5G NR is connected to the fourth-generation (4G) core network, whereas in SA the 5G NR is directly connected to the 5G core network. Global activities on 6G definition and standardisation have also been started [32,44].

1.2.3 Technology trends

A raft of new 5G and emerging wireless technologies is increasingly emerging in the arena of modern devices and system design. The following briefly discusses several key technology trends.

1.2.3.1 RF spectrum

Technology advancements together with spectrum scarcity has driven the exploration of new RF spectrum suitable for 5G and emerging wireless technologies in the millimetre wave (mm-wave), terahertz (THz) and even optical frequency bands. ITU-R is responsible for the global RF spectrum regulation for different radiocommunication services and applications [45] whereby new RF bands are defined continuously by 3GPP. Operating in higher frequency RF bands, which inherently offer larger data bandwidths, is a route to achieve the requirement on data capacity and to support a significantly increased user density to meet the 5G and future-generation wireless network promises. However, the propagation properties of these bands limit their usage for wide-area coverage. On the other hand, lower frequency RF bands are good for wide-area coverage deployments. There are currently two allocated operating RF bands for 5G NR: frequency range 1 (FR1) – 410 MHz to 7.125 GHz, and frequency range 2 (FR2) – 24.25–52.6 GHz. Testing at these bands presents different challenges both within the system and in free space.

1.2.3.2 Energy efficiency

Earlier mobile communication generations focused on spectral efficiency but starting from 5G the focus has shifted towards balancing this with energy efficiency. Achieving energy efficiency is a multifaceted problem. A variety of strategies have been considered for achieving the high levels of efficiency and areal information density required by 5G and emerging wireless systems. These include power-efficient amplifiers, complex signal coding and processing for multi-antenna transmission. However, ultimately, the linearity and the efficiency of the RF system will define the practical limits beyond which baseband processing will be ineffective, i.e. non-linearity inherently limits coexistence and ultimately the system capacity. It is normal practice to compensate for low levels of non-linearity by using predistortion, requiring additional baseband processing of the modulated waveform. As the bandwidth and number of concurrent signals increase, this solution attracts increased operational expenditure for such baseband processing.

1.2.3.3 Waveforms and modulation

Waveforms are a core physical layer component of any wireless technology, whereby modulation schemes with a large numbers of constellations points play an increasing role, especially in higher frequency RF bands. Similar to LTE [46], the 5G NR standard uses cyclic-prefix orthogonal frequency division multiplexing (CP-OFDM) waveforms due to their robustness to time dispersion and the ease of exploiting both the time and frequency domains when defining the structure for different signals and channels [47]. To support a wide range of deployment scenarios, NR supports adaptive modulation, including quadrature phase-shift keying (QPSK), quadrature amplitude modulation (QAM) and a flexible OFDM methodology with a proportional change in cyclic prefix duration and subcarrier spacings. Moreover, CP-OFDM waveforms can be used for either single-stream or multi-stream (i.e. multiple-input–multiple-output (MIMO)) channels and they are more robust to Doppler and oscillator phase noise than other multicarrier waveforms where robustness to phase noise is crucial for operation at mm-wave bands.

1.2.3.4 Beamforming

Beamforming is a key technology for 5G and future wireless communications. It is a signal processing procedure used with multi-antenna systems (e.g. MIMO), at the transmitter and/or receiver side to achieve dynamic spatial diversity and to overcome issues caused by multipath fading and co-channel interference. Together with a suitable beamforming technique, multi-antenna wireless systems can adaptively form directional beams towards the desired users while forming nulls towards the interference sources. Such technology has the capability to extend the communication range, reduce the transmit power, increase spectrum efficiency, improve the capacity of wireless communication networks and hence improve the performance and reliability of wireless communication links. The current tendency in multi-antenna systems used for mobile communications is to employ massive MIMO (mMIMO) at the base station and MIMO for the mobile handset, especially for the realisation of reliable communication links at mm-wave frequency bands and for compensation of their inherently high atmospheric and free-space path losses (PLs).

1.2.4 Metrological R&D

With the industrial exploitation and adoption of complex NR signals, energy efficient devices and large-scale multi-antenna beamforming technologies at different RF bands in 5G and emerging wireless systems, several worldwide industries, research communities and standard bodies are now facing new measurement challenges on efficient and accurate verification of NR products that meet desired performance parameters for fulfilling the diverse technical requirements set by ITU-R [48], especially, for high-volume beam-reconfigurable issues [49–51]. Moreover, measuring the performance of the devices/systems with adaptive antennas and characterising the propagation channel creates further challenges at mm-wave bands, where innovative test methods and novel measurement equipment are required throughout R&D and production.

Metrological support is essential for underpinning all aspects of 5G and beyond in response to the high demand for universal, dynamic, and data-rich wireless applications. The development of the metrology for 5G and emerging wireless technologies is being intensively pushed forward by means of metrological R&D activities carried out by collaboration between NMIs, academia, and industry, all of whom play key roles in international standard developments. Measurements of the NR waveforms, devices, antennas, system-level and RF exposure assessments are needed to develop, manufacture and deploy these new wireless communication technologies. The following describes an overview and highlights some global metrological R&D activities in key interlinked metrological areas for 5G and beyond.

1.2.4.1 Waveform metrology

Accurate electrical and optical waveform measurements play a crucial role in signal quality assessment of 5G and emerging wireless systems, especially during early research and standard definition stages. The relevant instruments, for example vector signal generator (VSG), arbitrary waveform generator (AWG), vector signal analyser (VSA), real-time digital oscilloscope (RTDO), are required to support a wide range of modulation bandwidths, a wide range of frequency bands, and be able to generate and analyse multi-format waveforms with high fidelity. Such instruments require calibration to obtain accurate results, especially, for the accurate measurement of a variety of standards-compliant and complex NR multi-carrier CP-OFDM waveforms.

Using electro-optic sampling (EOS) techniques as primary standards, NPL, NIST and PTB has individually established a traceable calibration chain between high-speed electrical waveform measurements and the SI [52]. Such techniques achieve accurate calibration of reference pulses from a photodetector and are capable of measurement bandwidths on the order of 1 THz. It enables traceable measurements of digital waveform parameters such as jitter and error vector magnitude (EVM). EVM is a parameter that is often used for waveform error assessment and is currently being studied by IEEE P1765 [53], and together with other system-level distortion metrics. It can be tied to the predicted bit-error-rate (BER) [54] and signal-to-interference-plus-noise ratio (SINR) [55]. To cross-validate the primary standards, several inter-comparison activities on waveform parameters such as on pulse generator aberrations [56], oscilloscope transition duration [57] and common-mode rejection ratio [58] have been carried out.

An international comparison has been carried out between measurement institutes from China (Beijing Institute of Radio Metrology and Measurement, BIRMM and National Institute of Metrology, NIM), the United States (NIST) and Germany (PTB) [59]. NIST has reviewed its developed waveform metrological work in [60]. Within Europe, many relevant metrological R&D collaborations between NMIs, industry and academia have been carried out with the aim of improving the traceability and uncertainty propagation for dynamic waveform generation and measurement. These include several collaborative European Association of National Metrology Institutes (EURAMET) projects such as 'Metrology for ultrafast electronics and high-speed communications' [61], 'Metrology for optical and RF communication systems' [62], 'Metrology for the photonics industry – optical fibres, waveguides and applications' [63] and 'Metrology for 5G communications' [64].

1.2.4.2 Device metrology

To support the wireless network promise on energy efficient implementation, the efficiency of operating devices located in 5G and emerging wireless systems must be greatly improved. For example, the complex NR modulation scheme for 5G and beyond has driven the development of complex power amplifier designs (e.g. Doherty and Envelope tracking) to service the increased Peak-to-Average Power Ratio (PAPR) and energy demand. To achieve this, such devices are operated in a non-linear regime. However, this causes undesired distortion to the waveform signals, resulting in degraded EVM and BER where the strength of this distortion is dependent on the degree of device non-linearity. Thereby, the challenge lies in undertaking accurate non-linear parameter device measurements to support the development of appropriate predistortion and filtering techniques [65–67] for correction of the relevant non-linear effects while keeping the efficiency benefits.

For mobile RF communications, it has always been important to be able to measure the complete RF performance, from individual components to subsystems. For linear RF components, this could traditionally be done with, for example, vector network analysers (VNAs) or VSAs. For the measurement of non-linear RF device-under-test (DUT), several non-linear vector measurement instruments have been proposed such as large signal network analysers (LSNA) [68], non-linear VNA

(NVNA) [69,70], load-pull systems [71–73]. To reduce measurement burden, behavioural models, such as X-parameters [74,75], have been developed that can encapsulate non-linear device characteristics in a more efficient manner. Using EOS and RF power primary standards, a traceable calibration chain between power and phase reference of these instruments used for non-linear measurement and the SI [52] can be established [76] for coaxial and on-wafer environments. Another important aspect to consider is the multiphysics characterisation of these RF devices considering both electromagnetic and thermal quantities and their distribution across the device which provides a fuller understanding of the internal device operation [77,78].

An inter-laboratory comparison of NVNA measurements has been carried out involving four organisations – NPL, the United Kingdom; University of Surrey, the United Kingdom; Chalmers University of Technology, Sweden; and Keysight Technologies, Denmark – under an EU collaborative project [64,79] to establish and demonstrate the quality of non-linear measurements made using commercially available NVNAs on several non-linear RF devices over a wide frequency range. Within Europe, the relevant metrological R&D collaborations between NMIs, industry and academia with the aim of improving the traceability and uncertainty propagation for device characterisation have been carried out for several EURAMET projects such as 'Metrology for new electrical measurement quantities in high-frequency circuits' [80], 'Metrology for 5G communications' [64] and 'Metrology for advanced energy-saving technology in next-generation electronics applications' [81].

1.2.4.3 Antenna metrology

Antennas for 5G and emerging wireless technologies are envisaged to have reconfigurable beams. They are typically formed by adaptive arrays, for example phased array antennas, digital beamforming array antennas, parasitic array antennas, MIMO and mMIMO [82–85]. However, conventional antenna testing schemes typically consider the antenna under test (AUT) as having a fixed radiation pattern. Hence, the practical evaluation of such antenna systems (i.e. with beam reconfigurability and smart connection functionality) presents significant cost and time burden issues [86]. The efficient testing of these antenna systems plays a key role in 5G and emerging wireless system deployment.

Typical antenna measurands are antenna factor, gain, axial ratio, efficiency, radiation pattern, etc., but as wireless technologies evolve and modern antennas are becoming highly integrated into wireless systems, there is a shift towards measuring some other integrated power pattern metrics such as total radiated power (TRP), total isotropic sensitivity (TIS), effective isotropic radiated power (EIRP), effective isotropic sensitivity (EIS) [87–89]. There are several different types of antenna measurement environment, for example, fully anechoic chamber, reverberation chamber, open area test site (OATS). In practice, the relevant antenna measurement environment is selected according to the relevant AUT measurands [90]. For example, a fully anechoic chamber is ideal for antenna radiation pattern and gain measurements, a reverberation chamber is ideal for antenna efficiency, TRP and MIMO diversity gain measurements, and an OATS is ideal for antenna factor measurements. With traceable calibrated instrumentation, and using primary standards such as field strength, RF power, reference antennas, a traceable calibration chain between the antenna measurements and the SI can be achieved [91].

An international comparison for antenna gain measurement of two Ku-band standard gain horn (SGH) antennas that involved 12 NMIs has been carried out [92] whereby a variety of measurement techniques such as extrapolation, far field, gain transfer and reciprocity were used. A similar international comparison for antenna pattern measurement of a reference SGH antenna that involved eight universities and industry was also performed [93] using different measurement techniques and antenna facilities, including spherical near-field, compact range, planar near-field and far-field facilities. Validation and accreditation of different facilities can be achieved with inter-comparisons and careful evaluation of error budgets and different measurement standards.

Several large-scale mm-wave beamforming testbeds have been established, for example, at Southeast University, China [94]; NPL and University of Surrey, the United Kingdom [83], to exploit the advantages of large-scale antenna arrays at mm-wave frequency bands. Using software-defined radio (SDR) technologies, NPL has developed several measurement capabilities and testbeds for beamforming antenna characterisations [95,96]. Work on antenna array near-field characterisation at mm-wave bands has been reported by Korea Research Institute of Standards and Science (KRISS) at 28 GHz [97] and Technical University of Delft at 60 and 94 GHz [98]. Within Europe, relevant metrological R&D collaborations between NMIs, industry and academia with the aim of improving the traceability and uncertainty propagation for antenna characterisations have been carried out under several EURAMET projects such as 'Metrology for optical and RF communication systems' [62] and 'Metrology for 5G communications' [64].

1.2.4.4 System OTA metrology

Conformance testing is essential to ensure that relevant radios comply with the requirements of standardisation. To understand the end-to-end reception performance of a wireless communication system or device, over-the-air (OTA) testing, which includes both antenna and propagation characteristics at the same time, is needed [99]. Throughput is often considered the key system-level link performance measurement metric. Furthermore, complex NR implementation has driven the thorough testing of 5G and emerging wireless systems, including RF parameters [100], array calibration and linearisation [101] and end-to-end performance testing, which is essential before their massive rollout in future networks [100,101]. For 5G NR FR1, traditional conducted testing will still be used for most compliance test requirements except TRP and the system-level link performance throughput test. For 5G NR FR2, all tests are required to be carried out by OTA radiated methods, due to the lack of antenna connectors for their compact and highly integrated designs [102–104].

This makes the current OTA testing methods impractical for wide-bandwidth product verification with advanced antenna technologies (e.g. high cost burden over large-scale array characterisation at mm-wave bands). Many alternate OTA test methods have been proposed, which vary widely with respect to their implementation, equipment used, chamber size, running cost and measurement time [105]. The typical measurement instruments for OTA testing are base station emulator (BSE), channel emulator (CE) and user equipment (UE) emulator, which often are based on using VSG, VSA, etc. Furthermore, within its test environment, the far-field plane wave condition is adopted as the reference for NR antenna array calibration and RF conformance radiated testing (i.e. the cost and complexity of existing OTA testing solutions such as anechoic chambers [106–109] and reverberation chambers [104] for end-to-end performance testing must be scaled up significantly). However, current state-of-the-art solutions to mimic the default far-field condition, e.g. direct far-field (DFF) method, compact antenna test range (CATR), plane wave generator (PWG) and near-field to far-field (NF–FF) transformation techniques [100], cannot meet the full requirements of 5G NR testing. These measurement parameters are mutually dependent, which makes OTA testing a problem of joint multidimensional parameter estimation. It is subject of much ongoing research to address such challenges and establish traceability to the SI.

A series of comparison activities for MIMO OTA measurements obtained from different measurement facilities/methods have been carried out since 4G, e.g., an inter-lab/inter-technique 3GPP/CTIA round robin test using RC [110], etc. Within Europe, relevant metrological R&D collaboration between NMIs, industry and academia for system OTA characterisation has been carried out under several EURAMET projects such as 'Metrology for optical and RF communication systems' [62] and 'Metrology for 5G communications' [64].

1.2.4.5 Propagation channel metrology

Radio wave propagation holds the key to understanding the transmitter power requirements, antenna requirements, interference levels, receiver design and expected distances for wireless communication links [111]. The different RF bands offer different propagation properties. As frequency increases, the channel characteristics in the mm-wave bands are significantly different from those in the sub-6 GHz bands in terms of large-scale and small-scale fading [112]. Signals operating in some mm-wave and THz bands experience severe inherent atmospheric and free-space PL and are more sensitive to physical blockages due to the relatively short wavelength. A sparse propagation channel is generally observed in mm-wave bands, where the line-of-sight (LoS) direct-path is the most dominant component. Many propagation paths that bring significant power at a lower frequency are either attenuated by obstacles or suffer from some diffraction and diffuse scattering. That is still true in the THz spectrum.

Typical propagation channel measurands are, for example, PL, power delay profile (PDP), power angle profile (PAP), angular spread (AS), root mean square delay spread (RMS DS), direction-of-departure (DoD), direction-of-arrival (DoA) and Doppler shift. These parameters enable detailed analysis of space-time channel properties in typical environments and network configurations. There are two major categories of channel measurement (also referred to as channel sounding) methods, frequency-domain channel sounding (FDCS) and time-domain channel sounding (TDCS). Typically, FDCS uses VNA-based instrumentation whereby

TDCS uses VSG-VSA-based instrumentation. The choice of method to be used depends on the relevant channel properties, i.e. FDCS is ideal for quasi-static channels, and TDCS is ideal for dynamic channels. However, the measurement parameters are mutually dependent, which makes channel sounding a problem of joint multidimensional parameter estimation [113–115]. It is the subject of ongoing research to optimise resolution, precision and accuracy and to establish traceability to the SI, which is a prerequisite for reliable quantitative extraction of channel parameters with known measurement uncertainty [116].

Several real-world trial activities have been carried out, including a sub-6 GHz mMIMO trial in Bristol under the UK 5G Testbeds and trials program [117], and cellular V2X trial in Japan [118]. Also a list of other trial activities can be found in [119]. Propagation channel models for sub-THz for in-city deployment are an active research topic and work has been reported at 90 and 200 GHz, however, only from a numerical and analytical modelling prospective with limited variations in both environmental settings and parametric studies [120]. The state of the art for THz communication in Europe is a single-input–single-output system at 237.5 GHz for transmitting data over 20 m at a data rate of 100 Gbit/s [121] showing that this technique has promise but the required technology is not widely available at present to validate or provide a comprehensive model and system for further adoption. The most recent advances, globally, in terms of sub-THz channel sounding are reported in [122], where an instrument-based time-domain THz channel sounder for the frequency range between 280 and 330 GHz was developed and assembled which offers excellent sensitivity and dynamic range, with a maximum measurable PL of 122.6 dB and a dynamic range of almost 70 dB. The first standard for wireless communications operating in the frequency range of 252–325 GHz has been published by IEEE Task Group on “100G Wireless” (TG100G) in 2017 [123]. Within Europe, relevant metrological R&D collaboration between NMIs, industries and academics on propagation within buildings have been carried out under a EURAMET project entitled ‘Metrology for ultrafast electronics and high-speed communications’ [124].

1.2.4.6 RF exposure metrology

The rollout of 5G and emerging wireless technologies has started since 2020, but in many countries the deployment has been delayed or even stopped by the local government. Two of the major reasons for this delays were, first, that questions still exist about the possible health impacts caused by the new frequency bands used (e.g. mm-wave), and, second, it is claimed that methods intended to quantify the exposure to NR are too complex, not independent enough from manufacturers or network providers and too time-consuming. Furthermore, public concern over potential health risks from radiation exposure has led to stringent RF exposure compliance regulation in some countries. The time-burden involved in employing current RF exposure assessment methods applicable to specific absorption rate (SAR) and power density as defined in the CENELEC [15] and IEC/IEEE International Standards [125] will need to be addressed, especially for high-volume beam reconfigurable product verification.

The principles of 5G exposure measurements have been presented in different contributions [126,127]. The methods are based on the measurement of the synchronization signal block (SSB) but the traceability with respect to the SI is not available. Moreover, there are concerns about the use of the maximum worst case exposure to quantify the exposure of 5G base stations. Different contributions tend to demonstrate that this would not be optimum [128–130]. This might be due to the use of complex adaptive antennas for which current metrological methods are not well established. Concerning the measurement of exposure in terms of SAR, it is not clear how to derive measurands to quantify the exposure due to mMIMO antennas.

Several spectrum regulators have reported RF exposure field trials in different regions, for example, UK Office of Communications (Ofcom) [131], France National Frequency Agency (ANFR) [132]. Also, METAS has reported a measurement method for 5G NR base stations [133]. Within Europe, the relevant metrological R&D collaborations between NMIs, industry and academia on RF exposure metrology have been carried out under a EURAMET project entitled 'Metrology for RF exposure from mMIMO 5G base station: Impact on 5G network deployment' [134]. The work focussed on an experimental RF exposure study of a stochastic nature of mMIMO systems at sub-6 GHz [135,136].

1.3 Structure of the book

This book is divided into six parts: (1) waveform metrology; (2) device metrology; (3) antenna metrology; (4) system OTA metrology; (5) propagation channel metrology; and (6) RF exposure metrology. The book is organised as follows.

Part 1: Waveform metrology

In Chapter 2, the authors discuss the prediction of SINR based on EVM. The subject matters include the definition of SINR, its relation to EVM, and a discussion of how to use SINR in an OTA environment for mobile terminal conformance testing.

Part 2: Device metrology

In Chapter 3, the authors present a selection of non-linear measurement techniques for 5G devices. The subject matter includes NVNA measurements, load-pull measurements, the *X*-parameter non-linear behavioural model and the practical limitations of RF measurements and techniques for evaluating their associated uncertainties.

In Chapter 4, the authors present a multiphysics measurement system used to characterise the distributed behaviour of devices with several examples showing the insights that can be gained through examining distributed electromagnetic and thermal measurements.

Part 3: Antenna metrology

In Chapter 5, the authors present a wide-bandwidth mm-wave fully connected hybrid beamforming metrological testbed with a large antenna array. Discussions include testbed design, calibration procedures, experimental evaluations, as well as the critical factors to consider for their practical implementation.

In Chapter 6, the author presents an assessment of the broadband sources and defocusing effects of a CATR for 5G antenna testing and discusses how common industry metrics for compact range performance metrics provide insights to the quality of the plane wave generated.

In Chapter 7, the authors present work carried out under several international intercomparison campaigns for mobile telecommunication antenna characterisation. These international comparison activities support interoperability of communication equipment in today's world market.

In Chapter 8, the authors present several fabrication methods for 5G wearable antennas, and the relevant methodologies and measurement techniques to characterise their performance in dynamic body-worn scenarios. Several examples of 5G wearable antennas operating at different frequencies are presented.

In Chapter 9, the authors present several methods and new metrics specified by 3GPP for the evaluation of 5G mm-wave mobile handset antenna systems. The subject matter includes a new near-field measurement approach, quantification of spatial coverage performance degradation and user blockage.

Part 4: System OTA metrology

In Chapter 10, the authors provide an overview of OTA test metrology of 5G NR systems and their current standardisation status. The subject matter includes a new mid-field-testing method, and two key 3GPP-accepted methods, the multi-probe anechoic chamber (MPAC) and radiated two-stage (RTS) methods.

In Chapter 11, the authors present work on OTA testing using synthetic-aperture techniques. The subject matter includes methods for design and the use of synthetic-aperture measurement systems that can be used for the characterisation of wireless channels at different 5G frequencies, with guidelines, examples and case studies.

In Chapter 12, the authors present an overview on using reverberation chambers for mm-wave OTA and electromagnetic compatibility (EMC) measurements. The relevant background, measurement procedures, uncertainty and implementation considerations are detailed as well as several verification examples.

In Chapter 13, the authors present work on OTA testing and channel sounding for V2X autonomous vehicles communications at FR1 and FR2 bands within university laboratory and campus environments. The subject matters include crucial parameters to investigate, influence of weather, air quality and traffic, etc.

In Chapter 14, the authors provide an overview of OTA performance evaluation and 3GPP/CTIA compliance testing of 5G/6G base stations, user devices and systems. The subject matter includes DFF, NF–FF, multi-probe systems, CATR, array-based PWG, etc. Several FR1 and FR2 measurement examples are presented.

Part 5: Propagation channel metrology

In Chapter 15, the authors present an overview on metrology for 5G channel sounding. The subject matter includes measurement methods, system architecture and data processing algorithm. Several real-world outdoor channel sounding measurement examples are presented, e.g. suburban, university campus.

In Chapter 16, the authors present work on empirical mm-wave channel measurements and analysis in several indoor and outdoor complex environments for 5G communications. The subject matter also includes mm-wave channel models, and current efforts towards modelling the mm-wave radio propagation channel.

In Chapter 17, the authors present work on the world's first real-time mMIMO testbed implementation and how it was used to evaluate the technology in real-world trials. These trials reveal insights into SINR estimation and a mechanism for predicting EVM of connected terminals in a multi-user configuration.

In Chapter 18, the authors present work on mm-wave radio propagation measurements towards 5G NR standardisation. The subject matter includes channel sounding methods and their implementation. Results from extensive channel measurement campaigns conducted in multiple indoor environments are analysed.

In Chapter 19, the authors present a comprehensive introduction to various real-world use cases in the THz band for 5G and beyond as well as their propagation channel characterisation and challenges. The subject matter includes, for example, IoT, railway and drone use cases, and the impact of weather in the THz band.

Part 6: RF exposure metrology

In Chapter 20, the author presents measurement methods to quantify the exposure from 5G NR base stations for the FR1 band. Both code-selective-based and test-mode-based approaches are considered, and measurement methods, procedures, and practical considerations are discussed.

In Chapter 21, the authors present the experimental assessment of RF exposure from a fully reconfigurable 5G NR FR1 mMIMO system. Several measurement campaigns have been carried out to determine the stochastic nature of mMIMO systems. Key factors considered include the number of users, their positions and date rates.

Finally, Chapter 22 provides a conclusion for the book as well as a brief discussion of future perspectives on metrology for 5G and emerging wireless technologies associated with multi-parameter characterisation of emerging wireless technologies, e.g. 3D hologram, autonomous vehicles and reconfigurable electromagnetic environments.

References

[1] European Council. *Why the EU is betting big on 5G – Research*eu Focus Magazine* [online]. 26 February 2015. Available: https://digital-strategy.ec.europa.eu/en/library/why-eu-betting-big-5g-researcheu-focus-magazine [Accessed October 2021].

[2] Ericsson. *Ericsson Mobility Report* [online]. June 2020. Available: https://www.ericsson.com/en/press-releases/2020/6/ericsson-mobility-report—june-2020 [Accessed October 2021].

[3] Information handling services (IHS). *The 5G Economy: How 5G Technology Will Contribute to the Global Economy* [online]. January 2017. Available: https://cdn.ihs.com/www/pdf/IHS-Technology-5G-Economic-Impact-Study.pdf [Accessed May 2021].

[4] UK Department for Digital, Culture, Media & Sport. *Next Generation Mobile Technologies: A 5G Strategy for the UK* [online]. March 2017. Available: https://assets.publishing.service.gov.uk/government/uploads/system/uploads/attachment_data/file/597421/07.03.17_5G_strategy_-_for_publication.pdf [Accessed May 2021].

[5] Research & Innovation Europe Magazine. *Objective 5G: The EU Is Reinventing 'Connectivity' for the Next Decade!* [online]. March 2015. Available: http://www.researchandinnovation.ie/objective-5g-eu-reinventing-connectivity-next-decade/ [Accessed May 2021].

[6] Bureau International des Piods et Mesures (BIPM). *What Is Metrology?* [online]. 2017. Available: https://web.archive.org/web/20170324081343/http://www.bipm.org/en/worldwide-metrology/ [Access September 2021]

[7] BIPM, IEC, IFCC, ILAC, IUPAC, IUPAP, ISO, OIML. *The international vocabulary of metrology—basic and general concepts and associated terms (VIM)*, 3rd edn, JCGM 200:2012 [online]. 2012. Available: https://www.bipm.org/documents/20126/2071204/JCGM_200_2012.pdf/f0e1ad45-d337-bbeb-53a6-15fe649d0ff1 [Accessed September 2021].

[8] Goldsmith M. *A Beginner's Guide to Measurement* [online]. March 2017. Available: https://www.npl.co.uk/special-pages/guides/gpg118_begguide2-measure [Accessed May 2021].

[9] Wikipedia. *Traceability*. 2021 [online]. Available: https://en.wikipedia.org/wiki/Traceability [Accessed September 2021].

[10] Wikipedia. *Measurement Traceability Pyramid*. 2020 [online]. Available: https://commons.wikimedia.org/wiki/File:Traceability_Pyramid.png [Accessed September 2021].

[11] Wikepdeia Commons. *Overview of the structure of a national measurement system*. 2020 [online]. Available: https://commons.wikimedia.org/wiki/File:National_Measurement_System_Overview.png [Accessed September 2021].

[12] European Commission. *International Cooperation on 5G*. September 2020 [online]. Available: https://ec.europa.eu/digital-single-market/en/5G-international-cooperation [Accessed September 2021].

[13] IEEE. *5G World Forum 2020: Multilateral MoU Signing-In Ceremony* [online]. November 2020. Available: https://ieeetv.ieee.org/5g-world-forum-2020-multilateral-mou-signing-in-ceremony [Accessed May 2021].

[14] 5G PPP. *5G Infrastructure Association (5GIA)*. 2017 [online]. Available: https://5g-ia.eu/ [Accessed September 2021].

[15] 5G Americas. *The Voice of 5G and LTE for the Americas*. 2021 [online]. Available: https://www.5gamericas.org/ [Accessed September 2021].

[16] IMT2020. *International Mobile Telecommunications-2020 (IMT-2020) 5G promotion group*. 2014 [online]. Available: http://www.imt2020.org.cn/en [Accessed September 2021].

[17] 5G Forum. *5G Forum Vision.* 2013 [online]. Available: http://www.5gforum.org/html/en/main.php [Accessed September 2021].

[18] 5GMF. *The Fifth Generation Mobile Communication Promotion Forum.* 2014 [online]. Available: https://5gmf.jp/en/ [Accessed September 2021].

[19] Netherlands Enterprise Agency. *5G in Brazil - the state of affairs.* 2019 [online]. Available: https://www.rvo.nl/sites/default/files/2019/09/5G-in-Brazil-the-state-of-affairs.pdf [Accessed September 2021].

[20] TSDSI. *Telecommunications Standards Development Society India (TSDSI)* [online]. April 2021. Available: https://tsdsi.in/ [Accessed May 2021].

[21] 5GPPP. *5GPPP (5G-Infrastructure Public Private Partnership Project) – A 1.4 bn € Joint Initiative Between European ICT Industry and European Commission* [online]. April 2021. Available: http://5g-ppp.eu/ [Accessed May 2021].

[22] METIS. *Mobile and Wireless Communications Enablers for the 2020 Information Society, EU 7th Framework Programme project.* 2012. [online]. Available: www.metis2020.com [Accessed May 2021].

[23] University of Oulu. *6G Flagship.* 2019 [online]. Available: https://www.oulu.fi/6gflagship/ [Accessed September 2021].

[24] Hexa-X. *Hexa-X vision on 6G and research challenges.* 2020 [online]. Available: https://hexa-x.eu/about/ [Accessed September 2021].

[25] Kagermann H., Wahlster W., and Helbig J. *Recommendations for implementing the strategic initiative Industrie 4.0: Final report of the Industrie 4.0 Working Group* [online]. 2013. Available: https://www.din.de/blob/76902/e8cac883f42bf28536e7e8165993f1fd/recommendations-for-implementing-industry-4-0-data.pdf [Accessed May 2021].

[26] University of Surrey. *5G Innovation Centre (5GIC).* 2014 [online]. Available: https://www.surrey.ac.uk/institute-communication-systems/5g-innovation-centre [Accessed September 2021].

[27] University of Surrey. *6G Innovation Centre (6GIC).* 2020 [online]. Available: https://www.surrey.ac.uk/news/university-surrey-unveils-its-6g-innovation-centre-and-distinctive-6g-vision [Accessed September 2021].

[28] UK Department for Digital, Culture, Media & Sport. *UK 5G Testbeds and Trials Programme.* March 2021 [online]. Available: https://www.gov.uk/guidance/5g-testbeds-and-trials-programme [Accessed September 2021].

[29] UK5G. *UK5G Innovation Network.* 2020 [online]. Available: https://uk5g.org/ [Accessed September 2021].

[30] Association of Radio Industries and Businesses (ARIB). *About ARIB.* 2001 [online]. Available: https://www.arib.or.jp/english/arib/about_arib.html [Accessed September 2021].

[31] New York University. *NYU Wireless research center.* 2012 [online]. Available: http://nyuwireless.com/ [Accessed September 2021].

[32] ATIS. *Next G Alliance.* 2020 [online]. Available: https://nextgalliance.org/ [Accessed September 2021].

[33] Object Management Group, Inc. *Industrial Internet Consortium (IIC).* 2020 [online] Available: https://www.iiconsortium.org/ [Accessed September 2021].

[34] ITU. *International Telecommunications Union radiocommunication sector (ITU-R)*. 1992 [online]. Available: https://www.itu.int/en/ITU-R/Pages/default.aspx [Accessed September 2021].

[35] ITU-R. *Detailed Specifications of the Terrestrial Radio Interfaces of International Mobile Telecommunications-2020 (IMT-2020)*. February 2021.

[36] 3GPP. *The 3rd Generation Partnership Project.* 2000 [online]. Available: https://www.3gpp.org/ [Accessed September 2021].

[37] ETSI. *European Telecommunications Standards Institute.* 1988 [online]. Available: http://www.etsi.org/ [Accessed September 2021].

[38] ATIS. *Alliance for Telecommunications Industry Solutions.* 2015 [online]. Available: https://www.atis.org/ [Accessed September 2021].

[39] CCSA. *China Communications Standards Association.* 2002 [online]. Available: http://www.ccsa.org.cn/ [Accessed September 2021].

[40] TTA. *Telecommunications Technology Association.* 1988 [online]. Available: http://www.tta.or.kr/eng/index.jsp [Accessed September 2021].

[41] TTC. *Telecommunication Technology Committee.* 1985 [online]. Available: https://www.ttc.or.jp/e [Accessed September 2021].

[42] CTIA. *International Association for the Wireless Telecommunication Industry.* 1984 [online]. Available: http://ctia.org/ [Accessed September 2021].

[43] NGMN. *Next Generation Mobile Networks*. 2006 [online]. Available: https://www.ngmn.org/ [Accessed September 2021].

[44] European Council. *Europe Moves Towards 6G* [online]. March 2021. Available: https://ec.europa.eu/digital-single-market/en/blogposts/europe-moves-towards-6g [Accessed May 2021].

[45] ITU. ITU-R Radio regulations 2016 Edition. 2016.

[46] Dahlman E., Parkvall S., and Skold J. *4G LTE-Advanced Pro and the Road to 5G*, 3rd edn. London: Elsevier Academic Press, 2016.

[47] 3GPP. *3GPP Release 15* [online]. April 2019. Available: https://www.3gpp.org/release-15 [Accessed May 2021].

[48] ITU-R. *Minimum Requirements, Related to Technical Performance for IMT-2020 Radio Interface(s). Report ITU-R M.2410*. November 2017.

[49] Frenzel L. *Major Trends in Wireless Test* [online]. March 2017. Available: https://www.electronicdesign.com/technologies/test-measurement/article/21804695/major-trends-in-wireless-test [Accessed May 2021].

[50] Remley K. A., Gordon J.A., Novotny D., *et al.* Measurement Challenges for 5G and Beyond: An Update from the National Institute of Standards and Technology. *IEEE Microwave Magazine*. 2017. 18, pp. 41–56.

[51] Foegelle M. D. 'New and Continuing Measurement Challenges for 5G mmWave and Beamforming Technologies'. *The 13th European Conference on Antennas and Propagation (EuCAP 2019)*, Krakow, Poland, 2019, pp. 1–5.

[52] Hale P. D., Williams D.F., Dienstfrey A., *et al.* 'Traceability of High-Speed Electrical Waveforms at NIST, NPL, and PTB'. *The 2012 Conference on Precision Electromagnetic Measurements*, Washington, DC, USA, 2012.

[53] IEEE. *IEEE P1765 - Recommended Practice for Estimating the Uncertainty in Error Vector Magnitude of Measureed Digitally Modulated Signals for Wireless Communications*. Active Project. Project Authorisation Request (PAR) approved by March 2021 [online]. Available: https://standards.ieee.org/project/1765.html [Accessed September 2021].

[54] Hudliþka M., Lundström C., Humphreys D. A., and Fatadin I. 'BER Estimation from EVM for QPSK and 16-QAM Coherent Optical Systems'. *The 2016 IEEE 6th International Conference on Photonics (ICP)*, Kuching, Malaysia, 2016.

[55] Brown T., Humphreys D., Hudlicka M., and Loh T. H. 'Prediction of SINR Using BER and EVM for Massive MIMO Applications'. *The 12th European Conference on Antennas and Propagation (EuCAP 2018)*, London, United Kingdom, 9–13 April 2018.

[56] Paulter N. G., Smith A. J. A., Larson D. R., Souders T. M., and Roddie A. G. NIST-NPL Interlaboratory Pulse Measurement Comparison. *IEEE Transactions on Instrumentation and Measurement*. 2003, 52, pp. 1825–1833.

[57] Bieler M., Seitz S., Spitzer M. *et al.* Rise-Time Calibration of 50-GHz Sampling Oscilloscopes: Intercomparison Between PTB and NPL. *IEEE Transactions on Instrumentation and Measurement*. 2007, 56, pp. 266–270.

[58] Struszewski P., Bieler M., Humphreys D., Bao H., Peccianti M., and Pasquazi A. Characterization of High-Speed Balanced Photodetectors. *IEEE Transactions on Instrumentation and Measurement*. 2017, 66, pp. 1613–1620.

[59] Bieler M., Struszewski P., Feldman A., *et al.* 'International Comparison on Ultrafast Waveform Metrology'. *The 2020 Conference on Precision Electromagnetic Measurements (CPEM)*, Denver, CO, USA, 2020.

[60] Hale P. D., Williams D. F., and Dienstfrey A. Waveform Metrology: Signal Measurements in a Modulated World. *Metrologia*. 2018, 55, pp. S135–S151.

[61] EURAMET. *EURAMET EMRP project IND16 "Metrology for ultrafast electronics and high-speed communications"*. 2011 [online]. Available: https://www.euramet.org/research-innovation/search-research-projects/details/project/metrology-for-ultrafast-electronics-and-high-speed-communications/ [Accessed September 2021].

[62] EURAMET. *EURAMET EMRP project IND51 "Metrology for optical and RF communication systems"*. 2013 [online]. Available: https://www.euramet.org/project-ind51/ [Accessed September 2021].

[63] EURAMET. *EURAMET EMPIR project 14IND13 "Metrology for the photonics industry - optical fibres, waveguides and applications"*. 2015 [online]. Available: https://www.euramet.org/research-innovation/search-research-projects/details/project/metrology-for-the-photonics-industry-optical-fibres-waveguides-and-applications/ [Accessed September 2021].

[64] EURAMET. *EURAMET EMPIR project 14IND10 "Metrology for 5G Communcations"*. 2014 [online]. Available: https://www.euramet.org/project-14ind10/ [Accessed September 2021].

[65] Jaraut P., Rawat M., and Roblin P. Digital Predistortion Technique for Low Resource Consumption Using Carrier Aggregated 4G/5G Signals. *IET MAP*. 2019, 13, pp. 197–207.

[66] Kim J., Roblin P., Chaillot D., and Xie Z. A Generalized Architecture for the Frequency – Selective Digital Predistortion Linearization Technique. *IEEE Transactions on Microwave Theory and Techniques*. 2013, 61, pp. 596–605.
[67] Abdelaziz M. Reduced-Complexity Digital Predistortion in Flexible Radio Spectrum Access. PhD Thesis, Tampere University of Technology, 2017.
[68] Moer W. V. and Gomme L. NVNA Versus LSNA: Enemies or Friends?. *IEEE Microwave Magazine*. 2010, pp. 97–103.
[69] Zhang Y., Nian F., Yuan G., *et al.* Precisely Synchronized NVNA Setup for Digitally Modulated Signal Generation and Measurement at 5G-Oriented Millimeter-Wave Test Bands. *IEEE Transactions on Microwave Theory and Techniques*. 2021, 69, pp. 833–845.
[70] Keysight. *Nonlinear Vector Network Analyzer (NVNA) - Breakthrough Technology for Nonlinear Vector Network Analysis from 10 MHz to 67 GHz*. June 2019 [online]. Available: https://www.keysight.com/gb/en/assets/7018-01822/brochures/5989-8575.pdf [Accessed September 2021].
[71] Chappidi C. R. and Sengupta K. 'A 26-42 GHz Broadband, Back-off Efficient and VSWR Tolerant CMOS Power Amplifier Architecture for 5G Applications'. *The 2019 Symposium on VLSI Circuits*, Kyoto, Japan, 2019.
[72] Chappidi C. R., Sharma T., and Sengupta K. Multi-port Active Load Pulling for mm-Wave 5G Power Amplifiers: Bandwidth, Back-Off Efficiency, and VSWR Tolerance. *IEEE Transactions on Microwave Theory and Techniques*. 2020, 68, pp. 2998–3016.
[73] Tsironis C. *WP64 - Load Pull Characterization, Focus Microwave Group*. March 2019 [online]. Available: https://www.aspen-electronics.com/uploads/3/7/1/2/37123419/wp-lpc-4.pdf [Accessed September 2021].
[74] Root D., Verspecht J., Horn J., and Marcu M. *X-Parameters: Characterization, Modeling, and Design of Nonlinear RF and Microwave Components (The Cambridge RF and Microwave Engineering Series)*. New York, NY: Cambridge University Press, 2013.
[75] Pedro J. C., Root D. E., Xu J., and Nunes L. C. *Nonlinear Circuit Simulation and Modeling: Fundamentals for Microwave Design (The Cambridge RF and Microwave Engineering Series)*, 1st edn. Cambridge, UK: Cambridge University Press, 2018.
[76] Rajabi M., Humphreys D.A., Avolio G., *et al.,* Design and Evaluation of Nonlinear Verification Device for Nonlinear Vector Network Analyzers. *IEEE Transactions on Microwave Theory and Techniques*. 2018, 66, pp. 1121–1130.
[77] Denis D., Snowden C. M., and Hunter I. C. Coupled Electrothermal, Electromagnetic, and Physical Modeling of Microwave Power FETs. *IEEE Transactions on Microwave Theory and Techniques*. 2006, 54, pp. 2465–2470.
[78] Urbonas J. *Multiphysics Measurement of High-Power Microwave Transistors and Amplifiers*. PhD Dissertation, University of Surrey. Guildford, Surrey, United Kingdom, 2019.

[79] Salter M., Stant L., Buisman K., and Nielsen T. ‘An Inter-Laboratory Comparison of NVNA Measurements’. *The 2018 International Workshop on Integrated Nonlinear Microwave and Millimetre-wave Circuits (INMMIC 2018)*, Brive-la-Gaillarde, France, July 2018.

[80] EURAMET. *EURAMET SI Broader Scope project SIB62 "Metrology for new electrical measurement quantities in high-frequency circuits"*. 2013 [online]. Available: https://www.euramet.org/research-innovation/search-research-projects/details/project/metrology-for-new-electrical-measurement-quantities-in-high-frequency-circuits/ [Accessed September 2021].

[81] EURAMET. *EURAMET EMPIR project 16ENG06 “Metrology for advanced energy-saving technology in next-generation electronics applications”*. 2017 [online]. Available: https://www.euramet.org/research-innovation/search-research-projects/details/project/metrology-for-advanced-energy-saving-technology-in-next-generation-electronics-applications/ [Accessed September 2021].

[82] Asplund H., Astely D., Butovitsch P.V., *et al. Advanced Antenna Systems for 5G Network Deployments: Bridging the Gap Between Theory and Practice*. London, UK: Academic Press, 2020.

[83] Payami S., Khalily M., Araghi A., *et al.* Developing the First mmWave Fully-Connected Hybrid Beamformer With a Large Antenna Array. *IEEE Access*. 2020, 8, pp. 141282–141291.

[84] Liu H., Gao S., and Loh T. H. Electrically Small and Low Cost Smart Antenna for Wireless Communications. *IEEE Transaction on Antennas and Propagation*. 2012, 60, pp. 1540–1549.

[85] Liu H., Gao S., and Loh T. H. Small Director Array for Low profile Smart Antennas Achieving Higher Gain. *IEEE Transaction on Antennas and Propagation*. 2013, 61, pp. 162–168.

[86] Fan W., Zhang F. and Wang Z. Over-the-Air Testing of 5G Communication Systems: Validation of the Test Environment in Simple-Sectored Multiprobe Anechoic Chamber Setups. *IEEE Antennas and Propagation Magazine*. Feb. 2021, 63, pp. 40–50.

[87] IEEE. *IEEE Standard Test Procedures for Antennas, ANSI/IEEE Std 149-1979*. 1979.

[88] CTIA. *CTIA OTA Test Plan v 3.8.2*. Apr. 2019 [online]. Available: https://api.ctia.org/wp-content/uploads/2019/04/CTIA_OTA_Test_Plan_3_8_2.pdf [Accessed September 2021].

[89] 3GPP. *3GPP TS 34.114 V12.2. 0 – User Equipment (UE)/Mobile Station (MS) Over the Air (OTA) Antenna Performance – Conformance Testing* [online]. October 2016. Available: https://portal.3gpp.org/desktopmodules/Specifications/SpecificationDetails.aspx?specificationId=2361 [Accessed May 2021].

[90] Zhou L.-L. and Zhao Y. ‘A Review of Antenna Radiation Measurement Environment’. *The 2016 Asia-Pacific International Symposium on Electromagnetic Compatibility (APEMC)*, 2016.

[91] NPL. *Antenna measurements*. 2021 [online]. Available: https://www.npl.co.uk/products-services/radiofrequency/antenna-measurements [Accessed September 2021].

[92] Guerrieri J. R., Coder J. B., Francis M. H., Gordon J., Novotny D. R., and Wittmann R. C. 'International Comparison of Ku-Band Standard Gain Horn Characterization'. *The 8th European Conference on Antennas and Propagation (EuCAP 2014)*, 2014.

[93] Pivnenko S., Pallesen J.E., Breinbjerg O., *et al.* Comparison of Antenna Measurement Facilities With the DTU-ESA 12 GHz Validation Standard Antenna Within the EU Antenna Centre of Excellence. *IEEE Transactions on Antennas and Propagation*. 2009, 57, pp. 1863–1878.

[94] Yang B., Yu Z., Lan J., Zhang R., Zhou J., and Hong W. Digital Beamforming-Based Massive MIMO Transceiver for 5G Millimeter-Wave Communications. *IEEE Transactions on Microwave Theory and Techniques*. 2018, 66, pp. 3403–3418.

[95] Loh T. H., Liu H., and Gao S. A Cost-Effective Direct Magnitude Measurement Methodology for Smart Antennas. *IEEE Transactions on Antennas and Propagation*. 2013, 61, pp. 2043–2050.

[96] Loh T. H. and Cheadle D. 'An Electronically Steerable Parasitic Array Radiator Antenna with Adaptive Beamforming Ability – A Radiation Pattern Assessment'. *The 2017 IEEE International Workshop on Electromagnetics (iWEM 2017)*, London, United Kingdom, May–June 2017.

[97] Yoo K., Chung T., Kim N., Seol B., Hong Y., and Lee D. 'Phase Measurement Techniques for Detecting the Defective Elements of an Array Antenna at 28 GHz'. *The 2019 International Symposium on Electromagnetic Compatibility (EMC EUROPE 2019)*, 2019.

[98] Alonso-delPino M., Rosa M. D., Simeoni M., Spella M., Martino C. D., and Spirito M. A Planar Near-Field Setup for Millimeter-Wave System-Embedded Antenna Testing. *IEEE Antennas and Wireless Propagation Letters*. 2017, 16, pp. 83–86.

[99] Loh T. H., Li C., Wang H., and Qin F. 'A Software-Defined-Radio Platform for Multiple-Input-Multiple-Output Over-the-Air Measurement'. *The 10th European Conference on Antennas and Propagation (EuCAP 2016)*, Davos, Switzerland, April 2016.

[100] Kong H., Wen Z., Jing Y., and Yau M. Midfield Over-the-Air Test: A New OTA RF Performance Test Method for 5G Massive MIMO Devices. *IEEE Transactions on Microwave Theory and Techniques*. 2019. 67, pp. 2873–2883.

[101] Long R., Ouyang J., Yang F., Han W., and Zhou L. Fast Amplitude-Only Measurement Method for Phased Array Calibration. *IEEE Transactions on Antennas and Propagation*. 2017, 65, pp. 1815–1822.

[102] Qi Y., Yang G., Liu L., *et al.* 5G Over-the-Air Measurement Challenges: Overview. *IEEE Transactions on Electromagnetic Compatibility*. 2017, 59, pp. 1661–1670.

[103] Foegelle M. D. The Future of MIMO Over-the-Air Testing. *IEEE Communications Magazine*. 2014, 52, pp. 134–142.

[104] Fan W., Kyosti P., Rumney M., Chen X., and Pedersen G. F. Over-the-Air Radiated Testing of Millimeter-Wave Beam-Steerable Devices in a Cost-Effective Measurement Setup. *IEEE Communications Magazine*. 2018, 56, pp. 64–71.

[105] Loh T. H. *OTA Test Methods and Candidates for 5G and Beyond, in Antennas and Propagation for 5G and Beyond*, Q. Abbasi, S. F. Jilani, A. Alomainy, and M. A. Imran, Eds., The Institution of Engineering and Technology, London, September 2020.

[106] Nilsson M. G., Hallbjörner P., Arabäck N., Bergqvist B., Abbas T., and Tufvesson F. Measurement Uncertainty, Channel Simulation, and Disturbance Characterization of an Over-the-Air Multiprobe Setup for Cars at 5.9 GHz. *IEEE Transactions on Industrial Electronics*. 2015, 62, pp. 7859–7869.

[107] Ji Y., Fan W., Nilsson M. G. *et al.* Virtual Drive Testing Over-the-Air for Vehicular Communications. *IEEE Transactions on Vehicular Technology*. 2000, 69, pp. 1203–1213.

[108] 3GPP. *Study on Radiated Metrics and Test Methodology for the Verification of Multi-Antenna Reception Performance of NR User Equipment (UE)*. June 2020.

[109] Fan W., Kyösti P., Hentilä L., and Pedersen G. F. A Flexible Millimeter-Wave Radio Channel Emulator Design With Experimental Validations. *IEEE Transactions on Antennas and Propagation*. 2018, 66, pp. 6446–6451.

[110] Mora-Andreu M. and Sánchez-Hernández D. A. 'Reverberation Chamber results on 3GPP/CTIA LTE MIMO OTA 2013 Round Robin Tests Using Different Channel Models—A Comparison of Performance'. *The 8th European Conference on Antennas and Propagation (EuCAP 2014)*, 2014.

[111] Rappaport T. S., Daniels R., Heath R.. Jr., and Murdock J. N. *Millimeter Wave Wireless Communications (Prentice Hall Communications Engineering and Emerging Technologies)*, 1st edn. New York, NY: Pearson, 2014.

[112] Shafi M., Zhang J., Tataria H., *et al.* Microwave vs. Millimeter-Wave Propagation Channels: Key Differences and Impact on 5G Cellular Systems. *IEEE Communications Magazine*. 2018, 56, pp. 14–20.

[113] Wang R., Renaudin O., Bas C. U., Sangodoyin S., and Molisch A. F. High-Resolution Parameter Estimation for Time-Varying Double Directional V2V Channel. *IEEE Transactions on Wireless Communications*. 2017, 16, pp. 7264–7275.

[114] Haardt M., Thomä R. S., Richter A. '*Multidimensional High-Resolution Parameter Estimation with Applications to Channel Sounding'*. in *High-Resolution and Robust Signal Processing*, Hua Y. Gershman A. Chen Q. (Eds.), CRC Press, New York, USA, 2003, pp. 253–337.

[115] Müller R., Häfner S., Dupleich D. *et al.* 'Simultaneous Multi-Band Channel Sounding at mm-Wave Frequencies'. *The EUCAP 2016*, Davos, Switzerland, April 2016.

[116] Salhi M. A., Kleine-Ostmann T., and Schrader T. Propagation Channel Measurements in the mm and Sub-mm Wave Range for Different Indoor

Communication Scenarios. *Journal of Infrared, Millimeter, and Terahertz Waves*. 2021, 42, pp. 357–370.

[117] University of Bristol, *World's first urban 5G experience hosted by University of Bristol and Watershed.* March 2018 [online]. Available: https://www.bristol.ac.uk/news/2018/march/5gexperience.html [Accessed September 2021].

[118] NTT Docomo. *Leading Automotive, Telecom and ITS Companies Unveil First Announced Cellular V2X Trials in Japan — Continental, Ericsson, Nissan, NTT DOCOMO, OKI and Qualcomm Technologies join forces to host C-V2X trials in Japan in 2018 to validate and demonstrate C-V2X benefits*. 2018 [online]. 2018. Available: https://www.nttdocomo.co.jp/english/info/media_center/pr/2018/0112_00.html [Accessed September 2021].

[119] European 5G Observatory. *Major International 5G Trials and Pilots. 2018 [online]*. Available: https://5gobservatory.eu/5g-trial/major-international-5g-trials-and-pilots/ [Accessed September 2021].

[120] Gougeon G., Corre Y., and Aslam M. Z. Sub-THz Channel Characterization. *ITU Journal: ICT Discoveries*. 2019, 2.

[121] Koenig S., Lopez-Diaz D., Antes J., *et al.* Wireless sub-THz Communication System With High Data Rate. *Nature Photonics*. 2013, 7, pp. 977–981.

[122] Schmieder M., Keusgen W., Peter M., *et al.* 'THz Channel Sounding: Design and Validation of a High Performance Channel Sounder at 300 GHz'. *2020 IEEE Wireless Communications and Networking Conference Workshops (WCNCW)*, Seoul, Korea (South), 2020, pp. 1–6.

[123] IEEE, *IEEE 802.15.3d-2017 – IEEE Standard for High Data Rate Wireless Multi-Media Networks Amendment 2: 100 Gb/s Wireless Switched Point-to-Point Physical Layer*. Oct. 2017.

[124] PTB. *Euramet EMRP project "IND16 Ultrafast"*. 2014 [online]. Available: https://www.ptb.de/emrp/ultrafast.html [Accessed September 2021].

[125] IEC. *IEC CDV IEC/IEEE 63195-1: Assessment of Power Density of Human Exposure to Radio Frequency Field Form Wireless devices in close proximity to the head and Body (Frequency range of 6 GHz to 300 GHz).* April 2018 [online]. Available: https://www.iec.ch/dyn/www/f?p=103:38:5094981455406::::FSP_ORG_ID,FSP_APEX_PAGE,FSP_PROJECT_ID:1303,23,101946 [Accessed September 2021].

[126] Keller H. On the Assessment of the Human Exposure to Electromagnetic Fields Transmitted by 5g Nr Base Stations. *Health Physics*. 2019, 117.

[127] METAS. *Technical Report: Measurement Method for 5G NR Base Stations up to 6 GHz*, METAS-Report 154.1-2020-5218-1016. February 2020.

[128] ANFR. *Evaluation de l'exposition du public aux ones électromagnétiques 5G: Volet 2: premiers résultats de mesures sur les pilotes 5G dans la bande 3400-3800 MHz, Rapport de l'agence nationale des fréquences*, April 2020 [online]. Available: https://www.anfr.fr/fileadmin/mediatheque/documents/

5G/20200410-ANFR-rapport-mesures-pilotes-5G.pdf [Accessed September 2021].

[129] Baracca P., Weber A., Wild T., and Grangeat C. 'A Statistical Approach for RF Exposure Compliance Boundary Assessment in Massive MIMO Systems'. *The 22nd International ITG Workshop on Smart Antennas (WSA2018)*, March 2018.

[130] Colombi D., Joshi P., Xu B., Ghasemifard F., Narasaraju V., Törnevik C. *et al.* Analysis of the Actual Power and EMF Exposure from Base Stations in a Commercial 5G Network. *Applied Sciences*. 2020, 10.

[131] Ofcom. *Electromagnetic Field (EMF) Measurements Near 5G Mobile Phone Base Stations – Summary of Results*, February 2020.

[132] ANFR. *Assessment of the Exposure of the General Public to 5G Electromagnetic Waves – Part 2: First Measurement Results on 5G Pilots in the 3,400-3,800 MHz Band* [online]. April 2020. Available: https://www.anfr.fr/fileadmin/mediatheque/documents/5G/20200410-ANFR-rapport-mesures-pilotes-5G-EN.pdf [Accessed May 2021].

[133] METAS. *Technical Report Measurement Method for 5G NR Base Stations*. February 2020.

[134] 5GRFEX. *EURAMET EMPIR SIP project 18SIP10 "Metrology for RF exposure from massive MIMO 5G base station: Impact on 5G network deployment"*. 2019 [online]. Available: http://empir.npl.co.uk/5grfex/ [Accessed September 2021].

[135] Loh T. H., Heliot F., Cheadle D., and Fielder T. 'An Assessment of the Radio Frequency Electromagnetic Field Exposure from A Massive MIMO 5G Testbed'. *The 14th European Conference on Antennas and Propagation (EuCAP 2020)*, Copenhagen, Denmark, March 2020.

[136] Loh T. H., Cheadle D., Heliot F., Sunday A., and Dieudonne M. 'A Study of Experiment-based Radio Frequency Electromagnetic Field Exposure Evidence on Stochastic Nature of A Massive MIMO System'. *The 15th European Conference on Antennas and Propagation (EuCAP 2021)*, Düsseldorf, Germany, March 2021.

Part I

Waveform metrology

Chapter 2
Metrology for 5G link adaptation and signal-to-interference-plus-noise ratio

Tim W.C. Brown[1], *Martin Hudlicka*[2], *David A. Humphreys*[3], *Min Wang*[4] *and Tian Hong Loh*[3]

Before a mobile terminal device transmits uplink data to or receives downlink data from the base station it has established a connection with, the quality of this connection/channel needs to be quantified, usually via the channel quality indicator (CQI). This is reported to the base station by the mobile terminal and it will determine the rate at which data can be transmitted as well as the type of modulation scheme that should be used. This will inherently depend on how high the signal-to-interference-plus-noise ratio (SINR) is at the mobile terminal such that a higher SINR will result in a higher CQI. Prior to the roll out of fifth-generation (5G) wireless mobile networks, the CQI was determined by the mobile terminal by finding the point at which higher data rates and more complex modulation schemes would begin to fail. This did not require actual measurement or estimation of the SINR as it was sufficient for the mobile terminal to know that it had found an achievable data rate and modulation scheme that works well. This is the principle of link adaptation whereby the data transmission has adapted to the quality of the wireless connection. In 5G systems, on the other hand, where massive multiple input–multiple output (mMIMO) is used, it is of vital importance for the mobile terminal to reliably estimate the SINR and report a corresponding CQI, not only to ensure that the highest possible data rate is delivered to the mobile terminal, but that the base station is delivering the best quality of service to other users as well. Therefore in 5G, the link adaptation is not an individual user problem anymore but a multi-user one (based on all the users within a cell served by a given base station), which relies upon the need for each mobile terminal to reliably evaluate its SINR and report back a correct corresponding CQI to the base station.

[1]Institute for Communication Systems, University of Surrey, Guildford, United Kingdom
[2]Czech Metrology Institute, Prague, Czech Republic
[3]5G and Future Communications Technology Group, National Physical Laboratory, Teddington, United Kingdom
[4]Purple Mountain Laboratories, Nanjing, P.R. China

This chapter first provides background information about the digital modulations used in fourth generation (4G) and 5G. It then defines the SINR and its relation to error vector magnitude (EVM) used as the means for a mobile terminal to predict its SINR where the sources of interference are unknown. Finally, it will discuss how mobile terminals are tested in an over-the-air (OTA) testing environment for purposes of conformance testing the estimation of SINR, which is an important aspect of standardisation within mobile terminals.

2.1 Definition of error vector magnitude and signal-to-interference-plus-noise ratio

Digital communications employ phase-shift keying (PSK) and quadrature amplitude modulation (QAM) schemes in order to send information over a radio frequency (RF) signal. The simplest form of such modulation is binary phase-shift keying (BPSK), which consists of two binary states: 0 and 1 [1]. An RF carrier wave sinusoidal signal can hold a phase shift of zero to correspond to state 0, while it changes to $\pm 180°$ phase shift to correspond to state 1. Therefore, on an Argand diagram or I and Q channel plot as it is known (corresponding to the real and imaginary axes respectively) state 0 would be at $+1$ on the I axis, while state 1 would be at -1 on the I axis. The two states correspond to one single digital bit.

A four-state, digital modulation scheme is commonly known as quadrature phase-shift keying (QPSK); it is illustrated in an I and Q channel plot on the left side of Figure 2.1. There are four states that are formed here based on two digital bits: 01, 10, 00 and 11. These four states can be made to arbitrarily correspond to four phase states: 45°, 135°, $-45°$ and $-135°$. Their corresponding I and Q channel coordinates are $(1/\sqrt{2}, 1/\sqrt{2})$, $(-1/\sqrt{2}, 1/\sqrt{2})$, $(1/\sqrt{2}, -1/\sqrt{2})$ and $(-1/\sqrt{2}, -1/\sqrt{2})$, respectively. It can be inferred that the four states in QPSK correspond to vectors, where the amplitude of each vector is constant. In the case of QAM schemes, the phase of the vector not only changes but also the amplitude. An example constellation of a 16-state QAM scheme is shown in the middle of Figure 2.1. A four-bit word is required to encode the 16-QAM vectors, each having

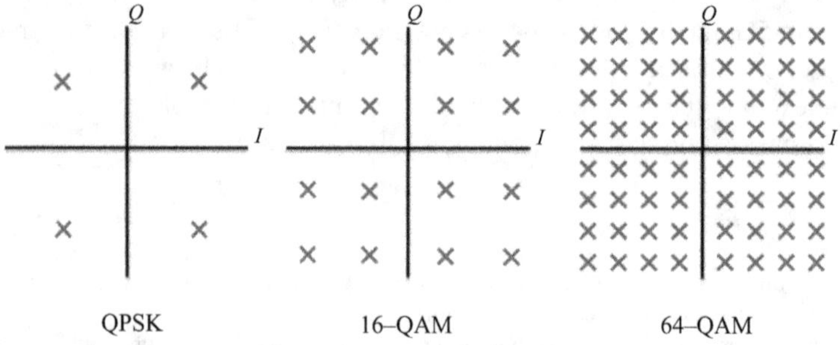

Figure 2.1 Illustration of QPSK, 16 and 64-state QAM constellations

a different combination of amplitude and phase. Other constellations could be achieved in 16-QAM by changing the position of the vector points corresponding to different states.

To carry more symbols in a constellation, modern communication systems use higher order modulation schemes such as 64-QAM, shown on the right side of Figure 2.1, encode six binary bits to achieve the 64 states. It is noteworthy that for an N-QAM scheme where $N=2^b$ and b is the number of bits. For even values of b the data can be Gray-coded to avoid multi-bit errors.

A modulated waveform will be effectively transmitted OTA by adjusting the amplitude and phase of the carrier wave and will be demodulated at the receiver. However, whilst transmitting OTA, the strength of the signal will decrease and at the same time additive white Gaussian noise (AWGN), dependent on the temperature and bandwidth of the environment, will be added at the receiver resulting in a receiver's signal-to-noise ratio (SNR). Where the SNR is weak, this will cause the additive noise to shift the transmitted vector to another position on the constellation plot. This change in Euclidian position constitutes the EVM and is illustrated in Figure 2.2. The bold dotted line on the I and Q channel plot indicates the intended vector, $\mathbf{v}_{\text{ref}}$, while due to the low SNR, the actual received vector is shown by the vector indicated in the dark line, $\mathbf{v}_{\text{received}}$. The change in Euclidian position is defined by the fine dotted line showing an error vector, $\mathbf{v}_{\text{error}}$, from which the EVM is derived. There is also a phase error angle between the two vectors ϕ_{EVM}. As depicted in Figure 2.1, it is clear that if the size of N increases, the Euclidian distance between two states gets smaller. If $|\mathbf{v}_{\text{error}}|$ becomes substantially large compared to the Euclidian separation, then higher order QAM modes will be more

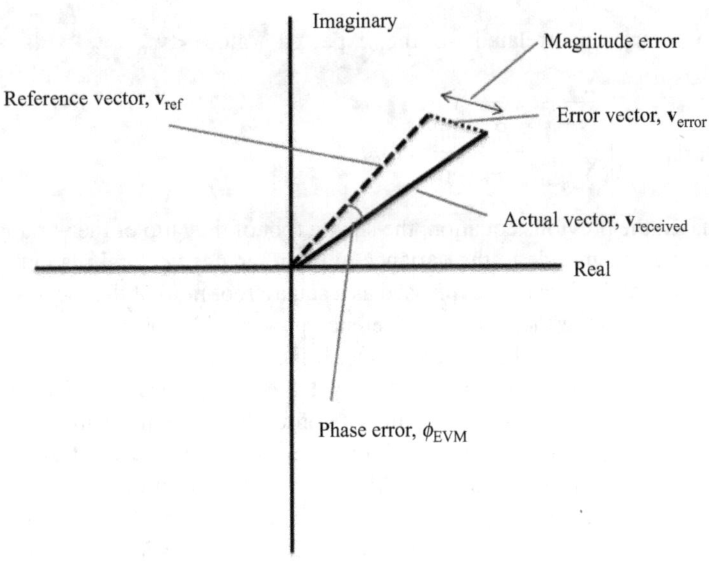

Figure 2.2 Illustration explaining the concept of error vector magnitude

vulnerable to detection errors, or symbol error, due to the EVM. Therefore, higher order QAM requires higher SNR to minimise symbol error.

$$\mathbf{v}_{\text{error}} = \mathbf{v}_{\text{ref}} - \mathbf{v}_{\text{received}}$$

$$\cos \phi_{\text{EVM}} = \frac{\mathbf{v}_{\text{ref}} \cdot \mathbf{v}_{\text{received}}}{|\mathbf{v}_{\text{ref}}||\mathbf{v}_{\text{received}}|}$$

Taking the vectors defined in Figure 2.2, the following well-known vector relations hold assuming that they are defined as vectors from the origin [2].

The EVM is defined by the proportion of $|\mathbf{v}_{\text{error}}|$ to $|\mathbf{v}_{\text{received}}|$ and is usually expressed as a percentage [3]:

$$\text{EVM}(\%) = \frac{|\mathbf{v}_{\text{error}}|}{|\mathbf{v}_{\text{ref}}|} \times 100 = \sqrt{\frac{|\mathbf{v}_{\text{error}}|^2}{|\mathbf{v}_{\text{ref}}|^2}} \times 100$$

Another understanding of EVM from the previous equation is that it can be found from a power error relative to the power of the transmitted vector. When using EVM to evaluate the average system performance, the instantaneous EVM of a single vector is not interesting. Rather it is of interest to determine the variance of the EVM over a given number of transmitted states or symbols, N_{Symbols}. Therefore, the root mean square (RMS) EVM is calculated as the standard deviation from comparing the transmitted and received vectors of N_{Symbols} in the following equation:

$$\text{EVM}_{\text{RMS}} = \sqrt{\frac{\sum_{n=1}^{N_{\text{Symbols}}} |\mathbf{v}_{\text{ref}}[n] - \mathbf{v}_{\text{received}}[n]|^2}{\sum_{n=1}^{N_{\text{Symbols}}} |\mathbf{v}_{\text{ref}}[n]|^2}} \times 100$$

$\mathbf{v}_{\text{error}}$ and $\mathbf{v}_{\text{ref}}$ are uncorrelated so the expected value $<\mathbf{v}_{\text{error}} \cdot \mathbf{v}_{\text{ref}}>$ is zero. Consequently,

$$\text{EVM}_{\text{RMS}} = \sqrt{\frac{\sum_{n=1}^{N_{\text{Symbols}}} P_{\text{error}}[n]}{\sum_{n=1}^{N_{\text{Symbols}}} P_{\text{ref}}[n]}} \times 100$$

Note that in the previous equation, the square root of the ratio of the variance in the error vectors' magnitude to the variance in the reference vectors' magnitude is calculated. Alternately, it can be expressed as a square root ratio of the mean square power error power to the mean square reference power. Either way they reach the same value and are expressed as a percentage. It is also worthy of noting that both the numerator and denominator are divided by $1/\sqrt{N_{\text{Symbols}}}$ because it is a square root ratio of variances but they cancel out. Therefore, the ratio also can be viewed as the ratio of total error power to the total reference power of N_{Symbols}. Calculating a reliable and repeatable value of the RMS EVM depends on both the value of N_{Symbols} but also the value of P_{ref} for each symbol, since for smaller values of P_{ref}, the relative magnitude of P_{error} will be larger and in vice versa. Higher order N-QAM schemes will require a higher value of N_{Symbols} as well as greater spread in P_{ref}. Further discussion on robust computation of RMS EVM can be found in [2].

The RMS EVM has a direct dependency on the SNR, which is defined as the ratio of the received power, P_{Rx}, to the noise power, N_0, of the AWGN at the receiver. The noise sources come from both thermal and shot noise effects [4] that are combined into defining a noise temperature from which the noise power can be found when the system bandwidth is known. The noise is Gaussian distributed such that it has a zero mean value and its power level is dependent on an increasing mean square or variance, σ^2. By assuming a Gaussian distribution, the SNR is related to the RMS EVM by the following equation [5]:

$$\text{EVM}_{\text{RMS}} \propto \frac{1}{\sqrt{\text{SNR}}}$$

The previous relation holds where it is assumed that the noise is purely AWGN and therefore the SNR can be readily predicted if necessary based on the RMS EVM at a mobile terminal based on receiving a known set of pilot symbols. Alternatively, just QPSK could be used and based on the low chance of symbol error, assume that the symbol is correct, yet the EVM could still be evaluated within each received symbol, which could reduce overhead. In a 5G communication system, it is necessary to predict the SINR at a mobile terminal and report a CQI to the base station based on it as opposed to a block error rate (BLER) that can be met as in 4G. This is due to the need to ensure that the mobile network is optimising its multiple downlink delivery reliably to all users with mMIMO pre-coding, while substantial co-channel interference could result if this was not sustained. In order for SINR to be predicted reliably from the EVM, it would require the interference having a Gaussian distribution and therefore the inverse square root proportionality between EVM and SINR would hold. The following equation defines the SINR by which the interference element is a sum of powers added to white Gaussian noise:

$$\text{SINR} = \frac{P_{\text{Rx}}}{\sum_{i=1}^{N_{\text{Int}}} P_{\text{Rx}|\ \text{Int}\,i} + N_0}$$

where $P_{\text{Rx}|\text{Int}\,i}$ is the received interference power from the source of interference i. Therefore, if the additive 'noise' created by each interferer has also a Gaussian distribution then the whole denominator will have a Gaussian distribution and therefore the SINR can be treated like the SNR and therefore predicted from the RMS EVM. Furthermore, the RMS EVM can be calculated at the receiver based on comparing the received vectors with the actual or expected vectors. It is also noteworthy that the maximum value of the SINR is the SNR so that the relation will hold where there is no interference present.

In order to ascertain if the interference is Gaussian, it is necessary to first identify the waveform and signalling that is used in 5G new radio. An orthogonal frequency division multiplexing (OFDM) [6] is used to transmit multiple carriers that overlap as shown in Figure 2.3 but are encoded such that they are orthogonal and do not interfere with each other. On each carrier, N-QAM data are sent and so the whole system provides substantial spectrum efficiency as otherwise the carriers would have to be separated by large guard bands if they were not orthogonally

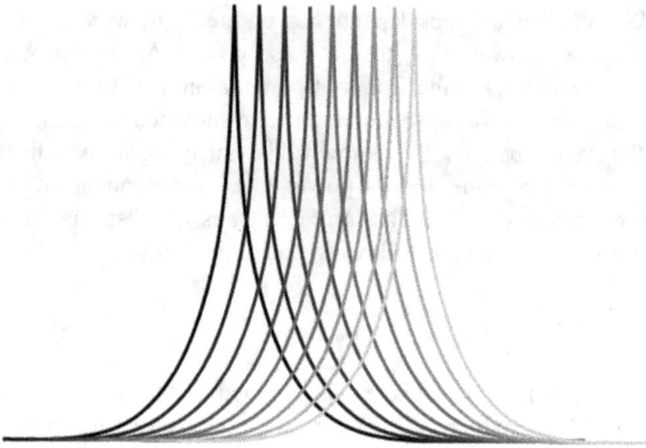

Figure 2.3 Illustration of multiple carriers in orthogonal frequency division multiplexing

encoded. One question therefore arises whether interference caused by a multi-carrier OFDM waveform employing *N*-QAM can be assumed to be Gaussian random noise. This will be discussed in Section 2.3 but first it is important to make a note of the impact of non-linearities on the SNR in this section while also the interference scenarios and link adaptation will be discussed in Section 2.2.

Up to this point, it has been assumed that the only cause of EVM is the AWGN at the receiver and that a high SNR or SINR will reduce it substantially. In a scenario with no interference, this would imply that if the transmit power at the desired base station is increased, then the SNR would also increase and therefore the RMS EVM would tend towards zero as the transmit power increases. This is not quite correct as the power amplifier in the base station has a finite capacity. Increasing the transmit power will drive the amplifier operating point into a non-linear regime, increasing the transmitted EVM. Whenever the OFDM signal operates in the non-linear region, the multiple carriers illustrated in Figure 2.3 will mix, or multiply together, which is known as intermodulation. The exact sampling of the OFDM waveform will only show the modulated subcarriers, but if the waveform is oversampled this extends the frequency range to show the out-of-band as well as in-band terms. The presence of non-zero out-of-band components is directly attributable to third-order and other odd-order distortion terms. When the carriers intermodulate with each other, they will produce new frequencies that are both in-band, which will be detected as increased EVM, and out-of-band of the OFDM signal, which may be detected from adjacent channels. Some of the new frequencies will be at the same frequencies as the carriers and cause interference to those carriers at the transmitter before the signal has even propagated OTA. Such interference will therefore cause a minimum RMS EVM that will eventually increase at the receiver, based on the SNR value. Therefore, based on this, the RMS EVM can be plotted against transmit power as shown in Figure 2.4(a) whereby as

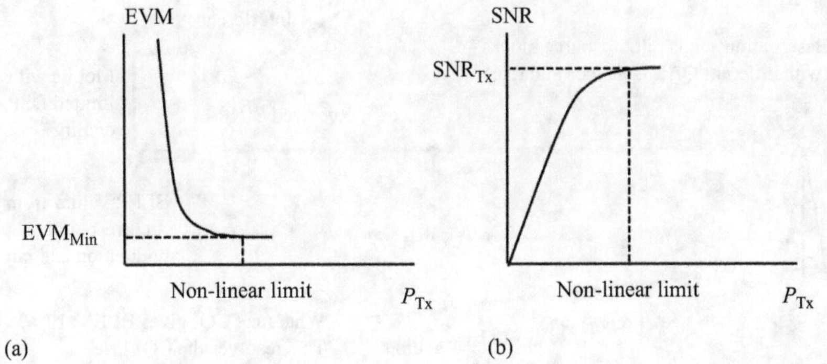

Figure 2.4 Illustration of (a) the received error vector magnitude reaching a minimum as the transmit power reaches the non-linear limit and (b) the corresponding maximum transmit signal-to-noise ratio

the transmit power, P_{Tx}, is increased, the EVM drops but as the non-linear limit point is reached a floor of minimum RMS EVM appears and at higher powers the power amplifier distortion will cause the EVM to increase (not shown in Figure 2.4(a)). At the same time in Figure 2.4(b), a corresponding SNR, which increases as the EVM decreases, reaches an effective saturation point and this could be termed the 'transmit SNR', which is an SNR limit at the receiver corresponding to the RMS EVM floor from the non-linear limit point. This transmit SNR limit is modelled in [7] by using a formula akin to two parallel resistors whereby the smaller resistor corresponds to the transmit SNR limit, SNR_{Tx} and the SNR at the Rx, SNR_{Rx} can increase infinitely but the overall resulting SNR will not go beyond SNR_{Tx}.

Like two resistors in parallel, the equation can be rearranged as follows:

$$\mathrm{SNR} = \frac{\mathrm{SNR_{Tx}SNR_{Rx}}}{\mathrm{SNR_{Tx}} + \mathrm{SNR_{Rx}}}$$

Noting that there is an SNR limit, the prediction of the SNR or SINR at the receiver based on the EVM will need to take into account any transmitter non-linearity. This can, however, be accounted for by ensuring that a suitable back-off is required such that when sending pilot signals, the peaks of the OFDM signals sent will then not reach the non-linear region or otherwise have minimum impact by which only the received noise and interference will be the dominant cause of EVM.

2.2 Link adaptation techniques and channel quality indicator

Link adaptation techniques are used in 4G communications to determine the highest order modulation scheme and code rate that can be used to send data to a mobile terminal before the communication link fails [8]. The link adaptation procedure is illustrated in Figure 2.5 where the base station will send out pilot resource

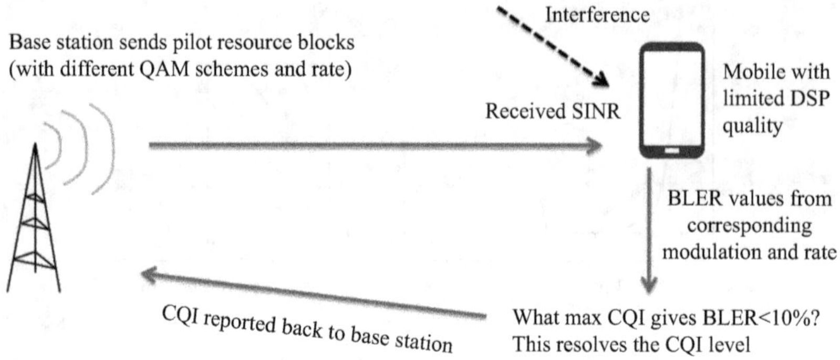

Figure 2.5 Diagram showing the process by which the channel quality indicator is calculated and reported in a 4G system

blocks to the mobile. A resource block is defined as a block of OFDM transmissions sent within a given time frame and over a specified frequency bandwidth. Initially the simplest modulation scheme used in 4G, being that of QPSK with the slowest code rate, is first sent to the mobile terminal which, based on its limited digital signal processing (DSP) quality, will evaluate a BLER for the resource blocks sent. The BLER itself will be dependent on the DSP quality, the strength of the signal arriving at the mobile and any interference that may be present. This procedure is repeated such that the next sets of resource blocks sent will have increasing code rates and also will move to higher order QAM schemes, including 16-QAM and 64-QAM. Depending on the quality of the channel, there will come a point where the BLER will increase above a threshold of 10% at which point it can be deemed that the lower level modulation scheme and code rate will have to be used as the best scheme that will not fail. The mobile terminal will return a CQI level and this can range anything from 1 to 15, or zero if the link is completely failed. When 4G evaluates the CQI, it will be implicitly due to a received SINR as well as the DSP quality of the mobile terminal. There is no need for the mobile terminal to reliably predict the SINR in 4G as link adaptation is formed based on what modulation and coding scheme the mobile terminal can receive data below a BLER threshold. Furthermore, the mobile terminal need not report an accurate CQI, whereby it could report a lower one than necessary, which would only be a disadvantage to the user as they will subsequently receive a slower data rate. In 5G however, there is a requirement to predict SINR well and report an accurate CQI for the benefit of all users within the cell as will be studied with regard to mMIMO.

The importance of a reliable SINR prediction and CQI reporting in 5G is due to the deployment/usage of mMIMO base stations. mMIMO systems have (>100) antenna elements so it is impractical to hold channel state information at the mobile terminal for so many antennas. Rather each mobile terminal will need to send pilot signals to the base station as illustrated in Figure 2.6 where the base station will be able to carry out pilot channel estimation for each antenna element from each mobile terminal. It is possible that pilot channels can also get contaminated, by for

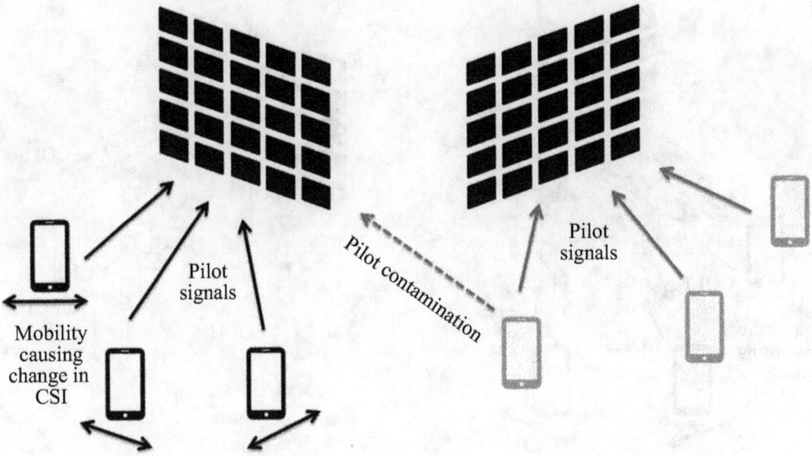

Figure 2.6 Illustration of pilot signals being sent from the served mobile terminals to a massive MIMO base station (left) and the pilot contamination caused by pilot signals going to the base station in the neighbouring cell (right)

instance pilot signals going to the base station in the neighbouring cell (see Figure 2.6), which will cause error in the pilot channel estimation and this may eventually degrade the SINR of the signal that is then sent to the mobile terminal.

Once the pilot channels are estimated, the base station will apply a pre-coder whereby a phase shift and amplitude weighting is applied to each antenna element at the mMIMO base station and it will form beams as illustrated in Figure 2.7 through an adaptive antenna array [1]. Adaptive antennas are able to radiate a beam in a given direction but also at the same time not radiate the beam or any side lobes as shown towards other mobile terminals so that it will help to suppress any interference to other mobile terminals. Such beamforming is also necessary to minimise any interference from one cell to the other.

Once the downlink beams are formed to each mobile terminal simultaneously and within the same frequency band, it is then necessary for each mobile terminal in Figure 2.7 to predict its SINR and then return a CQI to the base station which is found from a pre-specified range that the SINR falls within. Each mobile terminal must report an accurate CQI to the base station as it has to know that each mobile terminal has a sufficient SINR. If a mobile terminal wrongly reports a CQI that it is substantially lower than it actually is, then it will cause the base station to unnecessarily update the pilot channel estimations from each terminal and put more load on the network. Similarly, a mobile terminal may report a CQI that is higher than the actual, whereby it may cause the network not to update the pilot channel estimations regularly enough, thus affecting other users. A final point to note in Figure 2.7 is that the mobile terminals have mobility and this will dictate that there is a minimum requirement to update the pilot channel estimations based on the velocity at which the mobile terminals are moving, which will cause a subsequent change in channel state after a given time. This will be additional to the problems

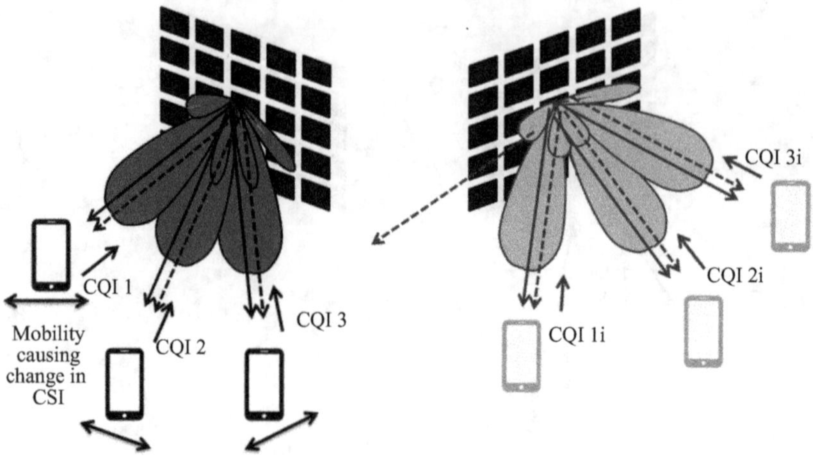

Figure 2.7 Illustration of CQI being reported from the served mobile terminals to a massive MIMO base station with pre-coded downlink channels (left) and interference caused by pre-coded channels from the neighbouring cell (right)

caused by pilot channel estimation which may cause the mobile terminals to report a weak SINR, thus requiring another round of pilot channel estimation.

Though mMIMO is the main use case of link adaptation in 5G cellular communications, it will have uses in other areas of the mobile network, including densified base stations, whereby it can be helpful if resource blocks have varying SINR across the band that is in use so that appropriate resource blocks can be chosen. Smaller densified cells are more vulnerable to interference given the relatively smaller reuse distance that will exist between cells. Another potential usage would be in millimetre-wave (mm-wave) small cell access such that the highly directive pencil beam from the base station may be weakly tracking the mobile terminal from which a CQI can indicate if it has been sufficiently updated.

2.3 Characterising Gaussian noise and interference

Having established the need to reliably predict SINR in 5G by using its relation to EVM, one factor now remains with regard to whether the interference caused by OFDM resource blocks from interferers can be assumed to be equivalent to Gaussian noise. This mainly depends upon whether the modulation scheme is complex enough to create sufficiently random signalling while there is also the need to de-correlate the signal and interference.

From [9], it is shown that with a long enough OFDM set of resource blocks (i.e. a sufficient number of transmitted frames in time and wide enough bandwidth or number of sub carriers) and a high enough QAM order, a signal that has Gaussian noise characteristics can be formed. Analyses of the probability density of simulated

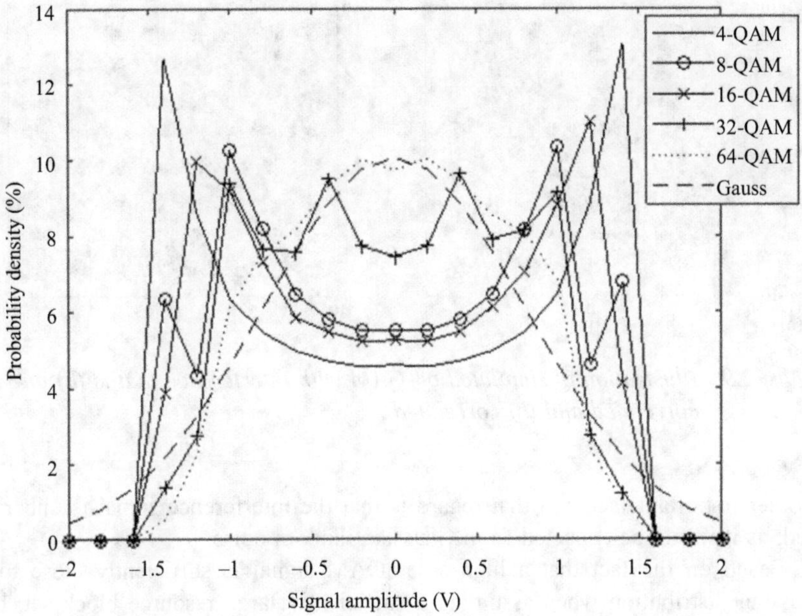

Figure 2.8 Comparison of Gaussian noise with histogram plots of signalling of different orders of QAM, which has a weak comparison for low-order 4-QAM but a strong comparison for 32-QAM and higher

4-QAM (which can be considered equivalent to QPSK in this analysis) through to 64-QAM schemes when using 20 frames of a 20-MHz OFDM with 1,200 orthogonal carriers are plotted in Figure 2.8, where the noise statistics are compared against a Gaussian theoretical curve. It is interesting to see that rising towards 32-QAM, the histogram becomes closer to that of the theoretical Gaussian curve because there are substantially more random changes in the phase and amplitude of the carrier that it becomes more like Gaussian noise. Going down to 4-QAM, it becomes more like a standard carrier signal that has a peak amplitude at fixed positive and negative positions. For higher order QAM, it is worthy of noting that in some instances, prior work has performed more finer detailed models using the α-stable distribution [10,11], which, unlike Gaussian distribution, can better model the two tails at the end of the 'bell', which therefore will enable a random process to better model the smaller components of the interference as noise. The more significant components are still appropriate to model as Gaussian, which can simplify the modelling of interference when the interfering source is an OFDM signal employing 32-QAM and above.

Despite the possibility to use the α-stable distribution, it can still be seen in Figure 2.8 that the main part of the probability density function is fitting well to the Gaussian curve for the 64-QAM. Ultimately it is important to determine if the EVM that is caused by such OFDM signals as interference does reflect that of Gaussian noise and, in turn, if the SINR becomes equivalent to a reduced SNR.

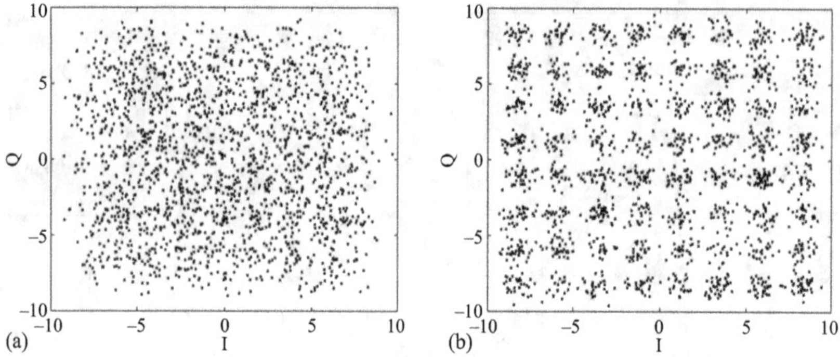

Figure 2.9 Illustration of simulated 64-QAM with interference that is (a) uncorrelated and (b) correlated

Another important factor in that regard is that the interference signal is uncorrelated, as it would be expected from Gaussian random noise.

Based on the fact that a high-order QAM signal is sufficiently close to a Gaussian distribution when assuming a sufficiently large resource block set, the correlation is evaluated by observing the characteristics of the signal with the addition of interference on a constellation diagram. A simulated *I* and *Q* channel constellation is plotted in Figure 2.9(a) and (b) for two experimental scenarios with an SINR of 15 dB, where the desired and interfering signals are de-correlated and correlated, respectively. Clearly in both cases their EVM statistics are going to be somewhat different since in the first case the EVM is much like what would be expected if the interference was Gaussian, while in the second case the interference is having substantially less effect. In order for correlated interference to occur, it will depend on if the desired and interfering signals have very comparable data being transmitted in downlink, along with very comparable channel state and time delay. All these factors for high-order QAM are unlikely to happen since mobile terminals are physically separated in most cases, i.e. a movement of more than half a wavelength is enough to cause channel state de-correlation, such that an instance of correlated interference is likely to be very rare.

2.4 Using error vector magnitude to predict signal-to-interference-plus-noise ratio

The RMS EVM has already been established as a strong candidate to predict SINR, where the sources of interference are not known, in other works with an SINR prediction error of less than 1 dB [12,13]. This is acceptable for returning a CQI value since each CQI value usually corresponds to an SINR range that is larger than 2–3 dB from the lowest to the highest. Therefore, any CQI inaccuracy is only likely to occur where the predicted SINR is close to the boundary, though the data rate would not be substantially decreased if a lower CQI level was chosen to avoid

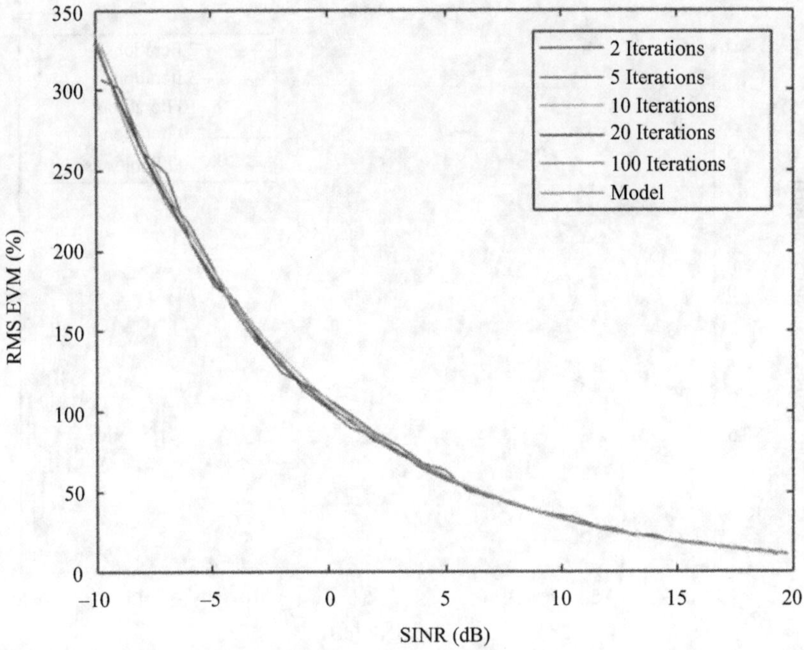

Figure 2.10 Plot of the RMS EVM against SINR when evaluated over a number of iterations of OFDM bursts (Source [15])

uncertainty in choosing the higher CQI. Other techniques have also been evaluated to predict SINR such as moment-based estimators in [14], though such a prediction method is likely to require a substantial number of iterations to arrive at an accurate prediction for this application due to the computational complexity.

Using EVM requires the need to send a number of different OFDM frames within a set of resource blocks. The RMS EVM is plotted for simulated OFDM data in Figure 2.10 by using a 20-MHz OFDM signal with 1,200 orthogonal carriers when employing 64-QAM on each carrier. The interferer also has the same OFDM configuration with 64-QAM and is generated with an independent set of random data in order to be de-correlated. As the SINR can also be directly computed in simulation, then the RMS EVM can be plotted directly against the actual SINR. This simulation was repeated when using 2, 5, 10, 20 and 100 iterations, of OFDM frames in order to evaluate the RMS EVM. Using more iterations inevitably means that a larger number of random vectors were transmitted from which the EVM was computed.

The plotted values of RMS EVM in Figure 2.10 all show good agreement with the model of RMS EVM against SINR. However, for low values of SINR, some error can be seen to be present with a small number of iterations particularly with SINR values below 0 dB. The level of error compared to the model is evaluated in Figure 2.11 where such deviations can be seen more clearly. For 20 iterations or more, the prediction error reduces to less than ±0.5 dB and this is likewise

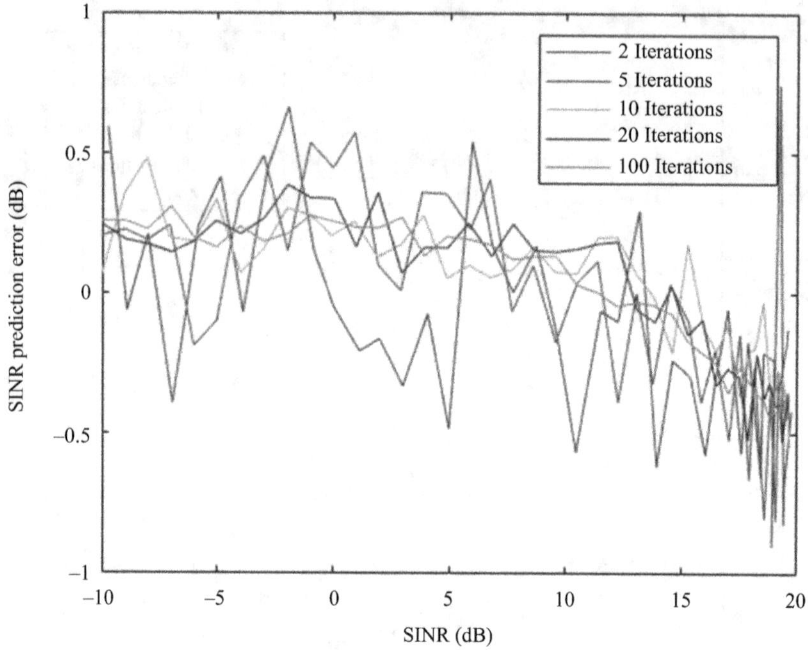

Figure 2.11 Plot of the RMS EVM error against SINR when evaluated over a number of iterations of OFDM bursts (Source [15])

acceptable for reliable prediction of CQI while the use of 20 iterations can expect to take in the order of just 200 ms to transmit OTA.

The final evaluation that needs to be carried out according to [15] is to determine the reliability of the SINR prediction within a real propagation environment. The designed signal will be subject to path loss as well as rapid fluctuations known as fast fading as the mobile terminal moves [1]. Similarly, the interferer will also be subject to path loss and fast fading. However, what is important here is that over a given bandwidth, the fading at each frequency will be substantially different and as such the channel is defined by a coherence bandwidth, i.e. the frequency span over which the fading is highly correlated. Beyond this coherence bandwidth, the fading is considered to be substantially different, i.e. de-correlated. An OFDM signal may have a bandwidth that is narrower or wider than the coherence bandwidth which will have substantial effect on the SINR.

The test procedure was carried out in [15] with an mMIMO base station of 32 elements communicating with three mobile terminals each with a single antenna. However, the first two antennas were stationary, while the third antenna (acting as the desired link) was moving and the SINR was calculated for that link by using the following steps:

1. Take the measured 32×3 MIMO channel from the base station to the three mobile terminals Rx1 to Rx3 as shown in Figure 2.12. It is assumed that perfect

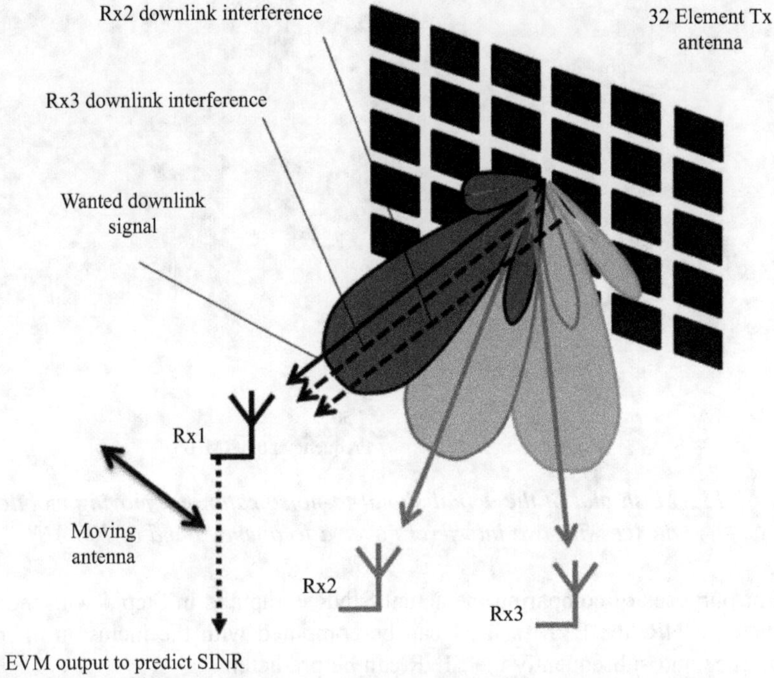

Figure 2.12 Diagram showing the calculation of SINR with signalling and the corresponding error vector magnitude used for the prediction of SINR

channel state information is known such that the actual channel measurements were therefore taken. The channel state is measured as Rx1 moves (where both Rx2 and Rx3 are stationary) by using a channel sounder. Note that in a real mobile network, this would be achieved by using pilot channels.

2. By applying a pre-coding technique known as zero forcing [1], the adaptive array beam can be formed towards the intended user by using the pre-coder vector applied to the 32 Tx elements. First of all, a pre-coder is applied such that the beam from the Tx to Rx1 forms a desired pre-coded link. This pre-coder is set for each data sample as Rx1 moves such that the maximum possible SINR in the desired downlink is obtained.
3. The maximum SINR of Rx1 depends also on the pre-coded downlinks to Rx2 and Rx3. These pre-coders can in turn form an Rx2 and Rx3 downlink interference (inter-beam interference), respectively, arriving at Rx1. At the same time, they are designed to maximise the SINR to Rx2 and Rx3, but for this work only, the SINR at Rx1 is considered.
4. A simulated OFDM signal is transmitted down the desired downlink to Rx1, while two independent OFDM signals are sent over the Rx2 and Rx3 interfering downlink and therefore the signal and interfering power levels can be computed such that the SINR with OFDM signalling is found. The AWGN is also taken into account here with an assumed SNR of 20 dB.

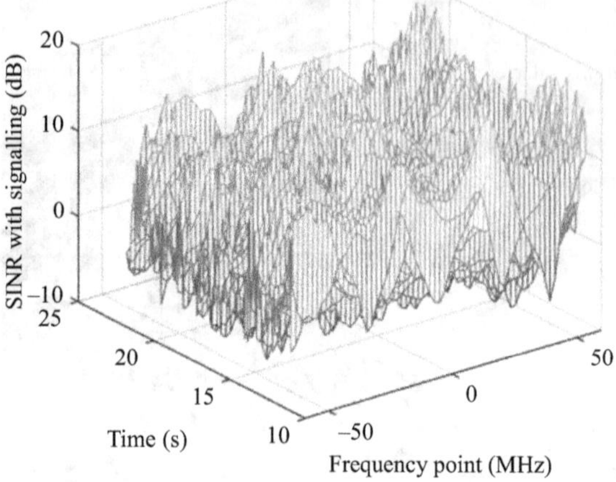

Figure 2.13 Mesh plot of the actual signal-to-noise ratio of a moving mobile device with two interferers over a frequency band of 100 MHz

5. For purposes of comparing the actual SINR computed in Step 4 with the predicted SINR, the EVM at Rx1 can be computed with the inclusion of interference and subsequently the SINR can be predicted.

In the computational steps to compute the SINR with signalling, it should be noted that the OFDM signal has a bandwidth of 20 MHz, and 20 iterations of frames are sent when using 64-QAM modulation over 1,200 orthogonal carriers.

The computed SINRs using OFDM signalling (in Step 4) are displayed in the mesh plot in Figure 2.13. For every frequency point sample space in time, an SINR value is computed. It can be clearly seen that even at one single time sample, the SINR fluctuates substantially since the channel characteristics of the desired and interfering signal both fluctuate substantially over frequency but are very independent in how they do so, thus resulting in such changes in SINR. The same happens for different time instances at the same frequency where the fading fluctuates substantially as the mobile terminal moves. It should be noted that in this measurement, the SINR does drop down to substantially low levels as low as −5 dB while it does not increase far beyond 15 dB. This is because the distance of the desired moving mobile terminal from the base station is relatively short and in a line of sight. In reality, this would not normally take place and as such the interference would be somewhat lower. However, for test purposes, this set-up helps to show a rapid change in SINR to help validate the prediction method.

For each data point where SINR with signalling is evaluated in Figure 2.13, the EVM was also calculated in Step 5 as both the transmit and receive vectors were known. Then, the EVM could be plotted in a mesh plot instead of SNR, though such a plot would lack clarity as RMS EVM reaches high values for low SINR. Therefore, the predicted SINR can be formed in a mesh plot, which can be resolved from the calculated RMS EVM. Taking the actual SINR with signalling and the

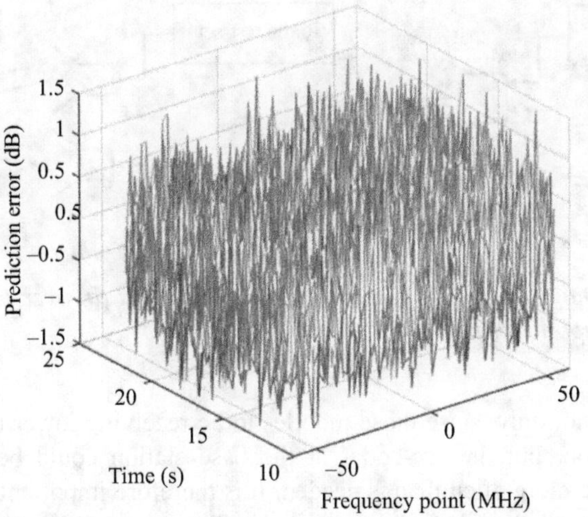

Figure 2.14 Mesh plot of the prediction error of signal-to-noise ratio of a moving mobile device with two interferers over a frequency band of 100 MHz

predicted SINR, it is then possible to compute the prediction error in decibels by taking the direct difference between the two, as it is plotted in Figure 2.14. As it can be seen over the whole measured frequency band and over all time-samples, the maximum prediction error does not exceed 1.5 dB. However, the mean error is found to be just under 1 dB. This is acceptable for reporting a reliable CQI. Reducing the OFDM bandwidth, and subsequently the number of orthogonal carriers, reduces the prediction error, though 20 MHz is typically the bandwidth that is adopted in current standards, while future standards with larger bandwidths may require sub-bands within which to predict the varying SINR over the whole band.

2.5 Application of interference in over the air testing

Having established that the SINR of a mobile terminal can be accurately evaluated via the EVM method, it is then necessary that any mobile terminal sold will meet a standard of compliance. This particular compliance will be that the mobile terminal must reliably show that it can produce an EVM within a given range for a given SINR. Failure to do so due to any signal-processing defects or otherwise within the mobile terminal may cause the predicted SINR to be substantially inaccurate and therefore return an inaccurate CQI. Returning an inaccurate CQI would not only cause problems for the mobile terminal itself, but it could potentially disrupt the mobile network. In the context of mMIMO, a CQI higher than the actual level would not only cause a modulation and coding scheme to be sent to the mobile with a high BLER, but also it may cause the base station to believe that its pre-coder does not need to update especially if several mobile users are wrongly reporting a high CQI. Similarly if one or more mobile terminals are reporting a wrongly low

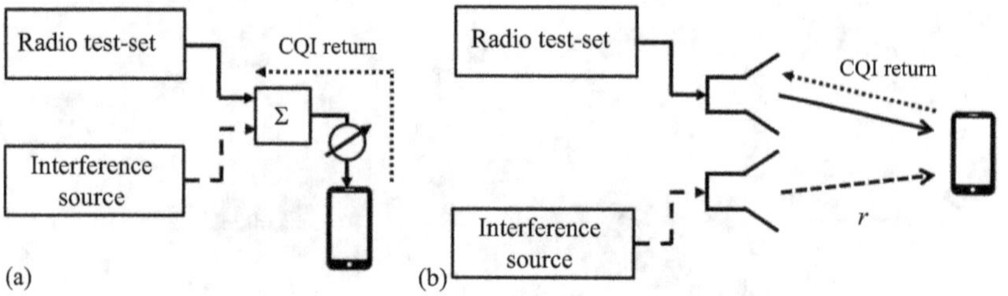

Figure 2.15 Diagram of a test set-up for (a) cabled and (b) over the air tests of a mobile terminal

CQI, they will not only cause those mobiles to be receiving lower rates and lower order modulation, but the pre-coder at the base station could be unnecessarily forced to update more often than is needed. It is therefore important for the mobile network to know that each mobile terminal will report a CQI within an allowed level of uncertainty.

We have shown that OTA testing has the capability to create a radio test environment, where the signal and interference are controlled so that the system can identify the expected SINR at the mobile terminal. The result will depend on the mobile terminal's hardware and implement of the standard to estimate the EVM of the known sample sequence, predict the SINR and report the correct CQI. The test set-up can progressively increase or decrease the SINR so that it will sweep across the range of SINR values and determine where CQI boundaries are crossed. There are two ways principally in which a test environment can be made for a mobile terminal. The first is in Figure 2.15(a) where it is a cabled test. A radio test-set, which will emulate the base station as well as its downlink signal, is connected to one input of a combiner and a source of interference is connected to the other. The interference can either be a Gaussian noise source if the interference is to be assumed as such, or an emulated OFDM signal with the appropriate bandwidth and number of carriers. It is important that the interference source is calibrated against the signal from the radio test-set to ensure that a reliably accurate SINR is delivered out of the combiner. Beyond this point, an attenuator may be used before connecting to the mobile terminal in order to ensure that the signal is attenuated to a level that the mobile can receive. With this set-up the radio test-set can increase its signal power in steps of 1 dB from the lowest level such that the SINR changes in such steps. At each step, the mobile will return a CQI level back through the attenuator and the splitter (which was the combiner in the downlink) to the radio test-set. It should be noted that the signal reporting the CQI in uplink will be attenuated by 3 dB, but this can be managed and the other half of the signal going to the interference source can be discarded.

Another important point in the test set-up is that it is necessary to know the noise temperature of the test environment as well as the noise figure of the receiver. These two physical parameters along with the system bandwidth are important in order to determine where the noise floor is and therefore a full SINR can be determined.

The cabled method allows the RF transceiver to be tested directly and all other interference sources are removed. However, in most practical cases for testing mobile terminals, especially in terms of testing an end product, it is not as desirable an option since the antennas cannot be easily removed in order to connect it to such a test set-up. Therefore, a more favourable option is to do so with an OTA set-up as shown in Figure 2.15(b).

The OTA set-up would conduct a test acquisition in much the same way as that of the cabled set-up. Clearly, the difference is that the radio test-set and the interference source are each connected to two directive antennas. The set-up and position of the two antennas would have to be in a shielded and anechoic environment. This will ensure that the downlink has an SINR incident on the mobile terminal, whereby the signal relative to the interference is determined by the ratio of electromagnetic power density incident on the mobile terminal. This electromagnetic power density has to be converted to a received power at the output flange of the mobile terminal antenna. This is estimated by using the effective area of the receiver antenna at the angle at which the signal and interference arrive. It is assumed that the angular separation between the signal and the interference is negligible if the distance r shown in Figure 2.15(b) is large enough and the separation between the two horn antennas is small. Therefore, in such a case, they are assumed to arrive at the same angle and there will be no difference in the effective area in the two cases. Measuring the effective area of the antenna directly is not possible, but it is possible that a link test can be first carried out by the radio test-set in order to determine the gain and the effective area of the mobile terminal based on sending a known signal and knowing the distance r, and using a received signal received power reported back by the terminal. From knowing this as well as the noise figure and ambient noise temperature, the SINR within the mobile terminal can be determined. Therefore, though the OTA has the advantage of not requiring direct connection to the terminal, it does require a fully calibrated test environment from which an accurate test SINR can be assumed.

Another important factor with OTA evaluation of SINR and reliability for CQI return is the impact of the downlink interference on the uplink and in vice versa. Though the downlink and uplink are normally separated in frequency by several MHz, it is still possible that the downlink can cause substantial interference on the uplink and vice versa when transmitting at the highest power levels. This is illustrated by results given in Figure 2.16 from work undertaken in [16]. A test set-up was carried out similar to that in Figure 2.15(b) though in this instance, the source of interference was fixed and as such the transmit power was adjusted in 1-dB steps from 12 dB below the interference through to 17 dB above the interference. Therefore, the SINR ranged from -12 up to 17 dB. A probe antenna was also used next to the mobile terminal in order to measure the incident signal and interference power levels. The measurement uncertainty for $k = 2$ (i.e. 95% confidence interval) has also been assessed and plotted as error bar as shown in Figure 2.16. It would be expected that for SINR below 0 dB, the downlink throughput would be weak and indeed as shown in Figure 2.16(a), it drops to 0 Mbps and in the region around 0 dB. Above 0 dB, the expected downlink throughput of over 50 Mbps is delivered.

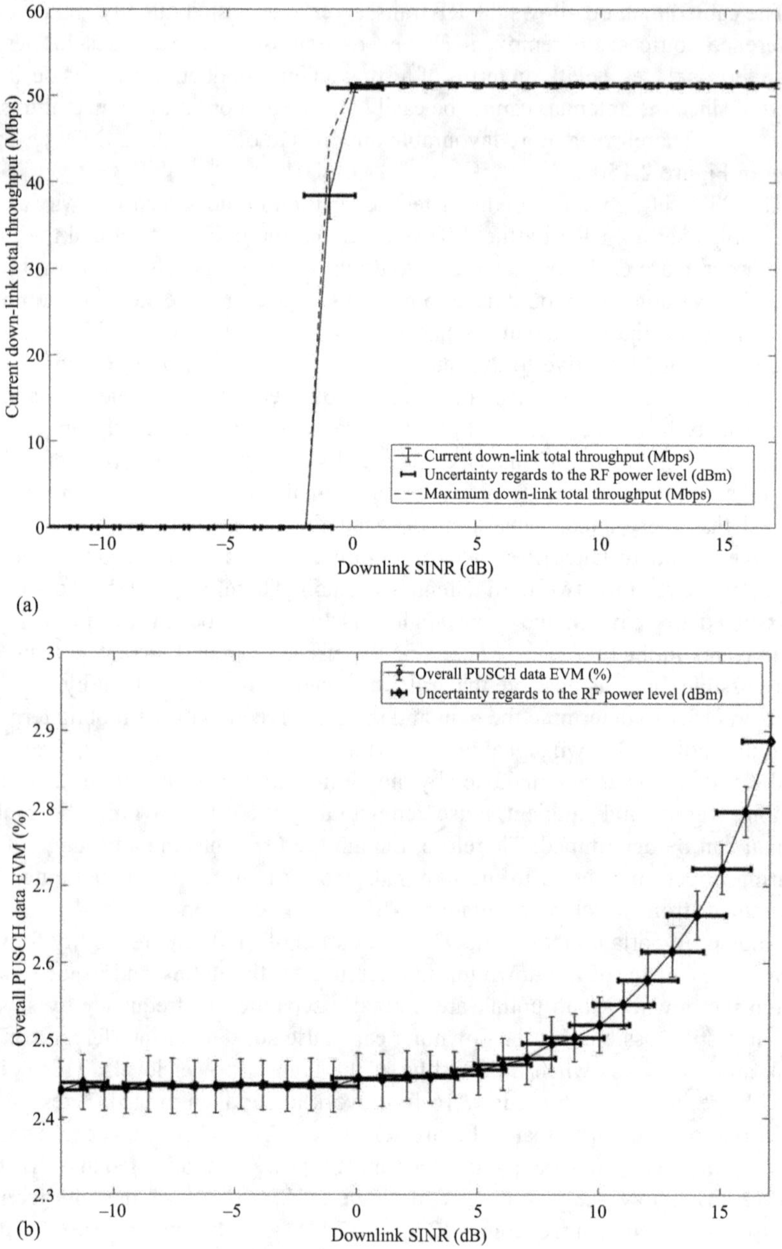

Figure 2.16 OTA test results for (a) downlink throughput vs SINR and (b) uplink EVM vs downlink SINR

However, at the same time the EVM in the uplink is also measured using the radio test-set, which has such capability. It would be expected that as the interfering signal is at a different frequency to the downlink (where interference is also present) that the uplink EVM would be at a constant low as the uplink SNR is constant. However, as seen in Figure 2.16(b), the uplink EVM begins to increase from around 2.4% to 2.9% as the downlink SINR rises above 0 dB. Though this is a small change in EVM, it clearly shows that as the downlink power is increasing, the uplink receiver in the same location as the downlink transmitter is being subjected to interference. Though the downlink frequency is at 2.14 GHz with only 20-MHz bandwidth in this case, while the uplink is at 1.95 GHz, the waveform for the downlink will produce spurious frequencies out of band and there will be potential effects from non-linearity of the transmitter. The isolation in the duplexer is also not sufficiently strong enough to prevent such interference when there are high transmit power levels, while also the out of band attenuation on the filter at the transmitter in the downlink will be limited. A final point to note in Figure 2.16(a) and (b) is that error bars are provided for SINR and EVM measurements which will carry uncertainty from the signalling used both in the desired signal and unwanted interference. Nonetheless the uncertainty in both cases is sufficiently low that SINR can be measured to within ±1-dB accuracy and EVM is within ±0.1%.

Having seen from experimental evidence of an OTA test that high levels of downlink SINR can have an effect on the measurement of EVM in uplink, and subsequently the prediction of SINR, the same issue can occur with regard to high-power levels in uplink impacting the measurement if EVM in downlink. Such errors in EVM could cause inaccuracies in measured SINR in the order of more than 3 dB especially where SINR levels in downlink exceed 20 dB corresponding to downlink EVM below 2%. To avoid this problem, it is possible that time domain duplexing can be used in link adaptation such that the uplink and downlink are given time slots. This will subsequently remove any such interferences. Alternatively, as the coupling level in a device is known for a given uplink transmit power, a calibration correction could be applied to the SINR prediction in downlink, where its effectiveness would be evaluated in an OTA conformance test.

2.6 Summary

Link adaptation in 4G communications is designed to determine the best modulation and coding scheme that a mobile terminal device can handle based on remaining below a threshold BLER. The ability of the device to handle such a scheme and corresponding data rate will depend on the quality of the wireless link leading to an SNR but also the quality of the DSP in the mobile and any co-channel interference. There is no requirement for the mobile terminal to give any measure of the SINR in order to return a CQI to the base station. Within 5G communications, including mMIMO, it is necessary for the mobile terminal to evaluate a reliable prediction of the SINR from which it can return an accurate CQI to the base station such that the base station can adequately serve all users at a given time instant. Inaccurate reporting could compromise the level of quality of service within the mobile network. An evaluation of the ability to predict SINR based on EVM has been presented in this chapter and also conformance testing of mobile terminals OTA has also been discussed. This will be invaluable for testing the standard of mobile terminals at present and beyond for applications requiring reliable prediction of SINR. In addition to mMIMO, this would be needed in full duplex and potentially small-cell-densified deployments which are both subject to substantial interference.

References

[1] Saunders S. R. and Aragón Zavala A. *Antennas and Propagation for Wireless Communication Systems*. Chichester, UK: Wiley, 2007.

[2] Jensen T. L. and Larsen T. 'Robust Computation of Error Vector Magnitude for Wireless Standards'. *IEEE Transactions on Communications*. 2013, 61 (2), pp. 648–657.

[3] Carvalho N. B. and Schreurs D. *Microwave and Wireless Measurement Techniques*. Cambridge, UK: Cambridge University Press, 2014.

[4] Poole C. and Darwazeh I. *Microwave Active Circuit Analysis and Design*. Amsterdam, Holland: Academic Press, 2015.

[5] Mahmoud H. A. and Arslan H. 'Error Vector Magnitude to SNR Conversion for Nondata-Aided Receivers'. *IEEE Transactions on Wireless Communications*. 2009, 8(5), pp. 2694–2704.

[6] Molisch A. F. *Wireless Communication Systems*. 2nd Edition, Chichester, UK: Wiley-IEEE Press, 2010.

[7] Tuovinen T., Tervo N., and Pärssinen A. 'Analyzing 5G RF System Performance and Relation to Link Budget for Directive MIMO'. *IEEE Transactions on Antennas and Propagation*. 2017, 65(12), pp. 6636–6645.

[8] Sesia S., Baker M., and Toufik I. *LTE—The UMTS Long Term Evolution, From Theory to Practice*. 2nd Edition, Chichester, UK: John Wiley & Sons, 2011.

[9] Feng C., Cui H., Ma M., and Jiao B. 'On Statistical Properties of Co-channel Interference in OFDM Systems'. *IEEE Communications Letters*. 2013, 17 (12), pp. 2328–2331.

[10] Cardieri P. 'Modeling Interference in Wireless Ad Hoc Networks'. *IEEE Communications Surveys and Tutorials*. 2010, 12(4), pp. 551–572.

[11] Middleton D. 'Non-Gaussian Noise Models in Signal Processing for Telecommunications: New Methods an Results for Class A and Class B Noise Models'. *IEEE Transactions on Information Theory*. 1999, 45(4), pp. 1129–1149.

[12] Hasan W. B., Doufexi A., Oikonomou G., and Beach M. 'EVM prediction for massive MIMO'. *The 30th IEEE Annual International Symposium on Personal, Indoor and Mobile Radio Communications (PIMRC 2019)*, September 2019, pp. 1–7.

[13] Hasan W. B., Harris P., Bromell H., Doufexi A., and Beach M. 'Adaptive User Grouping Based on EVM Prediction for Efficient & Robust Massive MIMO in TDD'. *IEEE Access*. 2019, 7, pp. 162683–162696.

[14] Chen Y. and Beaulieu N. C. 'NDA Estimation of SINR for QAM Signals'. *IEEE Communications Letters*. 2005, 9(8), pp. 688–690.
[15] Brown T. W. C., Humphreys D. A., Hudlicka M., and Loh T. H. 'Prediction of SINR using BER and EVM for massive MIMO applications'. *The 12th European Conference on Antennas and Propagation (EuCAP 2018)*, April 2018, London, UK, pp. 1–5.
[16] Wang M., Loh T. H., Cheadle D., Zhao Y., and Zhou Y. 'A rigorous link performance and measurement uncertainty assessment for MIMO OTA characterisation'. *2018 Loughborough Antennas & Propagation Conference (LAPC 2018)*, November 2018, pp. 1–6.

Part II

Device metrology

Chapter 3

Non-linear measurements for 5G devices and the associated uncertainties

Martin Salter[1], *Dilbagh Singh*[1] *and Laurence Stant*[2]

3.1 Introduction

Technological advancements for 5G New Radio present new challenges not just for digital systems and software, but also for the analogue radio frequency (RF) hardware involved. Increased bandwidths to accommodate higher data rates have raised the operating frequencies of spectrum channels, which, due to absorption and other frequency-dependent channel effects, reduce the usable distance between the base stations and user equipment. To achieve suitable coverage, the total number of base stations must therefore increase, potentially causing significant environmental impact unless the power consumption is sustainably sourced. A key opportunity to improve this situation is to increase the efficiency of base station and user equipment power amplifiers, which boost the RF signal before it is transmitted by the antennas.

This amplifier, of which at its heart is a power transistor as shown in Figure 3.1, is typically the most inefficient component in base stations and in handsets rivals the display and processor for consuming battery charge. For an amplifier the typical efficiency to compare against is the power-added efficiency (PAE), which considers the amount of DC power required for RF power gained. The PAE can be greatly increased if the device is operated in a non-linear regime, as opposed to a traditional linear regime where it is theoretically limited to 50 per cent or less. Whereas during linear operation the output of the amplifier is purely a scaled representation of the input signal, during non-linear operation the output amplitude becomes compressed (shown in Figure 3.2) and the spectrum will contain undesired signals at harmonics of the input frequency. The time-domain result of this effect, of interest to communications engineers, is that distortion is introduced to the signal which can increase the error vector magnitude and bit error rate. The strength of this distortion is dependent on the degree of nonlinearity of the amplifier. The challenge is then for communications engineers to develop

[1]National Physical Laboratory, Teddington, United Kingdom
[2]Diamond Light Source, Oxford, United Kingdom

Figure 3.1 A typical packaged RF power transistor (active device) as used in cellular base stations

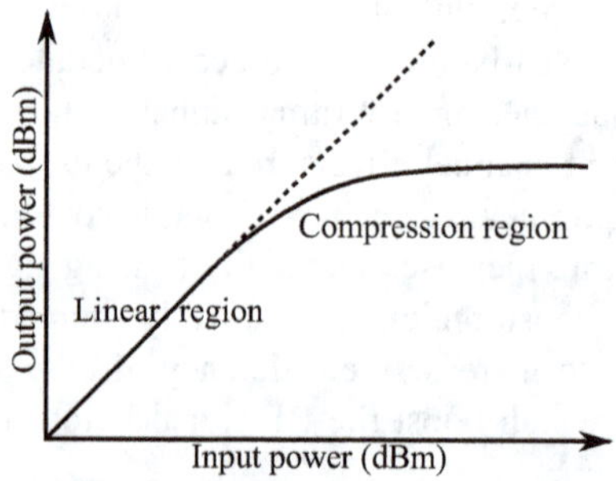

Figure 3.2 Gain compression occurs when an amplifier is driven into a non-linear operating regime [1]

techniques such as predistortion to correct for the non-linear effects and keep the efficiency benefits.

To support those efforts, the non-linear response of the device-under-test (DUT) must be accurately characterised through RF measurements so that it can be evaluated in system simulations. In addition, some classes of strongly non-linear amplifier (e.g. class F) require well-defined source and load impedances at harmonic frequencies to achieve their method of operation, for which accurate characterisation of the transistor aids in their design. Whilst the efforts outlined above address the distortion introduced by using non-linear amplifiers, continual improvement to baseband technology provides the opportunity for ever-denser modulation schemes with finer spacing between symbols. This puts an even stricter requirement on linearity, or at least the accurate characterisation of the complete analogue front-end, of which the amplifier is the main concern. Of course, some components of the RF front-end rely on non-linear effects for their functionality. Frequency mixers and multipliers are perhaps the main examples of these, which also require characterisation if they are to be included in standard computer-aided design simulations.

A further consequence of the higher operating frequencies of 5G RF systems is that accurate measurements become, at a fundamental level, more challenging. As will be discussed later in this chapter, unavoidable sources of error in RF measurements introduce uncertainties in the results, some of which scale with frequency. This is due to the physical nature of the electromagnetic waves propagating through the system and how they interact with mechanical components and transmission lines inside the front-end and devices therein. For example, tolerances of a few degrees of phase at traditional ultra-high frequency (UHF) frequencies become tens and even hundreds of degrees at microwave and millimetre-wave frequencies, respectively. This puts pressure on RF instrumentation and measurement techniques to provide the same level of confidence in their results at these much higher frequencies as was previously experienced at UHF and lower.

The number of measurements required to fully characterise a non-linear RF device is much greater than for a linear device. Many combinations of stimulus conditions must be given to the DUT and the response measured for each one. This often creates large data files which can be inconvenient for sharing with other designers and engineers or cause computing issues. To address this, behavioural models have been developed which can encapsulate non-linear device characteristics in a more efficient manner.

This chapter describes a selection of measurement techniques and considerations which address the above challenges presented by 5G RF requirements. The subject matter is divided into the following four sections:

1. RF instrumentation used for common device measurements and the calibration process used to remove their main sources of error;
2. the use of 'load-pull' measurements to characterise the response of power transistors across a range of source and load impedances;
3. *X*-parameters, a popular non-linear behavioural model used to capture the response of non-linear devices for use in design simulations; and
4. practical limitations of RF measurements and techniques for evaluating their associated uncertainties.

3.2 Non-linear vector network analyser measurements

3.2.1 Introduction to NVNA measurements

An RF device can be characterised by capturing its behaviour in a mathematical model whose parameters are found through measurement. For electromagnetic wave measurements, we are primarily interested in the travelling waves incident on and scattered by the DUT. These waves occur at all ports of the device and are termed a and b waves, respectively, and are illustrated in Figure 3.3.

For a linear device (where b depends linearly on a and which exhibits no frequency mixing), the de-facto model is scattering parameters (S-parameters). For

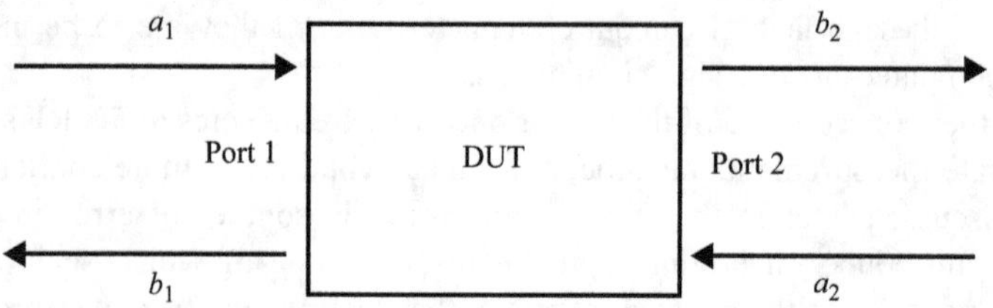

Figure 3.3 A diagram showing the relationship between a and b waves interacting with a two-port DUT

a two-port device, these can be written as

$$b_1 = S_{11}a_1 + S_{12}a_2$$
$$b_2 = S_{21}a_1 + S_{22}a_2$$

where a_i and b_i are the incident and scattered waves at port i, respectively, and the S terms are the S-parameters indexed against frequency. For a device with n ports, the equation can be written in matrix form:

$$\begin{bmatrix} b_1 \\ b_2 \\ \vdots \\ b_n \end{bmatrix} = \begin{bmatrix} S_{11} & S_{12} & \cdots & S_{1n} \\ S_{21} & S_{22} & \cdots & S_{2n} \\ \vdots & \vdots & \ddots & \vdots \\ S_{n1} & S_{n2} & \cdots & S_{nn} \end{bmatrix} \begin{bmatrix} a_1 \\ a_2 \\ \vdots \\ a_n \end{bmatrix}$$

$$\mathbf{b} = \mathbf{Sa}$$

Individual S-parameters can be defined as

$$S_{ij} = \left.\frac{b_i}{a_j}\right| a_k = 0 \text{ for } k \neq j$$

The measurement of electromagnetic waves is performed by an instrument known as a vector network analyser (VNA). Traditionally, the VNA has taken advantage of the fact that by measuring the ratio of these waves (i.e. S-parameters) in a process known as calibration, most of the systematic errors present in the system can be estimated and corrected for. These measurements are the mainstay of RF device characterisation used for both passive and active linear devices (filters, splitters, couplers, small-signal amplifiers, etc.). A thorough study of VNA theory and practice can be found in [2]. Once the S-parameters have been generated for a device, they may be used in a variety of circuit simulation and design environments to enable RF systems to be developed and optimised.

For non-linear devices, where the scattered waves may contain energy at harmonics of the excitation frequency and the ratio of scattered to incident waves is sensitive to input power, the S-parameters are no longer sufficient to accurately model the behaviour of the DUT. In this case, instead of measuring the ratios of incident and scattered waves, the absolute magnitude and phase of each wave must

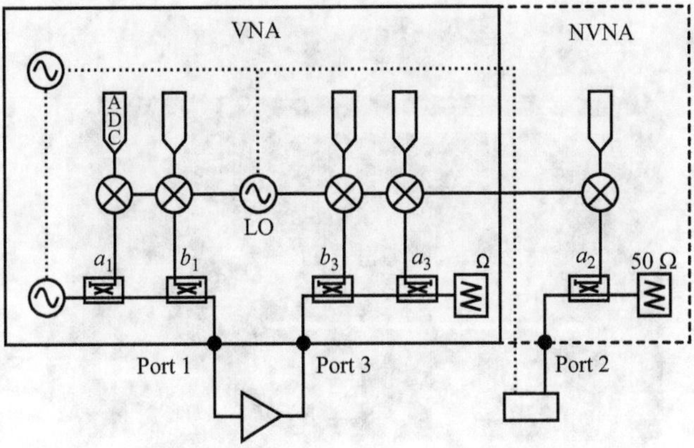

Figure 3.4 Schematic diagram of a modern mixer-based VNA or NVNA

be measured separately. With this information, the DUT can be simulated either by interpolating the measurements directly (a data-based model), or by taking advantage of non-linear behavioural models such as *X*-parameters which are explained later in this chapter.

To measure the *a* and *b* waves separately and to capture their harmonic frequency components, an upgraded VNA is required, known as a non-linear VNA, or NVNA. The internal structure of mixer-based versions of the two instruments is shown in Figure 3.4. The VNA and NVNA are very similar in terms of hardware, with the main difference being the calibration technique. However, for mixer-based NVNAs, an additional receiver is required to continuously measure a phase reference (comb generator); this ensures phase alignment when the instrument changes frequency to measure harmonic content. A high specification VNA may already contain enough receivers and so adding NVNA capability may only require a software upgrade. Figure 3.5 shows a typical NVNA with a DUT connected to the test ports via cables and a phase reference directly attached to a spare test port.

3.2.2 NVNA calibration

Because the VNA measures electromagnetic waves which are referenced to a characteristic impedance (typically 50 Ω), the raw measurements must be normalised such that this value is set at the test ports (the 'measurement planes'), between which the DUT is defined. In addition, any leakage or reflection of signals internal to the instrument configuration as shown in Figure 3.4 must be sufficiently corrected for or they will perturb DUT measurements as a function of impedance and frequency. This normalisation and error correction process, or calibration, has been subject to much development over the preceding decades and there are several popular mathematical models of the VNA signal path which are available to use in modern VNAs.

The two-port VNA calibration model typically used as a basis for NVNA calibration is the 8-term 'error box' model. This model, shown in Figure 3.6,

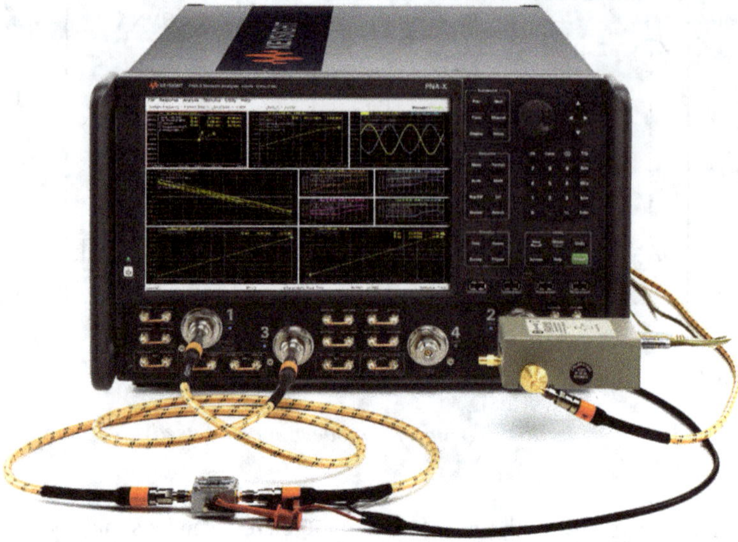

Figure 3.5 A Keysight PNA-X NVNA configured for measuring a two-port DUT, attached to test ports using cables. A phase reference is attached directly to a spare port.

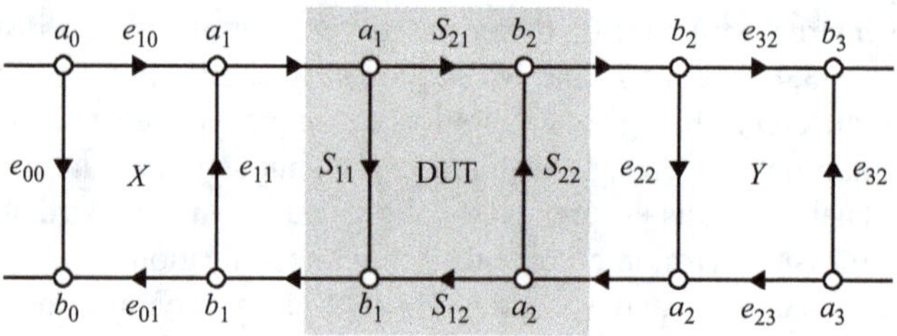

Figure 3.6 The 8-term 'error box' model used to characterise and de-embed systematic errors from VNA and NVNA measurements

defines two virtual two-port devices (error boxes), *X* and *Y*, between the DUT and the VNA ports. Crosstalk between ports is ignored, but for on-wafer measurements at high frequencies it can become significant and may require use of the more comprehensive 16-term model. The goal of the VNA calibration process, therefore, is to quantify the coefficients of the error box (effectively *S*-parameters), which are then de-embedded from raw DUT values measured by the VNA to produce an accurate DUT measurement result.

To perform the calibration procedure, a series of standards are connected to the VNA test ports. These are artefacts whose *S*-parameters are known to an appropriate degree of accuracy across the measured frequency range, and the uncertainty in that knowledge will typically be traceable to national standards to provide confidence in the calibration. The chosen standards will have *S*-parameters which are spread widely across the measurable *S*-parameter space to ensure that the determination of the coefficients of the calibration model is well conditioned. There are several popular sets of standards to use for this task, for example a short-circuit, open-circuit, matched load and straight through connection in a short-open-line-thru (SOLT) calibration, or a straight through, (high) reflect, and length of transmission line in a thru-reflect-line (TRL) calibration. The TRL calibration is popular for precision metrology and on-wafer measurements as the matched load used for the SOLT calibration is not easy to accurately define or fabricate on-wafer.

For *S*-parameter measurements the two transmission terms of each error box (the e_{10} and e_{01} terms and also the e_{32} and e_{23} terms) always occur as a product (either $e_{10}e_{01}$ or $e_{32}e_{23}$) and can only be solved as such using conventional VNA calibration methods. However, for NVNA measurements where the absolute values of the complex waves must be known, these terms must be individually quantified. To accomplish this, two additional measurements must be performed to upgrade the VNA calibration into an 'absolute calibration' suitable for measuring individual waves. These measurements need only be made on one port of the NVNA, from which the VNA calibration can be used to solve the terms for the other port.

The first additional measurement uses a calibrated power meter to find the amplitude of the loss through an error box from the NVNA signal source to the test port. The meter is an external device driven by the NVNA software to perform the measurements and the sensor head is attached to the test port. By analysing the signal path shown in Figure 3.7, the magnitude of the forward error term e_{01} can be found by

$$|e_{01}| = \frac{|a_0|}{\sqrt{P_{\text{meter}}}}\sqrt{|\Gamma_{\text{PM\_raw}}e_{11} - \Delta x|^2 - |\Gamma_{\text{PM\_raw}} - e_{00}|^2}$$

where a_0 is the uncalibrated a wave measurement, P_{meter} is the power meter reading, $\Gamma_{\text{PM\_raw}}$ is the uncalibrated power meter reflection coefficient (b_0/a_0) and $\Delta x = e_{00}e_{11} - e_{01}e_{10}$ [3].

The second additional measurement necessary for the absolute calibration requires another phase reference, separate to the phase reference already used on the mixer-based NVNA. This calibration phase reference must have a well-defined phase relationship at each measurement frequency which will be used to characterise the phase delay through the error box. This signal path is shown in

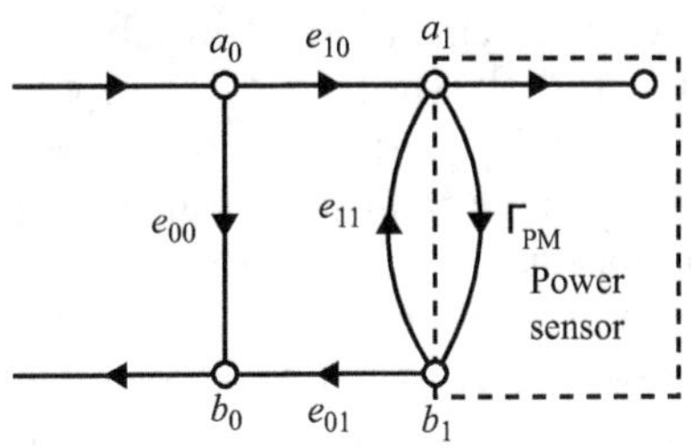

Figure 3.7 Signal path for absolute calibration power measurement

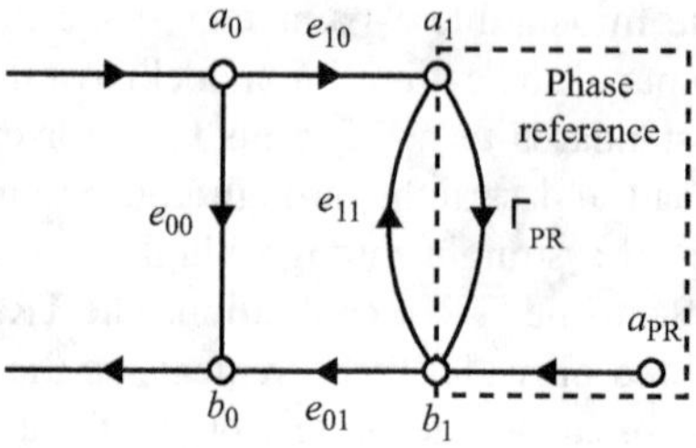

Figure 3.8 Signal path for absolute calibration phase measurement

Figure 3.8 and can be used to determine the phase of e_{01} using

$$\varphi(e_{01}) = \varphi(a_0) + \varphi(\Gamma_{PR\_raw} - \Gamma_{PR\_raw}\Gamma_{PR}e_{11} - e_{00} + \Delta x\Gamma_{PR}) - \varphi(a_{PR})$$

where Γ_{PR_raw} and Γ_{PR} are the uncalibrated and calibrated reflection coefficients of the phase reference and $\varphi(a_{PR})$ is the known phase value from the phase reference [3].

3.2.3 Time domain vs frequency domain

Throughout this section only mixer-based NVNAs have been considered. Although these instruments measure waves in the frequency domain, it is possible to convert their results into time-domain waveforms using an inverse Fourier transform. This capability can be built into modern NVNAs, or instead performed afterwards on a computer. Figure 3.9 shows an example of this conversion for a non-linear amplifier across a range of input powers.

The alternative type of NVNA uses a real-time sampling oscilloscope to sample the electromagnetic wave directly. These instruments are inherently time domain and so produce waveforms natively, requiring a Fourier transform to convert their results into the frequency domain [4].

3.3 Load-pull measurements

3.3.1 Introduction to load-pull measurements

A two-port RF device such as an amplifier is typically operated between a source and a load as shown in Figure 3.10. The amplifier accepts a signal from the source and delivers a scaled version of the signal to the load. There are four impedances of interest in this situation: the source and load impedances (Z_S and Z_L, respectively) and the input and output impedances of the device (Z_{IN} and Z_{OUT}, respectively).

The basic idea of load-pull is to systematically vary the source and load impedances presented to the device and also to sweep the source power and, for each impedance state and source power level, to measure the circuit parameters of interest (such as gain, output power, efficiency, voltage and current waveforms and dynamic load lines, etc.). An important application of load-pull measurements is to

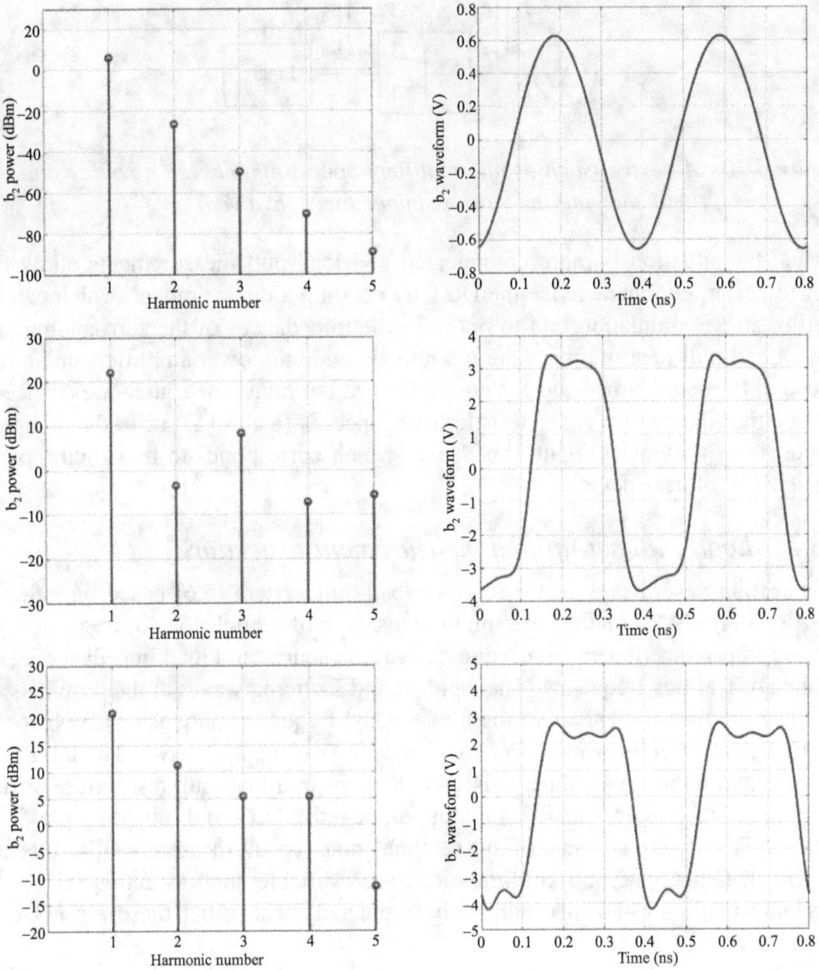

Figure 3.9 (Left) Amplitudes of the first five harmonics of the scattered wave from port 2 (b_2 wave) of an amplifier (frequency domain) for three different power levels of a 2-GHz incident wave on port 1 (a_1 wave) of the amplifier and (right) the corresponding b_2 waveforms (time domain). The a_2 wave power levels are (top) −15.4 dBm, (middle) +3.7 dBm and (bottom) +9.7 dBm.

determine the ideal matching network impedances to be used with a transistor in an amplifier circuit. The circuit designer can then design matching networks with these impedances to achieve an optimal amplifier design. If the device is operating linearly, then the behaviour of the circuit is fully determined by the DC bias voltages and currents, the *S*-parameters of the device and the reflection coefficients (or equivalently the impedances) of the source and load. However, for a non-linear

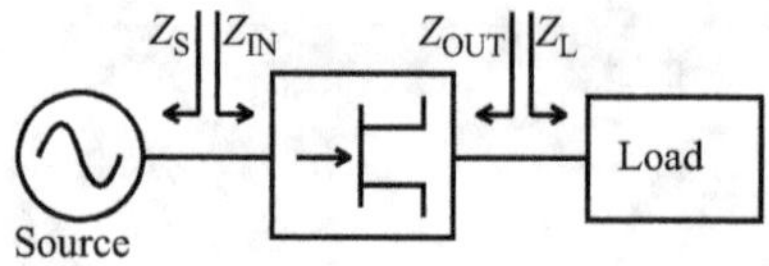

Figure 3.10 A device (such as an amplifier) operating between a source and a load showing the various impedances involved

device, the situation is more complicated and load-pull measurements allow the user to accurately characterise the DUT by creating a device model, which can be used in an RF simulation tool to optimise the impedances of the surrounding circuits. Load-pull measurements are often made on high-power amplifiers and a full sweep of the source impedance is not performed but rather the source impedance is set to either the system characteristic impedance Z_0 (e.g. 50 Ω) or to the complex conjugate of the device input impedance (which corresponds to maximum power transfer for a linear device).

3.3.2 Load-pull set-up and measurement procedure

This section describes a vector receiver load-pull system [5,6] based on either a VNA or an NVNA with the RF stimulus consisting of a single tone of frequency, f_0. Passive impedance tuners are used to control the source and load impedances seen by the DUT at this frequency. The incident and scattered waves at the terminals of the DUT are measured at either the fundamental frequency only using a VNA or at the fundamental and several harmonic frequencies using an NVNA. The DUT can be measured either in a connectorised package or in a load-pull test fixture or on-wafer. The DC bias and the RF stimulus can be pulsed to avoid self-heating effects in the DUT. Towards the end of this section, some typical measurement results are presented. Other load-pull configurations are available such as active load-pull, harmonic load-pull and load-pull with modulated signals, but these are not discussed here [7].

3.3.2.1 Load-pull test bench

Figure 3.11 shows a schematic diagram and Figure 3.12 a photograph of a typical load-pull test bench. The system shown in Figure 3.12 includes a probe station for the measurement of on-wafer DUTs. The load-pull set-up consists of the following components: (i) a source and pre-amplifier to supply a large RF signal to the DUT. The amplifier has a circulator at its output to protect it from signals reflected from the DUT. The signal consists of a single tone at frequency f_0 (RF stimulus); (ii) a DC measurement supply unit (MSU) and bias tees to supply and measure the DC voltage and current to the DUT (DC bias); (iii) circuitry to pulse the RF stimulus and the DC bias (if required) – not shown in Figure 3.11; (iv) low-loss couplers to separate the incident and scattered waves at the ports of the DUT (losses between the tuners and the DUT limit the range of impedances which can be presented at the DUT terminals); (v) a vector receiver (VNA or NVNA) to measure the complex amplitudes (i.e. the amplitudes and phases) of the incident and scattered waves; (vi)

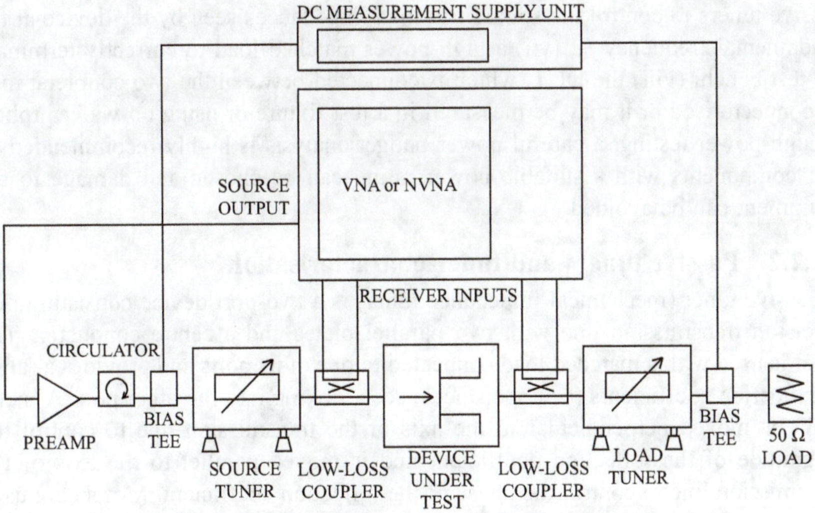

Figure 3.11 Vector receiver load-pull set-up showing the device under test, low-loss couplers, mechanical tuners and bias tees as well as the RF source, DC MSU and vector receiver (VNA or NVNA)

Figure 3.12 Load-pull test bench at the University of Surrey

passive tuners to control the source and load impedances seen by the device at the fundamental frequency f_0; (vii) a high-power matched load to correctly terminate the test bench; (viii) the DUT, which is connected between the two couplers, may be connectorised or it may be measured in a test fixture or using on-wafer probes. In high-power testing, a careful power budget analysis is highly recommended so that components with a suitable power rating can be chosen and damage to the equipment can be avoided.

3.3.2.2 Passive tuners and tuner characterisation

A passive tuner (mechanical impedance tuner) is a two-port device consisting of a precision transmission line with two parallel plates and a centre conductor. The tuner is used with a matched load connected to one of its ports and provides a range of reflection coefficients (or impedances) seen looking into the other port. A metal probe is moved perpendicular to the axis of the transmission line to control the magnitude of the reflection coefficient and is moved parallel to the axis of the transmission line to control the phase of the reflection coefficient. Motors are used to accurately set the position of the probe with respect to the two axes, which defines the state (impedance) of the tuner.

The source and load tuners are each characterised by measuring their two-port *S*-parameters on a grid of frequencies and tuner states using a calibrated VNA. This can be done periodically and does not need to be repeated prior to every load-pull measurement. The *S*-parameters of tuner states not on the measured grid can be obtained by interpolation, although extrapolation for frequencies and tuner states outside of the measured range is not recommended.

3.3.2.3 Vector receiver set-up and calibration

Directional couplers placed between the DUT and the tuners (as shown in Figure 3.11) separate the incident and scattered waves (*a* and *b* waves, respectively) at the input and output of the DUT. A vector receiver such as a VNA or an NVNA is used to measure the magnitudes and phases of the *a* and *b* waves at the measurement reference planes at the excitation (fundamental) frequency and also, in the case of an NVNA, at a chosen number of harmonics of the excitation frequency.

A calibration of the VNA or NVNA with the tuners set to the matched condition establishes the measurement reference planes at which the *a* and *b* waves are to be measured and the impedance to which the measured waves are to be normalised. During calibration, the RF source in the VNA or NVNA needs to be switched between the two ports of the load-pull test bench. Therefore, for calibration, the matched load is removed, and both ends of the test bench are connected to the VNA or NVNA. As described in Section 3.2, the calibration consists of three steps: (i) a vector calibration, (ii) a power calibration (for both VNA and NVNA load-pull) and (iii) a phase calibration (for NVNA calibration only). Depending on the nature of the DUT, the calibration may be carried out in either coaxial line or on-wafer. If the calibration is carried out on-wafer, the power sensor and phase reference, which are fitted with coaxial connectors, cannot be directly connected at the on-wafer measurement reference plane. Instead, the two on-wafer probes are landed on a thru and the power

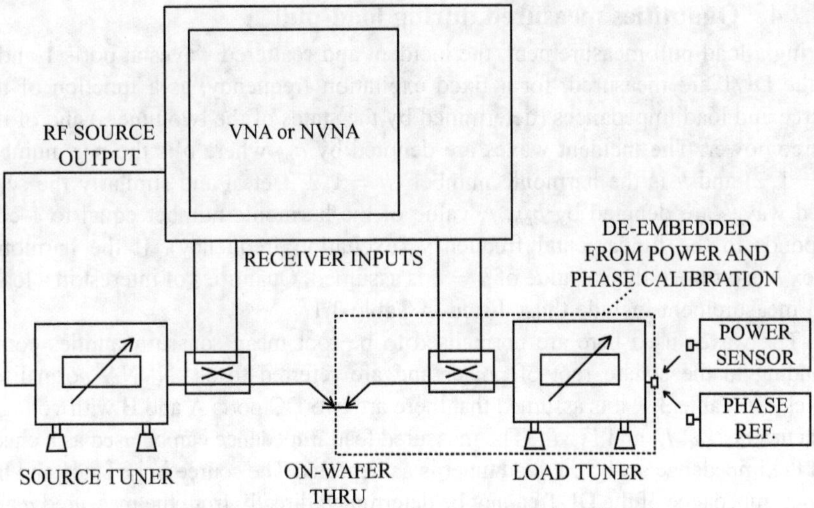

Figure 3.13 Connection of the power sensor and phase reference during calibration of the load-pull test bench for measurement of on-wafer DUTs

sensor and comb generator are connected at the output of the second tuner (see Figure 3.13). The combination of the on-wafer thru and the second tuner is characterised by measuring short, open and load calibration standards on the output of the second tuner with respect to the on-wafer calibration. This allows the thru and the second tuner to be de-embedded from the power sensor and phase reference measurements. The advantage of placing the tuners behind the couplers (i.e. further from the DUT than the couplers) is that, in this position, the states of the tuners can be changed without affecting the VNA or NVNA calibration.

The source or load impedance seen at the measurement reference plane depends on (i) the state of the tuner, (ii) the components between the reference plane and the tuner (couplers, cables, etc.) and (iii) the load terminating the tuner. Once the tuners have been characterised, the *S*-parameters of the components between the reference planes and the tuners and of the loads terminating the tuners (the output impedance of the pre-amplifier for the source tuner and the matched load for the load tuner) can be determined in situ by a de-embedding process. This is performed after the VNA or NVNA calibration to allow flexibility in the choice of components. This is done for both tuners by measuring the reflection coefficients at the reference plane looking towards the tuner for four different states of the tuner (three high reflection states to measure the coupler and cables in front of the tuner and one low reflection state to measure the load behind the tuner).

After the calibration and the tuner de-embedding, port 2 of the test bench is disconnected from the VNA or NVNA and the matched load is re-connected for the load-pull measurement as shown in Figure 3.11.

3.3.2.4 Quantities measured during load-pull

During a load-pull measurement, the incident and scattered waves at ports 1 and 2 of the DUT are measured, for a fixed excitation frequency, as a function of the source and load impedances (determined by the states of the two tuners) and of the source power. The incident waves are denoted by a_{ph} where p is the port number ($p = 1, 2$) and h is the harmonic number ($h = 1, 2, 3$, etc.) and similarly the scattered waves are denoted by b_{ph}. A value of the harmonic number equal to 1 corresponds to the fundamental frequency (excitation frequency). If the harmonic index h is omitted, then a value of $h = 1$ is assumed. Quantities of interest in a load-pull measurement include those listed in Table 3.1.

The waves used here are normalised to be root mean square quantities corresponding to the square root of power and are referred to as 'NVNA normalised waves'. In Table 3.1, it is assumed that there are two DC ports A and B with voltages and currents V_A, I_A and V_B, I_B. The measured load impedance can be used as a check that the impedance set by the load tuner is as expected The source impedance and the output impedance of the DUT cannot be determined directly from the measured a and b waves because the direction of signal flow is from the source to the load. If harmonics of the a and b waves are measured using an NVNA, then time-domain waveforms (e.g. voltage and current waveforms) and dynamic load lines can be obtained at the device terminals. For example, if n harmonics are measured, the corresponding voltage and current waveforms at port 2 of the DUT are given by

Table 3.1 Typical quantities of interest in a load-pull measurement

Quantity of interest	Expression in terms of measured waves and DC voltages and currents
Input reflection coefficient of the DUT	$\Gamma_{IN} = \frac{b_1}{a_1}$
Input impedance of the DUT	$Z_{IN} = Z_0 \frac{1+\Gamma_{IN}}{1-\Gamma_{IN}}$
Load reflection coefficient	$\Gamma_L = \frac{a_2}{b_2}$
Load impedance	$Z_L = Z_0 \frac{1+\Gamma_L}{1-\Gamma_L}$
Input power to the DUT	$P_{IN} = \lvert a_1\rvert^2 - \lvert b_1\rvert^2$
Available power from the source	$P_A = MP_{IN}$
Conjugate mismatch loss factor	$M = \frac{P_A}{P_{IN}} = \frac{(1-\Gamma_S\Gamma_{IN})}{\left(1-\lvert\Gamma_S\rvert^2\right)\left(1-\lvert\Gamma_{IN}\rvert^2\right)}$
Output power from the DUT	$P_{OUT} = \lvert b_2\rvert^2 - \lvert a_2\rvert^2$
DC power delivered to the DUT	$P_{DC} = V_A I_A + V_B I_B$
Transducer gain of the DUT	$G_T = \frac{P_{OUT}}{P_A}$
Operating gain of the DUT	$G_P = \frac{P_{OUT}}{P_{IN}}$
PAE of the DUT	$\eta_{PAE} = \frac{P_{OUT}-P_{IN}}{P_{DC}}$

$$V_2(t) = \sqrt{Z_0}\sum_{h=1}^{n}(a_{2h} + b_{2h})e^{jh\omega t},$$

$$I_2(t) = \frac{1}{\sqrt{Z_0}}\sum_{h=1}^{n}(a_{2h} - b_{2h})e^{jh\omega t}.$$

3.3.2.5 Fixture de-embedding

If the DUT is fitted with coaxial connectors, it can be connected directly to the load-pull test bench or if it is on-wafer, it can be measured using probes. Other DUTs, such as packaged transistors, can be measured by being embedded in a test fixture as represented in Figure 3.14 where 'FixA' and 'FixB' represent the left-hand and right-hand sides of the fixture, respectively [8].

A load-pull test fixture is shown in Figure 3.15. The input and output impedances of the transistor are about 5 Ω and so the fixture incorporates 10:1 impedance

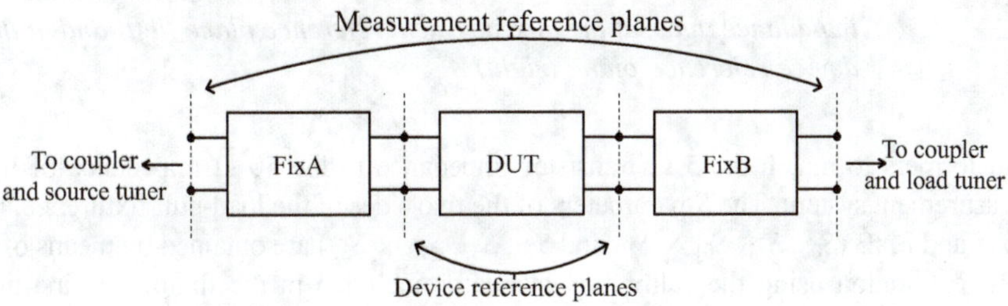

Figure 3.14 Schematic representation of a DUT embedded within a test fixture

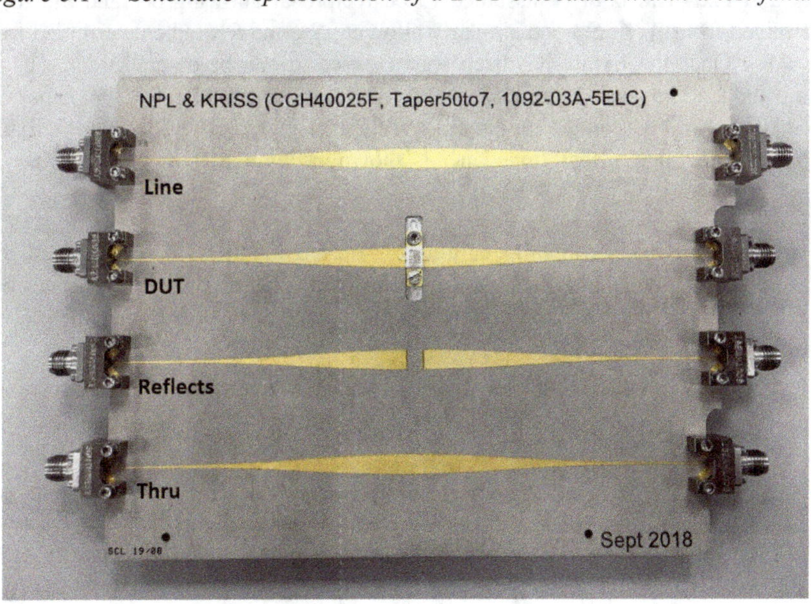

Figure 3.15 Load-pull test fixture with transistor under test

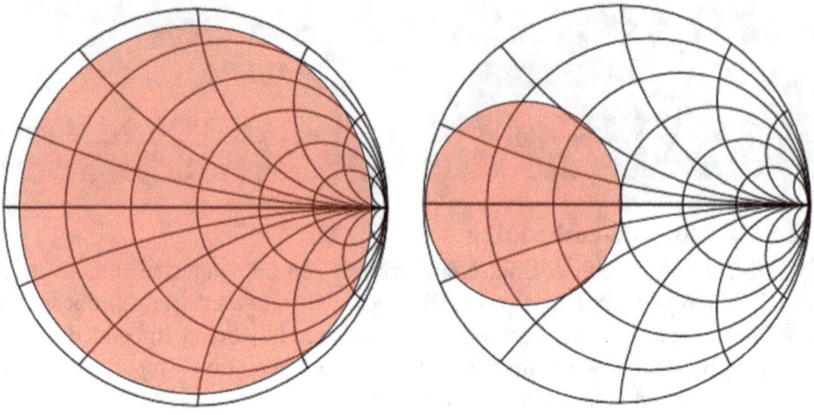

Figure 3.16 Smith chart illustration of the available tuning range of the load impedance tuner at the measurement reference plane (left) and at the device reference plane (right)

transformers to match the 5 Ω transistor impedance to the 50 Ω impedance of the measurement system. The *S*-parameters of the two sides of the load-pull fixture, i.e. of FixA and FixB (S_{00}, S_{11}, $S_{01} \times S_{10}$ and S_{22}, S_{33}, $S_{23} \times S_{32}$) are obtained by means of a TRL calibration using the calibration standards included in the fixture. A first-tier coaxial calibration is followed by a second-tier calibration using the fixture standards. FixA and FixB can both be assumed to be reciprocal allowing S_{01} and S_{10} to be determined from $S_{01} \times S_{10}$ and similarly allowing S_{23} and S_{32} to be determined from $S_{23} \times S_{32}$. Once the fixture has been characterised, it can be de-embedded, i.e. the source and load impedances and the *a* and *b* waves at the device reference planes can be deduced from those at the measurement reference planes. The effect of the fixture on the tunable range of load impedance is shown in Figure 3.16.

3.3.2.6 Thru check

As a check on the set-up and calibration of the load-pull bench, a load-pull measurement is made in which the operating gain of a coaxial thru is measured as a function of load impedance and source power. The operating gain (G_P) is defined by

$$G_P = \frac{P_{OUT}}{P_{IN}} = \frac{|b_2|^2 - |a_2|^2}{|a_1|^2 - |b_1|^2} = \frac{|b_2|^2\left(1 - |\Gamma_L|^2\right)}{|a_1|^2\left(1 - |\Gamma_{IN}|^2\right)}.$$

For a thru, this is expected to be close to 0 dB for all values of the load impedance (it can be calculated exactly in terms of the *S*-parameters of the thru and the load reflection coefficient). If all the measured operating gain values are within 0.2 dB (say) of 0 dB, the system is considered verified. This is referred to as a 'delta-G_P test'.

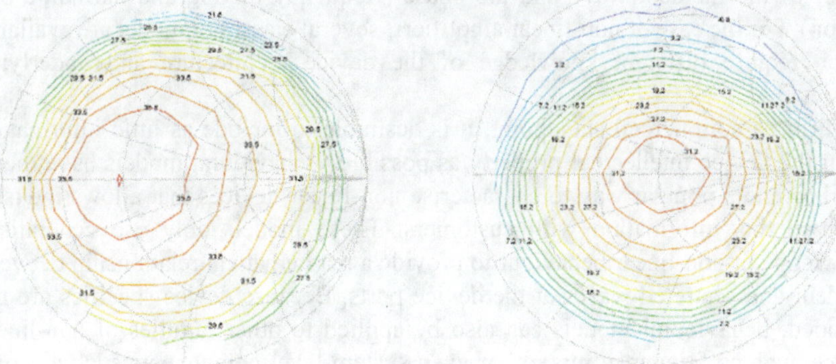

Figure 3.17 *Load-pull contours of output power (left) and PAE (right) for a 25 W GaN on SiC transistor measured in a load-pull test fixture. The contours are plotted on a Smith Chart representing the load impedance.*

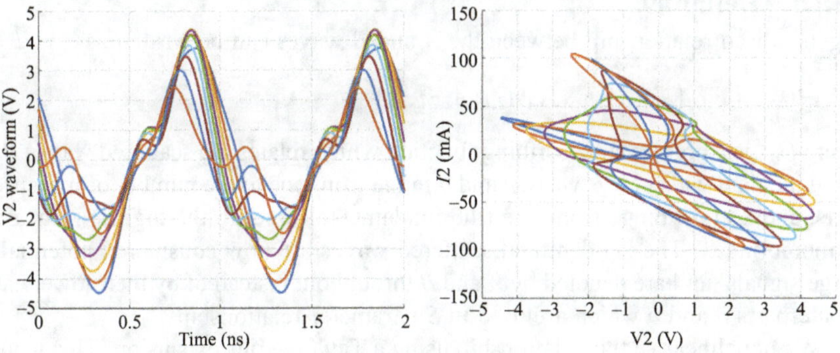

Figure 3.18 *Output voltage waveforms and dynamic load lines for a low-noise amplifier as a function of the phase of the load reflection coefficient measured using an NVNA load-pull system*

3.3.3 Typical load-pull results

Figures 3.17 and 3.18 show some typical load-pull results: contours of output power and PAE (Figure 3.17) and output voltage waveform and dynamic load lines (Figure 3.18).

3.4 *X*-parameter measurements

3.4.1 Basic X-parameters

Devices which exhibit non-linear behaviour cannot be characterised by a model as simple as *S*-parameters, with added complexities due to dependence on both the power of the incident wave and the phase (frequency mixing and harmonic distortion). For the case of non-linear amplifiers, several classes of model are available which require different knowledge of the device architecture and underlying physics.

From a commercial standpoint, it is desirable to include as little information about the device intellectual property as possible. Behavioural models have therefore become a popular way to characterise non-linear devices that allows industry to share this information with customers. These models only use generalised knowledge of non-linear behaviour to provide a mathematical relationship between incident and scattered waves at the device ports. Because device specifics are not included, behavioural models can also be applied to other families of non-linear device such as frequency mixers, or even system-level components such as complete front ends.

Arguably the most popular non-linear behavioural model since its release in 2009 has been the *X*-parameter model [9]. Two significant merits of this model are that it is included in the instrumentation and simulation solutions from a major manufacturer and that *X*-parameters reduce directly to *S*-parameters for linear devices.

3.4.1.1 Definition

A generalised relationship between the *A* and *B* waves can be written as

$$B_{p,k} = F_{p,k}(A_{1,1}, A_{1,2}, \ldots, A_{2,1}, A_{2,2}, \ldots)$$

where $F_{p,k}$ is a non-linear describing function which relates the scattered waves $B_{p,k}$ to the incident waves $A_{q,l}$, where p and q range from one to the number of measured ports, and k and l range from one (the fundamental tone) to the highest measured harmonic index. The incident and scattered waves are now considered potentially 'large signals' and are denoted by *A* and *B* throughout instead of by their lower case counterparts *a* and *b* which are used in *S*-parameter relationships.

We now linearise this relationship using a Taylor series expansion. This would be simple if it were not for the non-analyticity of the describing functions,[1] so Wirtinger calculus [10] must be used, which provides a solution of the form

$$B_{p,k} \approx F_{p,k}(A_{1,1}) + \sum_{q,l} S_{p,k;q,l}(A_{1,1})A_{q,l} + \sum_{q,l} T_{p,k;q,l}(A_{1,1})A^*_{q,l}$$

where the incident wave $A_{1,1}$ is large and defines the large-signal operating point (LSOP) of the DUT whilst the other incident waves $A_{q,l}$ with $q, l \neq (1, 1)$ are small and $A^*_{q,l}$ is the complex conjugate of $A_{q,l}$. The parameter, $F_{p,k}$, represents the DUT response at the LSOP, over which it is indexed. The parameters $S_{p,k}$ and $T_{p,k}$ represent the DUT response to small-signal perturbations to the LSOP with the conjugate terms being a result of the non-analyticity of the describing functions. In summary, the $F_{p,k}$ terms capture the large-signal non-linear response of the DUT as indexed values, while the other terms define how the small-signal behaviour of the

[1]The functions are analytic if using double-sided spectra, but this is non-standard [11].

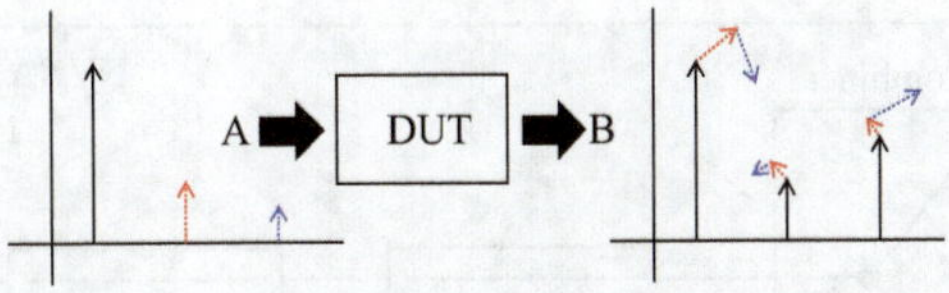

Figure 3.19 The harmonic superposition principle. If the input (a) contains a large-signal fundamental wave (solid phasor) and several small-signal harmonics (dotted phasors), the output (b) will comprise the large-signal response plus the sum of the individual responses due to each small-signal harmonic, at each frequency. The plots show complex phasors of increasing harmonic index along the horizontal axis.

DUT changes with the large-signal input. This model provides a better description of the DUT behaviour than *S*-parameters as the harmonic response is captured as well as phase interactions via the conjugate terms. The ability to sum the contributions of the different small-signal stimuli at each harmonic is known as the harmonic superposition principle [12] and is illustrated in Figure 3.19. An important assumption is that the LSOP is determined by a single large-signal $A_{1,1}$. To accommodate situations where this is not true, for example when an impedance mismatch at the output causes a large reflection, the model must be modified as will be described later for load-dependent *X*-parameters.

The behaviour of the device is considered to be time invariant. This property means that any delay to all of the incident waves will result in the same delay to all of the scattered waves, whilst maintaining the shape of the scattered signal waveforms. The previous relationship can now be normalised with respect to the phase of $A_{1,1}$ and with a slight change of notation the basic *X*-parameter model is defined as

$$B_{p,k} = X_{p,k}^{(F)}(|A_{1,1}|)P^k + \sum_{q,l\neq(1,1)} \left[X_{p,k;q,l}^{(S)}(|A_{1,1}|)A_{q,l}P^{k-l} + X_{p,k;q,l}^{(T)}(|A_{1,1}|)A_{q,l}^{*}P^{k+l} \right]$$

where $X^{(F)}$ are the large-signal terms, $X^{(S)}$ and $X^{(T)}$ are the small-signal terms and $P = e^{j\phi(A_{1,1})}$ is the $A_{1,1}$ phase term. In the case of linear operation when $|A_{1,1}|$ is small, all cross-frequency small-signal terms and all $X^{(T)}$ terms reduce to zero, and the remaining $X^{(F)}$ and $X^{(S)}$ terms reduce to conventional *S*-parameters.

Additional *X*-parameters which characterise the DC response of a device are also available and are defined in [9].

3.4.1.2 Extraction

X-parameters can be extracted from a DUT using a calibrated NVNA with a second signal source which can vary in frequency and phase when compared with the main source. This second source applies small-signal 'extraction' tones to the DUT whilst it is driven to a particular LSOP by the main source, capturing the linearised

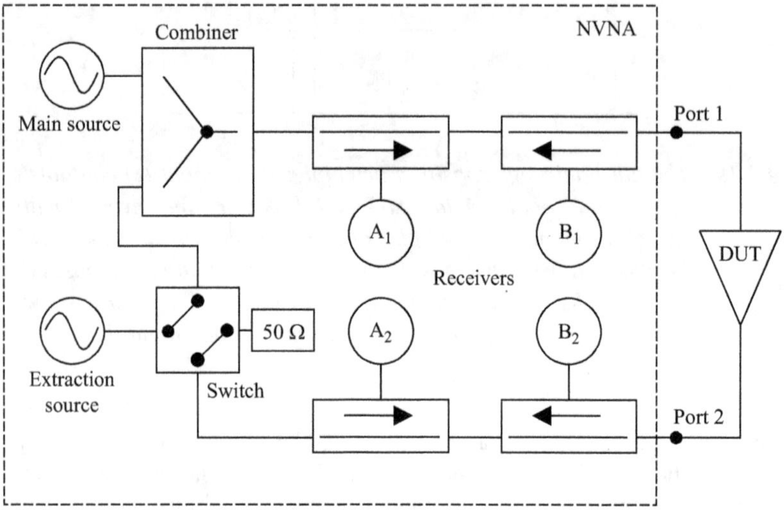

Figure 3.20 The NVNA set-up required for X-parameter extraction. The switch terminates the unused port with a 50 Ω load. The phase reference required for a mixer-based NVNA is connected to a spare receiver and is not shown [1].

response to incident small signals in the $X^{(S)}$ and $X^{(T)}$ terms. An example of the set-up configured in a standard NVNA is shown in Figure 3.20.

There are two methods for acquiring the measurements necessary for X-parameter extraction. The first is based around offsetting the extraction tone frequency from the particular small-signal frequency of interest (i.e. fundamental or harmonic) and using mixer theory to obtain values for $X^{(S)}$ and $X^{(T)}$. The second and more widely used method, which will be described here, uses extraction tones at phases offset from each other to characterise $X^{(S)}$ and $X^{(T)}$.

For extraction, we can simplify the previous equation by setting the phase of $A_{1,1}$ to zero, which eliminates the P terms from the equation. For each combination of small-signal stimulus port q and harmonic l and response port p and harmonic k, there are three X-parameters to solve for, $X^{(F)}$, $X^{(S)}$ and $X^{(T)}$:

$$B_{(p,k)_0} = X^{(F)}_{p,k} + X^{(S)}_{p,k,q,l}A_{(q,l)_0} + X^{(T)}_{p,k,q,l}A^*_{(q,l)_0}$$

$$B_{(p,k)_1} = X^{(F)}_{p,k} + X^{(S)}_{p,k,q,l}A_{(q,l)_1} + X^{(T)}_{p,k,q,l}A^*_{(q,l)_1}$$

$$B_{(p,k)_2} = X^{(F)}_{p,k} + X^{(S)}_{p,k,q,l}A_{(q,l)_2} + X^{(T)}_{p,k,q,l}A^*_{(q,l)_2}$$

Therefore, by performing three measurements at orthogonal excitations, we can obtain a good fit for the three terms. These excitations are as follows:

1. LSOP only, no small-signal extraction tones.

2. LSOP and extraction tone at phase θ.
3. LSOP and extraction tone at phase $\theta + 90°$ or $\theta - 90°$.

The initial phase θ can be of any value and the extraction tone amplitude must be low enough to only induce linear perturbations to the large-signal response (an amplitude 16 dB below the LSOP is typically used). More reliable values for the X-parameters may be obtained by taking further measurements at different extraction tone phases and solving by linear regression (least-squares method). We can define a matrix where, for each stimulus condition from zero (no extraction tone applied – LSOP only) to M, the columns contain the measured stimuli at each port q and harmonic l, besides their respective conjugates, such that

$$\boldsymbol{A} = \begin{bmatrix} 1 & 0 & A_{(2,1)_0} & A^*_{(2,1)_0} & A_{(1,2)_0} & A^*_{(1,2)_0} & A_{(2,2)_0} & A^*_{(2,2)_0} & \cdots \\ 1 & A_{(1,1)_1} - A_{(1,1)_0} & A_{(2,1)_1} & A^*_{(2,1)_1} & A_{(1,2)_1} & A^*_{(1,2)_1} & A_{(2,2)_1} & A^*_{(2,2)_1} & \cdots \\ 1 & A_{(1,1)_2} - A_{(1,1)_0} & A_{(2,1)_2} & A^*_{(2,1)_2} & A_{(1,2)_2} & A^*_{(1,2)_2} & A_{(2,2)_2} & A^*_{(2,2)_2} & \cdots \\ \vdots & \vdots & \vdots & \vdots & \vdots & \vdots & \vdots & \vdots & \cdots \\ 1 & A_{(1,1)_M} - A_{(1,1)_0} & A_{(2,1)_M} & A^*_{(2,1)_M} & A_{(1,2)_M} & A^*_{(1,2)_M} & A_{(2,2)_M} & A^*_{(2,2)_M} & \cdots \end{bmatrix}$$

As an exception, the first column is set to the unit vector to extract the large-signal term $X^{(F)}$ and the second column contains the small-signal perturbation to $A_{1,1}$; hence, it is set to zero for the first row and calculated as $A_{(1,1)_n} - A_{(1,1)_0}$ thereafter. Each row in the matrix corresponds to a different stimulus condition (with the large signal applied to port 1 and the extraction tone applied to a particular port, at a particular harmonic and with a particular phase) and typically only the terms on each row corresponding to the active stimuli will be significant. A column vector $\boldsymbol{b}_{p,k}$ of length $M+1$ is defined to contain the respective measurements of a single scattered wave $b_{p,k}$ for the different stimulus conditions:

$$\boldsymbol{b}_{p,k} = \begin{bmatrix} B_{(p,k)_0} & B_{(p,k)_1} & \cdots & B_{(p,k)_M} \end{bmatrix}^T$$

then all the X-parameters for p, k can be obtained by least squares as follows:

$$\widehat{\boldsymbol{X}}_{p,k} = \left(\boldsymbol{A}^T \boldsymbol{A}\right)^{-1} \boldsymbol{A}^T \boldsymbol{b}_{p,k}$$

with the column vector $\widehat{\boldsymbol{X}}_{p,k}$ containing X-parameters:

$$\widehat{\boldsymbol{X}}_{p,k} = \begin{bmatrix} X^{(F)}_{p,k} & X^{(S)}_{p,k,1,1} & X^{(S)}_{p,k,2,1} & X^{(T)}_{p,k,2,1} & X^{(S)}_{p,k,1,2} & X^{(T)}_{p,k,1,2} & \cdots \end{bmatrix}^T.$$

Note that $X^{(T)}_{p,k,1,1}$ is not extracted during measurement as the phase normalisation of $A_{1,1}$ by the NVNA causes its contribution to be combined with $X^{(S)}_{p,k,1,1}$. This will differ with X-parameters extracted from a simulation via harmonic balance where the two terms are correctly separated.

3.4.2 Load-dependent X-parameters

The basic X-parameters can accommodate small-signals incident upon all ports at all harmonics in addition to the LSOP. However, the accuracy of the model reduces significantly when for example a large signal appears at the output (port 2) of an amplifier, typically when a significant reflection is present. This is a common scenario when the load impedance at port 2 is mismatched. To improve the accuracy, the X-parameters are further indexed against the large signal at the fundamental frequency at the output, $A_{2,1}$, [9] to give

$$B_{p,k} = X_{p,k}^{(F)}\left(|A_{1,1}|, A_{2,1}\right)P^{k} + \sum_{q,l \neq (1,1)} \left[X_{p,k;q,l}^{(S)}\left(|A_{1,1}|, A_{2,1}\right)A_{q,l}P^{k-l} + X_{p,k;q,l}^{(T)}\left(|A_{1,1}|, A_{2,1}\right)A_{q,l}^{*}P^{k+l}\right]$$

The previous equation can be rewritten using the load reflection coefficient at the output $\Gamma_{2,1}$ to obtain

$$B_{p,k} = X_{p,k}^{(F)}\left(|A_{1,1}|, \Gamma_{2,1}\right)P^{k} + \sum_{q,l \neq (1,1)} \left[X_{p,k;q,l}^{(S)}\left(|A_{1,1}|, \Gamma_{2,1}\right)A_{q,l}P^{k-l} + X_{p,k;q,l}^{(T)}\left(|A_{1,1}|, \Gamma_{2,1}\right)A_{q,l}^{*}P^{k+l}\right]$$

In order to extract load-dependent X-parameters, it is necessary to perform load-pull measurements on the DUT as described in Section 3.3.

3.5 Uncertainty in non-linear measurements

3.5.1 Introduction to measurement uncertainty

All types of measurements have some sort of associated uncertainties, which are unavoidable due to limitations of the measurement instrument capabilities. The measurement results can only be relied upon with confidence when all the uncertainties associated with a measurement result are fully understood and known. Therefore, it is important to evaluate the measurement uncertainty in order to quantify the reliability of the measurement result.

In this section, the basic uncertainties associated with RF circuit measurements are given. The uncertainties with respect to the RF behaviour of circuit or system measurements are discussed. In measurements of active RF circuits such as amplifiers, which may involve DC components, only the RF uncertainties are treated and the uncertainties associated with the DC components are not covered in this section.

In S-parameter measurements of RF circuits, there are many potential sources of uncertainties or errors. Most of the measurement errors associated with the measurement set-up are greatly reduced by a calibration process such as TRL, SOLT, etc. However, there are further errors, which cannot be completely eliminated by the calibration. These sources of errors can be categorised as systematic or random errors. The systematic error sources are due to the following:

1. The uncertainties in the calibration standards due to physical imperfections of line length, characteristic impedance and non-zero resistivity of the conductors.

2. Imperfections in the instrumentation due to linearity and internal isolation.
3. Post calibration residual errors giving rise to mismatch factors (involving source- and load-match terms).
4. Drift of the VNA.

The random errors are due to the following:

1. DUT connector repeatability, including variations due to changing the orientation between the DUT and the reference test port.
2. Calibration standard connector repeatability, including variations due to changing the orientation between the calibration standard and the reference test port.
3. Laboratory environmental fluctuations (causing random errors in the assumed 'known' characteristics of the calibration items – such as mechanical dimensions of the conductors, permittivity of the dielectric, etc. – and in the behaviour of the instrumentation and the DUT's physical properties).
4. The receiver noise floor due to thermal noise and phase noise.
5. Changes in the characteristics of the RF cable due to any movements (flexing).

3.5.2 *Tools for uncertainty evaluation*

Uncertainties in *S*-parameter measurements can be evaluated by making several repeat measurements of calibration standards and DUTs to estimate the random errors and using knowledge of the measurement process to estimate systematic errors. The associated measurement uncertainties, which include error sources from calibration standards, calibration connector repeatability, DUT connector repeatability, cable flexing, the network analyser noise floor and the equipment linearity are usually incorporated in such measurements. Post-processing of the collected measured data is analysed to evaluate the measurement uncertainties. This method can be very time-consuming for both the user and the test equipment. These methods are used at national metrology institutes (NMIs) and at the equipment manufacturer's calibration laboratories.

The above-mentioned methods are suitable for evaluating the measurement uncertainties in *S*-parameters of DUTs whose characteristics are fixed and not expected to change during a time-consuming set of repeat measurements. For example, at the UK's National Physical Laboratory (NPL), when *S*-parameter measurements are supplied to customers, the measurement uncertainty has been evaluated using the NPL Primary Impedance Microwave Measurement System (PIMMS). The use of PIMMS can be quite onerous as multiple repeat calibrations and measurements are required to evaluate the measurement uncertainty, which is calculated during post-processing of the measurement data.

In the case of evaluating uncertainties in active devices, which may have characteristics which constantly change due to multi-physics properties such as DC supply, thermal self-heating, changes in active junction regions, electromagnetic influence due to proximity, environmental changes, etc., the traditional methods can potentially fail. At NPL, the measurement uncertainty in *S*-parameters in such DUTs is evaluated using the real-time VNA-DUO tool, which is discussed in more detail in the following section.

3.5.2.1 VNA-DUO

The commercial instrumentation manufacturers have often provided software tools such as the Keysight Uncertainty Calculator [13] as a method to evaluate *S*-parameter measurement uncertainty. These tools provide an approximate method of evaluating the measurement uncertainties based on the measurement settings, the connector types and the calibration kit used. Recently, for a more accurate evaluation of measurement uncertainty in *S*-parameters, Keysight Technologies Inc. has introduced the VNA dynamic uncertainty option (VNA-DUO) [14] installable on their current VNAs. VNA-DUO has recently been commissioned for use at NPL [15]. This enables the end-user to obtain a real-time display of *S*-parameter measurements and their associated measurement uncertainty at microwave frequencies.

The following three sources of uncertainty are characterised and accounted for with VNA-DUO for *S*-parameters:

1. **Calibration Standards** – The calibration standards, which have already been characterised by the manufacturer to a known degree of uncertainty, can be used. Alternatively, custom calibration kits with specified uncertainties can be used, e.g. calibration standards characterised at an NMI.
2. **Noise** – The noise at each test port is first characterised using 50 Ω load and short-circuit standards at measurement settings (frequencies, RF power, IF bandwidth, averaging factor) that include the ranges of later DUT measurements. These characterisation measurements are stored and included in subsequent uncertainty evaluations.
3. **Cable and connector repeatability** – The repeatability of test port cables is first characterised using 50 Ω load and short-circuit standards as terminations. These characterisation measurements are stored and included in subsequent uncertainty evaluations.

VNA-DUO makes full use of the capabilities of modern measurement instruments, which have state-of-the-art data processing within the instrument to enable such analysis. VNA-DUO provides accurate measurements and measurement uncertainties in real time. VNA-DUO is particularly important when evaluating active devices, which may have characteristics that vary as described earlier. Tools like VNA-DUO can be easily installed on a supported VNA and provide the end-user with the capability to easily evaluate the measurement uncertainty in real time.

3.5.2.2 MUF

The National Institute of Standards and Technology Microwave Uncertainty Framework (MUF) [16] provides a ‘drag-and-drop’ toolkit for managing the calculation of uncertainties in VNA and other measurements. The MUF makes it easy to construct models for calibration standards and automates the calculation of uncertainties with both a conventional linear error-propagation analysis and a Monte Carlo analysis capable of propagating uncertainties through non-linear models. A VNA Uncertainty Calculator tool is also included in the MUF for guiding the generation of uncertainties in *S*-parameters. The framework also includes many post-processors, including those which allow the uncertainties in measured *S*-parameters to be propagated to derived measurements such as transistor gain, output power, material parameters and mismatch-corrected temporal signals.

The MUF includes the option to drive an NVNA and therefore can measure harmonic performance of a non-linear device and associated uncertainties. The measurements of non-linear devices are more challenging compared to *S*-parameters as they involve further calibrations such as power and phase, etc., in addition to those required for standard *S*-parameter calibrations. These calibrations are not only performed at the fundamental frequency but also at the harmonic frequencies, depending on the measurement type. An example of a non-linear measurement using the MUF is given in Section 3.5.3.2.

3.5.3 Examples of uncertainty evaluation

3.5.3.1 VNA-DUO example

In the case of VNA-DUO, once the VNA measurement set-up has been characterised for noise, cable and connector repeatability, and calibrated with defined calibration standards, the VNA-DUO will display the measurement result and associated uncertainty on the VNA display. Figure 3.21 shows some of the formats that are used to plot the results and hence show the versatility of the tool. The user can select any combination of the three error sources for all display formats. The measurement uncertainty can also be displayed with different levels of confidence, e.g. 67, 95, 99 per cent, etc.

3.5.3.2 MUF example

To set up a VNA measurement using the MUF, the user must first define the calibration standards in use. Common standards are available, although the versatility of the software means that any combination of physical models of standards can be virtually assembled, with parameters of those (e.g. line lengths, diameters) able to have an associated uncertainty and probability distribution assigned to them. Alternatively, data-based calibration standards can be used, including uncertainty, based on previous MUF measurements.

The uncertainty evaluation is executed by a post-processor application, which again includes numerous RF metrology tasks. Each task has a mathematical model defined internally (custom tasks can be added or defined via MATLAB®) and the user can assign input files to the required parameters. The evaluation process is then executed and can utilise Monte Carlo or linear propagations of uncertainty at the discretion of the user. The linear propagation also provides a sensitivity analysis allowing individual contributions of each input quantity to be quantified, which can be useful when identifying potential areas to improve the total measurement uncertainty. An example of this analysis using the MUF is shown for a physical model of a coaxial short-circuit attached to a length of transmission line (an 'offset short') in Figure 3.22.

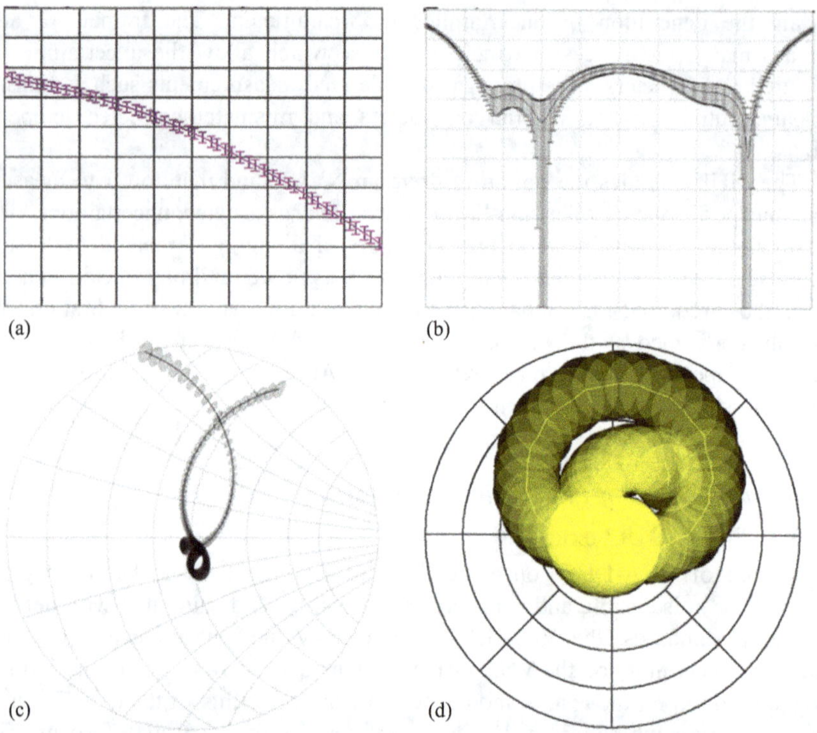

Figure 3.21 Typical S-parameter measurement results and measurement uncertainties using VNA-DUO in different formats: (a) linear magnitude of S_{11} and uncertainty bars, (b) S_{11} magnitude in dB and uncertainty bars, (c) magnitude and phase of S_{11} and uncertainty ellipses plotted on a Smith Chart and (d) magnitude and phase of S_{11} and uncertainty ellipses plotted in polar format

Because VNA calibrations are a very common use for the MUF, a dedicated graphical user interface is provided for the post-processor which concerns them. It shows a synoptic view of the instrument and allows the user to drag measurement files into the programme and allocate them to parameters such as cable repeatability, noise characterisation, calibration standards measurements. These are not the only parameters available and the user can choose what to include depending on their requirements. In contrast to the VNA-DUO, the MUF also includes the ability to calibrate NVNA measurements, by including measurements of the comb generator and power meter. Wave parameter (.wnp) files can be used instead of *S*-parameter files, which can be imported from MDIF files provided by NVNAs. The MUF also provides VNA and NVNA control software which can drive the instrument and keep track of the measurements. Once the calibration has been processed by the MUF, the user can then propagate the results, including associated uncertainties, through further measurement models in additional post-processor tasks. These cover a wide array of

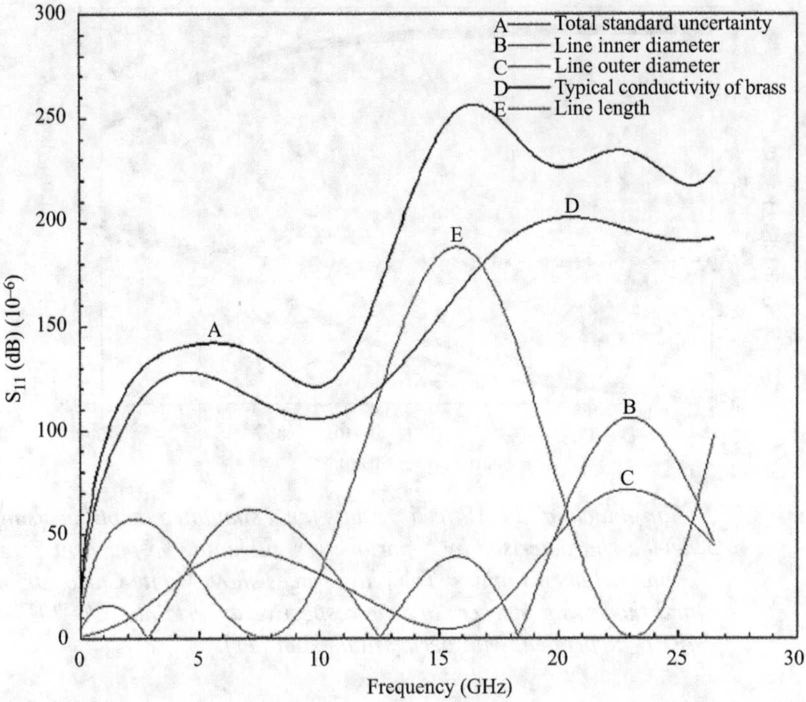

Figure 3.22 The MUF sensitivity analysis results for a 10-mm offset short-circuit in a 3.5-mm coaxial transmission line

applications such as antenna measurements, material characterisation and non-linear device characterisation such as *X*-parameters.

3.5.3.3 *X*-parameter example

By propagating NVNA measurement uncertainty through the *X*-parameter extraction process (e.g. by perturbing the measurements via the Monte Carlo method), confidence in the characterisation of the device can be captured in the *X*-parameter model. This model can then be loaded into design software to predict the expected device performance in a system, including information about the uncertainty of the predicted metrics by way of further Monte Carlo simulations. Figure 3.23 shows the results of this procedure applied to a modern millimetre-wave amplifier [17].

3.6 Summary

This chapter has presented some selected non-linear measurements for 5G devices and the associated uncertainties. Specific topics discussed have included NVNA measurements, load-pull measurements, *X*-parameter measurements, tools for uncertainty evaluation and examples of uncertainty evaluation.

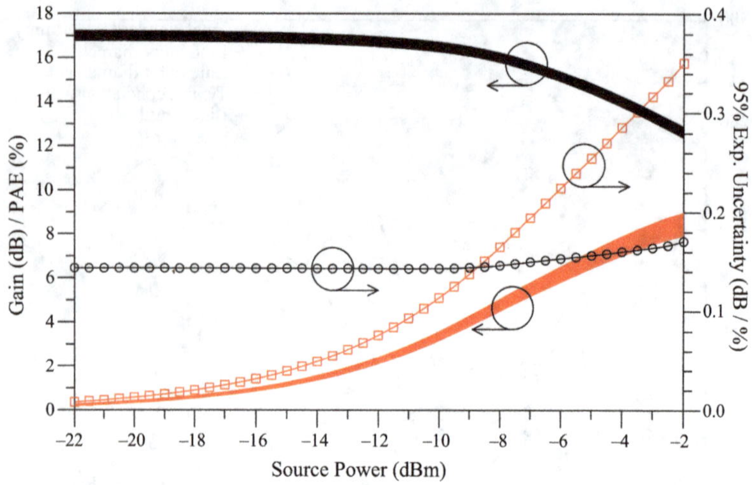

Figure 3.23 Gain (black) and PAE (red) results for a simulated amplifier using a device characterised by X-parameters, including 95 per cent expanded uncertainties. The plain traces show Monte Carlo samples and those with markers are the respective uncertainties. © 2019 IEEE. Reprinted, with permission, from [17].

Acknowledgements

Dr Hyunji Koo from the Korean Research Institute of Standards and Science (KRISS) designed the load-pull test fixture shown in Figure 3.15 whilst a visiting researcher at the National Physical Laboratory, UK (NPL). Clive Barnett from Keysight Technologies supplied Figure 3.21. The work described in this paper was partially funded by the UK Government's National Measurement System (NMS) programme and partially funded through the European Metrology Programme for Innovation and Research (EMPIR) project 16ENG06 'Metrology for advanced energy-saving technology in the next-generation electronic applications'. The EMPIR is jointly funded by the EMPIR participating countries within EURAMET and the European Union.

References

[1] Stant L. 'Measurement uncertainty in nonlinear behavioural models of microwave and millimetre-wave amplifiers', PhD Thesis, University of Surrey, 2020.

[2] Dunsmore J. P. *Handbook of Microwave Component Measurements: with Advanced VNA Techniques*. Hoboken, NJ: John Wiley and Sons, Inc., 2012, pp. 66–243.

[3] Lin M. and Zhang Y. 'Covariance-matrix-based uncertainty analysis for NVNA measurements'. *IEEE Transactions on Instrumentation and Measurement*. 2012, 61(1), pp. 93–102.

[4] Van Moer W. and Gomme L. 'NVNA versus LSNA: enemies or friends?'. *IEEE Microwave Magazine*. 2010, 11(1), pp. 97–103.

[5] Simpson G. Maury Microwave Application Note – A Beginner's Guide To All Things Load Pull OR Impedance Tuning 101 [online]. 2015. Available from: https://www.maurymw.com/ [Accessed March 2021].

[6] Dudkiewicz S. Maury Microwave Application Note – Vector Receiver Load-Pull Measurement [online]. 2011. Available from: https://www.maurymw.com/ [Accessed March 2021].

[7] Ghannouchi F. M. and Hashmi M. S. *Load-Pull Techniques with Applications to Power Amplifier Design*. Dordrecht: Springer, 2013.

[8] Aaen P., Pla J. A. and Wood J. *Modeling and Characterization of RF and Microwave Power FETs*. Cambridge: Cambridge University Press, 2011.

[9] Root, D., Verspecht J., Horn J. and Marcu M. *X-Parameters: Characterization, Modeling, and Design of Nonlinear RF and Microwave Components*. Cambridge: Cambridge University Press, 2013.

[10] Fischer R. F. H. *Precoding and Signal Shaping for Digital Transmission – Appendix A Wirtinger Calculus*. Hoboken, NJ: John Wiley and Sons Inc., 2002, pp. 405–413.

[11] Biernacki R. M., Marcu M., and Root D. E. 'Circuit Optimization with X-Parameter Models'. *IEEE International Microwave Symposium Digest*. June 2017, Honolulu, Hawaii.

[12] Verspecht J., and Root D. E. 'Polyharmonic distortion modeling'. *IEEE Microwave Magazine*. 2006, 7(3), pp. 44–57.

[13] Keysight Technologies. Vector Network Analyzer Uncertainty Calculator [online]. 2021. Available from: https://www.keysight.com/main/software.jspx?cc=GB&lc=eng&nid=-11143.0.00&id=1000000418:epsg:sud&pageMode=CV [Accessed March 2021].

[14] Keysight Technologies. Application Note 5992-3595 – VNA Measurement Uncertainty [online]. 2019. Available from https://www.keysight.com/gb/en/assets/7018-06463/application-notes/5992-3595.pdf [Accessed March 2021].

[15] Singh D., Salter M., Skinner J., and Ridler N. M. NPL Report TQE 16 – Commissioning of a VNA Dynamic Uncertainty Tool for Microwave *S*-Parameter Measurements [online]. 2021. https://doi.org/10.47120/npl.TQE16 [Accessed March 2021].

[16] National Institute for Standards and Technology, Microwave Uncertainty Framework website [online]. 2016. Available from: https://www.nist.gov/servicesresources/software/wafer-calibration-software [Accessed March 2021].

[17] Stant L. T., Salter M. J., Ridler N. M., Williams D. F., and Aaen P. H. 'Propagating measurement uncertainty to microwave amplifier nonlinear behavioral models'. *IEEE Transactions on Microwave Theory and Techniques*. 2019, 67(2), pp. 815–821.

Chapter 4
Multiphysics measurements of RF/microwave power amplifiers

Peter H. Aaen[1] *and Jonas Urbonas*[2]

When you can measure what you are speaking about, and express it in numbers, you know something about it, when you cannot express it in numbers, your knowledge is of a meager and unsatisfactory kind; it may be the beginning of knowledge, but you have scarcely, in your thoughts advanced to the stage of science, whatever the matter may be.

–Lord Kelvin

4.1 Introduction

As you will have seen so far in this book, obtaining precise measurements at RF, microwave, and mm-wave frequencies is a demanding task. With the adoption of 5G, and as we look toward 6G, increasingly complex communication standards pose additional challenges to power amplifier design, measurement, and modeling, as we have to develop techniques to support new challenging modulation schemes that have significantly wider bandwidths, high peak-to-average power ratios, and operation at much higher frequencies.

To achieve the required performance, the construction of a power transistor has also become increasingly more complex and over the past several decades, we have seen it transition from a single semiconductor die within a package to very sophisticated multi-chip modules. In spite of this increasing complexity, the measurement techniques and compact transistor models that are available are often only able to provide information at the transistor input and output ports. For measurements, this is usually restricted to the ports of the power amplifier assembly or reference planes at the edge of the packaged device leads. Apart from a few exceptions as noted in [1,2], the internal operation of the packaged transistor is described only coarsely, and the voltages and currents are only available at a few nodes of interest, effectively obscuring the detailed internal operation of the packaged device. Without fully understanding the internal transistor operation, we

[1]Colorado School of Mines, Department of Electrical Engineering, Golden, CO, USA
[2]Maury Microwave Corporation, Ontario, CA, USA

are often perplexed as we cannot explain poor performance scaling [3], as well as temperature distributions that are dependent on frequency, power, and load terminations [1].

The modeling community has begun to develop distributed coupled multiphysics approaches to transistor modeling, where the co-coupling between the device physics, electromagnetics, and thermodynamics are all solved simultaneously. Several of these approaches have been defined where electromagnetic simulation results (through S-parameters), thermal models (through a thermal impedance matrix), and nonlinear electrothermal transistor models are coupled in the netlist of a circuit simulator. This provides a comprehensive description of the transistor for computationally efficient simulation [1], where the voltages and currents at all nodes of interest within the packaged device are available, and they can be interrogated to develop a better understanding of its internal operation. However, while a circuit simulator can easily provide a circuit designer with the internal transistor current and voltage information, the measurement of these quantities is not a trivial task as contact probing methods, popular in low-frequency analog electronics, will significantly load and alter the performance of a circuit operating at microwave frequencies. Instead, non-contacting measurements of the electromagnetic fields and temperatures can be used to gain similar insights into device behavior.

The focus of this chapter is to motivate the need to treat high-power RF, microwave, and mm-wave transistors as distributed multiphysical systems and this in turn motivates the need to develop new measurement methodologies. As we will present, these devices can be measured as multiphysical systems by augmenting traditional measurement techniques, e.g., load–pull, with electromagnetic and thermal measurement capabilities. The chapter begins with a brief review of classical power amplifier design methodologies, presents the layout of a power transistor, and further motivates the needs for new measurement approaches. This is followed by a detailed discussion of the various methods of measuring distributed electromagnetic fields and temperature gradients as applied to electronic components. We then introduce a multiphysics measurement system used to characterize the distributed behavior of microwave power amplifiers. Finally, we conclude the chapter with several examples showing the insights that can be gained through examining distributed electromagnetic and thermal measurements of a high-power amplifier, measure and visualize an oscillation within a packaged transistor, and assess the performance of a Doherty amplifier. It is our hope that by presenting a detailed overview of recent work, readers will be able to make further advancements in the triad of design, measurement, and modeling of microwave power amplifiers.

4.2 Classical design techniques and transistor layout

In principle, designing a microwave power amplifier should be straight-forward as the shapes of the output voltage and current waveforms dictate the performance of the amplifier. By determining appropriate biasing and load terminations at the fundamental and harmonic frequencies, we can control the waveform shapes and achieve our desired output powers, efficiencies, etc. [4]. However, in practice it is significantly more complicated. Waveform engineering offers an insightful approach to amplifier design and, as noted by Tasker [5], the terminal waveforms provide the unifying theoretical link between the transistor technology, circuit design, and system performance. Over the past several decades, there has been significant effort to develop techniques to measure voltage and current waveforms at RF and microwave frequencies, which has been in long use at low frequencies, and to optimize the shape of the waveforms using harmonic load-pull [5]. In addition, since the waveform engineering technique requires voltage and current waveforms at the current generator plane, efforts to de-embed the package, in-package matching networks, and the nonlinear drain–source capacitance, have been developed [6]. The modeling community has also adapted large-signal modeling approaches by offering additional ports to provide access to these waveforms [7] through the development of embedding methods to include the package model [8]. Through these waveform engineering-enabled test methodologies and nonlinear models, we can systematically approach high-frequency power amplifier design.

An assumption inherent in the waveform engineering approach is that the waveforms of the transistor can be obtained at its drain terminal, as shown in Figure 4.1. This assumption is satisfied for low-power transistors, i.e., transistors with low total gate periphery, but significant complexities arise as the output power is increased. High output powers are achieved through the connection of many smaller transistors in parallel. The implicit assumption is that all individual transistors are performing the same, are excited equally, and contribute equally to the output.

To elaborate on this point, we need to examine the layout of a power transistor, as shown in Figure 4.2, where a photograph of a gallium nitride on silicon carbide (GaN-on-SiC) microwave power transistor with its protective ceramic lid removed is shown. The die is composed of an array of transistors with a unit-gate width (UGW) of 445 μm that are connected in parallel and is often referred to as a power bar. The layout of each transistor is identical and they are interconnected by metallization that is connected to a

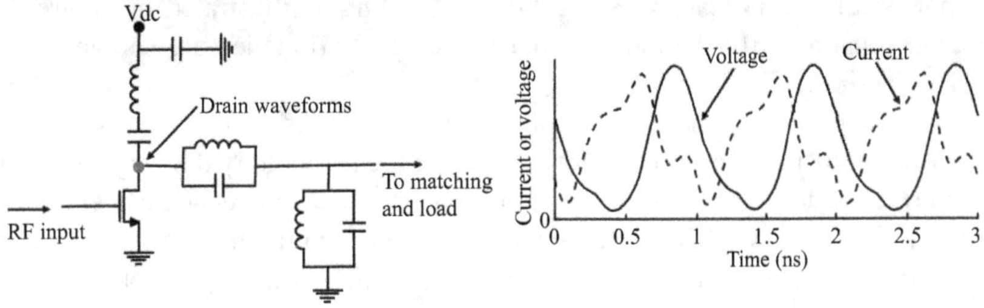

Figure 4.1 An illustration of the waveforms captured at the drain of the power amplifier. The impedances presented to the transistor are systematically varied to optimize the performance of the amplifier.

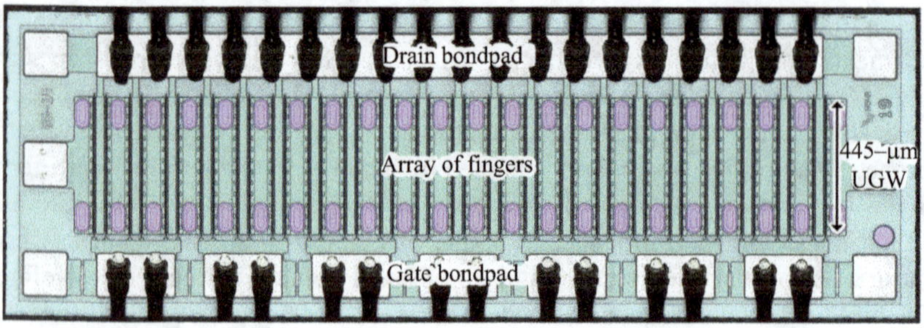

Figure 4.2 Photograph of a high-power gallium nitride (GaN) microwave transistor die showing an array of discrete transistor fingers. Metallization connects all of the transistors in parallel and bondwires are then attached to provide the connection to the package [43].

bondpad. For high-power applications, at frequencies usually lower than 6 GHz, the die may have additional matching circuitry adjacent to it that is composed of bondwires and discrete capacitors, with the entire assembly housed within a plastic or metal–ceramic package. For higher frequencies, more integrated versions with the matching circuitry on-chip are usually designed.

The individual transistors, or small groups of transistors, are called fingers and unit-cells, respectively. During the development of a semiconductor technology, there is a significant investment of resources to optimize the unit-cell for the intended application and this requires extensive collaboration between device, measurement, modeling, and design engineers. The unit-cell is extensively measured and many fabrication cycles will have occurred to optimize its performance and extensive measurement, including DC-IV and pulsed-IV, S-parameter, load–pull, etc. will have been performed. Once the unit-cell design is finalized, it will become a basic building block for power transistors to be designed using this new generation of the semiconductor process.

With the placement of many transistors in parallel, we would expect that the total performance scales linearly as a function of the number of transistors or equivalently the total gate periphery. In practice, what is seen is that linear scaling does not occur and instead we see sub-linear scaling as illustrated in Figure 4.3, where an illustration of the gain, output power, and efficiency are shown vs. the total gate periphery.

Unfortunately, a significant performance reduction is seen as compared with the unit-cell transistor that was optimized during the technology development. This is often referred to as the "scaling" issue for RF/microwave PAs and is well known within the industry. Modeling engineers are also very familiar with this issue and will develop arbitrary empirical, or nonlinear scaling, rules to account for this behavior so that their models can scale as a function of total gate periphery. The onset of sub-linear scaling begins to occur when the electrical phase difference across the die begins to get significantly large [9]. This results in each transistor being excited differently and no longer operating in unison with respect to its neighbors.

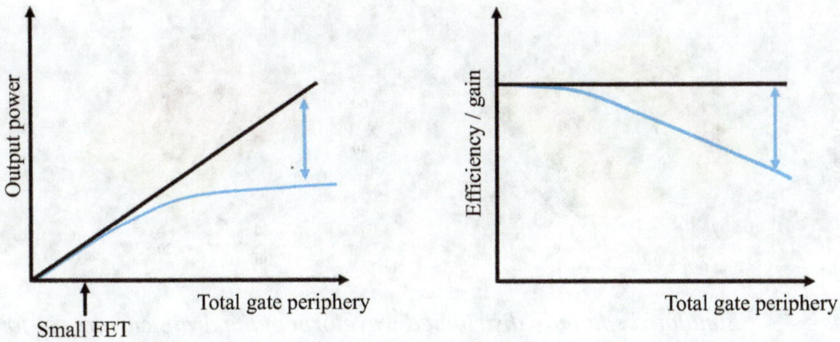

Figure 4.3 An illustration of the transistor performance as more fingers are connected in parallel, increasing the total gate periphery. The performance of the small FET does not scale linearly as desired and instead scales sub-linearly resulting in decreased output power, efficiency, and gain.

As noted at the beginning of the introduction, the transistor modeling community has been developing multiphysics modeling approaches that examine the performance of the individual transistors, and several publications have shown that complex thermal and electromagnetic interactions within the multifinger structure and nonuniform distributions at the transistor gates limit the transistor performance. These limitations manifest themselves as the reduced performance seen when plotting performance as a function of total gate periphery, as illustrated in Figure 4.3.

In a unique simulation, one of the authors here, Aaen *et al.*, presented the results from a multiphysics simulation of a 102-mm high-power laterally-diffused metal-oxide semiconductor (LDMOS) transistor operating at its 1-dB gain compression point (P_{1dB}) at 2.14 GHz [2]. Harmonic balance simulations of a packaged transistor were performed that contained an electrothermal nonlinear transistor model for each individual transistor finger and in total 204 transistor models were used to model the die in combination with a distributed thermal model, and *S*-parameters obtained from electromagnetic simulations for all of the metallization, bondwires, and packaging. From this simulation, the voltages, currents, and temperatures were extracted at DC, fundamental, and harmonic frequencies for each individual transistor within the die. Time-domain waveforms for each pair of unitary transistors that connect to the drain bondpad were computed, using the voltages and currents from the harmonic-balance simulation.

To analyze the performance of the large periphery transistor, the time-domain waveforms at the gates and drains of the individual transistors can be plotted vs. their position along the die. Although all unitary transistor models are identical, they are operating dramatically differently from one another as shown in the drain current and efficiency as plotted in Figure 4.4, for the 102-mm LDMOS die. This composite time-domain waveform allows for the comparison of the performance of each individual transistor relative to one another. While these waveforms vary

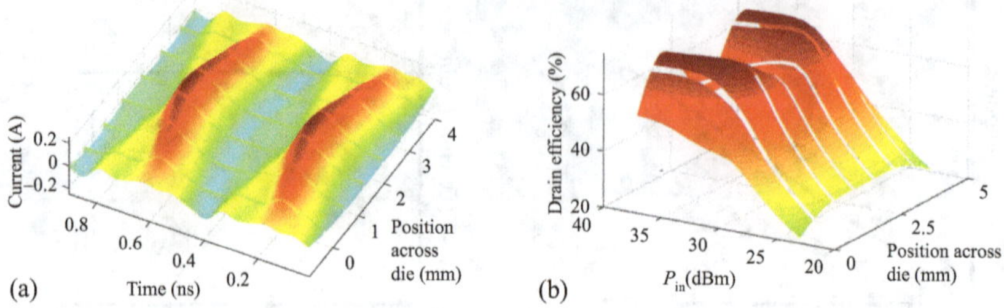

Figure 4.4 Simulated spatially distributed waveform of the drain current (a) for a 102-mm LDMOS packaged power transistor operating at 2.14 GHz, at P_{1dB}. *The waveform shows significant spatial and temporal variation across the die. This results in a significant change in drain efficiency across the die plotted in (b).*

across the die, they are an extension of the waveforms shown in Figure 4.1. With spatial and temporal variations evident, as the magnitude of the drain current is seen to change across the die and is shifted in time, the transistor drain can no longer be considered to be a single node.

It is this background that motivated us to develop new measurement techniques to examine the internal operation of the packaged transistor. In the next sections, we present an overview of electromagnetic and temperature sensing techniques that will be used to augment the conventional measurement methods for power transistors and amplifiers.

4.3 Thermal measurements

Power transistors generate lots of heat and a careful consideration of how to best dissipate it is required to ensure the optimal electrical device performance and reliability. Thermal measurements are used during the optimization of the semiconductor technology, engineering the packaging for best thermal transfer, and to optimize the spacing between transistor fingers. In addition, accurate temperature measurements are essential during the development of nonlinear electrothermal models. In this section, we review three of the most commonly used contactless thermal measurement techniques.

4.3.1 Infrared thermography

Infrared (IR) thermography is widely used and a measurement system comprises an IR-sensitive detector, a microscope with compatible, optics and sensor to measure the intensity of the thermal radiation emitted by the device under test. The thermal radiation from a blackbody is governed by the Stefan–Boltzmann law,

$$W = \sigma T^4, \tag{4.1}$$

where σ is the Stefan–Boltzmann constant and T is temperature in K. Due to the response of the materials used in the detector to a limited range of wavelengths near IR, only a portion of the total spectrum is actually measured [10]. Equation (4.1) is valid for ideal blackbodies with emissivity, ε , equal to 1. Emissivity of a material at a specified wavelength and temperature is defined as the ratio of the emitted thermal radiation by the material and the predicted radiation of an ideal blackbody. Real materials exhibit emissivity over the range of $0 < \varepsilon < 1$; therefore, (4.1) becomes

$$W = \varepsilon\sigma T^4. \tag{4.2}$$

A microwave transistor is constructed with many different materials and each has a different emissivity, i.e., emissivity of some semiconductors can vary between 0.3 and 0.8, and some metals can exhibit emissivities $\varepsilon < 0.1$ [11]. This variability complicates the thermal measurement procedure and requires a calibration scheme. One method to overcome this variability in thermal emissivity, especially when measuring high-power silicon-based devices that are translucent to IR, is to apply a uniform layer of an IR-opaque coating with high emissivity [12]. This ensures a fixed emissivity across the device and simplifies thermal measurements.

Overall, IR microscopy is a very versatile, widely adopted and very accurate measurement technique used to obtain high-temperature resolution measurements [13]. It offers spatial resolutions of ~3 μm and while transient measurements are possible, it is most often used for steady-state thermal imaging.

4.3.2 *Raman spectroscopy*

Another noncontacting method uses a laser that is focused on the material and by measuring the scattered photons the temperature of the material can be determined. The photons from the laser that are incident on the material will be either scattered or absorbed. The photons can be elastically scattered (Rayleigh scattering) or inelastically scattered (Raman scattering). In Raman scattering, the energy of the photons scattered from the sample is different from the incident photons due to the exchange of energy with lattice phonons (vibrations) in the material. This interaction causes the emitted photon to shift to a lower frequency or a higher frequency. As the temperature increases, the ratio between lower and higher frequency photon emissions changes and the absolute temperature can be calculated by analyzing this ratio.

The spatial resolution of the measurement is determined primarily by the focused laser spot size [14]. Spatial resolution offered by this method is significantly better than IR microscopy but the measurements can take longer to perform. The temperature resolution is approximately 0.5–1°C and the accuracy is dependent on the experimental calibration of the Raman spectra [15]. By modulating the laser, transient temperature measurements with sub-nano second temporal resolutions can be achieved. The measurement method is material selective so it excels at measuring the temperature in semiconductors. In addition to the excellent temporal resolution, a significant advantage this technique has over others is the ability to measure the temperature at different depths allowing 3D temperature mapping, although the method can be time-consuming to measure over large volumes or areas as the laser must be scanned across the sample.

4.3.3 Thermoreflectance imaging

Thermoreflectance (TR) has been widely used since the 1960s for semiconductor band structure and dielectric response function studies [16]. Recently it has been adapted for very high spatial resolution thermal imaging. A photograph of a typical TR measurement system is shown in Figure 4.5.

The TR imaging technique relies on the measurement of a thermally induced relative reflectivity change exhibited by the surface of a material. With the change in temperature, the refractive index of the material changes, which in turn changes the reflectivity. This is a complex relationship, but it can be approximated to a first-order equation as

$$\frac{\Delta R}{R} = \left(\frac{1}{R}\frac{\partial R}{\partial T}\right)\Delta T = C_{\text{th}}(\lambda)\Delta T, \tag{4.3}$$

where ΔT is the change in temperature and C_{th} is the TR coefficient. The C_{th} of a material depends on the wavelength of the light used, the angle of incidence, surface roughness, and the overall chemical composition of the sample [17].

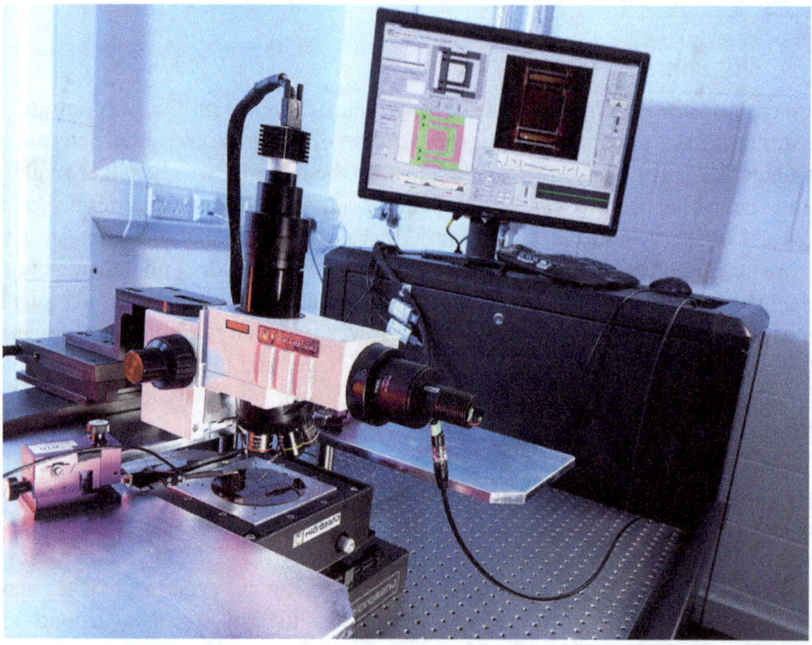

Figure 4.5 A thermoreflectance system from Microsanj LLC. Photograph taken of the installation at n3m-labs, University of Surrey (photograph courtesy of NPL).

The spatial resolution of the TR measurement system at high magnification is limited by diffraction

$$d = \frac{\lambda}{2(nsin\theta)}, \tag{4.4}$$

where d is the spatial resolution, λ is the optical wavelength used to illuminate the sample, θ is the half angle from the light source, and n is the refractive index. The TR approach uses visible or ultraviolet light for thermal measurements, which has shorter wavelengths than, for example, IR thermography, enabling a thermal measurement spatial resolution of 250–300 nm.

The thermoreflectance measurement approach is also well suited to perform very high temporal resolution measurements. This is achieved through constructing a time series of thermal images that are taken at different times in the thermal cycle. Instead of relying on the response time of the charge-coupled device (CCD) camera, modern thermoreflectance systems employ an interlocking transient imaging method, also known as pulsed boxcar averaging, where the measurement system varies the delay between the light emitting diode (LED) pulses used to illuminate the device under test and the device-under test (DUT) excitation pulses. The temporal resolution is limited by the LED pulse duration and is typically around 50 ns, although laser-based systems with a 200-fs temporal resolution have also been reported [18]. Due to the short duration of the LED pulse, the reflected sample illumination power received by the CCD is very low, and a large number of images are averaged to increase the signal-to-noise ratio of the measured signal. The temperature resolution of a TR-based measurement technique is approximately 0.5–1 K for commercially available systems with moderate per-point averaging. Systems having 10 mK thermal resolution have also been reported [17].

The high dynamic temperature measurement range and increased spatial and temporal resolution make thermoreflectance imaging a very appealing noninvasive contactless thermal measurement technique for the characterization of RF and microwave transistors.

4.4 Electromagnetic field measurements

Network parameter and load-pull measurements are extremely useful as they allow us to accurately characterize the behavior of microwave circuits at their ports. They are insufficient, however, for distributed *in situ* characterization of microwave transistors and PAs, which requires access to sub-circuit elements as noted in Section 4.1. Moreover, circuit designers are often in need of voltage, and current measurements and this cannot be easily achieved by using nonlinear de-embedding techniques or oscilloscopes due to the probe invasiveness and limited frequency ranges available. While we cannot measure the internal currents and voltages directly, we can measure the electromagnetic fields within and above the packaged power transistor.

There is a range of measurement techniques and the fields can be measured anywhere above the circuit starting from a few microns above the device (reactive near-field) to several meters away (far-field). The region where the fields are measured dictates the techniques and equipment needed to perform those measurements. Far-field and radiating near-field measurements are usually employed for antenna, antenna array, or over-the-air measurements and require the use of anechoic chambers, calibrated receiver antennas, and signal measurement instrumentation. Measurements of the electric and magnetic fields in the reactive near-field region, on the other hand, are often used to characterize the field behavior of various RF and microwave components as they can be representative of the voltage and current waveforms within DUTs. There are several techniques that have been developed and used to measure waveforms and we summarize several of these in the subsections that follow.

4.4.1 Metallic probes

Most metallic probes used for electric field (*E*-field) probing are based on coaxial assemblies [19–21], although more advanced resonance probes [22], dipole probes used in modulated scattering configuration [23], miniature dipole probes [24], and active miniaturized coaxial probes [25] have also been reported. A coaxial probe acts as a monopole antenna sensitive only to the normal *E*-field component, and it is usually electrically small to minimize the perturbations of the electric field. If the probe is positioned close to a device under test, the electric fields are sensed by capacitive coupling between the probe and the DUT metallizations, thus enabling excellent spatial resolution (sub 100 μm [25]) and very high sensitivity compared to other probing techniques. Similarly using small conductive loops, or arrangements of loops, the magnetic field can also be measured.

Although metallic probes can be used for *E*-field probing, the probes require extensive calibration in order to characterize its response. This is due to the height and the metallization-dependent capacitive coupling between the probe and the transistor. Unfortunately, this can make it complicated to map distributed *E*-fields above transistors as many parts of the transistor are not metallized resulting in a loss of the coupling capacitance. Moreover, the presence of a metallic probe can disturb the electromagnetic fields above the DUT. Care must be taken as the probe can induce strong localized electric fields that may not be observed as changes in measured port-based parameters, for example, S-parameters.

4.4.2 Electro-optic probing

Electro- and magneto-optic probing are some of the most well established RF and microwave electromagnetic field probing techniques for very accurate and minimally invasive field measurements [26]. The technique has been widely adopted for many applications ranging from monolithic microwave-integrated circuit diagnostics [27] to electromagnetic leakage inspection in RF multichip modules [28], and antenna and phased array measurements [29,30].

The electro-optic (EO) method is based on the linear electro-optic effect, also known as the Pockels effect, whereby an externally applied electric field changes the refractive index of a material, which in turn modifies the polarization of the light passing through that material, as illustrated in Figure 4.6.

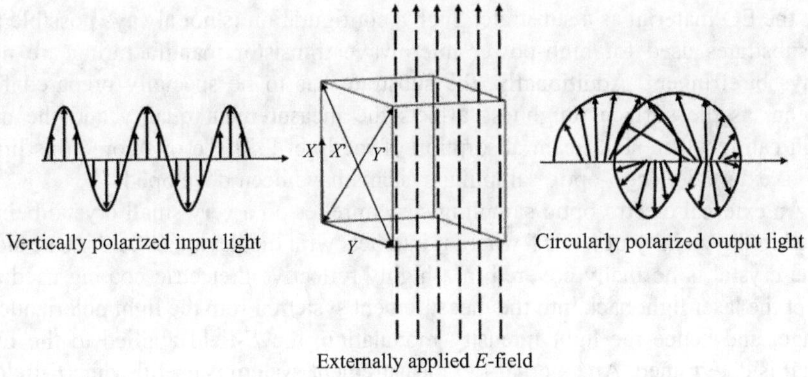

Figure 4.6 Light polarization change due to induced birefringence in an electro-optic crystal by an externally applied electric field [43]

The Pockels effect is exhibited by birefringent anisotropic crystals or other non-centrosymmetric media [31]. To use a birefringent crystal to deduce the strength of an externally applied E-field, the crystal is usually placed between two crossed polarizers together with a quarter wave plate forming a light intensity modulator. A quarter wave plate is used to eliminate the effects of a static material birefringence and bias the system for 50% light intensity transmission. If the crystal is exposed to a sinusoidally varying E-field, it will create a sinusoidally varying potential difference across it, resulting in the sinusoidal light intensity modulation as

$$\frac{I_o}{I_i} \simeq \frac{1}{2}(1 + \Gamma_m \sin\omega_m t) \tag{4.5}$$

for $\Gamma_m \ll 1$, where I_o is a modulated output beam intensity, I_i is the incident light beam intensity, Γ_m is the phase retardation incurred due to the externally applied E-field, and ω_m is the modulation frequency. By employing the intensity modulation configuration, electro-optic E-field probing systems can be manufactured, which deduce the strength of an externally applied electric-field to the EO crystal by measuring the change in the light intensity. Moreover, as the light intensity modulation is the direct replica of its modulating E-field, full time-domain information about the signal can be easily deduced.

There are two classifications of EO probing methodologies: internal and external. In an internal EO probing system, the substrate of the DUT is used as an EO medium and the laser light is shone through the substrate in a way that the light is reflected from the circuit metallizations. This enables the detection of a polarization change in the laser light and the measurement of the E-field inside the substrate [32]. Internal EO probing is completely noninvasive as no equipment is brought into the vicinity of the DUT and E-fields are measured optically without coupling away a portion of the signal. Internal EO measurement systems also boast excellent spatial resolution, down to 0.5 μm [33], as it is only dependent on the laser beam spot size. However, internal EO probing is limited only to the DUTs that have the EO material as a substrate. Such a configuration is not always possible as the substrates used for high-power microwave transistor manufacturing are not always birefringent. Additionally, the substrate has to be specially prepared for probing, as the surface roughness affects the measurement quality and the die attach can result in significant absorption of the light [32]. To overcome this limitation, external electro-optic sampling systems have been developed.

An external electro-optic sampling system relies on a very small crystal being placed in the vicinity of a DUT where it interacts with the device E-fields. One side of the crystal is normally covered in a highly reflective dielectric coating used to reflect the laser light back into the measurement system. From the light polarization change, and hence the light intensity modulation, the E-field applied to the EO crystal is determined. An external EO measurement system is useful when E-fields have to be measured above the device and for those built on non-EO substrates. To make the probe positioning easier, an all-dielectric optical fiber-coupled probe architecture has been developed [34]. In this probe configuration, the birefringent crystal is connected to the polarization-maintaining single-mode optical fiber, which is used to propagate the light into the crystal. Such a probe construction allows an easy placement of the probe anywhere around the DUT, without requiring any complicated probe holders. A typical external EO measurement system using fiber-coupled probes is shown in Figure 4.7.

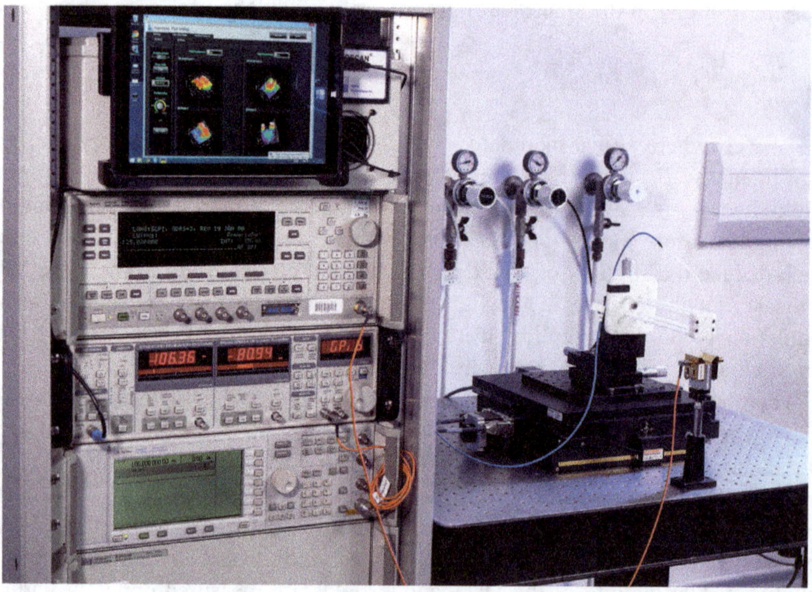

Figure 4.7 An electro-optic near-field scanning system from EMAG Technologies Inc. Photograph taken of the installation at n3m-labs, University of Surrey (photograph courtesy of NPL).

Due to the axial dependence of the birefringence in the electro-optic materials, the crystal used to probe the *E*-fields can be cut in a way that light intensity modulation is only produced by one *E*-field vector component. This property can be used to manufacture probes to sense E_x, E_y, and E_z field components separately. The *E*-field measurements are minimally invasive as the birefringent crystal and the fiber-optic cable to which the crystal is attached are made out of dielectric materials [34].

4.5 Multiphysics measurements

To measure the electric field and temperatures within a packaged microwave power transistor, we assembled a nonlinear vector network analyzer (NVNA) based load-pull system and augmented it to include electro-optic and thermoreflectance measurements as shown in Figure 4.8. The thermoreflectance technique was selected as it has a superior spatial and temporal resolution as compared with IR, but also allows the temperatures to be measured over a wide area. The electro-optic technique was selected to measure the electric fields as it allows full-vector (magnitude and phase) measurements of the x, y, and z components of *E*-field. As a result, the measurement setup allows nonlinear transistor electrical (port-based) characterization, in addition to transient and steady-state thermal, and distributed multi-harmonic vector *E*-field measurements. The NVNA-based measurement setup is used as it allows reconstruction of the large-signal time-domain current and voltage waveforms at the device reference planes via the inverse Fourier transform [35].

Load-pull measurements are enabled by the use of two passive impedance tuners. One tuner is used to control the impedance presented to the DUT on the source side and the other tuner is used to control the impedance presented to the DUT on the load side, although more tuners can be incorporated to perform harmonic impedance tuning. Once load-pull measurements have been completed and impedance states of interest identified, the TR system can perform transient and steady-state thermal measurements. Traditionally, the CCD-based transient TR measurements of microwave transistors are obtained under pulsed-DC operation and in this system this concept is extended to include the RF DUT excitation pulse. The bias and RF excitation are turned on and off synchronously with the CCD camera shutter and an LED illumination pulse, which acts as a gating signal for the thermal measurement. By offsetting the LED pulse with respect to the bias and RF excitation pulse, the temperature can be sampled, which results in the series of transient thermal images. The combination of TR measurements with load-pull enables a simultaneous interrogation of thermal and electrical performance at various source and load impedance states. A fully assembled TR and load-pull measurement bench is shown in Figure 4.9.

Finally, after both load-pull and TR measurements are performed, one of the VNA receivers, receiver A in Figure 4.8, is repurposed to perform multi-harmonic vector *E*-field measurements. The time-domain output signal from the EO system is measured directly by the NVNA, where, with the appropriate system calibration, accurate time-domain waveforms can be measured at any position above the DUT.

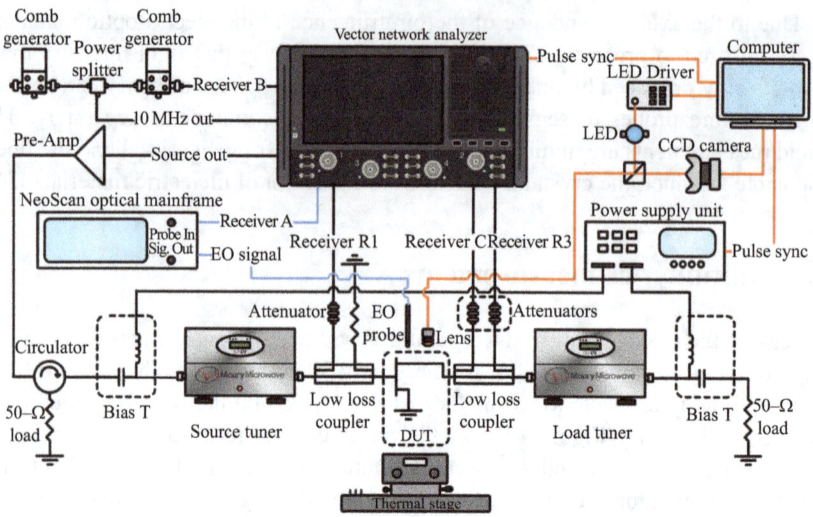

Figure 4.8 Schematic showing integrated multiphysics measurement system for transient and steady-state thermal (orange path), minimally invasive multi-harmonic vector E-field (blue path) and NVNA-based load-pull (black path) measurements combining commercially available EO, TR, and NVNA measurement systems

During the *E*-field measurement, the EO probe is placed into a dielectric holder that is mounted on a 2D translation stage. The translation stage is moved in fixed steps within the *x*–*y* plane above the DUT. The measurement is controlled by a custom-control software, which triggers the pulsed NVNA measurements and synchronously controls the movement of the 2D translation stage. The fully assembled measurement system is shown in Figure 4.10. This setup allows fully automated measurements of the distributed multi-harmonic vector *E*-fields with cross-frequency phase coherence under varying source and load terminations.

All three measurement methodologies: load-pull, thermoreflectance imaging, and EO *E*-field scanning requires calibration and system validation before measurements can be performed. The system calibration and validation for load-pull measurements was discussed in Chapter 2 and the calibrations for the TR and EO systems with the NVNA for waveform measurement are described next.

4.5.1 System calibration and validation

The Microsanj NT220B transient thermoreflectance imaging system used in the multiphysics measurement bench in Figure 4.8 is able to assess the DUT temperature by measuring the relative change in reflectivity from its surface. The reflectivity change with respect to the temperature is dependent on the sample material and wavelength of illumination. A TR calibration is performed by placing the transistor on a thermal chuck and varying its temperature. A monochromator is

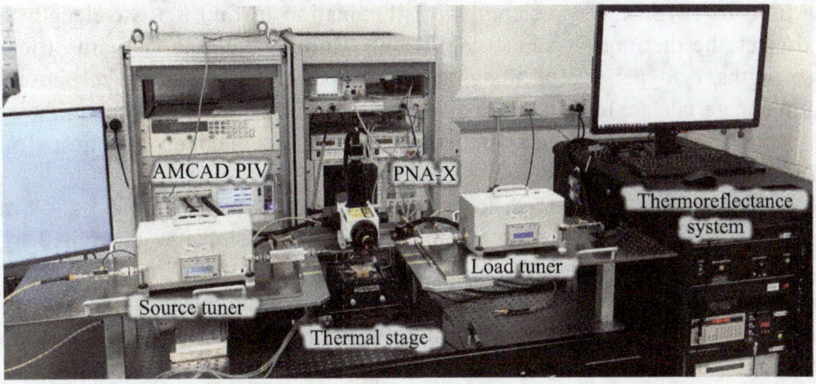

Figure 4.9 Photograph of the fully assembled thermoreflectance and NVNA load-pull measurement system at n3m-labs, University of Surrey. In this setup the Keysight PNA-X is used as an NVNA and AMCAD PIV system is used as a pulsed-DC power supply. © 2019 IEEE. Reprinted, with permission, from [41].

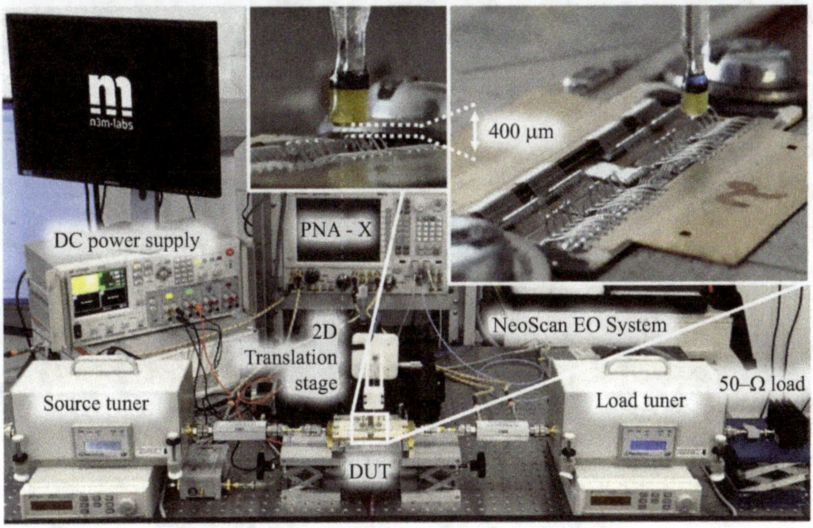

Figure 4.10 Fully assembled electro-optic and pulsed NVNA load-pull measurement system at n3m-labs, University of Surrey. In this setup the Keysight PNA-X is used as an NVNA. During the E-field scans of LDMOS transistors, the probe was positioned 400 μm above their tallest bondwires. © 2018 IEEE. Reprinted, with permission, from [37].

then used to sweep the wavelengths of illumination to find the wavelength that maximizes the thermoreflectance coefficient. Multiple calibrations with different wavelengths, or hyperspectral imaging, will be performed if thermal responses of multiple materials are to be measured. This is especially important when measuring microwave transistors and temperatures of both the semiconductor channel and circuit metallizations need to be obtained [36].

To calibrate the EO system for accurate distributed vector *E*-field and time-domain waveform measurements, a microstrip transmission line is excited by a multi-harmonic time-domain signal. This signal is reconstructed and converted to time-domain by first obtaining a calibrated NVNA measurement at the port reference planes, and then performing an inverse Fourier transform on the measured frequency-domain data. The EO system is then connected to the NVNA and a complex calibration coefficient is derived by comparing the raw EO signal measurement to the calibrated NVNA port reading. Application of the calibration coefficient to the raw EO measurement signal enables calibrated time-domain measurements above a DUT. Finally, the calibrated EO and NVNA system is used to measure a different time-domain signal, as illustrated in Figure 4.11, verifying the calibration and the capability of the system to perform the distributed time-domain *E*-field measurements. Full details of the calibration procedure and comparison of the measured time-domain *E*-field values are provided in [37].

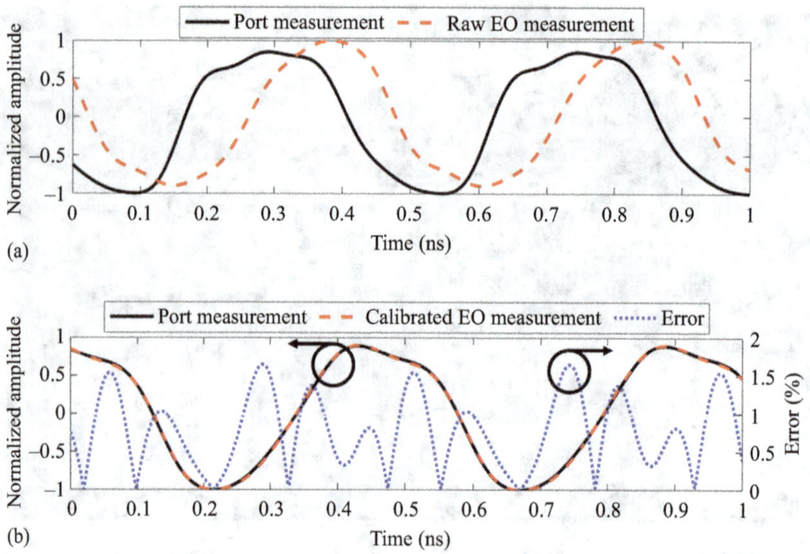

Figure 4.11 (a) Time-domain waveform used to obtain the correction factor, and (b) a calibrated waveform measurement of different multi-harmonic excitation using the same correction factor. Both waveforms are reconstructed using five harmonics. © 2018 IEEE. Reprinted, with permission, from [37].

During the time-domain E-field measurements, the NVNA uses an external harmonic phase reference to establish the cross-frequency phase coherence between measured fundamental and harmonic tones [35]. This enables the time-domain E-field reconstruction at all measurement points above the DUT. The time-domain E-field reconstruction from the measured data is computed by

$$\boldsymbol{E_{TD}} = \sum_{n=1}^{k} \Re\{\boldsymbol{\kappa}_n \boldsymbol{E}_n e^{jn\omega t}\}, \tag{4.6}$$

where $\boldsymbol{E_{TD}}$ is the time-domain distributed electric field above the DUT, $\boldsymbol{\kappa}$ is the complex EO signal calibration coefficient, ω is the radial frequency, t is time, n is the harmonic number, k is the number of measured harmonics, and $\boldsymbol{E}_n$ is the measured distributed electric field at nth harmonic as given by

$$\boldsymbol{E}_n = \begin{bmatrix} E_{n0,0} & E_{n1,0} & \cdots & E_{nx,0} \\ E_{n0,1} & \cdots & \cdots & \cdots \\ \cdots & \cdots & \cdots & \cdots \\ E_{n0,y} & \cdots & \cdots & E_{nx,y} \end{bmatrix}, \tag{4.7}$$

where x and y are the translation stage coordinates of x- and y-axes, respectively. Throughout this chapter, the distributed vector E-field results are normalized to their own maximum, as the E-field and light polarization change relationship inside the EO crystal is scalar and linear. The absolute E-field amplitude calibration is not necessary, albeit possible by employing μ-transverse electromagnetic cells that allow the determination of the scalar E-field amplitude calibration factor [38].

4.5.2 Measurements of high-power LDMOS transistors

4.5.2.1 Stable operation

The performance of large-periphery microwave transistors can be substantially degraded by nonuniform voltage and current distributions across the transistor dies [2], and E-field probing allows the measurement and visualization of that non-uniformity. As a demonstration, a commercially available A2T21S260W12N LDMOS 260-W power transistor was measured [37]. The packaged transistor had four parallel LDMOS dies, metal oxide–semiconductor (MOS) capacitors inter-connected with over 100 bondwires forming matching networks, with everything housed in a single air-cavity metal ceramic package as shown in Figure 4.12. The transistor package lid was removed during all measurements to allow access for the EO probe.

The NVNA load-pull measurements were performed under a pulsed continuous wave excitation with a pulse duration of 10 μs, and a 10% duty cycle. The load impedances presented to the DUT were controlled at 2.2-GHz fundamental while measuring three harmonics to find the impedance point at which the DUT output power is maximized at the 3-dB gain compression point (P_{3dB}). After this impedance point was identified, the EO system was connected to the receiver A of the NVNA and positioned 400 μm above the tallest bondwires of the DUT as

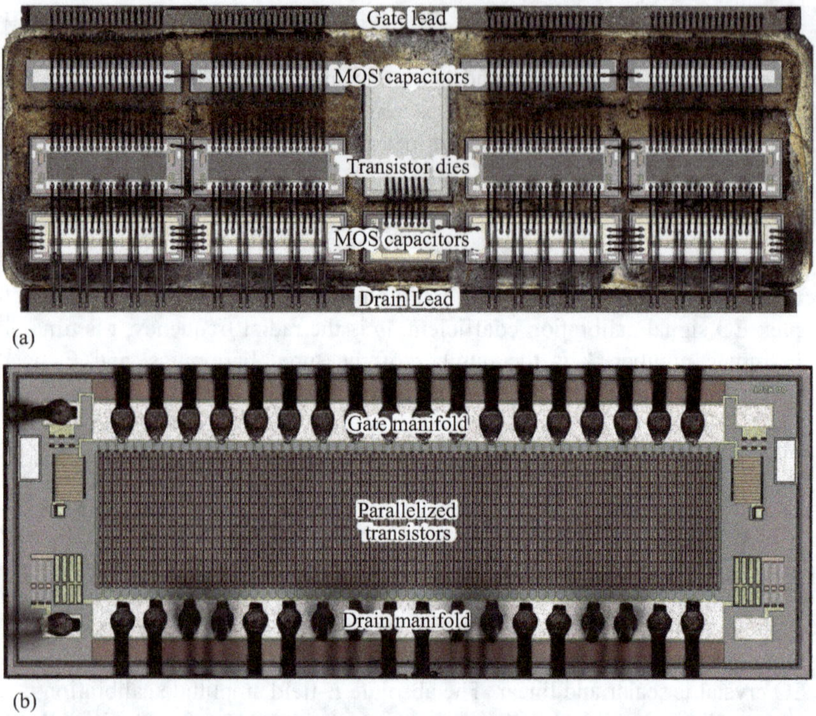

Figure 4.12 Large-periphery 260-W LDMOS transistor photographs: (a) image of the transistor with its protective lid removed revealing multiple transistor dies connected in parallel and arrays of bondwires and MOS capacitors forming pre- and post-matching networks; (b) a zoomed-in image of a single-transistor die, revealing 134 parallelized individual transistors required to achieve high output power

shown in the insert of Figure 4.10. The *E*-field scan area was 12 mm $\times$ 18 mm and all three vector components were measured. After the scan, plots of the distributed fields above the transistor can be generated using (4.7).

All three harmonics of the E_z field component are presented in Figure 4.13 along with the full time-domain-reconstructed E_x and E_y components. The measured E_z field distribution across the transistor is uniform at the fundamental frequency as can be seen in Figure 4.13(a), but the 2nd and 3rd harmonics in Figure 4.13(b) and 4.13(c), respectively, show non-uniform *E*-field distributions. This suggests that the DUT is well optimized at the fundamental frequency, but the harmonic terminations for the individual fingers may not be optimal. The E_x field measurements show a strong drain-to-gate coupling as can be seen in Figure 4.13(d), as a consequence of the gate-to-drain (C_{gd}) capacitance of the packaged transistor. The E_y field component shown in Figure 4.13(e), on the other hand,

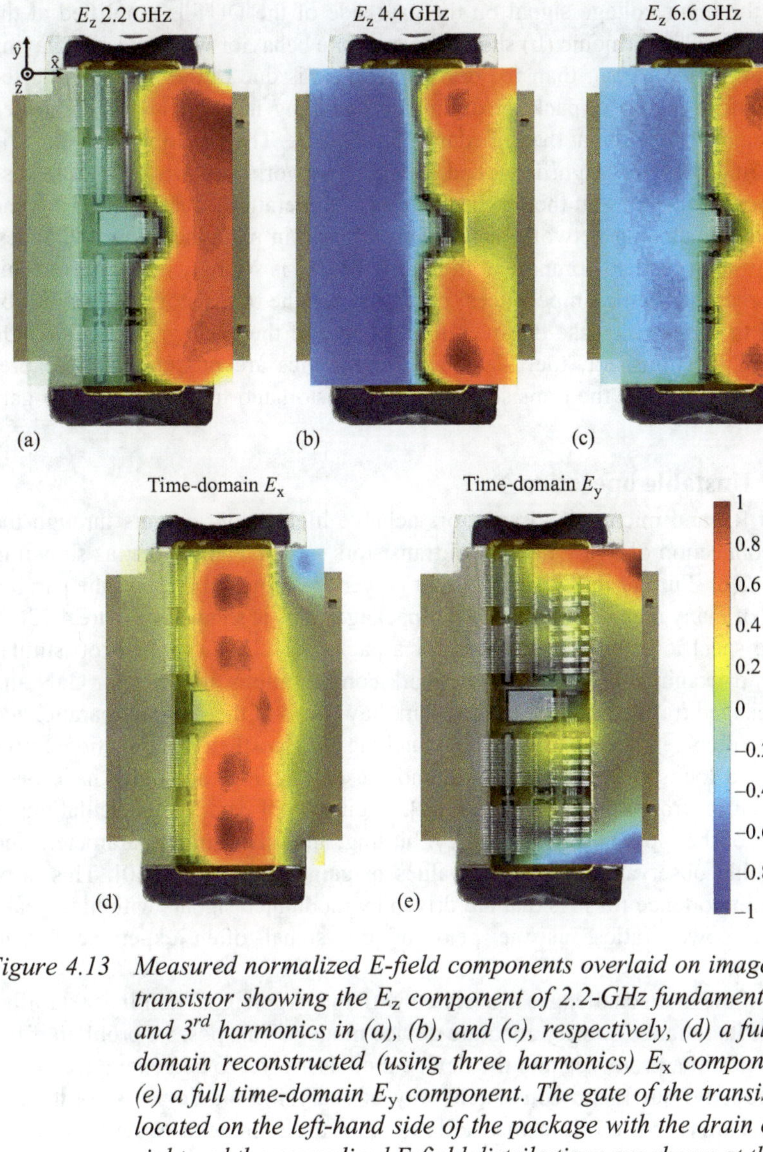

Figure 4.13 Measured normalized E-field components overlaid on images of the transistor showing the E_z *component of 2.2-GHz fundamental, 2nd and 3rd harmonics in (a), (b), and (c), respectively, (d) a full time-domain reconstructed (using three harmonics)* E_x *component, and (e) a full time-domain* E_y *component. The gate of the transistor is located on the left-hand side of the package with the drain on the right and the normalized E-field distributions are shown at their peak amplitudes. © 2018 IEEE. Reprinted, with permission, from [37].*

shows minimal coupling between the transistor dies as they all operate at the same potential.

As can be seen from the *E*-field plots in Figure 4.13, the fundamental (a) and the 3rd harmonic (c) of the E_z component are showing a typical amplifier behavior, whereby the input voltage signal on the gate side of the DUT is amplified at the drain side. The 2nd harmonic (b) shows the opposite behavior with gate-side having a larger amplitude signal, than the drain side. This is due to the partially shorted 2nd harmonic using an in-package drain-side matching network and uncontrolled 2nd harmonic termination at the impedance tuner plane. The variation of the field at the 2nd harmonic is very significant and matching networks need to be optimized so that all dies are working in their expected class of operation. Despite each die and its associated matching network being identical, within manufacturing tolerances, the 2nd harmonic *E*-field component in Figure 4.13(b) is strongly nonuniform. This suggests that harmonic impedances presented to the transistor dies could be improved to further increase the overall efficiency of the packaged transistor. On the gate of the transistor, the 2nd and 3rd harmonics are produced by the pre-amplifier used to drive the transistor into compression and transistor drain-to-gate coupling.

4.5.2.2 Unstable operation

Packaged RF and microwave transistors achieve high output powers through the parallel connection of many individual transistors within a single die, as shown in Figure 4.12(b). Further increases in output power are achieved through the parallel assembly of many dies within a transistor package. This is shown in Figure 4.12(a). While the specific assembly is shown for a packaged LDMOS transistor, similar assembly, packaging, and matching network configurations are used for GaN and GaAs packaged transistors. One of the main drawbacks of these highly parallelized networks, often referred to as cluster matching, is that their construction forms many closed loops within the package and these matching topologies have been shown to suffer from loop oscillations [39]. Excitation of parametric oscillations is a function of the input power, frequency, biasing, and other circuit parameters and they are only observable at specific values of gain compression [40]. This is of particular importance for PAs that are driven by modulated signals with high peak-to-average power ratios, as the peak of the signal often experiences gain compression.

One of the main challenges associated with the large-signal nonlinear stability analysis is its reliance on very accurate device models. This poses a problem if the transistor is still a prototype and the device models are yet to be generated. In this case, there is no quick and convenient way of analyzing the device stability and directly identifying the origin of oscillations. In a circuit with N active devices connected in parallel, there might be $N-1$ independent odd-mode oscillations; thus tests that are capable of identifying and visualizing oscillations *in situ* are needed.

Until now, the AM/AM power sweep curves obtained through load-pull measurements and the power spectrum analysis were the only measurement-based techniques to identify high-frequency parametric oscillations within microwave power transistors and PAs. However, these methods provide no information about the location of the oscillations, particularly the closed loops in the case of a parametric odd-mode oscillation. To overcome this, the multiphysics measurement

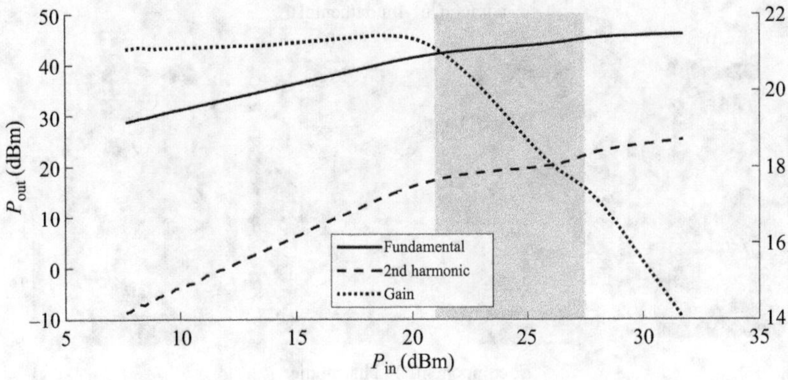

Figure 4.14 The P_{out} and gain curves vs. P_{in} of an LDMOS transistor at a load impedance where parametric odd-mode oscillation is observed. The shaded area indicates the approximate region with a parametric oscillation. © 2018 IEEE. Reprinted, with permission, from [44].

system was used to identify and visualize the closed loops within the power transistor *in situ*.

A modified prototype variant of the commercially available 260-W LDMOS transistor presented in the previous section was deliberately designed to exhibit parametric odd-mode oscillations within the transistor package. To obtain power sweep curves that reveal the parametric odd-mode instabilities, pulsed NVNA load-pull measurements were first performed on the transistor. This was needed as the parametric oscillations within transistors are dependent on the source and load impedances, drive power, gain compression, and other circuit parameters [40]. The NVNA was configured to measure three harmonics of a 2.11-, 2.14-, 2.17-, and 2.2-GHz fundamental tone with a 10-μs RF signal pulse width and 10% duty cycle.

The power sweep curves at several impedance points at a 2.2-GHz fundamental had a decrease in the P_{in} vs. P_{out} curve as shown in Figure 4.14, which is characteristic of a parametric oscillation. A change in the slope of the gain curve can also be observed, further denoting the presence of a parametric oscillation. No spurious spectral components were observed in the measured P_{out} spectrum suggesting that the closed loops formed at both fundamental and 2nd harmonics support odd-mode oscillations that lead to partial cancellation of waves inside the transistor.

After the load-pull measurements were completed and load impedance points at which the device exhibits parametric oscillations were identified, the measurement system was configured to measure the normal *E*-field component, E_z, perpendicular to the LDMOS transistor surface. The load impedance was varied to present specific impedance points where the oscillation is present. During the E_z field measurements, the external EO probe was positioned 300 μm above the tallest bondwires of the packaged transistor. The measurement area was 18 mm × 12 mm.

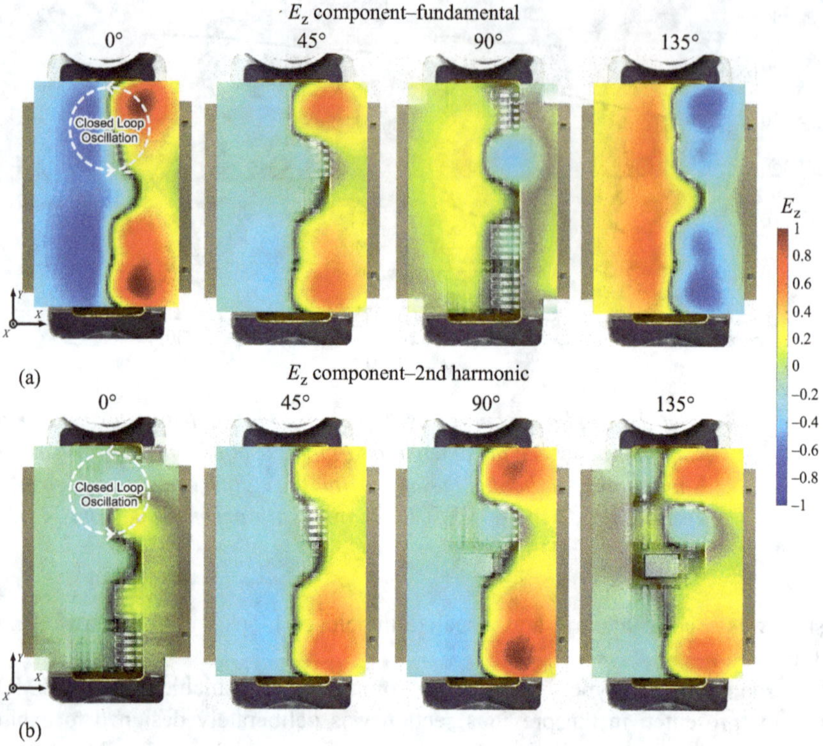

Figure 4.15 Normalized E_z-field measurement overlaid on a 4-die LDMOS transistor picture showing the locations of closed loops and dies oscillating in odd-mode. The measurement results are shown at (a) 2.2-GHz signal and (b) 4.4-GHz for 0°, 45°, 90°, and 135° phase of the measured signal. In both (a) and (b), only the top pair of transistor dies is oscillating in odd-mode resulting in an asymmetrical device operation. © 2018 IEEE. Reprinted, with permission, from [44].

The normalized *E*-field measurement results shown in Figure 4.15 reveal the bottom two dies operating in even-mode and the top two dies operating in odd-mode at both fundamental (Figure 4.15(a)) and 2nd harmonic (Figure 4.15(b)). This is seen by the out-of-phase normalized E_z-field values on top of the output matching networks. As the magnitude and phase of the *E*-field are measured at multiple harmonics for each pixel, time series of the measured *E*-fields are computed by employing an inverse Fourier transform as described in Section 4.5.1. These surfaces overlaid over the transistor image are shown in Figure 4.15 for 0°, 45°, 90°, and 135° of the measured signal and reveal the location of closed loops of the odd-mode oscillation, the signal coupling back to the input, and show the nonuniform device operation.

Having the ability to characterize such DUT behavior *in situ* enables PA designers to add oscillation suppression and performance optimization circuitry intentionally and only in the areas of the circuit where they are absolutely needed. This limits the degradation in device performance that results from the introduction of the stabilization circuitry.

4.5.3 *Measurements of GaN HEMTs*

To demonstrate the capability of the multiphysics measurement system to perform transient and steady-state thermal measurements, we performed a thermal characterization of a commercially available Cree (CGH40025) GaN-on-SiC 25-W 20-finger high electron mobility transistor (HEMT) operating at 3.5 GHz. Load-pull measurements were first performed to find the maximum output power and maximum drain efficiency impedance points. For each of these impedance points the maximum output power (max P_{out}) and drain efficiency were captured at the saturated output power, P_{sat}, which is defined when the gate current, $I_g = 720$ μA. At their respective impedances, the maximum output power was 45.4 dBm and the maximum drain efficiency was 73.9%. We note that if harmonic tuning was available, we would expect the performance of the transistor to be improved over these measurements.

After the optimum impedances were identified, the RF and DC biasing pulses were synchronized to the LED pulses and the CCD camera shutter to enable TR measurements. The TR measurements of the transistor were obtained at the quiescent DC operating point ($V_{ds} = 28$ V, $I_{dq} = 250$ mA), maximum efficiency impedance point, and maximum P_{out} impedance point. Two sets of thermal measurements were performed: transient and steady-state at 20× magnification using the ultraviolet LED with a 365-nm wavelength. The 365-nm LED was selected as the GaN channel of the device is found to have the highest TR coefficient at this wavelength.

Temperatures measured during the steady-state device operation, 10 s after the DUT excitation, are shown in Figures 4.16 and 4.17 revealing the thermal distributions for different device operating conditions. As expected, there is a significant increase in the DUT temperature at the maximum P_{out} impedance state. The measurement results shown in Figure 4.16 used a pixel-by-pixel TR calibration, which allows the extraction of the temperature for each individual pixel in the image, including the GaN channel. To examine the thermal distribution across the transistor, the channel temperatures were averaged along the transistor gate width and plotted in Figure 4.17 for each channel. It can be seen that there is a 70 °C difference between the maximum output power (Figure 4.17(a)) and DC (Figure 4.17(c)) peak temperatures. Only a small thermal difference can be observed in Figure 4.17(b) compared to the DC only measurements, which is due to a relatively small increase in the power dissipation when the RF drive is applied to the transistor and it is tuned for maximum efficiency. The thermal distributions in Figure 4.17 show a uniform temperature across the transistor die. This is especially apparent for the temperature measured at the maximum P_{out} impedance state.

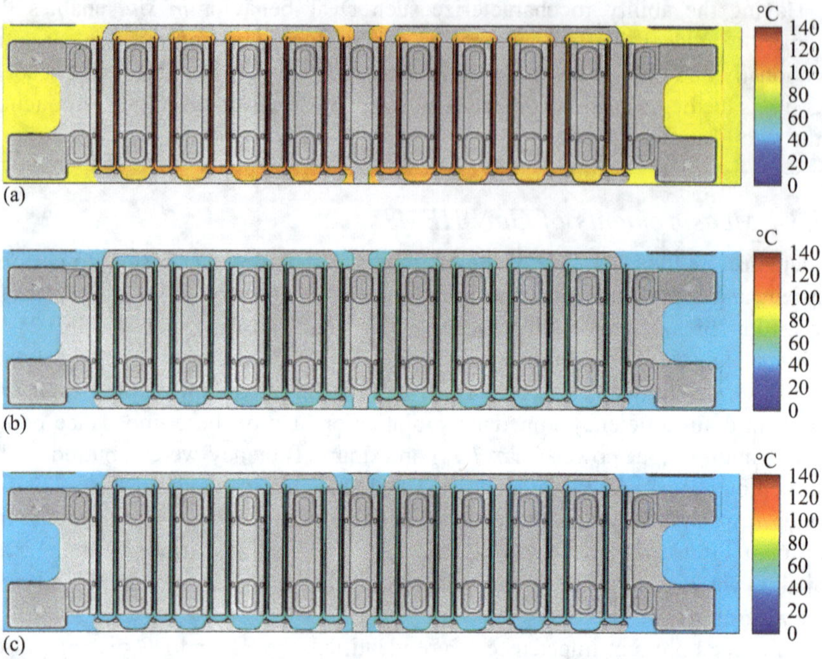

Figure 4.16 Thermal distribution overlay across the entire GaN transistor die as measured using the TR and load-pull measurement setup: (a) shows max P_{out}, (b) max efficiency, and (c) DC only thermal distributions with the TR coefficient obtained from the pixel-by-pixel calibration. © 2019 IEEE. Reprinted, with permission, from [41].

In all thermal distributions shown in Figure 4.17, thermal noise from the edges of the transistor vias can be observed. This is an artifact from the data post-processing as the TR coefficient of the transistor channel is very similar to that of the edges of the vias. Thus, when filtering out all of the TR coefficients except for the ones in the transistor channel, the edges of the vias are also left in the thermal image. This noise is highlighted in Figure 4.17(a).

The transient thermal measurement results of the first 200 μs over the whole transistor die operating at maximum P_{out} were also performed. It can be seen in Figure 4.18 that the individual transistor fingers within the finger pairs are heating asynchronously, with adjacent pairs of fingers heating at different rates until steady state is reached. In Figure 4.18 only fingers 9 and 10 are shown, where finger 9 is substantially hotter across the entire transistor gate width when compared to the finger 10, but all other pairs of transistor fingers exhibit the same behavior, which is a surprising result considering a symmetric, back-to-back structure of the transistors within the die. In steady state, the difference in the finger temperature becomes negligible as shown in Figure 4.17(a) [41].

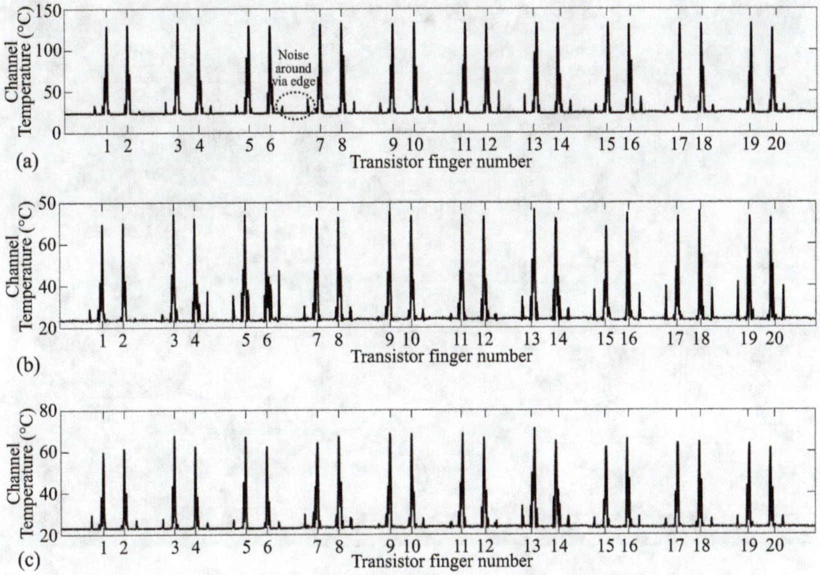

Figure 4.17 *GaN HEMT channel thermal distribution profiles obtained across the transistor die where the transistor fingers are numbered from left to right: (a) shows max.* P_{out}, *(b) max. drain efficiency, and (c) DC only thermal distributions. 2019 IEEE. Reprinted, with permission, from [41].*

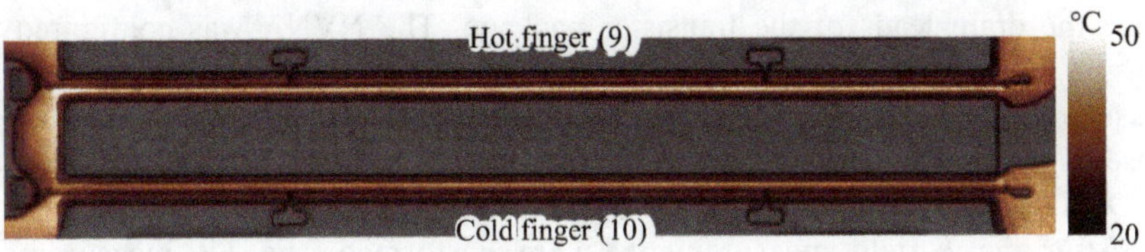

Figure 4.18 *Thermal measurement overlaid over the image of transistor fingers 9 and 10. The top finger (9) is 8 °C hotter than the bottom finger (10) across the entire length of the transistor gate width.*

4.5.4 *Measurements of Doherty power amplifiers*

Finally, we tie electrical, electromagnetic, and thermal measurements together by performing a full multiphysics characterization of a 250-W commercially available 2.496–2.69 GHz asymmetric Doherty power amplifier (DPA) HEMT in GaN technology from NXP Semiconductors The transistor was measured in its 50-Ω DPA test fixture as shown in Figure 4.19. During the electrical measurements, a pulsed CW excitation with a 10% duty cycle and 10-μs pulse width was used and was biased with the carrier set to $I_{dq} = 150$ mA, $V_{ds} = 48$ V (class AB operation) and the peaking set to $V_{gs} = -5.4V$, $V_{ds} = 48$ V (class C operation).

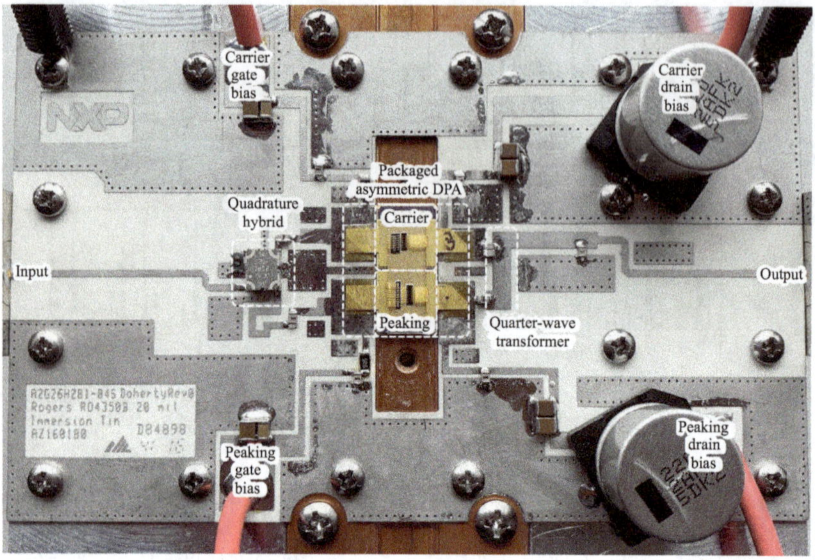

Figure 4.19 Asymmetric 250-W GaN HEMT DPA transistor in its 50-Ω test circuit. The 2-dB quadrature hybrid is used to split the input power and provide the required 90° phase difference between the peaking and the carrier PAs.

To measure the *E*-fields above the DPA, the EO probe was positioned 300 μm above the drain leads of the transistor package. The NVNA was configured to measure five harmonics of the 2.6 GHz fundamental. The E_z-field component was measured and the results over the different drive powers and measured *E*-field phase angles are shown in Figure 4.20, showing the detailed internal operation of the DPA.

After the *E*-field measurements were performed, the TR measurement system was configured to measure the GaN DPA thermal distributions. The thermal measurements were performed at 40 × magnification in the middle of the transistor dies, where the temperatures are expected to be at their most extreme. The DPA was operating in saturation with 1-ms excitation pulses and 10% duty cycle. As the gate width of the transistors in both dies is wider than the camera lens can capture at 40 × magnification, two separate thermal images of the transistor fingers in the middle of both peaking and carrier dies were obtained, one capturing the heating at the gate side of the transistor and the other at the drain side. The thermal measurement results of the carrier and peaking PAs are plotted in Figures 4.21 and 4.22, respectively. The TR calibration coefficients were only used for the transistor channel regions, and other coefficients in the thermal image were filtered out. This resulted in the accurate thermal measurements only for the transistor channels and the other areas within the thermal plots were set to the ambient temperature of 22 °C.

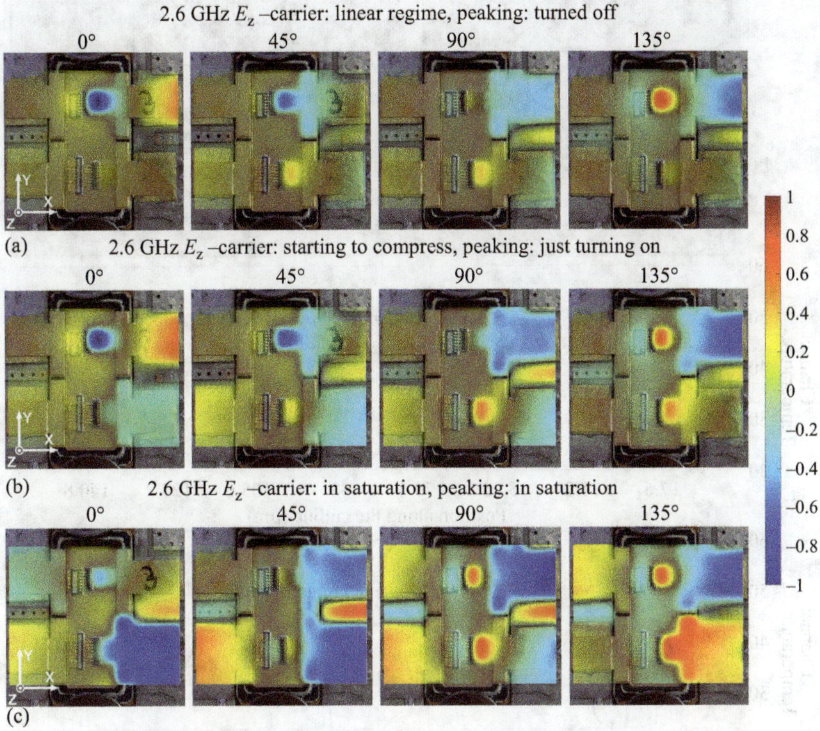

Figure 4.20 Normalized E_z-field component overlay over a 250-W asymmetric GaN HEMT DPA at different measured phase angles and under different input power drives: (a) carrier on, peaking off, (b) carrier close to saturation, peaking just turning on, and (c) both carrier and peaking in saturation

As can be seen from the measurement results for the carrier die in Figure 4.21, the temperatures at both ends of the transistor channels (at gate and drain sides), along the red and blue cutlines, are the same, averaging 44 °C. This is due to the lower output power of the carrier PA and shorter unit gate widths of the transistor. Moreover, the channel temperature may not have reached steady-state under the 1-ms DUT excitation pulse, at which the channel temperatures would be higher.

The thermal results of the peaking PA shown in Figure 4.22, however, display substantially increased transistor channel temperatures and a 14 °C temperature variation along the transistor channels. This temperature variation, with temperature decreasing toward the drain side, implies a voltage drop along the transistor gate width, resulting in non-uniform conduction along the full width of the transistor channel. The UGW of the peaking die is substantially larger than the carrier (445 μm

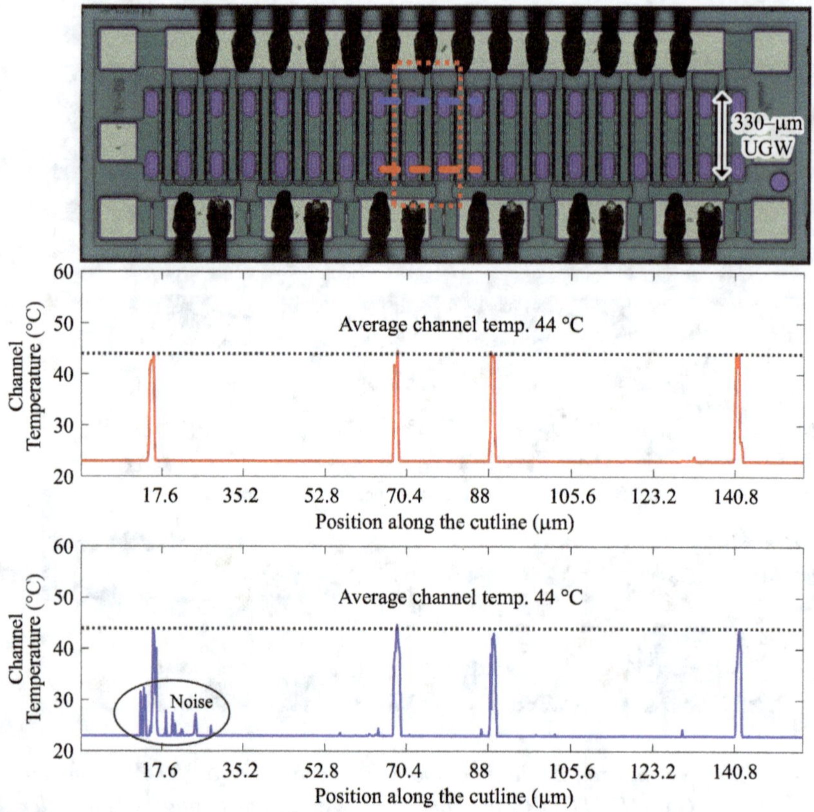

Figure 4.21 GaN channel temperatures of the carrier PA die, comprising 36 individual HEMTs, obtained using a TR measurement system. The temperatures are the same along the transistor channels at both the gate (red cutline) and the drain (blue cutline) sides of the die. Drain side measurements are slightly more noisy when compared to the gate side as indicated in the blue plot.

vs. 330 µm), as can be seen in Figures 4.21 and 4.22; thus, the voltage drop incurred along the gate width is also larger. This non-uniform thermal distribution along the gate width has been reported previously in literature [42] and also appears to be the case in this transistor.

4.6 Conclusion

In this chapter, we have attempted to show how multiphysics-enabled measurements can offer unprecedented insight into the internal operation of packaged RF

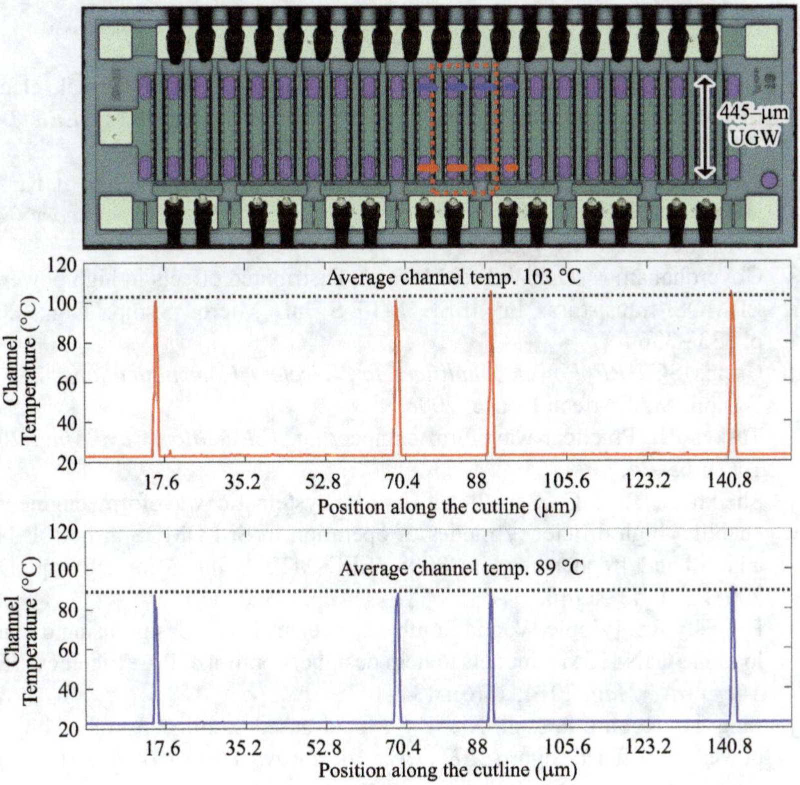

Figure 4.22 GaN channel temperatures of the peaking PA die, comprised of 42 individual HEMTs, obtained using a TR measurement system. The channel temperatures show a 14 °C change along the gate width when comparing the gate (red cutline) and the drain (blue cutline) sides of the die. This is a known phenomenon, when the conduction along the transistor gate width is non-uniform due to the incurred voltage drop along the gate.

and microwave transistors and how this can open significant new avenues for performance optimization, design troubleshooting, and model validation. The applications shown in this chapter are only a small subset of the measurements that are enabled by the multiphysics measurement methodology and this area of research is expected to grow in parallel to the ever-increasing multiphysics simulation capabilities, which more and more computer-aided engineering software companies are starting to offer. It is the vision of the authors that multiphysics device characterization methodologies will need to compliment traditional design techniques and laboratory measurements to develop future generations of wireless technologies.

References

[1] Denis D, Snowden CM, and Hunter IC., Coupled electrothermal, electromagnetic, and physical modeling of microwave power FETs. *IEEE Trans Microwave Theory Tech*. 2006; 54(6): 2465–2470.

[2] Aaen PH, Wood J, Bridges D, *et al.*, Multiphysics modeling of RF and microwave high-power transistors. *IEEE Trans Microwave Theory Tech*. 2012; 60(12): 4013–4023.

[3] Goverdhanam K, Dai W, Frei M, *et al.*, Distributed effects in high power RF LDMOS transistors. In: IEEE MTT-S Int. Micro. Symp. Dig.; 2005. p. 333–336.

[4] Cripps SC., *RF Power Amplifiers for Wireless Communications*. 2nd ed. Boston, MA: Artech House; 2006.

[5] Tasker PJ., Practical waveform engineering. *IEEE Microwave Mag*. 2009; 10(7): 65–76.

[6] Sheikh A, Roff C, Benedikt J, *et al.*, Systematic waveform engineering enabling high efficiency modes of operation in Si LDMOS at both L-band and S-band frequencies. In: 2008 IEEE MTT-S Int. Micro. Symp. Dig.; 2008. p. 1143–1146.

[7] Pengelly RS, Pribble W, and Smith T., Inverse class-F design using dynamic loadline GaN HEMT models to help designers optimize PA efficiency. *IEEE Microwave Mag*. 2014; 15(6): 134–147.

[8] Jang H, Roblin P, and Xie Z., Model-based nonlinear embedding for power-amplifier design. *IEEE Trans Microwave Theory Tech*. 2014; 62(9): 1986–2002.

[9] Jansen RH and Ghazinour A., Design-oriented EM-based distributed active device modeling. In: Workshop on Novel Methodologies for Device and Circuit CAD, at the 1999 IEEE MTT-S Int. Micro. Symp. Anaheim, CA, USA; 1999. p. 1–16.

[10] Blackburn DL., Temperature measurements of semiconductor devices - a review. In: 20th Ann. IEEE Semicond. Thermal Measurement Manage. Symp. San Jose, CA, USA; 2004. p. 70–80.

[11] Sarua A, Hangfeng J, Kuball M, *et al.*, Integrated micro-Raman/infrared thermography probe for monitoring of self-heating in AlGaN/GaN transistor structures. *IEEE Trans Electron Devices*. 2006; 53(10): 2438–2447.

[12] Mahalingam M and Mares E., Infrared temperature characterization of high-power RF devices. In: IEEE MTT-S Int. Microw. Symp. Dig. vol. 3. Phoenix, AZ, USA; 2001. p. 2199–2202.

[13] Zhao M, Liu X, Yingkui Z, *et al.*, Thermal analysis of AlGaN/GaN high-electron-mobility transistors by infrared microscopy. In: 19th IEEE Int. Symp. Physical Failure Anal. Integr. Circuits. Singapore; 2012. p. 1–4.

[14] Choi S, Heller ER, Dorsey D, *et al.*, Thermometry of AlGaN/GaN HEMTs using multispectral Raman features. *IEEE Trans Electron Devices*. 2013; 60 (6): 1898–1904.

[15] Kuball M, Hayes JM, Uren MJ, *et al.*, Measurement of temperature in active high-power AlGaN/GaN HFETs using Raman spectroscopy. *IEEE Electron Device Lett*. 2002; 23(1): 7–9.

[16] Matatagui E, Thompson AG, and Cardona M., Thermoreflectance in semiconductors. *Phys Rev*. 1968; 176(3): 950–960.

[17] Farzaneh M, Maize K, Lüerßen D, *et al.*, CCD-based thermoreflectance microscopy: principles and applications. *J Phys D: Appl Phys*. 2009; 42(14): 143001 (20pp).

[18] Stevens RJ, Smith AN, and Norris PM., Signal analysis and characterization of experimental setup for the transient thermoreflectance technique. *Rev Sci Instrum*. 2006; 77(8): 084901 (8pp).

[19] Yaghjian AD., An overview of near-field antenna measurements. *IEEE Trans Antennas Propag*. 1986; 34(1): 30–45.

[20] Yingjie G and Wolff I. Measurements of field distributions and scattering parameters in multiconductor structures using an electric field probe. In: IEEE MTT-S Int. Microw. Symp. Dig. vol. 3. Denver, CO, USA; 1997. p. 1741 – 1744.

[21] Hou R, Spirito M, Heeres R, *et al.*, Non-intrusive near-field characterization of distributed effects in large-periphery LDMOS RF power transistors. In: IEEE MTT-S Int. Microw. Symp. (IMS). Pheonix, AZ, USA; 2015. p. 1 – 3.

[22] Malyuskin O and Fusco VF., High-resolution microwave near-field surface imaging using resonance probes. *IEEE Trans Instrum Meas*. 2016; 65(1): 189–200.

[23] Budka TP, Waclawik SD, and Rebeiz GM., A coaxial 0.5-18 GHz near electric field measurement system for planar microwave circuits using integrated probes. *IEEE Trans Microwave Theory Tech*. 1996; 44(12): 2174–2184.

[24] Yingjie G and Wolff I., Miniature electric near-field probes for measuring 3-D fields in planar microwave circuits. *IEEE Trans Microwave Theory Tech*. 1998; 46(7): 907–913.

[25] Dehghan N, Porch A, Cripps SC, *et al.*, A novel high resolution *E*-field microscope system with applications in HPA diagnostics. In: 78th ARFTG Microw. Meas. Symp. Tempe, AZ, USA; 2011. p. 1 – 3.

[26] Valdmanis J and Mourou G., Subpicosecond electrooptic sampling: principles and applications. *IEEE J Quantum Electron*. 1986; 22(1): 69–78.

[27] Jäger D and David G., MMIC characterization using electro-optic field mapping. In: Int. Symp. Signals, Syst., Electron. Proc. San Francisco, CA, USA; 1995. p. 41 – 44.

[28] Rodenbeck CT, Peterson KA, Sandoval CE, *et al.*, Electrooptic inspection of vector leakage in radiofrequency multichip modules. *IEEE Trans Electromagn Compat*. 2013; 55(6): 1093–1099.

[29] Lee DJ, Kwon JY, Kang NW, *et al.*, Vector-stabilized reactive-near-field imaging system. *IEEE Trans Instrum Meas*. 2011; 60(7): 2702–2708.

[30] Sabet K, Darragh R, Sabet A, *et al.* Characterization and diagnostics of active phased array modules using non-invasive electro-optic field probes

with a CW laser source. In: IEEE MTT-S Int. Microw. Symp. Honolulu, HI, USA; 2017. p. 1617–1619.

[31] Yariv A and Yeh P., *Optical Waves in Crystals: Propagation and Control of Laser Radiation*. New York, NY: John Wiley & Sons; 1984.

[32] Hjelme DR, Yadlowsky MJ, and Mickelson AR., Two-dimensional mapping of the microwave potential on MMICs using electrooptic sampling. *IEEE Trans Microwave Theory Tech*. 1993; 41(6): 1149–1158.

[33] David G, Bussek P, Auer U, *et al.*, Electro-optic probing of RF signals in submicrometre MMIC devices. *Electron Lett*. 1995; 31(25): 2188–2189.

[34] Lee DJ and Whitaker JF., An optical-fiber-scale electro-optic probe for minimally invasive high-frequency field sensing. *Opt Express*. 2008; 16(26): 21587–21597.

[35] Root DE, Verspecht J, Horn J, *et al.*, *X-Parameters*. New York, NY: Cambridge University Press; 2013.

[36] Kendig D, Yazawa K, and Shakouri A., Hyperspectral thermoreflectance imaging for power devices. In: 2017 33rd Thermal Meas., Modeling & Manage. Symp. San Jose, CA, USA; 2017. p. 204–207.

[37] Urbonas J, Kim K, Vanaverbeke F, *et al.*, An electro-optic pulsed NVNA load-pull system for distributed E-field measurements. *IEEE Trans Microwave Theory Tech*. 2018; 66(6): 2896–2903.

[38] Lee DJ, Kwon JY, Kang NW, *et al.*, Calibrated 100-dB-dynamic-range electro-optic probe for high-power microwave applications. *Opt Express*. 2011; 19(15): 14437–14450.

[39] Tarui Y, Fujii K, and Itoh Y., Calculation of loop oscillations of microwave high-power amplifiers with several closed loop circuits and split-cell matching methods for high stability. *Electron Comm Jpn 2*. 2000; 83(8): 33–42.

[40] Mochizuki M, Nakayama M, Tarui Y, *et al.* Nonlinear analysis of f0/2 loop oscillation of high power amplifiers. In: IEEE MTT-S Int. Microw. Symp. Dig.. vol. 2. Orlando, FL, USA; 1995. p. 709–712.

[41] Urbonas J, Matei C, and Aaen PH. Transient and steady-state thermal measurements of GaN-on-SiC HEMT transistors under realistic microwave drive. In: 92nd Microw. Meas. Conf. (ARFTG). Orlando, FL, USA; 2019. p. 1–4.

[42] Pavlidis G, Kendig D, Heller ER, *et al.*, Transient thermal characterization of AlGaN/GaN HEMTs under pulsed biasing. *IEEE Trans Electron Devices*. 2018; 65(5): 1753–1758.

[43] Urbonas J., *Multiphysics measurement of high-power microwave transistors and amplifiers [PhD dissertation]*. University of Surrey. Guildford, Surrey, United Kingdom; 2019.

[44] Urbonas J, Kim K, and Aaen PH., Direct *E*-field. measurement and imaging of oscillations within power amplifiers. *IEEE Trans Instrum Meas*. 2019; 68 (8): 2971–2978.

Part III

Antenna metrology

Chapter 5

A metrological millimeter wave hybrid beamforming testbed with a large antenna array

Tian Hong Loh[1], David Cheadle[1], Mohsen Khalily[2], Sohail Payami[2], Konstantinos Nikitopoulos[2] and Rahim Tafazolli[2]

5.1 Introduction

Millimeter wave (mm-wave) frequency bands play an important role in fulfilling the high data-rate demands of current and future wireless communication systems. The rapid growth of the wireless communications market together with increasing requirements for functionality, energy efficiency, robustness, and reliability of communication systems has resulted in huge demands for new and advanced beamforming antenna technologies being developed for integration into these systems to improve performance and usage of the available spectrum in the most efficient way. In particular, these technologies are essential for mm-wave communications to be realized at both the user equipment and base station operating in real-world environment for the realization of reliable communication link and compensation of the inherent heavy atmospheric and free-space path loss. Such systems offer beam reconfigurability and smart connection functionality. However, their traceability and system-level testing present new measurement challenges as the associated measurement uncertainties at mm-wave frequencies become dramatically wider [1–3]. Furthermore, their practical evaluations and measurements are very time-consuming and often involve complicated procedures, leading to significant cost and time burden issues [4], whereby their time-efficient testing plays a key role for their applications in the fifth-generation (5G) communication technology, and beyond, applications that include the sixth generation (6G), Industry 4.0 and Internet of Things (IoT).

Current trends in developing cost-effective and energy-efficient wireless systems operating at mm-wave frequencies and with large-scale phased array antennas

[1]5G and Future Communications Technology Group, Electromagnetic and Electrochemical Technologies Department, National Physical Laboratory (NPL), Teddington, United Kingdom
[2]6G Innovation Centre, Institute for Communication Systems, University of Surrey, Guildford, United Kingdom

for fulfilling the high data-rate demands of 5G and beyond have driven the needs to explore the use of hybrid beamforming technologies. This chapter focuses on the development and experimental evaluations of a width-bandwidth mm-wave fully connected hybrid beamforming metrological testbed with large antenna array. Such testbed provides academics and industry a unique test platform with significantly more degrees of freedom on trial and the development of different robust hybrid beamforming algorithms for reaching higher data rate for mm-wave communications. It enables full exploitation over the advantages of large arrays as well as radically reduces the traceable metrological burden. Discussions include testbed design, calibration procedures, experimental evaluations, as well as the critical factors to consider for their practical implementation.

5.1.1 Beamforming architectures

Beamforming is a key technology for current and future wireless communications. Antenna systems based on such technology generally comprise multiple antenna elements, typically in the form of a uniformly spaced antenna array at the transmitting and/or receiving side of the wireless link. Signals are processed adaptively in order to overcome issues caused by multipath fading and co-channel interference and to achieve dynamic spatial diversity. Together with a suitable beamforming algorithm implemented using digital signal processing (DSP) units, antenna beamforming can be defined as the ability of an antenna array to steer, in space, its directional narrow beams toward specific directions of the desired users and nulls in the undesired directions, or conversely, the ability of the antenna array to estimate the direction-of-arrival (DOA) of the impinging signals [5,6]. Such technology is envisaged capable of improving the performance and reliability of wireless communication links, and consequently delivering a better end user experience for 5G and beyond by

- extending the communication range,
- reducing the required transmit power,
- increasing spectrum efficiency,
- improving network capacity,
- reducing interference,
- improving the signal-to-interference-plus-noise ratio (SINR).

Usually beamforming is achieved by controlling two variables at each antenna element, the amplitude and phase. The beamforming algorithm controls the array antennas and adjusts these variables to perform the complex phase shifting and weighted signal synthesis based on the calculated beamforming weights. The far-field radiation pattern is then formed by superposition of the radiation of all antenna elements in the array antennas. Thus, the radiation pattern (including beamforming and beam-nulling profile) of such antenna system can be controlled. Nevertheless, an important prerequisite for every beamforming architecture is phase coherency [7].

There are mainly three types of multi antenna beamforming architectures at the system level: analog beamforming, digital beamforming, and hybrid beamforming. Hybrid beamformers can further be divided into sub-connected and fully connected structures [8]. Figure 5.1 shows the basic implementation of these architectures at both the transmitter (Tx) and receiver (Rx) ends. Note that M_{TX}, M_{RX}, and N denote, respectively, the number of antennas at transmitting end and receiving end and the number of user data. A radio frequency (RF) chain generally consists of RF filters, a mixer and either a digital-to-analog converter (DAC) and power amplifier

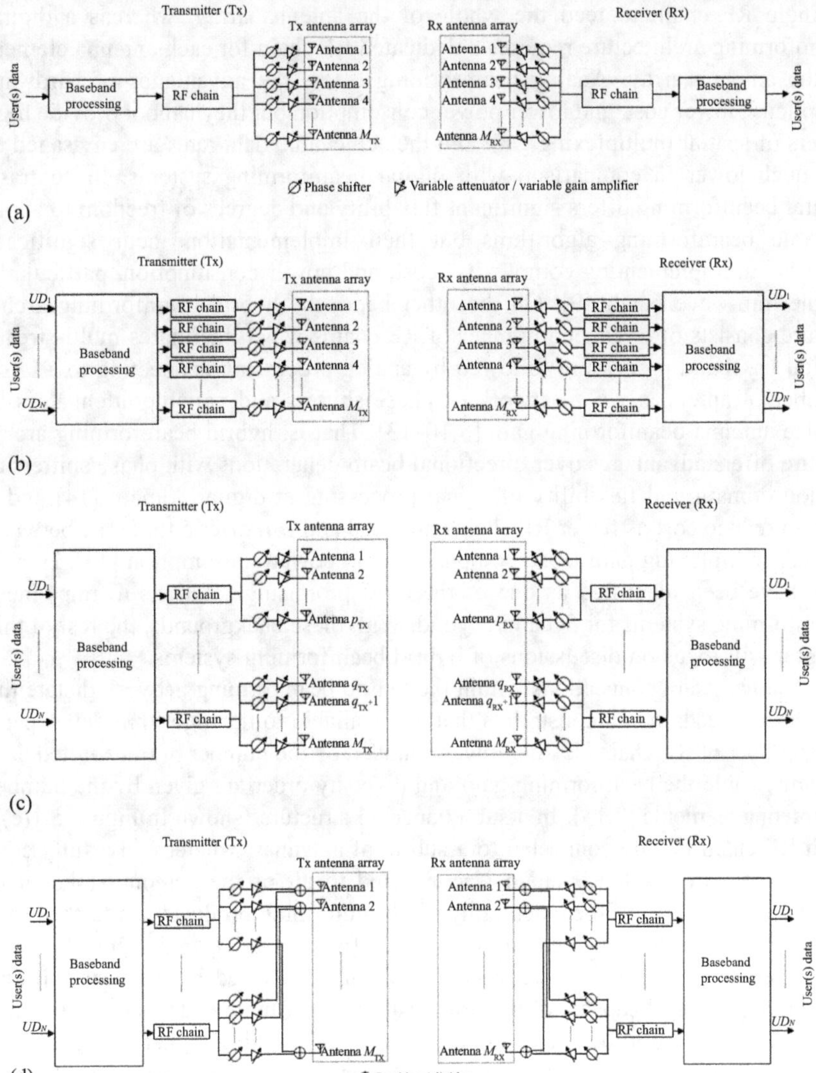

Figure 5.1 The basic implementation of beamforming architectures: (a) analog beamforming; (b) digital beamforming; (c) sub-connected hybrid beamforming; (d) fully connected hybrid beamforming

at the transmitter end or an analog-to-digital converter (ADC) and low noise amplifier (LNA) at receiver end. Each element of the antenna array is fed with a phase shifter and a variable attenuator (or variable gain amplifier). The RF signals received by each radiating element are down-converted to baseband signals and digitized individually. The digitized signals could then feed into a DSP module where the beamforming algorithms reside.

As depicted in Figure 5.1, the analog beamforming architecture requires only a single RF chain to feed the whole of the antenna array, whereas a digital beamforming architecture requires a dedicated RF chain for each antenna element in the antenna array. Analog beamforming offers the advantages of hardware simplicity, lower cost, and lower power consumption but they cannot provide high orders of spatial multiplexing gain and the achievable data rates are envisaged to be much lower in comparison with digital beamforming systems. In contrast, digital beamforming offers significant flexibility and degrees of freedom to apply suitable beamforming algorithms but their implementation incur significant increase in implementing complexity, cost, and power consumption, particularly in the mm-wave bands [9]. On the other hand, the hybrid beamforming architecture consists of a reduced number of RF chains which facilitates multi-stream digital baseband processing followed by analog processing connecting to a large number of antennas via a network of phase shifters and variable attenuators to realize antenna beamforming gain [8,10–13]. That is, hybrid beamforming architecture offers advantages over directional beam generations with phase shifters at analog domain and flexibility of signal processing at digital domain [14] and a way to reduce cost as fewer RF chains are used and can offer a trade-off between spatial multiplexing gain, cost, complexity, and power consumption [13]. Hence, they have been identified as one of the most promising solutions to implement beamforming systems for 5G and beyond. With these backgrounds, the rest of this chapter will focus on discussions of hybrid beamforming systems.

The hardware connections within the hybrid beamforming network dictate the degrees of freedom and constraints that are available to the algorithm developers. The number of RF chains is only lower bounded by the number of transmitted data streams, while the beamforming gain and diversity order are given by the number of antenna elements [8,15]. In a sub-connected structure (shown in Figure 5.1(c)), each RF chain is only connected to a subset of antennas, whereby in a fully connected structure, each RF chain is connected to all of the antennas (shown in Figure 5.1(d)). Most of the existing hybrid beamforming products and testbeds are based on the sub-connected architectures [16–19]. Compared with the sub-connected hybrid beamforming system, the fully connected hybrid beamforming system is envisaged capable of providing significantly more degrees of freedom to develop robust hybrid beamforming algorithms for reaching higher data rates and greater flexibility to control the phase and magnitude for the planar phased array. Nevertheless, hybrid beamforming systems based on fully connected architecture are more complex to build and evaluate. It is important to perform a rigorous evaluation of such systems so the advantages of large arrays could be fully exploited.

5.1.2 Challenges and state of the art

This subsection presents a brief overview of the challenges of hybrid beamformers from the algorithm, implementation, and measurement perspective as well as a brief discussion on current state-of-the-art hybrid beamforming systems. Achieving high data rates at mm-wave frequencies is very challenging [9,20,21]. The blockage phenomena are more severe in the mm-wave bands and the signals transmitted over mm-wave channels experience significant attenuation due to inherent heavy atmospheric and free-space path loss [22]. To overcome such challenges at mm-wave bands, an antenna array with a large number of elements (generally much greater than number of users) that enable the implementation of effective spatial multiplexing gain techniques and beamforming algorithms is needed to electronically steer the directional beams toward specific directions of the desired users and nulls in the undesired directions for establishing a reliable link between mm-wave terminals [23]. The recent studies on hybrid beamformers with a combination of switches and phase shifters indicate that such designs can reduce the cost and power consumption of massive multiple-input–multiple-output (mMIMO) systems but their hybrid beamforming algorithm implementation is challenging due to the nonconvex constant modulus combinatorial constraints that are imposed by the phase shifters during the optimization process of hybrid beamforming weights in which only the phase of the signal could be changed but not in magnitude [13]. Such constraints can significantly increase the computational complexity of the corresponding algorithms [23,24]. Note that for sub-connected architecture with M_{TX} transmit antennas and N_{RF} RF chains, the number of signal processing path is $M_{TX}\times N_{RF}$, whereas in fully connected architecture it is $M_{TX}\times N_{RF}^2$, i.e., the beamforming gain of the fully connected architecture is N_{RF} times greater than the sub-connected architecture [25,26]. Furthermore, the choice and performance of the suitable beamforming algorithms also depends on the performance metric, propagation channel, and the structure of the phase shifter network [13,27,28].

From implementation perspective, fabricating beamforming system based on fully connected architecture is much more challenging than the sub-connected architecture. In a typical fully connected beamforming system, a large number of electrical and RF components often need to be placed in proximity, which gives rise to mutual coupling effects between the RF paths, phase shifters, and antenna elements. Such issues become even more challenging with mm-wave operating frequency band and increase number of the antennas. Nevertheless, it is envisaged that beamforming system based on fully connected architecture can reach significantly higher data rates compared to sub-connected architecture [12,29]. Moreover, the larger number of phase shifters available in the fully connected architecture provides more degrees of freedom to design robust beamforming algorithms whereby the connection from each RF chain to all of the antennas provides higher array gains to boost the SINR (see Chapter 2 for detail definitions and measurement methods) for each data stream.

Several papers have reported on the overviews, developments, and challenges of hybrid beamforming techniques [8,10,13,20,30–32]. Different wireless communication applications impose different requirements on the design and operation of beamforming antenna system. The adoption of beamforming antenna techniques in future generation wireless systems often requires the phased array antennas to be an inherent part of the system, and it is envisaged that this would have a significant impact on the efficient use of the spectrum and the optimization of service quality [33]. Hence, from measurement perspective, when evaluating the performance of a beamforming system, one needs to rigorously consider the whole of the beamforming system from basebands, RF chains, to antenna arrays, also where possible, to consider the effects of the propagation channel and co-channel interferences.

Table 5.1 shows some state-of-the-art examples of the mm-wave hybrid beamforming testbeds, where f_c denotes the central frequency, BW denotes the bandwidth, and EIRP denotes the effective isotropic radiated power. Note that the dash symbol shown in the table indicates that the relevant information was unavailable. The hybrid beamforming systems shown in Table 5.1 are mostly established based on sub-connected hybrid beamforming architecture apart from the systems presented in [13,29,34]. Note that a sub-connected hybrid beamforming system with 32 antennas and 2 RF chains for 60-GHz applications presented in [16] shows that a sub-connected structure with two RF chains and phased arrays significantly outperforms two analog beamformers with the same array size, i.e., despite the directional nature of 60-GHz channels, the cooperation between the two RF chains provides a noticeable performance gain to improve the SINR for the users [13]. The 64-element (8×8) phased array presented in [17] operates at 28 GHz and provides ±50° scanning range in azimuth and can support EIRP of up

Table 5.1 Current state-of-the-art examples of mm-wave hybrid beamforming testbeds

	f_c (GHz)	BW (GHz)	Array size	Beam-forming range (°)	Phase resolution (bits)	Amplitude resolution (bits)	EIRP (dBm)	No. of RF chains
[35]	60	9	16×1	±20	5	–	53	1
[36]	60	9	6×6	±30	2	–	29	1
[37]	28	0.52	8×4	±30	–	–	49	1
[38]	111	6	4×4	±30	2	–	25	1
[39]	28	3	8×8	±50	–	–	–	1
[40]	28	2	16×16	±60	–	–	60	1
[41]	26	2	16×16	–	–	–	60	4
[16]	60	1.76	16×2	–	8	8	–	2
[17]	28	4	8×8	±50	6	–	52	1
[19]	28	0.8	8×1	±50	–	–	33	1
[18]	60	2	16×16	±50	5	9	42	1
[29]	27.5	5	8×1	±30	5	5	–	2
[34]	28/37	2.75/3.75	4×1	±30	5	–	–	2
[13]	26	2	16×8	±60	6	6	60	2

to 52 dBm. The 8-element (8×1) array of [19] provides a similar scanning range, and its EIRP at the saturation point is around 33 dBm. The 256-element transceiver operating at 60 GHz presented in [18] is also based on a sub-connected hybrid beamforming architecture. For mm-wave communications, a large number of antennas are required. However, most of the existing hybrid beamformers are based on sub-connected structures that cannot fully realize the benefits of large arrays.

The fully connected hybrid beamforming receiver systems operating at 28 and 37 GHz that are presented in [29,34] have 2 RF output ports, with 8 and 4 antennas, respectively. However, their array sizes are relatively small and may not be sufficient to compensate for the inherent high path loss at mm-wave bands in many practical operating scenarios. With this motivation, Payami *et al.* [13] have designed, built, and presented preliminary assessment of a wide bandwidth mm-wave fully connected hybrid beamforming system operating at 26 GHz with 2 RF input ports and 128-element phase array antennas. Compared with other hybrid beamformers, this system offers features that enable higher data rates (as compared to the sub-connected architectures) and increased flexibility to control the phase and magnitude for the planar phased array. Such system provides a unique platform for researchers to develop robust hybrid beamforming algorithms for mm-wave systems. However, it is important to perform rigorously the evaluation of such platform so the advantages of large arrays could be fully exploited. This chapter extends from earlier work in [13] to further evaluate experimentally the beamforming performance of the beamformer system together with the relevant RF chain and baseband systems with and without amplitude tapering as well as evaluating its linearity, phase, and attenuation performance. The following sections present the key steps that are required to design, implement, and verify such testbed system.

5.2 Testbed design and evaluation

This section presents the design of the aforementioned wide-bandwidth mm-wave fully connected hybrid beamforming testbed system. It offers 2-GHz instantaneous bandwidth, 6-bit phase control, 6-bit gain/attenuation control, 128-element phased array antenna (arranged as a 16×8 planar array), and two separate RF channels each capable of fully independent beamforming. If RF harmonics and spur issues are avoided with the use of suitable band pass filters (BPFs), the testbed could be setup to work between 25 and 30 GHz. Figure 5.2 shows the testbed system block diagram. It consists of a pair of 2×2 baseband systems with the relevant up- and down-converters, a 2×128 beamformer with the relevant digital input/output (DIO) control modules and power supply (established in [13]), and 128-element phased array and 2 standard gain horn (SGH) antennas. To evaluate the functionality of the testbed components, several pieces of RF equipment such as vector network analyzer (VNA), spectrum analyser (SA), and power meter have been employed. The following subsections provide their further descriptions and details.

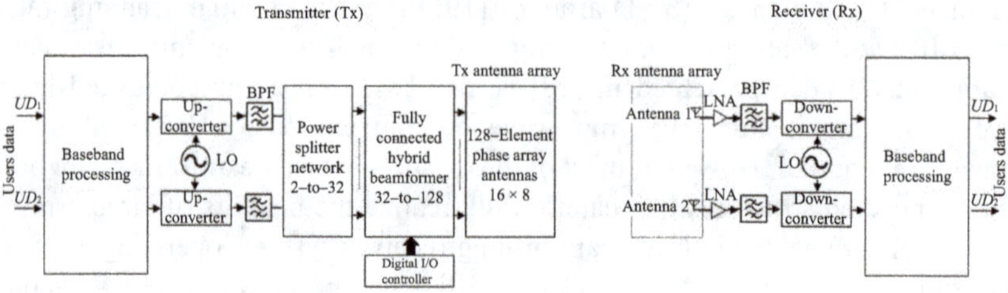

Figure 5.2 Block diagram of the wide-band mm-wave fully connected hybrid beamforming testbed system

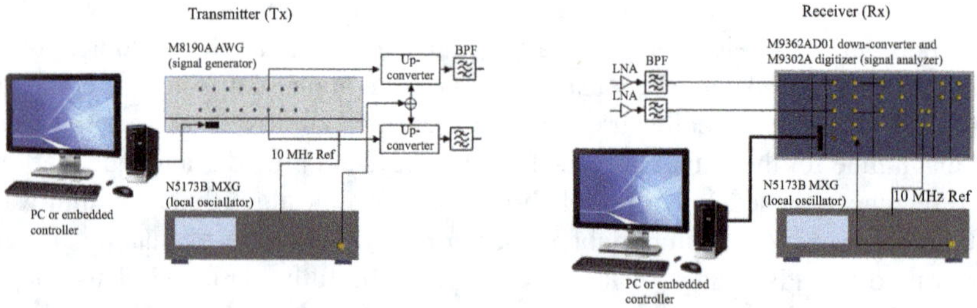

Figure 5.3 Illustrative block diagram of the two-channel wide-band baseband systems with mm-wave up- and down-converter

5.2.1 Baseband systems and up- and down-converters

The two-channel baseband systems are formed by a pair of Keysight M8190A arbitrary waveform generators (AWG) and Keysight M9203A digitizers, at the Tx and Rx ends, respectively. The system has a reference clock module (Keysight M9300A) that can generate either 10- or 100-MHz clock signals that are used to phase synchronize the separate modules at the receiver end. At the Tx end, the two-channel up-converter modules are composed of various commercially available component devices, whereby, at the Rx end, the baseband signal analyzer is preceded by RF front end with an LNA, BPF, and a down-converter (Keysight M9362AD01). The mm-wave up-converter consists of a power splitter (Marki PD-0140), a pair of mixers (Marki MM1-1140H), a pair of amplifiers (Erzia ERZ-HPA-1500-2700-29-E), and a pair of BPFs (Amwav AWBPF25.5-26.5GC2). At the front end of the mm-wave down-converter, a pair of LNAs (Erzia ERZ-LNA-2600-4000-30-2.5) and a pair of BPFs (AMWAV AWBPF25.5-26.5GC2) are employed. The up- and down-converters each employs a microwave analog signal generator, Keysight N5173B, to provide the local oscillator (LO) signal.

Figures 5.3 and 5.4 show the relevant illustrative block diagram of the two-channel baseband system with mm-wave up- and down-converters, respectively, and a photo setup of the system shown with two additional pairs of SGH antennas (Flann 20240-24NRL) at both the Tx and Rx RF front ends.

Figure 5.4 Photo setup of the two-channel wide-band baseband systems with mm-wave up- and down-converter and two pairs of standard gain horn antennas

With the use of a vector signal analyzer (VSA) software (Keysight 89601B) that can perform vector signal analysis to visualize signal quality with multiple domain traces in time, spectrum, and modulation, these systems can be used to analyze the user-defined modulated signals to test and investigate over the performance of new modulation schemes, for example, for 4G Long-Term Evolution and 5G new radio.

5.2.2 mm-Wave fully connected hybrid beamformer system

The wide-bandwidth mm-wave fully connected hybrid beamformer system consists of a user-programmable attenuator and phase shifter network with 2 RF inputs (Port 1 and Port 2) and 128 RF outputs, a cooling system, a power supply, and a DIO multifunction switch unit (Keysight 34980A) for controlling the attenuators and phase shifters digitally from a computer using via an Ethernet connection and under the control of custom-written software. Each RF input is split into 16 signals and then passed into 16 mm-wave fully connected beamforming modules, each with two RF inputs and 8 RF outputs. The outputs of the modules are then amplified and fed to the antennas. This results in a total of 128 RF outputs being available to support the phased array antenna of size 16×8. Together with the illustrative 128-element Tx antenna array, Figure 5.5 shows the detailed block diagram of the

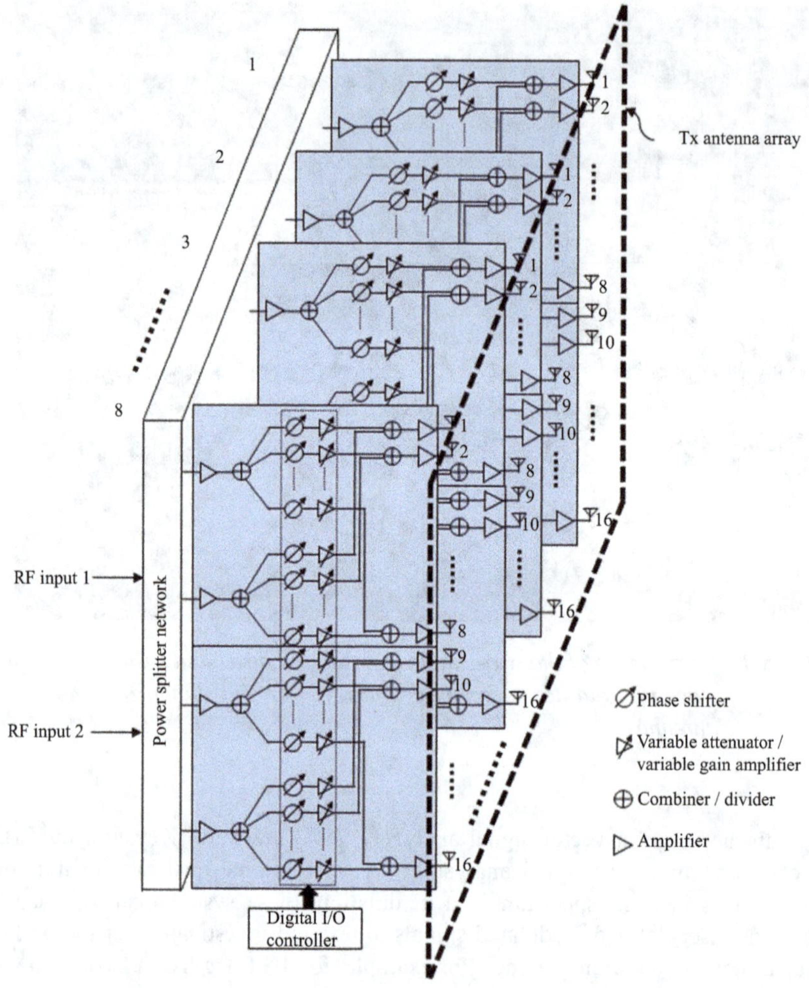

Figure 5.5 Detailed block diagram of the fully connected attenuator and phase shifter network modules within the wide-bandwidth mm-wave hybrid beamformer

fully connected network of phase shifters and adjustable attenuators. Cables with 78-pin D-sub connector are used for 5-V logic interfacing between the DIO multifunction switch unit and the mm-wave hybrid beamformer. The system includes a power supply (Keysight N5765) which is used to provide a DC voltage to the hybrid beamformer. Figure 5.6 shows the illustrative diagrams and a photo setup of the system shown with the DIO, power supply, and the 128-element microstrip phased array antenna. In the following subsections, the design and test procedures of different blocks/modules shown in Figure 5.5 are explained in detail.

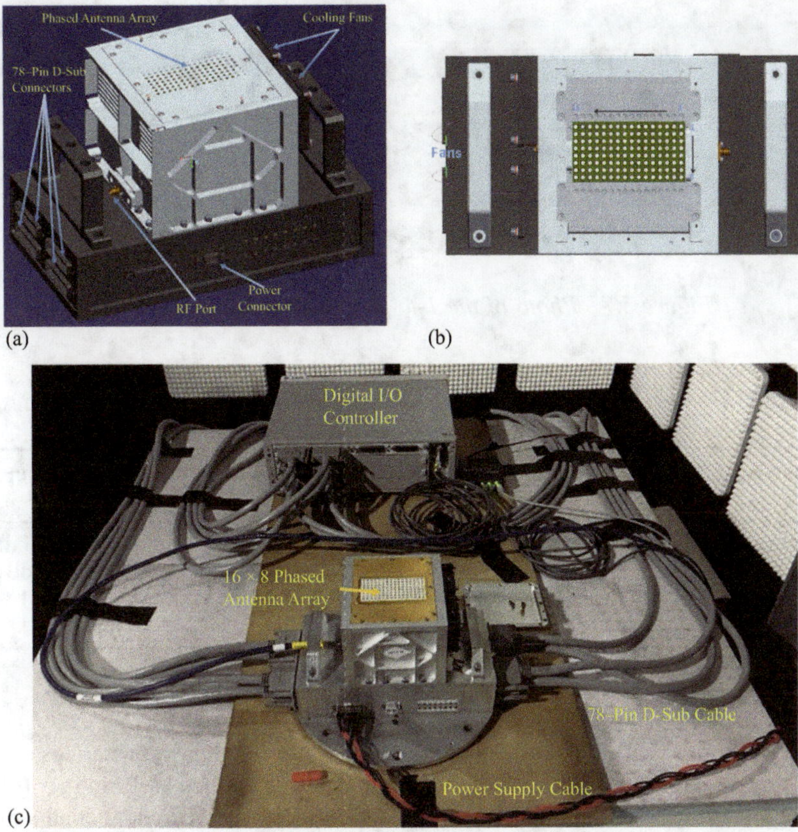

Figure 5.6 Illustrative diagrams and a photo setup of the system: (a) 3D view; (b) top view; (c) photo setup shown with the DIO, power supply, and the 128-element microstrip phased array antenna

5.2.2.1 Power splitter network

A compact 1-to-16 Wilkinson power splitter network has been designed and implemented (see photo in Figure 5.7) to divide each of the two RF input signals of the hybrid beamformer among the sixteen 2-to-8 beamforming modules. This passive power splitter network that is placed at the beginning stage of the hybrid beamformer (see Figure 5.5) and the traveling wave on the microstrip line is divided by two sub-paths for four times. As the signal attenuation is related to the length of the transmission lines, the compact size of network is an important factor to reduce the signal losses, especially in the mm-wave bands. Comparing to conventional network dividers, Wilkinson offers higher isolation between adjacent arms in a relatively compact size.

In general, apart from the fundamental power splitting functionality, the key factors that affect the signal attenuation between the signal level at the input and the

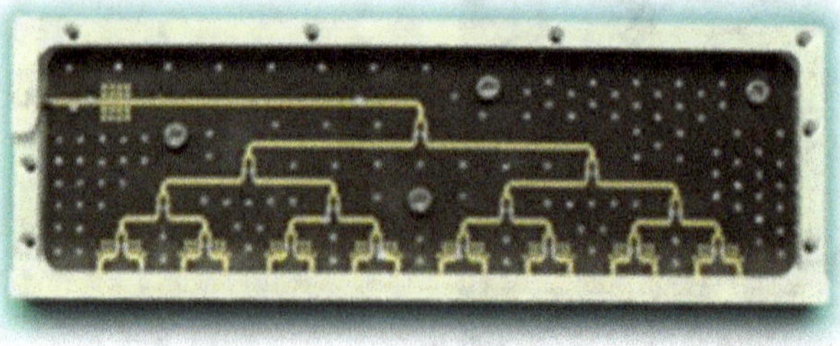

Figure 5.7 Photo of the 1-to-16 power splitter network

Table 5.2 Comparison between the theoretical and the actual measured values

Parameters	Theoretical targets in ideal scenarios	Actual measured values
Loss/channel	<18 dB	Max at 17.96 dB
Isolation between channels	>20 dB	Min at 21.9 dB
Input VSWR	<1.6:1	Max at 1.33:1
Output VSWR	<1.6:1	Max at 1.49:1

output of the power splitter network are the nominal loss and the insertion loss. For an N-way Wilkinson power splitter, the nominal loss is defined as $10\log_{10}N$ (dB) whereby the insertion loss is mainly originated from the dielectric and conductive loss of the transmission line, the fabrication and installation issues. In an ideal case (i.e., without any insertion loss), the signal level drops by $10\log_{10}16 = 12$ dB for the 1-to-16 Wilkinson power splitter network. Nevertheless, in practice, a well-designed and implemented power splitter can only minimize insertion loss, and not eliminate it. Compared with the theoretical targets in ideal scenarios, the actual measured values are summarized in Table 5.2, where VSWR denotes the voltage standing wave ratio. During the measurements, one observes that the mm-wave power divider offers high operational efficiency that meets the hybrid beamforming system requirements with the total loss per channel in the worst case scenario at 17.96 dB and the low splitter insertion loss at 5.96 dB.

5.2.2.2 Phase shifting and attenuation

The fully connected phase shifter and attenuator network consists of 256 user programmable phase shifters and attenuators that are equally distributed among the 16 mm-wave fully connected beamforming modules. Each module consists of 2 RF inputs and 8 RF outputs that are connected by 16 phase shifters and 16 attenuators (see the detailed block diagram in Figure 5.5 and an experimental setup photo in Figure 5.8). Each of these phase shifters and attenuators has six separate DIO control bits. During design stage, the user predefined phase value vector

Figure 5.8 Experimental setup photo for the evaluation of the 2-to-8 mm-wave fully connected beamforming module using a VNA (Figure 5 in [13], redistributed under CC BY 4.0)

{0°, 5.625°, 11.25°, 22.5°, 45°, 90°, 180°} and the attenuation value vector {0, 0.5, 1, 2, 4, 8, 12} have been chosen. Each phase shifter can select an arbitrary combination of the values in their corresponding user predefined vector by using their six relevant DIO control bits. The relationship between each phase shifter's control bits $\{b_0, b_1, b_2, \ldots, b_5\}$ and the corresponding phase value θ° is

$$\theta^\circ = b_0 \times 5.625^\circ + b_1 \times 11.25^\circ + b_2 \times 22.5^\circ + b_3 \times 45^\circ + b_4 \times 90^\circ + b_5 \times 180^\circ \tag{5.1}$$

From the experimental evaluation, the root mean square error of each of the modules is about 2.5° for the phase shifters and 0.3 dB for the attenuators. After assembling all the modules forming the whole phase shifting and attenuation network, the measured values of the phase-shifting and attenuation errors of the mm-wave fully connected beamforming modules are, respectively, within 5° and 0.5 dB, as compared with the theoretical targets. It is believed that this is due to the grounding performance of the modules and the matching state between the modules at mm-wave had slightly changed during the assembly process. These changes are envisaged to affect the precision of phase shifters and attenuators. Nevertheless, Section 5.3 shows that such deviations do not have a significant impact on the overall performance of the mm-wave hybrid beamforming testbed system. This is due to the fact that the random errors from a large number of phase shifters and attenuators are averaged out before the antennas.

5.2.2.3 Power amplifier

To apply the developed hybrid beamforming testbed system for both short-distance and long-distance mm-wave communication scenarios, it is necessary that the power amplifier of the system be capable of providing, linearly, a wide range of power levels. In practice, the trade-off between efficiency and linearity performance of RF power amplifiers will need to be considered. For beamforming applications, the linearity of the amplifiers has a higher priority as it guarantees the fidelity of the system's response versus different signal levels. A prototype of the power amplifier is shown in Figure 5.9. A VNA is used to test the functionality and performance of the power amplifiers. Compared to the design specifications, the results are reported as shown in Table 5.3, where IMD3 denotes the third-order intermodulation distortion.

Figure 5.9 Photo of the power amplifier prototype (Figure 6 in [13], redistributed under CC BY 4.0)

Table 5.3 Comparison between the theoretical and the actual measured values

Parameters	**Theoretical targets in ideal scenarios**	**Actual measured values**
Saturation point	>27 dBm	27.5 dBm
P1dB output power	>26 dBm	26.7 dBm
IMD3	>18 dBc	18.5–27 dBc

5.2.3 mm-Wave antennas

5.2.3.1 Transmitting antennas

At the transmitter end, a 128-element microstrip phased array antenna that can steer narrow beam(s) toward the angle of interest has been designed to operate at a center frequency of 26 GHz with 2 GHz operational bandwidth. The antenna is arranged as a 16×8 planar array with an individual exciting port per element. The spacing between elements is set to be $\lambda_0/2 = 5.8$ mm, where λ_0 is the wavelength of the center frequency. By manipulating the relative phase difference between the ports, the direction of constructed beam can be controlled, while the side-lobe-level (SLL) can be managed by controlling the amplitude of each excited port. The relevant theoretical details on deriving the relevant phase and amplitude values needed to steer the beam toward the angle of interest are given in [13,42]. Figure 5.10 shows the design and the photo of the phased array antenna.

As depicted in Figure 5.10(a) and (b), each element is a hat-shaped printed patch antenna excited by a single coaxial port. The optimized geometrical dimensions of each element at the presence of its corresponding connector are $d_1 = 3.3$ mm, $d_2 = 1$ mm, $d_3 = 1.78$ mm, and $d_4 = 0.68$ mm. One notes that the impedance matching is controlled by parameter d_4, while d_1, d_2, and d_3 impact the reactive loads of the patch [43,44] and, therefore, regulate the operating bandwidth. Each antenna element is directly attached to a standard mm-wave SMP connector. The substrate employed is Rogers RO4003C with $\varepsilon_r = 3.55$, tan$\delta = 0.0027$, and thickness, $h = 0.813$ mm. The array configuration is presented in Figure 5.10(c) with the overall substrate sheet dimensions of $d_5 = 95$ mm and $d_6 = 48.6$ mm. Figure 5.10(d) shows the fabricated prototype setup in an anechoic chamber.

The measurement results are validated with numerical simulation results using the commercial CST Microwave Studio time-domain software package [45], which utilizes the finite integration technique. The comparison between the measured and simulated scattering parameters (*S*-parameters) of the designed structure is presented in Figure 5.11. Note that the resonance frequency of each antenna element shifts down by around 300 MHz, when all the elements are placed adjacent to each other and form an array. This effect is being considered when designing the whole array with desired center frequency at 26 GHz. As shown in the results, the phased array elements resonate at the desired 26 GHz frequency with −10 dB bandwidth of more than 2 GHz and the maximum cross coupling level for the adjacent elements is less than −15 dB across the operating band. The measured S_{11} presented in Figure 5.11(a) clearly shows that the fabricated antenna is well matched at the frequency of interest. For operating frequency at 26 GHz, Figure 5.12 shows the comparison between the simulated and measured 2D radiation pattern for the single antenna element centrally located at #56 in the phased array (as shown in Figure 5.10(d)), whereby Figure 5.13 shows the simulated 3D radiation pattern for the single element and the phased array. The results clearly indicate that the array can produce narrow beams toward the angle of interest.

5.2.3.2 Receiving antennas

At the receiver end, two commercial off-the-shelf SGH antennas (Flann 20240-24NRL) are used. Figure 5.14 shows a photo setup of these SGH antennas being connected to the mm-wave receiver systems.

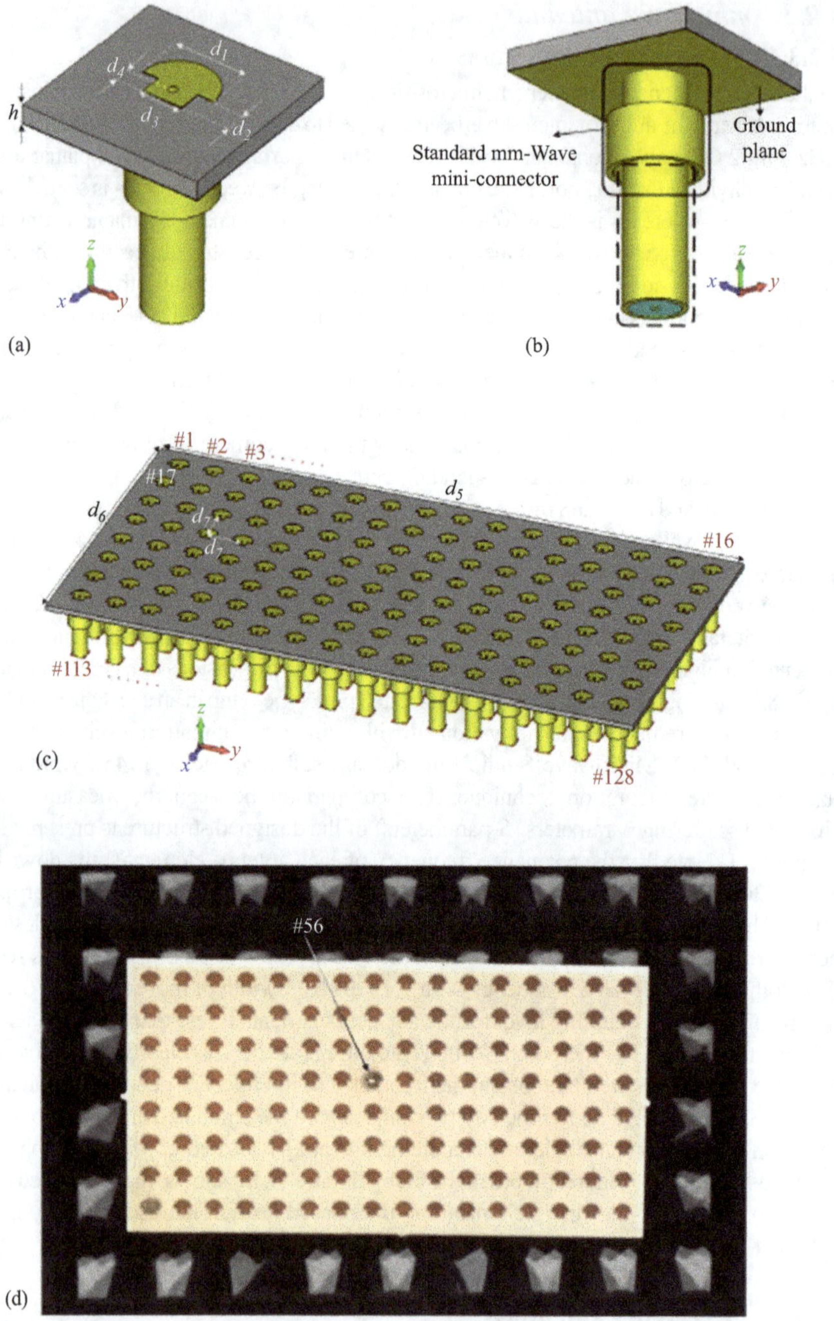

Figure 5.10 The phased array antenna (Figure 8 in [13], redistributed under CC BY 4.0): (a) single-element top view; (b) single-element back view; (c) illustrative diagram of the whole array; (d) fabricated prototype setup in an anechoic chamber

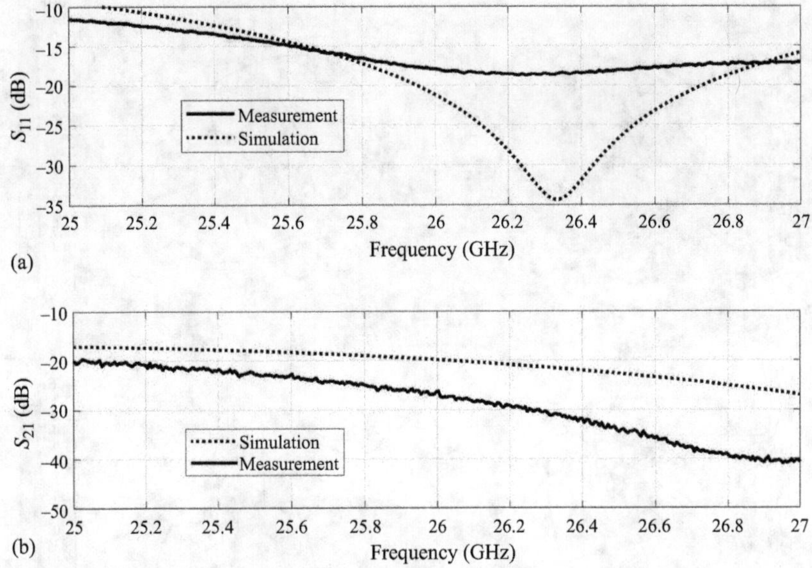

Figure 5.11 Comparison between the simulated and measured S-parameters: (a) S_{11}; (b) S_{21}

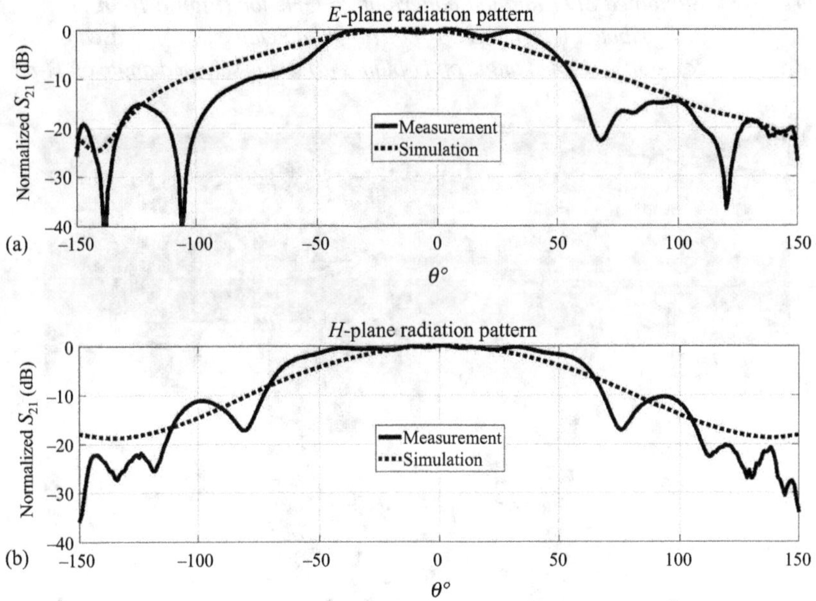

Figure 5.12 Comparison between the simulated and measured 2D radiation pattern for the single antenna element: (a) E-plane; (b) H-plane

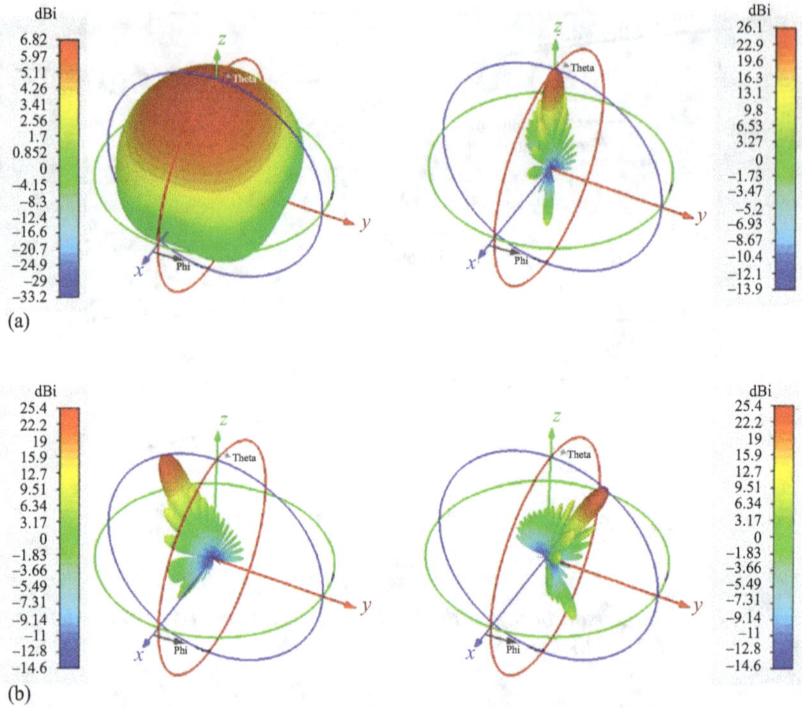

Figure 5.13 Simulated 3D radiation pattern at 26 GHz for (Figure 10 in [13], redistributed under CC BY 4.0) (a) single element; (b) phased array; (c) −30° scanned angle on H-plane; (d) +30° scanned angle on H-plane

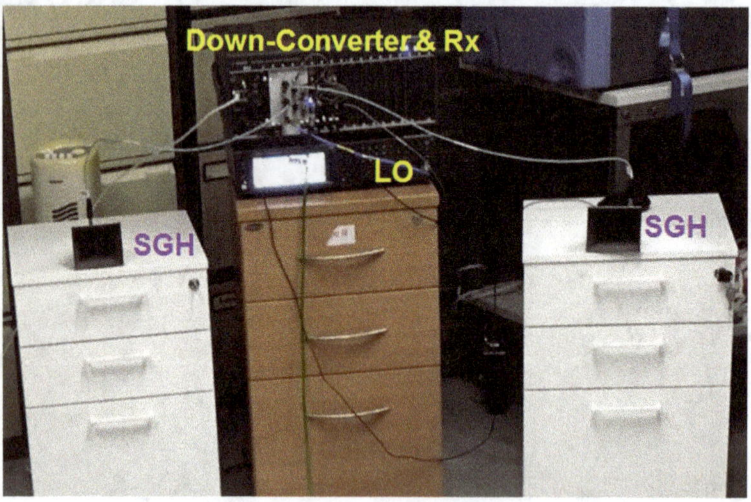

Figure 5.14 Photo showing experimental setup of the standard gain horn antennas being connected to the mm-wave receiver systems

5.3 Measurement metrologies

When using the traditional measurement methods [46,47], the phased array antenna is often measured together with a transceiver under particular circumstances such as multipath fading and co-channel interference where, for example, the bit error rate (BER) monitored by the DSP could be used as the critical parameter [33]. The beamforming control weights for amplitude and phase shift applied to each phased array antenna elements corresponding to the best BER are often recorded as the output parameters, which are then used by the DSP to control the phased array antennas when measuring its radiation pattern and SINR [46,48]. Nevertheless, quite often, the control weights obtained from traditional methods do not fully show the performance of the beamforming antenna system [33]. Also, if the interference is too large, the beamforming iteration cannot start; on the other hand, if the interference is not large enough, the beamforming algorithm may converge to produce a nonoptimal beam, and, consequently, the DSP unit will stop searching for a better solution which may be available. Furthermore, compared to sub-6-GHz communication systems, apart from the existing conventional hardware constraints such as phase noise, nonlinearities from power amplifiers, IQ-imbalance, etc., the practical implementation of mm-wave communication systems faces several exacerbated impact of hardware constraints, for example, ultrahigh data processing pressure and hardware imperfectness from the components of the wideband beamforming hardware [9].

To address these issues, it is important to perform rigorous calibration procedures and measurements for characterizing the whole of the beamforming system from basebands, RF chains, to antenna arrays, and where possible, to also consider the effects of propagation channel and co-channel interference. This section presents some experimental studies of the wide-bandwidth mm-wave fully connected hybrid beamformer metrological testbed system established in [13], together with the baseband and RF chains. The testbed was setup to operate at center frequency of 26 GHz with 128 active antenna elements arranged in the form of a 16×8 planar array and with 2 separate RF channels each capable of fully independent beamforming using the 256 user-programmable phase shifters and attenuators each with 6-bit control. Apart from describing the calibration procedures for the phase and amplitudes of the established fully connected hybrid beamformer system, the linearity, phase, and attenuation performance of the beamformer system between 25.5 and 26.5 GHz are evaluated as well as the beamforming and link performance of a 128-element planar phased array at 26 GHz where the measured radiation patterns with and without amplitude tapering are compared. The input power to the beamformer was −20 dBm for all tests except for the linearity test. The following shows some evaluation results of the testbed.

5.3.1 Calibration

After fabrication and assembly of all the mm-wave components such as splitters, beamforming modules, and power amplifiers for the wide bandwidth mm-wave

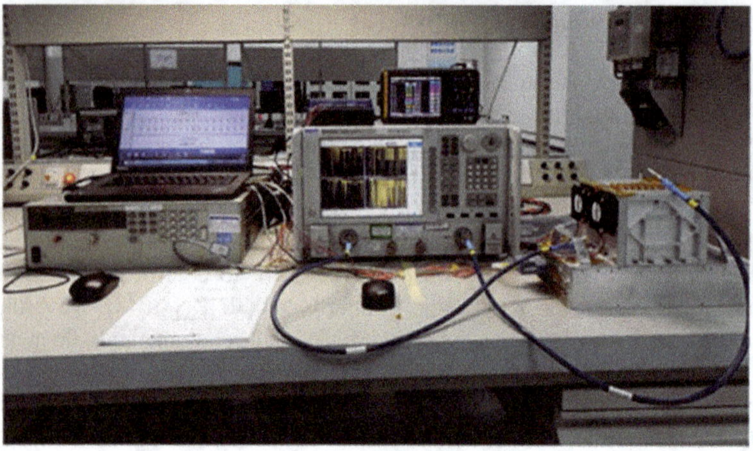

Figure 5.15 Photo showing experimental setup for the calibration of phase and amplitude of the mm-wave fully connected hybrid beamformer system using VNA

fully connected hybrid beamformer system, one needs to ensure performing calibration on the corresponding phases and amplitude at the output of the beamformer system. A calibrated VNA was employed for performing measurements of the phase and amplitude difference between the 2 RF inputs (Port 1 and Port 2) and 128 RF outputs of the hybrid beamformer (i.e., the output of the phase shifter and attenuator network). Using a DIO multifunction switch unit, the attenuators and phase shifters can be digitally programmed to set the envisaged values, with their six relevant DIO control bits and then a measurement can be made to confirm this. Figure 5.15 shows the relevant measurement setup. A calibration was performed before connecting the phased array antennas to the 128 RF output ports. Since the antennas are not connected at this stage, it is necessary to terminate all the ports, except the port under calibration, with a 50-Ω termination. It is important to ensure that the ports are terminated correctly or a condition could occur where signals are reflected back into the power amplifiers, potentially causing significant damage to the equipment.

This process needs to be repeated for all of the 256 phase shifters and attenuators for all the user predefined phase value vector {0°, 5.625°, 11.25°, 22.5°, 45°, 90°, 180°} and the attenuation value vector {0, 0.5, 1, 2, 4, 8, 12}. The phase calibration has two stages whereby two calibration tables are generated. The first table is used to make the phase differences at the antenna ports equal to 0°. In other words, this step alleviates the relative phase offsets between different RF ports and acts as the reference phase value for each phase shifter. Then, the second calibration table is applied to create the required phase shift with respect to the reference value for the corresponding phase shifter. To calibrate the amplitudes, one requires to evaluate the offset between the measured attenuation level and its envisaged value. Three calibration tables were created, for 25.5, 26.0, and 26.5 GHz, generated from measurements made at these frequencies.

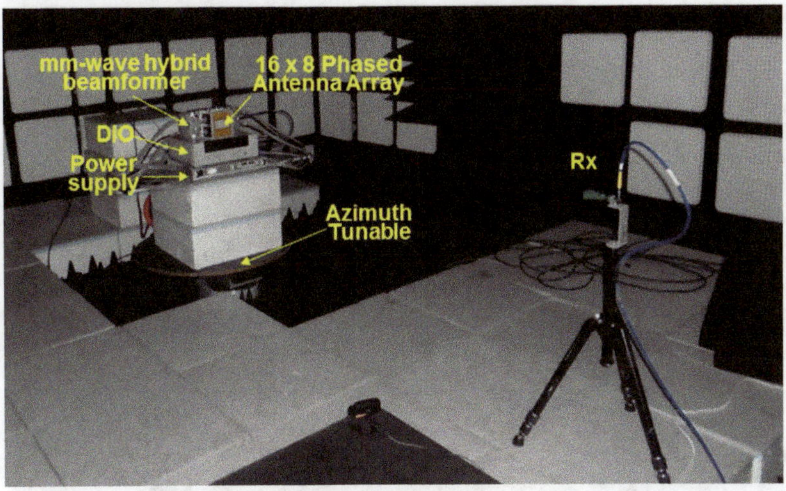

Figure 5.16 System performance measurement setup with the mm-wave hybrid beamformer mounted on the azimuth positioner in NPL's SMART anechoic chamber

Following the previous calibration procedures, a 128-element planar phased array is attached to the RF output front end of the mm-wave fully connected hybrid beamformer system. Apart from the link performance evaluation, all the measurements presented in the following sections were performed in the UK National Physical Laboratory's (NPL) small antenna-radiated test (SMART) range fully anechoic chamber [49]. Figure 5.16 shows the measurement setup where an azimuth positioner was employed to rotate the mm-wave hybrid beamformer system during the measurements. The DIO controller uses the calibration tables to improve the accuracy of the commanded phase and attenuation values. Figure 5.17 shows the front face of the planer phased array antenna that was aligned to the center of the azimuthal rotation as well as the azimuthal scanning coordinate. An SGH antenna (Flann 22240-20) was used as the receive antenna and was positioned at a distance of 2.35 m from the front of the planar array at $\theta = 0°$ (see Figure 5.16) and at the height of the center of the planar array. A Keysight PNA-X VNA was used to measure the radiation pattern of the array as it was rotated.

5.3.2 Linearity

The RF chains' linearity of the mm-wave fully connected hybrid beamformer system was assessed by varying the input power to the beamformer between -50 and -10 dBm with all phase shift and attenuation setting on the beamformer set to 0° and 0 dB, respectively. The VNA measures the S_{21} as the ratio between the transmitted and received power. In a system with good linearity, the measured S_{21} should not change significantly when the input power level is adjusted. Figure 5.18 shows the difference from the measured S_{21} result when the input power to the beamformer was

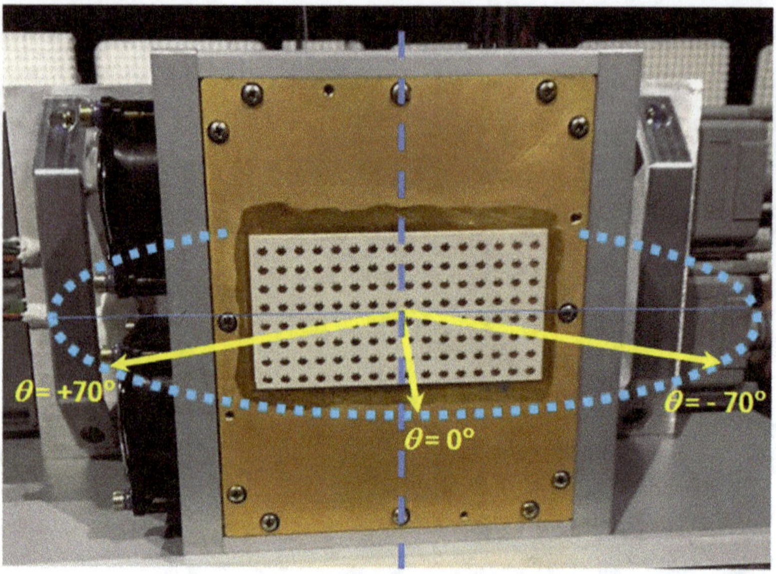

Figure 5.17 Photograph of the planar phased array antenna attached to the front of the mm-wave beamformer

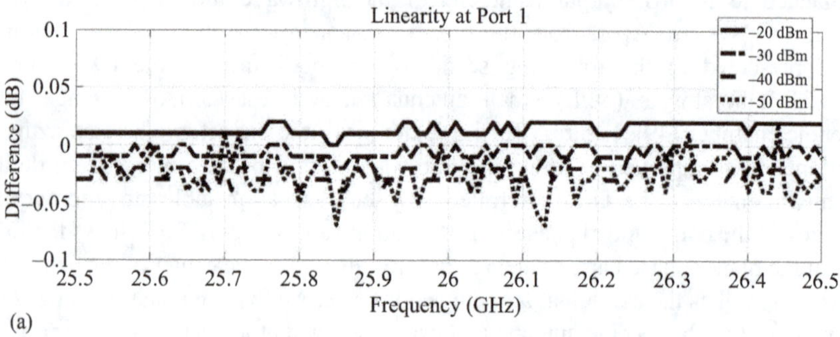

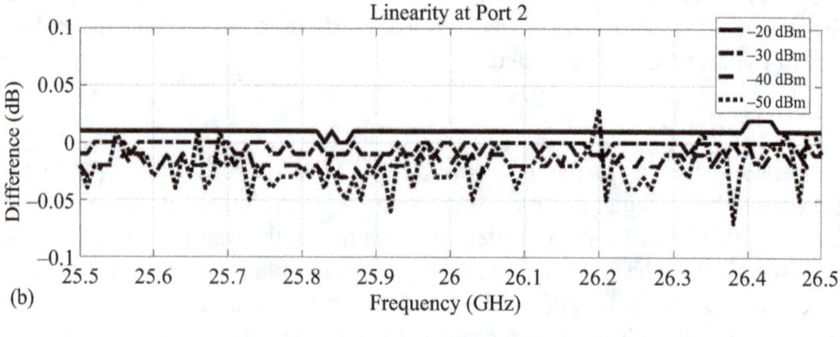

Figure 5.18 The linearity of the beamformer RF chains: (a) Port 1; (b) Port 2

set to −10 dBm for Port 1 and Port 2. The results show that the beamformer system has excellent linearity performance, with variation within ±0.06 dB.

5.3.3 Phase

The performances of the phase shifters of the mm-wave fully connected hybrid beamformer system were assessed by setting all phase shift setting to all RF outputs uniformly to 0, −5.625°, −11.25°, −22.5°, −45°, −90°, and −180° with all attenuation setting on the beamformer set to 0 dB and measuring the phase of the received signal and then normalizing the results to the 0° setting to access the relative phase differences. Figure 5.19 presents the difference between the measured S_{21} made with each uniform phase shift setting and the measurement made with uniform 0° phase shift. The results show that the mm-wave fully connected hybrid beamformer system could produce the reasonable phase shift to meet the desired target to within about ±6.5°.

5.3.4 Attenuation

The performance of the attenuators of the mm-wave fully connected hybrid beamformer system was assessed by setting all attenuation settings to all RF outputs uniformly to 0, 0.5, 1, 2, 4, and 6 dB with all phase shift setting on the beamformer set to 0°. Figure 5.20 presents the different between the measured S_{21} made with each uniform attenuation setting and the measurement made with uniform 0-dB attenuation. The results show that the beamformer system could produce the reasonable attenuation to meet the desired target to within about ±1 dB. Performance at the edges of the measured band could potentially be improved by selecting a calibration configuration appropriate for operating at these frequencies; this was not considered in this work as the focus here is on broadband operation centered at 26 GHz.

5.3.5 Beamforming

Using the established metrological mm-wave fully connected hybrid beamforming testbed system, flexible implementation of amplitude tapering techniques to reduce sidelobe levels could be explored. The beamforming capability of the mm-wave hybrid beamformer system was assessed by commanding the phase shifters to form beams at the azimuthal scanning coordinate (see Figure 5.17) from the boresight of the planar array with $\theta = -60°$, −45°, −30°, −22.5°, −15°, −7.5°, 0°, 7.5°, 15°, 22.5°, 30°, 45°, and 60° with and without amplitude tapering. The tapering weight along the 16-element array direction was set with attenuations of 8, 8, 6, 4, 2, 1, 0.5, 0, 0, 0.5, 1, 2, 4, 6, 8, and 8 dB (uniform at each 8-element array). Figures 5.21 and 5.22 show the normalized measured radiation pattern of the planar phased array between $\theta = -90°$ and $\theta = +90°$ with and without applying amplitude tapering, for Port 1 and Port 2, respectively. The results show that the mm-wave hybrid beamformer system could achieve good beamforming performance between ±60° where the SLL could be reduced by applying amplitude tapering.

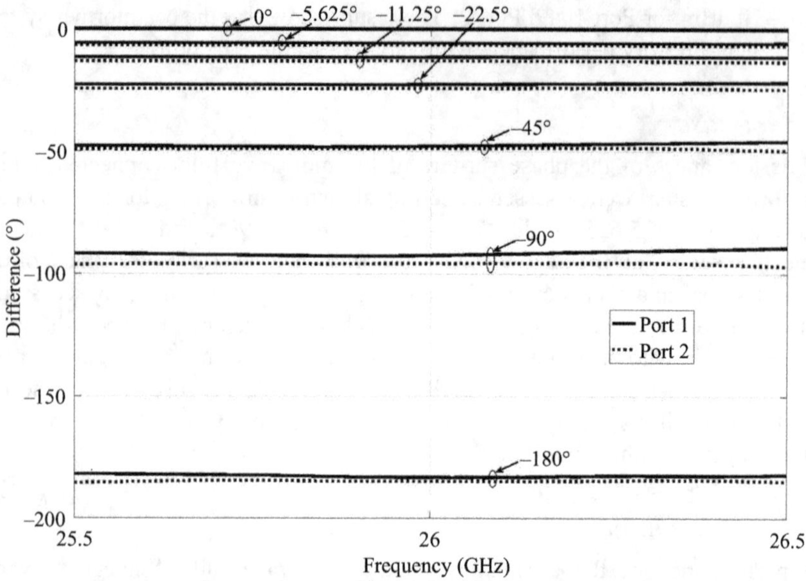

Figure 5.19 The phase performance of the beamformer for each RF chain

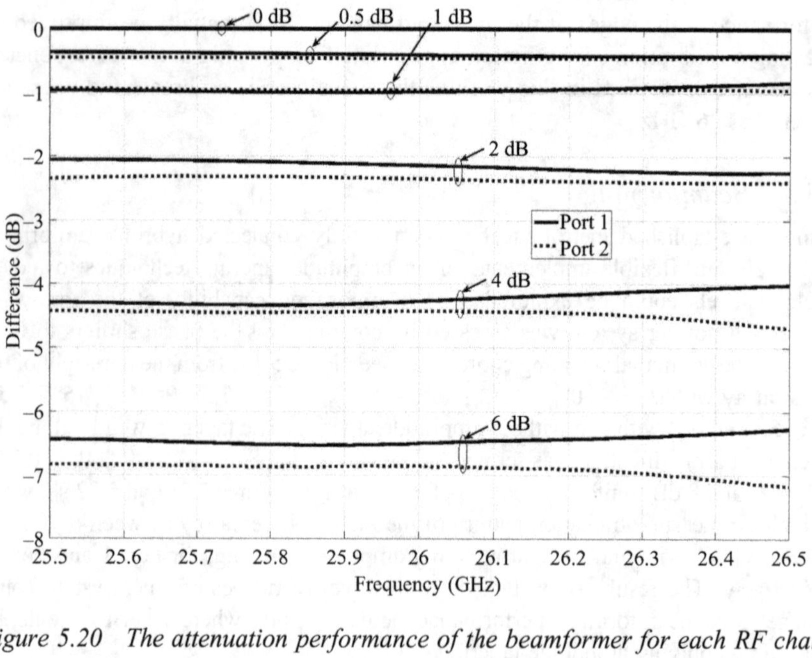

Figure 5.20 The attenuation performance of the beamformer for each RF chain

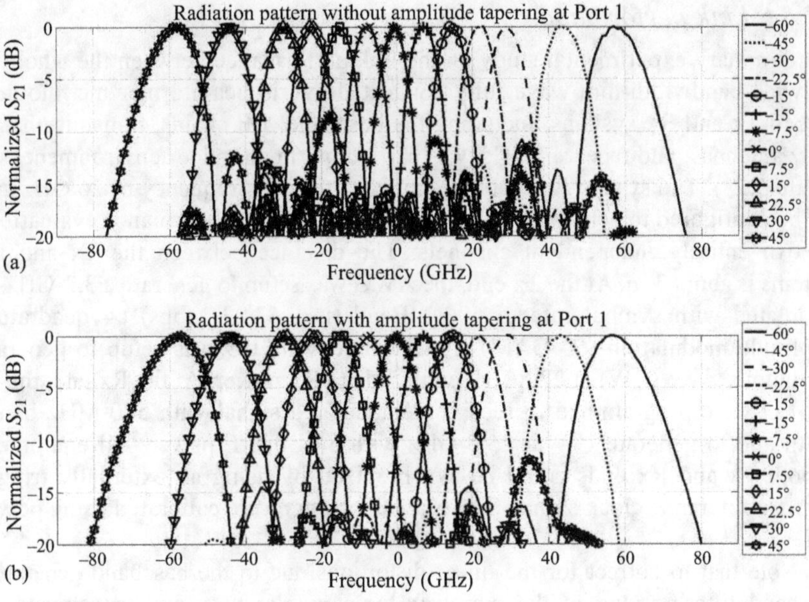

Figure 5.21 Beamforming capability of the beamformer for Port 1: (a) without tapering; (b) with tapering

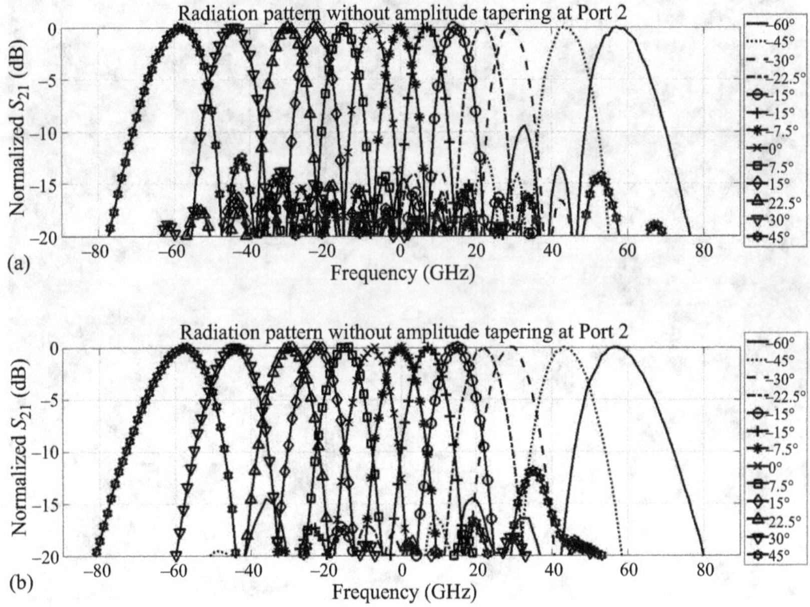

Figure 5.22 Beamforming capability of the beamformer for Port 2: (a) without tapering; (b) with tapering

5.3.6 Link performance

A preliminary experimental study on the link performance between the whole of the wide-bandwidth mm-wave fully connected hybrid beamformer metrological testbed Tx and Rx systems, including the baseband, RF chains, transmit phased array antennas, and receive SGHs, was carried out in an indoor environment (see Figure 5.23). The system was setup to operate at center frequency of 26 GHz and can be configured to perform the relevant mm-wave link performance evaluations for two entirely independent channels. The distance between the Tx and Rx systems is about 3 m. At the Tx end, the AWG was setup to generate a 3.2-GHz IF modulated with symbol rate of 500 MBaud (i.e., 524.3 Mbps) 64 quadrature amplitude modulation (64 QAM) signal whereby its LO was setup to generate continuous wave (CW) at 22.8 GHz with 5-dBm RF power. At the Rx end, the IF and LO were programmed to receive demodulated signal with 500-MHz bandwidth and to generate CW at 25.5 GHz with 5-dBm RF power. All equipment at both Tx and Rx ends could be synchronized by using an externally trigger 10-MHz reference clock signal. The channel powers were calibrated using power meters.

Note that to correct for the linear distortions due to the baseband generator, baseband filtering, and in the mm-wave up-converter unit, and to estimate an

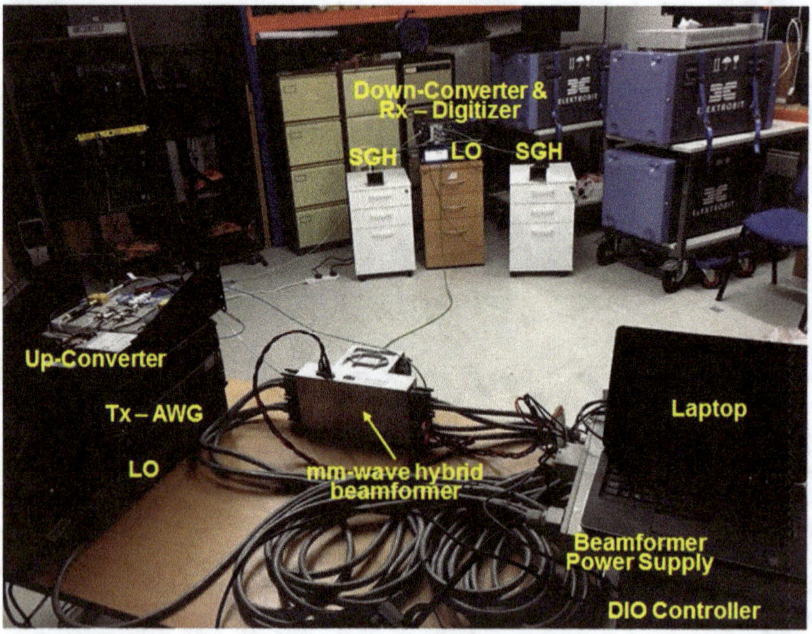

Figure 5.23 Photo of the measurement setup for mm-wave link performance evaluation in an indoor environment

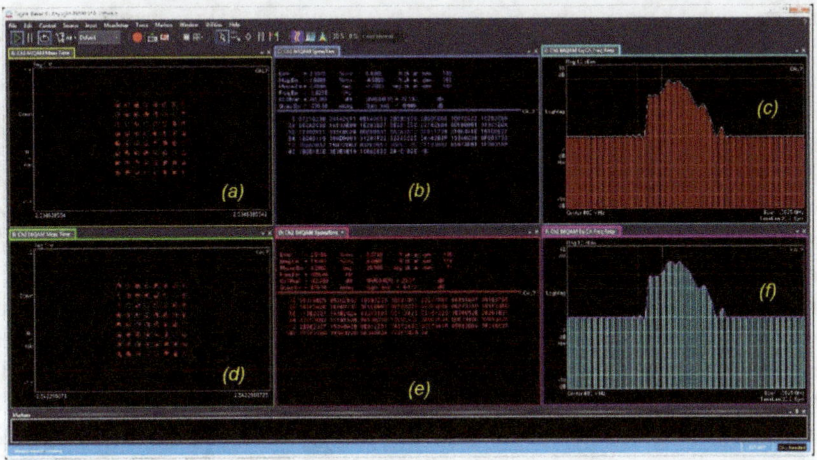

Figure 5.24 Link performance evaluation using VSA software tool: (a) constellation diagram for Channel 1; (b) system error for Channel 1; (c) frequency response for Channel 1; (d) constellation diagram for Channel 2; (e) system error for Channel 2; (f) frequency response for Channel 2

overall channel response for the signal path between baseband signal generation at the Tx end to the VSA demodulator input at the Rx end, flatness calibration for both the channels can be performed using the Least Mean Squared adaptive equalizer built into the Keysight 89601B VSA software tool. Such calibration requires direct connection between the RF output of Tx end to the RF input of Rx end and the use of the VSA software tool to demodulate a test signal. The calibration results could then be applied as a correction to the modulated signal generated by the AWG. Using the VSA software tool, Figure 5.24 shows the visualization of measured link performance results by means of parameters such as error vector magnitude (EVM), signal to noise ratio (SNR), IQ offset, channel frequency response, etc. The measured channel frequency response could then be used for the flatness calibration to generate the correction vector for preequalization of the test signal generated by the AWG.

5.4 Conclusion and future directions

This chapter has presented the insight methodologies on how to design, implement, and evaluate a width-bandwidth mm-wave fully connected hybrid beamforming metrological testbed with a large antenna array. The focus has been given on discussions include testbed design, calibration procedures, experimental evaluations, as well as the critical factors to consider for their practical implementation. If RF harmonics and spurious signal issues are avoided, one envisages that the testbed could be setup to work between 25 and 30 GHz with 2-GHz instantaneous bandwidth. Each of the phase shifters and attenuators in the mm-wave fully connected hybrid beamformer has six separate DIO control bits. Apart from describing the calibration procedures for the phase and amplitudes of the established fully connected hybrid beamformer system, the linearity, phase, and attenuation performance of the beamformer system between 25.5 and 26.5 GHz have been evaluated as well as the beamforming and link performance of a 128-element planar phased array at 26 GHz where the measured radiation patterns with and without amplitude tapering are compared.

This testbed offers significant advantages and increased flexibility compared to state-of-the-art hybrid beamformers, offering precise independent phase and attenuation control at each antenna element, arbitrary waveform generation, and a very flexible bespoke control software stack. It will enable full exploitation over the advantages of large arrays as well as radically reduce the traceable metrological burden by utilizing the advanced software-defined radios and signal processing techniques. The 2×128 beamformer allows for two RF channels being independently beamformed in both azimuth and elevation. Also, the ability to control the amplitude of the signals at each antenna element provides better control over the sidelobe levels. Despite the abundant literature on designing algorithms for hybrid beamformers [8,15], most of the existing work relies on theoretical assumptions about the hardware constraints and the propagation channels. The performance of such algorithms may significantly vary according to the propagation scenario, and the hardware design. The fully connected structure of the beamformer paves the path to develop, implement, and test a plethora of hybrid beamforming algorithms, which are be realizable with a conventional sub-connected architecture.

With the fully connected features that offer increased flexibility to control the phase and magnitude for the planar phased array, the beamforming system can create semi-omnidirectional and narrow directional beams. As a result, the high EIRP dynamic range of this testbed enables its deployment in various propagation scenarios. It provides researchers in academics and industry a unique test platform with significantly more degrees of freedom on trial and the development of different robust hybrid beamforming algorithms for reaching higher data rate for mm-wave communications. Furthermore, the linearity, high bandwidth, and dynamic range of the system enable the beamformer used to analyze various user-defined modulated signals to test and investigate over the performance of new modulation schemes for 5G and beyond. The designed testbed can also be used to further optimize the transmission over frequency-selective channels [50,51] and provide trade-offs between diversity and spatial multiplexing gains. Such studies may even motivate the deployment of hybrid beamformers in lower frequency bands, which are generally more frequency selective. Another research direction that has recently received significant interest is applying machine-learning-based techniques for channel estimation, beam alignment, and calculating the hybrid beamforming weights for mm-wave systems [52,53] operating over realistic mm-wave channels.

Acknowledgement

The work of T. H. Loh and D. Cheadle was supported by the 2017–2020 National Measurement System Programme of the UK government's Department for Business, Energy and Industrial Strategy (BEIS), under Science Theme Reference EMT20 of that Programme. The work of Sohail Payami was supported by NPL, through the NPL-Surrey 2017–2020 Collaboration Programme. The authors would additionally like to thank Will Pontin from NPL and Juan Carlos De Luna Ducoing from University of Surrey for their assistance during the measurements.

References

[1] Martens J. 'Meeting New RF Measurement Challenges: Wide Intermediate-Frequency Millimeter-Wave Modulated Device and Subsystem Characterization and Evolving Measurement Architectures'. *IEEE Microwave Magazine*. 19, April 2018, pp. 35–48.

[2] Foegelle M. D. '5G and mmwave device measurement challenges'. *The 12th European Conference on Antennas and Propagation (EuCAP 2018)*, London, United Kingdom, April 2018. pp. 1–5

[3] Kang N., Kwon J., Cho C., and Park J. 'Measurement System for Millimeter-Wave Antennas With Distributed External Local Oscillators and Mixers'. *IEEE Transactions on Instrumentation and Measurement*. 2019, 68(6), pp. 1967–1972.

[4] Fan W., Zhang F., and Wang Z. 'Over-the-Air Testing of 5G Communication Systems: Validation of the Test Environment in Simple-Sectored Multiprobe Anechoic Chamber Setups'. *IEEE Antennas and Propagation Magazine*. 2021, 63(1), pp. 40–50.

[5] Sun C., Cheng J., and Ohira T. (eds.). *Handbook on Advancements in Smart Antenna Technologies for Wireless Networks*. London, UK: IGI Global, 2008.

[6] Dahlman E., Parkvall S., and Skold J. *5G NR: The Next Generation Wireless Access Technology*. 2nd edn. London, UK: Academic Press, 2020.

[7] Reil M. and Lloyd G. *White Paper – Millimeter-Wave Beamforming: Antenna Array Design Choices & Characterization* [online]. Rohde & Schwarz; October 2016. Available from: https://scdn.rohde-schwarz.com/ur/pws/dl_downloads/dl_application/application_notes/1ma276/1MA276_2e_Beamform_mmW_AntArr.pdf [Accessed March 2021].

[8] Ahmed I., Khammari H., Shahid A., *et al.* 'A Survey on Hybrid Beamforming Techniques in 5G: Architecture and System Model Perspectives'. *IEEE Communications Surveys & Tutorials*. 2018, 20(4), pp. 3060–3097.

[9] Yang X., Matthaiou M., Yang J., Wen C., Gao F., and Jin S. 'Hardware-Constrained Millimeter-Wave Systems for 5G: Challenges, Opportunities, and Solutions'. *IEEE Communications Magazine*. 2019, 57(1), pp. 44–50.

[10] Molu M. M., Xiao P., Khalily M., Cumanan K., Zhang L., and Tafazolli R. 'Low-Complexity and Robust Hybrid Beamforming Design for Multi-Antenna Communication Systems'. *IEEE Transactions on Wireless Communications*. 2018, 17(3), pp. 1445–1459.

[11] Payami S., Khalily M., Loh T. H., and Nikitopoulos K. 'Hybrid Beamforming With Switches and Phase Shifters Over Frequency-Selective Channels'. *IEEE Wireless Communications Letters*. 2020, 9(8), pp. 1305–1308.

[12] Payami S., Ghoraishi M., Dianati M., and Sellathurai M. 'Hybrid Beamforming With a Reduced Number of Phase Shifters for Massive MIMO Systems'. *IEEE Transactions on Vehicular Technology*. 2018, 67(6), pp. 4843–4851.

[13] Payami S., Khalily M., Araghi A., *et al.* 'Developing the First mmWave Fully-Connected Hybrid Beamformer With a Large Antenna Array'. *IEEE Access*. 2020, 8, pp. 141282–141291.

[14] Jiang D. and Natarajan B. 'Hybrid Precoding With Compressive Sensing Based Limited Feedback in Massive MIMO Systems'. *Transactions on Emerging Telecommunications Technologies*. 2016, 27(12), pp. 1672–1678.

[15] Molisch A. F., Ratnam V. V., Han S., *et al.* 'Hybrid Beamforming for Massive MIMO: A Survey'. *IEEE Communications Magazine*. 2017, 55(9), pp. 134–141.

[16] Blandino S., Mangraviti G., Desset C., Bourdoux A., Wambacq P., and Pollin S. 'Multi-User Hybrid MIMO at 60 GHz Using 16-Antenna Transmitters'. *IEEE Transactions on Circuits and Systems I: Regular Papers*. 2019, 66(2), pp. 848–858.

[17] Kibaroglu K., Sayginer M., Phelps T., and Rebeiz G. M. 'A 64-Element 28-GHz Phased-Array Transceiver With 52-dBm EIRP and 8–12-Gb/s 5G Link at 300 Meters Without Any Calibration'. *IEEE Transactions on Microwave Theory and Techniques*. 2018, 66(12), pp. 5796–5811.

[18] Kodak U., Rupakula B., Zihir S., and Rebeiz G. M. '60-GHz 64- and 256-Element Dual-Polarized Dual-Beam Wafer-Scale Phased-Array Transceivers With Reticle-to-Reticle Stitching'. *IEEE Transactions on Microwave Theory and Techniques*. 2020, 68(7), pp. 2745–2767.

[19] Pang J., Wu R., Wang Y., *et al.* 'A 28-GHz CMOS Phased-Array Transceiver Based on LO Phase-Shifting Architecture With Gain Invariant Phase Tuning for 5G New Radio'. *IEEE Journal of Solid-State Circuits*. 2019, 54(5), pp. 1228–1242.

[20] Abbasi M. A. B. and Abbasi Q. H. 'Development Challenges of Millimeter-Wave 5G Beamformers' in Chatzimisios P., Tafazolli R., and Wang C.-L. (eds.). *Wiley 5G Ref: The Essential 5G Reference Online*. New York, NY: John Wiley & Sons, Ltd., 2020, pp. 1–25.

[21] Al-Ogaili F. and Shubair R. M. 'Millimeter-wave mobile communications for 5G: Challenges and opportunities'. *The 2016 IEEE International Symposium on Antennas and Propagation (APSURSI 2016)*, Fajardo, PR, USA, 2016. pp. 1003–1004.

[22] Rappaport T. S., MacCartney G. R., Samimi M. K., and Sun S. 'Wideband Millimeter-Wave Propagation Measurements and Channel Models for

Future Wireless Communication System Design'. *IEEE Transactions on Communications*. 2015, 63(9), pp. 3029–3056.

[23] Heath R. W., González-Prelcic N., Rangan S., Roh W., and Sayeed A. M. 'An Overview of Signal Processing Techniques for Millimeter Wave MIMO Systems'. *IEEE Journal of Selected Topics in Signal Processing*. 2016, 10 (3), pp. 436–453.

[24] Payami S., Shariat M., Ghoraishi M., and Dianati M. 'Effective RF codebook design and channel estimation for millimeter wave communication systems'. *The 2015 IEEE International Conference on Communication Workshop (ICCW 2015)*, London, United Kingdom, 2015. pp. 1226–1231.

[25] Kutty S. and Sen D. 'Beamforming for Millimeter Wave Communications: An Inclusive Survey'. *IEEE Communications Surveys & Tutorials*. 2016, 18 (2), pp. 949–973.

[26] Han S., I C.-L., Xu Z., and Rowell C. 'Large-Scale Antenna Systems With Hybrid Analog and Digital Beamforming for Millimeter Wave 5G'. *IEEE Communications Magazine*. 2015, 53(1), pp. 186–194.

[27] Payami S., Ghoraishi M., and Dianati M. 'Hybrid Beamforming for Large Antenna Arrays With Phase Shifter Selection'. *IEEE Transactions on Wireless Communications*. 2016, 15(11), pp. 7258–7271.

[28] Ayach O. E., Rajagopal S., Abu-Surra S., Pi Z., and Heath R. W. 'Spatially Sparse Precoding in Millimeter Wave MIMO Systems'. *IEEE Transactions on Wireless Communications*. 2014, 13(3), pp. 1499–1513.

[29] Mondal S., Singh R., Hussein A. I., and Paramesh J. 'A 25–30 GHz Fully-Connected Hybrid Beamforming Receiver for MIMO Communication'. *IEEE Journal of Solid-State Circuits*. 2018, 53(5), pp. 1275–1287.

[30] Mohammed S. L., Alsharif M. H., Gharghan S. K., Khan I., and Albreem M. 'Robust Hybrid Beamforming Scheme for Millimeter-Wave Massive-MIMO 5G Wireless Networks'. *Symmetry*. 2019, 11(11), pp. 1424.

[31] Chen R., Xu H., Li C., Zhu L., and Li J. 'Hybrid beamforming for broadband millimeter wave massive MIMO systems'. *The 2018 IEEE 87th Vehicular Technology Conference (VTC Spring)*, Porto, Portugal, 2018. pp. 1–5.

[32] Sun S., Rappaport T. S., and Shaft M. 'Hybrid beamforming for 5G millimeter-wave multi-cell networks'. *The 2018 IEEE Conference on Computer Communications Workshops (INFOCOM WKSHPS)*, Honolulu, HI, USA, 2018. pp. 589–596.

[33] Loh T. H., Liu H., and Gao S. 'A Cost-Effective Direct Magnitude Measurement Methodology for Smart Antennas'. *IEEE Transactions on Antennas and Propagation*. 2013, 61(4), pp. 2043–2050.

[34] Mondal S. and Paramesh J. 'A Reconfigurable 28-/37-GHz MMSE-Adaptive Hybrid-Beamforming Receiver for Carrier Aggregation and Multi-Standard MIMO Communication'. *IEEE Journal of Solid-State Circuits*. 2019, 54(5), pp. 1391–1406.

[35] Natarajan A., Reynolds S. K., Tsai M.-D., *et al.* 'A Fully-Integrated 16-Element Phased-Array Receiver in SiGe BiCMOS for 60-GHz Communications'. *IEEE Journal of Solid-State Circuits*. 2011, 46(5), pp. 1059–1075.

[36] Cohen E., Ruberto M., Cohen M., Degani O., Ravid S., and Ritter D. 'A CMOS Bidirectional 32-Element Phased-Array Transceiver at 60 GHz With LTCC Antenna'. *IEEE Transactions on Microwave Theory and Techniques*. 2013, 61(3), pp. 1359–1375.

[37] Roh W., Seol J.-Y., Park J., *et al.* 'Millimeter-Wave Beamforming as an Enabling Technology for 5G Cellular Communications: Theoretical Feasibility and Prototype Results'. *IEEE Communications Magazine*. 2014, 52(2), pp. 106–113.

[38] Rebeiz G. M., Kim S.-Y., Inac O., *et al.* 'Millimeter-wave large-scale phased-arrays for 5G systems'. *The 2015 IEEE MTT-S International Microwave Symposium*. Phoenix, AZ, USA, 2015. pp. 1–3.

[39] Labios L. *Keysight Technologies, UC San Diego Demonstrate the World's Fastest 28 GHz 5G Band, Bidirectional Phased-Array* [online]. UC San Diego; Jacobs School of Engineering; June 2017. Available from: https://jacobsschool.ucsd.edu/news/release?id=2226 [Accessed March 2021].

[40] Everything RF. *NI Demonstrates World's First Real-Time Over-the-Air Prototype for Verizon 5G at 28 GHz*. March 2017 [online]. Available from: https://www.everythingrf.com/news/details/3818-NI-Demonstrates-World-s-First-Real-Time-Over-the-Air-Prototype-for-Verizon-5G-at-28-GHz [Accessed September 2021].

[41] Keysight Technologies. *Keysight Technologies' mmWave Testbed Enables Ball Aerospace, Anokiwave to Validate First Commercial 5G NR Active Array Antenna Designs at 26 GHz* [online]. 2018. Available from: https://about.keysight.com/en/newsroom/pr/2018/20feb-nr18016.shtml [Accessed March 2021].

[42] Balanis C. A. *Antenna Theory: Analysis and Design*. Hoboken, NJ, USA: Wiley-Blackwell, 2005.

[43] Araghi A. and Dadashzadeh G. 'Detail-Oriented Design of a Dual-Mode Antenna With Orthogonal Radiation Patterns Utilizing Theory of Characteristic Modes'. *Applied Computational Electromagnetics Society (ACES) Journal*. 2013, 28(10), pp. 952–959.

[44] Araghi A. and Dadashzadeh G. 'Oriented Design of an Antenna for MIMO Applications Using Theory of Characteristic Modes'. *IEEE Antennas and Wireless Propagation Letters*. 2012, 11, pp. 1040–1043.

[45] Simulia. *CST Studio Suite Electromagnetic Field Simulation Software*. 2016 [online]. Available from: https://www.3ds.com/products-services/simulia/products/cst-studio-suite/?utm_source=cst.com&utm_medium=301&utm_campaign=cst [Accessed September 2021].

[46] Tyler N., Allen B., and Aghvami A. H. 'Calibration of Smart Antenna Systems: Measurements and Results'. IET Microwaves, *Antennas and Propagation*. 2007, 1(3), pp. 629–638.

[47] Alexiou A. and Haardt M. 'Smart Antenna Technologies for Future Wireless Systems: Trends and Challenges'. *IEEE Communications Magazine*. 2004, 42(9), pp. 90–97.

[48] Nasimuddin N., Chen Z. N., Qing X., and See T. S. P. 'Sectorised Antenna Array and Measurement Methodology for Indoor Ultra-Wideband Applications'. IET Microwaves, Antennas and Propagation. 2009, 3(4), pp. 621–629.

[49] National Physical Laboratory. *Small antenna radiated test range*. 2010 [online]. Available from: https://www.npl.co.uk/products-services/radio-frequency/smart-chamber [Accessed September 2021].

[50] Payami S., Sellathurai M., and Nikitopoulos K. 'Low-Complexity Hybrid Beamforming for Massive MIMO Systems in Frequency-Selective Channels'. *IEEE Access*. 2019, 7, pp. 36195–36206.

[51] Sohrabi F. and Yu W. 'Hybrid Analog and Digital Beamforming for mmWave OFDM Large-Scale Antenna Arrays'. *IEEE Journal on Selected Areas in Communications*. 2017, 35(7), pp. 1432–1443.

[52] Scalabrin M., Bielsa G., Loch A., Rossi M., and Widmer J. 'Machine Learning Based Network Analysis Using Millimeter-Wave Narrow-Band Energy Traces'. *IEEE Transactions on Mobile Computing*. 2020, 19(5), pp. 1138–1155.

[53] Huang H., Song Y., Yang J., Gui G., and Adachi F. 'Deep-Learning-Based Millimeter-Wave Massive MIMO for Hybrid Precoding'. *IEEE Transactions on Vehicular Technology*. 2019, 68(3), pp. 3027–3032.

Chapter 6

Assessment of broadband sources and defocusing effects of compact antenna test range for 5G antenna testing

Vince Rodriguez[1]

Compact range techniques are ideal for fifth-generation (5G) applications where electrically large arrays are tested and where a simple 16 cm by 8 cm handset becomes an 18-wavelength-sized device that requires a 6.25-m long test distance to be in the far field [1]. The compact range is the best approach or solution. This is especially true, if these arrays are fed internally and there is no means of tapping a reference for measuring the phase. The phase is necessary for near-to-far-field transformation measurements, which are the other potential technique for testing these devices. While there are other methods to reduce the test distance by generating a plane wave at the location of the device under test (DUT), compact ranges using parabolic reflectors are the simplest and more widely used approach [1]. Dielectric lenses are typically heavy and more complex to manufacture. There are new approaches to the manufacturing of dielectric lenses by using artificial materials. Some of these lenses have been used to increase the quiet zone (QZ) of far-field anechoic chambers [2]. These "meta-material" lenses are much lighter, but the materials only operate to an upper frequency of 8 GHz. Transmit- and reflect-arrays can also be used to generate near-field plane waves; however, these approaches are fairly narrow banded. Antenna arrays [3,4] tend to be limited in frequency and potentially more complex to manufacture [5]. However, plane wave generators are being developed for 5G testing and they are commercially available.

The complexity required in plane wave generators based on antenna arrays, or the issues related to dielectric lenses, does not mean that paraboloid compact ranges are simple. The concept is simple, and the approach has been around for over 50 years, yet these systems are complex and require tight tolerances in machining to get accurate solutions. Once standard committees identified the compact range as a valid approach for testing antennas and systems for 5G applications, several companies with little history in the manufacturing and implementation of compact ranges started to market compact range systems for antenna measurements in the Ku, K and Ka bands. In some cases, these systems were fitted with broadband

[1]NSI-MI Ametek, Suwanee, GA, USA

antennas to make them easier to use across the entire frequency band of interest. The use of these antennas that are not specifically designed to be compact range feeds needs to be assessed. Consideration should be paid to the feed antenna since the feed antenna determines a lot of the performance parameters of the compact antenna test range. The edge termination is also another important parameter as it determines the absorber treatment necessary in the range as was discussed by Rodriguez in [1].

In this chapter, the problems with a far-field measurement of 5G steerable arrays is shown by calculating the lower boundary of the far field for a typical handheld device at the Ka band. The results show that a physically very large antenna range in excess of 10 m is required to measure the pattern of a handheld device. Once the physical size issue of the far-field measurement approach has been shown, the compact range is presented as the best possible approach to measure these electrically large arrays used in 5G applications.

The industry-accepted compact range performance metrics are presented. Several solutions involving compact ranges are available in the market; however, in many cases the traditional metrics to ensure the quality of the plane wave created by the compact range are not given for these solutions. These solutions have broadband antennas to illuminate the compact range but the feed is a critical part of the compact range system. The effects of the feed antenna on the compact range metrics is shown through measurements performed on a compact antenna test range. The reader is given the information to understand the importance of qualifying the performance of the compact range for the feed antennas being used and understand what the potential trade-offs are of using some broadband antennas as feeds.

In this chapter, some broadband feeds like dual ridge horns and Vivaldi-type printed antennas are analyzed numerically to understand the phase center variation as well as the 1-dB beamwidth. Another topic, covered at the end of the chapter, to bear in mind when using compact ranges, is the potential for high fields to be present at the feed with the potential to damage the electronics of the receive system used in the measurement. The same metrics used in compact ranges can be used for other plane wave generation devices.

6.1 The compact range

In reference [1], the author looks at how to choose a testing approach for measuring radiated parameters of an antenna. Typically, we are interested in the antenna performance in the far field. Historically the far-field boundary was arbitrarily set to a point where the phase variation across the aperture of the antenna under test (AUT) of an incident wave radiated from a point source at a distance R was less than 22.5 degrees. This led to the known equation $R = 2D^2/\lambda$, where λ is the wavelength at the frequency of interest, and D is the largest dimension of the antenna. Assuming that the antenna is a number n of wavelengths in aperture, the test distance, R, is given by $R = 2n^2\lambda$. A 5λ antenna will require a test distance that is 50λ.

With the 5G devices operating potentially at the Ka band, even those physically small antennas may be placed on a handset. Such a handset may be 20 cm by 10 cm, and while the antenna or antennas on that device are not the entire size of the device, the entire device becomes part of the radiating structure. Hence, at the middle of the Ka band, the device is at its largest dimension, the diagonal, 22.4 cm or 24.8λ; hence, the far field starts at 11.07 m away. This is a significant distance and would require a very large test facility. Additionally, it should be noted that the $R = 2D^2/\lambda$ distance is a minimum and there are potential errors related to testing at that distance. In the classical blue book of microwave antenna measurements [6], it is shown that testing at the lower boundary of the far field will give significant errors on the first null and first sidelobe of the radiation pattern. The results of the analysis done in [6] are shown in Figure 6.1. The figure shows that at least for highly directive antennas, a distance of $4D^2/\lambda$ (equivalent to a phase variation of 11.25 degrees across the aperture of the antenna) will reduce the error on the first sidelobe and null significantly. But that means that for the case mentioned earlier,

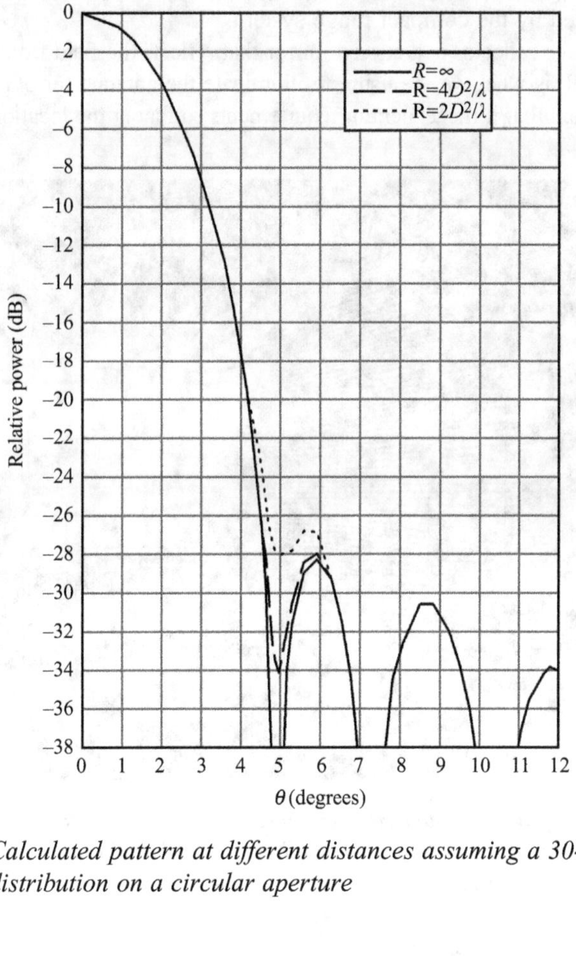

Figure 6.1 Calculated pattern at different distances assuming a 30-dB Taylor distribution on a circular aperture

the test distance would be doubled to 22 m. As Rodriguez mentions in [1], testing indoors in the far field would require a range that is potentially 25-m long and 23-m wide by 23 m-tall. To overcome this problem is the reason that compact ranges and near-to-far-field methods were developed during the last decades of the last century.

The compact range was invented by Richard C. Johnson in the middle of the 1960s. His patent [7] was filed in 1964 and awarded in 1967. As reported by Rodriguez on his book on anechoic ranges [1], even before the patent award, Johnson and his colleagues were already implementing compact range solutions [8] and reporting their findings prior to the publication of Johnson's classic paper [9] published in 1969. In Figure 6.2, a typical compact range is shown. The parabolic reflector is in this case terminated in serrations to reduce the effect of edge diffraction on the QZ. The feed and the measurement instrumentation are located on the floor between the reflector and the antenna positioner. The reflector shown supports a 90-cm QZ that is 101λ at the middle of the Ka band. A far field supporting that size QZ will require a test distance in excess of 18 km. The reflector frequency range covers from 2 up to 100 GHz. The parent anechoic room is 6 m in length by 3.6 m in width by 3.6 m in height. This provides an idea of the space savings provided by the compact range system.

The parabolic reflector is a curved mirror that reflects the field from the feed. The reflection, which is what we are using to illuminate the antenna (or 5G device) under test, must be such that it meets certain requirements so that at the location of the DUT,

Figure 6.2 A typical serrated-edge compact range reflector for a 90-cm QZ. The entire room is 6 m by 3.6 m by 3.6 m (photo courtesy of NSI-MI Ametek, used with permission).

the electromagnetic environment mimics a plane wave in the far field of its source. Thus, the source is a critical part of the compact range system. However, before we look at the source, let us understand how the performance of a compact range is quantified.

6.2 Compact range performance

In the compact range, the illuminating antenna located at the focal point of the paraboloid reflector illuminates the range. The illumination reflected from the reflector should arrive to the QZ or test zone, where the antenna being measured is located or positioned, as a plane wave. However, this will assume a uniform illumination of the reflector by the feed. In Figure 6.3, the radiation pattern of the feed is superimposed on a compact range sketch. The area of the pattern that is of interest to us is approximately from −15 to 15 degrees (see Figure 6.3). The pattern is not uniform for that angular range of 15 to −15 degrees; hence if the field was measured along a vertical line in the QZ, the amplitude will not be constant but related to the pattern of the feed across that angular range that illuminates the QZ. The 30-degree span is related to the focal length. Indeed, a longer focal length will mean that a smaller angle of the radiation pattern of the feed antenna will affect the QZ performance. But a longer length will also mean additional path losses in the measurement system. A shorter path length will reduce the losses but increase the angle of the feed and that increases the cross-polarization of the compact range as discussed by Stutzman and Thiele [10]. Hence the 30 degrees come from a compromise between losses and cross-polarization.

Ideally that will be the only variation; however, it was discussed in [1], there is reflected and diffracted energy entering the QZ that interferes with this direct reflection from the reflector causing a ripple on the illumination. A properly designed

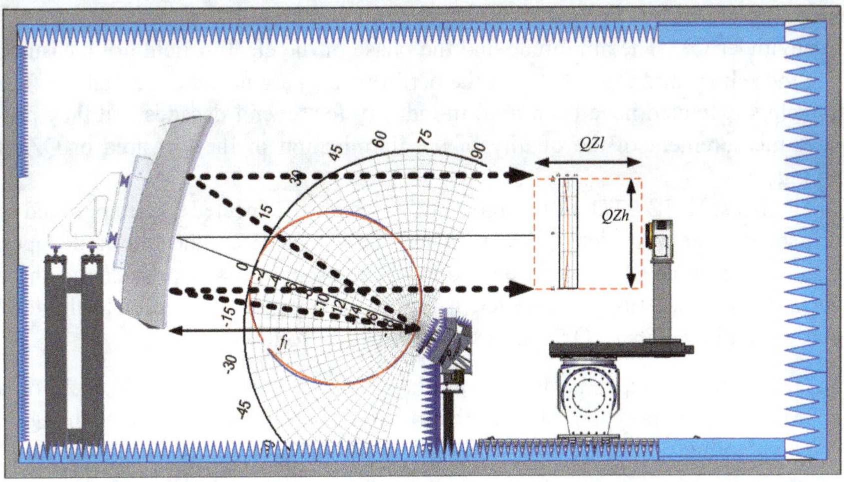

Figure 6.3 The feed pattern superimposed on a sketch of a compact range notice the field distribution on the QZ created by that illumination

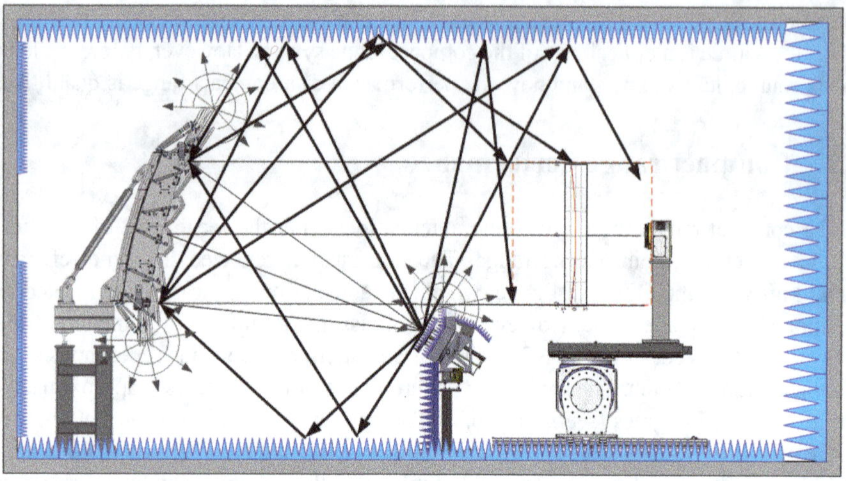

Figure 6.4 Potential reflected energy and diffracted energy that may interact with the direct illumination causing ripple on the QZ

range should minimize reflections from the lateral walls and also minimize direct interaction between the feed and the QZ volume. The edge treatment of the reflector will minimize the edge diffractions [11] or redirect them away from the QZ. Figure 6.4 shows all these potential sources of reflected energy. To assess the quality of the compact range and how do the reflector, the feed and feed support and the anechoic environment affect the plane wave illumination, the QZ is measured.

The compact range performance is typically measured along a vertical line and a horizontal line that form the main axes of the cylindrical QZ. Figure 6.5 shows the typical QZ geometry. Typically, the height and width are the same making a circular cylinder QZ. The amplitude and the phase of the electric field are measured along those lines and from that data the performance parameters are extracted. The parameters extracted have been used in industry for several decades and they provide a measurement of the quality of the illumination of the test area or QZ of the range.

The latest IEEE STD 149 standard [12], which is expected to be approved in 2021, states what the typical QZ metrics are. These metrics are the amplitude taper, the amplitude ripple and the phase variation. Some suppliers also provide the phase taper and the phase ripple; however, it is more common to just refer to the total variation in phase. These QZ metrics are defined as follows:

- The amplitude taper is defined as the variation expressed in decibels of a second-degree polynomial fit of the measured amplitude data across the QZ.
- The amplitude ripple, expressed in dB, is measured by determining the variation of the amplitude about the second-degree polynomial fit.
- The phase variation, which is given in degrees, is the difference between the highest and lowest phase values across the QZ.

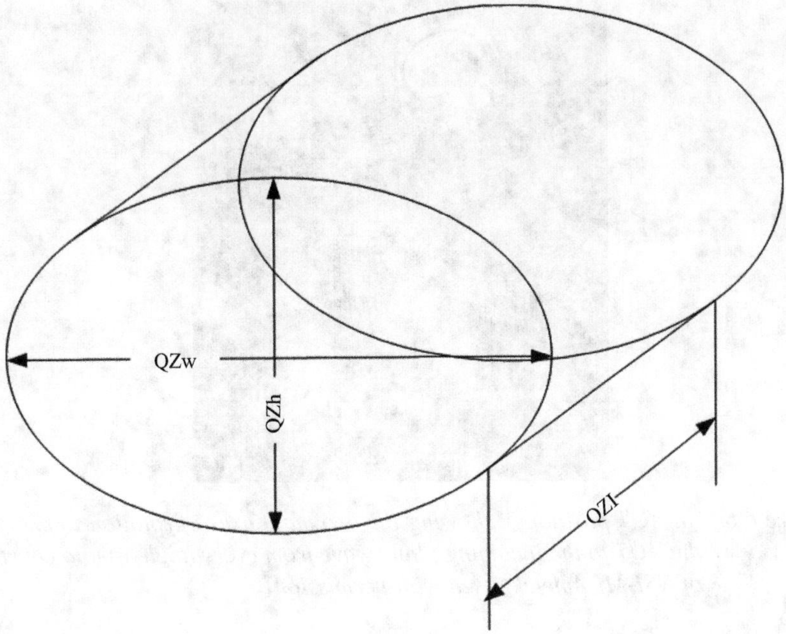

Figure 6.5 General geometry for a compact range QZ. The shape is typically a horizontal elliptical cylinder.

The phase taper and phase ripple are obtained in the same matter as the amplitude taper and amplitude ripple, but the phase data is the one being processed.

The measurement is carried out by a linear positioner that is mounted onto the AUT positioner in the chamber. The AUT positioner allows the AUT to be rotated around two axes as shown on Figure 6.6. This is a minimum, and positioners may have more axes of motion to help with the alignment of the AUT. The linear positioner has a rotational positioner that allows it to polarize the antenna used to probe the field as needed. The linear positioner and rotational stage are known as a "phase probe" in the industry, even that both magnitude and phase are measured. The upper rotational stage on the AUT positioner can rotate the phase probe as needed to scan in the vertical or horizontal directions across the QZ. Figure 6.7 shows a phase probe mounted to an AUT positioner. The probe antenna is a standard gain horn and the positioner is set for a horizontal scan across the QZ.

As the probe is moved along the line, the amplitude and phase of the field are measured. In Figure 6.8, measured results for one such a scan are shown. From these measured results, the QZ metrics are extracted.

Figure 6.8 shows one of the many datasets obtained during a qualifying test. Typically, the QZ is done for the vertical and horizontal lines at the front, center and back of the QZ, for each polarization. Hence, at a given frequency, there will be 12 datasets. Some of them are horizontal scan, horizontal polarization (as shown

Figure 6.6 An AUT positioner showing the two main axes of rotation to present the AUT to the incoming plane wave from every angle (photo courtesy of NSI-MI Ametek, used with permission)

Figure 6.7 A phase probe mounted not an AUT positioner (photo courtesy of NSI-MI Ametek, used with permission)

in Figure 6.8), horizontal scan vertical polarization, vertical scan horizontal polarization, vertical scan vertical polarization, etc.

Figure 6.9 shows the amplitude taper for the data in Figure 6.8 to be about 0.95 dB, while the phase variation is 2.53 degrees. Analysis of the data determined the amplitude ripple to be 1.6-dB peak to peak.

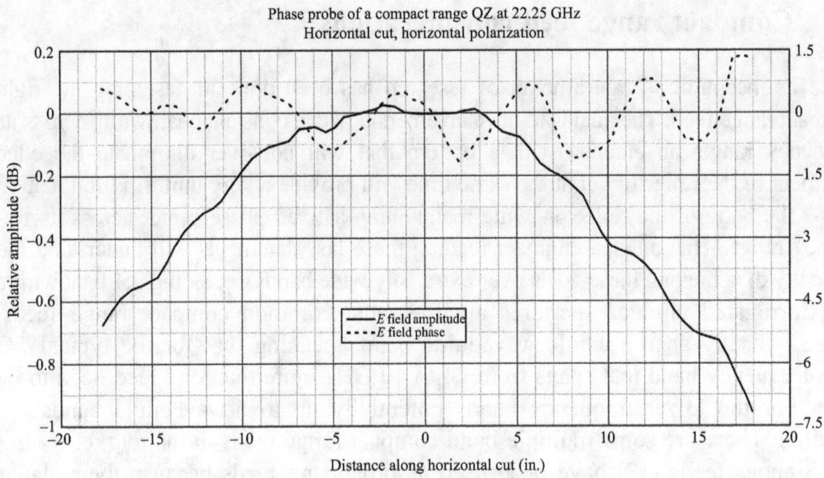

Figure 6.8 Measured results for amplitude and phase of a scan along a horizontal line for horizontal polarization on a 36-in. (91.4-cm) wide QZ at 22.25 GHz

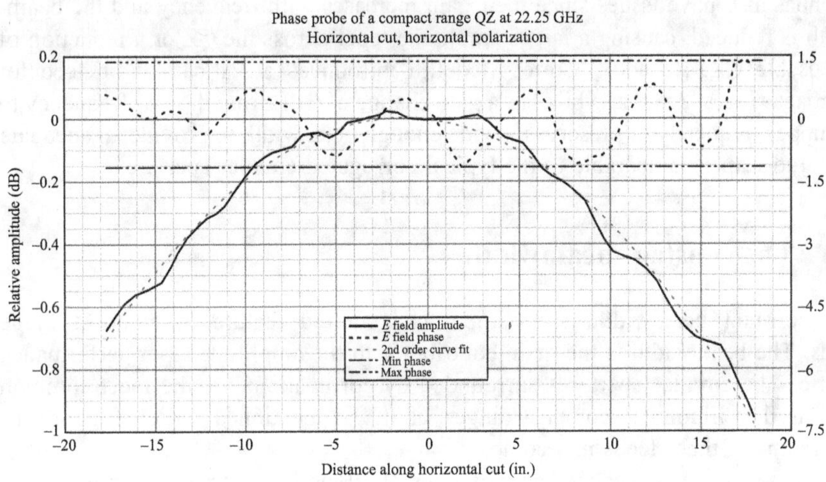

Figure 6.9 Amplitude taper and phase variation across the 36-in. (91.4-cm) QZ for the horizontal cut and horizontal polarization at 22.25 GHz

The amplitude taper is mainly dependent on the pattern of the illuminating feed. The phase variation is dependent mainly on how well focused is the compact range. The ripple is mainly dependent on edge effects and surface finish. If properly designed, the anechoic room effects on the ripple are smaller than the effects caused by the surface and edge of the reflector.

6.3 Compact range feed considerations

Since its inception a large amount of research has been done in designing the right compact range feed. The goal is to have an antenna with a constant beamwidth across its frequency range; an antenna with a pattern that will not over illuminate the edge treatment of the reflector yet at the same time will provide a very uniform illumination across the body of the reflector while maintaining a fixed phase center across its frequency range. The antenna engineer reading these last statements will understand the difficulty of achieving these goals over extremely wide bands exceeding 3:1 bandwidth.

Corrugated circular apertures are the most common compact range feeds. These typically span a single waveguide band, meaning that for the typical 5G high-frequency band that spans from 24 to 40 GHz more than one feed need to be used. For the lower frequency bands potentially up to 5-waveguide bands are required. There are some multiple band compact range feeds in the market.

Sinuous feeds [13] have been used as broadband feeds because their planar geometry causes the phase center not to change in the direction of propagation. They are, however, lossy as they are typically a loaded cavity-backed antenna. The other problem that sinuous antennas exhibit is the polarization "wobble." This "wobble" phenomena means that as frequency changes, the polarization tilt angle can shift as much as 17 degrees as reported by Walkenhorst [14]. Broadband horn antennas also have issues since their gain increases with frequency and the beamwidth is reduced, causing a larger amplitude taper across the QZ or a reduction of the usable QZ. Recently, Cortes-Medellin introduced a 3:1 bandwidth feed for compact ranges [15] that while has a gain change of 6 dB from the low frequency to the upper frequency, it preserves a wide enough beamwidth to provide an adequate amplitude taper comparable to the typical corrugated circular aperture.

6.4 The constant beamwidth

The constant beamwidth is an important required feature of compact range feeds. The beamwidth is the main contributor to the amplitude taper performance metric. The circular apertures surrounded by corrugations are the most common feed used to illuminate compact ranges, as it was mentioned earlier. Figure 6.10 shows one of these feeds in a compact range.

These feeds have realized gain of about 10 dB across its frequency range and a 1-dB beamwidth of about 30 degrees or larger across the entire frequency range. Figure 6.11 shows computed results for a corrugated circular aperture feed for the 26.5–40 GHz range fed by a WR-28 rectangular waveguide. The model has the RF absorber commonly used to treat the feed. The gain is shown to change only 1 dB across the entire band from 9.5 to 10.5 dBi. The stable gain also shows in a very constant illumination across the band with the 1-dB beamwidth in both principal planes. The change in beamwidth is less than 4 degrees across the entire frequency range and it is always larger than 30 degrees, which is the required beamwidth given the focal distance and intended QZ size for the reflector.

Figure 6.10 A corrugated circular aperture used as a feed antenna in compact ranges (photo courtesy of NSI-MI Ametek, used with permission)

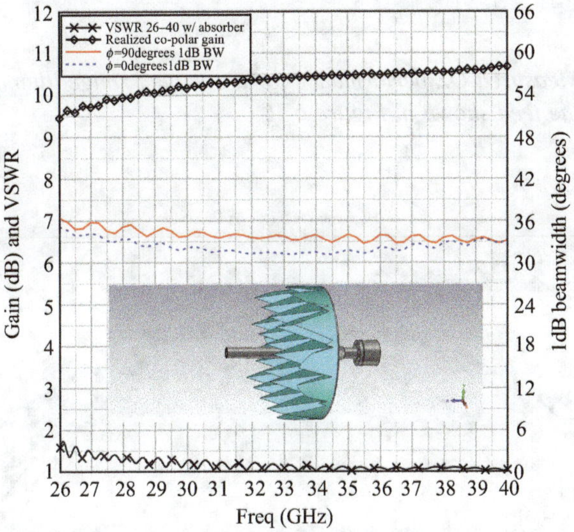

Figure 6.11 Computed performance of a circular aperture corrugated feed used to illuminate compact range reflectors in the Ka band

If we look at the amplitude probe across the band of the feed antenna simulated with the results shown in Figure 6.11, we see that the illumination of the QZ is very constant across the range. The measured data is shown in Figure 6.12 for one of the many cuts done on the QZ. Less than 0.1 dB of difference in the illumination across the band of the antenna is observed. The QZ performance remains unchanged across the entire Ka band.

Let us now compare the performance of the circular corrugated aperture feed to the performance of a dual ridge horn typically used in electromagnetic compatibility measurement applications. This particular dual-ridge horn is the one analyzed by Rodriguez in [16] to improve its radiation pattern. While the dual-ridge

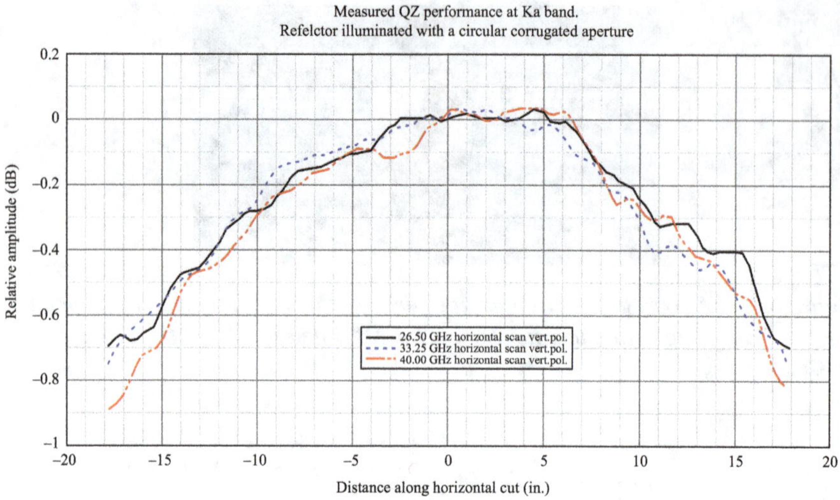

Figure 6.12 Measured QZ performance for a compact range illuminated with the feed shown in Figure 6.10

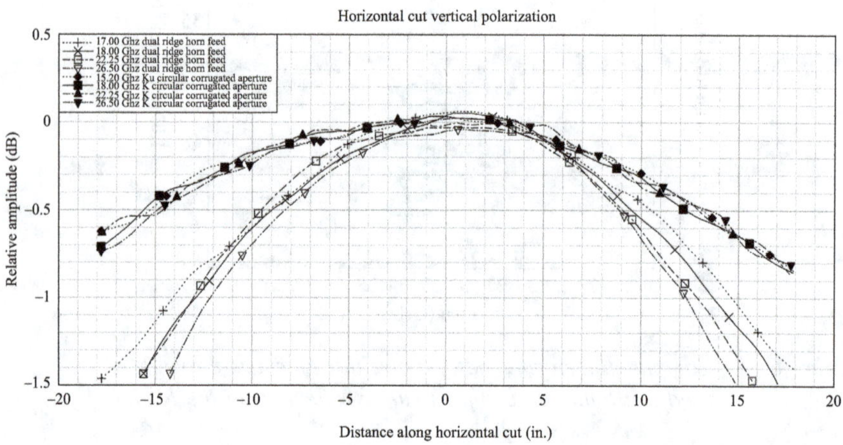

Figure 6.13 Horizontal cut scan vertical polarization in the same compact range with two different feeds

horn operates from 10 to 40 GHz, the comparison was done in the K band (18–26.5 GHz). The measured results are for the same reflector, in the same chamber and the same phase probe fixture. The only difference is the feed used to illuminate the reflector.

Figure 6.13 shows that the dual-ridge horn, although only having a couple of dB higher gain than the circular aperture, gives a larger aperture taper. From 17 to

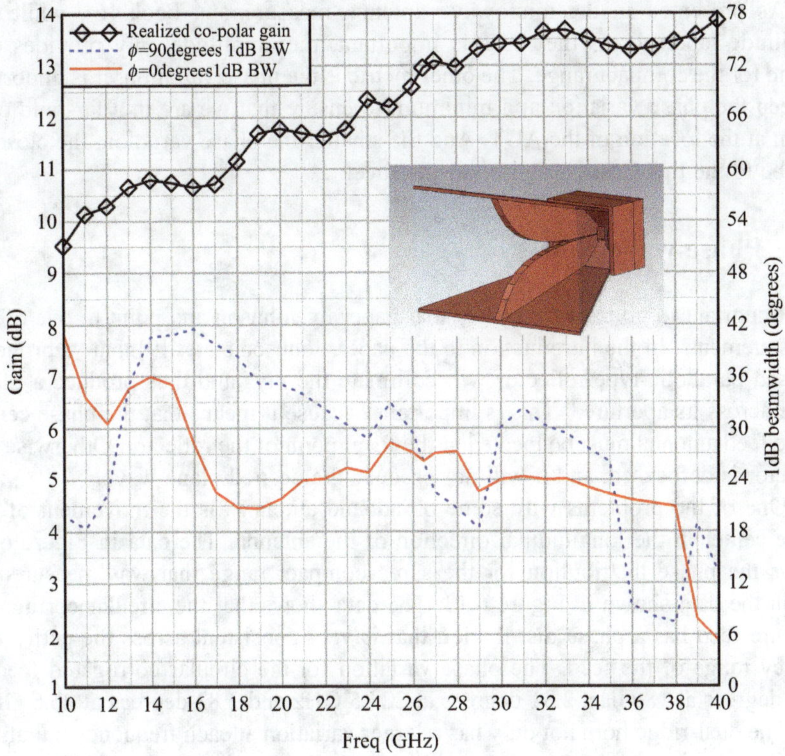

Figure 6.14 Computed performance of the dual ridge horn used as feed in the measurements

26.5 GHz, the 1.5-dB amplitude taper has been reduced in span by about 20 cm (8 in.). To achieve the same 0.75 amplitude taper that the circular corrugated aperture provides over the entire 91-cm QZ, the dual-ridge horn will provide that amplitude taper only across a 54.05-cm (21.28-in.) diameter QZ at the highest frequency that is a 40% reduction in QZ. The measured data illustrates the importance of the feed in a compact range system and shows how a simple increase in gain of 2 dB can significantly reduce the QZ.

More important is the beamwidth of the antenna. In Figure 6.14, the 1-dB beamwidth of the dual-ridge horn used is computed with the realized gain. The results show that the beamwidth at the frequencies of interest is much smaller than 30 degrees for the $\varphi = 0$ degree cut, which corresponds to the *H*-plane of the antenna and to the horizontal plane when the antenna is vertically polarized as was the case in the data shown in Figure 6.13, which showed a horizontal scan with the antennas vertically polarized. Notice also that the beamwidth is very different for both principal planes in contrast to the circular corrugate aperture feed that exhibits a very similar beamwidth for both principal planes of the radiation pattern.

As presented in the microwave antenna measurement book [6], while the amplitude taper across the AUT is important, the amplitude only provides one metric for the compact range. The other metric is the phase variation. It is important to keep the phase variation at a minimum to ensure that we are in a far-field condition at the location of the AUT. And the smaller the phase variation, the closer it will be to the theoretical infinite test distance.

6.5 Phase wavefront

It is important to go back to the original paper by Johnson and point to one of the key statements. In the introduction to the paper, Johnson states [9] that "a properly focused parabolic-type reflector will collimate the rays and thus produce a plane wave across its aperture." This is important because it means that the phase center of the feed antenna must be located at the focal point of the reflector. Otherwise the reflector is defocused, and it will not produce the desired plane wave.

One of the problems with some broadband antennas is the movement of the phase center in the longitudinal direction of the antenna. The data in Figure 6.15 shows the phase distribution for the same compact range that was measured to obtain the data shown in Figure 6.13. The data shows that the circular corrugated aperture feed has a phase distribution that is very consistent across the entire frequency range of the feed. The phase variation for the circular corrugated feed is 1.98 degrees at 18 GHz, 2.31 degrees at 22.25 GHz and 2.83 degrees at 26.5 GHz.

The dual-ridge horn not only has a larger variation at each frequency but also a larger change across the frequency range. At the lowest frequency of 18 GHz, the phase variation is 6.8 degrees. At 22.25 GHz, the phase variation is 4.72 degrees and at 26.5 GHz the variation is 4.02 degrees. These phase variations are still better than those across the QZ of a typical far-field range using the test distance equal to

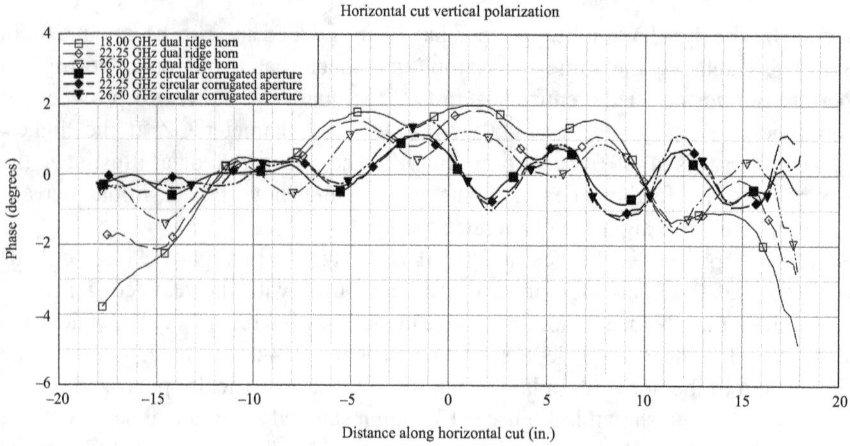

Figure 6.15 Horizontal phase scan across the QZ for the vertical polarization using two different antenna feeds

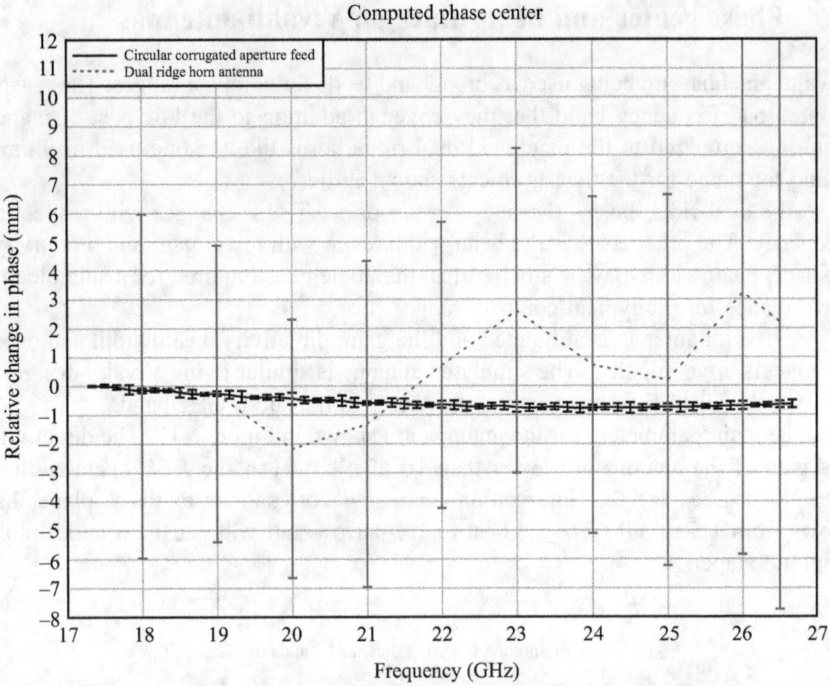

Figure 6.16 Computed phase center for the dual ridge horn and for the circular corrugated aperture

$2D^2/\lambda$ (where D is the QZ diameter) or $4D^2/\lambda$ [1] that provides a theoretical 22.5 or 11.25 degrees phase variation, respectively.

This example illustrates the issue with an antenna being defocused as the frequency changes and the phase center moves away from the focal point of the reflector. It should be noted that the data presented earlier uses a fraction of the broadband dual-ridge horns frequency range of 10–40 GHz. For the range of frequencies used, the phase center only moves a small amount. In Figure 6.16, the computed phase center for the circular corrugated aperture and the dual-ridge horn are compared. The phase center is computed for the boresight using the 30-degree beamwidth that is used to illuminate the QZ. The error bars are also plotted in Figure 6.16. The plot shows that the change in the location of the phase center for the circular corrugated aperture is very small and that the error in the calculation is also very small indicating a very well behaved pattern. The dual ridge horn shows a larger variation with the phase center moving over 5 mm across the frequency band of interest. More important is the large error bars denoting a difficulty in computing a clear phase center for that antenna. The measured and computed data in Figures 1.15 and 1.16 show that even a small change in the location of the phase center (2 mm between 18 and 26.5 GHz) versus the location of the focal point caused a change in the phase variation of 2.8 degrees between those two frequency points.

6.6 Phase center and beamwidth on Vivaldi antennas

Vivaldi antennas are being used as broadband feeds for compact ranges. The reason is the broad frequency band that they cover in addition to the low cost of manufacturing compared to the machined dual ridge horns like the one used in illuminating the range for the measurements shown earlier.

Like dual ridge horns, the phase center on a Vivaldi changes with respect to frequency. The other issue is the beamwidth; even with lower gain and directivity, the 1-dB beamwidth may be smaller than the 30 degrees required for a suitable QZ performance for the typical compact range.

A Vivaldi antenna is simulated and the gain, directivity, beamwidth and other parameters are computed. The simulated antenna is similar to the Vivaldi designed for over the air [17] measurements of systems in the sub-6 GHz band.

The main parameters for the antenna are shown in Figure 6.17. The directivity and gain of the antenna are lower than 10 dBi; however, the 1-dB beamwidth at $\varphi = 90$ degrees for the polarization computed corresponds to the *E*-plane that shows from 1.5 to 3.9 GHz the beam is to narrow and will cause an undesirable amplitude taper.

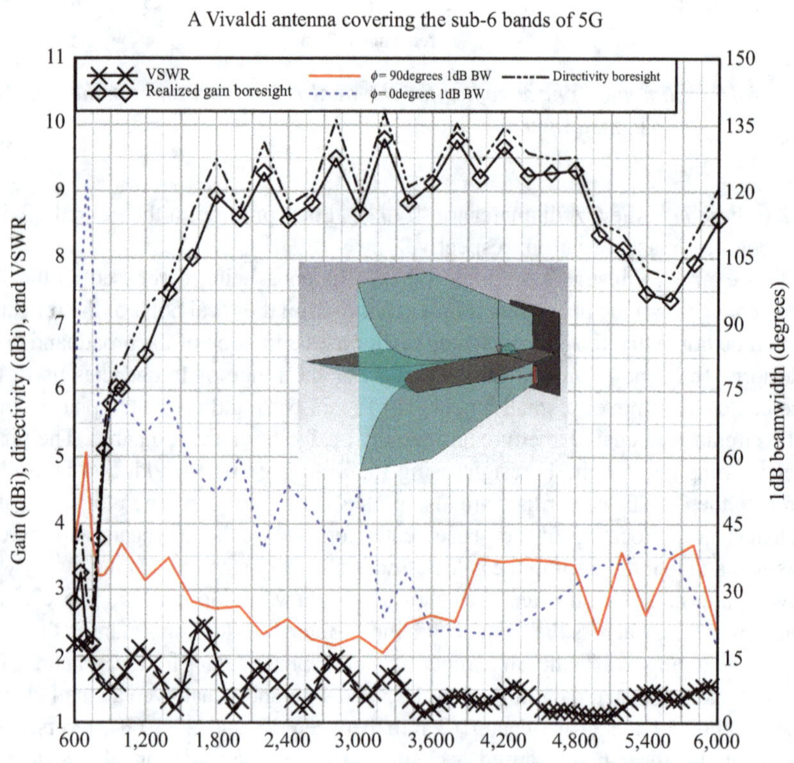

Figure 6.17 A Vivaldi antenna computed performance

Additionally, the beamwidth is very different between the two principal planes which will translate into a very different performance between the vertical and horizontal polarizations with regards to the amplitude taper. For example, this antenna will provide an adequate amplitude taper when performing a horizontal scan with the antenna vertically polarized at 2.4 GHz, but the horizontal scan with the antenna horizontally polarized will give a much higher amplitude taper across the QZ. This is not desirable and is shown for the 1-dB beamwidth of circular corrugate aperture feeds introduced by Cortes-Medellin [15] which exhibits a 1-dB beamwidth larger than 30 degrees for the full band and for both principal planes.

As shown earlier the beamwidth will mainly affect the amplitude taper across the QZ. It should be pointed out that the Vivaldi having such a wide beam will cause an additional illumination of the edge treatment of the reflector; this will typically cause larger diffraction on the serrations and potentially a larger ripple on the amplitude. Also illuminating more of the lateral surfaces will require a thicker treatment of absorber on the lateral surfaces of the range as Rodriguez discusses in [1].

The other issue is the potential defocusing of the reflector related to the movement of the phase center. The computed data shown in Figure 6.18 shows a significant movement of the phase center, given that the antenna is 20 cm in length

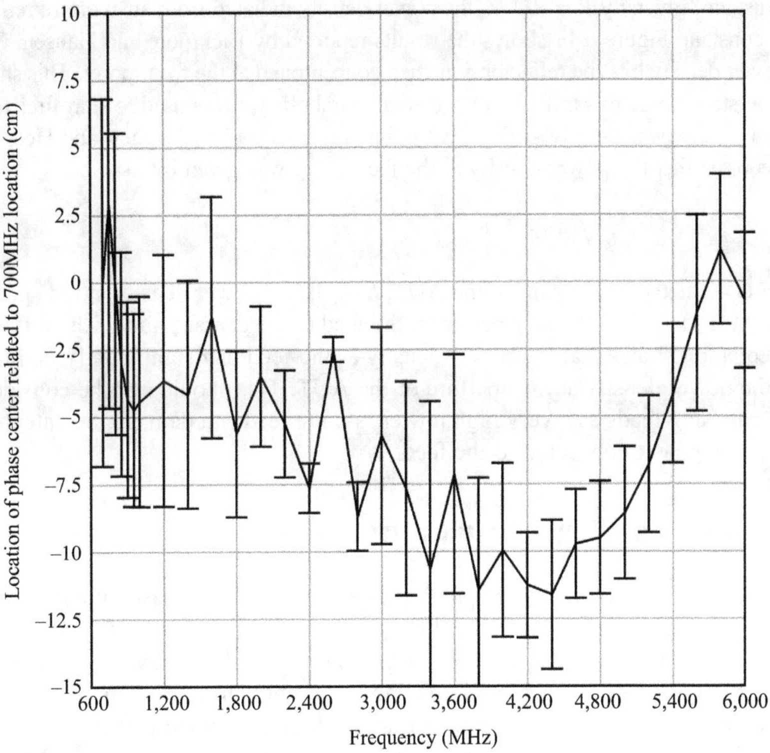

Figure 6.18 Phase center computed for the Vivaldi antenna

from the feed point to the end of the radiating elements, the phase center changes a total distance of 14.5 cm. The large, 5-cm phase center change between the 2.4 and the 2.6 GHz is of interest. These variations will cause the compact range system to be defocused and increase the phase variation across the QZ. Increasing the error on the nulls and sidelobes. This exercise, in addition to the previous measured data, shows how important the feed antenna is to the performance of the range.

6.7 Power handling of the feed

While in the previous examples the compact range has been analyzed with the feed radiating, the reader must remember that the compact range is a passive device and as such reciprocity holds. The plane weave component radiated from an AUT in the QZ will be focused at the feeds focal point. This is because as stated in [1], the compact range is an angular filter.

The power radiated from the feed spreads over and area of the reflector and then propagates to the QZ as a plane wave, without attenuating. Similarly, the power radiated by the AUT propagates without attenuation, to the reflector, since we are in the near-field of the AUT. The power density in the near-field of an aperture antenna was given by Bickmore and Hansen [18], and it is shown that for distances that are 0.01 of the far-field distance given by $R = 2D^2/\lambda$, the power density radiated from an aperture remains fairly constant. Figure 6.19 shows the results reported by Bickmore and Hansen. Once the power density hits the reflector, it is then concentrated at the focal point. This should be understood since this is analogous to small parabolic mirrors used to start fires using the sun. The power density at the focal point was analyzed by Baggett and Hess [19] and reported that the power density at the feed, S_{feed}, was given by

$$S_{feed} = \left[\lambda^2 G_{AUT}^2/(4\pi R_o)^2\right] S_{AUT}$$

where G_{AUT} is the linear gain of the AUT, S_{AUT} is the power density at the aperture of the AUT, and R_o is the distance from the feed to the center point of the reflector. This equation shows that the power density at the feed can significantly be higher than the power density at the aperture of the AUT. This should also be considered when choosing feeds as a very high power into the feed can damage the antenna or the RF equipment connected to the feed.

6.8 Understanding your measurement range

Regardless of the test approach, it is imperative that the measurement range is properly characterized. This does not only apply to compact range systems, but to plane wave generators, and far-field ranges. For both plane wave generators and compact ranges not only should the range be measured for amplitude taper, amplitude ripple, but also for phase variation. It is very important not to assume that a plane wave exists, but to verify it exists, and to check how pure the plane wave illumination is and perform an uncertainty analysis.

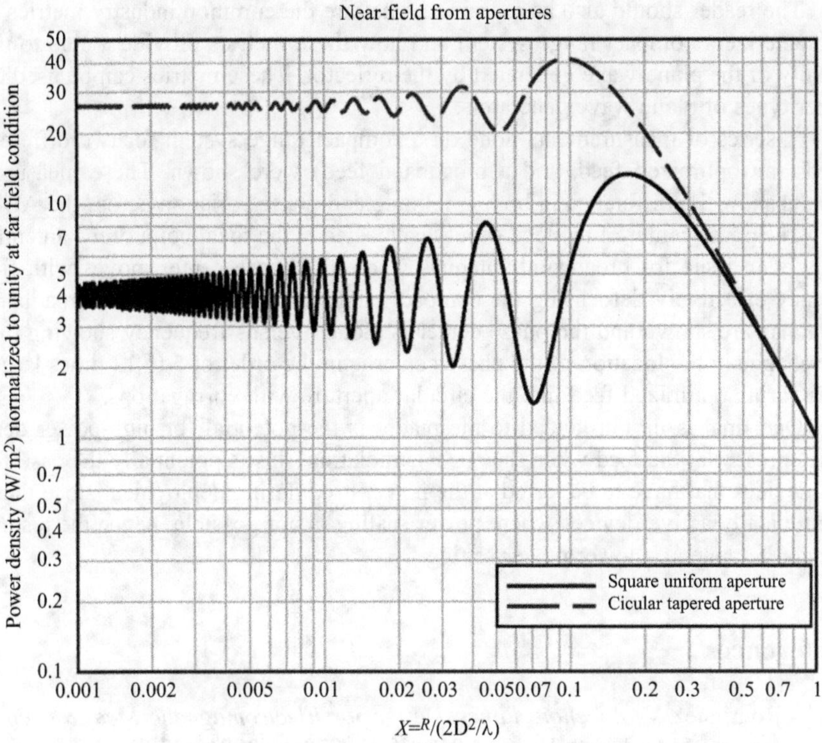

Figure 6.19 Power density from apertures in the near-field

The user should beware of so-called standard products with typical performance specifications and request the system be properly characterized with a phase probe of the QZ. Then use that information to determine how large the test AUT can be, and at which frequencies.

It is not the intent to demonize broadband feeds such as Vivaldi's and ridged horns but also to understand the trade-off between performance and convenience. This trade-off is critical. A broadband feed will save money since it saves time, but that savings is paid in terms of increased uncertainty or reduced QZ size.

Also remember to consider the focusing effect of the compact range reflector when doing link budgets since high-power densities can be present at the feed point that can damage said feed or the RF equipment connected to said feed.

6.9 Conclusion

In this chapter, the reader should have gotten a better understanding of the compact range as an antenna measurement tool. It should be clear that given the electrical size of the arrays used in 5G communications, a far-field measurement will require a very large physical area to get a low error measurement of the sidelobes and nulls of the array pattern.

The reader should also have learned what are the common industry metrics to characterize a compact range system and how these metrics provide a clue to the quality of the plane wave generated by the reflector. These metrics can be used on other types of plane wave generators.

A series of measurements done on a compact range system for two different feeds (an optimized feed, and a broadband feed) were shown. These measured results show the difference in both amplitude and phase variation across the AUT.

The phase center of the feed should be located at the focal point of the antenna. This is an issue for broadband antennas where the phase center moves with frequency effectively defocusing the compact range. Simulations for two broadband antennas are shown and the phase center is plotted versus frequency showing that the change in the location of the phase center is in the order of 5 to 13 times larger than for an optimized feed like the circular aperture with corrugations.

One final issue introduced to the reader is the potential for high-power densities present at the feed when the AUT is radiating, this is potentially the case for 5G devices that have to be tested at the power used during deployment since these are typically active devices where power scaling is not possible, hence the testing cannot be done at a lower power setting.

References

[1] Rodriguez V. *Anechoic Range Design for Electromagnetic Measurements*. Boston, MA: Artech House, 2019 (ISBN 978-1-63081-537-0).

[2] Rodriguez V., Matitisine S., Chia T. T., *et al.* 'A Cone Shaped Tapered Chamber for Antenna Measurements Both in Near Field and Far Field in the 200 MHz to 18 GHz Frequency Range and Extension of the Quiet Zone using an RF Lens'. *Journal of the Applied Computational Electromagnetics Society*. 2013, 28(12), pp. 1162–1170.

[3] Haupt R. 'Generating a Plane Wave in the Near Field with a Planar Array Antenna'. *Microwave Journal*. 2003, pp. 152–166.

[4] Courtney C. C., Voss D. E., Haupt R., and LeDuc L. 'The Measured Performance of a Planewave Generator Prototype'. *The 24th Annual Meeting of the Antenna Measurement Techniques Association (AMTA 2002)*, Cleveland, OH, 2002.

[5] Tankelium A., Belkacem A., Akinci M., Celik M., Bartko H., and Derat B. 'Quiet Zone Verification of Plane Wave Synthesizer Using Polar Near-Field Scanner'. *The 2020 14th European Conference on Antennas and Propagation (EuCAP 2020)*, Copenhagen, Denmark, 2020.

[6] Hollis J. S., Lyon T. J., and Clayton J. L., Eds. *Microwave Antenna Measurements*. 2nd ed., Atlanta, GA: Scientific Atlanta, Inc., 1970.

[7] Johnson R. C. *Antenna Range for Providing a Plane Wave for Antenna Measurements*. United States Patent 3,302,205, 31 January 1967.

[8] Johnson R. C. and Poinsett R. J. 'Compact Antenna Range Techniques'. N0. RADC-TR-66-15, Griffiss Air Force Base, New York, April 1966.

[9] Johnson R. C., Ecker H. A., and Moore R. A. 'Compact Antenna Range Techniques'. *IEEE Transactions on Antennas and Propagation*. 1969, AP-17(5), pp. 568–576.

[10] Stutzman W. L. and Thiele G. A. 'Cross-polarization and Scanning Properties of Reflector Antennas' in *Antenna Theory and Design*. 2nd ed., New York, NY, John Wiley & Sons, Inc., 1998, pp. 82–93.

[11] Lee T. H. and Burnside W. D. 'Performance Trade-off Between Serrated Edge and Blended Rolled Edge Compact Range Reflectors'. *IEEE Transactions on Antennas and Propagation*. 1996, 44(1), pp. 87–96.

[12] IEEE, 'P149™/D8 Draft Recommended Practice for Antenna Measurements'. Unapproved Draft, New York, NY, 2020.

[13] DuHamel R. H. *Dual Polarized Sinuous Antennas*. United States Patent 4,658,262, 14 April 1987.

[14] Walkenhorst B. T. and Tammen D. 'Correcting Polarization Distortion in a Compact Range Feed'. *Proceedings of the 38th Annual Symposium of the Antenna Measurement Techniques Association (AMTA 2016)*, Austin, TX, 2016.

[15] Cortes-Medellin G. and Walkenhorst B. T. '3:1 Bandwidth Dual Polarized Feeds for Compact Range and Near-Field Probes'. *The 2019 IEEE International Symposium on Antennas and Propagation and USNC-URSI Radio Science Meeting*, Atlanta, GA, USA, July 7–12 2019.

[16] Rodriguez V. 'Recent Improvements to Dual Ridge Horn Antennas: The 200MHz to 2GHz and 18GHz to 40GHz Models'. *The 2009 IEEE International Symposium on EMC*, Austin, TX, August 17–21, 2009.

[17] Sonkki M., Sánchez-Escuderos D., Hovinen V., Salonen E. T., and Ferrando-Bataller M. 'Wideband Dual Polarized Cross-Shaped Vivaldi Antenna'. *IEEE Transactions on Antennas and Propagation*. 2015, 63(6), pp. 2813–2819.

[18] Bickmore R. W. and Hansen R. C. 'Antenna Power Densities in the Fresnel Region'. *Proceedings of the IRE*. 1959, 47(12), pp. 2119–2120.

[19] Baggett M. and Hess D. 'Power Handling Considerations in a Compact Range'. *The 35th Annual Meeting and Symposium of the Antenna Measurement Techniques Association (AMTA 2013)*, Columbus, OH, October 2013.

Chapter 7

International intercomparison campaigns for mobile communication antenna characterizations towards 5G and beyond

Maria Alberica Saporetti[1], Lars Foged[1], Sergiy Pivnenko[2], Dirk Heberling[3], Rasmus Cornelius[3], Martin Böttcher[4], Stefan Weitz[4] and Manuel Sierra Castañer[5]

In recent years, many measurement facilities have been involved in international comparison campaigns led and supported by the European Association on Antennas and Propagation (EurAAP). These activities play an important role for the documentation and validation of laboratory proficiency and competence, helping to improve the antenna measurement procedures/protocols in facilities and standards. The analysis and data elaboration have promoted discussions among the antenna measurement community experts and have led to modernization of comparison techniques. A very large set of measured information is collected during such campaigns representing a valuable database of information for exploitation in further studies and analysis.

Mobile telecommunication antennas have been included in the set of reference antennas selected for the EurAAP intercomparison campaigns that have been performed starting from 2005. The know-how acquired, thanks to this long-term experience, will prove to be extremely useful for future intercomparison campaigns dedicated to 5G applications. The exponential increase of 5G devices and applications in the future years will require quality improvements of antenna test facilities. In this sense, these types of campaigns are important not only for official certification programs (e.g. ISO 17025), but also for the validation of antenna measurement set-ups.

This chapter provides an overview of the international intercomparison campaigns, mainly in EurAAPs, detailing those involving mobile telecommunication

[1]Microwave Vision Italy (MVI), Pomezia, Italy
[2]Antenna Systems Solutions (Asysol), Santander, Spain
[3]Institute of High Frequency Technology RWTH Aachen University, Aachen, Germany
[4]IMST GmbH, Kamp-Lintfort, Germany
[5]Information Processing and Telecommunications Center, ETSI Telecomunicacion, Universidad Politecnica de Madrid, Madrid, Spain

antenna characterizations where the gained experience are exploited towards characterization of emerging antenna systems for 5G and beyond. In particular, Section 7.1 gives a brief overview of the interlaboratory evaluations and on the background, history and management of EurAAP intercomparison campaigns. Section 7.2 describes the past and present EurAAP campaigns and also future campaigns with millimetre-wave (mm-wave) chip reference antennas for 5G applications. The procedure, reported in Section 7.3, to correlate the measurements from different laboratories has been reviewed in the last years. Terminal and infrastructure mobile telecommunication antennas, like base transceiver station (BTS) and multiple-input–multiple-output (MIMO) antennas, have been the reference antennas of EurAAP intercomparison campaigns concluded in recent years as detailed in Section 7.4. The data collected in the frame of these campaigns will be used in the next future constituting a benchmark for new facility self-evaluation EurAAP activities. Finally, in Section 7.5, the future trends for intercomparison campaigns for the characterization of emerging antennas for 5G and beyond are outlined. Conclusions are drawn in Section 7.6.

7.1 Interlaboratory evaluations

In this section, some examples of intercomparison activities conducted from institutional and non-institutional organizations all over the world during the years are presented. Such descriptions aim to prove the recognition of the relevance and associated benefits of interlaboratory comparisons.

7.1.1 General overview

The measurement of any antenna performance parameter is considered to be incomplete without information of the measurement accuracy. It is therefore important for a facility to evaluate and document measurement error estimates [1–3].

Independently of the technology and measurement approach, most measurement facilities follow internal procedures for uncertainty budgets and quality policy, based on reference gain antennas and a certain measurement experience. Certifications like ISO 17025 [1] prove the formal recognition of that competence. However, the measurement uncertainty or error estimation for a specific antenna, in a specific measurement range, remains an approximation until this estimate has been successfully validated against reference data. This reference data might be gained by averaging measurement data from many different laboratories and measurements.

Several interlaboratory evaluations have been performed for a long time among the National Measurement Institutes (NMIs), where the National Institute of Standards and Technology (NIST) served as the pilot laboratory. For example, in 1996, an international intercomparison of horn antenna measurements (gain and polarization) at X-band was completed with seven participating laboratories [4]. From 2008 to 2013, an intercomparison measurement of two Ku-band standard gain horns was conducted involving 12 NMIs, from Australia, China, Czech Republic, France, Japan, Russia, South Korea, Sweden, Switzerland, Turkey, the

United Kingdom and the United States [5]. All of the NMIs performed on-axis gain measurements, return loss, S_{11}, parameters, measurements and uncertainty analyses in the Ku-band. The comparison was conducted with the NIST measuring at the beginning and end of the campaign. Averages for the gain and uncertainties reported by the NMIs were calculated as weighted averages of the (linear) gain measured by the participants. To this purpose, NIST has investigated [6,7] on how to determine the combined result and its associated uncertainty in the measurement of a common measurand by a group of competent laboratories to establish the degree of equivalence of national measurement standards maintained by NMIs. International comparisons are not limited to antennas: for example, [6] presents the analysis of a comparison of cryogenic radiometers that involved 17 NMIs.

To name other similar activities, the Swiss Foundation for Research on Information Technologies in Society (IT'IS) [8] has promoted in 2018–19 a intercomparison campaign with a mm-wave antenna, including a total of eight devices (antennas and sources), five of which corresponded to the specifications in the latest standard draft, and three of which were modified from those specifications. None of the participating parties knew which devices were modified, nor the radiated power. The round-robin devices had been fully characterized numerically by the IT'IS Foundation and validated experimentally. IT'IS collected the results and performed the evaluation, together with an independent third.

In order to ensure quality control, the participants in the Eurolab France (a professional association of laboratory) dedicated to electromagnetic compatibility regularly organize interlaboratory comparison scheme [9]. For each scheme, a protocol is defined for a specific measure and according to ISO standards.

7.1.2 EurAAP intercomparison campaigns

The EurAAP intercomparison campaigns rely on the know-how acquired during the sixth EU Framework network 'Antenna Centre of Excellence (ACE)' [10] as reported in [11–18]. The antenna measurement activity of the ACE, in the period 2004–07, started to define some reference antennas: a high directive reflector antenna, Technical University of Denmark (DTU) – European Space Agency (ESA) 12 GHz VAST12 [19,20] and two dual-ridge horn antennas, Microwave Vision Group (MVI)/SATIMO SH800 [21] in L/S/C-band and MVI/SATIMO SH2000 in Ku/Ka-band [22]. Afterwards, the different activities related to this topic have been evolved in the frame of the European Cooperation in Science and Technology (COST) ASSIST (IC0603) [23] and COST VISTA (IC1102) [24] and were then included in the EurAAP [25] measurement Working Group (WG) 5 [26] on antenna measurements which constitutes a framework for cooperation to advance research and development of antenna measurements in support of the ever-increasing use of wireless technology in modern society. These campaigns have proved that comparative measurements based on high-accuracy reference antennas and involving different antenna measurement systems are important instruments in the evaluation, benchmarking and calibration of the measurement facilities.

As a complementary activity to formal certifications, measurement ranges shall regularly perform activities such as an intercomparison, to properly validate the measurement procedures. The main aim of EurAAP facility comparison activities is to provide a formal opportunity for the participants to validate and document their achieved measurement accuracy and procedures by comparison with other facilities. A positive outcome has been the stimulation of the investigation and evaluation of possible improvements in measurement set-ups and procedures. Further activities have concerned the improvement of standard procedures/protocols and tools for verification, fostering a fruitful discussion on the harmonization of the techniques to determine intercomparison performance parameters. As a further benefit, the campaigns have initiated a dialogue among different laboratories throughout Europe and the United States.

The EurAAP facility comparison campaigns are conducted by a campaign leader or institution. The management of these intercomparison activities has been complex and has required corresponding regulation. The comparison leader collects the measured and processed data for comparison processing, following a strict procedure. Data of each participant is 'incognito' to other participants before officially released. Each participant can investigate his results against a set of 'reference' data provided by the comparison manager based on other measurements. If there are no evident errors and the participant agrees, the measured data is released and accepted in the campaign. This activity has been supported by the EurAAP WG5 on antenna measurements that covered the transportation costs and insurance for shipment from one laboratory to another.

As different laboratories are interested in measuring different antenna performance figures, in order to meet everyone's need, the 'typical' antenna performance figures common to most measurement situations are required:

1. Peak gain (IEEE definition) at discrete frequencies.
2. Directivity/gain patterns.
3. *S*-parameters.
4. Description of measurement facility and measurement procedure.
5. Description of mechanical and electrical set-up and alignment
6. Uncertainty budget.

7.2 EurAAP measurement campaign reference antennas

A key element of the success of an intercomparison campaign is the stability of the antenna that has to be measured. To these purposes, an exhaustive set of highly stable reference antennas in terms of type and bands were selected:

- L-band BTS antenna with directive elevation beam and wide azimuth beam (MVI/SATIMO BTS 1940);
- X/Ku/Ka-band high gain reflector antenna (MVI/SATIMO SR40-A fed by SH4000 dual-ridge horn) [27–31];
- L/C-band medium gain dual-ridge horn (MVI/SATIMO SH800) with absorbers plate [31–33];
- Cellular Telecommunications and Internet Association (CTIA) MIMO LTE antennas.

All the antenna designs are characterized by high stability, ensure excellent repeatability thanks to the alignment accuracy and, in some cases, have been enhanced to improve the correlation in different facilities.

The campaign with BTS and CTIA MIMO LTE antennas will be described in detail in Section 7.4. Figures 7.1 and 7.2 show the SR40 reflector antenna and the map of the facilities involved in the intercomparison campaign. Figures 7.3 and 7.4 show the SH800 dual-ridge horn and the map of the facilities who joined the campaign.

The measurements collected during EurAAP campaigns have been also exploited to investigate the achievable agreement between antenna measurement and simulation tools, as was done for the high gain reflector antenna [29,30,34] and for the CTIA MIMO antennas (see references in the following sections).

Figure 7.1 Reflector SR 40-A fed by SH4000 dual-ridge horn antenna

Figure 7.2 Map of the facilities involved in the Reflector SR 40-A fed by SH4000 dual-ridge horn antenna campaign

Figure 7.3 SH800 medium gain dual-ridge horn antenna with absorber plate

Figure 7.4 Map of the involved facilities in the SH800 medium gain dual-ridge horn antenna with absorber plate

The above-mentioned campaigns have been completed between 2017 and 2021 and results have been published since 2005 in more than 20 IEEE reference papers. The set of collected measurements will also be exploited in the frame of a new task, approved in 2019 within the EurAAP WG5, about laboratory/facility self-evaluation from intercomparison results [35]. The objective is to establish a process to self-evaluate antenna measurement systems. For this service, the reference parameters, including their uncertainties, for each gold standard are available to be compared with the measurements provided by other antenna measurement facilities. Therefore, those facilities do not need to wait for another intercomparison campaign to validate their measurement systems.

A new campaign, started in 2019, has as test object the ESA-DTU mm-VAST antenna [36], shown in Figure 7.5.

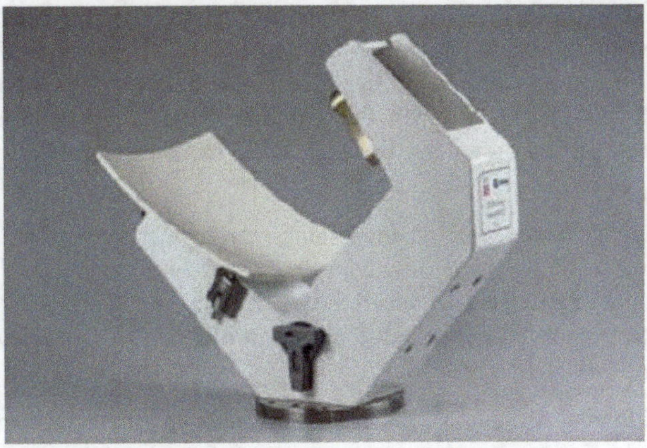

Figure 7.5 ESA-DTU mm-VAST antenna

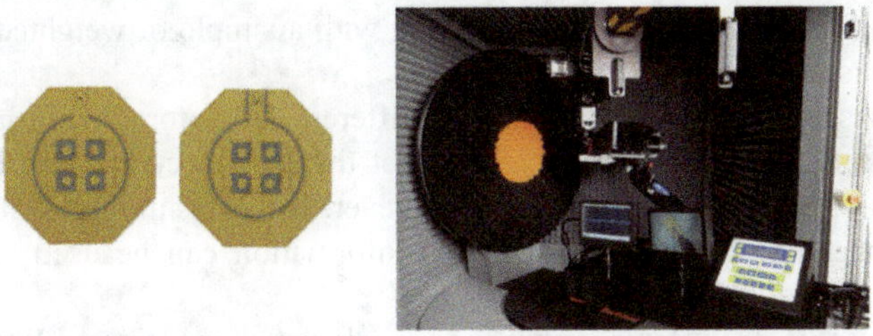

Figure 7.6 mm-wave chip antennas for 5G applications and microprobing measurement set-up

Mm-wave chip reference antennas [37], shown in Figure 7.6, will serve as 'gold-standard' antennas for microprobed antenna measurements. They are suitable for 5G wireless applications and have been selected for a future intercomparison campaign that will start in 2021.

7.3 Procedures for EurAAP measurements comparison

To evaluate the results of any comparative analysis, the following figures of merits are deemed as relevant:

- the reference value and its uncertainty computed from several independent measurements,
- the correlation between the reference value and each measurement expressed through the equivalent noise level (ENL).

To evaluate the results of a comparative analysis, many methods exist and have been described in literature, especially by the NIST, resulting from national measurement laboratories of a Ku-band standard gain horn [38–40].

Research activities, relying on the available large amount of measurement data, have been carried on among EurAAP WG5 members [15,16,22,41,42]. The outcome has led to the definition of:

- The reference value and its uncertainty through a proper combination of measurements and weights, involving the uncertainty budgets of the results;
- Consistency of the declared uncertainty budgets;
- ENL between the reference value and each measured value to express their correlation.

In the following paragraphs, the achieved results and the agreed methodology chosen to evaluate the measurement data coming from EurAAP intercomparison campaigns [42] are shown.

7.3.1 Reference value

The reference value can be defined as the common best value among different measures and can be obtained by averaging, with a simple or weighted mean, the different measurements.

The measurements are performed in different laboratories and, hence, have different uncertainties. A simple mean is not the most accurate solution since it gives the same weight to all samples. However, the measurement values have a specified uncertainty estimate. Thus, this information can be used as a weight factor.

Based on that, the reference value, μ, is calculated as weighted linear mean:

$$\mu_{Lin} = \frac{\sum_{i=1}^{n} w_i x_i Lin}{\sum_{i=1}^{n} w_{iLin}} \tag{7.1}$$

where n is the total number of measurements, i is the measurement of the ith participant to the campaign, and x_{iLin} is the measurement value (linear).

The weight w_i associated to the ith measurement is computed as:

$$w_{iLin} = \frac{1}{\sigma_{i\ Lin}^2}. \tag{7.2}$$

In (7.2), $\sigma_{i\ Lin}$ is the uncertainty obtained from $\sigma_{i\ \mathrm{dB}}$ that is the calculated uncertainty (in dB) by the ith facility taking part in the intercomparison campaign. $\sigma_{i\ \mathrm{dB}}$ is calculated as root mean square of different uncertainty sources. $\sigma_{i\ Lin}$ is obtained as:

$$\sigma_{i\ Lin} = \left(10^{\left(\frac{\sigma_{\mathrm{dB}}}{20}\right)} - 1\right) x_{Lin}. \tag{7.3}$$

The uncertainty associated to the weighted mean of (7.1) is given by

$$\sigma_{\mu Lin} = \sqrt{\frac{1}{\sum_{i=1}^{n} 1/\sigma_{iLin}^2}}. \tag{7.4}$$

7.3.2 *Equivalent noise level determination*

The ENL is a figure of merit that helps to understand the level of correlation between each measurement and the reference value. All the deviations of each measurement with respect to the reference value are converted into an equivalent 'noise' level. To express all deviations in a single value is useful, for example in the case of an antenna radiation pattern, where measurement values for many different angles exist. The expression for the ENL for antenna pattern, evaluated on the entire θ circle or part of it (e.g. $\pm 45°$ or $\pm 60°$), is the following:

$$ENL = \left[\sqrt{\frac{1}{n-1} \sum_{i=\theta_1}^{n} \left(\frac{Dir_{i\_co,cx} - Dir_{i\_ref\_co,cx}}{Dir_{co,ref,boresight}} \right)^2} \right] \tag{7.5}$$

where n is the total number of points comprised in the θ circle (e.g. 91 if the circle is $\pm 45°$, 121 if the circle is $\pm 60°$ and 1° resolution), i is the point of the θ circle: $\theta_i = -45°$ or $-60°$ and $\theta_n = 45°$ or $60°$, $Dir_{i\_co,cx}$ is the directivity (co or cx) of the measured pattern at $\theta = \theta_i$, $Dir_{i\_ref\_ co,cx}$ is the directivity of the reference pattern (co or cx) at $\theta = \theta_i$, $Dir_{co,ref,boresight}$ is the directivity of the co-polar component of the measured pattern in boresight.

Equation (7.5) can be also expressed as root mean square error (RMSE):

$$ENL = RMSE \left(\frac{Dir_{i\_co,cx} - Dir_{i\_ref\_co,cx}}{Dir_{co,ref,boresight}} \right) \tag{7.6}$$

In (7.5) and (7.6), directivity is the measurement value. In the case of gain, in order to compute a value that gives information on the pattern envelope error without being affected by the error in the gain measurement, this measured gain data must be replaced by a normalized value (*GainOffset*), which is obtained by subtracting a constant offset (in dB) all over the measured gain pattern: In the case of gain values, the measured radiation pattern should be normalized. Otherwise, errors in the gain measurement would affect the radiation pattern comparison. The offset can be computed as the co-polar gain difference in boresight direction:

$$Offset_{dB} = Gain_{dB,co\_boresight} - Gain_{dB,co,ref\_boresight} \tag{7.7}$$

This offset must be applied to all measured gain values:

$$GainOffset_{dB} = Gain_{dB} - Offset_{dB}. \tag{7.8}$$

Hence, the subtraction of the offset corresponds to a normalization with respect to the boresight reference pattern. The ENL in the case of gain can be computed as:

$$ENL = \left[RMSE\left(\frac{GainOffset_{co,xp} - Gain_{ref\_co,xp}}{Gain_{co,ref\_boresight}} \right) \right] \tag{7.9}$$

7.3.3 Uncertainty budget consistency check

The weights for the computation of the reference pattern (Equations (7.1) and (7.2)) depend on the uncertainty budget. Facilities participating in the intercomparison campaign must provide the standard deviation σ of the measurement. 2σ expresses the 95.4% confidence that the measurements error is within this level (99.7% is 3σ) in the case of normal distribution. The uncertainty is calculated by each facility following its own established protocols, some of them under the ISO 17025 standard.

A commonly used strategy in facility comparisons is to exclude one or more results that are not conformal to the general consensus. The aim of this approach is to avoid having a bad result that 'pollutes' the overall result. However, since these campaigns have statistically few participants (it is complex to find laboratories with measurement capability for these antennas), such a procedure is not reliable to assure convergence. A further drawback of this approach is that the situation for the excluded participants gets delicate. Human or component failure can lead to wrong measurement results. The purpose of the facility comparison is not to investigate such errors but to validate and confirm uncertainty budgets based on correctly executed procedures and equipment.

The methods agreed by EurAAP WG5 members to check the consistency of the declared uncertainties are the Birge ratio and the *E score* (see (7.11)).

The Birge ratio [43] is a classic method to check the consistency of a set of intercomparison results. The Birge ratio, R_B, can be computed both for directivity and for gain and is defined as:

$$R_B = \sqrt{\frac{\sum_i^n w_{iLin}\left(x_{iLin} - \mu_{lin}\right)^2}{n-1}} \tag{7.10}$$

where w_{iLin} is the weight, x_{iLin} are the measurements (gain or directivity), μ_{Lin} is the weighted mean (gain or directivity), and n is the number of involved facilities.

Values of R_B are much larger than 1 indicates that the measurements $x_1, \ldots, x_n$ are inconsistent. Instead, if RB is close to 1 or less, the measurements $x_1, \ldots, x_n$ are consistent.

An alternative metric to check the consistency of a set of data is provided by the *E score* that, as the Birge ratio, can be computed both for directivity and for gain. It verifies the accuracy of each declared uncertainty:

$$E\ score_i = \frac{|x_{iLin} - \mu_{lin}|}{\sqrt{U_i^2 + U^2}} \tag{7.11}$$

where $x_{iLin}, \ldots, x_{nLin}$ are the measurements (gain or directivity), μ_{Lin} is the weighted mean (gain or directivity), $U_i = 2\alpha_i$, $\alpha_i = (\sigma_{iLin} - 1)x_{iDir/Gain\_Boresight}$, where σ_{iLin} is the uncertainty declared by the ith facility that performed the ith facility (computed by (7.3)) and $U = 2\alpha$ with $\alpha = \sqrt{\frac{1}{\frac{1}{(\alpha_1)^2}+\frac{1}{(\alpha_2)^2}+\frac{1}{(\alpha_3)^2}+\cdots}}$.

An *E score* value smaller than 1 means that the declared uncertainty is most probably correct, otherwise the uncertainty is underestimated.

The use of both factors allows the determination of a coherent set of measurements and the possibility to request additional information about the uncertainty calculation of a specific facility.

7.4 Mobile communication antennas EurAAP intercomparison campaigns

This section shows the results of the concluded EurAAP intercomparison campaigns that have involved mobile telecommunication antennas, namely:

- MVI/SATIMO BTS 1940
- CTIA MIMO LTE antennas.

The BTS campaign is described to show an example of the achievable results of an intercomparison campaign using a standard antenna for mobile communication infrastructures. The results are obtained following the procedures described in Section 7.3.

Regarding the CTIA MIMO antennas campaign, as only few measurements were performed, it was not reliable to apply the procedures described in Section 7.3. The performed measurements were exploited for a comparison with simulations from different software vendors and the results are shown to provide an example for terminal antennas.

7.4.1 MVI BTS 1940 campaign

The reference antenna for this campaign was the L-band base station antenna MVI/SATIMO BTS 1940 (Figure 7.7). The antenna is a linear array with dual slant ±45°or H/V polarized elements working in GSM1800 and UMTS bands (1,710–2,170 MHz). The linear array antenna performances allow for excellent cross-polar discrimination and maintain a well-defined radiation pattern in the direction of the boresight axis throughout the working bandwidth. The BTS1940 antenna is equipped with high precision female N-type connectors to ensure superior repeatability and durability. The nominal impedance is 50 Ω with return loss values better than 15 dB.

Figure 7.7 MVI/SATIMO BTS 1940 high-accuracy reference antenna

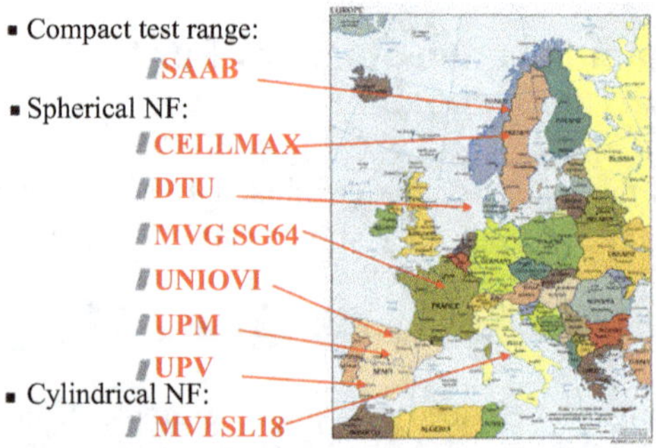

Figure 7.8 European location of the facilities involved in the BTS campaign

7.4.1.1 Campaign

The campaign started in 2009 and ended in 2016 [22,31,41,42,44]. The BTS has been measured in all the facilities shown in Figure 7.8.

The test plan and type of measurements are reported in Table 7.1.

The set of measurements gathered during the BTS campaign are employed to evaluate the measurements of the participants of the laboratory/facility self-evaluation activity from intercomparison results described in Section 7.2.

7.4.1.2 Results

The results that will be shown hereafter include the following facilities: DTU, UPM, SAAB, MVG SG64, MVI-SL, Cellmax. The weighted reference pattern has been computed, including all the facilities except Cellmax and MVI-SL. The declared uncertainties are reported in Table 7.2.

The comparison is shown in the following Sections 7.4.1.2.1 and 7.4.1.2.2 in terms of radiation patterns and ENL. As it can be inferred, the measurements are quite in agreement with each other, except some discrepancies due to positioner

Table 7.1 Required measurements for MVI/SATIMO BTS1940 campaign

	Frequency range	ϕ	θ
Full 3D gain measurement	1,500–2,400 MHz 5 MHz step	From 0° to 135° (45° step)	From −180° to 180° (1° step)

Table 7.2 Facilities and uncertainties for the reference pattern computation

Facility	Uncertainty on directivity (3σ)	Uncertainty on gain (3σ)
DTU	0.240	0.240
UPM	0.273	0.313
SAAB	0.345	0.525
MVG SG/SL	0.200	0.500

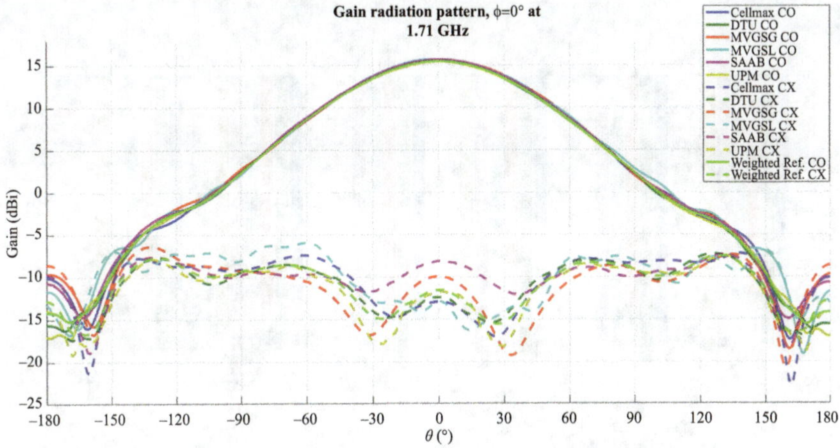

Figure 7.9 BTS gain radiation pattern – azimuth cut, ϕ=0°@1.71 GHz

effect, that affects only some angular ranges, and is due to the noise when the signal level is too low. The comparison results have led some of the participants to improve their measurement set-ups and procedures, to investigate their measurement further or perform a new measurement to correct for errors. Similar considerations hold for the uncertainty declared by each facility which, if not accurately computed, has been reviewed based on the computed consistency checks.

7.4.1.2.1 Radiation patterns

Measured co-polar and cross-polar antenna radiation pattern at ϕ=0° and 90° at 1.71 are compared with the weighted reference pattern (computed with Equation 1) in Figures 7.9 and 7.10.

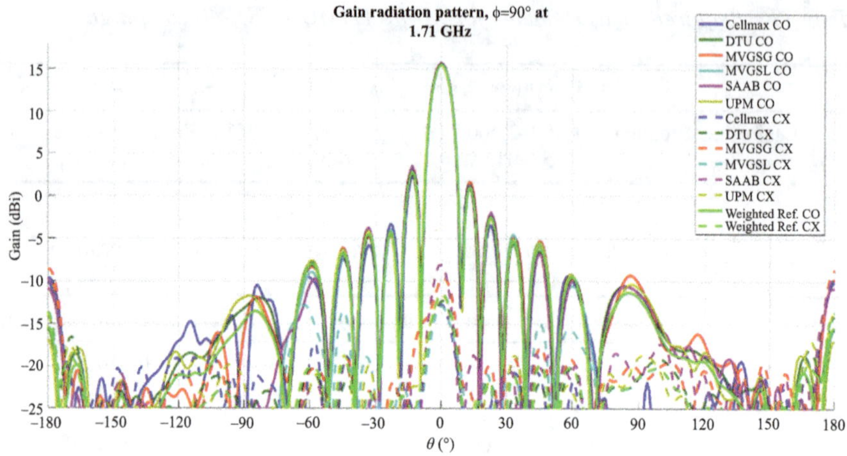

Figure 7.10 *BTS gain radiation pattern – elevation cut, ϕ=90°@1.71 GHz*

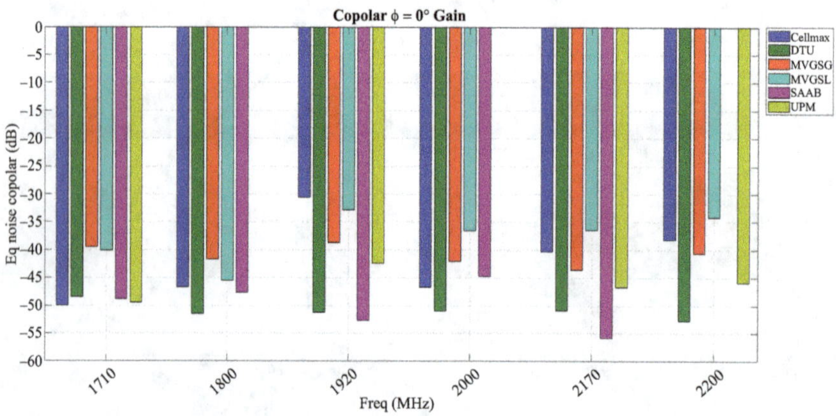

Figure 7.11 *BTS Equivalent noise level in the azimuth plane, for the co-polar BTS antenna radiation pattern (azimuth plane) (no data provided for SAAB at 2.2 GHz)*

7.4.1.2.2 Equivalent noise level

The ENL, @1.71, 1.92, 2.17 and 2.20 GHz, has been computed using the gain (Equation (7.9)) considering a θ cone of ±45°. Figures 7.11 and 7.12 show the noise level in the azimuth and in the elevation planes for the co-polar component. The peak gain results for the reference and for each facility are shown in Table 7.3. Figures 7.13 and 7.14 show the equivalent error level as a function of θ at 1.71 GHz for the co-polar components in the azimuth and elevation planes for all facilities.

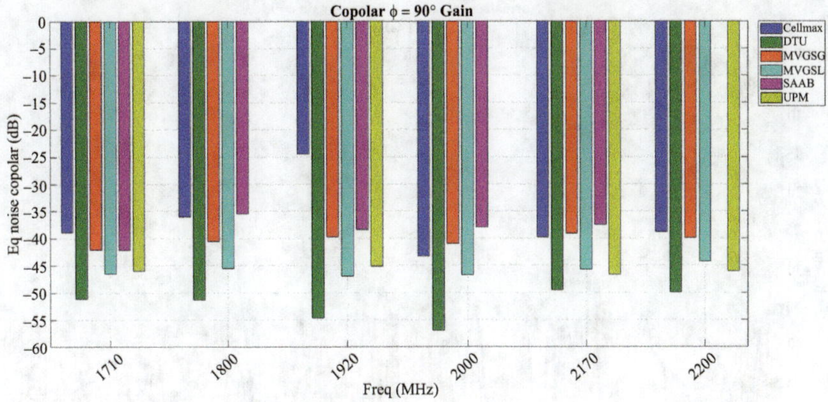

Figure 7.12 Equivalent noise level for the co-polar BTS antenna radiation pattern (elevation plane) (no data provided for SAAB at 2.2 GHz)

Table 7.3 Peak gain results for the BTS campaign

	Peak Gain						
Freq. (GHz)	**REF**	**DTU**	**UPM**	**MVG SG64**	**MVG SL**	**SAAB**	**Cellmax**
1.71	15.54	15.51	15.48	15.64	15.69	15.72	15.57
1.92	16.07	16.15	16.00	15.75	15.60	16.24	16.23
2.17	16.60	16.82	16.39	16.43	16.03	16.41	16.54
2.20	16.58	16.73	16.43	16.37	15.95	–	16.50

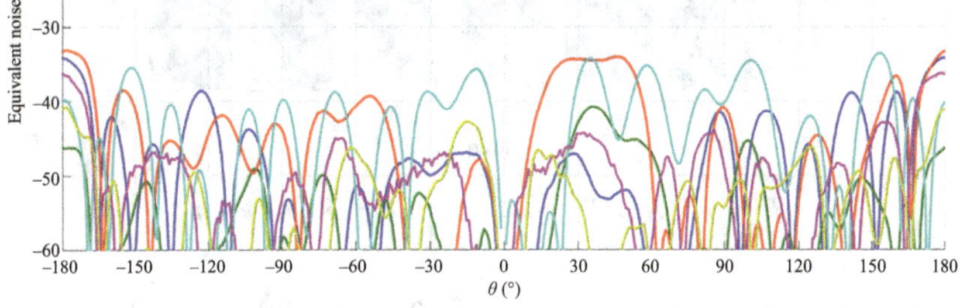

Figure 7.13 Normalized absolute difference for the co-polar BTS antenna gain radiation pattern (azimuth plane) at 1.71 GHz in the cut θ=0°

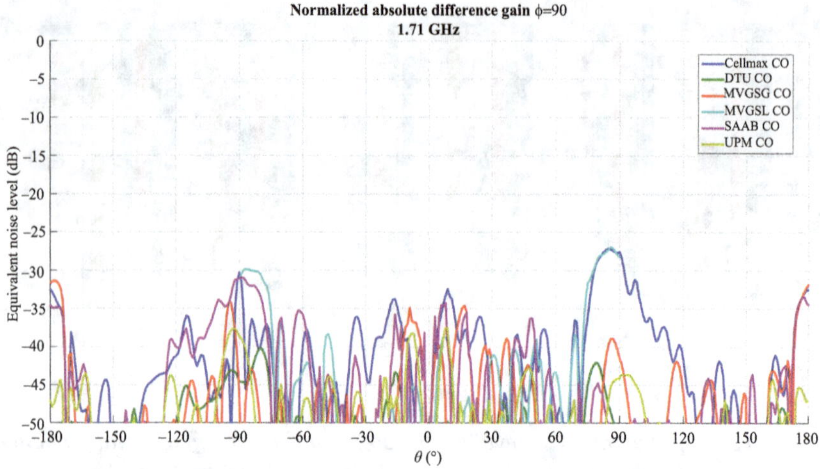

Figure 7.14 Normalized absolute difference level for the co-polar BTS antenna gain radiation pattern (elevation plane) at 1.71 GHz (limited validity of angular range: −120°/+60°): for the co-polar component in the cut θ=90°

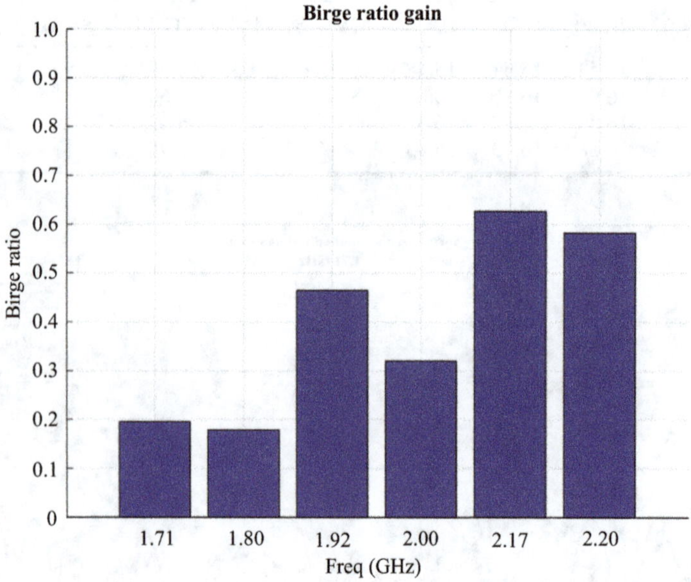

Figure 7.15 Birge ratio for the BTS campaign

The consistency checks computed with (7.9) and (7.10) can be both shown as a function of θ or as a function of frequency. Figures 7.15 and 7.16 show the Birge ratio and *E score*, respectively, over frequency and over θ. Both consistency checks

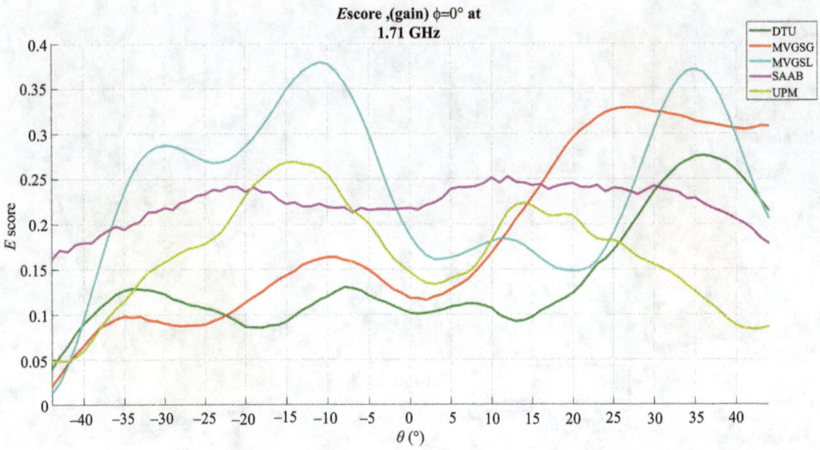

Figure 7.16 E score for the BTS campaign in the azimuth plane @1.71 GHz

are smaller than 1 meaning a consistent data set: the declared uncertainties are coherent.

7.4.2 CTIA MIMO LTE antennas campaign

The reference antenna of this campaign was the CTIA 2×2 LTE MIMO antenna. The antenna concept was conceived in 2011 during the first CTIA MIMO OTA Subgroup meetings [45–47]. In that period, different standardization organizations were working on test methodologies and procedures to determine accurately MIMO performance. This task had become difficult since each test methodology was characterized by several implementation details [48]. The CTIA LTE MIMO 2×2 reference antennas were designed with the objective to remove the uncertainties related to the unknown antenna performance of the few LTE MIMO 2×2 reference devices or golden standards that were available in those years for evaluating radiated data throughput measurement methodologies and test facilities [49]. By knowing the radiated performances of the reference MIMO 2×2 antennas, plus the conducted performances of each device under test (DUT), it is possible to predict the data throughput of each DUT. The predicted performances can be compared with measured OTA radiated throughput values. The adoption of simple reference antennas with well-known performance also eliminates part of the measurement uncertainty and increases the correlation among different MIMO OTA test methodologies and test facilities.

7.4.2.1 Design and challenges of the antenna

The antenna design, shown in Figure 7.17, was simple, low-cost and with high manufacturing repeatability to enable widespread use as a reference device. The antenna concept was designed by Motorola and developed for series production by SATIMO/MVI.

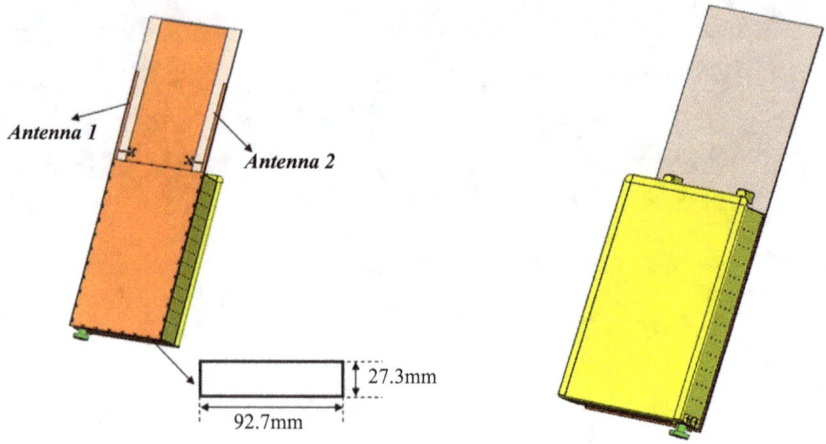

Figure 7.17 MIMO 2×2 reference antenna schematic

In order to properly test and verify the correlation and accuracy of different MIMO data throughput measurement systems, the set of MIMO reference antennas covers three LTE bands: 2, 7 and 13.

For each band, three antennas were designed to emulate a 'good' MIMO antenna system, i.e. low correlation coefficient ($\rho<0.1$), high total antenna efficiency (>90%) and low gain imbalance ($\Delta G \cong 0$ dB). Respectively, the 'nominal' MIMO antenna system has moderate correlation coefficient ($\rho \leq 0.5$), moderate total antenna efficiency ($\geq$ 50%) and low gain imbalance ($\Delta G \cong 0$ dB). Finally, the 'bad' MIMO antenna system, having poor correlation coefficient ($\rho \geq 0.9$), moderate-to-poor total antenna efficiency ($\leq$ 50%) and low gain imbalance ($\Delta G \cong 0$ dB).

Other than the electrical performance constrains, the reference antenna also needed to solve the potential problem with connecting any external antenna into a portable device. The connection between the portable device radio frequency (RF) port and the external antenna, normally a coaxial cable, can potentially carry current in the outer conductor [50]. The associated radiation perturbs the antenna system radiation and influence system parameters like correlation coefficient, absolute gain and gain imbalance. For this reason, the antenna was conceived by attaching MIMO 2×2 external antennas to an RF enclosure, where the DUT and its RF connections are located (Figure 7.18).

The antennas are self-resonant based on the Inverted-F antenna topology. The control of magnitude of the complex correlation coefficient is achieved through the reference ground plane placement between antennas radiators. Since the antennas are absolutely symmetric, the gain imbalance between the antennas is nearly 0 dB, which is important in an environment where the fields impinge from all directions. With this feature, the gain imbalance between antennas can be artificially controlled through discrete RF attenuators placed between the DUT and reference antenna RF port, inside the RF enclosure.

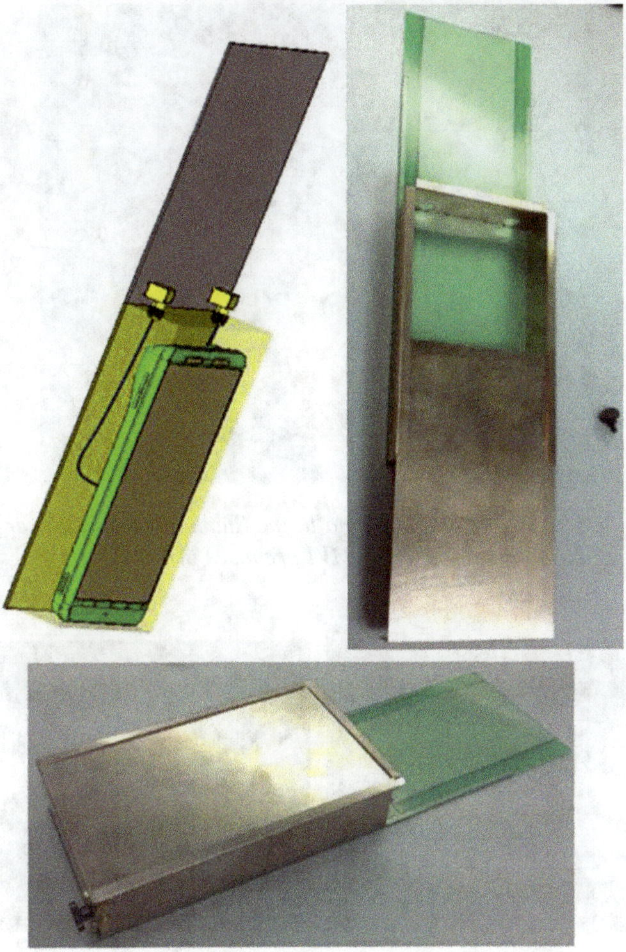

Figure 7.18 Band 13 'Good' MIMO 2×2 antenna

This allows for the possibility of both chamber/test methodology validation and comparative DUT design baseline testing against known references. In particular, CTIA 2×2 MIMO antennas were included in the EurAAP measurement facility comparison campaign in collaboration with the 'Small Antenna Measurements Group'.

7.4.2.2 Campaign

In the frame of this campaign, the MIMO antennas were measured with the MVG StarLab spherical near-field system at MVI premises (Figure 7.19) in Italy and in the far-field system at IMST premises in Germany (Figure 7.20). The test plan and type of measurements are reported in Table 7.4.

Figure 7.19 CTIA MIMO 2×2 antenna in the StarLab spherical near-field measurement system at MVI premises in Italy

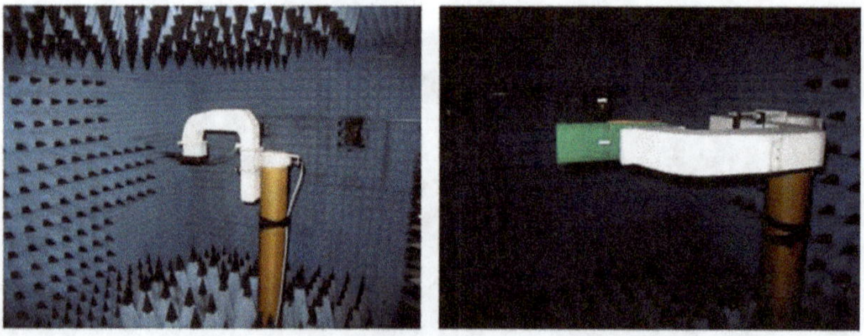

Figure 7.20 CTIA MIMO 2×2 antenna in the far-field system at IMST premises in Germany. The photos show the same roll-over-azimuth positioner rotated in two different orientations.

7.4.2.3 Results: comparison with simulations

As already pointed out, since only two participants were involved, it is not suggested to apply the statistical approach described in Section 7.3. The measurements performed within the intercomparison campaign have been exploited for comparison with simulations from software vendors [51,52]. These antennas have been selected to 'bridge the simulation-measurement gap' in the frame of a convened session in cooperation between the EurAAP WGs on software and measurements at the 2016 European Conference on Antennas and Propagation.

Simulations were performed by software vendors with different numerical methods and compared with the measurements obtained in the intercomparison

Table 7.4 Test plan of the CTIA 2×2 MIMO antennas intercomparison campaign

Full 3D gain measurement	**Frequency range**	ϕ	θ	**Ports**
	LTE Band 7 [2,620–2,690 MHz] Good antenna	From 0° to 135° (45° step)	From −180° to 180° (1° step)	Antennas 1 and 2
	LTE Band 13 [746–756 MHz] Good antenna Bad antenna	From 0° to 135° (45° step)	From −180° to 180° (1° step)	Antennas 1 and 2

campaign. The involved software vendors and numerical methods employed were as follows:

- EMPIRE:
  Finite-Difference Time-Domain method (FDTD)
- WIPL-D:
  Method of Moments (MoM)
- FEKO:
  Finite Element Method (FEM) and MoM
- ANSYS:
  FEM
- FASANT:

Multilevel Fast Multiple (MLFMM), Macro basis functions, Geometrical Theory of Diffraction (GTD), Physical Optics (PO).

The full computer-aided design (CAD) model of the RF enclosure module with and without cover (Figure 7.21) was provided together with details on the printed circuit boards, including drawings and datasheet of dielectric substrate. No details were provided on the feeding point, considered as second-order effect.

The results are shown in terms of matching, isolation, pattern, gain and directivity, gain imbalance and envelope correlation. Since no information about the uncertainty of the simulations was available, the shown average data is not weighted, but just a simple mean. Results for the band 7 good antenna are shown in the following figures comparing measurements and simulations. The matching is shown in Figure 7.22, while the isolation is shown in Figure 7.23. The antenna reference system is shown in Figure 7.24.

Even if not realistic, in a reverberation and controlled anechoic test environment, the received power is uniformly distributed in both θ and ϕ directions. The propagation models approved by 3rd Generation Partnership Project standardization body, when the MIMO antennas were developed, had information of the power distribution in azimuth while there was no information on the power distribution in elevation. Figure 7.25 shows the antenna radiation pattern (E_θ) at 2,656 MHz for antennas 1 and 2 (antennas 1 and 2 are visible in Figure 7.17). The averaged

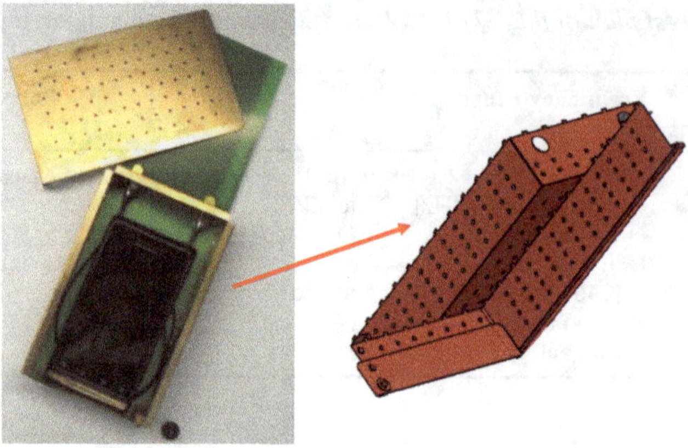

Figure 7.21 CAD model of the RF enclosure of the 2×2 MIMO reference antenna without the cover

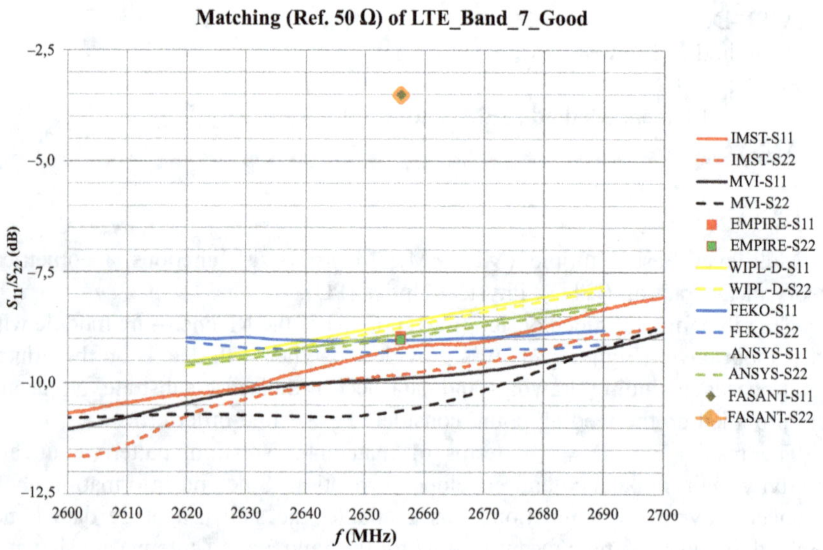

Figure 7.22 Measured (IMST and MVI) and simulated matching (EMPIRE, WIPL-D, FEKO, ANSYS, FASANT) for the band 7 good antenna

patterns (obtained as simple not weighted mean) are shown in Figure 7.26 while the peak directivity and peak gain are shown, respectively, in Figures 7.27 and 7.28.

The gain imbalance and envelope correlation are shown in Figure 7.29. The envelope correlation is a parameter to characterize the suitability of an antenna for diversity or MIMO applications. For high diversity or MIMO performance, low absolute correlation values are required. The envelope correlation of simulations

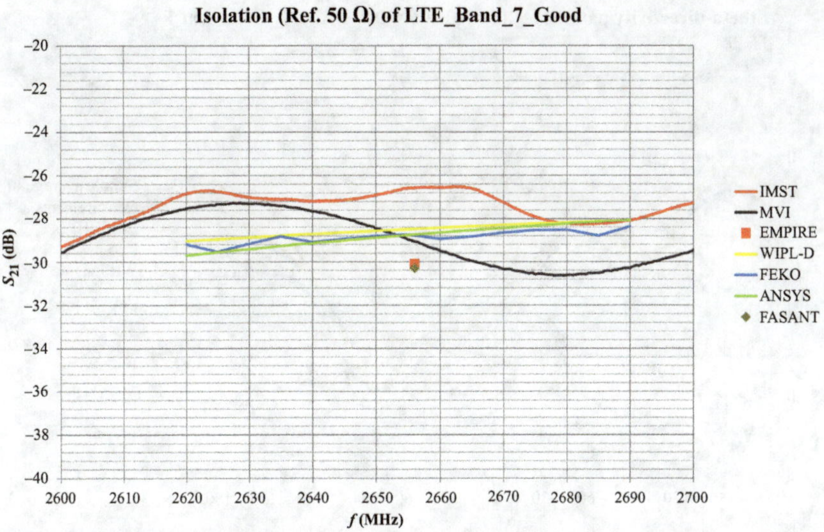

Figure 7.23 Measured (IMST and MVI) and simulated (EMPIRE, WIPL-D, FEKO, ANSYS, FASANT) isolation for the band 7 good antenna

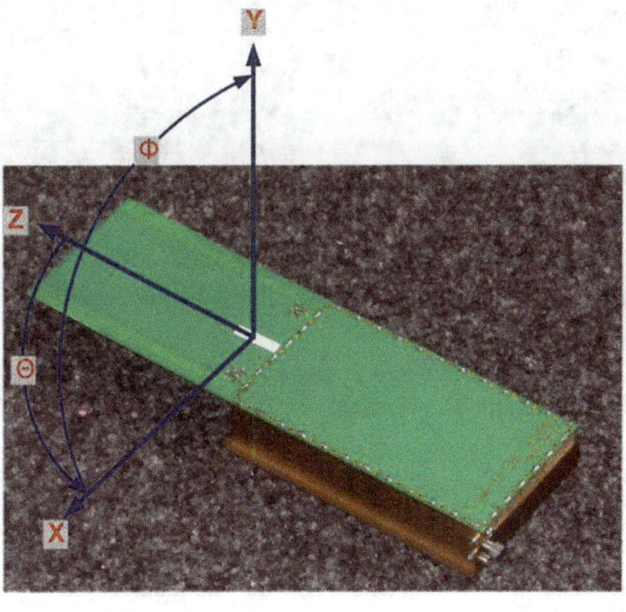

Figure 7.24 Antenna reference system for measurements and simulations

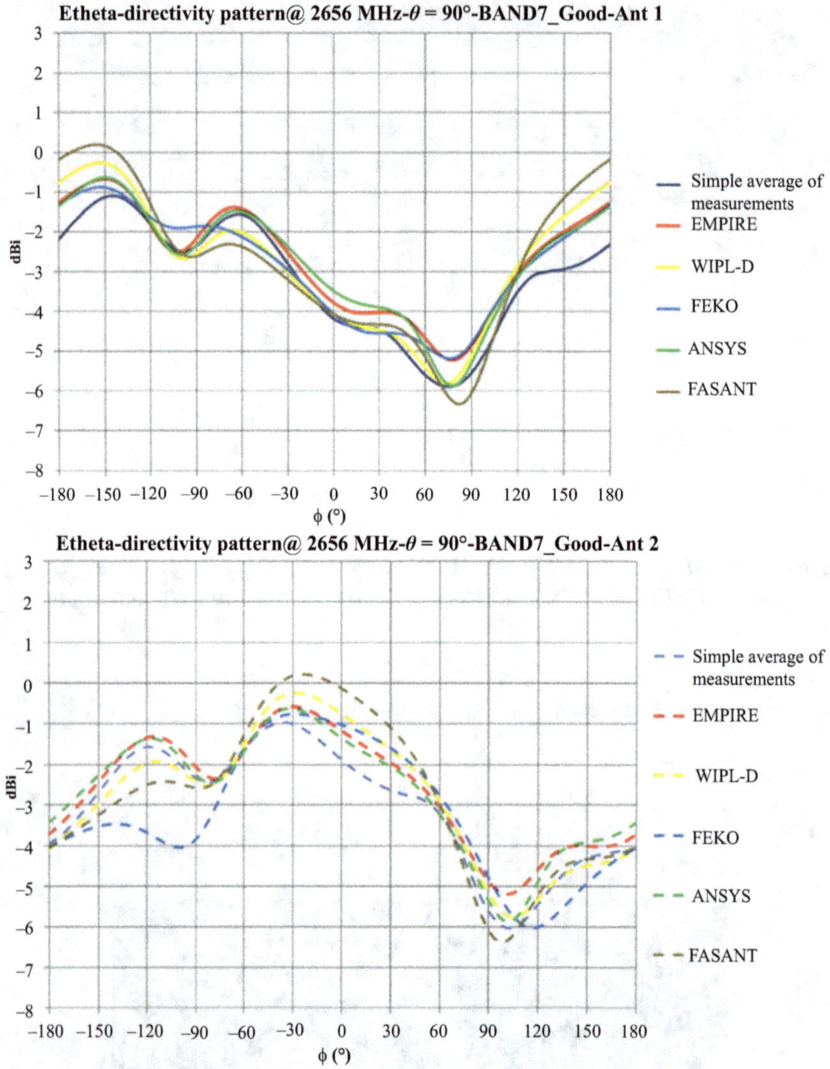

Figure 7.25 Measured and simulated antenna radiation pattern (E_θ) of antenna 1 (top) and antenna 2 (bottom)

has been computed considering line of sight and isotropic channel module to emulate measurement conditions.

As visible, a reasonable agreement among the simulations and measurements was obtained for band 7 good antenna, both for the radiation pattern and for the *S*-parameters.

Considering the other bands, for example in band 13, even though measured *S*-parameters agreed very well between each other, the same correspondence with

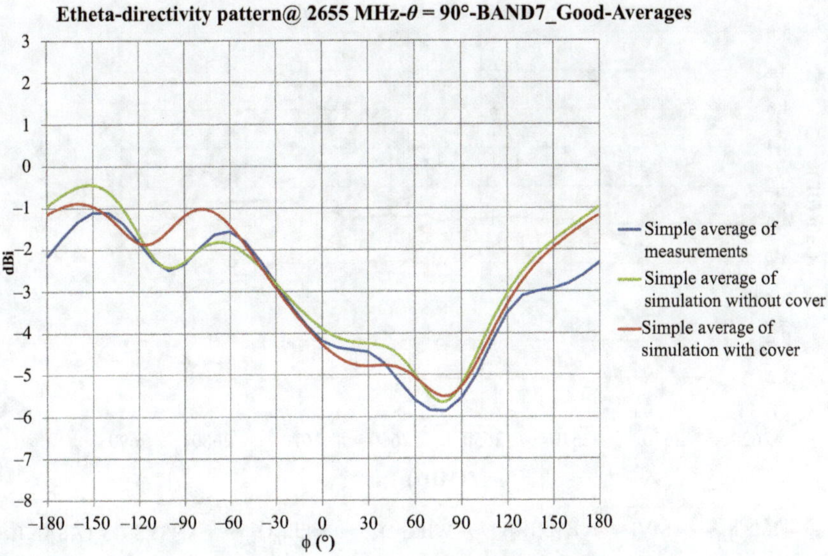

Figure 7.26 Averaged simulated and measured antenna radiation pattern (E_θ)

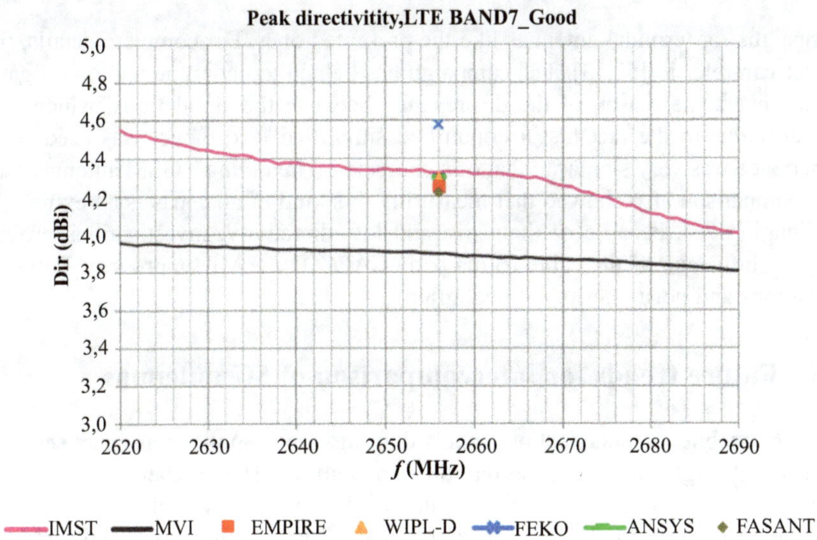

Figure 7.27 Measured and simulated peak directivity

the simulations was not found. Additionally, the simulations did not match among one another. A similar situation was visible for the radiation patterns. The plots for the band 13 antenna are not shown here since the aim of the section was to provide some examples to show the benefits that can arise from the intercomparison

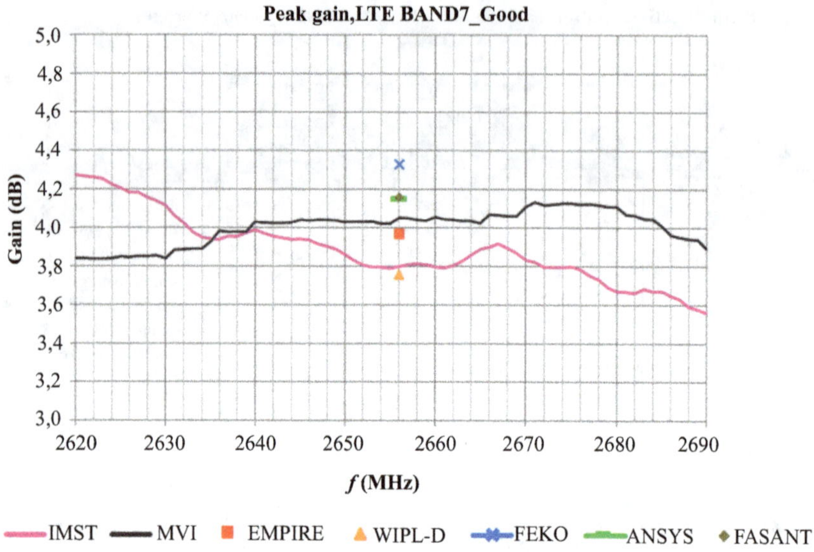

Figure 7.28 Measured and simulated peak gain

campaigns for terminal antennas like the presented ones. For complete details, the reader can refer to [52,53]. The campaign has helped to underline and investigate, in this band, the reason of the discrepancy between the simulations, which was probably due to the fact that a nonunified simplified feed model was used. This experience was very similar to that of a comparison involving a GSM antenna [54]. The comparison also showed that measuring this kind of antennas is an extremely challenging task because of their size and low directivity: investigations proved that the influence of the cable (6 dB pad+CAPCON® EMI suppressant Tubing), transitions and positioner is not negligible.

7.5 Future trends for intercomparison of 5G antennas

The 5G mobile communication standard brings new applications and services thanks to its high bandwidth and real-time capabilities. The number of use cases is incredibly large; Internet of things, autonomous driving, vehicle to everything communications, virtual and augmented reality, new multimedia, industrial or medical applications among others. This means that during the next years, there will be a high increase of the number of devices for user terminals and for infrastructure. Therefore, more antennas, and probably in a faster way, will have to be measured making necessary an evolution in antenna measurement facilities, reducing measurement time and maintaining high quality in order to test those terminals.

The services provided by EurAAP, the intercomparison campaigns presented here, and the self-evaluation campaigns introduced in Section 7.2, are two different

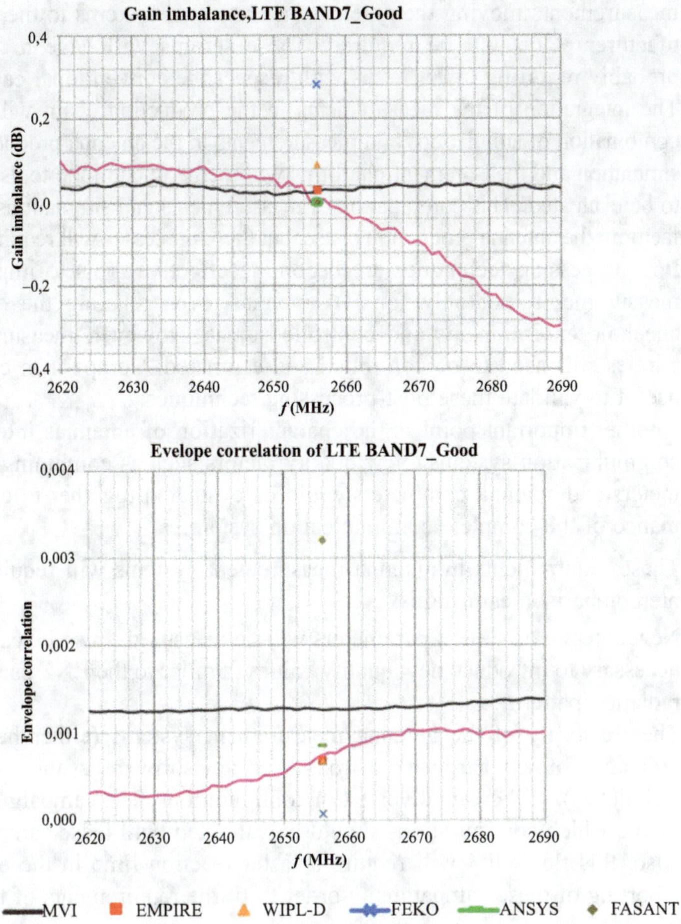

Figure 7.29 Measurements and simulations of band 7 good antenna: Gain imbalance (up) and envelope correlation (bottom)

tools to assure the quality of those test facilities and to improve the technical capabilities of the antenna test laboratories.

In any case, it is clear that these campaigns have to evolve to consider new antenna parameters, integration of antenna with RF components, or with the communication system itself, over the air measurements, and other new trends in communication and antenna domain.

Let us summarize first some of the requirements for the future 5G antenna measurements [55], some of them already in 4G:

- The antenna industry will demand in situ, real-time, accurate and cost-effective measurements, using for instance commercial robots with more complicated acquisition schemes for reducing antenna manufacturing time. New test facilities integrated in the manufacturing chain will appear. For instance, in situ measurements moving the antenna measurement systems to the antenna manufacturer place will be required. These set-ups will have to be evaluated, probably requiring lower times with respect the conventional campaigns.

- The integration of the measurements in the production chain will facilitate the combination of simulations and measurements in the antenna production. Antenna simulation and measurement platforms will merge in common tools that will have to be evaluated. This combination of measurements and simulations will be a key factor in the antenna production phase, but these processes will require a validation.
- Post-processing techniques are becoming effective tools to improve antenna measurements, not only for 5G antennas, even relaxing the quality of the anechoic environments and controlled areas. However, measurement uncertainties still need to be controlled and estimated. Comparison campaigns are useful to validate these post-processing techniques.
- Another important point is the characterization of antennas integrated in the communication systems. New quality factors such as combining system parameters and antenna parameters could be evaluated together to correct performance of the complete communication system.

These requirements in antenna measurement systems will require changes in the intercomparison campaigns:

- New antenna or system parameters will be measured. In some cases, it will be necessary to introduce new quality values (similar to the ENL used here for the radiation pattern).
- The flexibility of the antenna measurement systems (either because of the utilization of new hardware, as robots, or new software, as the post-processing techniques), will create diverse test configurations. The campaigns should also define which configurations should be validated with these campaigns.
- Also, this flexibility will require a faster reaction time in the execution and reporting of these campaigns in order to fit the requirements of the industry.

These are only some examples of the future requirements for the intercomparison campaigns. It seems clear that they will be more complex in the future, but they will be still an excellent tool to validate the antenna measurement systems. A variety of well-characterized test antennas is a valuable result of the campaigns and can be used as benchmark for new measurement and post-processing techniques.

7.6 Conclusions

This chapter has discussed intercomparison campaigns and showed that they are an important tool for the validation of test facilities. The work done under the EurAAP umbrella has allowed to establish a set of criteria to evaluate the capabilities and quality of the antenna test facilities, to discover errors in the measurement process and to check the consistency of the uncertainty budget estimation. One of the most important conclusions is that the different antenna test facilities participating in these campaigns are able to measure with extraordinary high quality, improving their capabilities during the last years. Also, some of the participants are certified by ISO 17025 standards, and these campaigns are used to show their measurement capabilities.

The measurement comparison procedures have been discussed for two different campaigns using mobile communication antennas: one base station antenna and one MIMO antenna for a user terminal. It was emphasized that the antenna selection and the organization of the campaign are key factors for the success of the campaign. Furthermore, it was shown that, besides measurement results, simulation results can be included in comparison campaigns.

Finally, we analysed the challenges and trends for the future. The expected rapid deployment of 5G services will bring new devices, either for terminals or infrastructure, and new requirements in RF measurements. In addition, the industry demand for in situ, real-time, accurate and cost-effective measurements requires modifications of current measurement techniques. Post-processing techniques play an important role here. However, all measurement procedures must still provide uncertainty estimations and must be validated. Comparison campaigns and the availability of well-characterized test antennas are an excellent method for this validation.

References

[1] ISO/IEC 17025 International Standard [online]. Available from: www.iso-iec17025.com [Accessed March 2021].

[2] ANSI/IEEE Standard 149-1979. *IEEE Standard Test Procedures for Antennas*, 1979.

[3] IEEE Standard 1720-2012. *IEEE Recommended Practices for Near-Field Antenna Measurements*, December 2012.

[4] Stubenrauch C. F., Newell A. C., Repjar A. G., *et al.* 'International intercomparison of horn gain at X-band'. *IEEE Transactions on Antennas and Propagation*. 1996, 44(10), pp. 1367–1374.

[5] Gentle D., Beardmore A., Guerrieri J., *et al. Measurement Techniques and Results of an Intercomparison of Horn Antenna Gain at Frequencies of 12.4, 15.0, 18.0 GHz.* CCEN Key Comparison CCEN.RF-K23.F, May 2016 [online]. Available from: https://www.bipm.org/utils/common/pdf/final_reports/EM/RF-K23/CCEM.RF-K23.F.pdf [Accessed March 2021].

[6] Kacker R., Datla R., and Parr A. 'Combined result and associated uncertainty from interlaboratory evaluations based on the ISO Guide'. *Metrologia*. 2002, 39(3), pp. 279.

[7] Guerrieri J. R., Francis M., and Wittmann R. 'Consensus value methods to compile on-axis gain measurement results'. *Proceedings of the 2015 Antenna Measurement Techniques Association*, Long Beach, CA, USA, October 2015, pp. 255–259.

[8] The Foundation for Research on Information Technologies in Society (IT'IS) [online]. Available from: https://itis.swiss/who-we-are/ [Accessed March 2021].

[9] Allard A., and Amarouche S. 'Analysis of interlaboratory comparison when the measurements are not normally distributed'. *18th International Congress of Metrology*, September 2017, 12003, pp. 1–14.

[10] Antenna Centre of Excellence (ACE) [online]. Available from: http://www.ist-ace.org/ [Accessed March 2021].

[11] Foged L. J., Garreau P., Breinbjerg O., Pivnenko S., Castañer M., and Zackrisson J. 'Facility comparison and evaluation using dual ridge horn'. *Proceedings of the 2005 Antenna Measurement Techniques Association*, Newport, RI, USA, 2005, pp. 161–165.

[12] Foged L. J., Iversen P. O., Duchesne L., Breinbjerg O., and Pivnenko S. 'Comparative Measurement of standard gain horns'. *The 28th ESA Antenna Workshop on Space Antenna Systems and Technologies (ESTEC)*, Noordwijk, the Netherlands, June 2005.

[13] Foged L. J., Breinbjerg O., Pivnenko S., Di Massa G., and Sabatier C. 'Antenna measurement facility comparison within the European Antenna Centre of Excellence'. *2005 European Microwave Conference*, October 2005.

[14] Giacomini A., Bencivenga B., Duchesne L., and Foged L. J. 'Determination of high accuracy performance data for dipole reference antennas'. *Proceeding of the First Antenna Measurements Techniques Association Europe (AMTA Europe) Symposium*, Munich, Germany, May 2006.

[15] Foged L. J., Bencivenga B., Pivnenko S., *et al.* 'Error calculation techniques and their application to the Antenna Measurement Facility Comparison within the European Antenna Centre of Excellence'. *2nd European Conference on Antennas and Propagation (EuCAP 2007)*, Edinburgh, UK, November 2007.

[16] Foged L. J., Bencivenga B., Breinbjerg O., Pivnenko S., and Sierra-Castañer M. 'Measurement facility comparisons within the European antenna centre of excellence'. *ISAP2008, International symposium on Antennas and propagation*, Taipei, Taiwan, October 2008.

[17] Pivnenko S., Breinbjerg O., Burgos S., Sierra-Castañer M., and Eriksson H. 'Dedicated measurement campaign for definition of accurate reference pattern of the VAST12 antenna'. *Proceedings of the 2008 Antenna Measurement Techniques Association*. Boston, MA, USA, 2008, pp. 249–253.

[18] Foged L. J., Bencivenga B., Sierra Castañer M., *et al.* 'Facility comparison and evaluation using dual ridge horns'. *3th European Conference on Antennas and Propagation (EuCAP 2009)*, Berlin, Germany, March 2009.

[19] Hansen J. E. *Definition, Design, Manufacture, Test and Use of a 12 GHz Validation Standard Antenna, Executive Summary*. Technical University of Denmark, Lyngby, Denmark, 1997, Tech. ReportR672.

[20] Pivnenko S., Olav Breinbjerg O., Sierra Castañer M., *et al.* 'Comparison of antenna measurement facilities with the DTU-ESA 12 GHz validation standard antenna within the EU Antenna Centre of Excellence'. *IEEE Transactions on Antennas and Propagation*. 2009, 57(7), pp. 1863–1878.

[21] Foged L. J., Giacomini A., Duchesne L., Leroux E., Sassi L., and Mollet J. 'Advanced modelling and measurements of wideband horn antennas'. *11th International Symposium on Antenna Technology and Applied Electromagnetics (ANTEM 2005)*, Saint-Malo, France, June 2005.

[22] Foged L. J., Sierra Castañer M., and Scialacqua L. 'Facility comparison campaigns within EurAAP'. *5th European Conference on Antennas and Propagation (EuCAP 2011)*, Rome, Italy, April 2011.

[23] COST ASSIST IC0603. *Antenna Systems and Sensors for Information Society Technologies* [online]. Available from: www.cost-ic0603.org [Accessed March 2021].

[24] COST VISTA IC1102. *Action on Versatile, Integrated and Signal-aware Technologies for Antennas* [online]. Available from: www.cost-vista.eu [Accessed March 2021].

[25] EurAAP (European Association on Antennas and Propagation) [online]. Available from: www.euraap.org [Accessed March 2021].

[26] EurAAP Activities. *Measurements* [online]. Available from: https://www.euraap.org/Activities/wg/detailed-presentation-for-each-wg/measurements [Accessed March 2021].

[27] Microwave Vision Group. High Precision Offset Parabolic Reflector [online]. https://www.mvg-world.com/es/products/antennas/reference-antennas/high-precision-offset-parabolic-reflector [Accessed March 2021].

[28] Foged L. J., Saporetti M. A., and Sierra-Castañer M., *et al.* 'Measurement and simulation of reflector antenna'. *9th European Conference on Antennas and Propagation (EuCAP 2015)*, Lisbon, Portugal, April 2015.

[29] Sierra Castañer M., Foged L. J., Saporetti M.A., *et al.* 'Comparison of reflector antenna, measurements and simulations'. *37th Antenna Measurement Techniques Association Annual Meeting & Symposium (AMTA 2015)*, Long Beach, CA, USA, October 2015.

[30] Saporetti M. A., Foged L. J., Sierra Castañer M., *et al.* 'Measurements and simulations correlation of high reliability reflector antenna'. *10th European Conference on Antennas and Propagation (EuCAP 2016)*, Davos, Switzerland, April 2016.

[31] Saporetti M. A., Foged L. J., Sierra Castañer M., *et al.* 'Recent developments in international facility comparison campaigns', *41st Antenna Measurement Techniques Association Annual Meeting & Symposium (AMTA 2019)*, San Diego, CA, USA, October 2019.

[32] Saporetti M. A., Foged L. J., Sierra Castañer M., *et al.* International facility comparison campaign at L/C band frequencies. *39th Antenna Measurement Techniques Association Annual Meeting & Symposium (AMTA* 2017), Atlanta, GA, USA, October 2017.

[33] Saporetti M. A., Foged L. J., Sierra Castañer M., *et al.* 'L/S & C band medium gain ridge horn intercomparison campaign results'. *14th European Conference on Antennas and Propagation (EuCAP 2020)*. Copenhagen, Denmark, March 2020.

[34] Saporetti M. A., Foged L. J., Sierra Castañer M., *et al.* 'Measurement and simulation of reflector antenna'. *9th European Conference on Antennas and Propagation (EuCAP 2015)*, Lisbon, Portugal, April 2015.
[35] EurAAP Activities – Measurements [online]. Available from: https://www.euraap.org/Activities/wg/detailed-presentation-for-each-wg/measurements [Accessed March 2021].
[36] Pivnenko S., Oleksiy S. K., Breinbjerg O., *et al.* 'DTU-ESA millimeter-wave validation standard antenna (Mm-Vast) – performance verification'. *36th ESA Antenna Workshop on Antennas and RF Systems for Space Science*, Noordwijk, Netherland, 2015.
[37] Microwave Vision Group. Mm-Wave Chip Reference Antenna [online]. Available from: https://www.mvg-world.com/en/resources/mm-wave-chip-reference-antenna [Accessed March 2021].
[38] Ludwig A. C. 'The definition of cross polarization'. *IEEE Transactions on Antennas and Propagation*. 1973, AP-21, pp. 116–119.
[39] Guerrieri J., Coder J., Francis M., Gordon J., and Novotny D., Wittmann R. 'International comparison of Ku-band – standard gain horn characterization'. *The 8th European Conference on Antennas and Propagation (EuCAP 2014)*, The Hague, the Netherlands, 2014, pp. 2653–2654.
[40] Guerrieri J., Francis M., and Wittmann R. 'Consensus value methods to compile on-axis gain measurement results'. *The 10th European Conference on Antennas and Propagation (EuCAP 2016)*, Davos, Switzerland, 2016, pp. 1–4.
[41] Scialacqua L., Mioc F., Zhang J., Foged L. J., and Sierra-Castañer M. 'Antenna Measurement Intercomparison Campaigns in the framework of the European Association of Antennas and Propagation'. *2013 International Symposium on Antennas and Propagation (ISAP 2013)*. Nanjing, China, October 2013.
[42] Saporetti M. A., Foged L. J., Sierra Castañer M., Pivnenko S., Cornelius R., and Herberling D. 'Description and preliminary results of antenna measurement facility comparisons in EurAAP'. *IEEE Antennas and Propagation Magazine*. 2017, 59(3), pp. 108–116.
[43] Birge R. T. 'The calculation of errors by the method of least squares'. *Physical Review*. 1932, 40, pp. 207–227.
[44] Saporetti M. A., Foged L. J., Scialacqua L., and Saccardi F. 'Antenna measurement system verification by correlation with facility-comparison campaign in the framework of the European Association of Antennas and Propagation'. *2014 International Wireless Symposium (IWS 2014)*, Xi'an, China, March 2014.
[45] Motorola Mobility. *Reference Antennas proposal for MIMO OTA, 3GPP TSG RAN WG4#59, R4-113032*, Barcelona, Spain, May 2011.
[46] Szini I., Pedersen G. F., Scannavini A., and Foged L. J. 'MIMO 2X2 reference antennas concept'. *The 6th European conference on Antennas and propagation (EuCAP2012)*, Prague, Czech Republic, March 2012.
[47] Szini I., Pedersen G. F., Scannavini A., and Foged L. J. 'Design and verification of MIMO 2x2 reference antennas'. *2012 IEEE International*

Symposium on Antennas and Propagation and USNC-URSI National Radio Science Meeting (APS2012), Chicago, IL, USA, July 2012.

[48] Agilent Technologies. *Analysis of Initial Measurement Campaign Results*. 3GPP TSG RAN WG4#59, R4-113097, Barcelona, May 2011.

[49] Elektrobit, Nokia. *Comparison of full Anechoic Chamber Measurement Results*. 3GPP TSG RAN WG4#61, R4-115927, San Francisco, November 2011.

[50] Yanakiev B., Nielsen J. O., and Pedersen G. F. 'On small antenna measurements in a realistic MIMO scenario'. *The 4th European Conference on Antennas and Propagation (EuCAP 2010)*, Barcelona, Spain, April 2010.

[51] Saporetti M. A., Foged L. J., Scannavini A., Weitz S., and Kunisch J. 'Measurements and simulations of MIMO 2x2 reference antennas', *10th European Conference on Antennas and Propagation (EuCAP 2016)*, Davos, Switzerland, April 2016.

[52] Vandenbosch G. A. E., Mioc F., Saporetti M., and Foged L. J. 'Bridging the simulations-measurements gap. State of the art'. *IEEE Antennas and Propagation Magazine*. 2016, 58(6), pp. 12–14.

[53] Vandenbosch G. A. E. and Gillard R. 'Benchmarking of optimally used commercial software tools for challenging antenna topologies'. *IEEE Antennas and Propagation Magazine*. 2013, 55(3), pp. 281–292.

[54] Vandenbosch G. A. E. 'State-of-the-art in antenna software benchmarking—'Are we there, yet?'. *IEEE Antennas and Propagation Magazine*. 2014, 56(4), pp. 300–308.

[55] Sierra-Castañer M. 'Review of recent advances and future challenges in antenna measurement'. *Applied Computational Electromagnetics Society Express Journal*. 2016, 1(2), pp. 64–67.

Chapter 8

Numerical and experimental analyses of wearable antennas, including novel fabrication and metrology techniques

Isidoro Ibanez-Labiano[1], Syeda Fizzah Jilani[2], Elif Ozden-Yenigun[3] and Akram Alomainy[1]

Most fifth-generation (5G) mobile network applications require wearable antennas to be unobtrusive, low-profile, low-power and electrically small. Such antennas are a crucial element in wearable body-centric wireless system designs for delivering 5G user's experience. Wearable antennas can be employed in a wide range of applications from communicating, harvesting energy to sensing capabilities. For this purpose, fabrics and novel materials such as graphene are been explored in order to cope with the wearable device demands in terms of flexibility, conformability and lightweight. Similarly, novel fabrication techniques for wearable antenna prototyping such as screen printing, inkjet printing, embroidery and cutters are been investigated to exploit the unique characteristics of various materials. These innovative fabrication methods allow a high degree of fabrication precision enabling their uptake for 5G applications. Due to power absorption by lossy human body tissues, a distorted radiation pattern and lower radiation efficiency are envisaged when they are worn on and at proximity to the body. Furthermore, when designing the antenna, the body proximity effects must be considered to prevent significant antenna detuning and the consequent mismatch. Numerical and experimental human body phantoms are used with a view to simulate its impact. This chapter presents an analysis of novel fabrication methods for wearable antennas, methodologies and measurement techniques to characterise their performance in a dynamic body-worn communication environment. This chapter delivers a review of different novel fabrication techniques for wearable antennas such as various printing processes, machine embroidery and laser methods. Follow by a metrology section, where numerical and experimental human phantoms are explained and durability tests described. Three examples of 5G wearable antennas

[1]School of Electronic Engineering and Computer Science, Queen Mary University of London, London, United Kingdom
[2]Department of Physics, Aberystwyth University, Ceredigion, United Kingdom
[3]School of Design, Textiles, Royal College of Art, London, United Kingdom

for different frequencies and their on-body performance characterisation are provided. Finally, it is closed with a summary of the outcomes achieved in this chapter and future prospects of the research.

8.1 Introduction

The idea of application-specific user-body-centric wireless networks built around wearable devices and nodes placed at certain locations has significantly evolved in the past few years with them becoming integrated aspects of our daily activities for healthcare, well-being and immersive information and communication technology and entertainment experience. A significant progress has been made particularly for biometry and communication at microwave frequencies [1,2]; however, the migration of developing wireless systems towards the millimetre-wave (mm-wave) spectrum for fifth-generation (5G) and beyond is driving various research trends, including design, fabrication and measurement techniques of wearable antennas towards higher frequency bands due to the attraction over their inherent advantages on high data rates, secured links and lower probability of interference [3–6].

A number of commercial wearable devices have been introduced and available in the market; however, they are restricted to certain applications and demand more versatility, compactness and diversity [7]. A wearable antenna is often regarded as an integral part of a wireless wearable device and its performance should guarantee a satisfactory wireless communication link between the communicating nodes positioned on different parts of the human body and also with off-body units to ensure the required power transfer. The antenna solution in such application should be low-profile, compact and conformal enough to adapt to the shape and curvature of the human body, as well as easily integrate-able within garments and everyday clothing. In practice, wearable antennas may experience a resonance shift or distortion in radiation characteristics when placed closer to the human body which requires high robustness to reduce the impact of the body on antenna performance [8]. It is important that the proposed antenna solutions at mm-wave should be insensitive to the effect of the body presence or at least include such effect in the design and analysis procedure to ensure efficient link performance. They must ensure high-gain characteristics to reduce node-to-node path-loss [9], which can be avoided or minimised through the use of high-gain and directional antennas. Pattern reconfiguration would provide a suitable solution, as the cognitive control on the direction of radiation can configure the shortest distance between the transmitter and receiver and enhance the link quality significantly. It is also important to ensure adaptability of radiation performance to the required communication path; with the direction of maximum radiation of the antenna being tangential to the body surface for on-body links, while being orthogonal for off-body links [10,11].

A number of versatile mm-wave antennas consisting of conventional or novel patch geometries provided with the broadside radiation response suitable for off-body communication, or end-fire radiation for on-body links have been reported on

60 GHz. For instance, a four-element patch array is proposed on a 127-μm thin RT/duroid film with a gain of approximately 12 dBi, and a human phantom is also used for the characterisation of body effects [12]. It is observed that the effects of the phantom on the antenna performance are negligible due to broadside radiation being shielded from the body by the back-layer ground plane. The four-element antenna array is also designed and implemented on a textile substrate for off-body scenario [13]. In addition, a 60-GHz textile-based Yagi–Uda end-fire antenna fabricated by ad hoc manufacturing process is suggested for body area networks and also realised on thin substrates to demonstrated on-body communications [14,15]. The authors have led research activities to design and experimentally validate an inkjet-printed flexible leaky-wave wideband antenna array at mm-wave with the capability of limited beam-steering [16,17] and also mm-wave wearable antenna solution with significant size reduction [18], respectively. However, the major drawback of reported and presented designs in the open literature is the limited flexibility in terms of functionality and reconfiguration and also the lack of adaptability in a cognitive manner to the radio link.

Considering all the progress that has been made in the area, there are still critical challenges remain to be solved. High free-space path loss and oxygen-induced atmospheric attenuations are major challenges in the realisation of a reliable and efficient mm-wave communication links, which require innovations and original contributions in terms of radio frequency (RF) front-end that could compensate for such losses and enhance the overall system performance. Although these issues might provide difficulties and obstacles in the realisation on wearable technologies at mm-wave, they could be addressed and minimised (in some cases eliminated) in efficiently designed antenna solutions. In addition, experimental characterisation of suitable arrays at mm-wave need to adhere to specific thresholds and end-user informed requirements. As an example, the human body presents a dynamic environment, specifically for wearable applications, which leads to increased signal losses due to the misalignment of communicating nodes that is, in some cases, greater than the absorption loss.

8.2 Advanced fabrication techniques for wearable antennas

Wearable antennas are a key component of wearable wireless communication systems. Such antennas are often unobtrusive, low-profile, electrically small and with low-power consumption for delivering the desired user's experience. Wearable antennas can be used for communication, energy harvesting or as sensors. This section presents the advances in fabrication techniques for flexible wearable antennas. Such antennas, design to be worn, have specific requirements such as flexibility, compactness, reconfigurability, conformability and durability. Therefore, the materials used and production methods in wearable antennas are quite unconventional and need to be evaluated.

Recently, textile-based antennas and electronic circuitries integrated as a part of a garment or a clothing have recently gathered huge attention [19,20]. Flexible antennas are anticipated to have a substantial impact in the development of a numerous wireless applications for 5G and beyond. Printed devices on flexible substrates combine the benefits of low-cost fabrication, light-weight, thin form factor, conformity and compliance with the contours of non-planar surfaces, rollability and robustness against cracks due to stretching, bending and folding [21]. These features along with the smart packaging techniques are necessary for wearable electronics, body-centric sensors and many other future wireless applications. A high-frequency spectrum has gained substantial attention to increase the bandwidth resources for future wireless networks to accommodate a rapidly increasing number of devices and applications. This transition implies a significant reduction in the size of the antenna and other electronic components and thus requires advanced manufacturing techniques to fulfil the high-accuracy demands associated with the shorter wavelength.

Several efficient methods such as photolithography, inkjet printing [22–25], flexography [26] and screen printing [27,28] are famous for inexpensive mass-production of printed flexible electronics. These fabrication processes are additive growth methods and involve an automated layer-by-layer deposition or growth sequence on a substrate [29,30]. Photolithography offers high accuracy, although it involves highly sophisticated clean-room facilities to transfer a pattern on a substrate by means of light [31]. Flexible substrates with laminated metallic sheets of copper or aluminium can be prototyped with high accuracy by means of a laser milling equipment [32,33]. Inkjet printing is another method to achieve good accuracy by using a computer-aided inkjet printer and can be extended to a reel-to-reel process for a time-efficient bulk-fabrication [34,35]. Conductive embroidered patterning with conductive yarns on fabrics for antennas is also a reasonable alternative especially in wearable antennas integrated in clothing. In addition, several conductive woven fabrics can be sewed on the textile substrate as a radiating area of the antennas as well as other circuitries. Among the fast, high-resolution and cost-effective printing techniques, screen printing is promising due to simplicity, affordability, time-efficiency and reduced material wastage [36]. When compared with the expensive instruments required for the inkjet printing, laser prototyping or cleanroom conditions, screen printing is an appropriate choice with reasonable fabrication accuracy on a variety of substrates like plastics, fabrics, polymers, etc. Further discussion on these techniques is presented as follows.

8.2.1 Screen printing

Screen printing is a process of transferring a pattern by means of a screen or an ink-blocking stencil that defines the area to be patterned on the substrate using a paste. Conductive pastes and metallic epoxies are widely used in this regard to pattern conductive lines, pads and metallic structures for the printable electronics and antennas by using the screen printing process. A number of variants are commercially available such as micro/nanoparticle conductive inks, pastes and epoxies based on variation in the composition, wettability, adhesion and conductivity. The conductivity of these pastes is mainly dependent on the concentration of the metal particles in the solvent, while the uniformity, homogeneity and thickness of the printed pattern depend on the particle size.

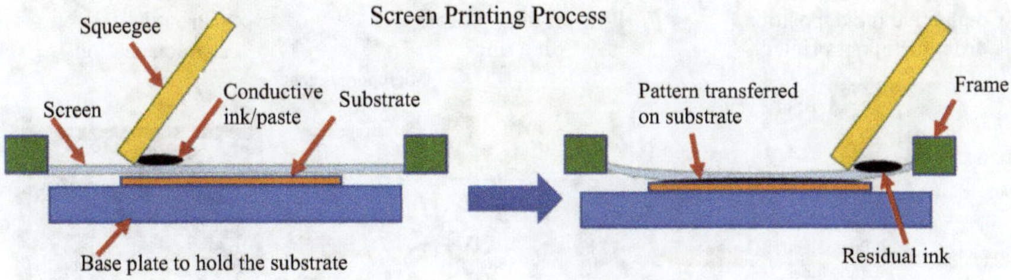

Figure 8.1 Screen printing process for flexible antennas and electronics

The screen is usually made with a fine net sheet coated with a photo emulsion with an engraved pattern designed to transfer the paste to the substrate placed under the screen frame. The complete screen printing process is presented in Figure 8.1. A reasonable level of prototyping accuracy can be achieved by a careful deployment of the process. After transferring of the pattern, the next step involves the sintering of the printed surface. The desired level of conductivity is achieved after the curing and sintering of the printed prototype. Several methods have been developed for sintering by using heat, laser, ultraviolet (UV) beam exposure, etc. For instance, one method involves exposure of laser beam on the printed surface by using formic acid in a nitrogen atmosphere [36,37]. Belt furnace with formic acid dosing system is used where laser sintering is carried out on the FASTLAPS roll-fed prototype system [36,37].

8.2.2 Inkjet printing

Another additive manufacturing technique is inkjet printing which can successfully be used for an accurate fabrication of several printed circuits and antennas. For instance, the inkjet printing by using silver nanoparticle ink has been used on a flexible polydimethylsiloxane substrate in [38] and on a polyethylene terephthalate (PET) substrate in [39] for a wearable antenna fabrication. Recently, carbon-based compounds such as graphene enhanced nano-composite materials and carbon nanotubes (CNTs) have gained attention in the printable and flexible electronics to replace conductive metal inks. These biodegradable materials have many environment-friendly attributes, reduced cost as compared to metal-based inks and suitability to be used in antennas. For instance, a flexible and stretchable antenna on a stretchable single-walled CNTs/Lycra fabric is reported in [40]. Moreover, there are many stretchable conductive ink solutions compatible with the body-centric wearable antennas commercially available [41].

Advanced inkjet printers are designed with the drop-on-demand technology, where a piezo or thermo component is incorporated in the printer's printhead to generate pressurised pulses to control the ink discharge from the nozzle [35]. The entire inkjet printing process involves three essential stages: printing, sintering and characterisation as shown in Figure 8.2. Printing is performed by a set of nozzles integrated in the inkjet printer, conductive ink and an automated system, where the user controls the speed/jetting of ink, feature size and resolution, thickness of a

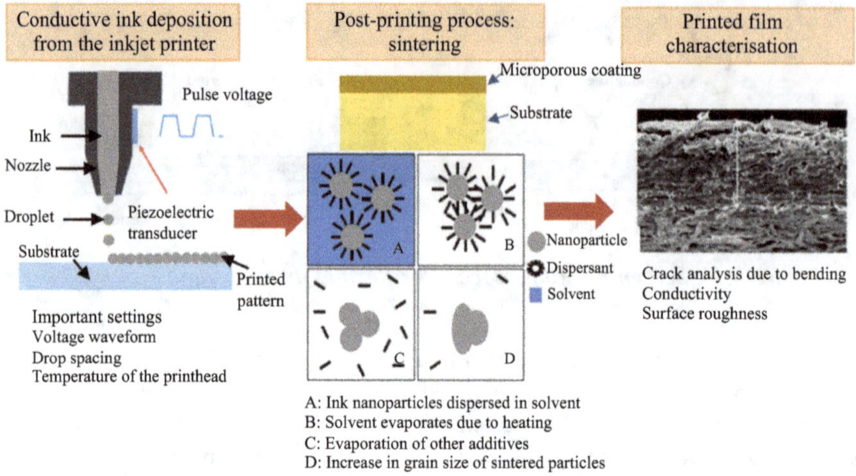

Figure 8.2 Inkjet printing process for antenna fabrication

single printed layer, drop spacing and thus the overall accuracy of the printed pattern. The next step is to sinter the printed pattern to improve the conductivity of the printed layer. In metal-based inks, the conductive particles are usually encapsulated in a sheath of polymer to avoid oxidation due to solvent. Sintering involves breaking the sheath to release the encapsulated conductive nanoparticles. Laser sintering and heat sintering are two most used methods, where high-energy laser beam or heating at high temperature melts the polymer coating as well as the ink particles. The molten ink particles combine to form a continuous layer of metal. Laser printing is time efficient with a risk of defects if the intensity and time of exposure is not properly controlled. Some commercially available substrates are coated with a microporous layer that offers rapid drying of the ink and chemical sintering.

In general, the conductivity of a silver nanoparticle single layer is $0.3–0.7\times10^7$ S/m and can vary up to $0.4–2.5\times10^7$ S/m depending on the number of successive printing iterations, layer thickness, temperature, and curing and sintering durations [29,33]. Thicker layers provide better conductivity; however, thicker metal layers are more brittle and can be easily cracked by successive bending and folding in the case of flexible and wearable antennas. With an optimisation of these constraints, inkjet printing can be a reliable choice for wearable antennas.

8.2.3 Embroidery with conductive yarns

Embroidered textile-based antennas can be potentially used in a variety of wearables and can be easily integrated along with other woven electronic components as a part of daily garments. The fabrication of embroidered antennas is performed by using conductive threads and computer-aided embroidery machines to construct the antenna pattern on a textile substrate. Besides metal-based conductive threads, many other polymer fibres with metallic coating have been introduced. These fibres comprise approximately 10-μm thick robust and flexible polymer core with a metallic coating of 2-μm thickness [42]. In order to achieve good surface uniformity and high conductivity, fully automated embroidery machinery and high-density stitching are deployed. This method is suitable for the fabrication of a variety of textile-based embroidered wearable antennas, RF sensors and the associated electronic circuitries required for many body-centric applications such as medical monitoring, and RF identification and tracking. These fabric antennas have shown great potential for the 5G networks and beyond for future RF functionalised fashionable garments and body-worn communications due to excellent RF performance compatible to conventional rigid metallic antennas.

8.2.4 Woven and non-woven antennas

This method utilises the conductive sheet either sewed or glued on textile substrates. For instance, graphene sheets, copper and aluminium sheets, etc. are cut in a designed pattern by means of blade cutting or laser cutting and installed on the substrate with adhesives. Wearable antennas designed with metallic fabrics sewn on cotton, denim and other similar textiles have also been reported. Besides sewing the metallic sheet on a textile, the conductive fabric is often woven in a defined pattern by weaving. The common weaving techniques in this regard are basic plain weave which is strong and durable, and a rip stop weave that is effective to avoid tearing. Threads used for weaving are usually metal based such as silver, copper, nickel and tin in various combinations. The performance of these textiles-based antennas is usually characterised by repeated stress cycles that involve cold water washing, hot water soaking, outdoor drying, ironing, etc. and impact of these successive washing cycles on the conductivity, durability and adhesion and overall robustness of these antennas [43]. The use of graphene-based materials for flexible non-woven antennas is well known due to good conductivity and thin-film characteristics suitable of bending and folding. Graphene has an anisotropic two-dimensional (2D) shape and good processability. A few layers assemble to form a flexible 'graphene paper' which offers good conductivity and can be deployed on various rigid and flexible substrates and even on textiles for antennas as demonstrated in recent works.

8.2.5 Vinyl cutters and plotters

Vinyl cutters and plotters have been widely used in several industries, for example within graphic design. Currently is being explored and applied for rapid prototyping in label-type designs, mainly for UHF RFID tags. The method consists of shaping the antenna pattern on different films and substrates, employing cutter plotters to mechanically cut the structure according to the virtual design. Flexible adhesive copper sheets or other conductive films like graphene are used for this purpose, followed by either subtractive by removing the excesses or additive via transferring the final patterns on target soft substrates [44] (Figure 8.3).

This fabrication method is particularly suitable for cost-effective and time-saving antenna designs. No chemicals or heat is involved in this process, turning it

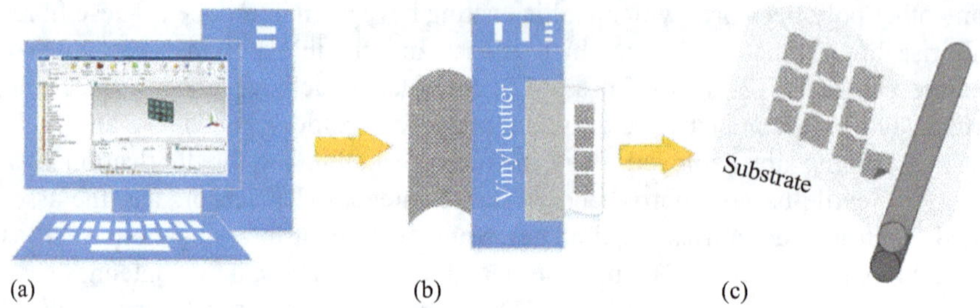

Figure 8.3 Flow chart of the procedure: (a) computational modelling; (b) cutting process; and (c) lamination on the substrate

into a suitable alternative for more complex techniques. These characteristics perfectly match the requirements of soft electronics devices such as being planar, flexibility, stretchability, conformability, lightweight, as well as low cost. There is an ever-growing interest in this method for the digital manufacturing of flexible devices that is compatible with several materials such as sheets of metals [45] and even 2D materials like graphene [46].

Shape and size of the antenna will be constrained by the resolution of the cutting equipment. In the case of simple antenna patterns, 0.5-mm resolution is readily achievable, while for more complex shapes, recent studies claimed that 1 mm would be the smallest line completed [44]. These characteristics cope with the requirements for 5G antenna models in the frequency band of sub-6 GHz in [45] antennas operating around 2.45 GHz and in [46] from 3 to 9 GHz, have been manufactured following this approach. The improvement in the features of cutting equipment allows some plotters to for resolutions up to 24.5 μm [47], allowing us to reach the K band (i.e. 18–26.5 GHz). In Figure 8.4, some prototypes fabricated with this method are depicted even within at mm-wave frequencies Figure 8.4(g).

8.2.6 Automated laser prototyping on flexible substrates

Automated laser micromachining is one of the methods for flexible antenna fabrication [33]. Several rigid and flexible substrates laminated with copper and aluminium are available that can be subjected to laser micromachining to etch out the designed pattern. The working principle of these laser-based machines is similar to basic printed circuit boards milling for electronic circuits and antennas, though these advanced machines use an extremely precise UV laser beam instead of metallic tools like mill/drill/router bits to remove metal in a specified pattern from the surface of the cladded flexible substrate. The laser beam prevents hazardous chemicals usage for etching, requires no masks or photolithographic patterning, time-efficient and holds minimum manufacturing cost. The laser probe operates in a contact-free environment and maintains an appropriate height above the substrate and thus can be easily used on delicate materials. These laser-milling machines are highly efficient in prototyping on both rigid and flexible substrates and provide fast and reliable fabrication with high accuracy.

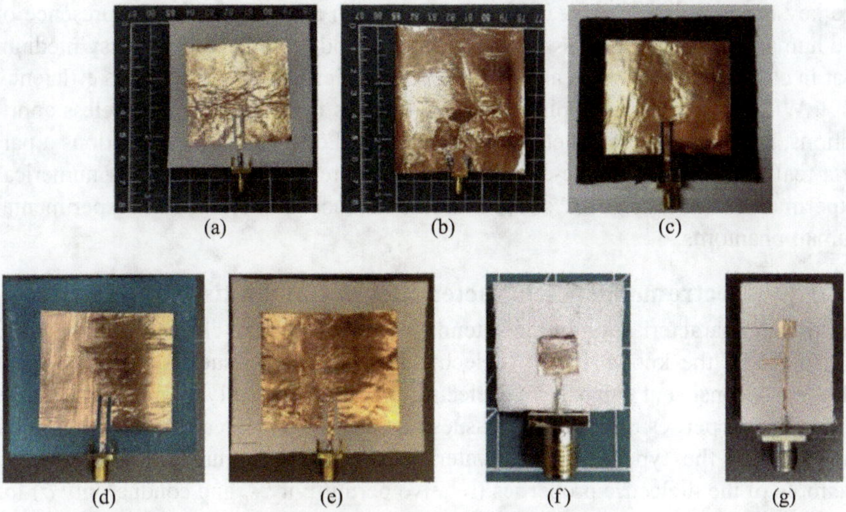

Figure 8.4 Photos of the antenna prototypes: (a) cotton top-view; (b) cotton bottom-view; (c) jeans; (d) viscose; (e) Lycra; (f) cotton at 9.45 GHz; and (g) cotton at 38 GHz [45]. Reproduced, with permission, from Isidoro Ibanez-Labiano, MDPI Materials; published by MDPI, 2020.

8.3 Metrology techniques for wearable antennas

This section of the chapter describes measurement techniques of 5G wearable devices and antennas for on-body scenarios. Different set-ups are presented to evaluate the key features of wearable antennas.

For any type of antenna, measurement parameters such as scattering parameters (*S*-parameters), gain and radiation patterns can be evaluated within a far-field screened anechoic chamber using vector network analyser (VNA) to assess and analyse their basic radiation characteristics and features, where VNA is typical equipment used for classical antenna metrology. For the specific case of wearable antennas, measurements are often performed with either real human or human tissue equivalent phantom in order to take into account the impact of the human presence. The effect of the human body on the antenna's performance, as well as the consequences on human tissues as a result of absorbing RF energy, specific absorption rate (SAR). In reality, wearable systems often perform in a dynamic and constantly changing environment, where they should sustain their performance under physical changes for example bending and crumpling conditions. Environmental factors, including moisture and temperature variations, will modify the electromagnetic properties of the materials used in these wearable antennas, thus altering their radiation and link performance.

8.3.1 Phantom numerical and experimental evaluation

The effects of a numerical and physical phantom for different wearable antenna systems are presented later. Impedance bandwidth, efficiency and radiation pattern are key characteristics of wearable antenna which are affected by the presence of the human body or phantoms [48]. The human body is inherently a lossy medium that in close proximity of an antenna has detuning effects and reduces its efficiency [49]. With a view to developing wearable devices for body-centric wireless applications, it is crucial to characterise these effects. For these characterisations a part of a real human body, tissue-equivalent options are commonly used for numerical experimental analysis. This section describes such numerical and experimental human phantoms.

8.3.1.1 Electromagnetic characteristics of human tissues

For proper characterisation of the antenna's behaviour in the close proximity of the human body, the knowledge of dielectric properties of tissues is very important [48,49]. A consistent database of dielectric properties is really complex to achieve due to the properties of biological tissues that change with frequency, temperature, position and the type of tissue (water content) [49]. Figure 8.5 illustrates an example of the dielectric properties (relative permittivity ε_r and conductivity σ) for two types of human tissues that differ from water-content with fat having lower values than muscle. It can be seen that the dielectric properties are dependent on the frequency and water content.

Open databases of dielectric data, such as the one of the Institute for Applied Physics, are available for all tissues at single frequencies or for a single tissue in a full span between 10 Hz and 100 GHz [50]. Previous research studies have calculated these characteristics: conductivity, σ, relative permittivity, ε_r, loss tangent, tanδ, and penetration depth for a number of different human body tissues at the Industrial Scientific Medical (ISM) band around 2.45 GHz [10]. In the case of

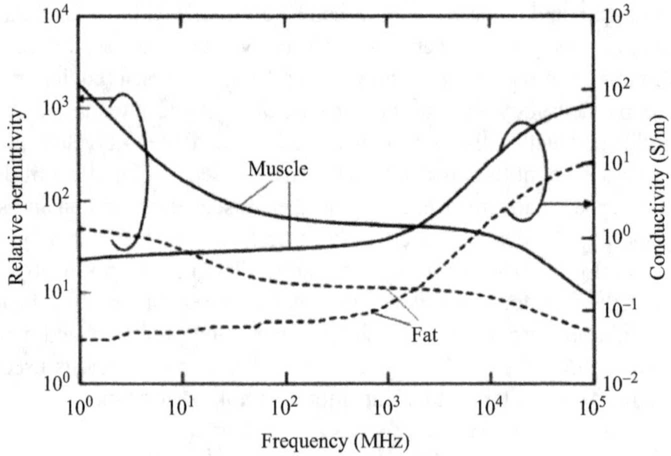

Figure 8.5 Dielectric properties of muscle and fat from 1 MHz to 100 GHz at 37°C [49]

Table 8.1 Dielectric properties of human body tissues at ISM Band – 2.45 GHz [50]

Tissue	ε_r	tanδ	σ (S/m)
Dry skin	38.007	0.28262	1.464
Fat	5.2801	0.14524	0.1045
Muscle	52.729	0.24194	1.7388

Table 8.2 Fully details of the seven models of the CST Voxel family [53]

Model	Age/Sex	Size (cm)	Weight (kg)	Resolution (mm)
Baby	8-week female	57	4.2	0.85×0.85×4.0
Child	7-year female	115	21.7	1.54×1.54×8.0
Donna	40-year female	176	79	1.875×1.875×10
Emma	26-year female	170	81	0.98×0.98×10
Gustav	38-year male	176	69	2.08×2.08×8.0
Laura	43-year female	163	51	1.875×1.875×7.0
Katja	43-year pregnant 24 weeks	163	62	1.775×1.775×4.84

body-worn devices, the main influential tissues are skin, fat and muscle date is depicted in Table 8.1.

8.3.1.2 Numerical phantoms

Theoretical homogeneous or layered flat phantoms are the most basic theoretical phantoms in simple shapes, which are used for running computational simulations and numerical analyses [46,51]. Thanks to the development of different technologies such as X-ray, ultrasound, computed tomography and magnetic resonance imaging (MRI) [52] more complex models have been developed, Voxel models. Among commercial options, Computer Simulation Technology (CST) Studio Suite has a biological database that provides different solutions [53]. CST Voxel family with seven models represent seven persons of different gender, age and stature, Table 8.2. The three-dimensional computer-aided design models of the seven types of individuals are shown in Figure 8.6.

CST also provides a visible human model dataset with more layers and a specific anthropomorphic mannequin, which is for standardised certification purposes. Also, there are available different CTIA (Cellular Telecommunications and Internet Association) head and hand models to simulate different types of devices in several holding situations and again for certification purposes [53], see Figure 8.7.

8.3.1.3 Experimental phantoms

Usually, it is not recommended the use of a real human body for test campaigns due to the complexity of the set-up and health concerns [54]. For that purpose the vast range of experimental phantoms is employed, where they replicate a specific part of

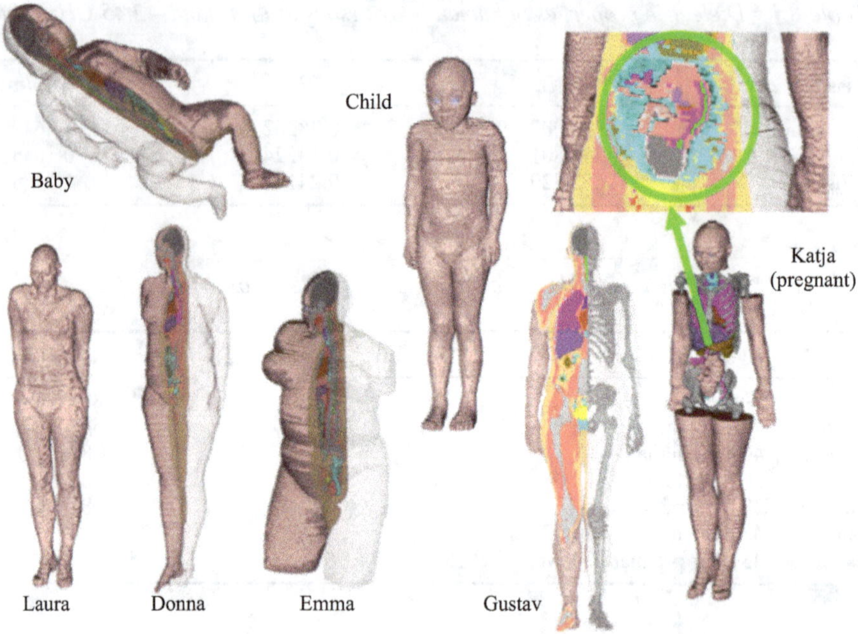

Figure 8.6 CST Voxel family options: baby, child, Gustav, Laura, Katja, Donna and Emma [53]

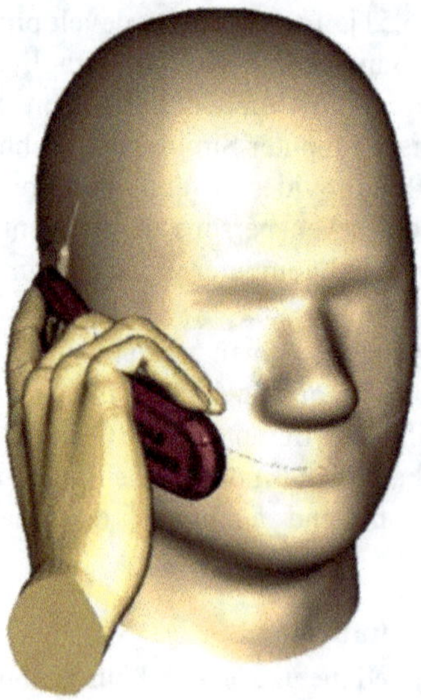

Figure 8.7 Phantom head representation holding a device [53]

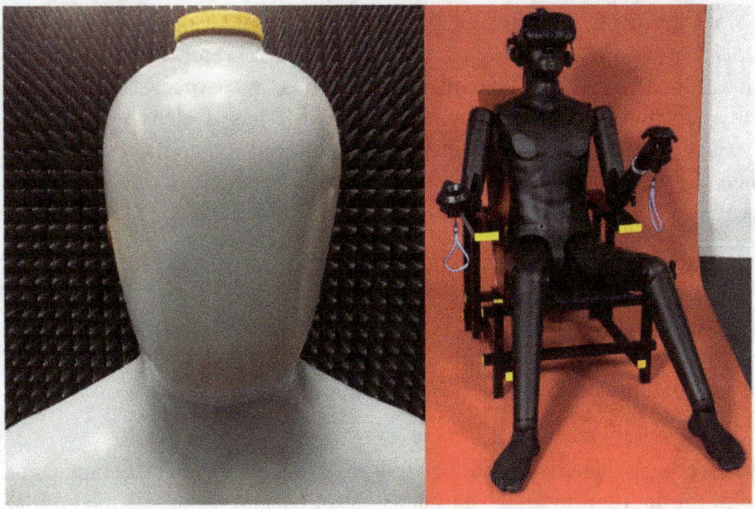

(a) (b)

Figure 8.8 (a) Human bust mannequin and (b) SPEAG model [63]

the human anatomy or the whole body. One type of experimental phantoms are liquid ones [55,56], which usually are homogeneous but they cannot perform over a wide frequency range. They are typically containers filled with a mixture that replicates the dielectric properties of the human tissues that wanted to be mimic [46]. There is not a unique recipe, so different mixtures of solutions were proposed in the research literature [57–59]. In [46] a solution consisted of 79.7% of deionised water 0.25% of sodium chloride, 16% of Triton X-100 (polyethylene glycol monophenyl ether), 4% of diethylene glycol butyl ether (DGBE) and 0.05% of boric acid, was used to fill the phantom depicted in Figure 8.8(a). Other phantoms can be classed as a solid and can be cast into different sizes and shapes. They commonly used deionised water, sodium chloride, agar, sodium dehydroacetate and polyethylene powder TX-150 and TX-151, in various quantities for different solutions [60–62].

Their commercial solutions are available from the Swedish company SPEAG, one of the leaders regarding this area. One of its full model phantoms can be seen in Figure 8.8(b).

8.3.2 Mechanical durability of wearable devices

The fabrics are subjected to multi-directional stress and strain components; thus they behave non-linearly due to the mechanical interaction between human body and garment that depends on the garment style, garment space allowance, dynamic wearing and the posture of the human body [64]. Standard textile testing methods over current 2000 ISO, BS and ASTM standards are very well defined to assess mechanical, thermal and chemical stability of textile structures varying from fibres to the final garments. Due to the lack of mechanical testing standard, specially narrated for wearable technologies, quite often the researchers have followed the methods proposed for the standard textiles. Thus, we expect a similar grade of durability and testing configuration in any textile-based electronic systems that might cause misinterpretation when the electronic integration has drastically influence the mechanical stability. Nevertheless, current textile testing standards include the following:

- Breaking Force and Elongation of Textile Fabrics (ASTM D5035-11:2019).
- Determination of the elasticity of fabrics, Multiaxial test (BS EN ISO 20932-2:2020).
- Determination of the elasticity of fabrics, Strip test (ISO 20932-1:2018).
- Determination of bending length and flexural rigidity of fabrics (BS 3356:1990).

It is very challenging to do quantitative analysis; thus the appearance of creases in the fabric after cleansing is assessed based on the test method BS ISO 7769:2009. Coupled with the electronic component, the textiles' behaviour when subjected to bending, folding and abrasion dominated the durability of e-textile, including smart fabric sensor or transducers, as stated in [65]. Hence, the antenna requires to match draping and haptic qualities of the textiles, especially if it would be worn on the body. References [66,67] have demonstrated that antenna under bending could widen the radiation pattern along the bending plane and amend the impedance matching. These results emphasised the necessity of user case predictions to adapt the communicating element along with the textiles and the integration method, especially on body applications. How body medium impacted the absorption of EM radiation and the SAR distribution as a function of the antenna curvature sizes and different on-body locations and distances was pointed out in [68]. Hence, the bending rigidity of the antenna associated with the textile integration method, including embroidery, printing and lamination, also have an impact on the antenna performance. Thus, stiff and robust embroidered antennas are less affected from the deviations in the bending curve.

The researchers could tune the antenna operational frequency by predicting the bending and tensile performance of textiles, whereas the durability in terms of abrasion, especially in varying temperature and humidity conditions, is more unpredictable. Different test configurations described in the following ASTM D4970/D4970M – 16e3Standard Test Method for Pilling Resistance and Other Related Surface Changes of Textile Fabrics: Martindale Tester, ASTM D3885 – 07a(2019) Standard Test Method for Abrasion Resistance of Textile Fabrics (Flexing and Abrasion Method) and ASTM D4157 – 13(2017) Standard Test Method for Abrasion Resistance of Textile Fabrics (Oscillatory Cylinder Method), and other abrasion standards can be followed to monitor the changes in the post- and pre-antenna performances. A similar approach was conducted in [69] to understand the effect of abrasion on the surface conductivity of nickel–copper yarn-coated nylon ripstop and silver yarn-coated cotton/polyester jersey knit and reported the damage on the surface associated with the increased resistivity.

8.3.3 Specific absorption rate

SAR is a measure widely used to verify how much energy is absorbed by a human body when this is exposed to an electromagnetic field of transmitters operating at frequencies from 300 kHz to 100 GHz. Mobile phone operations and MRI scans are two of the most common areas where this measure is required. It is commonly defined as the power absorbed per mass of human tissue, units of watts per kilogram (W/kg):

$$SAR = \frac{1}{V}\int_{Sample} \frac{\sigma(\boldsymbol{r})|\boldsymbol{E}(\boldsymbol{r})|^2}{\rho(\boldsymbol{r})} d\boldsymbol{r},$$

where σ is the electrical conductivity, $\boldsymbol{E}$ is the electric field, ρ is the sample density and V its volume.

The SAR testing is normally carried out using electric field probes [70], usually dipole antennas, submerged in a phantom that mimics the human tissue to be analysed. Several factors have to be taken into account, such as part of the body that is exposed to RF energy plus its shape, and intended position of RF source. Tests must replicate as accurate as possible the operational environment of the wearable system.

In some cases, temperature probes are placed to quantify temperature increase. Although temperature distributions do not always correlate well with SAR distributions, and that regulatory thresholds on local temperature values may not be surpassed as readily as those on local SAR [71].

Numerous measurement standards and directions, such as the IEC 62209-1 (head) and the IEC 62209-2 (body-worn, separation distance of up to 25 mm can apply), have been released to cover the standardisation of these types of measurements. IEEE STD 1528 IEEE is a recommended practice for measurements and computations with respect to human exposure on electromagnetic fields, from 100 kHz to 300 GHz. The Federal Communications Commission (FCC) proposes the FCC OET 65 Supplement C to evaluate the compliance of human exposure to RF energy. Most of the guidelines are cover in the International Commission on Non-Ionizing Radiation Protection (ICNIRP) 2020 Guidelines [72]. ICNIRP has been recommended by the International Telecommunication Union and World Health Organization.

8.3.4 Durability and washing tests

Sufficient knowledge must be possessed about the performance of the communication element under deformation and chemical agitation to strategically planning out the positioning and the antenna design itself. Textiles especially worn on the body are prone to abrasion and cyclic deformation that is associated with the dynamic wearing and environmental conditions and later care, including laundering. Coupled with the anisotropic textiles, it could be challenging to predict the durability of the communicating element. Reference [69] investigated the effects of pilling, wrinkling, abrasion and laundering of e-textiles by examining the surface resistivity performance from both direct current and radiofrequency perspective.

Proposed micro-strip patch antenna design made of copper yarn-coated nylon ripstop on durable denim fabric was tested by following ASTM D3512 to reveal the effect of pilling on reflection coefficient, S_{11}, and realised gain. The results demonstrated that the gain dropped almost 20 dB due to severe damage on the conductive coating of the fabric. However, wrinkling did not exhibit such a dramatic effect on the conductivity. As revealed by [73], printed conductive tags suffered more than conductive coated fabrics due to the inherent discontinuity along the surface when subjected to abrasion and pilling. The protective measures against environmental effects that included preshrinking of new fabrics to prevent the resonance shift of fabricated antennas over time and textile protective layers to shield water absorption and salt contamination related to the human bodily fluids and the harsh environments and abrasion were discussed in [74]. Particularly in conductive fabrics and inks made of different metals and their derivatives, the antenna performances were drastically influenced by the environmental conditions such as humidity, water/salt absorption and temperature due to oxidation and changes in electrical conductivity. They emphasised the importance of the chemical nature of water repellent coatings and additional security measures about how strong salt solution was capable of dissolving the protective barrier and reached the conductive metallic layer. At cryo temperatures, it was also reported that the ice significantly changed the antenna impedance matching [75]. Thus, we are expecting the same material failures associated with both printed and coated nanomaterial based wearable antennas that might become vulnerable due to external factors, including the temperature changes and humidity.

Despite all the recent efforts in the literature [76,77], from the perspectives of electronics, 'care' of antenna systems when placed on the body is still very intriguing and also highlights the practical limitations of antenna design. Lacking e-textile durability standards that include laundering or pilling makes the design impractical to compare different textile-based systems either made of conductive inks, yarns or fabrics. Nevertheless, we had clues how various textile forms ensured washing durability. On the first attempts on revealing the durability of laundering, Scarpello *et al.* [78] imposed a breathable thermoplastic polyurethane coating to protect the screen printed microstrip inset-fed patch antenna from water absorption and corrosion. They introduced the protective layer by ironing and explored the two competing parameters, reduced conductivity due to melted polymer and diminished roughness that had a positive contribution by reducing the surface resistivity. It is evident that applied thermal post-processes, including the ironing, led to more compact structures, especially working with conductive ink-based surfaces [78]. Either using metal-deposited fabrics or conductive inks, the mainstream approach is to apply superhydrophobic coatings to inhibit water absorption even after multiple wash cycles [79]. In particular, nanomaterial-based inks or screen-printed surfaces are susceptible to water migration and contamination through the cracks, where the pores are unavoidable due to the textile manufacturing method. Thus, washing durability of the textile antenna is strongly dependent on the physical and chemical absorption mechanism and surface quality, including cracks and uniformity. Even it is not inevitable, the antenna performance could be predicted to some extent. For instance, Carla *et al.* [80] revealed how increasing moisture could increase the permittivity and loss tangent of the textile material that led to a decrease in antenna efficiency by changing the resonance frequency. In [46] was reported the washing durability of graphene-based ultra-wideband (UWB) textile antennas by following the standard protocol in BS EN ISO 6330.

Reference [74] pointed out the uncertainties in the antenna modelling associated with the textile substrates and suggested the following protocol (i) modelling of conductive fabrics/patches or surfaces using impedance models to predict non-ideal conductivity, (ii) understanding the changes in permittivity either by washing or other external factors that could change the air gaps or the volume of air trapped in textile structure, (iii) assembly of the wearable antenna system, for instance attaching conductive layers to the textile. These points highlighted the heterogeneous medium along with the challenges in standardisation of e-textiles, including the antennas. Nevertheless, the current efforts on the standardisation of wearable electronic devices and technologies include a broad scope of testing methods such as BS EN IEC 63203-201-1: measurement methods for basic properties of conductive yarns, BS EN IEC 63203-201-2: measurement methods for basic properties of conductive fabric and insulation materials, BS EN IEC 63203-201-3: determination of electrical resistance of conductive textiles under simulated microclimate, BS EN 63203-204-1: washable durability test method for leisure and sportswear e-textile system and a standardisation in the terminology, BS EN 63203-101-1. Wearable electronic devices and technologies. Part 101-1. Terminology. 'Wearable electronics' coupled with the textiles is a still-developing field that has the potential to evolve more mature systems and technologies as in PD IEC/TR 62899-250:2016 Printed electronics. Material technologies required in printed electronics for wearable smart devices. More remarkably, when planned out to position the nanomaterial-based antenna systems either on the body or in the proximity of the human, it is highly recommended to follow the guidance on performing tests to simulate nanoparticle release, PD CEN/TR 17222:2019S: textile products and nanotechnologies: skin exposure and addressed the prospective safety issues for wearable applications.

8.3.5 Environmental factors (humidity and thermal tests)

Moisture and temperature are the two most common environmental factors to be analysed within wearable systems with respect to their impact on antenna's performance. The effect of these two factors can be measured following two different approaches – first by characterising the variations of the parameter itself, and second by measuring the alteration of the antenna's performance due to the change of the parameter – in both cases under a controlled environment.

Regarding the first case, dedicated methods previously described in the literature [81–83] are used to measure specific constitutive material parameters of the antenna, such as conductivity, σ, permeability, μ, and permittivity, ε, under variation of the environmental conditions. In the case of electrical conductivity, the standard procedure is to use a multimeter or a more precise instrument like a four-probe set-up and considering the geometry of sample [81]. For material characterisation in terms of dielectric properties, several methods are available, from microwave theory, they can be split up into two complementary types: non-resonant and resonant [82,83]. Within these two kinds, there are several sub-categories with their respective specific tools and apparatus. Each method has its pros and cons, so a trade-off is necessary for each case, but a general rule is to use non-resonant methods for wideband measurements and resonant ones for concrete frequencies. The second mechanism is based on measuring the changes of the antenna performances under certain environmental conditions, seeing the indirect impact of the changes from the parameters previously mentioned.

For instance, relative humidity (RH) has an impact on the antenna's behaviour by increasing its dimensions (substrate's thickness) [84], as well as rising the dielectric constant value due to moisture absorption [85]. These characteristics have motivated several research studies regarding the use of antennas as humidity sensors, mainly focusing on wearable technology [86,87]. In the case of temperature, generally, conductivity increases as materials get colder, as there is less atomic activity bouncing and the electrons can move more freely. It will affect the ε_r as well and consequently will influence the resonant frequency (f_r) of the device. Rising temperature in solid materials permits the dipoles of the molecules more free movement (like the case of electrons), conveying to an increase in the permittivity value. In cases with less restrains, such as liquids and gasses, this has an adverse effect [45].

For all cases, environmental factors can be taken into consideration during numerical analysis by performing parameters sweep in commercially available full-wave electromagnetic simulation software (CST Studio Suite). Other programs such as COMSOL Multiphysics can run multi-physics analysis, where an environmental factor coefficient can be inserted. For the purpose of test campaigns, several commercial solutions of environmental chambers, also known as climatic or climate chambers, are available.

These chambers are capable of artificially replicating different environmental conditions for a device under test (DUT). This environmental test equipment is able to accelerate the changes in the conditions or extreme situations in a controlled manner recording all the data. Several conditions can be manipulated such as temperature, moisture, UV and/or vacuum. Figure 8.9 depicts the diagram of a set-up to measure *S*-parameters of an antenna, where the DUT is placed inside the climatic chamber and then connected to the analysing equipment (in this case a VNA), and Figure 8.9(b) shows one of the commercial options. Depending of the test to be carried out, more bespoke solutions can be set [45].

8.4 Wearable antennas for 5G and beyond

8.4.1 Ultra-wideband frequency-reconfigurable antenna

A UWB-wearable reconfigurable antenna is presented in this section. The proposed configuration combines frequency and pattern reconfiguration on textile materials

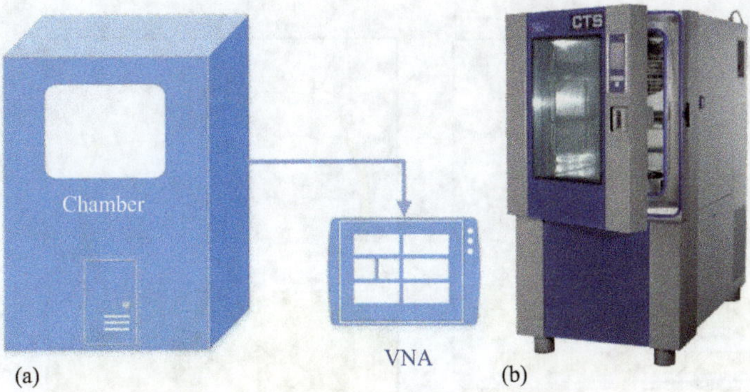

Figure 8.9 (a) Diagram of a thermal set-up and (b) CTS commercial ambient chamber [88]

and suitable for integration as a part of a military uniform or any clothing. Two different methods have been adopted to fabricate the patch and ground on a textile substrate. One method is conductive thread embroidery that was to fabricate the antenna on fabrics such as leather and cotton as a substrate. However, in the second method, a thin copper sheet was cut by using a laser cutter and mounted on the substrate. Commercial self-adhesive copper textile sheets were used.

The geometry of the antenna is based on the design proposed in [89,90]. Figure 8.10 presents the design proposed in [89,90], though the antenna dimensions have been optimised according to the dielectric properties of the textile substrates [91]. The proposed antenna offers both frequency and pattern reconfiguration, where the stubs length from the front part reconfigured the frequency while pattern reconfiguration was done by the two semi-elliptical stubs located on the back side. Connecting the ground plane with any of the stubs changes the radiation pattern of the antenna. The connected stub acts as a reflector and shifts the beam towards the opposite direction. As depicted in Figure 8.10, by using these two stubs, the radiation pattern can be reconfigured to the $+x$-axis and $-x$-axis of the antenna geometry. The dimensions of the proposed textile antenna are tabulated in Table 8.3 [91].

The 50-Ω matched feedline is designed for all the proposed configurations. In the first configuration, no stub was connected to the feedline and the antenna was tuned in UWB configuration that covers approximately 3–8 GHz. The proposed design in [89–91] reconfigured the frequency in six different bands. Table 8.3 depicts the designed dimensions and Table 8.4 presents the obtained bandwidth with respect to the proposed switch configuration as shown in Figure 8.10.

The simulation was done in CST Studio Suite. The pattern reconfiguration is performed with the semi-elliptical stubs included in the ground geometry. Each stub acts like a reflector to switch the radiation beam of the antenna to the opposite side when connected to the ground. Therefore, three modes are generated: (i) Mode 0: when no stub attached with the ground plane; (ii) Mode 1: the switch on left side

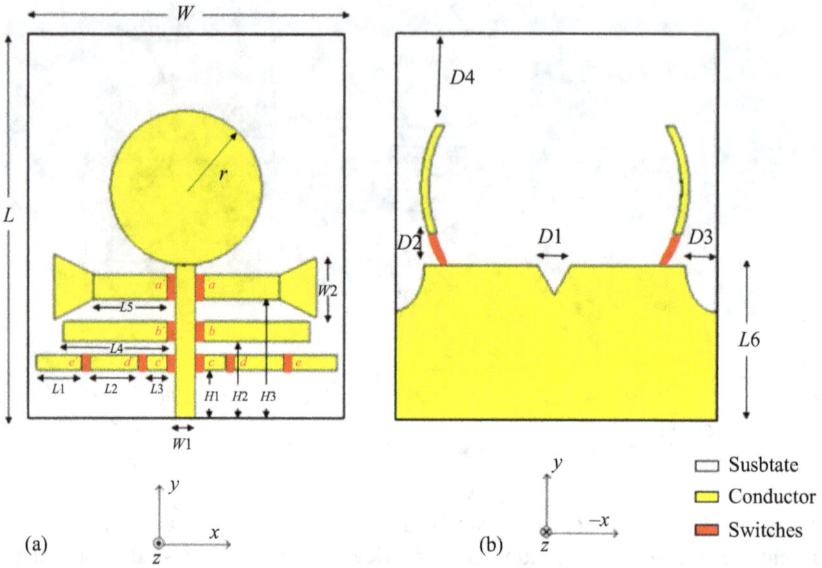

Figure 8.10 The proposed multi-mode reconfigurable ultra-wideband antenna: (a) front view, (b) back view [91]. Reproduced, with permission, from Syeda Fizzah Jilani, EuCAP 2016; published by IEEE, 2020.

Table 8.3 Dimensions of the proposed textile-based antenna

Dimensions	mm	Dimensions	mm
Radius of the patch, *r*	10	Ground length, *L*6	20
Width of the antenna, *W*	42	First distance from bottom edge, *H*1	6.3
Width of the patch, *W*1	3	Second distance from bottom edge, *H*2	10
Stub width, *W*2	8	Third distance from bottom edge, *H*3	15
Length of the antenna, *L*	8	Length of V-cut, *D*1	4.5
First stub length, *L*1	8	Distance of reflector, *D*2	4
Second stub length, *L*2	4	Distance of reflector, *D*3	4
Third stub length, *L*3	18	Distance, *D*4	8
Fourth stub length, *L*4	10	Substrate thickness	1
Fifth stub length, *L*5	10	Conductive path thickness	0.08

is connected (ON) and the switch on right side is OFF (not connected); and (iii) Mode 2: the switch on left side is OFF and the switch on right side is ON, which results in $\pm 90°$ beam shift from broadside to the $+x$-axis and $-x$-axis.

Five textile-based antenna prototypes were fabricated and the effect of materials on the antenna performance was investigated. In first two antennas, the conductive sewing thread embroidery using an embroidering machine was employed as shown in Figure 8.11. Next three antennas were fabricated by using VLS6.60 laser cutter to cut the metallic pattern from thin conductive textile sheets

Table 8.4 Switch configuration and respective applications [91]

Switch configuration	Switches ON	Bands (GHz)	Application
Ultra-wideband	None	3–8	Cognitive radio
1	*c*, *c′*	2.4	ISM (industrial, scientific and medical) [10]
2	*c*, *c′*, *d*, *e*	3.2–4.2	Wi-Max (Worldwide Interoperability for Microwave Access) [11]
3	*a*, *a′*	4.1–5.4	Some fixed mobile and satellites applications, Wi-Fi (Band A)
4	All	5	Wi-Fi
5	*b*, *b′*	6.5–7.5	Some fixed mobile and satellites applications
6	*c*, *c′*, *d*	2.4; 5	ISM and Wi-Fi

(a) (b)

Figure 8.11 Front and back view of two embroidered: (a) antenna on leather and (b) antenna on cotton [91]. Reproduced, with permission, from Syeda Fizzah Jilani, EuCAP 2016; published by IEEE, 2020.

Table 8.5 Details of the fabricated prototypes of antennas [91]

Fabricated antenna	Substrate used	Material used
Antenna 1	Cotton denim (ε_r=1.2)	Conductive sewing thread
Antenna 2	Cotton calico (ε_r=1.2)	Conductive sewing thread
Antenna 3	Cotton denim	Copper fabric
Antenna 4	Cotton calico	Copper fabric
Antenna 5	Leather (ε_r=2.3)	Copper fabric

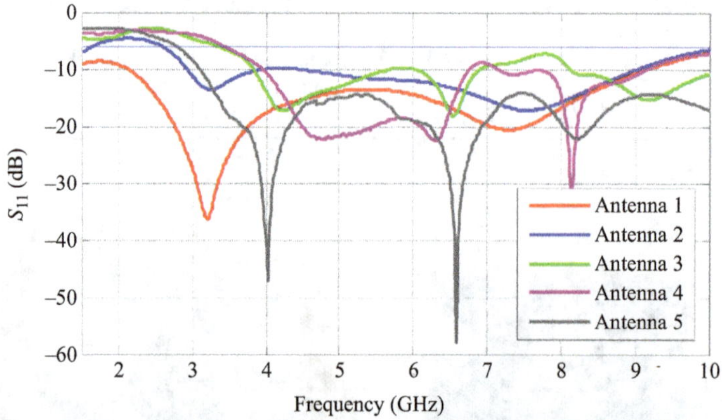

Figure 8.12 Measured S_{11} plots of the proposed textile-based antennas at the ultra-wideband configuration [91]. Reproduced, with permission, from Syeda Fizzah Jilani, EuCAP 2016; published by IEEE, 2020.

and sticking it to the fabric with conductive glue. The fabrication details of the antennas are provided in Table 8.5. The conductive epoxy is used to connect the antenna stubs by creating a short circuit as an ideal switch.

The reflection coefficients, S_{11}, of the proposed antennas are measured by using a VNA. The measured S_{11} for antennas 1–5 is shown in Figure 8.12 in UWB configuration.

Based on the testing of the fabricated antenna, Table 8.6 was developed to tabulate the obtained corresponding active bands. It is observed that antennas 1, 2 and 5 cover maximum number of bands in different switch configurations.

8.4.2 Millimetre-wave flexible antenna design and metrology

In this work, a flexible beam scanning mm-wave array has been introduced. The design methodology of the proposed antenna has been derived from the series arrays; widely reported as fixed-beam arrays, frequency scanning and leaky-wave antennas (LWAs) [92]. The collinear principle, by the Franklin's collinear arrays of patch antennas demonstrates that by using non-radiating quarter-wave stubs for

Table 8.6 The operating bands obtained from the fabricated antennas in different switch configurations [91]

Antenna	Operating bands at different switch configurations
Antenna 1	Ultra-wideband, 3.2–4.2, 4.2–5.4, 6.5–7.5 GHz
Antenna 2	Ultra-wideband, 3.2–4.2, 4.2–5.4, 6.5–7.5 GHz
Antenna 3	Ultra-wideband, 2.4 GHz
Antenna 4	Ultra-wideband, 4.2–5.4 GHz
Antenna 5	Ultra-wideband, 2.4, 4.2–5.4, 6.5–7.5 GHz

180° phase shift in a series of patch antennas, the original out-phase current distribution can be converted into an in-phased current on each collinear segment that provides one major radiating beam in a broadside direction [93–95]. Several modifications have been reported in Franklin arrays concept in order to achieve high-gain characteristics [96,97]. Besides, these arrays typically offer narrow bandwidth, as the design geometry is particularly based on the quarter-wavelength of the selected centre/resonant frequency. This work improved this limitation and presented a flexible and wideband LWA with frequency scanning. The proposed flexible antenna array retains the high gain of conventional Franklin array with the additional features of beam steering over a wide bandwidth.

In typical Franklin array assembly, each unit cell consists of a patch and a transmission line or a stub designed with the radiating length of quarter-wavelength ($\lambda/4$) (i.e. $f=c/\lambda$, where f=resonant frequency, c=speed of light and λ=wavelength). These unit cells are repeated to construct a linear array. In this work, array bandwidth is improved by changing the design lengths of the geometry with respect to the lower and upper cut-off frequencies of the intended bandwidth. Thus, the patch length is $\lambda_L/4$ and stub length is $\lambda_U/4$ in order to generate two individual resonant bands at lower cut-off (f_L) and upper cut-off frequency (f_U). In addition, the operating bandwidth between the two designed resonances is adjusted by means of parametric analysis of the patch width, stub width and the gap between the folded stubs [16].

The optimised dimensions of antenna design shown in Figure 8.13 are provided in Table 8.7. Simulation of the proposed mm-wave antenna array is carried out using the CST Microwave Studio software. The six-element array is designed on the PET substrate of $12\times28.2\times0.4$ mm^3 with a continuous metallic ground at the bottom. Height of substrate (h) is 0.4 mm for good return loss magnitudes while retaining the desired flexibility of the substrate. Designed array is fabricated on a surface-treated PET film (ε_r=3.2, tanδ=0.022 at 10 GHz) from Mitsubishi Paper Mills with silver nanoparticle ink by inkjet printing process. The drop-spacing of 15 μm (i.e. 1,693.33 dpi), firing voltage of 15 V and jetting frequency of 5 kHz have been set on the Dimatix Material Printer. Three PET sheets each of height of 135$\pm$12 μm are bonded together by means of double-sided tape to attain h=0.4 mm. The nanoparticle silver ink was used for printing and a conductive layer with a thickness of 0.5 μm was deposited on a PET substrate with a pattern resolution of $\pm$20 μm. Post-printing drying, curing and sintering processes were performed to improve the conductivity of

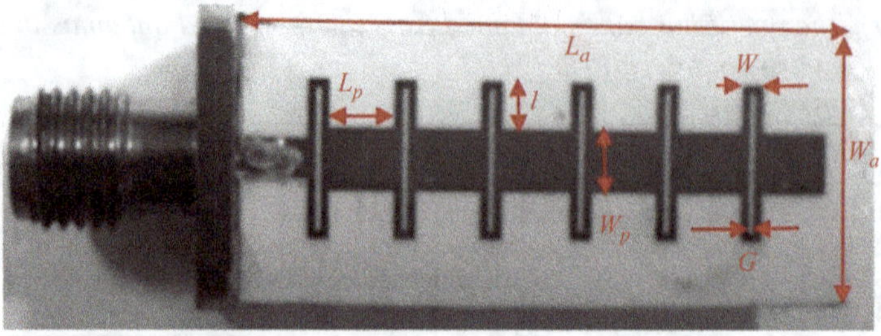

Figure 8.13 Proposed mm-wave leaky-wave antenna (LWA) mm-wave frequency-scanning array fabricated prototype [16]. Reproduced, with permission, from Syeda Fizzah Jilani, IEEE AP-S Symposium; published by IEEE, 2020.

Table 8.7 Optimised dimensions of antenna design

Design parameters	mm	Design parameters	mm
Patch length, $L_p=\lambda_L/4$, $f_L=24$ GHz, $\lambda_L=12.5$ mm	3.1	Length of the folded stub, $2l+w=2\times(\lambda_U/4)$, $f_U=30$ GHz, $\lambda_U=10$ mm	5.2
Patch width, W_p	2.7	Gap, G	0.3
Length of the stub, $l\approx\lambda_U/4$	2.15	Feed width, w	0.9

the printed layer. The selected PET substrates are coated with a chemically active microporous adhesive that provides rapid drying and chemical sintering of the ink. This chemical method avoids a need of laser exposure or high-temperature sintering treatments that may cause bending, shrinking, decolourisation or melting, if the exposure energy or temperature is not appropriately handled.

The S_{11} plot of Figure 8.14 shows a good agreement between the measured and simulated results, hence validating the proposed findings of the designed antenna array. Measurements have shown that a bandwidth of 24.6–30 GHz has been achieved. Figure 8.15 depicts that the directive beam scans the spherical quadrant from the broadside to end-fire direction, as the frequency sweeps from 24.6 to 30 GHz. Figure 8.16 shows the numerically evaluated realised gain and efficiency vs. frequency of proposed array. The peak gain is 8.3 dBi at 28 GHz, and efficiency is approximately 60% in the overall operating range of the antenna. This flexible antenna is a good contribution in the development of the body-centric devices for future medical, security and many other communication applications because of its distinctive features.

8.4.3 On-body measurements on wearable antennas

In [46] a graphene textile-based UWB antenna was designed to cover a wide bandwidth ranging from 3 to 9 GHz. In this study, a numerical and experimental

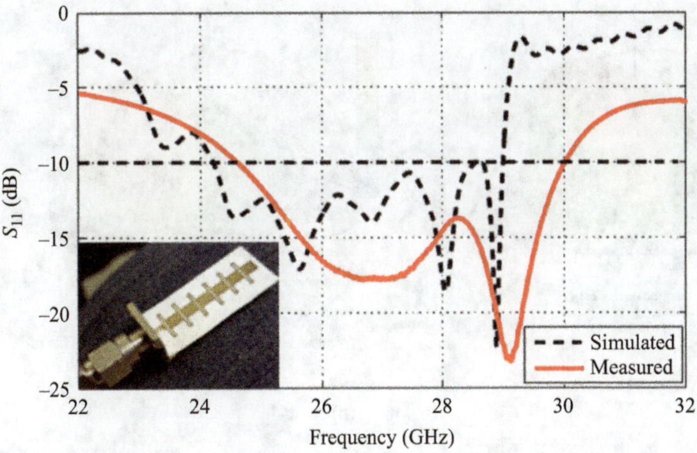

Figure 8.14 Simulated and measured S_{11} plots of proposed mm-wave frequency-scanning array [16]. Reproduced, with permission, from Syeda Fizzah Jilani, IEEE AP-S Symposium; published by IEEE, 2020.

analysis of the antenna performance under the presence of a phantom model of human body parts was carried out.

For the simulations, a block of four layers in total, 44-mm thick (the thickness of body skin, fat and muscle layers were 1, 3 and 40 mm, respectively), was used. The electromagnetic properties employed of the different tissues – dielectric constant, ε_r, loss tangent, tanδ, conductivity, σ, and resistivity, ρ – are listed in Table 8.1 [50,98]. A 2D representation of the model is depicted in Figure 8.17(c). In order to look into the effects of the human body experimentally, a torso phantom loaded with a liquid to replicate human tissues was used. The solution selected is explained in Section 8.3.1.3; 79.7% of deionised water, 0.25% of sodium chloride, 16% of Triton X-100, 4% of DGBE and 0.05% of boric acid, which stuffed the torso phantom shown in Figure 8.17(b).

A test campaign was carried out at the Antenna Measurement Laboratory at Queen Mary University of London [99] with the aim to evaluate the performance of the antenna in the vicinity of a human phantom. For measuring the return losses, a VNA, PNA-L Agilent N5230C, was connected to the antenna under test (AUT). This set-up was calibrated with a standard electronic calibration kit to the end of the coaxial cable to have the same initial reference for all the measurements, for reproducibility purposes. For calculating the radiation properties of electromagnetic compatibility (EMC)-screened anechoic chamber equipped with two open boundary quad-ridge horn antennas operating from 400 MHz to 6 GHz (ETS-Lindgren 3164-06) and from 0.8 to 12 GHz (Satimo QH800), granting measurements in both vertical and horizontal linear polarisations were used, Figure 8.17(a).

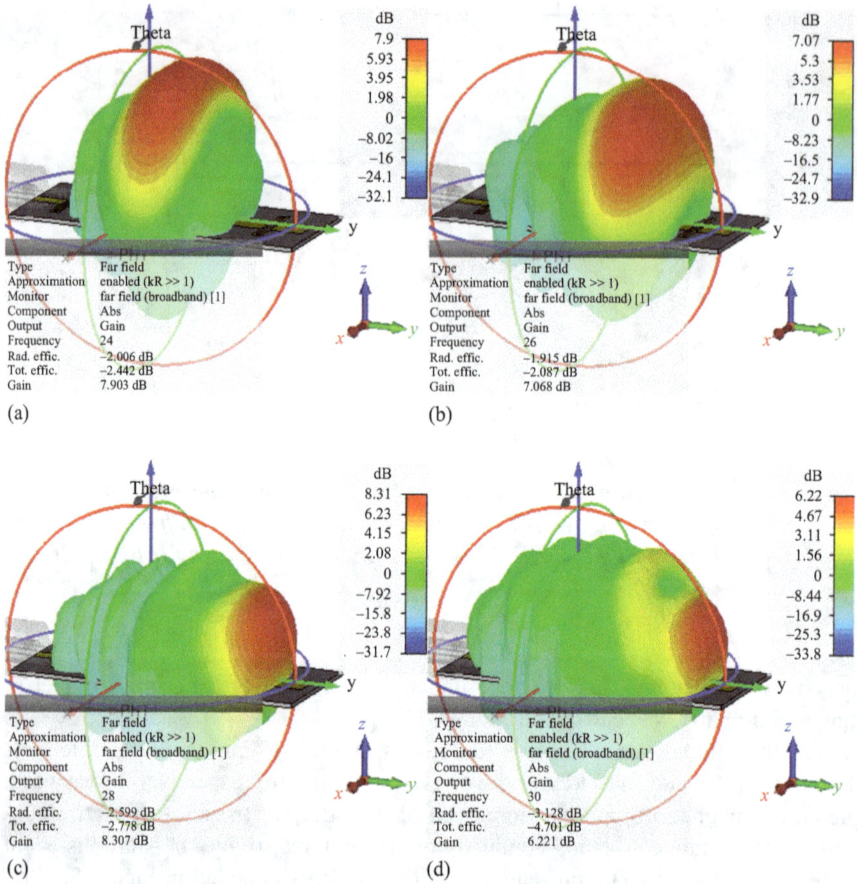

Figure 8.15 The simulated EM radiation patterns of proposed mm-wave frequency-scanning array: (a) at 24 GHz; (b) at 26 GHz; (c) at 28 GHz; (d) at 30 GHz [16]. Reproduced, with permission, from Syeda Fizzah Jilani, IEEE AP-S Symposium; published by IEEE, 2020.

Figure 8.18 shows that the simulated on-body scenario exhibited a return loss of less than 10 dB in a wider range (2–12 GHz) in comparison with the off-body case (3–11.5 GHz). The slight frequency detuning and this increase in bandwidth could be due to the dissipative nature of the medium, the human body. It can be seen that the correlation between simulated and measured results, thus validating the model used.

Figure 8.19(a)–(f) presents the measured and simulated radiation patterns of the AUT for the on-body scenario, for frequencies: 3, 5 and 7 GHz. As well as for the return losses, the far-field measurement results agreed significantly with the simulated ones. However, due to the absence of a background plane that shields the antenna from the body, the antenna radiation properties were being affected.

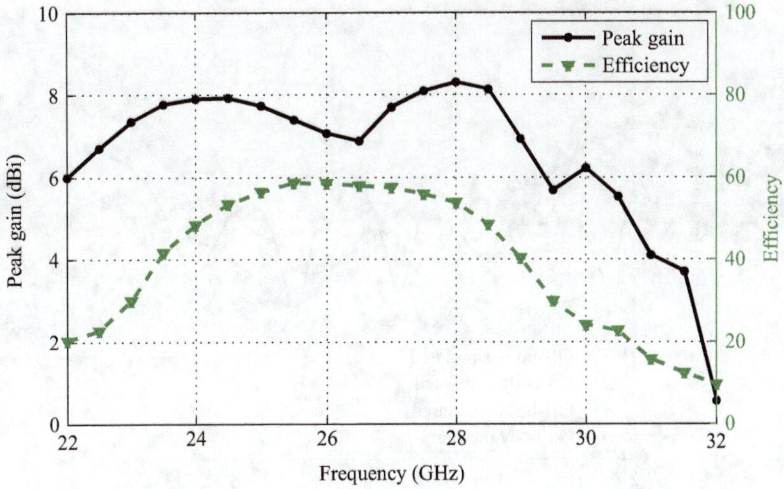

Figure 8.16 Peak realised gain and efficiency vs. frequency of proposed mm-wave frequency-scanning array [16]. Reproduced, with permission, from Syeda Fizzah Jilani, IEEE AP-S Symposium; published by IEEE, 2020.

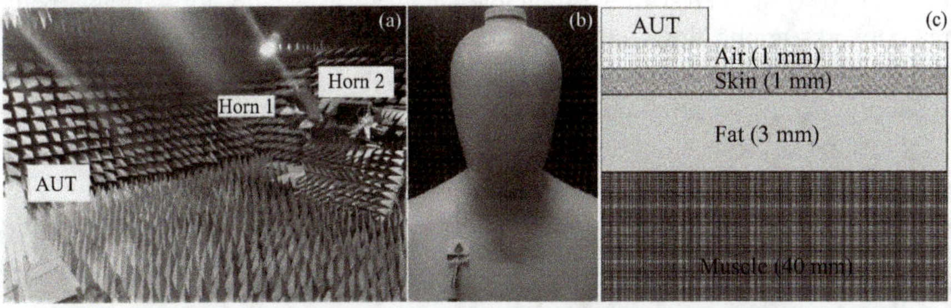

Figure 8.17 (a) EMC chamber set-up used in this study, (b) an image showing an AUT on a phantom and (c) four-layer phantom model used for on-body simulations [46]. Reproduced, with permission, from IIL, Applied Materials Today; published by Elsevier, 2020.

The front-to-back ratio (*F*/*B* ratio) rises as the phantom absorbed the reflected waves that accompanied radiation at the front furthermore, which can be noticed that the experimental phantom absorbed more energy than the numerical model. This deviation in the expected behaviour is due to the complexity of exactly replicating the human model with all its variables.

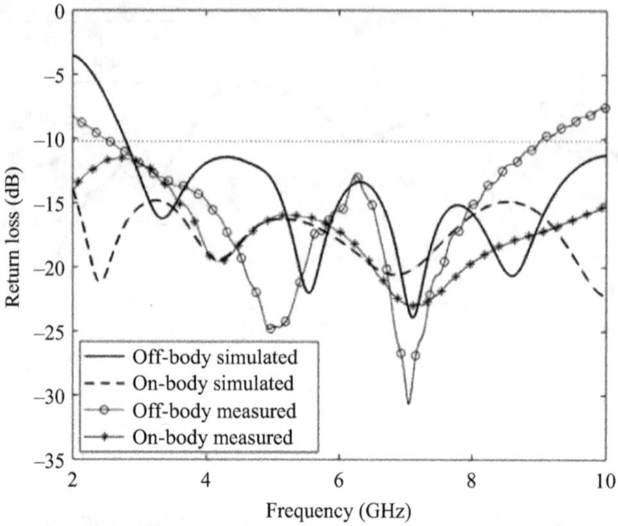

Figure 8.18 Simulated vs. measured S_{11} of graphene-based antennas in off- and on-body settings [46]. Reproduced, with permission, from IIL, Applied Materials Today; published by Elsevier, 2020.

8.5 Conclusions

With the growing interest in 5G wearable antennas, especially for wearable body area networks applications, novel materials and fabrication techniques are being investigated and they were discussed in this chapter. Different methods such as screen printing, inkjet printing, embroidery, used conductive yarns and fabrics, and cutters were explained and evaluated. Each manufacturing technique was analysed in detail considering materials that can be used and the properties obtained for the antennas fabricated by that process. Every method has been supported by examples and references to facilitate the comparison between techniques and to highlight their advantages and features.

Current developments in the field focus on improving the efficiency and precision of the manufacturing techniques, employing new flexible materials as a substrate, orientated towards higher frequency bands like the mm-wave band that demands high accuracy and precision. For this purpose, additive manufacturing techniques such as inkjet printing stands out and besides reduce the waste of material. Although in certain cases post-processing is required and that proposes a challenge for some materials like textiles. Novel conductive materials such as yarns or cloths and carbon-based ones like graphene, with a high conductivity, can be applied for the new fabrication and printing techniques. Graphene and other carbon-based solutions are promising alternatives to conventional metallic materials on the market, possess various properties, including electrical and thermal conductivity, optical transmittance and high strength/stiffness.

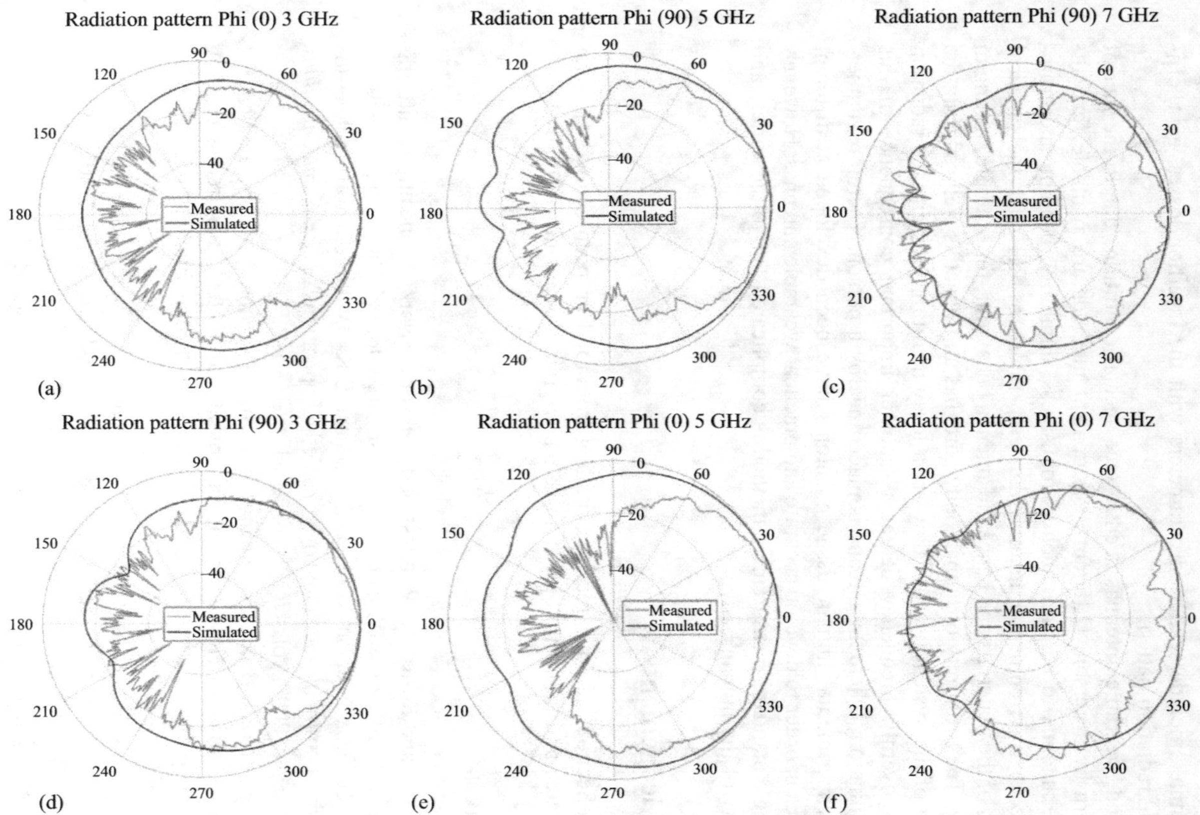

Figure 8.19 *Measured vs. simulated radiation pattern of the proposed graphene-based antenna on the phantom at: E-plane cut, at φ=90°: (a) 3 GHz, (b) 5 GHz and (c) 7 GHz; H-plane cut, at φ=0°: (d) 3 GHz, (e) 5 GHz and (f) 7 GHz [46]. Reproduced, with permission, from IIL, Applied Materials Today; published by Elsevier, 2020.*

Metrology and characterisation of wearable antennas require two validation approaches due to the dynamic environment of being worn. The first one is the same as for conventional antennas that cover fundamental parameters in order to verify the performance correlation between fabricated and simulated designs. These features include scattering parameters with the insertion losses, gain, efficiency and radiation patterns.

Second, wearable antennas need to face special conditions due to the fact of being worn by a person. Different measurement methods are available for validating the antenna to be worn such as robustness, durability, bending and crumpling effects. Environmental factors like temperature or RH alter the physical properties of the materials and hence the antenna behaviour, for example shifting in frequency. The proximity of human body parts has an impact as well reducing the efficiency of the system because they are a lossy medium, affecting the radiation properties overall. Frequency detune is another important consequence of this specific situation, where an antenna is placed nearby a person. Therefore, numerical and experimental human–tissue phantoms have been developed at different frequencies in order to develop accurate and practical wearable antennas. However, developing a standard measurement method is complicated due to the variety and complexity of the human body.

Acknowledgements

Project in Section 8.4.3 has received funding from the European Union's Horizon 2020 research and innovation programme under grant agreement no. 796640.

References

[1] Hao Y. and Foster R. 'Wireless body sensor networks for health-monitoring applications'. *Physiological Measurement*, 2008, vol. 28(11), pp. 27–56.

[2] Cotton S.L., Scanlon W.G., and Madahar B.K. 'Millimeter-wave soldier-to-soldier communications for covert battlefield operations'. *IEEE Communications Magazine*, 2009, vol. 47(10), pp. 72–81.

[3] Pellegrini A., Brizzi A., Zhang L., *et al.* 'Antennas and propagation for body-centric wireless communications at millimeter-wave frequencies: A review [wireless corner]'. *IEEE Antennas and Propagation Magazine*, 2013, vol. 55(4), pp. 262–287.

[4] Rappaport T.S., Sun S., Mayzus R., *et al.* 'Millimeter wave mobile communications for 5G cellular: It will work!'. *IEEE Access*, 2013, vol. 1, pp. 335–349.

[5] Yang Z., Pathak P.H., Zeng Y., Liran X., and Mohapatra P. 'Monitoring vital signs using millimeter wave'. *Proceedings of the IEEE MobiHoc'16*, 2016, pp. 211–220.

[6] Zeng Y., Pathak P.H., Yang Z., and Mohapatra P. 'Human tracking and activity monitoring using 60 GHz mm-wave'. *15th ACM/IEEE International*

Conference on Information Processing in Sensor Networks (IPSN), Vienna, 2016, pp. 1–2.

[7] Chahat N., Tang A., Guraliuc A., Zhadobov M., Sauleau R., and Valerio G. 'Antennas, phantoms, and body-centric propagation at millimeter-waves' in Werner D.H. and Jiang Z.H. (eds.). *Electromagnetics of Body Area Networks: Antennas, Propagation, and RF Systems*, Hoboken, NJ, Wiley, 2016, pp. 205–259.

[8] Leduc C. and Zhadobov M. 'Impact of antenna topology and feeding technique on coupling with human body: Application to 60-GHz antenna arrays'. *IEEE Transactions on Antennas and Propagation*, 2017, vol. 65(12), pp. 6779–6787.

[9] Nechayev Y.I., Wu X., Constantinou C., and Hall P.S. 'Millimetre-wave path loss variability between two body-mounted monopole antennas'. *IET Microwaves, Antennas & Propagation*, 2013, vol. 7(1), pp. 1–7.

[10] Hall P.S. and Hao Y. *Antennas and Propagation for Body-Centric Wireless Communications*, Norwood, MA, Artech House, 2006.

[11] Conway G.A. and Scanlon W.G. 'Antennas for over-body-surface communication at 2.45 GHz'. *IEEE Transactions on Antennas and Propagation*, 2009, vol. 57(4), pp. 844–855.

[12] Chahat N., Zhadobov M., Le Coq L., Alekseev S.I., and Sauleau R. 'Characterization of the interactions between a 60-GHz antenna and the human body in an off-body scenario'. *IEEE Transactions on Antennas and Propagation*, 2012, vol. 60(12) pp. 5958–5965.

[13] Chahat N., Zhadobov M., Muhammad S.A., Le Coq L., and Sauleau R. '60-GHz textile antenna array for body-centric communications'. *IEEE Transactions on Antennas and Propagation*, 2013, vol. 61(4), pp. 1816–1824.

[14] Chahat N., Zhadobov M., Le Coq L., and Sauleau R. 'Wearable end-fire textile antenna for on-body communications at 60 GHz'. *IEEE Antennas and Wireless Propagation Letters*, 2012, vol. 11, pp. 799–802.

[15] Wu X.Y., Akhoondzadeh L., and Hall P.S. 'Printed Yagi-Uda array for on-body communication channels at 60 GHz'. *Microwave and Optical Technology Letters*, 2011, vol. 53(12), pp. 2728–2730.

[16] Jilani S.F. and Alomainy A. 'Millimeter-wave conformal antenna array for 5G wireless applications'. *IEEE AP-S Symposium*, San Diego, CA, USA, 2017.

[17] Jilani S.F. and Alomainy A. 'An inkjet-printed MMW frequency reconfigurable antenna on a flexible PET substrate for 5G wireless systems'. *LAPC IET*, Loughborough, UK, 2017, pp. 13–14.

[18] Ur-Rehman M., Malik N.A., Yang X., Abbasi Q.H., Zhang Z., and Zhao N. 'A low profile antenna for millimeter-wave body-centric applications'. *IEEE Transactions on Antennas and Propagation*, 2017, vol. 65(12), pp. 6329–6337.

[19] Simorangkir R.B.V.B., Yang Y., and Esselle K.P. 'Performance of embroidered higher-order mode antennas with different stitching patterns'. *11th European Conference on Antennas and Propagation (EUCAP)*, Paris, 2017, pp. 177–180.

[20] Yoshikawa T. and Maeda T. 'Embroidered wideband wearable slot antennas using conductive yarn with different yarn-spacing densities'. *International Workshop on Electromagnetics: Applications and Student Innovation Competition (iWEM)*, Hsinchu, 2015, pp. 1–2.

[21] Elobaid H.A.E., Abdul Rahim S.K., Himdi M., Castel X., and Abedian Kasgari M. 'A transparent and flexible polymer-fabric tissue UWB antenna for future wireless networks'. *IEEE Antennas and Wireless Propagation Letters*, 2017, vol. 16, pp. 1333–1336.

[22] Khan S., Lorenzelli L., and Dahiya R.S. 'Technologies for printing sensors and electronics over large flexible substrates: A review'. *IEEE Sensors Journal*, 2015, vol. 15(6), pp. 3164–3185.

[23] Shamim A. '3D inkjet printed flexible and wearable antenna systems'. *International Symposium on Antennas and Propagation (ISAP)*, Phuket, 2017, pp. 1–2.

[24] Hettak K., Ross T., James R., Momciu A., and Wight J. 'Flexible polyethylene terephthalate-based inkjet printed CPW-fed monopole antenna for 60 GHz ISM applications'. *European Microwave Conference*, Nuremberg, 2013, pp. 1447–1450.

[25] Maza A.R., Cook B., Jabbour G., and Shamim A. 'Paper-based inkjet-printed ultra-wideband fractal antennas'. *IET Microwaves, Antennas & Propagation*, 2012, vol. 6(12), pp. 1366–1373.

[26] Olkkonen J., Lehtinen K., and Erho T. 'Flexographically printed fluidic structures in paper'. *Analytical Chemistry*, 2010, vol. 82(24), pp. 10246–10250.

[27] Eshkeiti A., Reddy A.S.G., Emamian S., *et al.* 'Screen printing of multilayered hybrid printed circuit boards on different substrates'. *IEEE Transactions on Components, Packaging and Manufacturing Technology*, 2015, vol. 5(3), pp. 415–421.

[28] Xiao G.G., Zhang Z, Lang S., and Tao Y.'Screen printing RF antennas'. *17th International Symposium on Antenna Technology and Applied Electromagnetics (ANTEM)*, Montreal, QC, 2016, pp. 1–2.

[29] Jilani S.F., Rahimian A., Alfadhl Y., and Alomainy A. 'Low-profile flexible frequency-reconfigurable millimetre-wave antenna for 5G applications'. *Flexible and Printed Electronics – IOPscience*, 2018, vol. 3(3), 5003.

[30] Fang Y., Hester J.G.D., Su W., Chow J.H., Sitaraman S.K., and Tentzeris M. M. 'A bio-enabled maximally mild layer-by-layer Kapton surface modification approach for the fabrication of all-inkjet-printed flexible electronic devices'. *Nature Scientific Reports*, 2016, vol. 6, 39909.

[31] Watson D.E., Ng, J.H., and Desmulliez M.P.Y. 'Additive photolithography based process for metal patterning using chemical reduction on surface modified polyimide'. *18th European Microelectronics & Packaging Conference*, Brighton, 2011, pp. 1–7.

[32] Rabbani M.S. and Ghafouri-Shiraz H. 'Liquid crystalline polymer substrate-based THz microstrip antenna arrays for medical applications'. *IEEE Antennas and Wireless Propagation Letters*, 2017, vol. 16, pp. 1533–1536.

[33] Jilani S.F., Munoz M.O., Abbasi Q.H., and Alomainy A. 'Millimeter-wave liquid crystal polymer based antenna array for conformal 5G applications'. *IEEE Antennas and Wireless Propagation Letters*, 2019, vol. 18(1), pp. 84–88.

[34] Jilani S.F., Aziz A.K., Abbasi Q.H., and Alomainy A. 'Ka-band flexible Koch fractal antenna with defected ground structure for 5G wearable and conformal applications'. *IEEE 29th Annual International Symposium on Personal, Indoor and Mobile Radio Communications (PIMRC 2018)*, 2018, pp. 361–364.

[35] Khaleel H.R. 'Design and fabrication of compact inkjet printed antennas for integration within flexible and wearable electronics'. *IEEE Transactions on Components, Packaging and Manufacturing Technology*, 2014, vol. 4(10), pp. 1722–1728.

[36] Jilani S.F., Falade O.P., Wildsmith T., Reip P., and Alomainy A. 'A 60-GHz ultra-thin and flexible metasurface for frequency-selective wireless applications'. *Applied Sciences, Special Issue: Frequency Selective Surfaces*, 2019, vol. 9(5), 945.

[37] Falade O.P., Jilani S.F., Ahmed A.Y., *et al.* 'Design and characterisation of a screen-printed millimetre-wave flexible metasurface using copper ink for communication applications'. *Flexible and Printed Electronics–IOPscience*, 2018, vol. 3(4), 5005.

[38] Guo X., Ying H., Weidong P., Wenging K., Leidong M., and Yugang Z. 'Flexible and deformable monopole antenna based on silver nanoparticles for wearable electronics'. *Nanoscience Nanotechnology Letters*, 2017, vol. 9 (11), pp. 1632–1638.

[39] Guo X., Hang Y., Xie Z., Wu C., Gao L., and Liu C. 'Flexible and wearable 2.45 GHz CPW-fed antenna using inkjet-printing of silver nanoparticles on PET substrate'. *Microwave Optical Technology Letters*, 2017, vol. 59(1), pp. 204–208.

[40] Guo X., Huang Y., Wu C., *et al.* 'Flexible and reversibly deformable radio-frequency antenna based on stretchable SWCNTs/PANI/Lycra conductive fabric'. *Smart Materials and Structures-IOPscience*, 2017, vol. 26(10), 5036.

[41] Kumar A., Saghlatoon H., La T.G., *et al.* 'A highly deformable conducting traces for printed antennas and interconnects: Silver/fluoropolymer composite amalgamated by triethanolamine'. *Flexible and Printed Electronics – IOPscience*, 2017, vol. 2(4), 5001.

[42] Wang Z., Volakis J.L., and Kiourti A. 'Chapter 10: Embroidered antennas for communication systems' in *Electronic Textiles*, Amsterdam, Elsevier, 2015, pp. 201–237.

[43] Corchia L., Monti G., and Tarricone L. 'Durability of wearable antennas based on nonwoven conductive fabrics: Experimental study on resistance to washing and ironing'. *International Journal of Antennas and Propagation*, 2018, pp. 1–8, Article ID 2340293.

[44] Yang X., Huang Y., Dai Z., Barber J., Wang P., and Lu N. 'Cut-and-paste: Method for the rapid prototyping of soft electronics'. *Science China Technological Sciences*, 2019, vol. 62(2), pp. 199–208.

[45] Ibanez-Labiano I. and Alomainy A. 'Dielectric characterization of non-conductive fabrics for temperature sensing through resonating antenna structures'. *MDPI Materials*, 2020, vol. 13(6), pp. 1271.

[46] Ibanez-Labiano I., Ergoktas M.S., Kocabas C., Toomey A., Alomainy A., and Ozden-Yenigun E. 'Graphene-based soft wearable antennas'. *Applied Materials Today*, 2020, vol. 20, 100727.

[47] Mazzinghi A., Blaschi L., and Freni A. '3D printed low cost high gain antennas'. *IEEE Antennas and Wireless Propagation Letters*, 2018, vol. 1(1).

[48] Pethig R. 'Dielectric properties of body tissues'. *Clinical Physics and Physiological Measurement*, 1987, vol. 8(4A), pp. 5–12.

[49] Gabriel C. 'Compilation of the dielectric properties of body tissues at RF and microwave frequencies'. *Physics*, 1996.

[50] Institute for Applied Physics. Dielectric properties of body tissues in the frequency range 10 Hz–100 GHz Italian National Research Council [online]. Available from: http://niremf.ifac.cnr.it/tissprop/ [Accessed 4 March 2021].

[51] Tell R.A. 'Microwave energy absorption in tissue'. *EPA, Twinbrook Research Laboratory Technical Report*, 1972.

[52] Ito, K. 'Numerical and experimental human body phantoms'. *IET Seminar on Antennas and Propagation for Body-Centric Wireless Communications*, 2007, pp. 6–12.

[53] CST Microwave Studio [online]. Available from: https://www.3ds.com/products-services/simulia/products/cst-studio-suite/ [Accessed 4 March 2021].

[54] Abd Rahman N.H., Yamada Y., and Amin Nordin M.S. 'Analysis on the effects of the human body on the performance of electro-textile antennas for wearable monitoring and tracking application'. *MDPI Materials*, 2019, vol. 12(10), 1636.

[55] Federal Communications Commission. *Evaluating compliance with FCC guidelines for human exposure to radiofrequency electromagnetic fields*. Supplement C to bulletin 65, 2001.

[56] Kanda M.Y., Flallen M., Chou C.K., and Balzano Q. 'Formulation and characterization of tissue equivalent liquids used for RF densitometry and dosimetry measurements'. *IEEE Transactions on Microwave Theory and Techniques*, 2004, vol. 52(8), pp. 2046–2056.

[57] Takimoto T., Onishi T., Saito K., Takahashi M., Uebayashi S., and Ito K. 'Characteristics of biological tissue equivalent phantoms applied to UWB communications'. *Electronics Communications in Japan (Part I: Communications)*, 2007, vol. 90(5), pp. 48–55.

[58] Garrett J. and Fear E. 'Stable and flexible materials to mimic the dielectric properties of human soft tissues'. *IEEE Antennas Wireless and Propagation Letters*, 2014, vol. 13, pp. 599–602.

[59] Conway G.A. and Scanlon W.G. 'Antennas for over-body-surface communication at 2.45 GHz'. *IEEE Transactions on Antennas Propagation*, 2009, vol. 57(4), pp. 844–855, 10560828.

[60] Giddens H., Paul D.L., Hilton G., and McGeehan J.P. 'Numerical and experimental evaluation of phantoms for off-body wireless communications'. *Loughborough Antennas & Propagation Conference*, 2011, pp. 1–4.

[61] Guy A.W. 'Analysis of electromagnetic fields induced in biological tissues by the thermographic studies on equivalent phantom models'. *IEEE Transactions on Microwave Theory and Techniques*, 1971, vol. 19(2), pp. 205–214.

[62] Okano Y., Ito K., Ida I., and Takahashi M. 'The SAR evaluation method by a combination of thermographic experiments and biological tissue-equivalent phantoms'. *IEEE Transactions on Microwave Theory and Techniques*, 2000, vol. 48(11), pp. 2094–2103.

[63] SPEAG [online]. Available from: https://speag.swiss/ [Accessed 4 March 2021].

[64] Li Y. and Dai X.Q. *Biomechanical Engineering of Textiles and Clothing*, 1st edn. Cambridge: Woodhead Publishing, 2006, p. 428.

[65] Castano L. and Flatau A. 'Smart fabric sensors and e-textile technologies: A review'. *Smart Materials Structures*, 2014, vol. 23, 053001.

[66] Pekka S., Mikko K., and Yahya S. 'Textile antennas: Effects of antenna bending on radiation pattern and efficiency'. *Antennas and Propagation Society International Symposium*, San Diego, CA, 5–11 July 2008, 10219545.

[67] Salonen P. and Rahmat-Samii Y. 'Textile antennas: Effects of antenna bending on input matching and impedance bandwidth'. *First European Conference on Antennas and Propagation*, Nice, 2006, pp. 1–5.

[68] Elias N.A., Samurai N.A., and Rahim M.K.A. 'Analytical study of energy absorption in human body due to bendable behaviour of textile antenna' in Zainal H. (ed.). *Theory and Applications of Applied Electromagnetics*. Lecture Notes in Electrical Engineering Series, no. 344, 2015, pp. 329–337.

[69] Xu B., Eike R.J., Cliett A., Ni L., Cloud R., and Li Y. 'Durability testing of electronic textile surface resistivity and textile antenna performance'. *Textile Research Journal*, 2019, vol. 89(18), pp. 3708–3721.

[70] Fernandes D.E.D., Marcos J.S., Maslovski S.I., and Silveirinha, M.G. 'Measurement of the specific absorption rate using a single electric field sensor'. *IEEE EUROCON – International Conference on Computer as a Tool*, 2011, pp. 1–4.

[71] Wang Z., Lin J.C., Mao W., Liu W., Smith M.B., and Collins C.M. 'SAR and temperature: Simulations and comparison to regulatory limits for MRI'. *Journal of Magnetic Resonance Imaging*, 2007, vol. 26(2), pp. 437–441.

[72] ICNIRP. 'Guidelines for limiting exposure to electromagnetic fields (100 kHz to 300 GHz)'. *Health Physics*, 2020, vol. 18(5), pp. 483–524.

[73] Klancnik M., Rijavec B., Pivar M., and Muck D. 'Influence of printing material and printing ink layer on RFID antenna operation'. *Tekstilec*, 2018, vol. 61(3), pp. 162–170.

[74] Lilja J., Salonen P., Kaija T., and Maagt P. 'Design and manufacturing of robust textile antennas for harsh environments'. *IEEE Transactions on Antennas and Propagation*, 2012, vol. 60(9), pp. 4130–4140.

[75] Branimir I., Goran G., and Davor B. 'Performance of wearable antenna exposed to adverse environmental conditions'. *21st International Conference on Applied Electromagnetics and Communications*, Dubrovnik, Croatia, 14–16 October 2013, paper no. 13990440.

[76] Moraru A., Ursachi C., and Helerea E. 'A new washable UHF RFID tag: Design, fabrication, and assessment'. *MDPI Sensors*, 2010, vol. 20(12), 3451.

[77] Simorangkir R.B.V.B., Le D., Björninen T., Sayem A.S.M., Zhadobov M., and Sauleau R. 'Washing durability of PDMS-conductive fabric composite: Realizing washable UHF RFID tags'. *IEEE Antennas and Wireless Propagation Letters*, 2019, vol. 18(12), pp. 2572–2576.

[78] Scarpello M.L., Kazani I., Hertleer C., Rogier H., and Ginste D.V. 'Stability and efficiency of screen-printed wearable and washable antennas'. *IEEE Antennas and Wireless Propagation Letters*, 2012, vol. 11, pp. 838–841.

[79] Gorgutsa S., Bachus K., LaRochelle S., Oleschuk R.D., and Messaddeq Y. 'Washable hydrophobic smart textiles and multi-material fibers for wireless communication'. *Smart Materials and Structures*, 2016, vol. 25(11), 115027.

[80] Carla H., Annelien L., Hendrik R., and Van Langenhove L. 'Influence of relative humidity on textile antenna performance'. *Textile Research Journal*, 2009, vol. 80(2), pp. 177–183.

[81] Yilmaz S. 'The geometric resistivity correction factor for several geometrical samples'. *Journal of Semiconductors*, 2015, vol. 36(8), 082001.

[82] Sankaralingam S. and Gupta B. 'Determination of dielectric constant of fabric materials and their use as substrates for design and development of antennas for wearable applications'. *IEEE Transactions on Instrumentation and Measurement*, 2012, vol. 59(12), pp. 3122–3130.

[83] Lesnikowski J. 'Dielectric permittivity measurement methods of textile substrate of textile transmission lines'. *Przeglad Elektrotechniczny (Electrical Review)*, 2012, R.88 NR-3a.

[84] Morton W.E. and Hearle J.W.S. *Physical Properties of Textile Fibres*, 4th edn. Cambridge: Woodhead Publishing Limited and CRC Press LLC, 2008.

[85] Lin X., Seet B.C., and Joseph F. 'Wearable humidity sensing antenna for BAN applications over 5G networks'. *IEEE 19th Wireless and Microwave Technology Conference (WAMICON)*, Sand Key, FL, 2018, pp. 1–4.

[86] Jang J.H. and Han J.I. 'Cylindrical relative humidity sensor based on polyvinyl alcohol (PVA) for wearable computing devices with enhanced sensitivity'. *Sensors Actuators A: Physical*, 2017, vol. 261, pp. 268–273.

[87] Chang K., Kim Y.H., Kim Y.J., and Yoon Y.J. 'Functional antenna integrated with relative humidity sensor using synthesised polyimide for passive RFID sensing'. *Electronics Letters*, 2007, vol. 43(5), pp. 7–8.

[88] Controltecnica-CTS. Climatic test chambers [online]. Available from: https://www.cts-clima.com/en/climatic-testing/climatic-test-chambers.html [Accessed 4 March 2021].

[89] Aboufoul T., Parini C., Chen X., and Alomainy A. 'Pattern-reconfigurable planar circular ultra-wideband monopole antenna'. *IEEE Transactions on Antennas and Propagation*, 2013, vol. 61(10), pp. 4973–4980.

[90] Aboufoul T., Ali K., Alomainy A., and Parini C. 'Combined pattern and frequency reconfiguration of single-element ultra-wideband monopole antenna for cognitive radio devices'. *7th European Conference on Antennas and Propagation (EuCAP)*, 2013, pp. 932–936.

[91] Andrade A.d.C., Fonseca I.P., Jilani S.F., and Alomainy A. 'Reconfigurable textile-based ultra-wideband antenna for wearable applications'. *10th European Conference on Antennas and Propagation (EuCAP)*, 2016, pp. 1–4.
[92] Singh J., Rani E.S., and Singh E.M. 'Design of microstrip patch antenna array for traffic speed detector application'. *5th International Conf. on Advanced Computing & Communication Technologies (ACCT)*, 2015, pp. 4–7.
[93] Holub A. and Polivka M. 'Collinear microstrip patch antenna' in *Passive Microwave Components and Antennas*. London: InTech, 2010, Ch. 24, pp. 513–530.
[94] Polivka M., Holub A., and Mazanek M. 'Collinear microstrip patch antenna'. *Radioengineering*, 2005, vol. 14(4), pp. 40–42.
[95] Solbach K. 'Microstrip-Franklin antenna'. *IEEE Transactions on Antennas and Propagation*, 1982, vol. 30(4), pp. 773–775.
[96] Wang P.P., Antoniades M.A., and Eleftheriades G.V. 'An investigation of printed Franklin antennas at X-band using artificial (metamaterial) phase-shifting lines'. *IEEE Transactions on Antennas and Propagation*, 2008, vol. 56(10), pp. 3118–3128.
[97] Chang S.H., Liao W.J., Peng K.W., and Hsieh C.Y. 'A Franklin array antenna for wireless charging applications'. *PIERS*, 2010, vol. 6(4), pp. 340–344.
[98] Azlan A.A., Ali M.T., Awang M.Z., and Jamlos M.F. 'A simulation of single patch antenna with phantom tissue for ISM band'. *National Conference on Medical at Center of Medical Electronic Politeknik Shah Alam*, 2014.
[99] Antennas & Electromagnetics Research Group, Queen Mary University of London. Available from: http://antennas.eecs.qmul.ac.uk/facilities/antenna-measurement-laboratory/ [Accessed 4 March 2021].

Chapter 9

5G mm-wave mobile handset antenna system, measurements and evaluations

Carla Di Paola[1], Bo Xu[2], Kun Zhao[1,3], Shuai Zhang[1], Zhinong Ying[1,3] and Gert Froelund Pedersen[1]

Nowadays, the large number of services provided by mobile phones drives the usage of the millimetre-wave (mm-wave) frequency range for the upcoming fifth-generation (5G) mobile communication system, in order to provide wider bandwidths that support high-throughput data transfer. However, the propagation environment becomes hostile at higher frequencies, due to larger path loss, so higher gain antenna arrays, that are able to achieve wide beam steering range, are proposed to be embedded in 5G mobile terminals. In order to assess the radio frequency (RF) performance of the mm-wave mobile handset antenna systems, new figures of merit are specified by the Third-Generation Partnership Project (3GPP). The RF requirements take into account multiple factors that affect the performance of the mm-wave antenna, such as the challenging electromagnetic environment inside the phone-housing and the high mobility. In addition, the blockage from the user limits the antenna coverage range. Since it is difficult for mm-wave signal to propagate through the human body, an accurate model of the user shadowing is required to be included in the mm-wave cellular network planning. In this chapter, the new metrics to evaluate the performance of mm-wave antenna systems are presented, and both traditional and novel approaches are used for the measurements of mm-wave mobile handsets mockups. The user blockage is investigated in different scenarios and a new model is drawn.

9.1 Introduction

The large number of services provided by mobile phones, e.g. video-calling, browsing, localization, gaming and so forth, results in a significant increase of data traffic. In order to satisfy the demand for higher data throughput, more spectrum

[1]Antennas, Propagation and Millimeter Wave Systems (APMS) Section, Department of Electronic Systems, Aalborg University, Aalborg, Denmark
[2]Ericsson Research, Ericsson AB, Stockholm, Sweden,
[3]Research Center Lund, Sony Corporation, Lund, Sweden

resources are needed. For the 5G mobile communication system two different frequency ranges are considered: frequency range 1 (FR1) [1] that overlaps some of the 4G bands, and frequency range 2 (FR2) [2–4], including bands in the mm-wave spectrum. On the one hand, FR1 benefits from the advantages of inherent physics properties at low-frequency bands over building penetration and radio coverage but suffers from reduced bandwidth availability. On the other hand, mm-wave bands offer wider bandwidths than bands in FR1 [5,6] but are affected by a more hostile propagation environment, due to the shorter wavelength. In fact, the free-space path loss results enlarged and the blockage from the interfering objects is more significant, as well as the penetration loss in conventional materials and the diffraction loss [7].

In order to mitigate the propagation loss in mm-wave bands, resulting from the above-mentioned free-space path loss, transmission loss, shadowing loss and user effect [8], higher gain antenna arrays are required. Since high gain leads to narrow radiation beamwidth, the issue of spatial coverage becomes more critical than at the FR1 bands. Thus, the demand for beam-steering antenna systems that, thanks to the ability to form multiple beams in different directions, allow to achieve a large coverage range [9,10] leads to the development of new phased antenna arrays [11], as well as lens antenna [12–18], butler matrix [19–21], pattern reconfigurable antenna [22–24] and beam-switchable antenna array [25].

For the assessment of the RF performance of the novel mobile antenna systems in FR2, the conventional metrics used for mobile devices in FR1, being omnidirectional parameters, result inadequate to evaluate beam-steering architectures. Therefore, proper figures of merit are specified by the 3GPP [26,27], to address the requirements that the mm-wave mobile handset needs to fulfil, before being on the market.

However, the coverage performance of the mobile device is highly impacted by the small mobile device form factor, the phone-case material, the presence of other dielectrics and electronic components, e.g. battery, speaker, case, screen and so forth, as well as the number and placement of antenna modules. Therefore, when defining the RF requirements, these factors must be taken into account.

Moreover, the impact of the integration on the radiation performance, mentioned earlier, must be considered also when designing mm-wave antennas [28]. This factor, though, cannot be predicted with high accuracy, since too many components are involved. To this end, an analysis a posteriori is offered by the simulation-assisted approach, developed with the near-field measurement system that, based on the reconstruction of the surface currents on the device, allows a more insight understanding of the interactions inside the phone-housing and to optimize the mm-wave antenna design accordingly.

Furthermore, the user blockage, responsible of spatial coverage degradation at FR1 bands, becomes more serious at mm-wave frequencies [29]. Therefore, when assessing the performance of the mm-wave mobile communication link, the body shadowing, causing significant attenuation of the transmitted/received signals, should be considered. The current models of the user shadowing, proposed by the 3GPP community [30], are though extremely simplified. Thus, the need for an accurate model of the user blockage at mm-wave, which takes into account the different shapes of the user body shadow, since the diffractions, creeping waves and surface waves change according to the body shape.

In this chapter, metrics and methods for the evaluation of 5G mm-wave mobile handset antenna systems are presented and discussed. In Section 9.2, the new figures of merit specified by 3GPP, to assess the RF performance in FR2, are provided and the factors influencing the requirements are analysed. The novel simulation-assisted approach for near-field measurements is described in Section 9.3, where its application to a mm-wave mockup is also included. In Section 9.4, the performance evaluation of a mm-wave antenna system is conducted, from the design and experimental validation of the antenna mockup, to the test with active measurements in a quasi-real scenario, where the antenna is integrated in the phone-housing. Finally, the body shadowing is investigated with statistical measurements in Section 5 and a new model of the user blockage is proposed.

9.2 3GPP RF requirements and implications on antenna design and integration in FR2

9.2.1 FR2 frequency spectrum

As previously stated in Section 9.1, the limited spectrum resources, available in FR1 bands for cellular communications, drive people's attention towards the (mm-wave) bands, which can provide significant data throughput, to enable the feature of enhanced mobile broadband. To date, five frequency bands in FR2 are defined in the 3GPP technical specification (TS) [2] for 5G new radio (NR) and are shown in Table 9.1. In addition to the bands listed, a new frequency band from 47.2 to 48.2 GHz is under discussion [3] and studies on frequency range above 52 GHz are kicked off in 3GPP as well [4].

Though rich spectrum resources are available in FR2, the higher operation frequencies demand high-gain antenna systems on both the base station (BS) side and the user equipment (UE) side to provide an acceptable link budget for communications. On the other hand, a high-gain antenna radiation pattern naturally leads to a narrow beamwidth, with limited spatial coverage. Since UEs, especially handheld mobile devices, are expected to operate under a highly random and varied

Table 9.1 NR operating bands in FR2

Operating band	Frequency (MHz)
n257	26,500–29,500
n258	24,250–27,500
n259	39,500–43,500
n260	37,000–40,000
n261	27,500–28,350

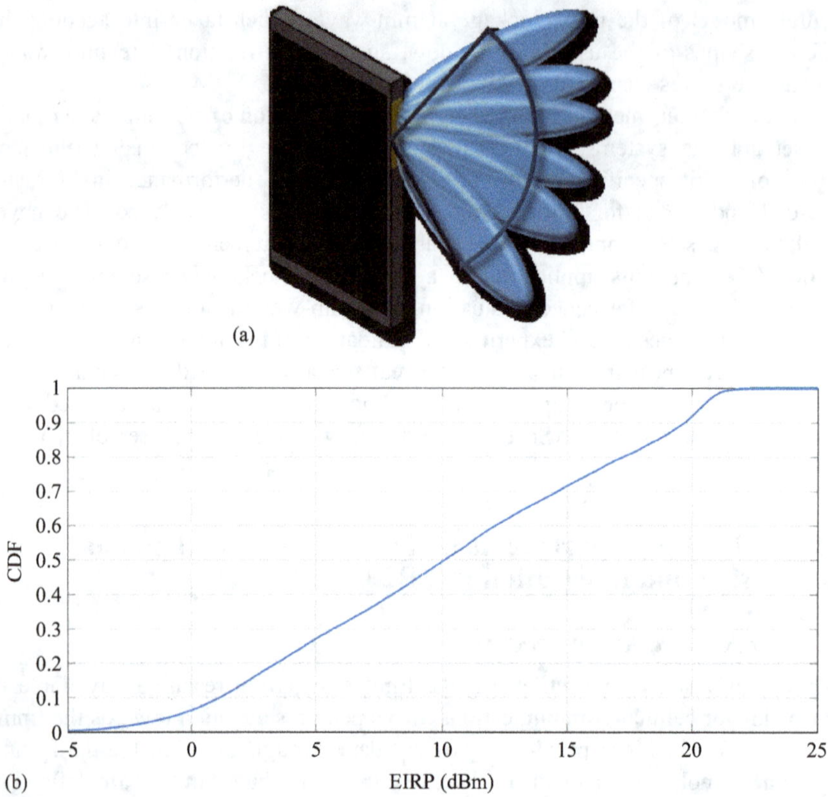

Figure 9.1 (a) Illustration of the spherical coverage of a mobile handset and (b) curve representing the CDF of EIRP

propagation environment, a sufficiently large spatial coverage is necessary to ensure stable performance, given the random angles of arrival (AoA) of wireless signals. Therefore, the beam-steering function is also necessary for antenna systems in mobile devices (see Figure 9.1(a)).

9.2.2 UE power class and RF requirements in FR2

Utilizing the FR2 spectrum does not require only novel antenna systems, but also proper figures of merit to assess the RF performance. Compared to the antennas in FR1, antenna systems in FR2 are usually highly integrated with the RF front end, making it difficult to be tested over a conducted set-up, so over the air (OTA) test becomes necessary. Moreover, conventional OTA metrics used for mobile devices in FR1, e.g. total radiated power (TRP) and total isotropic sensitivity, are omnidirectional parameters so are not sufficient to evaluate a device with a beam-steering antenna system.

A mobile antenna system in FR2 must provide both high gains and a large spherical coverage. To give a full picture, the cumulative distribution function (CDF) of the effective isotropic radiated power (EIRP) and the complementary CDF (CCDF) of the effective isotropic sensitivity (EIS) are used as the fundamental metrics to evaluate the UE OTA performance in FR2. EIRP and EIS are the measured radiated power and sensitivity in a single direction, respectively. The CDF is obtained by measuring the EIRP/EIS with uniform surface density over the whole sphere around the device. An example of CDF of EIRP is plotted in Figure 9.1(b). The CDF value at $Y\%$ means that $Y\%$ of the total sphere has an EIRP lower than the corresponding EIRP value on the X-axis. For example, 90% of the total sphere has EIRP values below 20 dBm, according to the CDF curve in Figure 9.1(b).

To guarantee that the UE antenna system in FR2 can achieve high gain, while keeping a sufficiently large spatial coverage, the minimum requirements on the EIRP/EIS values at the CDF/CCDF curve's peak point and at a lower percentile point are defined in the 3GPP specification. The EIRP/EIS requirement at the lower percentile point is also known as the spherical coverage requirement. The minimum requirement means that the UE is not allowed to transmit EIRP lower than the defined EIRP limit at the required percentile.

The percentile point of the spherical coverage depends on the power class (PC). Unlike the FR1 PC that is mainly determined by the maximum output power that the UE can reach, the PC in FR2 is specified as a package of multiple parameters, including the maximum allowed TRP/EIRP and the minimum peak EIRP/spherical coverage. More importantly, the PC in FR2 is also associated with different UE types. To date, four PCs in FR2 have been specified in 3GPP, which are shown in Table 9.2.

Obviously, different UE types result in a different form factor, that can significantly impact the RF performance. A front-end module (FEM), including the antenna system, in handheld UEs (PC3), e.g., mobile phones, usually faces a more challenging electromagnetic environment, due to the crowded space inside the phone housing. Therefore, the performance degradation due to the phone integration must be considered. In addition, the different types of UE are meant to operate in different scenarios. For example, PC1 UEs are fixed wireless access (FWA) devices, which operate in a fixed location. On the other hand, PC3 handheld UEs are usually moved around, so the spherical coverage requirement of PC3 UE is defined with a lower percentile on the CDF curve than PC1, to ensure the

Table 9.2 Power class in FR2 and the associated UE types

UE power class	UE type
PC1	Fixed wireless access (FWA) UE
PC2	Vehicular UE
PC3	Handheld UE
PC4	High-power non-handheld UE

Table 9.3 UE minimum peak EIRP and spherical coverage for power class 3

Operating band	Min peak EIRP (dBm)	Min EIRP at 50%-tile CDF (dBm)
n257	22.4	11.5
n258	22.4	11.5
n259	18.7	5.8
n260	20.6	8
n261	22.4	11.5

performance over a larger part of the spatial coverage. These factors need to be taken into account, when defining the RF requirements.

Considering the PC3 UE (handheld UE) as an example, its uplink (UL) spherical coverage requirement is defined at 50% of the CDF curve. The peak EIRP and the spherical coverage value are derived based on the assumption that each antenna panel consists of a four-element array [26]. The number of antenna panels per UE are considered for deriving the spherical coverage requirement [26]. The influence of the phone integration, mentioned previously, is also taken into account. The corresponding values are listed in Table 9.3. Note that those values are applicable only to single band design. For devices supporting carrier aggregation (CA) and multiple bands operation, performance relaxations are allowed. The detailed values for each scenario can be found in the 3GPP TS 38.101-2 [2].

On the other hand, the PC1 UE (FWA) includes a much larger number of antenna elements than PC3 UEs, as it is allowed by the form factor. As the PC1 UEs are not subject to high mobility, the array for PC1 UEs only needs to cover a small fraction of the sphere. Consequently, the spherical coverage requirement of PC1 is defined at a much higher percentile value compared to the PC3 UEs, which is 85%.

Apart from the minimum requirements, the maximum allowed TRP and EIRP are also captured in the 3GPP specification for each FR2 PC, to limit UL co-channel interference, as well as to meet corresponding regulatory requirements. Moreover, the RF requirement in FR2 also includes various other parameters in addition to the spherical coverage, such as transmitter power dynamic and signal quality. These FR2 RF requirements are specified in the 3GPP TS 38.101-2 [2].

9.2.3 The impact of the beam-based operation on the RF requirement

Since the UE adopts beamforming technology in FR2, the 5G NR protocol is designed to support beam-based operation, which has an undeniable impact on the achievable spherical coverage. An overview of NR beam management can be found in [31]. The FR2 network operates in time division duplex mode, where the UL and downlink (DL) transmission share the same channel. By exploiting the DL signals direction, FR2 UEs can autonomously select their UL beam according to the channel's reciprocity. 3GPP has also developed requirements and tests to verify this UE capability, known as beam correspondence. Beam correspondence is the ability of the UE to select a suitable beam for UL transmission, based on DL measurements, with or without relying on UL beam sweeping [2]. For UEs that are able to select a suitable UL beam, based on DL measurements, without relying on UL beam sweeping, the resulting EIRP should meet the UE minimum peak EIRP and UE spherical coverage requirements, as mentioned previously. On the other hand, a beam correspondence tolerance requirement should be fulfilled, if the UE needs the UL beam sweeping assistant to meet the spherical coverage requirement.

Another beam management aspect that influences the RF requirement concerns the inter-band CA operation, since BSs for different frequency bands may or may not be deployed at the same locations. Therefore, FR2 UEs sometimes need to flexibly beamform in either the same direction or in different directions, which requires independent beam management on each carrier. In this case, having an overlapped spatial coverage region between each carrier can help the UE to support both co-located and non-co-located deployment for inter-band carrier operation. Requirements for UEs supporting independent beam management with common spatial coverage are developed in TS 38.101-2 for the inter-band CA operation in DL [2].

9.2.4 Implication on the antenna integration in mobile handheld devices

The spherical coverage requirement for the FR2 UEs implicates different antenna designs and integration methods, compared to the conventional FR1 antenna system for handheld mobile devices. To reduce the insertion loss, antenna systems in FR2 are usually highly integrated with the RF FEM. To fully exploit the advantage of rich spectrum in mm-wave bands, antenna arrays need to cover a large bandwidth with multiple contiguous and non-contiguous frequency bands. As previously shown, multiple bands have been defined in FR2 and more bands would be added on in the future. Such a bandwidth demand particularly challenges a compact antenna array with limited thickness and clearance. Moreover, the dual-polarized antenna array is also highly preferred in the FR2 network, to support polarization-based MIMO transmission and obtain the polarization diversity gain.

Antenna arrays in mobile handheld devices typically have a semi-hemisphere spatial coverage, due to the phone ground plane existence. Therefore, multiple arrays covering different faces would be needed, to reach a decent spherical coverage. A simple comparison of single- and multi-antenna arrays in a mobile phone prototype is shown in Figure 9.2. It can be seen that, though the peak EIPR is not affected, the spherical coverage can be significantly improved by adding an additional panel face towards a different direction.

The limited space available in a smartphone requires the arrays to occupy an area as small as possible, challenging thus the mm-wave array design. Therefore, it is crucial to select a proper placement which maximizes the array performance. With the state-of-the-art antenna FEM designs, it is possible to place a planar antenna array, e.g. patch array, either parallel to the phone-printed circuit board (PCB), facing the front/backside of the phone, or perpendicular to the phone PCB,

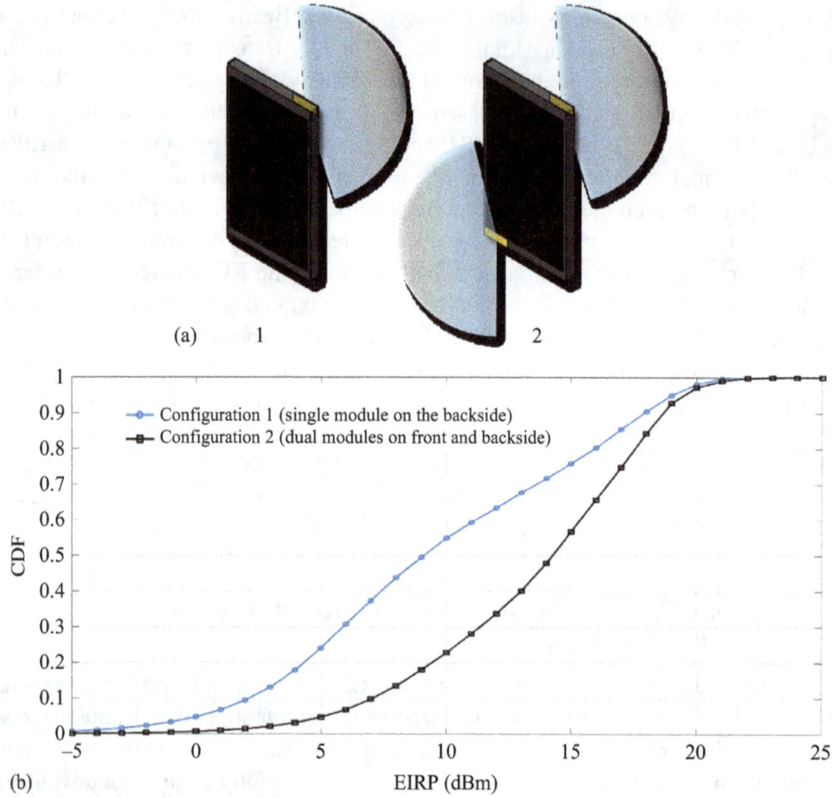

Figure 9.2 (a) UE configured with a single-antenna panel on the backside and two-antenna panels on the front and backside, and (b) the corresponding CDF of EIRP

towards the edge of the phone (see Figure 9.3). Comparing the two possible placements, the edge-facing antenna module usually shows better spherical coverage but less-peak EIRP, due to the lower directivity of the radiation pattern. From the integration aspect, the phone bezel design needs to be optimized, so that the radiation from the edge-mounted antenna module is not blocked. Multiple works have been published to solve this issue [32–35]. On the other hand, the front/backside facing design can incur in increased manufacturing costs and mounting problems, due to the presence of the display.

In addition, phone-housing materials significantly impact the spherical coverage performance of FR2 UEs (see Figure 9.4). For example, glass-based back covers are popularly used in commercial phones today, but the signal attenuation through glass covers is larger than plastic covers. A comparison of spherical coverage between a glass back covered phone vs. a plastic back covered phone is shown in Figure 9.5. Both phones have identical array configurations, but more than 2-dB difference can be observed at 50% of the CDF. To provide the same

Figure 9.3 Possible placements of the FR2 antenna module in a mobile phone

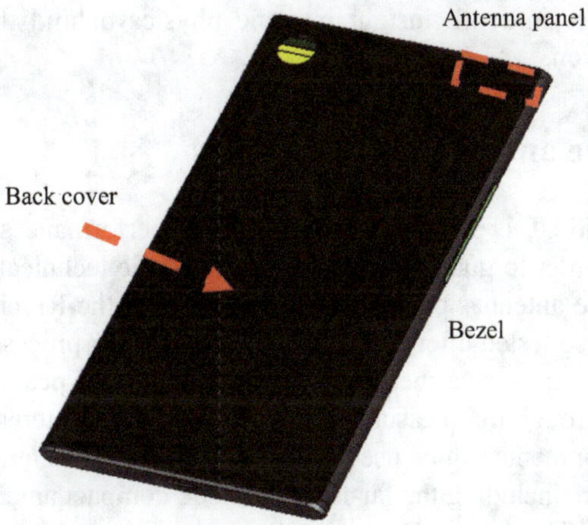

Figure 9.4 Mobile phone prototype with integrated mm-wave antenna panel

EIRP in the desired direction, more power is needed to be transmitted through the glass back cover than the plastic back cover. Besides, metallized structures (e.g. metal bezel or metal back cover) can easily block mm-wave signals radiation. Thus, the edge-mounted antenna array may also face a high impact from the smartphone's metal bezel, as mentioned previously.

As a remark, the spherical coverage of the mm-wave mobile handset is impacted by multiple factors, including the number and placement of antenna panels and the phone-housing materials. As the spherical coverage requirement on

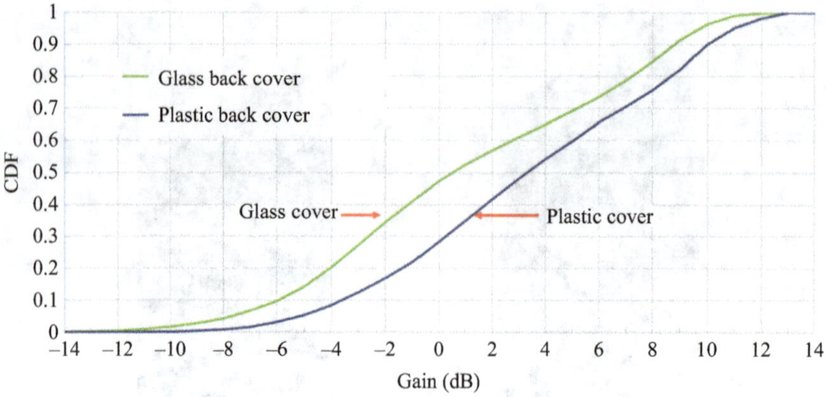

Figure 9.5 Spherical coverage comparison between a UE with glass back cover and plastic back cover

EIRP/EIS is one of the most fundamental requirements for FR2 UEs, all the related parameters must be carefully examined and selected. A more comprehensive and detailed study on UE spherical coverage, including various factors, can be found in [27], where a more in-depth insight into the physics behind the observed phenomena is also given.

9.3 5G mobile antenna measurements

Before the handheld UE is put on the market, the RF performances are evaluated via measurements in order to guarantee the fulfilment of the technical requirements for certification. As the antennas play a fundamental role in the RF chain, the antennas performance are also tested after the design and fabrication processes. The radiation performances in the mm-wave spectrum, assessed in terms of peak EIRP and CDF of EIRP, are tested through the measurement of the radiation patterns. Typically, three types of test environments can be used to measure the beam pattern in the mm-wave frequency spectrum, including the far-field chamber, compact antenna test range and near-field chamber. Thanks to the compact size and relatively less expensive cost, the near-field chambers and consequently the near-field measurement techniques have attracted a lot of attention, particularly in the industry. Initially, the near-field measurement methods have been used only to perform near-field to far-field transformation to obtain the antenna radiation patterns.

In recent years, the simulation-assisted approach has been developed with the near-field measurement system [36–39]. It not only allows to obtain the far-field patterns but also to reconstruct the approximate surface currents on the device under test (DUT), making it possible for additional measurement applications, e.g. antenna diagnoses [36], suppression of unwanted radiation in the measurements [37] and EMF exposure assessments [38,39]. This approach is very useful for mm-wave mobile antenna measurements, because the mm-wave antenna performance can be significantly impacted by the housing and test environments. In fact, although the surrounding components critically affect the radiation performance, especially at mm-wave frequencies, the information about their mechanical and electronic characteristics is unavailable for full-wave simulations, so it cannot be included in the antenna design stage. Even if radiation patterns measurements are required for validation, far-field radiation patterns may not provide enough information for adjustment.

The simulation-assisted approach aims to derive the surface equivalent currents (EQCs) (equivalent electric currents J_{eq}, equivalent magnetic currents M_{eq}, or both) on a reconstruction surface Σ_R (see Figure 9.3), based on the *Equivalence Theorem*. Instead of measuring the fields on Σ_M and perform near-field to far-field transformation (see Figure 9.10), the inverse source problem solves the surface EQCs reconstructed on a closed surface, denoted as Σ_R (see Figure 9.6(b)). According to the Equivalence Theorem, the fields radiated outside Σ_R from the source in Ω can be treated as those radiated from J_{eq} and M_{eq} on Σ_R. This allows to replace the DUT and any other radiating structure with a closed surface, which fully encloses them, whereon equivalent sources are placed. The inverse source problem can be formulated with the integral equation and then solved using the method of moments. This problem is known as ill-posed, because restrictions conditions or regularizations need to be applied, in order to acquire a unique solution. In [40], the M_{eq} is computed from the electric field on Σ_M by expanding the M_{eq} into Hertzian dipoles. In [41], the sources are represented by J_{eq}, and the restriction is applied using the method of least square with singular value decomposition. In [42], both the J_{eq} and M_{eq} are solved using the regularization technique through a conjugated gradient method. However, the methods presented in [40–42] give non-zero fields in Ω, leading to a larger reconstruction error at close distances outside Σ_R. In [43–45], the reconstruction algorithm has been further improved by enforcing Love's equivalence [46] on Σ_R (see Figure 9.6(c)), i.e. the field inside Σ_R is enforced as zero.

On Σ_M the integral equation can be formulated by considering the enforcement that J_{eq} and M_{eq} radiate the same tangential components on Σ_M:

$$\hat{n} \times E(r) = \hat{n} \times \left[-\eta_0 \mathscr{L}(J_{eq}; r) + \mathscr{K}(M_{eq}; r)\right] \qquad r \in \Sigma_M \tag{9.1}$$

Where

$$\mathscr{L}(J_{eq}; r) = jk_0 \int_{\Sigma_R} \left[J_{eq}(r') + \frac{1}{k_0^2} \nabla \nabla_s{}' \cdot J_{eq}(r')\right] g(r, r') ds'$$

$$\mathscr{K}(M_{eq}; r) = \int_{\Sigma_R} M_{eq}(r') \times \nabla g(r, r') ds'$$

$$g(r, r') = \frac{e^{(-jk_0|r - r'|)}}{4\pi|r - r'|}$$

$\eta_0 = \sqrt{\mu_0/\varepsilon_0}$, $k_0 = \omega\sqrt{\mu_0\varepsilon_0}$ is the surface divergence operator, r and r' are position vectors, $\hat{n}$ is the unit vector normal to the surface. As mentioned previously, solving (9.1) alone would give large near-field reconstruction error. This can be improved

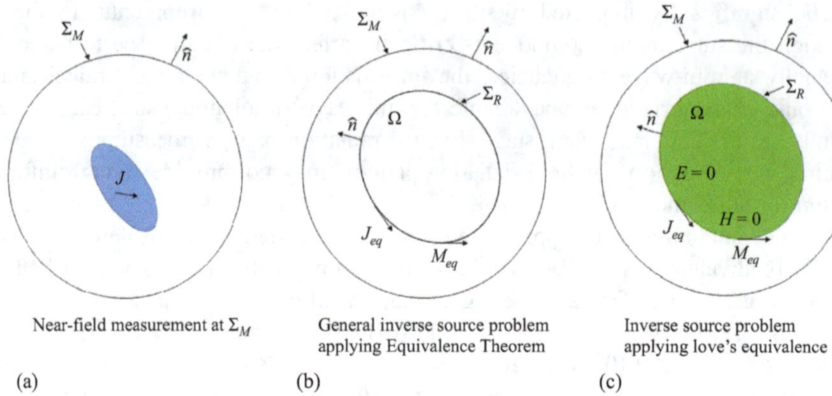

Figure 9.6 (a) Traditional near-field measurements of a DUT (source currents represented by J), (b) general inverse source problem applying Equivalence Theorem in near-field measurements, and (c) inverse source problem applying Love's equivalence

by enforcing Love's equivalence and results in the following boundary integral identities [43–45]:

$$\hat{n} \times \left[-\eta_0 \mathscr{L}(J_{eq}; r) + \mathscr{K}(M_{eq}; r)\right] = -\frac{1}{2} M_{eq}(r) \qquad r \in \Sigma_R \tag{9.2a}$$

$$\hat{n} \times \left[-\frac{1}{\eta_0} \mathscr{L}(M_{eq}; r) + \mathscr{K}(J_{eq}; r)\right] = \frac{1}{2} J_{eq}(r) \qquad r \in \Sigma_R \tag{9.2b}$$

As previously mentioned, solving the system of Equations 9.1a–9.2b gives a more accurate near-field reconstruction.

This approach is applied to a mm-wave mockup containing different types of elements, as shown in Figure 9.7. The dimensions of the mockup are 165 mm × 83 mm × 8 mm and it is equipped with chassis, case and display. All the elements are fed through 50-Ω microstrip lines connected to 50-Ω micro-miniature coaxial connectors.

The fields are measured using *MVG StarLab 50 GHz*, which is a portable spherical multi-probe system with an array of probes electronically scanning from 6 to 50 GHz, as shown in Figure 9.8. The measured data allow to obtain the far-field patterns through near- to far-field transformation and are also used for source reconstruction, implemented in the MVG software *INSIGHT* [48].

Figure 9.9 presents the simulated and reconstructed EQC results. Generally, the reconstructed J_{eq} and M_{eq} agree well with the simulated ones. The slot and the patch elements have wider J_{eq} and M_{eq} distributions, while the dipole and the notch have more limited J_{eq} and M_{eq} distributions. This is explained by the more severe pattern distortion that the slot and the patch suffer in the realistic housing environment. For the dipole and the notch elements, a portion of reconstructed J_{eq} and M_{eq} can be observed in the corner, stronger than the simulated EQCs in the same

Figure 9.7 Fabricated 28-GHz mockup: back view, front view and back view with the back cover removed. © IEEE. Reproduced, with permission, from [47].

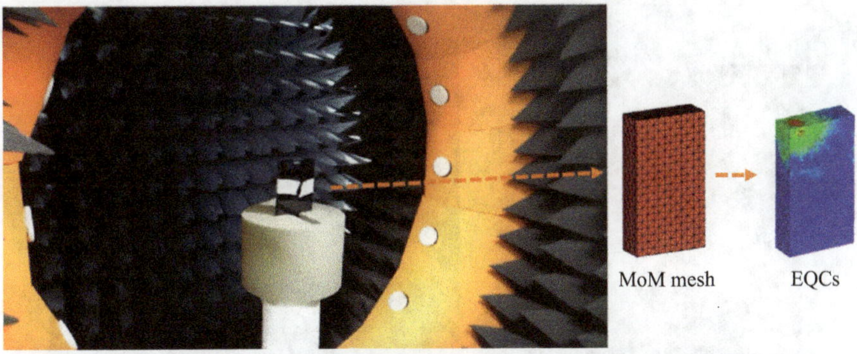

Figure 9.8 Antenna measurements and diagnosis using MVG StarLab 50 GHz near-field system and post-processed using INSIGHT [48]. © IEEE. Reproduced, with permission, from [47].

location. This may be attributed to the cable effects, which can possibly be eliminated by applying spatial filtering [37].

In conclusion, as the impact of housing is hard to be precisely predicted in practice, the simulation-assisted approach, based on the solution of the inverse source problem, can provide intuitive physical insight and extra diagnostic information about 5G mobile antenna design. The reconstructed EQC can facilitate further adjustment from a brand new perspective.

9.4 Passive and active measurements of a 5G beam-switchable antenna system with near-field chamber

As previously explained in Section 9.1, the hostile propagation environment at mm-wave requires to implement beam-forming antenna systems that achieve large

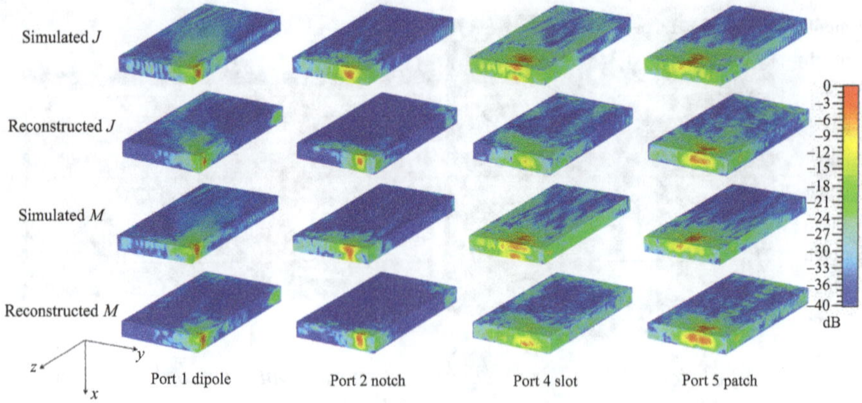

Figure 9.9 Simulated and reconstructed equivalent currents. © IEEE. Reproduced, with permission, from [47].

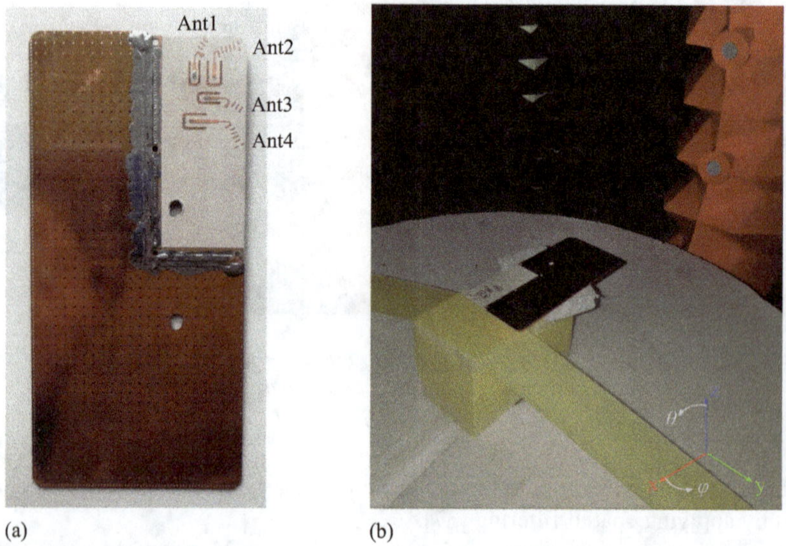

Figure 9.10 (a) Fabricated antenna array mounted to the ground plane and (b) passive measurements set-up in the MVG StarLab 50 GHz

coverage and high gain in order to face the severe propagation loss. Different beam-steering technologies have been developed in recent years from the most common phased antenna arrays [11] to lens antenna [12–18], butler matrix [19–21], pattern reconfigurable antennas [22–24] and beam-switchable antenna array [25]. In this section, the beam-switchable antenna system is investigated. In particular, the focus is on the difference between the performance achieved in the full-wave simulations, carried out with CST Microwave Studio [49], and the results obtained from both passive measurements, run over the antenna mockup, and active measurements, conducted on the array integrated into the phone-case with a glass, to test the feasibility for mobile devices. The spatial coverage performance, assessed in terms of CDF of EIRP, according to the 3GPP specification [26,27], evaluated for each stage, are compared and discussed.

9.4.1 Corner beam-switchable array simulated performance

The architecture consists of four microstrip-fed quasi-Yagi antennas printed on both sides of the corner of the PCB, pointing different directions, and a switch to feed each antenna and steer the beam. Yagi–Uda antennas are selected as array elements for the high directivity, high gain and simple structure, but any other directive element can be used instead. In fact, the antenna system is very flexible and different topologies and locations can be deployed in a mobile handset, as demonstrated in [25]. However, as previously stated in Section 9.2.4, the limited space available in a smartphone requires the antenna arrays to occupy an area as small as possible so, among the possible configurations, the corner placement represents the right trade-off between compactness and performance. To achieve the scan angle of 180° by simply switching to different antennas, it is necessary for each element to achieve ideally 45° beamwidth in different portions of the semi-hemisphere. For this reason, the Yagi–Uda elements are bent to point to different directions.

The simulated antenna array exhibits wideband performance, confirmed by the −10-dB impedance bandwidth from 24 to 42 GHz, with port-to-port isolation over 20 dB. The 3D-radiation patterns of the array elements, simulated at 28 GHz, are reported in Figure 9.11(a), 9.11(c), 9.11(e), and 9.11(g). Each antenna can steer the beam wider than 45° in a specific direction, allowing the overall scan angle of 180° with maximum gain of 8.2 dBi. Therefore, printing the same structure in the opposite corner allows to cover the total sphere. The optimized corner array of four elements is then fabricated (Figure 9.10(a)) and passive measurements are performed to validate the radiation performance of the beam-switchable antenna system.

9.4.2 Experimental validation with passive measurements

The set-up for the passive measurements is shown in Figure 9.10(b). The fabricated component is measured with the MVG StarLab 50 GHz, which allows to evaluate the realized gain of the four antenna elements for each value of ϕ, while θ varies from 0° to 180°. The 3D far-field radiation patterns at 28 GHz, obtained through near-field to far-field transformation, are reported in Figure 9.11(b), 9.11(d), 9.11(f), and 9.11(h), for comparison with the corresponding simulated ones. The measurements results are in good agreement with the simulations in terms of radiation properties. In fact, the desired scan angle of 180° is achieved, despite a decrease in gain of approximately 1 dB. However, due to limitations of the measurements set-up that were not included in the simulated environment, the measured gain is significantly low for $\theta > 160°$.

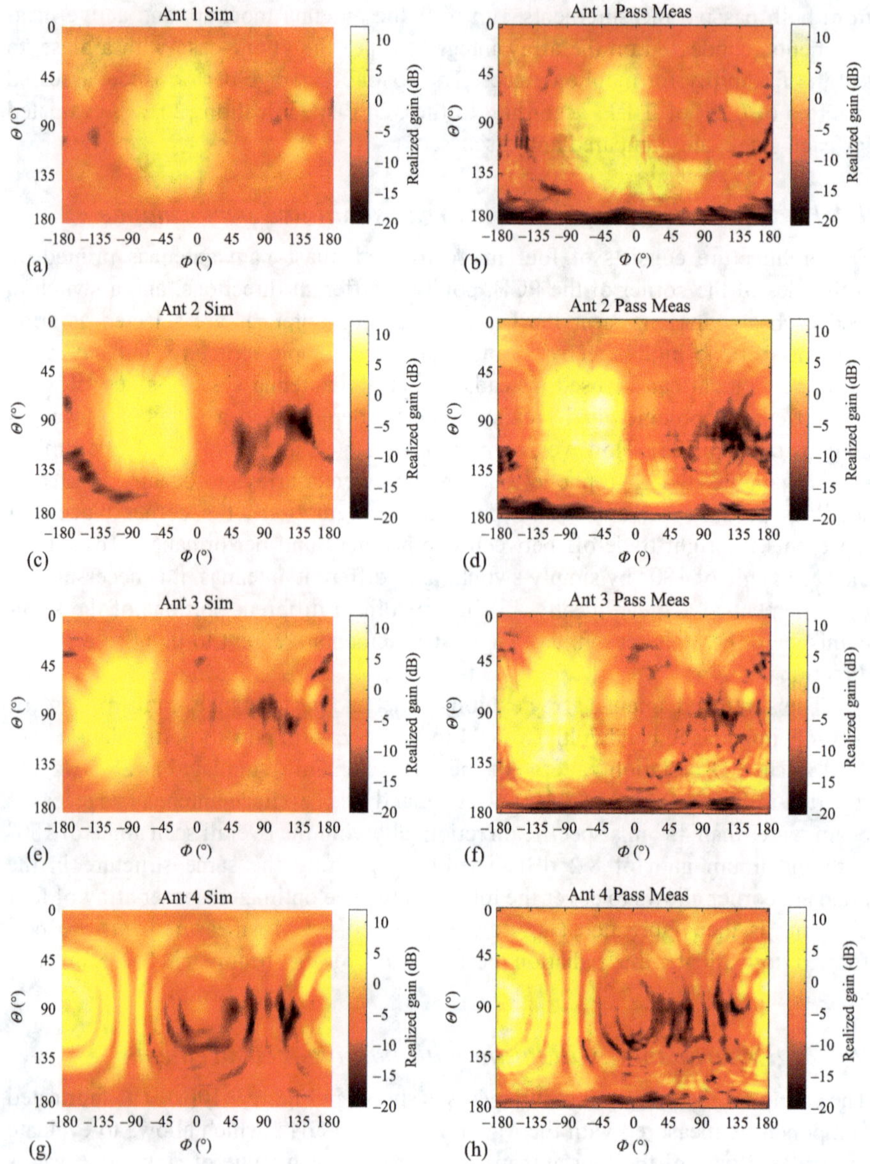

Figure 9.11 3D-radiation pattern of (a) antenna 1, (c) antenna 2, (e) antenna 3, (g) antenna 4 simulated and (b) antenna 1, (d) antenna 2, (f) antenna 3, and (h) antenna 4 measured at 28 GHz. For ϕ varying from $-180°$ to $180°$ the corresponding value along θ has been plotted

The beam-steering capability of the beam-switchable antenna array, shown by the simulations and confirmed by the measurements, is then further tested in the real scenario, where the antenna array is integrated in a phone-case with a glass. It is important to point out that, in the simulations stage, the impact of the housing and the other electronic components has not been taken into account; therefore the antenna array design has not been optimized considering the performance degradation caused by the integration.

9.4.3 Integration impact evaluation with active measurements

The performances of the prototype in the real scenario are tested through active measurements, where the array connected to the FEM is integrated in a polycarbonate phone-case, 0.7-mm distant from the cover, as presented in Figure 9.12(a) and 9.12(b). The FEM, realized by Sony, consists of a power amplifier (PA), a low-noise amplifier (LNA) and two single-pole four-throw switches, one switching between PA and LNA and the other selecting the array elements. The information regarding the feeding network and the gain of each antenna at 28 GHz, required for the evaluation of the link budget, are available in Table 9.4.

The active measurements set-up, shown in Figure 9.12(c), includes a spectrum analyser connected to the probe array, a power supplier, a signal generator with −10-dBm input power and the PA in the FEM. By rotating the set-up along ϕ, while θ varies from 0° to 162°, the 3D-radiation pattern of the four Yagi–Uda antennas at 28 GHz is measured and the results are presented in Figure 9.13(a), 9.13(c), 9.13(e) and 9.13(g). The desired coverage is achieved with EIRP of 16.7 dBm on average, due to limitations of the measurement system, which allows only −10-dBm input power. However, a back radiation, with around 19.5-dBm magnitude, is observed in the bottom right corner, caused by the presence of the switch, that is absent in simulations results, where the switch is not included. However, the most critical effect on the radiation properties of the antennas is given by the display. The highly conductive material, interacting with the electric currents on the surface of the antenna, highly affects the performance. In fact, a significant decrease in gain is observed in Figure 9.13(b), 9.13(d), 9.13(f), and 9.13(h), when the glass is mounted on the phone-case, whereas the unwanted back radiation results stronger and larger.

In order to evaluate the impact of the phone-case and the glass on the spatial performance of the antenna array at 28 GHz, in Figure 9.14 the corresponding CDF of EIRP is compared with the CDF obtained from simulations and passive measurements. To align the results, the same input power of −10 dBm is added to the gain obtained from simulations and passive measurements. The comparison shows that, with −10-dBm input power, the 3GPP spherical coverage requirements are not fulfilled. In particular, in order to meet both requirements, i.e. EIRP at 50%-tile CDF higher than 11.5 dBm and peak EIRP above 22.4 dBm, input power of minimum 12 dBm should be added to simulations, and 15 dBm to the other cases. Moreover, the plot offers a general view of the difference between simulated and

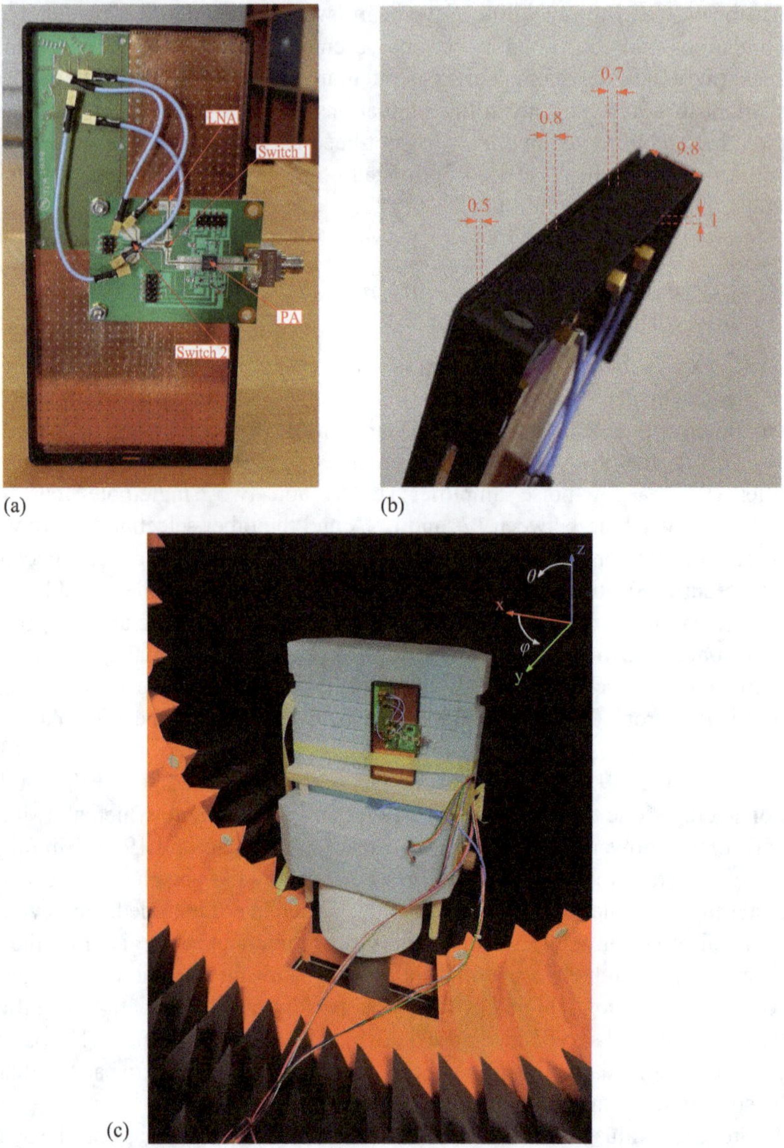

(a) (b) (c)

Figure 9.12 (a) Antenna array connected to the Sony FEM and integrated in the Sony phone-case prototype, (b) dimensions of case and glass in mm and (c) active measurements set-up in the MVG StarLab 50 GHz

fabricated prototype, and the performance degradation caused by the impact of the phone-case and the glass. In fact, at CDF = 50% the gain of the measured structure is approximately 3.5 dB lower than the simulated one and can be ascribed to poorer matching of the prototype, possibly due to the installation of the connectors, or to

Table 9.4 Array elements gain and feeding network characteristics at 28 GHz for the link budget

Component	**Parameter and value (dB)**
Antenna 1	Gain = 8
Antenna 2	Gain = 7.6
Antenna 3	Gain = 7.1
Antenna 4	Gain = 7.8
Switch	Insertion-loss = 1.6
FEM	Gain = 14 – 15*
Cables	Insertion loss = 1

*The value of the gain of the FEM includes the insertion loss of the switch.

fabrication issues. The gain decreases 0.5 dB, when the antenna array is integrated in the phone-case and further 0.5 dB, with the glass mounted.

An investigation a posteriori on the impact of the phone-housing and the glass on the array elements performance is conducted through the evaluation of the surface currents distribution in free space, after integration in the polycarbonate case and when the glass is added. The results for antenna 1 are reported in Figure 9.15. Compared to the free-space environment (Figure 9.15(a)), where the current generated by antenna 1 flows only to antenna 2, when adding the phone-case (Figure 9.15(b)), the current is spread along the PCB. Moreover, with the glass mounted, the surface currents are also directed to the ground plane, as shown in Figure 9.15(c). This analysis further corroborates the radiation performance degradation observed in the active measurements results and highlights the importance of including the integration impact in the design and optimization stage.

As stated in Section 9.2.4, phone-housing materials, such has polycarbonate, ceramic, glass, metal have a different impact on the spherical coverage performance of FR2 UEs. In fact, comparing the radiation pattern of antenna 1 placed inside polycarbonate and ceramic phone-cases, the signal attenuation through the ceramic cover is larger than the polycarbonate cover, as observed in Figure 9.16(a) and 9.16(b). Apart from high attenuation, the metal cover is also responsible of a significant distortion of the radiation characteristics of the antenna, decreasing the gain of the main beam, that is mainly reflected towards the bottom of the device (Figure 9.16(c)). However, the blockage of the main beam from the metal frame can be solved by adopting the solutions in [32–34].

In conclusion, the mm-wave antenna performances are highly impacted by the challenging electromagnetic environment encountered inside the mobile device, due to the small form factor and the interactions with the other components. Even though the model of phone-case and glass can be included in the simulations stage, the impact of the integration is hard to be predicted a priori. To this end, the simulation-assisted approach, presented in Section 9.3, through the reconstruction of the equivalent currents distribution over the phone-case, could represent a valid tool for a further investigation of the integrated beam-switchable array performance.

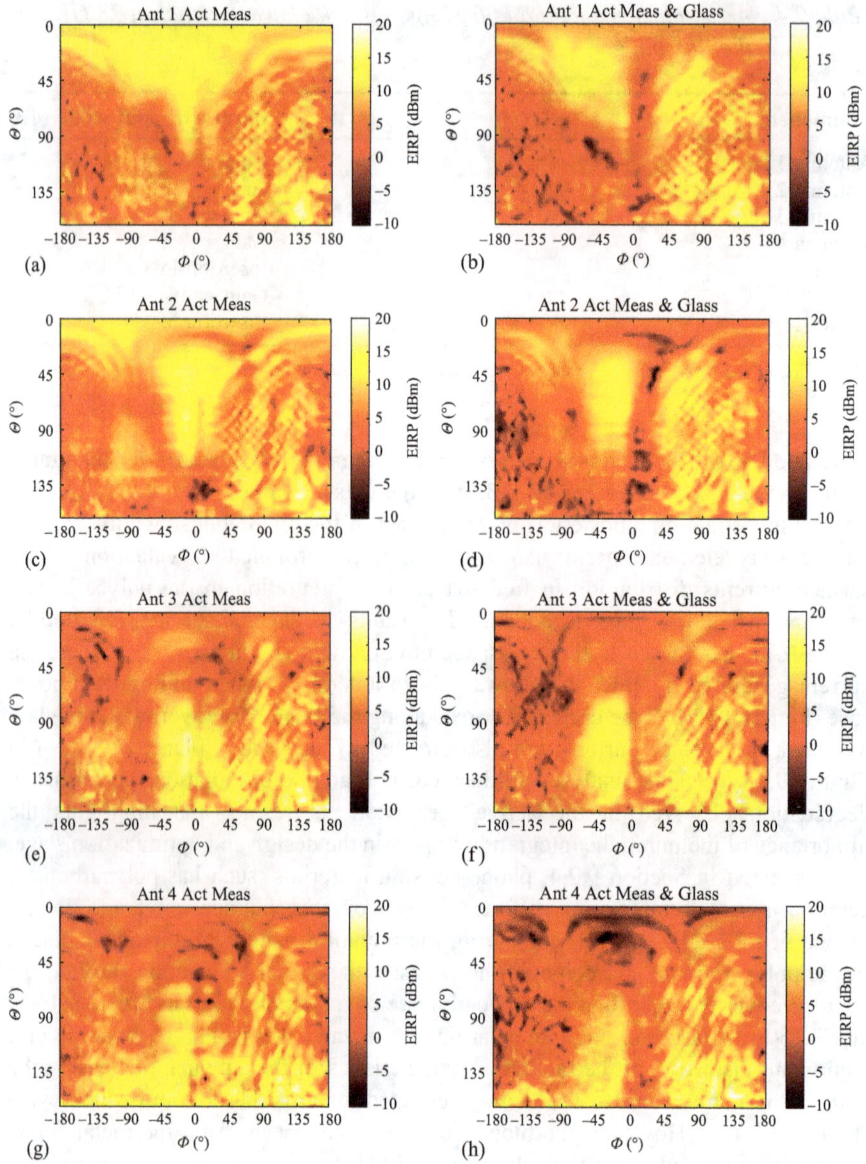

Figure 9.13 Measured radiation pattern at 28 GHz of (a) antenna 1, (c) antenna 2, (e) antenna 3, (g) antenna 4 inside the phone-case and (b) antenna 1, (d) antenna 2, (f) antenna 3, and (h) antenna 4 with the glass, respectively. For ϕ varying from $-180°$ to $180°$ the corresponding value along θ varying from $0°$ to $162°$ has been plotted

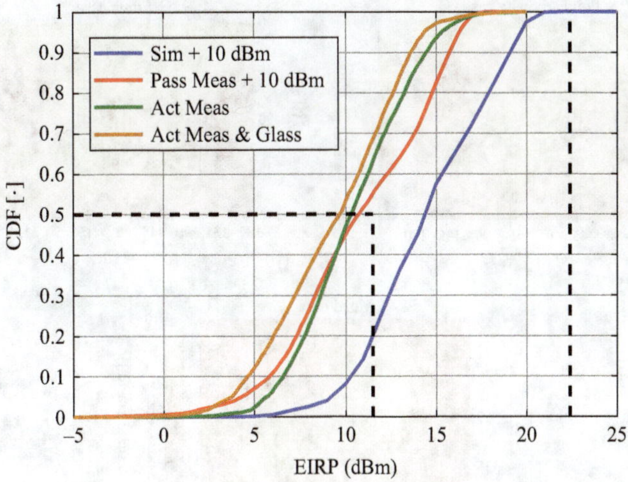

Figure 9.14 *CDF of the beam-switchable antenna array, simulated and measured at 28 GHz, including 12-dBm input power*

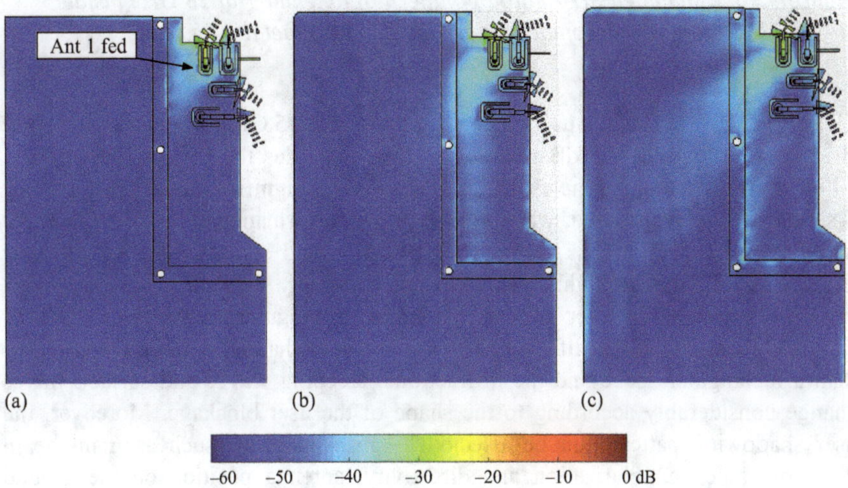

Figure 9.15 *Surface currents distribution generated by antenna 1 at 28 GHz: (a) in free space, (b) when the array is mounted inside the phone-case only and (c) with the glass mounted*

9.5 Effects and modelling of the user blockage on millimetre-wave mobile terminal antennas

Apart from the phone-housing integration, the spatial coverage is affected by the user blockage that becomes more serious at mm-wave frequencies, compared to

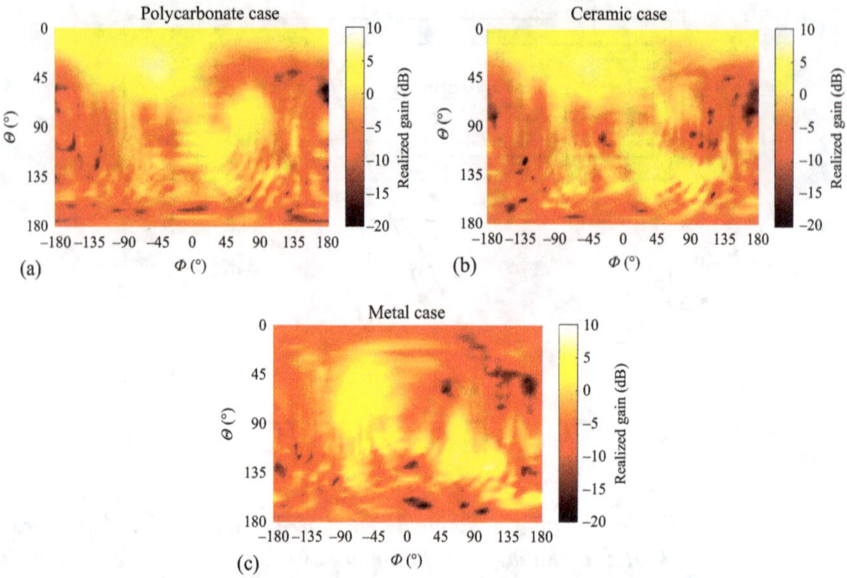

Figure 9.16 Simulated 3D-radiation pattern of antenna 1 at 28 GHz inside: (a) polycarbonate, (b) ceramic and (c) metal case

FR1 bands [29]. In [50] the user blockage is studied at 15 GHz, where a strong body shadow of around 20–25 dB is observed, that degrades the coverage efficiency. Although the blockage of the user at 28 GHz is not as significant as at 60 GHz [51], it should still be considered, when assessing the performance of the 5G mm-wave mobile communication link. The 3GPP has proposed some models of the user shadowing [30], though highly inaccurate and extremely simplified. For example, a model represents the user blockage with a rectangular space with −30-dB attenuation inside. In real life, the shape of the user shadow is more complicated than a rectangular space and the diffractions, creeping waves and surface waves change considerably according to the shape of the user blockage. Moreover, the user shadowing pattern can be affected by many aspects, such as main beam direction, antenna polarization and directivity, antenna position on the ground plane, user's grip and many others. For the above-mentioned reason, the user blockage is investigated in the following to draw an accurate model for mm-wave mobile communication.

9.5.1 Critical gesture investigation

Before modelling the user blockage pattern, measurements are conducted to find out how the different user gesture affects the shadow. Like in optics, the shape and intensity of the user shadow is different as the antenna–user distance and position changes. The measurements are carried out in both single- and dual-hand mode. The measurements set-up is shown in Figure 9.17(a), where the user holding the

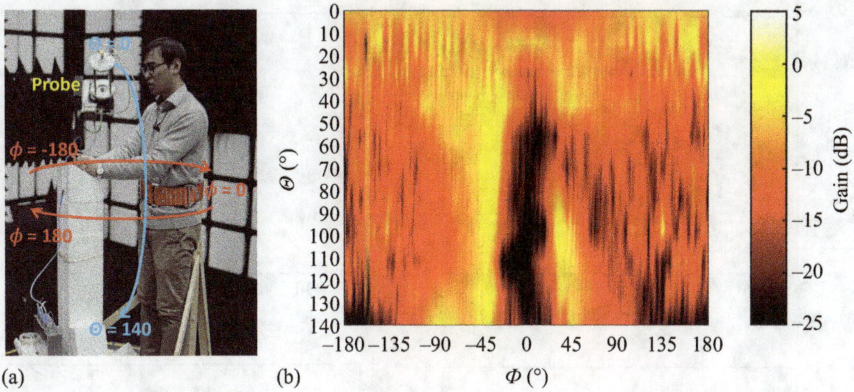

Figure 9.17 (a) Measurement set-up and (b) shadow pattern with the user in single-hand mode

antenna under test (AUT) stands on a platform, which rotates continuously in the ϕ-axis, while a probe mounted on a robot arm scans in the θ-axis, to obtain a 3D radiation pattern. The available measurement range is 0° to 140° in θ-axis and −180° to 180° in the ϕ-axis. Figure 9.17(b) shows the radiation pattern measured, while the user is holding the AUT in single-hand mode. The dark shape in the middle is the shadow of the user's body, and the dark parts to the right are caused by the user's arm and hand.

Two metrics are used to evaluate the user shadow:

- The shadowed antenna power ratio (SAPR) is used to assess the shadow intensity, which is first defined in [41]:

$$SAPR(\delta\theta, \delta\phi) = \frac{P_{shadow}}{P_{total}}$$

$$= \frac{\Delta\theta\Delta\phi \sum_{\theta_{shadow\_min}}^{\theta_{shadow\_max}} \sum_{\phi_{shadow\_min}}^{\phi_{shadow\_max}} (P_{ant,V}(\theta,\phi) + P_{ant,H}(\theta,\phi))\sin\theta}{\Delta\theta\Delta\phi \sum_{\theta_{min}}^{\theta_{max}} \sum_{\phi_{min}}^{\phi_{max}} (P_{ant,V}(\theta,\phi) + P_{ant,H}(\theta,\phi))\sin\theta}$$

  In the formula, P_{shadow} is the integrated power in the target shadow region, recorded by the probe for the two polarizations ($P_{ant,V}(\theta,\phi)$ and $P_{ant,H}(\theta,\phi)$), and P_{total} is the TRP measured in the whole space. SAPR is small when the user blockage is strong in the target region.
- The shadowed cumulative distribution function (SCDF) is introduced to indicate how the power intensity in the target shadow region is distributed and is defined as:

$$F(x) = P(X \leq x) \tag{9.4}$$

Where P is the probability that the power X in a point of the shadowing region has value less than or equal to a certain power level x.

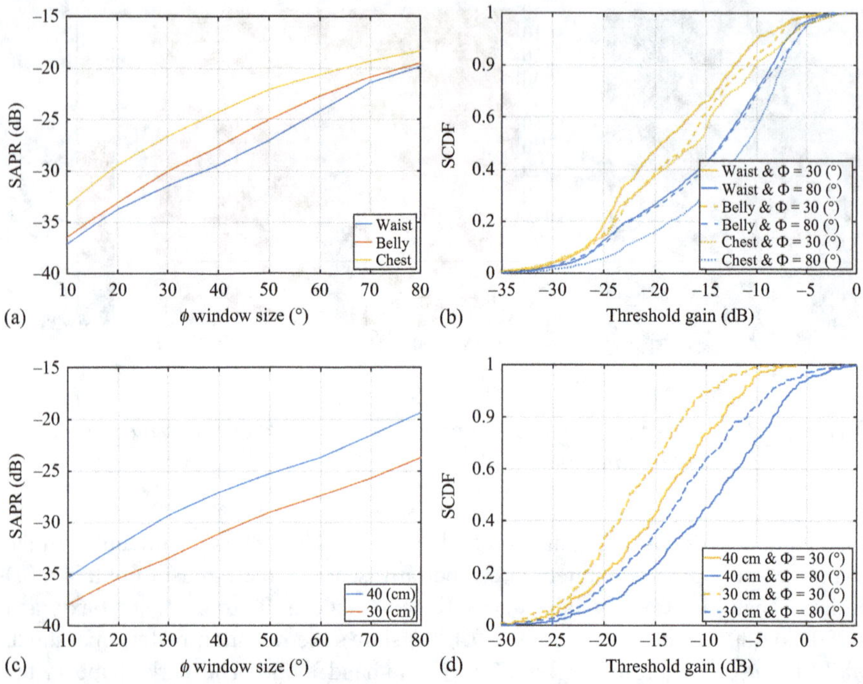

Figure 9.18 (a) SAPR and (b) SCDF evaluated at different heights and (c) SAPR and (d) SCDF calculated at different distances

In the investigation, the user shadow is sensitive only to the antenna–user height and the antenna–user distance. In the antenna–user height study, the user holds the AUT at three different heights which correspond to waist, belly and chest. In the antenna–user distance analysis, the AUT is positioned 30 and 40 cm away from the user. The calculated SAPRs and SCDFs for each configuration are shown in Figure 9.18. From the SAPR plots (Figure 9.18(a) and 9.18(c)), it can be observed that, as antenna height and distance to the user increase, the shadow intensity becomes smaller. In Figure 9.18(b) and 9.18(d), ϕ indicates the window size, meaning that the spatial points are integrated within $\pm\phi$ in the ϕ-axis and 0° to 140° in the θ-axis, which correspond to the main body shadow region. The SCDF distributions show the same trend as SAPR. Therefore, height and distance are the parameters considered for the following statistical measurements.

9.5.2 Statistical measurements

In order to obtain statistical results, multiple measurements with 18 different users are conducted in both single- and dual-hand mode. Based on the results previously discussed, two antenna positions are chosen. One position is at the user's waist height, while the user is watching movies or sending high dimensions files, the other is at the user's chest height, while the user is recording a video. In addition,

Table 9.5 Number of subjects for each height range

Height (cm)	<155	155–162	162–170	170–175	175–180	180–188	>188
No. of subjects	1	2	5	2	5	2	1

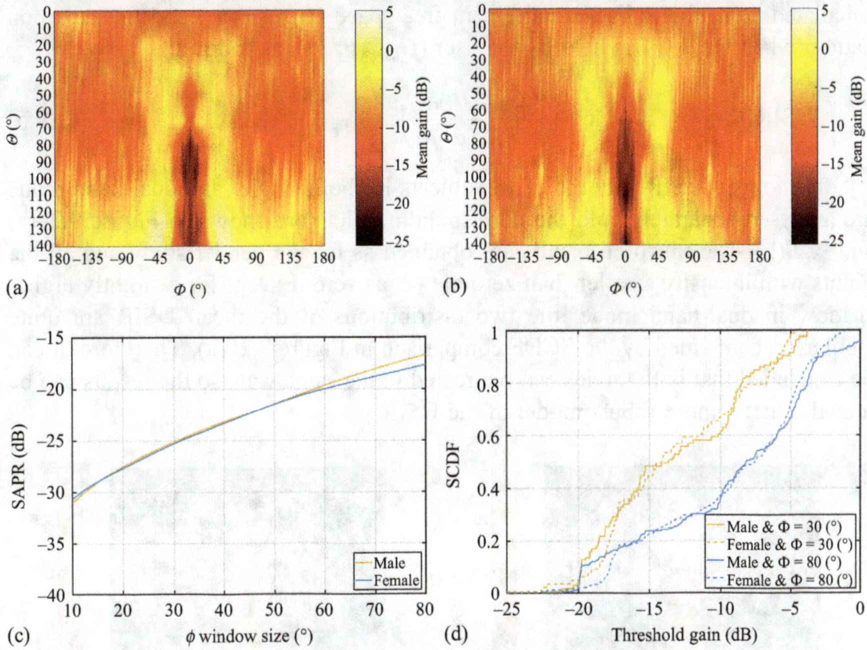

Figure 9.19 Mean gain radiation pattern of (a) male subject and (b) female subject with the same height. Comparison of the (c) SAPR and (d) SCDF of male and female subjects

the users are asked to choose their most comfortable position. According to the investigation in [52], the height distribution of 18 subjects is chosen and shown in Table 9.5.

9.5.2.1 User-shadowed investigation of different genders

In the first study, the difference between male and female subjects in terms of radiation pattern, SAPR and SCDF, is investigated. Two subjects of different gender with the same height are measured in both single- and dual-hand modes, then two extra results are obtained by mirroring at the centre ($\phi = 0$) to compensate the non-symmetrical difference. Mean gain radiation patterns for all the measured and mirrored results are shown in Figure 9.19(a) and 9.19(b). The patterns look quite similar, except from the wider shadow of the female subject in ϕ near the shoulders area, since the female body shape is not the same as the male. Concerning

SAPR and SCDF comparison, reported in Figure 9.19(c) and 9.19(d), respectively, the two male and female subjects show no significant difference; therefore the 18 users are selected with different gender for the statistical measurements.

9.5.2.2 User-shadowed investigation of single- and dual-hand mode

To assess the real shadowed intensity, the user-shadowed intensity ratio (USIR) is calculated from the radiation pattern in free space ($G_{free}(\theta, \phi)$) and the radiation pattern when the AUT is held by the user ($G_{user}(\theta, \phi)$), as follows:

$$\mathrm{USIR}(\theta, \phi) = 10 \times \log_{10}\left(\frac{G_{free}(\theta, \phi)}{G_{user}(\theta, \phi)}\right). \tag{9.5}$$

Then mean USIR patterns of 18 subjects for both single- and dual-hand mode are analysed separately, and the corresponding plots are shown in Figure 9.20(a) and 9.20(b). The mirrored results are obtained as for the gender study and all the points with intensity smaller than zero are set as zero. Except for a slightly higher shadow in dual-hand mode, the two distributions of the mean USIR are quite similar, as confirmed by the SCDF comparison in Figure 9.20(a). Therefore, it can be concluded that both modes can be treated in the same way, so the results can be mixed to get a more robust model of the USIR.

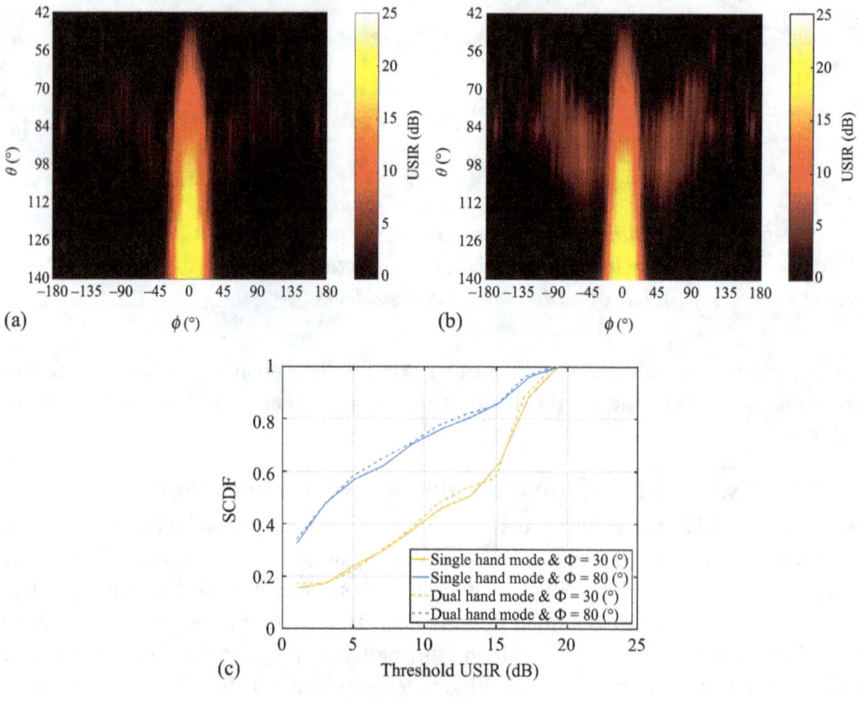

Figure 9.20 Mean USIR of (a) single-hand mode, (b) dual-hand mode and (c) corresponding SCDF comparison

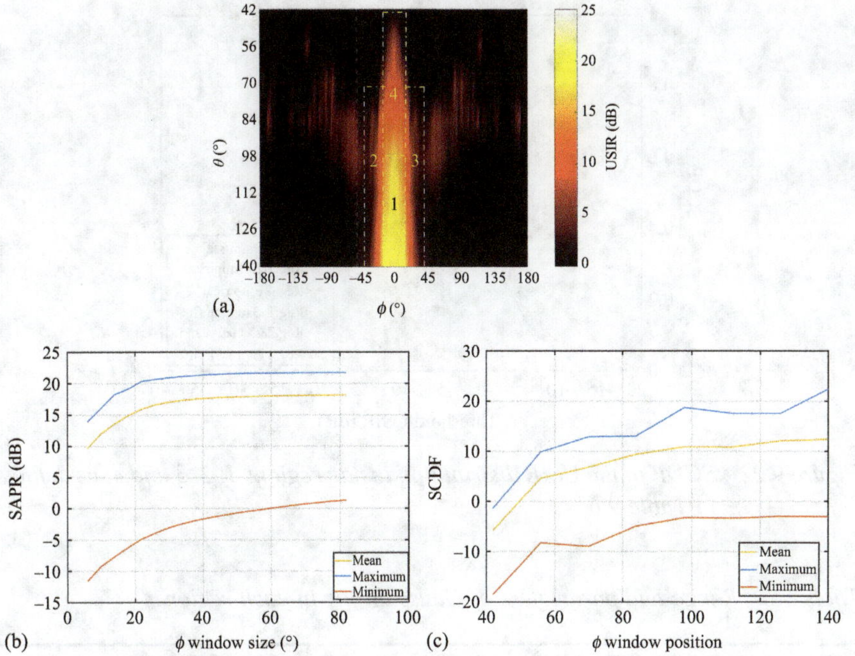

Figure 9.21 (a) Mean USIR of all subjects, (b) SAPR calculated in the ϕ window and (c) in the θ window

9.5.3 User shadowing pattern modelling

9.5.3.1 USIR modelling

The USIR is modelled as a random process. First, a mean pattern of all USIR is plotted in Figure 9.21(a). It can be noticed that the main shadow is within $\phi = \pm 45°$ and is much stronger than that of arms and hands at approximately $45° < \phi < 135°$ and $-135° < \phi < -45°$. However, in order to define these regions more accurately, SAPR is evaluated in ϕ and θ directions separately. First, SAPR is calculated for the variable window size from $6° < \phi < 90°$, where the size of θ window is from 42° to 140°. The SAPR calculated in the ϕ direction is shown in Figure 9.21(b), where two tendencies for all the curves can be observed. From 6° to approximately 32°, the curve increases steeply, then after 32° all three curves flatten. Then, it has been chosen to use two regions for the model: $\Delta\phi = \pm 16°$ and $\Delta\phi = \pm 16°$ to $\pm 45°$. Concerning the evaluation of SAPR in the θ direction (reported in Figure 9.21(c)), sliding window of 14° is used and SAPR is calculated at each position of the window within $\Delta\phi = \pm 16°$. In this case, it is quite difficult to identify head, shoulders and body regions. However, three tendencies for the mean curve can be observed. From 42° to 56°, the curve increases steeply, since the users head is detected, from 56° to 98° the curve is less steep, so it can be defined as transitional region, where diffractions, creeping waves and surface waves from the

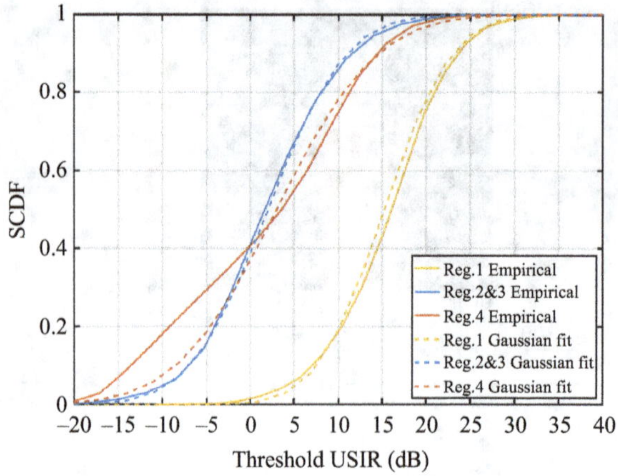

Figure 9.22 SCDF of the USIR distribution inside regions 1, 2, 3 and 4, as defined in Table 9.6

Table 9.6 Range and parameters of Gaussian fits in each region

Region number	**μ**	**σ**	**Range in ϕ**	**Range in θ**
1	16	6	−16° to 16°	98° to 140°
2	2	7	16° to 45°	70° to 140°
3	2	7	−45° to −16°	70° to 140°
4	3	9	−16° to 16°	42° to 98°

head occur together to differing degrees, according to users height. After 98° the curve flattens, representing the main body shadowing region. The four regions are shown together in Figure 9.22, where region 1 corresponds to the body shadow and regions 2, 3 and 4 are the transitional regions of body edge and head, respectively.

The USIR regions are based on the mean shadowing pattern from the measurements in single- and dual-hand modes. In order to model the USIR as a random process, the USIR in each spatial point within the region should follow a specific distribution. Since regions 2 and 3 have similar distributions, only SCDFs for regions 1, 2 and 4 are considered. With the aim of simplifying the model, it is chosen to fit the Gaussian distribution to the empirical curves. For the following Gaussian distribution, the standard deviation (σ), mean (μ) and size of the four regions are given in Table 9.6:

$$F(x) = \frac{1}{2}\left(1 + erf\left(\frac{x-\mu}{\sigma\sqrt{2}}\right)\right). \tag{9.6}$$

In Figure 9.22, the empirical curves correspond to the SCDFs of the USIRs of the 18 users with four different gestures and their mirrors, while the Gaussian fit curves are the proposed distributions. Empirical and Gaussian curves show high agreement. In region one, most of the spatial points are above 0 dB, which agrees with the main body shadowing region. Due to the diffractions and creeping waves in regions 2 and 3, about 38% of the total points are below 0 dB. Concerning region 4, the Gaussian curve fits the empirical one only above 0 dB, because of the impact of the different users height, apart from diffractions and creeping waves affecting also regions 2 and 3. As the main interest of the investigation is the users shadow, not the reflections, then Gaussian distribution is chosen to fit the empirical curve above 0 dB.

9.5.3.2 Model verification

To check if the proposed model is similar to the mean measured results, a MATLAB model is created consisting in running the random process 36 times (corresponding to 18 measurements in single-hand mode and 18 in dual-hand mode), according to the four regions Gaussian fit and sorting method mentioned before, that yields to 36 different USIRs. The mean USIR corresponding to the 36 times random process is plotted in Figure 9.23(a), where a good agreement with the results in Figure 9.21(a) can be noticed. However, the modelled figure is cleaner, due to the fact that random processes result is sorted, whereas the mean measured result is affected by the randomness given by diffractions, reflections and the different shadow shape of the subjects involved. Finally, the 3GPP self-blocking model in portrait mode [39] is shown in Figure 9.23(b). Compared to that, the proposed model gives a more realistic shape and distribution in the shadowing region. Moreover, it can be considered accurate and reliable, thanks to the different users and scenarios included in the statistical measurements.

9.5.4 User body effect on the propagation channel

The blockage from the user body limits the spherical coverage of mobile antennas at mm-wave frequencies and negatively impacts the mobile network's performance. The signal from the BS can be propagated through a multipath propagation channel and reaches the mobile terminal at one or more random AoA. In the

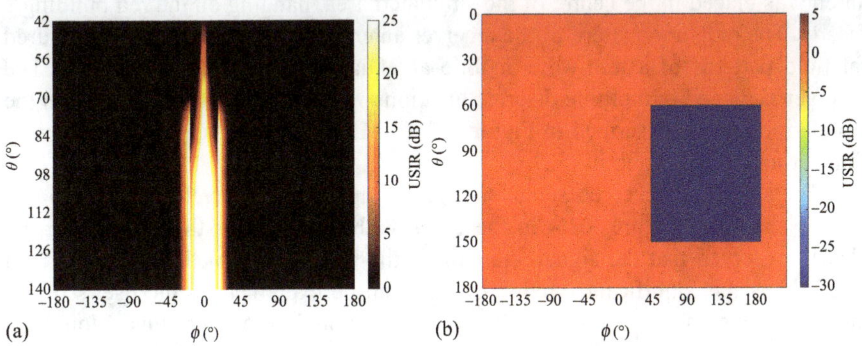

Figure 9.23 (a) Proposed model from 36 random processes and (b) 3GPP self-blocking model in portrait mode

Figure 9.24 An illustration of the user body shadowing the wireless signal from the BS

mm-wave spectrum, due to the significant blockage from the user's body, the signals received from the AoAs behind the user body can be severely shadowed (see Figure 9.24). Therefore, it is critical to assess the body effect on the propagation of the wireless signals to obtain a more intuitive result on the system-level impact.

In [52], a study about the wireless propagation, including the impact of user body at 28 GHz, is carried out based on 3D-raytracing simulations. The raytracing simulation model is shown in Figure 9.25(a), where a suburban environment with multiple buildings, high 15– 20 m, is studied. A transmitter with an omnidirectional antenna is placed in the centre of the simulated area, hanging on the red building's facade. The radiation pattern of the receiver antenna is obtained from the measured far-field patterns of a mm-wave mobile antenna with and without the user. The receiver is placed in multiple discrete locations within the simulation area, and the simulation flow is described in Figure 9.25(b). Detailed parameters of the simulation set-up can be found in [52].

The simulated received signal strength along the main street of the simulated area is shown in Figure 9.26, in the cases with and without the user. It can be clearly observed that the signal strength on the bottom part of the simulated area with the user is significantly reduced. This can be explained by the user's orientation with the back to the transmitter that blocks the line-of-sight link. More than 10– 20 dB signal attenuation can be observed, due to the shadowing effect of the user body, which points out the necessity of taking the user body effect into account, when planning the mm-wave cellular networks.

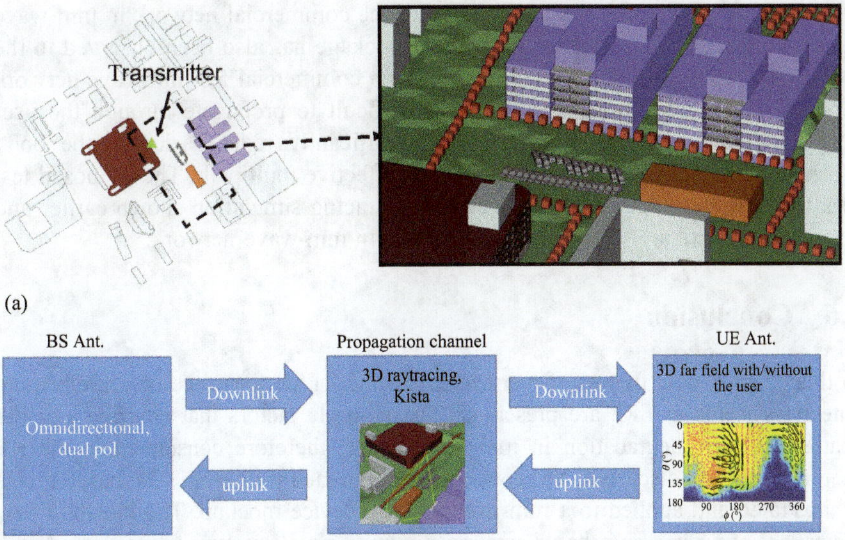

Figure 9.25 (a) The 3D raytracing model in a suburban environment and (b) the flow of the raytracing simulation with user body effect

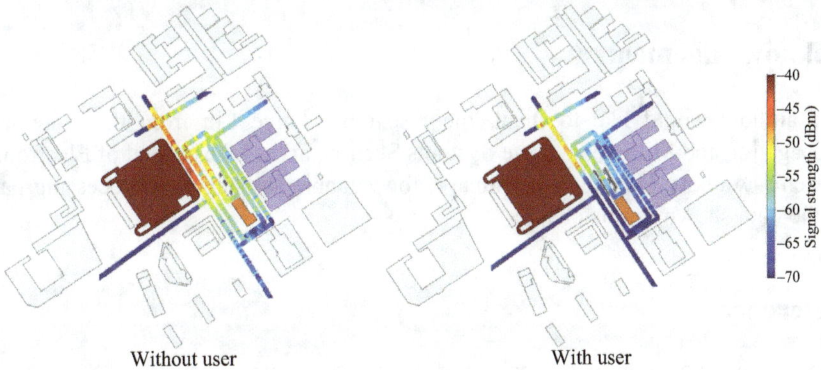

Figure 9.26 The simulated received signal strength at 28 GHz in the sub-urban environment with and without the user effect

It is important to mention that the above consideration is valid for a propagation environment with a strong waveguide effect, like the one of the buildings along the main street in this case. The AoAs are limited to a few directions with a small angular spread, which makes the user body blockage effect become more pronounced. On the other hand, if the channel contains rich multipath components with largely spread AoAs, the user body's blockage effect can be mitigated to a certain extent [52].

In recent years, with the deployment of the commercial network in mm-wave bands, the severe impact of the user body blockage has also been observed in the real scenario. According to the test results of a commercial 5G mm-wave network reported in [53], the mm-wave signal is difficult to propagate through the user body, and the data throughput can be significantly reduced due to the body blockage, especially in the locations with ineffective multipath. The proposed test result provides similar conclusions as the raytracing simulation shown earlier and confirms the critical effect of the user's body in mm-wave network.

9.6 Conclusion

In this chapter, new figures of merit to assess the RF performance of beamforming antenna systems in FR2 are presented. The multiple factors that contribute to the spatial coverage degradation in mm-wave bands, therefore considered in the RF requirements, are discussed. Moreover, a new approach for near-field measurements is illustrated and applied to a mm-wave mobile device mockup. The radiation performances of a beam-switchable antenna system in a real scenario are evaluated with active measurements to quantify the spatial coverage performance degradation caused by the antenna integration. Finally, the blockage from the user body is investigated in different scenarios with statistical measurements. A new model of the user shadowing, which takes into account different users gender, height, gestures, is proposed.

Acknowledgements

The authors would like to thank the researcher Peiye Liu from the Antennas Propagation and Millimeter-wave Systems Section, at the Department of Electronic Systems, Aalborg University, Denmark, for valuable assistance in processing part of the data shown in this chapter.

References

[1] 3GPP TS, 38 101-1. NR. User equipment (UE) radio transmission and reception; Part 1: Range 1 Standalone (16.6.0 ed.); 2021 January.

[2] 3GPP TS, 38 101-2. NR. User equipment (UE) radio transmission and reception; Part 2: Range 2 Standalone (16.6.0 ed.); 2020 December.

[3] 3GPP TR, 38 847. New frequency range for NR (47.2 48.2 GHz) (0.1.0 ed.); 2021 January.

[4] 3GPP TR, 38 808. Study on supporting NR from 52.6 GHz to 71 GHz (1.0.0 ed.); 202 November.

[5] Pi Z and Khan F., An introduction to millimeter-wave mobile broadband systems. *IEEE Communications Magazine*. 2011; 49(6).

[6] Rappaport TS, Sun S, Mayzus R, *et al.*, Millimeter wave mobile communications for 5G cellular: It will work! *IEEE Access*. 2013; 1: 335–349.

[7] Rodriguez I, Nguyen HC, Sorensen TB, *et al.*, Analysis of 38 GHz mmWave propagation characteristics of urban scenarios. In: *Proceedings of European Wireless 2015; 21th European Wireless Conference*. VDE; 2015. p. 1–8.

[8] Rajagopal S, Abu-Surra S, Pi Z, *et al.* Antenna array design for multi-GBPS mmwave mobile broadband communication. In: *Global Telecommunications Conference (GLOBECOM)*. IEEE; 2011. p. 1–6.

[9] Hong W, Jiang ZH, Yu C, *et al.*, Multibeam antenna technologies for 5G wireless communications. *IEEE Transactions on Antennas and Propagation*. 2017; 65(12): 6231–6249.

[10] Roh W, Seol JY, Park J, *et al.*, Millimeter-wave beamforming as an enabling technology for 5G cellular communications: Theoretical feasibility and prototype results. *IEEE Communications Magazine*. 2014; 52(2): 106–113.

[11] Hong W, Baek KH, Lee Y, *et al.*, Study and prototyping of practically large-scale mmWave antenna systems for 5G cellular devices. *IEEE Communications Magazine*. 2014; 52(9): 63–69.

[12] Lian JW, Ban YL, Chen Z, *et al.*, SIW folded Cassegrain lens for millimeter-wave multibeam application. *IEEE Antennas and Wireless Propagation Letters*. 2018; 17(4): 583–586.

[13] Cheng Y, Hong W, and Wu K., Design of a substrate integrated waveguide modified R-KR lens for millimetre-wave application. *IET Microwaves, Antennas & Propagation*. 2010; 4(4): 484–491.

[14] Tekkouk K, Ettorre M, Le Coq L, *et al.*, Multibeam SIW slotted waveguide antenna system fed by a compact dual-layer Rotman lens. *IEEE Transactions on Antennas and Propagation*. 2016; 64(2): 504–514.

[15] Zhang YS and Hong W., A millimeter-wave gain enhanced multi-beam antenna based on a coplanar cylindrical dielectric lens. *IEEE Transactions on Antennas and Propagation*. 2012; 60(7): 3485–3488.

[16] Di Paola C, Zhao K, Zhang S, *et al.*, Wideband SIW horn antenna with phase correction for new generation beam steerable arrays. In: *2019 IEEE 90th Vehicular Technology Conference (VTC2019-Fall)*. IEEE; 2019. p. 1–4.

[17] Di Paola C, Zhao K, Zhang S, *et al.*, A novel lens antenna design based on a bed of nails metasurface for new generation mobile devices. In: *The 14th European Conference on Antennas and Propagation*. IEEE; 2020.

[18] Quevedo-Teruel O, Ebrahimpouri M, and Ghasemifard F., Lens antennas for 5G communications systems. *IEEE Communications Magazine*. 2018; 56(7): 36–41.

[19] Lian JW, Ban YL, Xiao C, *et al.*, Compact substrate-integrated 4× 8 Butler matrix with sidelobe suppression for millimeter-wave multibeam application. *IEEE Antennas and Wireless Propagation Letters*. 2018; 17(5): 928–932.

[20] Cao Y, Chin KS, Che W, *et al.*, A compact 38 GHz multibeam antenna array with multifolded butler matrix for 5G applications. *IEEE Antennas and Wireless Propagation Letters*. 2017; 16: 2996–2999.

[21] Zhong LH, Ban YL, Lian JW, *et al.*, Miniaturized SIW multibeam antenna array fed by dual-layer 8× 8 Butler matrix. *IEEE Antennas and Wireless Propagation Letters*. 2017; 16: 3018–3021.

[22] Zhang S, Huff G, Feng J, *et al.*, A pattern reconfigurable microstrip parasitic array. *IEEE Transactions on Antennas and Propagation*. 2004; 52(10): 2773–2776.

[23] Trad IB, Floc'h JM, Rmili H, *et al.*, Design of reconfigurable radiation pattern dipole antenna with director and reflector elements for telecommunication systems. In: *2013 Loughborough Antennas & Propagation Conference (LAPC)*. IEEE; 2013. p. 117–121.

[24] Jusoh M, Aboufoul T, Sabapathy T, *et al.*, Pattern-reconfigurable microstrip patch antenna with multidirectional beam for WiMAX application. *IEEE Antennas and Wireless Propagation Letters*. 2014; 13: 860–863.

[25] Di Paola C, Zhang S, Zhao K, *et al.*, Wideband beam-switchable 28 GHz quasi-Yagi array for mobile devices. *IEEE Transactions on Antennas and Propagation*. 2019; 67(11): 6870–6882.

[26] 3GPP TR 38 817-01. General aspects for user equipment (UE) radio frequency (RF) for NR (16.2.0 ed.); 2020 October.

[27] Zhao K, Zhang S, Ho Z, *et al.*, Spherical coverage characterization of 5G millimeter wave user equipment with 3GPP specifications. *IEEE Access*. 2018; 7: 4442–4452.

[28] Hong W, Baek KH, and Ko S., Millimeter-wave 5G antennas for smartphones: Overview and experimental demonstration. *IEEE Transactions on Antennas and Propagation*. 2017; 65(12): 6250–6261.

[29] Krogerus J, Icheln C, and Vainikainen P., Effect of human body on 3-D radiation pattern and efficiency of mobile handsets. In: *2005 IEEE Instrumentation and Measurement Technology Conference Proceedings*. vol. 1. IEEE; 2005. p. 271–276.

[30] 3GPP TR 38 901. Technical specification group radio access network; study on channel model for frequencies from 0.5 to 100 GHz; 2017 July.

[31] Giordani M, Polese M, Roy A, *et al.*, Standalone and non-standalone beam management for 3GPP NR at mmWaves. *IEEE Communications Magazine*. 2019; 57(4): 123–129.

[32] Rodriguez-Cano R, Zhang S, Zhao K, *et al.*, Reduction of main beam-blockage in an integrated 5G array with a metal-frame antenna. *IEEE Transactions on Antennas and Propagation*. 2019; 67(5): 3161–3170.

[33] Rodriguez-Cano R, Zhang S, Zhao K, *et al.*, mm-Wave beam-steerable endfire array embedded in a slotted metal-frame LTE antenna. *IEEE Transactions on Antennas and Propagation*. 2020; 68(5): 3685–3694.

[34] Rodriguez-Cano R, Zhao K, Zhang S, *et al.*, Handset frame blockage reduction of 5G mm-wave phased arrays using hard surface inspired structure. *IEEE Transactions on Vehicular Technology*. 2020; 69(8): 8132–8139.

[35] Kurvinen J, Kähkönen H, Lehtovuori A, *et al.*, Co-designed mm-wave and LTE handset antennas. *IEEE Transactions on Antennas and Propagation*. 2018; 67(3): 1545–1553.

[36] Jørgensen E, Meincke P, Cappellin C, *et al.*, *Improved source reconstruction technique for antenna diagnostics*. In: Proceedings of the 32nd ESA Antenna Workshop; 2010.

[37] Quijano JLA, Scialacqua L, Zackrisson J, *et al.*, Suppression of undesired radiated fields based on equivalent currents reconstruction from measured data. *IEEE Antennas and Wireless Propagation Letters*. 2011; 10: 314–317.

[38] Lundgren J, Helander J, Gustafsson M, *et al.*, *A near-field measurement and calibration technique: Radio-frequency electromagnetic field exposure assessment of millimeter-wave 5G devices*. *IEEE Antennas and Propagation Magazine*. 2020.

[39] He W, Xu B, Scialacqua L, *et al.*, Fast power density assessment of 5G mobile handset using equivalent currents method. *IEEE Transactions on Antennas and Propagation*. 2021; 69(10): 6857–6869.

[40] Taaghol A and Sarkar TK., Near-field to near/far-field transformation for arbitrary near-field geometry, utilizing an equivalent magnetic current. *IEEE Transactions on Electromagnetic Compatibility*. 1996; 38(3): 536–542.

[41] Sarkar TK and Taaghol A., Near-field to near/far-field transformation for arbitrary near-field geometry utilizing an equivalent electric current and MoM. *IEEE Transactions on Antennas and Propagation*. 1999; 47(3): 566–573.

[42] Álvarez Y, Las-Heras F, and Pino MR., Reconstruction of equivalent currents distribution over arbitrary three-dimensional surfaces based on integral equation algorithms. *IEEE Transactions on Antennas and Propagation*. 2007; 55(12): 3460–3468.

[43] Quijano JLA and Vecchi G., Improved-accuracy source reconstruction on arbitrary 3-D surfaces. *IEEE Antennas and Wireless Propagation Letters*. 2009; 8: 1046–1049.

[44] Quijano JLA and Vecchi G., Field and source equivalence in source reconstruction on 3D surfaces. *Progress in Electromagnetics Research*. 2010; 103: 67–100.

[45] Quijano JLA and Vecchi G., Near-and very near-field accuracy in 3-D source reconstruction. *IEEE Antennas and Wireless Propagation Letters*. 2010; 9: 634–637.

[46] Harrington R, *Fields THE*. Hoboken. Wiley-IEEE Press; 2001.

[47] Xu B, Ying Z, Scialacqua L, *et al.*, Radiation performance analysis of 28 GHz antennas integrated in 5G mobile terminal housing. *IEEE Access*. 2018; 6: 48088–48101.

[48] Microwave Vision Group. Available from: https://www.mvg-world.com/en/products/antenna-measurement/software/insight.

[49] Dassault Systèmes. Available from: https://www.3ds.com/products-services/simulia/products/cst-studio-suite/.

[50] Zhao K, Helander J, Sjöberg D, *et al.*, User body effect on phased array in user equipment for the 5G mmWave communication system. *IEEE Antennas and Wireless Propagation Letters*. 2016; 16: 864–867.

[51] Wang T, Umehira M, Otsu H, *et al.*, A twin cylinder model for moving human body shadowing in 60GHz WLAN. In: *2015 21st Asia-Pacific Conference on Communications (APCC)*. IEEE; 2015. p. 188–192.

[52] Zhao K, Gustafson C, Liao Q, *et al.*, Channel characteristics and user body effects in an outdoor urban scenario at 15 and 28 GHz. *IEEE Transactions on Antennas and Propagation*. 2017; 65(12): 6534–6548.
[53] Narayanan A, Ramadan E, Carpenter J, *et al.*, *A first look at commercial 5G performance on smartphones*. In: Proceedings of The Web Conference 2020; 2020. p. 894–905.

Part IV

System OTA metrology

Chapter 10

Over-the-air testing metrology of 5G radios

Wei Fan[1], Pekka Kyosti[2,3], Ya Jing[4] and Zhiqin Wang[5]

10.1 Introduction

5G new radio (NR) (i.e., the global 5G standard) is introduced to significantly improve wireless system performance in terms of spectral efficiency, coverage, data-rate, latency, number of connected devices, and energy efficiency. As 5G matures, now is the time for its large-scale deployments and commercialization. Several key radio technologies have been introduced to enable the 5G NR, e.g., high operating frequency (e.g., millimeter-wave (mm-wave) frequency), wider system bandwidth as compared with 4G, large-scale antenna configuration, and integrated radio frequency (RF) front-end design. Thorough testing of 5G NR radios is essential before their massive rollout. Testing is required in the stage of device troubleshooting, regression, validation, and performance evaluation. However, these NR technology advances have posed significant challenges on testing methods. Test complexity has significantly increased due to higher operating frequencies, wider system bandwidth, increased minimum number of test cases, increased overall test time, electrically larger antenna under test (AUT), and necessity of over-the-air (OTA) testing.

Traditional cable conducted testing is getting problematic due to complexity introduced in advances RF and antenna technologies. Components of radio devices, including the modem, RF circuits, and antennas, were individually designed and separately tested in the traditional conducted test setup. However, integrated RF front-end design is expected in the 5G NR due to cost, size, and design consideration. If RF coaxial cables and connectors, which are not part of the final wireless product, are employed during the testing, they introduce losses and mismatches and may also intrusively affect the radiation pattern profile of the wireless product under test. Antenna ports, if accessible, become massively bulky due to large-scale antenna array configuration, which makes traditional cable conducted

[1]Department of Electronic Systems, Aalborg University, Aalborg, Denmark
[2]Centre for Wireless Communications, Oulu University and Keysight Technologies, Oulu, Finland
[3]Keysight Technologies, Oulu, Finland
[4]Keysight Technologies, Beijing, China
[5]China Academy of Information and Communications Technology, Beijing, China

testing complicated, expensive, and error-prone. Furthermore, many benefits of 5G NR systems are introduced by the spatial discrimination capability of their employed antenna arrays (e.g., beamforming toward desired signal directions and null steering toward undesired directions), which cannot be measured accurately at each antenna element port. Therefore, radiated OTA testing, where built-in antennas are used directly as the interface to receive/transmit test signals, will be the main-stream method for 5G radios, especially for radios operating at mm-wave bands and above.

In this chapter, the focus is given on OTA testing methodologies of 5G radios. The first part focuses on the wireless system performance testing, where the objective is to evaluate 5G radio performance, e.g., throughput, under realistic spatial channel conditions. Two standard OTA testing methods, namely, the multi-probe anechoic chamber (MPAC) and radiated two-stage (RTS) method, are detailed. The second part focuses on the RF testing, where the goal is to test 5G RF parameters under ideal plane wave conditions. The focus is given on a novel midfield testing method. Other mature RF testing methods, e.g., direct far-field (DFF), compact antenna testing range (CATR), plane wave generator (PWG), and near- to far-field (NF-FF) transformation method, can be found in many references in the literature and therefore are not detailed in this chapter. The third part introduces the current status of OTA testing of 5G radios in standardization, where both wireless system performance testing and RF testing are discussed.

10.2 Performance testing

10.2.1 Introduction

When wireless communication devices, e.g., mobile phones, are under development or being manufactured, it is typically mandatory to test if their radio performance meets the requirement or specification under realistic propagation conditions. A natural and intuitive way to test it is to measure the performance of the device-under-test (DUT) in the live network of different scenarios, e.g., urban or suburban scenarios. This is the so-called field trials. Field trials do tell the real-world performance of DUTs, but they often suffer from the uncertainty introduced from the environment. The test results may not be repetitive due to, e.g., various weather conditions and propagation phenomenon. The variation of the test results makes it difficult to judge if a DUT is qualified in regard to the designed specification. Moreover, it is also unfair to compare the test results of different DUTs in field trials due to any potential changes in the propagation channel.

To make the test results both representative of real-world performance and comparable between different DUTs, an alternative, i.e., virtual drive testing, is often adopted in the industry. Virtual mobility testing can be thought of as doing the aforementioned field trials in a more controllable, repeatable, and reproducible manner by "emulating" the propagation channel to laboratories. The propagation channel can be emulated according to the standard channel models that are widely adopted for testing. With virtual mobility testing, different DUTs can be tested under exactly the same channel, and hence the test results are more repeatable, and the comparison of the test results is fairer.

Conventionally, virtual mobility testing has been done in a conducted manner, where radio signals are sent over coaxial cables. This is called the conducted testing. A basic setup of the conducted testing consists of three components, including a signal generator that generates the test signal, a channel emulator (CE) that generates the wanted propagation channel and performs convolution between the test signal and the channel, and lastly the DUT that receives and demodulates the faded test signal. All three components are wired up with coaxial cables.

Though the setup is very simple and straightforward, there are a few disadvantages to it. First, the DUT is fed with the test signals via cables, which is done at the antenna ports in the DUT. This means the original enclosure of the DUT has to be opened, and the antennas inside the DUT need to be unmounted to give the port-to-cable connections. The manual handling could be tedious and time-consuming, especially with mass volume testing. Moreover, when the antenna module is directly integrated to the logic board, which means there are no antenna ports available for cable connection, the conducted testing becomes impractical. Second, the effect of the antenna radiation pattern is accounted for in the CE when generating the wanted channel coefficients. Nevertheless, the radiation pattern would need to be measured beforehand. If the radiation pattern is measured with the antenna module alone, the resulting radiation pattern may not reflect the influence of, e.g., the chassis and other components of the DUT, which may cause discrepancy on the antenna radiation pattern in reality, and hence, compromise the effectiveness of the test results. Some DUTs may support the so-called antenna test functions (ATFs), which allows us to measure the antenna radiation pattern onboard. In this case, the discrepancy of the measured antenna radiation pattern to the reality can be minimized.

Due to these reasons, the testing community has been moving to an OTA manner. However, since the desired radio signals are not guided to the DUT over cables as in the conducted testing, methods to achieve similar effects are needed for the OTA testing. How to emulate a channel OTA according to standard channel models with high fidelity in the test environment becomes the research focus of multiple-input–multiple-output (MIMO) OTA testing. Many OTA techniques have been proposed in the literature and selected in the standard covering different real-world operating scenarios. Though OTA testing is more complicated than that for the conducted testing, they offer some appealing features such as being non-intrusive to DUTs and providing realistic radio interaction with the DUT as if in real life. Next, the discussion is focused on two standard OTA testing methods for system performance testing, i.e., the MPAC solution and the RTS solution.

10.2.2 MPAC solution

10.2.2.1 Background

Current MPAC OTA test methods have their roots in the measurement specification for single-antenna system by the International Association for the Wireless

Telecommunication Industry (CTIA) [1]. Two test metrics, namely, Total radiated power (TRP) and total isotropic sensitivity (TIS), were defined for the transmitter and receiver performance evaluation of user equipment (UE) in the active mode, respectively. The relation to MPAC methods is two-fold. First, both measurements can be done using multiple probes, though only one of them is active at a time for single-input-single-output (SISO) OTA testing, while all probes are activated simultaneously in the MPAC setup for emulating the spatial channels. Second, the DUT is operating in its normal (active) mode and the intermediate metric in TIS is the link throughput, which is not a direct RF parameter. TRP measures how much power the DUT, together with its RF parts and antennas, can radiate overall. TIS measures the receiver sensitivity of the DUT, i.e., the minimum power level, when a certain throughput can still be achieved. Especially, TIS evaluates the DUT as a whole, with its baseband, RF, and antenna capability. These tests can be performed either in an anechoic chamber (AC) with single or multiple probes, or in a reverberation chamber (RC). In AC, there were similar research questions as in current MPAC systems, e.g., what the sufficient range length is, i.e., the distance between DUT and probe(s), and how many DUT orientations are to be measured to capture the isotropic condition.

With the introduction of multi-antenna UEs and use of MIMO techniques, it became evident that good or bad antenna designs cannot be assessed with the existing SISO OTA performance evaluation. For example, a UE with two high efficiency antennas, but with similar polarizations and short inter-element distance, is not optimal for diversity or spatial multiplexing schemes, due to the possible high antenna correlation at the two antennas. By MIMO techniques, we do not mean here only the spatial multiplexing, but any diversity or beamforming schemes. The incapability of previous OTA methods to test such simple MIMO operation as diversity switching was recognized and consequently the need for new test concept has been identified [2]. Moreover, it was remarked that antennas and other functional blocks of MIMO transceivers cannot be tested separately, since MIMO communications is highly dependent on the radio channel conditions. A static line-of-sight channel with averaging over many orientations (as done for the SISO OTA testing) does not give insights to the overall performance of multi-antenna terminals. Temporally varying radio channel with polarimetric and angular dispersion is needed for the evaluation of multiple antennas and MIMO processing. The concept of MPAC with ACs, CEs, and multiple dual polarized probes in different angles with respect to the DUT was sketched in [3]. A similar concept was published a bit earlier in [4].

The aforementioned is one perspective to the development of MIMO OTA testing. This perspective considers the evolution of antenna tests and measurement practices. Another perspective is related to the development of traditional conductive testing of baseband functionalities of MIMO capable devices. Radio channel models had been defined for MIMO evaluations in fading channel conditions. The question was how to extend these tests to incorporate antennas and corresponding RF designs. The question became unavoidable at latest in the advent of mm-wave frequency transceivers and disappearing of physical antenna connectors. MIMO capable devices had been tested with channel models and fading emulators in a conductive mode, i.e., not OTA, connecting the test equipment to DUT with RF cables. The traditional conductive MIMO emulation is briefly introduced in the following subsection. After that, we move to discussing how the same emulation functionality is pursued OTA.

10.2.2.2 System function of conductive MIMO emulation

The system model in [5] is

$$\mathbf{Y}(t,f) = \mathbf{H}'(t,f)\mathbf{X}(t,f) + \mathbf{N}(t,f), \tag{10.1}$$

where $\mathbf{Y}(t,f) \in \mathbb{C}^{(N1)}$ is the received signal vector, $\mathbf{H}'(t,f) \in \mathbb{C}^{(NM)}$ is the MIMO channel transfer function, $\mathbf{X}(t,f) \in \mathbb{C}^{(M1)}$ is the transmitted signal vector, $\mathbf{N}(t,f) \in \mathbb{C}^{(N1)}$ is the noise vector at receiver inputs, and N, M denote the number of receiver (Rx) and transmitter (Tx) antenna ports, respectively. MIMO channel model with L paths can be defined as

$$\mathbf{H}'(t,f) = \sum_{l=1}^{L} G_{rx}(t,\Omega_l^{rx}(t)) \begin{bmatrix} \alpha_l^{\theta\theta}(t,f) & \alpha_l^{\theta\phi}(t,f) \\ \alpha_l^{\phi\theta}(t,f) & \alpha_l^{\phi\phi}(t,f) \end{bmatrix} G_{tx}(t,\Omega_l^{tx}(t))^T, \tag{10.2}$$

where $\mathbf{G}_{rx} \in \mathbb{C}^{(N2)}$ and $\mathbf{G}_{tx} \in \mathbb{C}^{(M2)}$ are the antenna pattern matrices of θ and ϕ polarizations for Rx and Tx antenna arrays, respectively. Furthermore, α_l^{ab} are time t and frequency f selective fading complex channel gains of path l for received polarization a and transmitted polarization b, Ω_l^{rx} and Ω_l^{tx} are the arrival and departure solid angle of path l, respectively. The formulation of (10.2) enables polarimetric and spatially selective fading MIMO channels. Each row of an antenna pattern matrix $\mathbf{G}$ is composed of one or several physical antenna elements that are fed from baseband unit only with a single RF chain. Antenna patterns are time variant, e.g., in the case of analog beamforming subarrays with time-variant weights or adaptive switching of antennas.

Equation (10.1) is defined in frequency domain for mathematical convenience. Normally standard channel models are determined and fading emulation performed in delay (-time) domain, which is achievable by Fourier transformation over the frequency dimension. Transmitted signal vector, originated from a test equipment, is directed to fading CE through M RF ports with coaxial cables. CE performs convolution with vector $\mathbf{X}$ and channel matrix $\mathbf{H}'$ and outputs signal vector $\mathbf{Y}$ through N coaxial cables for the RF input ports of the DUT. With properly calibrated test system, it is possible to emulate $\mathbf{H}'$ shown in (10.2) on high precision, i.e., to test the DUT in channel conditions defined by the target channel model. Even though DUT antennas are bypassed by RF cables, their approximate model can be included in the emulation by defining $\mathbf{G}_{rx}$. In practice, it is difficult to include all coupling and other RF effects perfectly in the model.

10.2.2.3 Principle of MPAC

The MPAC test setup is composed of a digital fading emulator, an AC and a set of OTA probe antennas. It is initially described in [6] and thoroughly in [7]. The intention

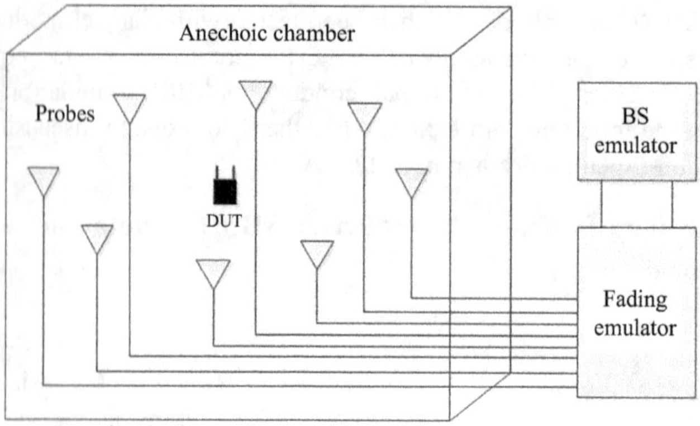

Figure 10.1 MPAC MIMO OTA test setup containing fading emulator, anechoic chamber, and multiple probes (Figure 1 in [7])

in MPAC OTA test setup is to generate time-variant electro-magnetic fields with specified angular, polarimetric, and delay dispersion characteristics around the DUT. The characteristics follow target channel models that are, typically, defined to reflect realistic propagation conditions. Figure 10.1 illustrates elements of the setup. The DUT is in the center of the AC, in an area called test zone, and the probes are arranged, e.g., uniformly around the DUT in a two-dimensional (2D) or three-dimensional (3D) constellation. Each probe is connected to an output port of the CE, possibly thorough a power amplifier. Amplification may be required between CE outputs and probes to compensate for the path loss between probes and the DUT. A communication tester generates the test signal that is fed to the CE. The CE performs the convolution of the channel model impulse responses and Tx signals. Thus, it constructs the multipath environment, including path delays, Doppler spread, and fast fading. The channel model, containing polarization characteristics and directions of departures and arrivals, is mapped to the CE such that the model allocation corresponds to the physical probe installation in the chamber. The DUT is assumed to be in the FF region of each probe, and vice versa. In the actual OTA test, an appropriate performance metric, e.g., throughput, is collected as a figure of merit (FoM). Target channel models define the temporal/Doppler, frequency/delay, space/angle, and polarimetric parameters of fading MIMO radio channel conditions. Standard channel models do not define, however, the instantaneous impulse or frequency responses of the channel.

The reconstruction of target field for the DUT is based on the following fact of the antenna measurement and EM theories. According to [8], a closed boundary, e.g., a spherical or cylindrical surface in the 3D case or a circle in the 2D case, determines the field within the boundary, when no sources are present inside the boundary. The intended target field results inside the test zone, if such a field is generated, where the components of both the electric and magnetic fields tangential to the surface of the test zone are equal to those of the target field [9].

Generally, both downlink and uplink transmissions and their fading emulation are required in active mode testing. In the simplest case, no fading or multipath effects are emulated in the uplink. In this case, an uplink communication antenna, connected to the communication tester with a coaxial cable, is located inside the AC such that it does not distract any probe–test zone clearance. In time division duplex (TDD) systems the uplink radio channel should be reciprocal with the downlink and thus both up- and downlink fading must be emulated. This can be achieved by receiving the uplink signals by probes and performing fading also for them within the CE.

Probes must be dual polarized when the polarization dimension of radio channel is considered. Both orthogonally polarized elements of a single probe must be connected to different CE outputs. This will enable independent fading on different polarizations, which is required by most standard channel models. It is possible to generate any polarization states, e.g., circular, linear, or elliptical with orthogonally polarized, colocated, and phase-calibrated probe elements.

10.2.2.4 System function of MPAC OTA emulation

Assume an MPAC setup with K probes located around the test zone with a certain polarization, distance, and directions from the center of the test zone. Assume that the bore sight of probe antennas is pointing toward the test zone. We can construct similar MIMO system function as (10.1) from the Tx antenna ports to the Rx, i.e., the DUT,* antenna ports. Now the MIMO channel is not determined only by a calibrated CE running the target channel model containing specific Tx and Rx antenna models. Instead, the transfer function is composed jointly by the CE with a channel model in it, cables, probes, propagation environment within the AC, and DUT antennas. Some of these components are typically partly or completely unknown and there is no intention (or means) to fully determine the unknowns by any calibration measurement.

In MPAC setup, the matrix $\mathbf{H}'$ in (10.1) is substituted by

$$\mathbf{H}(t,f) = \mathbf{F}(t,f)\mathbf{W}(t,f), \tag{10.3}$$

where the first term $\mathbf{F} \in \mathbb{C}^{(NK)}$ contains DUT antennas, the AC, and probes, i.e., the physical environment within the AC from K probes to N DUT antennas. Entries of matrix $\mathbf{F}$ are

$$\Psi(t,f) = \mathbf{G}_{rx,n}(t, -\vec{\beta}_{n,k})\mathbf{G}_{o,k}(\vec{\beta}_{n,k})^T \frac{\lambda}{4\pi d_{n,k}} e^{j\beta d_{n,k}}, \tag{10.4}$$

where $\mathbf{G}_{rx,n}$ and $\mathbf{G}_{o,k} \in \mathbb{C}^{(12)}$ are the polarimetric antenna pattern vectors of DUT antenna n and probe k, respectively. The direction and frequency are specified by the wave vector $\vec{\beta}_{n,k}$ from probe k to DUT antenna n. The corresponding distance is denoted by $d_{(n,k)}$. Finally, λ and β are the wavelength and the wave number, respectively. As mentioned previously, DUT antennas may have time-dependent radiation patterns due to possible analog beamforming or antenna switching.

*In this section, we label DUT as the receiver and the other link end, the transmitter for notational clarity. In practice, however, both link ends transmit and receive signals in the active mode.

The second term $\mathbf{W} \in \mathbb{C}^{(KM)}$ of (10.3) is a specific OTA channel model executed in the CE. It contains the time and frequency selective fading propagation coefficient $\alpha^{ab}_{(l,k)}$, Tx antenna radiation patterns $\mathbf{G}_{tx}$, and specific OTA probe weights $\gamma^{a}_{(l,k)}$. The kth row, $k = 1,\ldots,K$ of matrix $\mathbf{W}(t,f)$ is

$$\mathbf{W}(t,f) = \sum_{l=1}^{L} \begin{bmatrix} \gamma^{\theta}_{l,k} & \gamma^{\phi}_{l,k} \end{bmatrix} \begin{bmatrix} \alpha^{\theta\theta}_{l}(t,f) & \alpha^{\theta\phi}_{l}(t,f) \\ \alpha^{\phi\theta}_{l}(t,f) & \alpha^{\phi\phi}_{l}(t,f) \end{bmatrix} G_{tx}\left(t, \Omega^{tx}_{l}(t)\right)^{T}. \qquad (10.5)$$

Now the challenge is to determine weights $\gamma^{a}_{l,k}$ and fading coefficients $\alpha^{ab}_{(l,k)}$ such that DUT within the test zone experiences characteristics of the target channel model. These targeted characteristics are, e.g., the power–delay profile (PDP), Doppler spectrum, Ricean K-factor, amplitude distribution, cross-polarization power ratio, and power angular distribution as observed by the DUT.

10.2.2.5 Field synthesis methods

Two alternative methods can be used for reproducing fading radio channel conditions within a test zone. They are called the plane wave synthesis (PWS) and the prefaded synthesis (PFS). These methods set slightly different requirements for the setup and provide different capabilities, while in many cases they lead to statistically similar channel emulation.

10.2.2.5.1 Plane wave synthesis

The static target field inside a sufficient small test zone can be controlled by tuning the phases and amplitudes of fields radiated by the probes [7]. The MPAC setup must be phase and amplitude calibrated and the transfer functions to the center of the test zone must be known. Then different field shapes, i.e., polarizations, amplitudes, and phase-fronts can be controlled by setting complex excitation coefficients $\gamma^{\theta}_{l,k}$ and $\gamma^{\phi}_{l,k}$ in the fading emulator for signals routed to the probes. Any field structure can be decomposed to the contribution of plane waves with different directions, phases, and amplitudes. Moreover, standard channel models typically define geometric channels with rays, which can be interpreted as plane waves. Hence, the basic field component in PWS is the plane wave. The principle of plane wave generation in the 2D space is depicted in Figure 10.2.

A Doppler frequency can be introduced to plane waves by introducing a time-variant phase term. The polarization state is set by using dual polarized probes and controlling relative phases and amplitudes of orthogonal polarization components. Any number of plane waves with different directions and temporal characteristics can be generated without extra complexity. This enables the reconstruction of any instantaneous radio channel conditions and arbitrary power angular spectra (PAS) [11]. Path delays and time variability are generated by the fading emulator.

10.2.2.5.2 Prefaded synthesis

The approach of the PFS is different, though it uses the same HW resources and results in statistically equivalent emulation in the case of many standard channel models. The PFS does not aim to control instantaneous EM fields exactly and does not require phase calibration [7]. Instead, it controls the power angular distribution

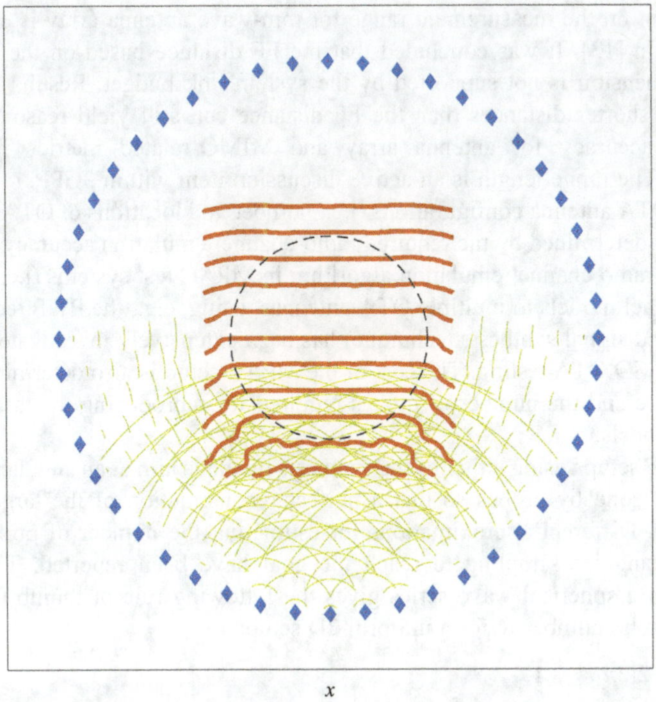

Figure 10.2 Construction of a 2D plane wave with PWS (Figure 2 in [10])

and the polarization power ratios as observable by the DUT. Time/Doppler and delay/frequency domain effects are generated in the fading emulator, similarly to the PWS method. Weight coefficients $\gamma_{l,k}^{\theta}$ and $\gamma_{l,k}^{\phi}$ of (10.5) are real-valued amplitudes in PFS [12]. The PFS method suits well for channel models with the cluster definition of propagation parameters, such as the third-generation partnership project (3GPP) models for 4G and 5G evaluations.

10.2.3 Research/design questions

The design of an MPAC test system must balance between the system precision and system complexity (and cost). Many research questions rise from the specification and designing of an MPAC system. What is the minimum number of probes needed for a certain DUT category? What is the minimum range length and consequently the minimum AC size? Where, i.e., to which directions, to place the limited number of probes? How to determine weights $\gamma_{(l,k)}$ for a specific channel model? These questions are briefly discussed in the following subsections.

10.2.3.1 MPAC configuration

The measurement range is an important aspect for system design, as it determines the chamber size (which is a major cost-determining factor of the setup). The requirement on the measurement range for mm-wave antenna array is extensively discussed in [13]. It was concluded that the FF distance based on the maximum device dimension is not supported by the system link budget. Results have indicated that shorter distances than the FF distance can still yield reasonable measurement accuracy for antenna array and MIMO-related metrics for testing purposes. The range length is an active discussion item within 3GPP.

The OTA antenna configurations (i.e., number and locations of OTA antennas) are mainly determined by the required radio-channel emulation accuracy in the test zone. The radio channel emulation algorithm in MPAC test systems (i.e., mapping radio channel models to multiple OTA antennas, using, e.g., the PWS technique or the prefaded signal synthesis technique) has been extensively investigated for sub-6-GHz MIMO OTA testing. The cost of the MPAC setup is also determined by the CE resource and the number of probes required for reproducing the target spatial channel models.

MPAC setups using either of the synthesis methods aim at an angular sampling of the test zone by probes so that a perfect reconstruction of the target PAS is achievable. Different approximations for estimating the number of probes or the minimum angular sampling to reach the goal have been reported. The cut-off property of a spherical wave series gives the following rule of thumb [9] for the required probe number K for a uniform 2D setup:

$$K = 2\lceil 2\pi r \rceil + 1, \tag{10.6}$$

where $\lceil \cdot \rceil$ is the integer ceiling operator and r is the radius of the cylindrical test zone in wavelengths. The maximum increment of the angular sampling by probes is consequently

$$\theta_{\Delta} = 360^{\circ}/K. \tag{10.7}$$

It has been demonstrated that a large number of probes are required to create a large test zone. Setting up a 3D MPAC configuration for 5G radio devices is costly, so finding ways to limit the number of probes while still approximating the target channels sufficiently accurately could make the test system both cheaper and simpler to implement. Inspired by the fact that radio channel models are generally directional in real-world scenarios, probe selection framework is generally adopted for MPAC setup for 5G radios [14–16]. A uniform configuration of the OTA probes is not ideal and a probe selection mechanism can save cost, via reducing the required number of fading channels. This way, the associated RF chains can be effectively reduced to save cost. This concept is also adopted for the MPAC setup for 5G frequency range 2 (FR 2) testing in the standardization.

For any OTA test system, we need to define testing metrics to evaluate how well the desired propagation channels are reconstructed. We can directly compare, e.g., the continuous target PAS and the discrete emulated PAS implementation using a limited number of probes in the MPAC setup. For sub-6-GHz systems, the PAS deviation is indirectly evaluated via the spatial correlation. That is, the spatial correlation calculated from the target spatial channel models is used as reference to judge the accuracy of emulated spatial correlation in the MPAC setup. For 5G beam-steerable devices, beam probability, which is adopted to characterize the beam selection performance of the mm-wave antenna system under spatial fading channels, can be utilized [5,14]. Another metric for 5G radios is the beamforming power spectra similarity, which is the similarity of the target PAS and the discrete PAS reproduced in the OTA test system as seen by the DUT [5,14].

10.2.3.2 Channel emulation method

The determination of complex excitation coefficients in the PWS method is discussed, e.g., in [7,11]. The principle is to define the target field in a set of sample points within or around the test zone, to specify the transfer function from K to all sample points, and to solve excitation coefficient from the determined set of linear equations. This is typically done for a set of plane waves, apparently arriving to the DUT from different directions.

In the case of the PFS, the probe weights are determined to reconstruct the power angular spectrum of the target channel model. In 4G systems, this was usually done by using the spatial correlation, or more specifically the error between the target and the reconstructed spatial correlation function within the test zone, as the metric to be minimized [7]. In 4G devices the MIMO techniques were mainly spatial multiplexing and diversity. This made the spatial correlation between, possibly, widely spaced DUT antenna locations a meaningful metric. Recall the Fourier transformation link between the power angular spectrum and the spatial correlation function. In 5G mm-wave systems the primary use of multiple antennas is beamforming. Hence, a potential metric for weight determination could be a deviation between the target and the reconstructed power angular spectrum, as observable by an ideal antenna array. This metric is called the PAS similarity percentage (PSP) and it is defined in [5].

The comparison between the PWS and PFS is performed for implementing geometry-based stochastic channel models (GSCMs) [16]. It shows the cluster-wise channel emulated by the PFS method is Kronecker structured, which is different from the general definition of GSCMs. In contrast, the channel emulated with the PWS method is consistent with GSCMs. Moreover, the emulation accuracies for the two methods are compared under different target channel settings, i.e., different cluster angular spreads. The simulation results demonstrate the advantage of the PWS method over the PFS method, especially when cluster angular spreads are small.

10.2.3.3 Channel validation

The objective of the channel validation measurement is to ensure that target channel models are correctly implemented inside the test area in the MPAC setup. Several aspects of the emulated channel models are analyzed in the channel validation measurements, including the fading distribution, PDP, power Doppler spectrum/temporal autocorrelation function, spatial correlation, and cross-polarization ratio. The spatial correlation, which is a statistical measure of the similarity between received signals at different spatial locations, has been used to represent the channel spatial characteristics at the receiver side and is selected as the FoM for OTA testing in conventional MPAC setups. The spatial correlation might be less relevant to determining OTA system performance for beam-steerable devices; the PAS of the emulated channels are more interesting. Therefore, PSP is used in the FR 2 to measure the similarity of the PAS produced by the MPAC system and the reference PAS in the target channel model. Research works have also been reported to validate the joint-angle delay power profile of the emulated channels in the MPAC setup [17].

Validation of spatial fading channel emulation in 5G frequency range 1 (FR 1) has been extensively discussed in the literature and a preliminary work has been reported on validation in the FR 2 chamber [18]. There are some practical differences of validation at the FR 1 and FR 2 chamber. The probe antennas and measurement antennas are more directive at the FR 2. Virtual array, which is used to estimate the emulated channel spatial profile in the chamber, will be more difficult to form with the mechanical positioner, as the wavelength at the FR 2 is much smaller. The channel emulation at FR 2 is also more complicated, which involves frequency up-and-down conversion for the sub- 6-GHz CE. Spatial correlation and power angle spectrum were investigated for the simple MPAC setup in the FR 2 chamber in [18]. An excellent agreement between the measured results and the target is achieved in terms of spatial correlation and power angle spectrum, which validates the MPAC method in the mm-wave bands in the FR 2 band.

10.3 Two-stage MIMO OTA test method

10.3.1 Principle of the radiated two-stage method

The basic concept of the two-stage MIMO OTA test method is to divide the OTA test into two stages: the first stage is to acquire the antenna pattern of DUT, the second stage is to perform the throughput measurements using a downlink signal generated by convolving the device antenna pattern with the desired spatial channel model through a conducted or radiated connection. This innovative idea was first proposed to 3GPP RAN 4 meeting in 2009 in [19] with the conducted second stage throughput test, and more analysis and test results are summarized in [20]. The radiated version of the RTS was proposed to 3GPP in [21] to overcome the shortcomings of cable-conduct throughput test and finally accepted by 3GPP and CTIA as one of the standardized MIMO OTA conformance test methods [22–24]. Several papers have already been published to introduce two-stage MIMO OTA method from different aspects [25–28].

In the first stage, the DUT's antenna pattern can be acquired, depending on the test context, by simulation, design, or measurement of the actual DUT antenna pattern. To overcome the shortcomings of the presence of the connecting cables, for example, the integrity of the device is altered and changes in the impedance seen by the antennas, two-stage method employs a nonintrusive antenna measurement method that does not require the connection of cables. This nonintrusive antenna measurement approach uses the ability of the receiver in the DUT to measure the amplitude and relative phase of known signals received by DUT antennas.

The amplitude and phase information can be collected in two ways: Layer 3 (L3) messages or user datagram protocol (UDP) packets. For L3 message method when base station (BS) emulator sends the antenna measurement request in downlink L3 message, the DUT reports the received signal's power and phase information in uplink L3 message following the format described in [29]. For the UDP method, one antenna measurement app should be preinstalled on DUT and launched after the call connection is set up. When the app receives the antenna measurement request from test PC, the received signal power and phase information are assembled into UDP packets and sent out to test PC with specified IP address and port number. By rotating the DUT relative to the known incident signal, it is possible to construct the 3D antenna patterns with amplitude and phase responses. To fully characterize the antennas, measurements are made at two orthogonal probe antenna orientations, typically vertical and horizontal. This can be done by switching between two separate antennas or by rotating a single antenna.

In the second stage, the desired antenna pattern is loaded into a CE and convolved with the spatial channel model being used to evaluate the DUT performance. How the antenna pattern is embedded into the channel model will be introduced in the following section. This process generates the signals at the DUT receiver that would have been received had the DUT been placed in the same 2D or 3D spatial field used for the convolution. During the second stage, it is not necessary to alter the device orientation relative to the probe antennas since the rotation of the DUT relative to the chosen channel model is performed electrically within the CE. In this stage, there are two options for connecting the signal generated by the CE to the device. The simplest way is to use cables connected to the device's temporary antenna connectors should they be available shown as Figure 10.3. This is usually not a problem for traditional RF test but is becoming increasingly more difficult on mobile devices where space is at a premium. When goes to FR 2 mm-wave frequency even there are no available RF connectors. The conducted second stage does not require the use of an AC for the throughput measurements although some form of RF shielding is desirable to prevent any ambient interference from affecting the results. A consequence, however, of using the cabled connection is that any radiated interference generated by the device that would otherwise have desensitized the receiver through its own antenna is no longer measured. When the device is operating at low transmit powers, this is not an issue but at high power, there may be differences in performance compared to radiated throughput measurements.

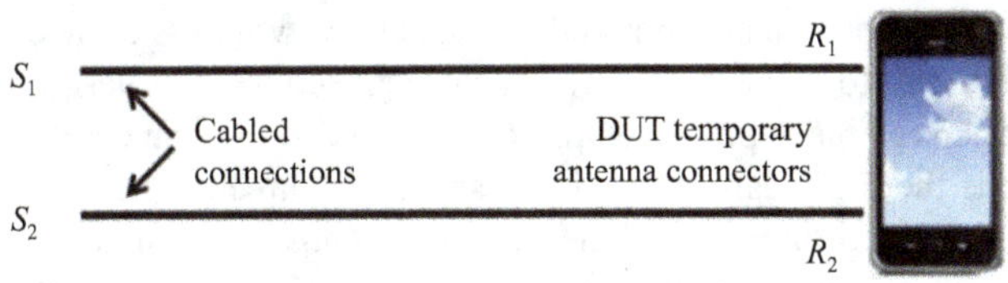

Figure 10.3 Cable-conducted second-stage throughput test

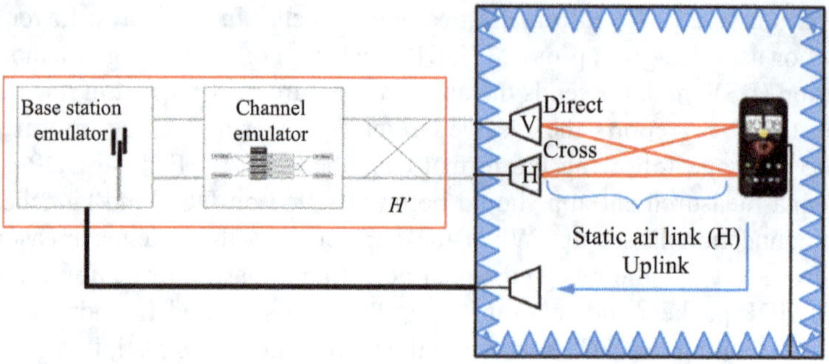

Figure 10.4 Radiated second-stage throughput test

The alternative to a conducted second-stage connection is to use a radiated "cable replacement" connection shown as Figure 10.4, which is known as RTS method. This approach does require the use of an AC for throughput measurements but has the benefit that any radiated interference generated by the device is fully considered in the measurements. The purpose of the radiated connection is to enable the signals generated by the CE, which are already conditioned to include the effect of the device antennas, to be directly connected to the device receiver. However, this can only be done by calibrating out the impact of the signal propagation in the AC and the impact of the device antenna.

Many strategies have been reported in the literature to obtain the radiated channel matrix between the probe antenna ports and the DUT antenna ports, depending on the available output information from the DUT. As discussed in [26], the channel matrix can be directly calculated based on the knowledge of complex antenna patterns of the DUT. This method, however, typically requires a large AC for FF antenna pattern measurement and support from a special chip-set to report DUT complex antenna patterns in a nonintrusive manner. This idea is detailed in the original RTS method. The channel matrix can also be directly estimated via channel estimation algorithms, e.g., utilizing pilot sequence. This idea, however, might only be supported by BS-type DUT where transmitted signals can be designed and therefore known. In [30], based on the received reference signal received power value per DUT antenna port, a calibration method is proposed to determine the channel matrix. The method is highly attractive since the testing can be performed in a small RF-shielded enclosure. The method was then extended for high-order MIMO DUT in [31], with a closed-form calibration method. The unique features of mm-wave antennas also open up new possibilities to achieve wireless cable connection. A straightforward way to achieve wireless cable connection is to design the radiated channel matrix in the multi-probe setup, saving the need to compensate it in the CE. For example, if the DUT can form widely separated directive beam patterns, each toward a directive and direction aligned probe antenna, we can achieve cable-like connection OTA via antenna pattern discrimination. Alternatively, polarization discrimination is another way to achieve virtual cable connection for 2×2 MIMO systems. The wireless cable can be directly achieved via adopting orthogonality between two cross-polarized components, where two wireless cable connections can be achieved. In [18], two proposed schemes, i.e., polarization discrimination and antenna pattern discrimination, are two examples to approximate wireless cable connection. This strategy is not generic for any DUT and in any multi-probe setup. For small mm-wave UEs, it might not be practical to design many widely separated beams due to design limitations. As for polarization discrimination case, it is only limited to 2×2 MIMO systems and MIMO handsets with two orthogonal linear polarized antenna designs. The more general calibration method, which can determine the radiated channel matrix **A** without relying on DUT antenna radiation properties and multi-probe setup, is therefore more generic to achieve wireless cable connection. Later, we focus on the methods introduced in the original RTS method to achieve the wireless cable concept.

The advantage of this two-stage approach is that radio conditions representing arbitrarily complex 2D or 3D spatial channels can be completely emulated in CE using a simple AC with high accuracy. The required probe antenna number depends on the measured MIMO order. For example, if 2×2 MIMO is tested, only two probe antennas are required. Another advantage is because of acquiring antenna pattern in first stage, more diagonal information can be got for performance debug. Except for antenna-only metrics such as gain, branch power imbalance, envelope correlation coefficient, etc., when the device antenna patter is convolved with the spatial channel model, including the Tx antenna assumptions, that it is possible to see the actual signal that will be presented to the receiver for demodulation. Since the two-stage method has access to all the elements in the Tx antenna, channel emulation, and Rx antenna, it is possible to compute a wide variety of signal characteristics that can help predict and explain the variation in DUT performance.

10.3.2 Applying measured antenna patterns into channel models

Spatial channel models can be described using either a geometric (sum of sinusoids) approach or using a correlation-based approach. DUT antenna patterns can be convolved with these two descriptions in different approaches.

Geometry-based modeling of the radio channel enables a separation of propagation parameters and antennas properties. The time-variant impulse response matrix of the $U \times S$ (with S and U the number of transmit and receive antenna, respectively) MIMO channel is given as

$$\mathbf{H}(t;\tau) = \sum_{n=1}^{N} \mathbf{H}_n(t;\tau), \tag{10.8}$$

where t is the time, τ is the delay, N is the number of clusters, and n is the cluster index. Equation (10.8) uses 2D geometry-based model implementation (only considering the angle spread in azimuth direction) to demonstrate how the antenna pattern is introduced in the channel model emulation. The impulse response matrix, composing Tx antenna array response matrices F_{tx}, Rx antenna array response matrices F_{rx}, respectively, and the channel from Tx antenna element s to Rx element u for cluster n, is expressed as

$$H_{u,s,n}(t;\tau) = \sum_{m=1}^{M} \begin{bmatrix} F_{rx,u,V}(\phi_{n,m}) \\ F_{rx,u,H}(\phi_{n,m}) \end{bmatrix}^T \begin{bmatrix} \alpha_{n,m,VV} & \alpha_{n,m,VH} \\ \alpha_{n,m,HV} & \alpha_{n,m,HH} \end{bmatrix} \begin{bmatrix} F_{tx,u,V}(\phi_{n,m}) \\ F_{tx,u,H}(\phi_{n,m}) \end{bmatrix}$$
$$\times \exp(j2\pi\lambda_0^{-1}(\bar{\phi}_{n,m} \cdot \bar{r}_{rx,u}))\exp(j2\pi\lambda_0^{-1}(\bar{\phi}_{n,m} \cdot \bar{r}_{tx,u}))$$
$$\times \exp(j2\pi\nu_{n,m}t)\delta(\tau - \tau_{n,m}), \tag{10.9}$$

where $F_{rx,u,V}$ and $F_{rx,u,H}$ are the antenna element u field patterns for vertical and horizontal polarizations, respectively, $\alpha_{n,m,VV}$, $\alpha_{n,m,VH}$, $\alpha_{n,m,HV}$, and $\alpha_{n,m,HH}$ are the complex gains of vertical-to-vertical, vertical-to-horizontal, horizontal-to-vertical, and horizontal-to-vertical polarizations of ray n, m, respectively, λ is the wave length of the carrier frequency, $\bar{\phi}_{n,m}$ is the AoD unit vector, $\bar{\phi}_{n,m}$ is the AoA unit vector, $\bar{r}_{tx,u}$ and $\bar{r}_{rx,u}$ are the location vectors of element s and u, respectively, and $\nu_{n,m}$ is the Doppler frequency component of ray n, m. For the RTS method, $F_{rx,u,V}$ and $F_{rx,u,H}$ are the measured UE receiver antenna patterns from the first stage, and $F_{tx,u,V}$ and $F_{rx,u,H}$ are the predefined patterns for the BS. In the second-stage throughput test, the CE rotates the antenna pattern data in (10.9) against the channel model that is the same as physically rotating the device against the fixed probes in the MPAC method. For correlation-based channel model, MIMO channel correlation and power imbalance property are introduced by explicitly multiplexing the corresponding covariance matrix on the independent MIMO channel coefficients. This covariance matrix can be discomposed to correlation matrix and power imbalance matrix, which is determined by the Tx, Rx antenna patterns, and the specified channel model. How to derive the covariance matrix for arbitrary antenna patterns under multipath channel conditions can refer to [32]. Both geometry-based and correlation-based models are accepted in [22] for MIMO OTA test.

10.3.3 Wireless cable principle in RTS

To achieve the wireless cable connection in second stage, it is necessary to measure the propagation conditions inside the AC and modify the transmitted signals in such a way to remove the crosstalk between probe antennas and DUT antennas.

To simplify the expression, a 2 × 2 MIMO configuration is illustrated to show how the wireless cable principle works. Assume that x_1 and x_2 are the transmitted signals from the BS emulator, after applying the desired multipath fading channel and convolving with the complex antenna pattern, we get $f(x_1)$ and $f(x_2)$. Without signal conditioning, the first spatial stream $f(x_1)$ transmitted from probe antenna 1 will be received by both DUT antennas. Similarly, the second stream $f(x_2)$ transmitted from probe antenna 2 will also be received by both receivers. In order to create the desired situation where $f(x_1)$ is received only by receiver 1 and $f(x_2)$ is received only by receiver 2, it is necessary to calculate the transmission properties of the AC then compute the inverse transmission matrix and multiply this by the two streams. Assume the radiated channel matrix between the probe antennas and the device antennas is

$$\mathbf{H} = \begin{pmatrix} h_{11} & h_{12} \\ h_{21} & h_{22} \end{pmatrix}. \tag{10.10}$$

This channel matrix is composed of the probe antenna patterns, the signal propagation path in the AC, and the DUT receive antenna patterns. The inverse of the radiated channel matrix can be tuned in different ways as

$$\mathbf{H} = \begin{pmatrix} \alpha & \beta \\ \gamma & \delta \end{pmatrix}. \tag{10.11}$$

After applying the inverse of the radiated channel matrix to $f(x_1)$ and $f(x_2)$, the signal received at the device antennas is the same as in the cable-conducted method being

$$\begin{pmatrix} y_1 \\ y_2 \end{pmatrix} = \begin{pmatrix} h_{11} & h_{12} \\ h_{21} & h_{22} \end{pmatrix} \begin{pmatrix} \alpha & \beta \\ \gamma & \delta \end{pmatrix} \begin{pmatrix} f(x_1) \\ f(x_2) \end{pmatrix}$$
$$= \begin{pmatrix} 1 & 0 \\ 0 & 1 \end{pmatrix} \begin{pmatrix} \alpha & \beta \\ \gamma & \delta \end{pmatrix} \begin{pmatrix} f(x_1) \\ f(x_2) \end{pmatrix}. \tag{10.12}$$

The process works as both probes transmit copies of $f(x_1)$ and $f(x_2)$ at different amplitudes and phases such that $f(x_1)$ is received only at receiver 1 and $f(x_2)$ is received only at receiver 2. The process to achieve this is very similar to the precoding used to optimize signal reception for spatial multiplexing MIMO gain, while the difference is, no predefined precoding matrix is available, the inverse matrix of propagation matrix is calculated as the precoding matrix for each test case, and this precoding matrix needs be applied after the channel model implementation.

This precoding process is not exact, so the ideal situation of complete isolation between the signals and receivers is not possible. In the real test case, the received signals look like there is some signal coupling between two branches as the following equation:

$$\begin{pmatrix} y_1 \\ y_2 \end{pmatrix} = \begin{pmatrix} 1 & \rho \\ \rho & 1 \end{pmatrix} \begin{pmatrix} f(x_1) \\ f(x_2) \end{pmatrix} = \begin{pmatrix} f(x_1) + \rho f(x_2) \\ \rho f(x_1) + f(x_2) \end{pmatrix}. \tag{10.13}$$

Define the power ratio between the desired signal and undesired signal as the isolation level $20log10(1/\rho)$ in dB low-level isolation (e.g., lower than 10 dB) will increase the received signal's correlation level, which may bring the throughput performance degradation. However, measurements have shown that above 15 dB, isolation is achievable for most devices, and this is sufficient isolation not to impact the throughput test results [27].

It should be noted that not arbitrary radiated channel matrix in (10.10) is proper to be used to generate wireless cable effect because of ill-conditioned properties. With the measured antenna pattern information the radiated channel matrixes can be constructed and analyzed with condition number and propagation loss for all orientations, the high isolation can be achieved, which is optimized by selecting DUT orientations with a smaller condition number and propagation loss.

10.3.4 RTS solution for 4G/ 5G system

The key instruments required for MIMO OTA test system is BS emulator and CE. Figure 10.5 shows one RTS solution setup using Keysight BS emulator UXM 7515B and CE Propsim, which can support both 4G LTE and 5G NR MIMO OTA test. UXM 7515B can support flexible signal format configuration, while all kinds of standardized and user-defined channel models can be emulated by Propsim, including options of inverse matrix implementation and device orientation rotation in electrical way required by RTS method.

Except for MIMO OTA performance test, wireless cable method is also approved in 3GPP TR 38. 810 for FR 2 5G NR UE Demod and Channel State Information characteristic testing under a virtually cabled scenario. For FR 2 Demod test the first-stage antenna pattern measurement is skipped, but the wireless cable principle of second stage of RTS is applied without including DUT's real antenna pattern into the channel model emulation and the throughput performance evaluation. The system setup in Figure 10.5 still can work by adding radio heads to match to proper frequencies.

One possible challenge for wireless cable solution is because standard UE antenna models are used in channel model generation instead of measured patterns, no available information can be utilized to optimize the orientation selection for inverse matrix tuning, using probe antennas with orthogonal polarizations is one

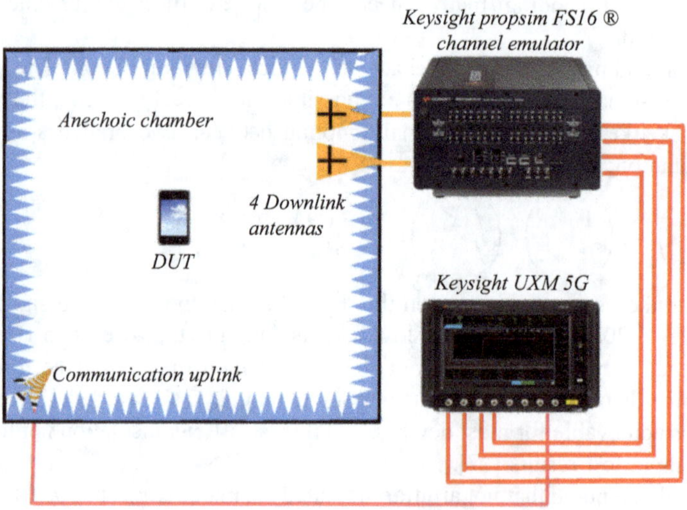

Figure 10.5 RTS MIMO OTA test setup with BS emulator and channel emulator

method to get high isolation by utilizing cross-polarization ratio, but this method may face challenges for DUTs with more than two antennas.

10.4 RF testing

The default reference testing condition for characterizing AUT radiation properties is plane wave field within the quiet zone, which is defined by the uniform amplitude and phase distribution over the AUT. Plane wave condition is required to accurately measure the AUT radiation characteristics (e.g., main beam, null depth, and sidelobe), RF receiver performance (e.g., sensitivity level and dynamic range), RF transmit performance (e.g., transmit power, signal quality, unwanted emissions), electromagnetic-susceptibility testing, scattering measurement in radar cross section, antenna array calibration and array faulty mode diagnosis, etc.

However, ideal plane wave condition does not exist in practice, yet it can be approximated using several strategies, namely, the DFF method, the CATR method, the PWG method, and the NF–FF method. The pros and cons of different approaches are well discussed in the literature and only are summarized later.

- For the DFF method, the basic idea is that a plane wave at the probe antenna can be approximated if the measurement range is no smaller than the Fraunhofer FF distance. The requirement for measurement range might lead to large AC and introduced excessive path-loss problem, especially for large-scale DUTs and for mm-wave radios.
- The basic idea of the CATR is that a plane wave can be generated using transformation with a parabolic reflector in a much shorter distance than the DFF method. The CATR offers good dynamic range due to reduced measurement range. Furthermore, it supports wide-band operation and dual-polarized measurements. However, CATR requires dedicated attention in design, e.g., dedicated edge treatment and fabrication of smooth surface for the reflector antenna, dedicated feed antenna alignment, etc. The size of the reflector antenna is typically 1.5D.
- The NF–FF method computes the performance metrics defined for FF by using mathematical NF–FF transformations. We need to sample the phase and amplitude of the electrical field in the near-region over a surface, satisfying Nyquist sampling theorem. The measurement range can be rather small (e.g., typically 3λ–5λ in planar NF measurement). However, there are many drawbacks of the method. First, both phase and amplitude measurements are needed for the NF measurement. The measurement time might be massive due to sampling requirement. It is required that antenna feed port should be accessible with a signal fed to the antenna that is used as phase reference; therefore, it might not support phase-less measurement. Furthermore, the supported RF signal is a continuous wave signal, while modulated signal 5G NR signal might not be supported.
- PWG can also enable OTA testing directly in the FF at a reduced distance, by exciting the PWG array elements by suitably optimized complex coefficients.

Compared to CATR, the PWG has several advantages, e.g., smaller measurement distance and smaller PWG aperture. It also avoids the problem of direct illumination of the quiet zone by the feed as in CATR systems. However, it is currently still not mature technology, e.g., its application to wideband and large scale DUT is not well investigated. Furthermore, the main application has been on sub-6-GHz frequency band, and few works have been reported on mm-wave bands.

The focus of this chapter is on a recently developed midfield OTA testing solution of 5G radios, as detailed in the following section.

10.4.1 Midfield solution

With the development of wireless communication systems, beamforming is one key technology adopted in 5G BS for both FR 1 and FR 2 and UE for FR 2. To fully measure DUT's RF performance utilizing beamforming technology, RF testing needs to be performed by OTA. On the other hand, the joint integration of the antenna array and the RF frontend are common practice for mm-wave devices, which makes the RF connector unavailable for the RF performance test, making OTA testing the only choice.

A lot of discussions focusing on 5G NR OTA test are proceeding in 3GPP standards [33–35]. Among these 3GPP Technical Specifications (TSs) and Technical Reports (TRs), all RF metrics are defined for FF condition and FF solutions are suggested as reference for both BS and UE testing, including DFF and indirect FF (IFF), which is also known as CATR. To satisfy the required test distance criteria, the chamber size for DFF will be very large and high propagation loss makes the measurement of signal with low power density impossible. CATR has a more compact test distance compared with DFF, but in order to generate enough test zone to accommodate large device, the required distance between feed probe antenna and reflector also makes CATR systems facing large path loss issue.

For the previous reasons, it is often desirable to do OTA measurement in a distance that is much shorter than the Fraunhofer distance but can still get the FF RF parametric results. Although NF-FF transformation is well accepted as a reliable technique to measure FF patterns, it requires the whole NF phase and amplitude contributions data to derive the FF patterns, which means that even the FF result in one direction can only be got after the surface scanning is finished. This test method is test time prohibitive for effective isotropic-radiated power (EIRP)-based test cases and currently not applicable to effective isotropic sensitivity (EIS).

To overcome the previous test challenges a middle-field (MF) test method is proposed with asymptotic expansion transform, and prototype test systems are developed in [36–38] to achieve OTA RF measurement with compact size. The MF testing method can reduce the test distance to $1/8$ of the Fraunhofer distance, and the MF transformation proposed for this method is significantly less complex than the traditional NFTF. The transformation does not require phase information and 3D scan for EIRP- and EIS-based test cases but use measured power results on two different radiated NF test distances to extrapolate the FF results.

10.4.2 MF measurement distance

The well-known FF Fraunhofer distance is defined as

$$R = \frac{2D^2}{\lambda}, \tag{10.14}$$

where D is the largest dimension of the AUT, and λ is the wavelength. Due to the dimensions of the massive MIMO (M-MIMO) array utilized in 5G NR, the FF requirements require a very large test distance and thus a very large OTA chamber. An MF test method is proposed to achieve accurate OTA RF measurement with compact size. This section will focus on the MF test method with known antenna array center positioning, but antenna array scale and beamforming weights information are not required for this method.

By assuming that DUTs utilize antenna arrays and MF measurement distances are well in the FF of the single-element antennas, the overall antenna array's radiation pattern respected to test distance d can be constructed using the following formula as linear superposition (Figure 10.6):

$$y(\theta,\phi,d) = \sum_{k=1}^{N} x_k G_k(\theta_k,\phi_k) \frac{\lambda d}{4\pi d_k} \exp\left(j2\pi \frac{d_k}{\lambda}\right) G_P(\alpha_k,\beta_k), \tag{10.15}$$

where (θ, ϕ) are the azimuth and elevation angles of the probe antennas relative to the AUT phase center, d is the distance between the probe antenna and the AUT phase center (please note d and the following d_k are different variables), k is the index for the antenna element k, N is the number of antenna elements, x_k is the beamforming weight for antenna element k, G_k is the FF complex antenna pattern for antenna element k, (θ_k, ϕ_k) are the azimuth and elevation angles of antenna element k with respect to the probe antenna, λ is the wavelength, d_k is the distance between antenna element k and the probe antenna, G_P is the FF complex antenna pattern for the probe antenna, (α_k, β_k) are the azimuth and elevation angles of the probe antenna with respect to the antenna element k, $\lambda/(4\pi d_k)$ is the path loss, and $2\pi d_k/\lambda$ is the phase variation at distance d_k. Using the mathematic model in (10.15) the antenna array's patterns can be calculated at different test distances. Figure 10.7 shows the antenna gain patterns normalized by test distance d for a sample 2×8 antenna array with $\lambda/2$ as the antenna element separation. The pattern shape of traditional FF test distance (blue line) is selected as the reference, by shortening the

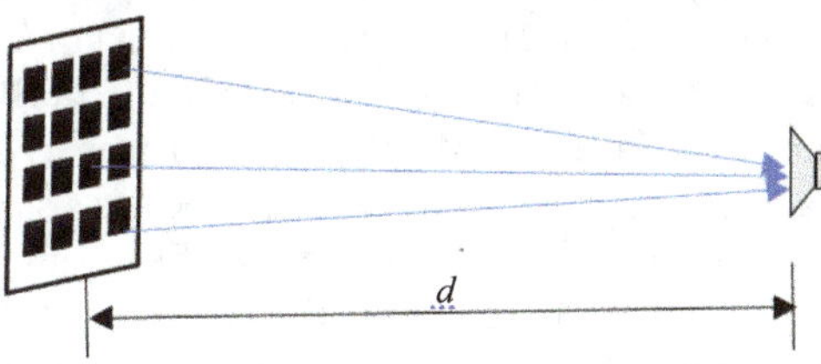

Figure 10.6 Antenna array measurement in midfield (Figure 1 in [38])

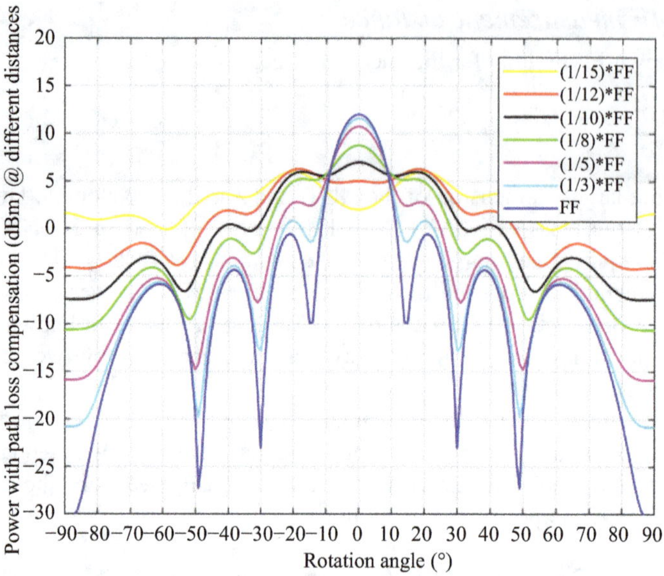

Figure 10.7 Beam gain pattern at different angles and different distances (Figure 1 in [39])

test distance to 1/3, 1/5, 1/8, 1/10, 1/12, and 1/15 of FF distance, we can observe the pattern shape's change tendency. It is noticed that with a reasonable distance, for example, larger than 1/8 of FF, even though the FF antenna pattern is not fully formed, the antenna pattern's main lobe shape is very similar to that of the antenna pattern in FF.

Further simulation results on the relationship of the received power with respect to the test distance and the probe antenna location direction within the 4-dB beam width are shown in Figure 10.8. The spatial intervals corresponding to the 4-dB beam width of the main lobe are shown in Figure 10.8 by four selected angles: the beam peak direction and directions which are 1-, 3-, and 4-dB gain lower than the peak direction, respectively. The power variation at different measurement distances is simulated using (10.15) with dashed lines. The solid lines in Figure 10.8 are reference lines that the power changes linearly using FF assumption. For this assumed antenna array, the traditional FF test distance is around 34λ, and in the FF region, there is no difference between results followed FF theory and simulated results using (10.15). In region MF (I) of larger than 1/3 of FF, the power offset between the solid lines and dash lines is less than 0.5 dB for all directions within the 4-dB beam width. In region MF (II) of larger than 1/8 of FF, the beam direction is still the same as that in FF, and the main lobe has a similar shape to that of FF with no sort order change and no cross, but the offset between the solid lines and dash lines is larger than 0.5 dB. Regions MF (I) and MF (II) of larger than 1/8 of FF are suggested for the new MF OTA test system, and MF-to-FF transformation is used to derive correct FF results.

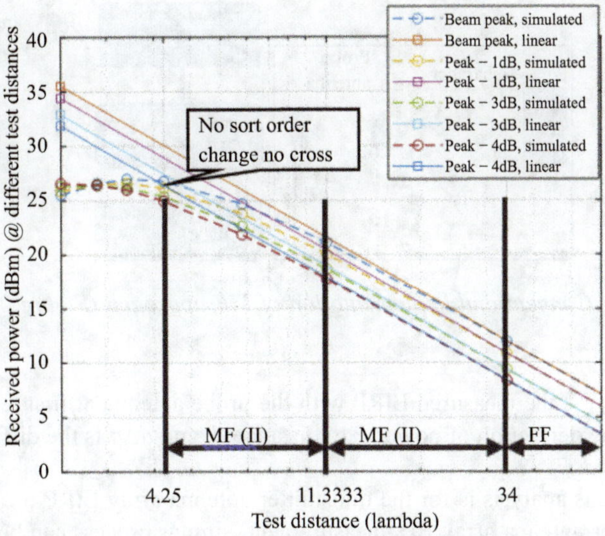

Figure 10.8 Received power at different directions within 4-dB beam width w.r.t. test distance (Figure 2 in [39])

10.4.3 Basic principle for MF test method

Given the observation outlined in the previous section, NF testing would require a transform of some sort, not necessarily an NF–FF transform utilizing a 3D scan in the vicinity of the DUT. In this section, we are outlining our basic principle for MF methodology utilizing an MF transform that allows highly accurate EIRP/EIS measurements in the NF.

The power pattern $P(\theta, \phi, d)$ of the array at test distance d can be further written as

$$P(\theta, \phi, d) = |y(\theta, \phi, d)|^2. \tag{10.16}$$

By putting (10.15) into (10.16), it turns out that for arbitrary specified orientation (θ, ϕ), the gradient of power P can be simplified as a function of variable d with unknown coefficient a.

$$\frac{\partial p}{\partial d} = ad^{-2} + \Delta(d), \tag{10.17}$$

where d is the measurement distance, a is a coefficient of expansion to be determined depending on (θ, ϕ, d), and $\Delta(d)$ is a redundant term consisting of terms having a lower order than d^{-2}, which can be ignored. Measurement results at two different MF distances in the same direction are sufficient to accurately estimate coefficient a. With the estimated coefficient a, the EIRP of the DUT at an arbitrary FF distance d_f from the DUT may be determined according to

$$EIRP(d_f) = EIRP(d_1) + \int_{d_1}^{d_f} \frac{\partial p}{\partial d} d\Delta d, \tag{10.18}$$

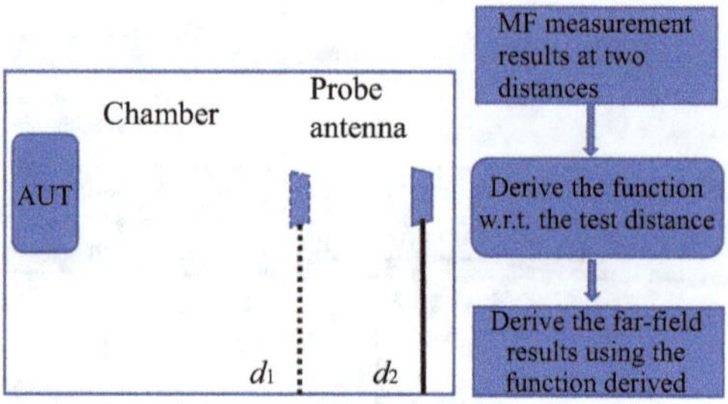

Figure 10.9 Concept and implementation of MF approach (Figure 2 in [38])

where $EIRP(d_1)$ is the measured EIRP with the probe antenna at first test distance d_1, $\partial p/\partial d$ is the derivation of power P to distance d, and $d\Delta d$ is the differentiation of the distance d.

The previous analysis is for the transmitter antenna array EIRP measurement. For the receiver antenna array EIS measurement, similar process can be followed. By measuring the power P required to achieve the given BLER at two different distances d_1 and d_2 ($d_1 < d_2$), the unknown coefficient of expansion can be determined, then the power needed to achieve the same BLER for FF can be derived. Based on the previous analysis for EIRP and EIS measurement approach, MF measurements at two different distances are needed to derive the function and the corresponding FF measurement results. With that MF concept the system implementation is designed as in Figure 10.9. It should be pointed out that this MF transform is not based on an NF–FF transformation utilizing a 3D scan of complex (magnitude and phase) fields, phase measurements of the field components are not required at all, i.e., the MF transform is based on the magnitude measurements (EIRP and EIS) only.

10.4.4 MF system and validation test

One MF OTA prototype system is set up in OTA lab shown in Figure 10.10 that can provide adjustable test range from 1 to maximum 2.5 m inside a chamber with external dimension 5 m $\times$3.5 m $\times$3.5 m. One CATR system with a 7 m $\times$7 m parabolic reflector is selected as the reference system to provide the cross-comparison results.

The DUT used for the performance comparison is a commercial 5G NR BS in FR 1. The size of the DUT is about 0.8 m $\times$0.6 m, and its antenna array is about the same size. The DUT's operating frequency is around 2.6 GHz. Its FF distance is 16.35 m, and the corresponding MF distance is about 2.2 m. We chose 2 and 2.5 m as the two MF distances to derive the FF results. In the comparison test, different broadcast beams and data traffic beams with different beam directions are measured. All the comparison results show that the test results in the two chambers are comparable. To save the space, only one beam's measurement results are shown in Figure 10.11 as one example, more comparison results can be found in [33,34].

Figure 10.10 Five-axis mechanical subsystem inside the chamber. 3D rotator for DUT orientation control (left). Three-axis positioner for probe location control (right) (Figure 8 in [38]).

10.5 5G testing standardization

10.5.1 Performance testing

10.5.1.1 Background

The standardization of the 5G NR has been led by 3GPP, among which the technical specification group (TSG) RAN Working Group 4 (RAN 4) deals with NR MIMO OTA testing. Two frequency ranges are defined for the study, namely, the FR1 covering 450–6, 000 MHz and the FR2 covering 24, 250–52, 600 MHz. Up to the RAN 4 97e meeting, study items on the FR1 MIMO OTA testing have almost been accomplished, and related items about the FR 2 are still under investigation. In addition, work items on MIMO OTA performance requirements for NR UEs are still ongoing within 3GPP RAN 4.

Performance metrics, measurement methodologies, channel models, and channel model validation procedure are mainly defined in [23]. For FR 1 MIMO OTA testing, MIMO throughput is adopted as the baseline FoM. Three DUT orientations, i.e., free-space data mode portrait, free-space data mode landscape, and free-space data mode screen up, are defined for FR1 MIMO OTA testing. For each orientation, a throughput measurement is conducted at every 30° after rotating the DUT in the azimuth plane. In total, 36 measurements are collected, and the measured sensitivity values are averaged to get the final throughput result. As for measurement methodologies, MPAC and RTS test methods are adopted for FR1 MIMO OTA testing. Furthermore, 3GPP spatial channel model extended urban micro-cell (UMi) clustered delay line A (CDL-A) and urban macro-cell (UMa)

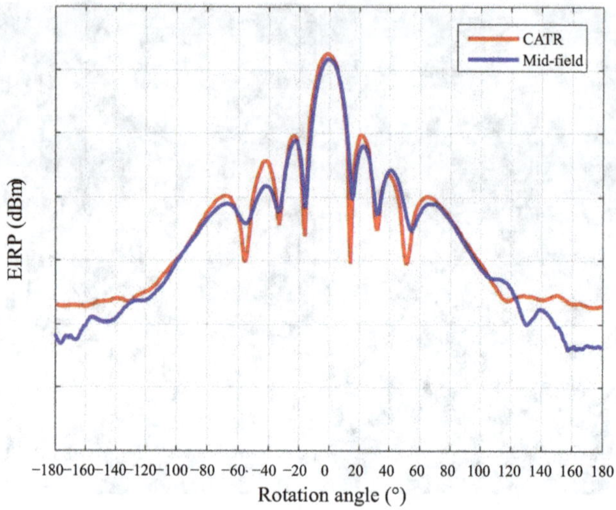

Figure 10.11 2D antenna pattern horizontal cut comparison (traffic beam H 0V 19) (Figure 9 in [38])

CDL-C channel models are required to be measured for FR 1. In order to ensure that the channel models are correctly implemented in the quiet zone, measurements, including Doppler/temporal correlation, PDP, spatial correlation, cross-polarization, and power validation, need to be validated in the test zone before the actual throughput testing.

For FR 2 MIMO OTA testing, similar as FR 1, MIMO throughput is adopted as the performance metric. Different from the 2D scan in the azimuth plane for FR1, a 3D scan consisting of 36 test points on a sphere is specified for FR 2 MIMO OTA testing. In terms of measurement methodologies, 3D MPAC is selected for FR 2 MIMO OTA testing, and the 3D channel model indoors (InO) CDL-A and UMi CDL-C are selected for testing. Regarding channel model validation, spatial correlation that is used for FR 1 is replaced with PSP.

10.5.1.2 Measurement methodologies

For FR 1, 2D MPAC and RTS test methods are adopted as the reference methods, while the 3D MPAC test method is selected in FR 2.

2*D MPAC for FR*1

2D MPAC test method is regarded as a reference for FR 1 MIMO OTA testing. By utilizing multiple probes around the DUT, the spatial profile of the target channel can be produced in the test zone [7,41]. Considering the tradeoff between the cost of CE resources and the possible desired test zone size [42], 16 uniformly spaced probes on a ring are adopted as the final probe layout in FR1 MIMO OTA testing by 3GPP RAN 4 as shown in Figure 10.12 and a photo of practical setup is shown in Figure 10.13, respectively.

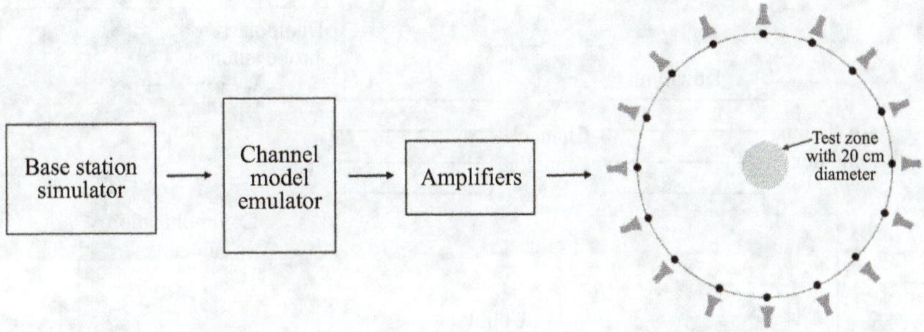

Figure 10.12 2D MPAC system layout for NR FR 1 MIMO OTA testing

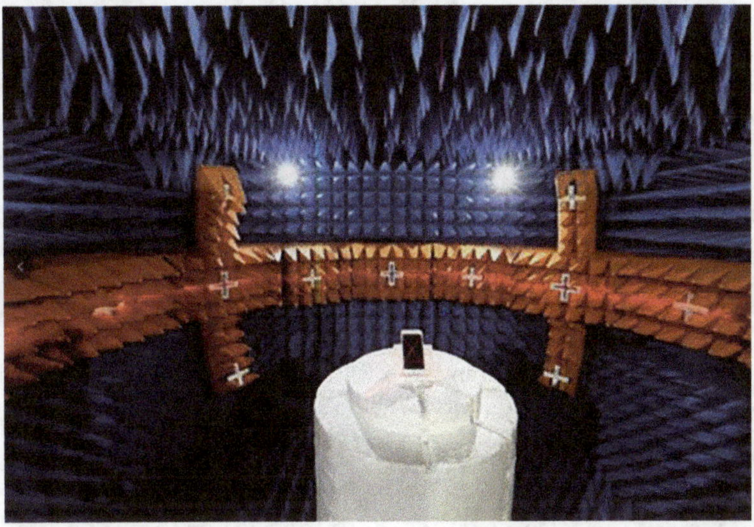

Figure 10.13 MPAC anechoic chamber for FR 1 MIMO OTA testing

***RTS for FR*1**

The RTS test method [43] is defined as the harmonized method for FR 1 MIMO OTA testing. An example of the RTS system layout for 4 × 4 NR FR 1 MIMO OTA testing is illustrated in Figure 10.14. The BS emulator sends downlink signals to the CE. Specified channel models are implemented in the CE, and the output signals from the CE are fed to the probes. The DUT is placed in the center of the AC, and a separate communication antenna is used for the uplink connection with the BS emulator. Depending on the chamber size and path loss, amplifiers may be needed for downlink and uplink.

The first stage of RTS is to acquire the DUT's complex antenna pattern, including both amplitude and phase. One way to obtain this is through the so-called ATF. In the second stage of RTS, the transfer function between the probes and the DUT antennas can be obtained through the ATF. A weighting matrix that is the

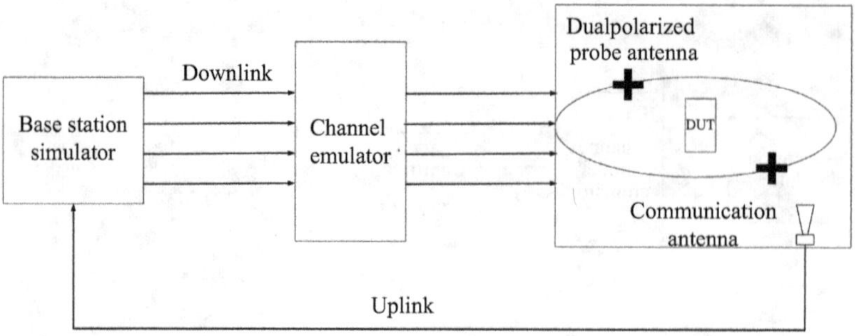

Figure 10.14 Example RTS system layout for 4 × 4 NR FR 1 MIMO OTA testing

inverse of the transfer function can be applied in the CE so that the isolation between the desired link and crosstalk link is maximized. The DUT antenna pattern measured in the first stage can be added to the emulated channel in the CE. Finally, throughput tests can be performed with different channel models.

The internal structure of the actual AC for the RTS method is depicted in Figure 10.15. The DUT is placed in the center of the AC and surrounded by multiple probes, and 2D or 3D antenna patterns can be measured by rotating the DUT with the ATF. During the throughput measurement, the action of physically rotating the DUT as with the MPAC method is, instead, realized by applying the resulting DUT antenna pattern that is mathematically rotated based on the measured DUT antenna pattern according to the desired angle in the CE.

The advantages and disadvantages of the MPAC and RTS methods for NR FR1 MIMO OTA testing are given in Table 10.1.

3*D MPAC for FR*2

Different methods for FR 2 MIMO OTA testing have been proposed in 3GPP and discussed extensively in [34,44,45]. As a result, 3D MPAC [5,17] is adopted as the final test method by 3GPP RAN 4, and the system setup is illustrated in Figure 10.16. Note that due to the much higher frequency and corresponding channel models for FR 2, the testing operation is different from 2D MPAC for FR 1, as listed in the following aspects:

- From the perspective of system setups, amplifiers used in FR 1 are replaced by radio heads with the functions of frequency conversion and power amplification.
- Compared with the spatial correlation used in FR 1, PSP is used in FR 2 to measure the similarity of the PAS produced by OTA system and the reference PAS [46].
- To emulate the behavior of the channel models in FR 2, six dual-polarized probes are placed on a 3D sector with minimum radius of 0.75 m from the center of the test zone. The exact probe locations over the channel model coordinate system are shown in [23]. The six probe locations are determined following the discussed probe-selection principle.

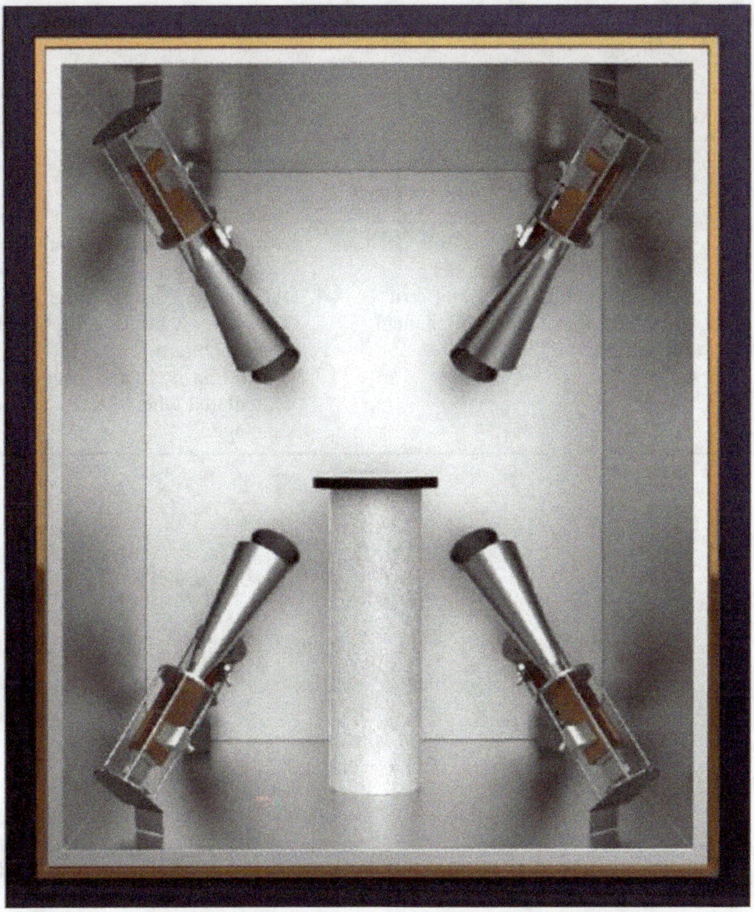

Figure 10.15 FR 1 RTS anechoic chamber

- In comparison to the 2D scan in FR 1 MIMO OTA testing, 3D scan is adopted in FR 2 MIMO OTA testing. Thirty-six evenly spaced test points determined using the charged particle approach for FR 2 MIMO OTA testing are listed in Table 6.2.3.2-1 of [23].

An exemplary environment inside the AC for FR 2 MIMO OTA testing is shown in Figure 10.17. For the channel models to be tested and the corresponding probe placement positions to be extendable in the future, a setup with six dual-polarized probes being placed in the 3D sector is designed, which can be moved freely, and the 2D positioner can be controlled with the software automatically. Furthermore, the addition of radio head equipment allows CEs that originally only support FR 1 testing to support for FR 2 as well. As an important component, both the capability and performance of the radio head need to be considered in the actual

Table 10.1 Advantages and disadvantages of MPAC and RTS in FR 1 MIMO OTA testing

Methods	Advantages	Disadvantages
MPAC	1. A real physical channel spatial profile on the DUT side is realized 2. The same fading sequence can be reused for different DUTs under the same target channel model	1. High setup complexity and cost 2. Channel model validation is needed prior to throughput testing
RTS	1. Low setup complexity and cost 2. Capable of any arbitrary channel models	1. DUTs need to support the ATF 2. Only support DUTs of nonadaptive antenna patterns 3. The weighting matrix needs to be recalculated when the DUT condition is changed

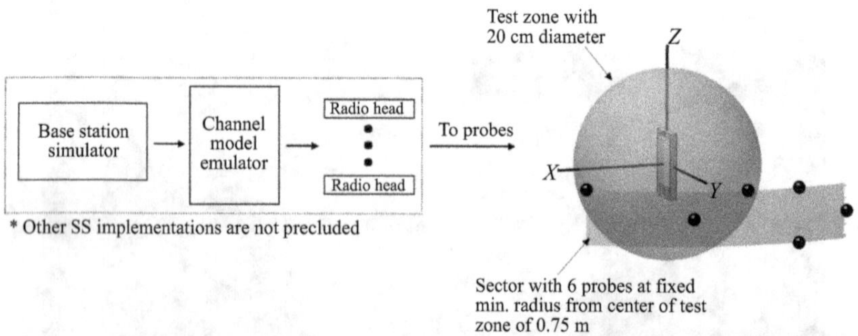

Figure 10.16 3D MPAC system layout for NR FR 2 MIMO OTA testing

test system. The radio head needs to have stable frequency conversion and accurate power amplification to avoid distortion of the signal by the component itself. Second, the single-port bidirectional transmission capability can halve the number of required CE ports, while supporting bidirectional testing.

10.5.1.3 Channel model validation

In order to ensure that the channel models are correctly emulated in the test zone, 3GPP RAN 4 also defined the parameters and procedure of channel model validation. Table 10.2 illustrates the key performance indicators and instruments needed to accomplish the procedure.

For FR 1 MIMO OTA channel models, the same verification parameters with 4G MIMO OTA testing, i.e., PDP, Doppler/temporal correlation, spatial correlation, cross-polarization, and power validation, are adopted for the channel model validation. For FR 2 MIMO OTA, PSP is used instead of the spatial correlation to validate how well the channel model is emulated compared with the target one. The validation procedures have been described in detail for different parameters in [23].

Figure 10.17 FR 2 3D MPAC anechoic chamber in labs

Table 10.2 Channel model validation for FR 1 and FR 2

FR 1	FR 2	Validation instrument
Power validation		BS simulator and spectrum analyzer
Doppler/temporal correlation		Signal generator and spectrum analyzer
Cross-polarization		Vector network analyzer (VNA)
PDP		
Spatial correlation	PSP	

Detailed parameters for time and frequency samples have not been defined by 3GPP RAN 4. Moreover, the procedure of the PSP validation in FR 2 is still under discussion within 3GPP RAN 4.

10.5.2 RF testing

10.5.2.1 5G NR RF BS standards

3GPP started 5G standard research in 2013, and the RF specifications for both BS and UE are discussed in the TSG RAN 4. After 5 years, the first edition of 5G NR standards was frozen in 2018. For the BS, there are four major RF TSs, i.e., BS transmission and reception characteristics [47], electromagnetic compatibility (EMC) [48], conducted conformance testing [49], and radiated conformance testing. Furthermore, three TRs [50–52] are recommended as the supplements to the previous TS. In [50], transmitter and receiver characteristics of the active antenna system (AAS) BS deployment scenarios are studied, and the EMC requirements are proposed as well. In [51], the detailed AAS BS RF OTA test methods are described, e.g., FF, NF, CATR, PWS, and RC.

The test cases can be divided into two parts, i.e., transmitter characteristics and receiver characteristics:

- Transmitter test cases can be summarized as three categories. The first is the power class test, e.g., TRP and EIRP. The second is the radiated signal quality test, e.g., error vector magnitude (EVM), frequency error and time alignment. The third is the radiated unwanted signal test, including unwanted emission and intermodulation. These previous cases measure the BS coverage ability and transmission signal quality.
- Receiver test cases can be divided into four kinds. The first is the OTA receiver power ability test cases, e.g., sensitivity, reference sensitivity level, and dynamic range. The second is the in- and out-band blocking test cases. The third is the unwanted signal test cases, including spurious emission (SE) and intermodulation. The last is the in-channel selectivity. These test cases verify the BS receive signals quality and anti-interference performance.

The basic specifications and capabilities of BS are given in Tables 10.3 and 10.4. The general commercial macro-BS specifications are shown in Table 10.3, which are based on the BS manufacture's declaration.

As shown in Table 10.3, the number of antenna elements of the FR 2 case is significantly larger than that of the FR 1, whereas the physical size of the BS is

Table 10.3 5G NR BS basic specifications

Specifications	FR 1 5G NR BS	FR 2 5G NR BS
Typical size	About 1 m ×0.5 m ×0.3 m	About 0.5 m ×0.3 m ×0.25 m
Frequency band	2.6, 3.5, 4.9 GHz	26, 28, 39 GHz
Working bandwidth	100 MHz	400 MHz
Duplex mode	TDD	TDD
Number of RF channels	8 Tx/ 8 Rx, 4 Tx/ 4 Rx	4 Tx/ 4 Rx, 2 Tx/ 2 Rx
Number of antenna elements	At least 96	512 / 1,024
Beam coverage	Horizontal maximum ±60° Vertical +6° to −17°	Horizontal maximum ±60° Vertical maximum ±17°

Table 10.4 5G NR BS typical radiated capabilities

Specifications	FR 1 5G NR BS	FR 2 5G NR BS
Test cases	FR 1 5G NR BS	FR 2 5G NR BS
Output power	≥ 53 dBm	≥ 30 dBm
Maximum EIRP	≥ 76 dBm	≥ 60 dBm
Antenna gain	≥ 23 dBi	≥ 30 dBi
Occupied bandwidth	≤ 100 MHz	≤ 400 MHz
EVM (64QAM)	≤ 8%	< 9%
Receiver sensitivity	≤ −116 dBm	≤ −113 dBm
Unwanted emissions	Meet the requirements	≤ −39 dBW/ 200 MHz

Figure 10.18 Plane wave synthesizer

smaller. When the frequency increases, the antenna element can be designed to be smaller; thus the mm-wave BS can accommodate more antenna elements. Moreover, the FR 2 BS has a larger working bandwidth than the FR 1 BS. In addition, the beam coverages of FR 1 BS and FR 2 BS are almost the same except it goes up to $+6°$ in the vertical direction in FR 1 and $+17°$ in FR 2.

As shown in Table 10.4, the output power of FR 1 BSs is much higher than that of FR 2 BSs, because the power amplifier has a better performance at the low frequency. Since the array aperture of FR 2 BSs in electric size is much bigger than that of FR 1 BSs, the beam width of FR 2 BS is smaller.

Traditional OTA test methods, e.g., FF, CATR, have been widely used. These test systems are usually large, heavy, and costly for BS measurement. To cope with that, the PWS method was developed in 2018 and approved as a BS test method in 2019 by 3GPP. Compared with the CATR test system, the PWS test system requires a smaller chamber size, and the weight is lighter. A practical PWS system is shown in Figure 10.18. The typical size of a PWS system is 6 m $\times$4 m $\times$4 m for the FR 1 BS measurement, and it can provide at least a test quiet zone of 1 m $\times$1 m.

It is noted that the measurement time is very important. At present, many OTA test cases are performed by directly sampling on the entire spherical surface,

Figure 10.19 Reverberation chamber

e.g., the equal angle spherical acquisition method for TRP measurements. For example, in FR1 BS TRP testing, when the sampling interval is 6°, a total of 1, 801 points on the 4π solid angle need to be measured. As the sampling interval decreases with the increase of the frequency, more data need to be collected in FR 2 TRP testing. In order to save time, a few other data acquisition methods have been included in [49], e.g., the wave vector space, the two-cut method, and the three-cut method. However, it may deteriorate in the test accuracy.

The RC test method is adopted in [51] for the BS OTA testing. In some laboratories, as illustrated in Figure 10.19, in an Reverberation chamber, the RC test method is performed for FR 2 BSs TRP testing. The difference of the TRP test value between CATR and RC is only about 0.3 dB, but the test time of RC is hundreds of times lower than that of CATR. Moreover, the chamber size of the RC is only about 1.5 m ×1.5 m ×2 m, and the cost of an RC test system is about one-third of the CATR test system. The RC test method can also be used for some testing items in FR 1 BS testing.

10.5.2.2 5G NR RF UE standards

The standards of 5G NR UE mainly include protocol conformance specifications, radio resource management (RRM) conformance specifications, radio transmission, and reception conformance specifications. The protocol conformance test requirement is introduced in [53], which specifies protocol stack development verification requirements. The RRM conformance test requirement is studied in [54], which specifies mobility management and performance testing requirements. The RF conformance test requirement is studied in [55], which specifies the transmitter and receiver requirement.

UE RF standards mainly include the technical requirement, test procedure, and test method. The RF TSs of FR 1 devices and FR 2 devices are introduced in [55,56], respectively, which specify the minimum RF transmission and reception requirements of standalone devices. The TS for FR 1 and FR 2 interworking operation with other radios is introduced in [57], which specifies the minimum RF transmission and reception requirements of NSA devices. The UE RF conformance specifications for FR 1 and FR 2 devices are introduced in [58,59] respectively, which specify the RF measurement procedures of the conformance testing. Moreover, the RF conformance specification for FR 1 and FR 2 interworking operation with other radios is introduced in [60], which specifies the measurement procedures of the conformance testing for carrier aggregation between FR 1 and FR 2, and additional requirements due to the NR NSA operation mode with evolved universal terrestrial radio access. Besides, the test method for FR 2 is studied in [61], which details the OTA test methodology, the associated measurement uncertainty budget, and the related test tolerance. FR1 devices are measured through conducted methods and FR 2 devices are measured through OTA methods. The enhanced test method for FR 2 UE is introduced in [62], which enhances the FR 2 RF testing methodology and quantifies the impact of the enhancement on the measured UE performance, as related to the polarization basis mismatch between the test equipment and UE. Moreover, it also adds support for testing under extreme temperature conditions.

In the 5G FR 1 frequency band, the characterization of RF performance is evaluated through cabled connections at the temporary antenna connectors. This bypasses the device antenna. In the 5G FR 2 mm-wave frequency band, however, RF units and antenna interfaces cannot be separated, and hence all the RF characteristics have to be tested through the OTA test method.

RF test cases are studied in the following three aspects: transmitter (Tx) and receiver (Rx) in-band test cases, Tx and Rx spurious test cases, and Rx blocking test cases. Tx and Rx in-band test cases mainly include the power class test, modulation and demodulation test, and adjacent channel leakage power test, e.g., EIRP, TRP, EIS, EVM, and frequency error. EIRP and TRP are the two key technical indicators to measure the power class of the DUT. A very high EIRP or TRP may cause interference to other frequency bands and systems. On the other hand, a low EIRP or TRP may decrease the coverage. In [59], the EIRP value of FR2 UE with power level 3 in band n 258 is limited between 22.4 and 43 dBm, and the maximum TRP value is less than 23 dBm. EIS is a key technical parameter of receiver performance test, which measures the minimum power with which the DUT can still maintain a predefined throughput level averaged over different directions. The EIS affects the effective coverage of BS. The reference EIS value of FR 2 UE with power level 3 supporting 100 MHz bandwidth in band n 258 is set lower than −85.3 dBm in [59]. The EVM is a key technical parameter for evaluating modulation and demodulation capability of FR 2 UEs, which represents the difference between the reference waveform and the measured waveform. The frequency error is to verify the accuracy of a UE's receiver and transmitter detecting and generating the correct carrier frequency with respect to the stimulus signal offered by the BS under ideal propagation conditions and low power level.

Tx and Rx spurious test cases mainly include the transmitter SE and receiver SE. SEs are caused by unwanted transmitter effects, e.g., harmonics emissions, parasitic emissions, intermodulation products, and frequency conversion products. Extremely high SEs may cause interference to other frequency bands and systems, which will affect the frequency planning.

Rx blocking test cases mainly include the adjacent channel selectivity (ACS) and in-band blocking, which represent the UE's anti-interference ability. ACS tests the UE's ability to receive data with a given average throughput for a specified reference measurement channel, in the presence of an adjacent channel signal at a given frequency offset from the central frequency of the assigned channel, under conditions of ideal propagation and no added noise. In-band blocking is defined for an unwanted interfering signal falling into the UE receive band or into the spectrum equivalent to twice the channel bandwidth below or above the UE receive band at which the relative throughput shall meet or exceed the minimum requirement for the specified measurement channels.

In order to test these cases in a reliable environment, many solutions for the OTA testing are proposed in [61], including DFF, IFF, and NF. Currently, many instrument manufacturers have proposed FR2 CATR solutions for the RF conformance testing. In this chapter, an CATR chamber will be introduced as a standard IFF test system to evaluate the FR2 UE's RF performance.

The 5G NR FR 2 UE RF conformance testing system is an integrated solution consisting of the CATR, combined axes UE positioner, multi-feed antenna array, link antenna, system simulator, signal analyzer, signal generator, and chamber control software, as illustrated in Figure 10.20, which could offer an effective interface for test case execution, analysis, and reporting. As shown in Figure 10.20, the CATR chamber is used to form a plane wave at the UE position in which the LTE and FR 2 link antenna are used to establish the call from the UE to the system simulator. Tx and Rx in-band test cases are executed by the system simulator and the combined axes UE positioner. The signal analyzer and multi-feed antenna array are used for the Tx and Rx spurious test. The signal generator is proposed to perform the Rx blocking test. With this chamber, the RF conformance testing specified by both 3GPP and global certification forum, including beamforming in 3D space at mm-wave frequencies, predefined and customizable FR 2 RF performance tests, extreme temperature and humidity condition tests, and in-band and out-of-band emission tests, can be performed efficiently. To meet the wide test frequency range and to reduce the test time in SE measurements, based on the CATR system, a multi-feed test system can be utilized, which is composed of one feed working from 24 to 44 GHz, one feed working from 6 to 24 GHz, one feed working from 40 to 60 GHz and one feed working from 60 to 90 GHz, which can meet the requirements of the SE test from 6 GHz to the second harmonic. This solution greatly reduces the test time, and multiple feeds are placed near the parabolic focus in comparison to the original single-feed system. The offset feed can form a plane wave with a certain angle within the original area in the CATR, so as to realize multi-band simultaneous measurements.

To sum up, in this section, the protocol conformance specification, RRM conformance specification, and RF conformance specification of 5G NR UE are

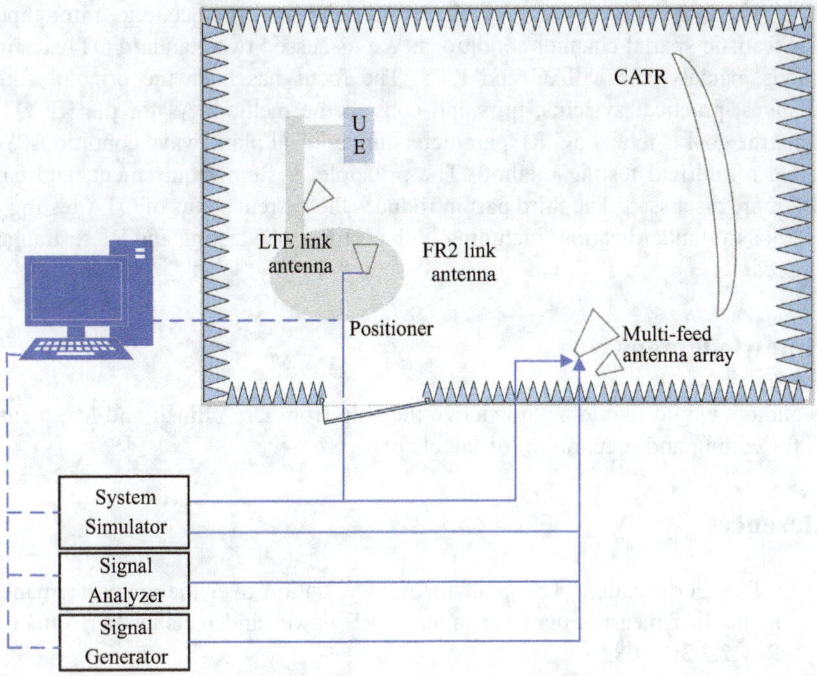

Figure 10.20 RF test system layout for 5G NR FR 2 UE RF test

reviewed. The RF-related standards are introduced in detail, including the TSs, test procedures, and test methods. The Tx and Rx in-band test cases, SE tests, and blocking test cases to evaluate the RF performance of 5G FR 2 UE are discussed. A multi-feed CATR RF test chamber is proposed to execute these RF test cases with which the efficiency and dynamic range is improved through three effective methods. In the future, low-cost solutions and more efficient systems will be proposed for FR 2 UE RF tests.

10.6 Conclusion

The 5G systems promise higher spectral efficiency and energy efficiency, lower latency, and more reliable communications. These advantages are supported by mm-wave and/or M-MIMO techniques. Cable conducted testing has been the dominant testing method for sub-6-GHz conventional communication systems, where antenna ports are mostly accessible for conducted testing. In the conducted testing, antenna characteristics are omitted completely by testing from antenna ports. However, for M-MIMO antenna systems with hundreds of antenna elements, conducted testing obviously becomes infeasible. Moreover, it is likely that mm-wave systems will not have standard antenna ports, rendering OTA the only testing solution. In this chapter, we first discussed the OTA solutions for the performance testing, where the objective is to evaluate 5G radio performance, e.g., throughput, under realistic spatial channel conditions. We discussed two standard OTA testing methods, namely, the MPAC and RTS. The focus has been the principle, key challenges, practical systems, pros and cons of the method. As for the RF OTA testing, the goal is to test 5G RF parameters under ideal plane wave conditions. We detailed a midfield testing method. The principle, system requirement, and capabilities are discussed. The third part introduces the current status of OTA testing of 5G radios in standardization, including both performance testing and RF testing are discussed.

Acknowledgment

The authors would like to acknowledge the help from Dr. Yilin Ji and Mr. Yejian Lyu for editing and preparation of the chapter.

References

[1] CTIA Certification. Test plan for mobile station over the air performance, method of measurement for radiated RF power and receiver performance. Rev 2.2.2; 2008.

[2] Foegelle M. Over-the-air performance testing of wireless devices with multiple antennas. *RF Design*. 2006 02; 2012.

[3] Foegelle M. Systems and methods for over the air performance testing of wireless devices with multiple antennas; 2012. US Patent 8,331,869.

[4] Iwai H, Yamamoto A, Sakata T, *et al.*, *Spatial fading emulator for handset antennas*. In: 2005 IEEE Antennas and Propagation Society International Symposium. vol. 1A; 2005. p. 218–221.

[5] Kyösti P, Hentilä L, Fan W, *et al.*, On Radiated Performance Evaluation of Massive MIMO Devices in Multiprobe Anechoic Chamber OTA Setups. *IEEE Transactions on Antennas and Propagation*. 2018; 66(10): 5485–5497.

[6] Kyösti P, Nuutinen J, Kolu J, *et al.*, Channel modelling for radiated testing of MIMO capable terminals. In: Proc. of ICT-Mobile Summit; 2009.

[7] Kyösti P, Jämsä T, and Nuutinen J., Channel Modelling for Multiprobe Over-the-Air MIMO Testing. *International Journal of Antennas and Propagation*. 2012; 2012.

[8] Kotterman WAT, Heuberger A, and ThomÃ RS., *On the accuracy of synthesised wave-fields in MIMO-OTA set-ups*. In: Proceedings of the 5th European Conference on Antennas and Propagation (EUCAP); 2011. p. 2560–2564.

[9] Laitinen T, Kyösti P, Nuutinen J, *et al.*, *On the number of OTA antenna elements for plane-wave synthesis in a MIMO-OTA test system involving a circular antenna array*. In: Proceedings of the Fourth European Conference on Antennas and Propagation; 2010. p. 1–5.

[10] Kotterman W., *Increasing the volume of test zones in anechoic chamber MIMO over-the-air test set-ups*. In: 2012 International Symposium on Antennas and Propagation (ISAP); 2012. p. 786–789.

[11] Fan W, Carton I, Kyösti P, *et al.*, Emulating Ray-Tracing Channels in Multiprobe Anechoic Chamber Setups for Virtual Drive Testing. *IEEE Transactions on Antennas and Propagation*. 2016; 64(2): 730–739.

[12] Fan W, Kyösti P, Fan S, *et al.*, *3D channel model emulation in a MIMO OTA setup*. In: 2013 IEEE 78th Vehicular Technology Conference (VTC Fall); 2013. p. 1–5.

[13] Kyösti P, Fan W, Pedersen GF, *et al.*, On Dimensions of OTA Setups for Massive MIMO Base Stations Radiated Testing. *IEEE Access.* 2016; 4: 5971–5981.

[14] Fan W, Kyösti P, Rumney M, *et al.*, Over-the-Air Radiated Testing of Millimeter-Wave Beam-Steerable Devices in a Cost-Effective Measurement Setup. *IEEE Communications Magazine*. 2018; 56(7): 64–71.

[15] Fan W, Carton I, Kyösti P, *et al.*, A Step Toward 5G in 2020: Low-Cost OTA Performance Evaluation of Massive MIMO Base Stations. *IEEE Antennas and Propagation Magazine*. 2017; 59(1): 38–47.

[16] Ji Y, Fan W, Pedersen GF, *et al.*, On Channel Emulation Methods in Multiprobe Anechoic Chamber Setups for Over-the-Air Testing. *IEEE Transactions on Vehicular Technology*. 2018; 67(8): 6740–6751.

[17] Fan W, Zhang F, and Wang Z., Over-the-Air Testing of 5G Communication Systems: Validation of the Test Environment in Simple-Sectored Multiprobe Anechoic Chamber Setups. *IEEE Antennas and Propagation Magazine*. 2021; 63(1): 40–50.

[18] Fan W, Hentilä L, and Kyösti P., Spatial Fading Channel Emulation for Over-the-air Testing of mmWave Radios: Concepts and Experimental Validations. *Frontiers of Information Technology & Electronic Engineering*. 2021; 22: 548–559.

[19] R4-091361. MIMO OTA test methodology proposal. Agilent Technologies, 3GPP TSG-RAN WG4 Meeting #bis. Rev 2.2.2; 2009.

[20] Jing Y, Zhao X, Kong H, *et al.*, Two-Stage Over-the-Air (OTA) Test Method for LTE MIMO Device Performance Evaluation. *International Journal of Antennas and Propagation*. 2012; 2012.

[21] R4-66AH-0012. Incorporating self-interference into the two-stage method. Agilent Technologies, Tri-L Solutions, 3GPP RAN4#66AH. Rev 2.2.2; 2013.

[22] 977 T. Verification of radiated multi-antenna reception performance of user equipment (UE). V14.5.0; 2018.

[23] 827 T. Study on radiated metrics and test methodology for the verification of multi-antenna reception performance of NR User Equipment (UE). V1.3.1; 2020.

[24] Certification C., Test methodology. MIMO radiated two stage; 2020.

[25] Rumney M, Kong H, and Jing Y., Practical active antenna evaluation using the two-stage MIMO OTA measurement method. In: The 8th European

Conference on Antennas and Propagation (EuCAP 2014); 2014. p. 3500–3503.

[26] Jing Y, Kong H, and Rumney M., MIMO OTA Test for a Mobile Station Performance Evaluation. *IEEE Instrumentation Measurement Magazine*. 2016; 19(3): 43–50.

[27] Rumney M, Kong H, Jing Y, *et al.*, *Recent advances in the radiated two-stage MIMO OTA test method and its value for antenna design optimization*. In: 2016 10th European Conference on Antennas and Propagation (EuCAP); 2016. p. 1–5.

[28] Jing Y, Hertel T, Kong H, *et al.*, *Recent developments in radiated two-stage MIMO OTA test method*. In: 2020 14th European Conference on Antennas and Propagation (EuCAP); 2020. p. 1–5.

[29] 978 T. E-UTRA UE Antenna test function definition for two-stage MIMO OTA test method. v13.0.0, 3GPP. https://portal.3gpp.org/desktopmodules/Specifications/SpecificationDetails.aspx?specificationId=2594.

[30] Fan W, Kyösti P, Hentilä L, *et al.*, MIMO Terminal Performance Evaluation With a Novel Wireless Cable Method. *IEEE Transactions on Antennas and Propagation*. 2017; 65(9): 4803–4814.

[31] Zhang F, Fan W, and Wang Z., Achieving Wireless Cable Testing of High-Order MIMO Devices With a Novel Closed-Form Calibration Method. *IEEE Transactions on Antennas and Propagation*. 2021; 69(1): 478–487.

[32] Dao M, Nguyen V, Im Y, *et al.*, 3D Polarized Channel Modeling and Performance Comparison of MIMO Antenna Configurations With Different Polarizations. *IEEE Transactions on Antennas and Propagation*. 2011; 59 (7): 2672–2682.

[33] NR. Base station (BS) conformance testing Part 2: Radiated conformance testing (Release 15). 3GPP TS 38141-2; 2018.

[34] NR. User equipment (UE) conformance specification; radio transmission and reception. 3GPP TS 38521 v1640; 2020.

[35] NR. Study on test methods. 3GPP TR 38810 v1650; 2019.

[36] Kong H, Wen Z, Jing Y, *et al.*, *A compact millimeter wave (mmWave) mid-field over the air (OTA) RF performance test system for 5G massive MIMO devices*. In: 2018 IEEE MTT-S International Wireless Symposium (IWS); 2018. p. 1–4.

[37] Kong H, Wen Z, Jing Y, *et al.*, Midfield Over-the-Air Test: A New OTA RF Performance Test Method for 5G Massive MIMO Devices. *IEEE Transactions on Microwave Theory and Techniques*. 2019; 67(7): 2873–2883.

[38] Kong H, Jing Y, Wen Z, *et al.*, *Mid-field OTA RF test method: new developments and performance comparison with the compact antenna test range (CATR)*. In: 2020 14th European Conference on Antennas and Propagation (EuCAP); 2020. p. 1–5.

[39] Jing Y, Wen Z, Hertel T, *et al.*, *Near field test methodology for 5G UE FR2 radiated performance measurements*. In: 2021 15th European Conference on Antennas and Propagation (EuCAP); 2021. p. 1–5.

[40] R4-2016213. On test methodology for high DL power and low UL power test cases. 3GPP TSG-RAN WG4 Meeting #97-e; 2020.

[41] Fan W, Carreo Bautista de Lisbona XCB, Sun F, *et al.*, Emulating Spatial Characteristics of MIMO Channels for OTA Testing. *IEEE Transactions on Antennas and Propagation*. 2013; 61(8): 4306–4314.

[42] R4-1910809. Further discussion on NR FR1 MPAC Probes Layout. VIVO, 3GPP RAN4 #92bis; 2019.

[43] Shen P, Qi Y, Yu W, *et al.*, Notice of Retraction: Inverse Matrix Auto-Search Technique for the RTS MIMO OTA TestPart 1: Theory. *IEEE Transactions on Electromagnetic Compatibility.* 2017; 59(6): 1716–1723.

[44] R4-1711501. Further analysis of RTS applicability to BS demodulation performance testing. Keysight Technologies, 3GPP RAN4 #84bis; 2017.

[45] R4-170669. SS MPAC for RRM/Demod. Keysight, RAN4 AH#2; 2017.

[46] R4-1705831. Metrics for evaluating RRM/Demodulation Measurement Setup. Keysight, RAN4 #83; 2017.

[47] NR. Base station (BS) radio transmission and reception. 3GPP TS 38104, Tech Spec Rel 16, V1650; 2020.

[48] NR. Base station (BS) electromagnetic compatibility (EMC). 3GPP TS 38113, Tech Spec Rel 16, V1650; 2020.

[49] NR. Base station (BS) conformance testing; Part 1: Conducted conformance testing. 3GPP TS 38141-1, Tech Spec Rel 16, V1650; 2020.

[50] Study of radio frequency (RF) and electromagnetic compatibility (EMC) requirements for active antenna array system (AAS) base station. 3GPP TR 37840, Tech Repo Rel 12, V1210; 2014.

[51] Evolved universal terrestrial radio access (E-UTRA) and universal terrestrial radio access (UTRA); radio frequency (RF) requirement background for active antenna system (AAS) base station (BS) radiated requirements. 3GPP TR 37843, Tech Repo Rel 15, V1570; 2020. https://portal.3gpp.org/desktopmodules/Specifications/SpecificationDetails.aspx?specificationId=3131.

[52] Radio frequency (RF) conformance testing background for radiated base station (BS) requirements. 3GPP TR 37941, Tech Repo Rel 16, V1661; 2020. https://portal.3gpp.org/desktopmodules/Specifications/SpecificationDetails.aspx?specificationId=3737

[53] 5GS. User equipment (UE) conformance specification; Part 1: Protocol. 3GPP TS 38523-1, Tech Spec Rel 16, V1650; 2020.

[54] NR., User equipment (UE) conformance specification; Radio resource management (RRM). 3GPP TS 38533, Tech Spec Rel 16, V1650; 2020.

[55] NR., User equipment (UE) radio transmission and reception; Part 1: Range 1 Standalone. 3GPP TS 38101-1, Tech Spec Rel 16, V1650; 2020.

[56] NR., User equipment (UE) radio transmission and reception; Part 2: Range 2 Standalone. 3GPP TS 38101-2, Tech Spec Rel 16, V1650; 2020.

[57] NR., User equipment (UE) radio transmission and reception; Part 3: Range 1 and Range 2 Interworking operation with other radios. 3GPP TS 38101-3, Tech Spec Rel 16, V1650; 2020.

[58] NR., User equipment (UE) conformance specification; radio transmission and reception; Part 1: Range 1 Standalone. 3GPP TS 38521-1, Tech Spec Rel 16, V1650; 2020.
[59] NR., User equipment (UE) conformance specification; radio transmission and reception; Part 2: Range 2 Standalone. 3GPP TS 38521-2, Tech Spec Rel 16, V1650; 2020.
[60] NR., User equipment (UE) conformance specification; Radio transmission and reception; Part 3: Range 1 and Range 2 interworking operation with other radios. 3GPP TS 38521-3, Tech Spec Rel 16, V1650; 2020.
[61] Study on test methods for New Radio. 3GPP TR 38810, Tech Repo Rel 16, V1661; 2020. https://portal.3gpp.org/desktopmodules/Specifications/SpecificationDetails.aspx?specificationId=3218.
[62] Study on enhanced test methods for FR2 NR Use. 3GPP TR 38884: Tech Repo Rel 17, V010; 2020. https://portal.3gpp.org/desktopmodules/Specifications/SpecificationDetails.aspx?specificationId=3707.

Chapter 11
Over-the-air testing with synthetic-aperture techniques

Peter Vouras[1], Kate A. Remley[2], Benjamin Jamroz[2], Maria Becker[3], Dylan F. Williams[2], Jeanne Quimby[2], Rodney Leonhardt[2], Robert D. Horansky[2], Alec Weiss[4], Robert D. Jones[4], Joshua Kast[4], Atef Elsherbeni[4], Paritosh Manurkar[5], Damla Guven[1], Jack Chuang[1], and Camillo Gentile[1]

This chapter provides an overview of synthetic-aperture systems for applications related to wireless communications. By sampling the fields propagating through a spatial volume at many locations, synthetic-aperture-system measurements allow the reconstruction of the temporal and spatial characteristics of the channel in post-processing. While often used to characterize spatially diverse multipath environments, synthetic-aperture systems can also provide reference measurements for assessing the over-the-air (OTA) performance of a wireless device under test (DUT).

The chapter covers the relationships between system parameters, such as aperture size and angular resolution, and includes guidelines for optimal configuration of synthetic-aperture systems for various types of measurements. The chapter concludes with examples that illustrate synthetic-aperture measurements of spatial multipath channels in both a repeatable, laboratory-based wireless communications test bed operated in microwave cellular-communication bands and in an industrial setting at millimeter-wave (mmWave) frequencies.

The application of synthetic-aperture systems has enabled the estimation of multipath temporal and spatial characteristics of a wireless channel to an unprecedented level of accuracy [1]. The temporal behavior of multipath signals at the receiver created by numerous propagation paths between the transmit and receive antennas can significantly degrade data-transfer performance of a wireless system.

[1]NIST Communications Technology Laboratory, Gaithersburg, MD, USA
[2]NIST Communications Technology Laboratory, Boulder, CO, USA
[3]Department of Physics, Nebraska Wesleyan University, Lincoln, NE, USA
[4]Department of Electrical Engineering and Computer Science, Colorado School of Mines, Golden, CO, USA
[5]Department of Electrical, Computer, and Energy Engineering, University of Colorado, Boulder, CO, USA

In the spatial domain, the active antennas and amplifiers in commercially available devices may be susceptible to inter-symbol interference from spatially diverse multipath channels, imperfectly modeled coupling between the device's antenna elements, and/or nonideal beamforming due to imperfect channel estimation, for example. To identify these issues, an understanding of both the environment and the response to the environment of the DUT is essential [2,3].

A typical synthetic-aperture channel-measurement system consists of a transmit antenna, which is typically held fixed, and a receive antenna mounted on a mechanical positioner. Typical positioners include translation stages, rotating stages, or robotic arms with multiple degrees-of-freedom that move the antenna through space and stop precisely at points along a predetermined sampling lattice. A receiver sequentially samples the received signal at each sampling-lattice location. An example of such a system is shown in Figure 11.1, where the robotic arm could perform planar, cylindrical, or spherical tests. In post-processing, the impulse response of a wireless channel can be reconstructed from the samples collected at each location in the lattice as if the data had been sampled simultaneously at each location with built-in receiver hardware such as a digital multichannel phased array. Note that the synthetic-aperture hardware implementation can support the characterization of channels in both near- and far-field conditions. However, most analysis software, including the examples shown in this chapter, make far-field assumptions [4].

In many implementations, a vector network analyzer (VNA) is used to excite and sample the OTA signal because of the potential for corrected, high-dynamic-

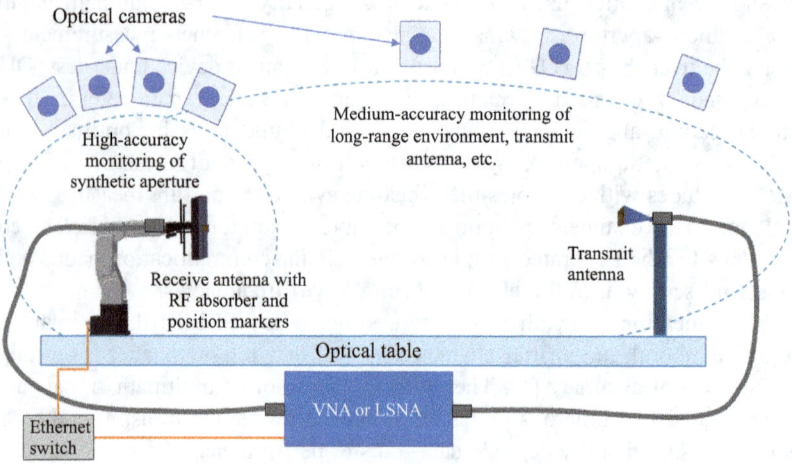

Figure 11.1 One type of synthetic-aperture system in which a robotic arm scans over discrete points in the aperture. At each position in the synthetic-aperture lattice, the vector network analyzer steps over a wide frequency range. The spatial channel is reconstructed in post-processing.

range measurements. VNAs have well-established calibrations that can be used to de-embed cables and connectors to easily correct for measurement-system imperfections that are not part of the channel being characterized. From the metrologist's point of view, another advantage is that VNA uncertainty analyses can be made traceable to fundamental physical quantities using well-established techniques [5,6]. However, a vector signal generator and vector receiver could be used as well, providing a potentially faster and more-compact measurement system at the expense of a narrower bandwidth and, thus, lower delay resolution [7]. Such a setup could also be made traceable, although this is typically less common.

The accuracy of the calculations can be further improved by leveraging an optical position-measurement system to better locate actual antenna locations during the measurement. The system shown in Figure 11.1 represents an mmWave synthetic-aperture system implemented at the National Institute of Standards and Technology (NIST). The optical table minimizes vibrations to reduce position uncertainty in the measurement. An external camera-based position measurement system is used to determine both the location of the receive-antenna element in the sampling lattice and the location of scatterers in the channel if desired. Typically, high accuracy, and a relatively high number of cameras, are desired for the former, whereas the latter may not require such accuracy. The position and rotation of each object in the channel may be determined by the use of three or more small infrared reflective markers mounted on the object and tracked by the camera system. The camera-based position measurements improve positioning accuracy and can be used to incorporate positional uncertainties into the synthetic-aperture measurements.

The measurement of the characteristics of realistic, static OTA spatial-channel conditions is critical for the design, test, and standardization of wireless technology, in particular, for devices and systems that use multiple antennas. Such antenna arrays are used to efficiently transmit data in reflective and dynamically changing environments and are also used to compensate for higher losses experienced at mmWave frequencies. For example, Fifth-Generation, New Radio (5G NR) base stations typically use multiple-element antenna arrays with beamforming capabilities to transmit multiple narrow beams that produce high antenna gain [8,9]. Multiple-input–multiple-output (MIMO) technology is also used [9] to leverage multiple transmission paths simultaneously.

The success of such wireless technology depends, in part, on the antenna array performance and the ability of user equipment to find and track these multiple beams [10]. Understanding the characteristics of spatial-channel test environments is also crucial for product development, performance evaluation, and conformance testing purposes [11]. Because of the critical role that the antenna array plays in device performance, many conformance tests must now be completed using OTA measurements [12]. For all of these cases, an independent characterization of the spatial channel can be provided by synthetic-aperture systems due to their relatively small footprint, especially at mmWave frequencies. Methods for design and use of synthetic-aperture measurement systems that can be used for the characterization of wireless channels or test chambers used for OTA channel emulation are described next, with guidelines, examples, and case studies.

11.1 Synthetic-aperture methods

Synthetic-aperture systems are capable of making very precise and high-resolution measurements of multiple reflections in wireless communication channels. They can characterize the signal-propagation environment with a high degree of fidelity. The resulting estimate of the wireless channel's impulse response provides information on the source of multipath signals caused by reflections, the extent of spatial noise created by diffuse scattering and diffraction, and shadowing or signal blockage created by stationary objects or by people and vehicles moving through the environment.

11.1.1 Prior work

Synthetic-aperture systems have been utilized in different indoor or outdoor configurations with a VNA measuring the channel's impulse response for a variety of channel-characterization applications using a variety of sampling geometries. For example, authors in [13] describe a three-dimensional virtual array with spatial samples collected inside the volume of a cube. This sampling lattice was utilized to measure indoor propagation channels at 60 GHz. Also, in an indoor environment, authors in [14] describe a linear virtual array used to estimate the propagation channel at 17 GHz. This virtual array was constructed by moving the receive horn along a straight track. A circular virtual array was constructed in [15] for channel measurements by rotating the receive antenna on a turntable. A virtual aperture in space and time was described in [16] to extend the performance of a MIMO radar. The impact of position errors was analyzed in [17] to determine the impact on angle estimation accuracy. Wideband channel measurements using a VNA are discussed in [18,19], while [20–22] describe extended range wideband channel sounders that also use a VNA. A linear virtual array described in [4] was used to measure the propagation channel at 1.8 GHz.

A novel approach for constructing linear virtual arrays by moving a sparse antenna array is analyzed in [23]. A cubic synthetic aperture with a VNA is utilized in [24] to estimate line-of-sight (LOS) and non-line-of-sight (NLOS) propagation paths in an indoor environment. Linear virtual arrays constructed by mounting the receive antenna on a moving train are described in [25,26]. A wideband channel sounder using a VNA with a cylindrical synthetic aperture is described in [27] for characterizing an emulated 3D spatial channel within a test chamber. This system, developed by NIST, will be described in Section 11.4, providing an example microwave-frequency band configuration and uncertainty analysis. A 35-by-35-element, planar-grid, mmWave synthetic-aperture array, also implemented at NIST [28,29], provides illustrative examples throughout this chapter. Measurements and analyses from this system are described further in Section 11.5.

11.1.2 The role of the spatial Fourier transform in synthetic-aperture processing

Many synthetic-aperture measurement systems are based on VNA measurements due to the VNA's ability to collect data over a wide frequency range with narrow frequency spacing and high dynamic range. Because the scans can be time-consuming, this high resolution comes at the expense of measurement time, typically requiring that channel conditions remain stationary during the measurement sweep.

For VNA-based synthetic-aperture systems, the S_{21} scattering parameter, which represents the channel's transfer function $H(f,\mathbf{r})$ between the transmit and receive antennas, is measured as a function of frequency, f. The receive antenna position, $\mathbf{r}$, may be defined in various coordinate systems, with the most common being rectangular [$\mathbf{r}=(x,y,z)$], cylindrical [$\mathbf{r}=(r,\theta,z)$] with $0\le\theta\le 2\pi$, or spherical [$\mathbf{r}=(R,\theta,\phi)$] coordinates with $0\le\theta\le\pi$ and with $-\pi\le\varphi\le\pi$. For the case of hemispherical coverage $0\le\theta\le\pi$ in cylindrical coordinates and $0\le\theta\le\pi/2$ with $-\pi\le\varphi\le\pi$ for spherical coordinates. The array of transfer-function measurements is then used to determine the angles and times of arrival of the static spatial wireless channels at the synthetic aperture using Fourier analysis. The resulting power-delay-angle profiles provide the distribution of received power as a function of temporal delay, τ, elevation angle, θ, and azimuth angle, ϕ [30].

The magnitude-squared of the temporal signal derived from an inverse two-dimensional Fourier transform of the channel transfer function generates the power-delay-wave-vector profile as

$$p(\tau,\mathbf{k}) = \left|\iint H(f,\mathbf{r})e^{j2\pi\tau f}e^{j\mathbf{k}\cdot\mathbf{r}}dfd\mathbf{r}\right|^2, \tag{11.1}$$

where f varies over the measurement bandwidth and $\mathbf{r}$ varies over the spatial sample locations. In spherical coordinates, for example, the wave vector $\mathbf{k}$ is given by

$$\mathbf{k} = (2\pi/\lambda)[\sin\theta\cos\phi, \sin\theta\sin\phi, \cos\theta].$$

The wave vector, with units of inverse length, represents the projection of the wavelength of a propagating plane wave onto the three Cartesian x–y–z coordinate axes containing the synthetic aperture. The angle of arrival (AoA) of the plane wave is given in spherical coordinates θ and ϕ. The projection terms onto the x–y–z axes are given by $\sin\theta\cos\phi$, $\sin\theta\sin\phi$, $\cos\theta$.

For ease of analysis, prior to taking the Fourier transform in (11.1), cylindrical or spherical synthetic-aperture measurements of the transfer function are interpolated to a rectilinear grid of positions $\mathbf{r}=(x,y,z)$. Often [31], a three-dimensional Hamming window is applied to minimize aperture sidelobes. Or, more involved frequency dependent methods may be used for wide measurement bandwidths [32].

To capture the power represented by the three spatial dimensions of the power-delay wave vector while reducing the results to a function of azimuth and elevation angles, the power-delay-wave-vector profile may then be interpolated to a spherical coordinate system of positions $\mathbf{r}=(r,\theta,\phi)$ and integrated along the radial component of the function to obtain the power-delay-angle profile

$$p(\tau,\theta,\phi) = \int_0^\infty p(\tau,r,\theta,\phi)dr. \tag{11.2}$$

Other implementations may be used as well. Integrating the power-delay-angle profile over all delay times gives the power-angle profile, which is the distribution of the total received power as a function of elevation angle, θ, and azimuth angle, ϕ [30]. This provides one valuable visual representation of the multipath channel.

11.2 Beamforming and measurement attributes of synthetic apertures for wireless communications applications

Synthetic-aperture systems can take many physical forms, with, for example, a receive-antenna element sampling the received signal along a linear translation stage, a planar array, or a cylindrical or spherical volume or surface. As long as the sampling locations of the antenna in the aperture are precisely known, the signals that were incident on the aperture can be reconstructed by the use of appropriate coordinate transformations and analysis algorithms, some of which are discussed later.

However, no matter which geometry is utilized, it is important to understand the interaction between the physical parameters of the scan (e.g., spatial extent, number of points, point spacing), the frequency of operation, and the effects of the receive-antenna-element pattern relative to the features to be resolved. The following sections explain these interactions, with equations and guidance that are intended to be useful in implementing synthetic-aperture systems.

11.2.1 Array coordinate systems

Figure 11.2(a) shows the angles used to describe beam scanning directions for the three common array coordinate systems defined earlier. In this illustration, the array lies in the x–y plane and the z-axis points along the normal to the plane of the

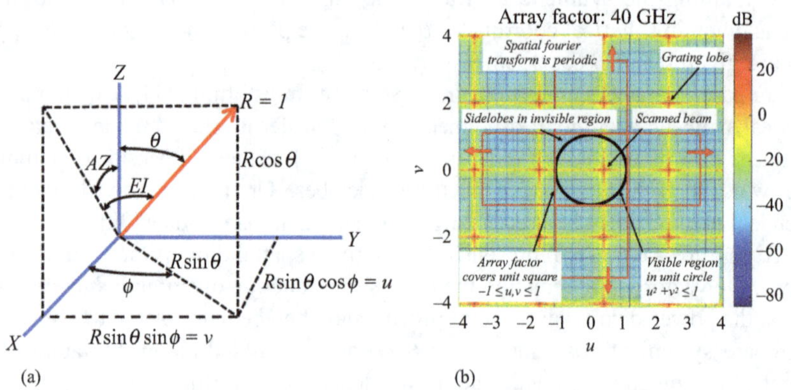

Figure 11.2 (a) Array coordinate system and (b) spatial Fourier transform output and array factor for a 35-by-35-element planar synthetic aperture at 40 GHz. © 2020 IEEE. Reprinted, with permission, from [32].

array, also known as the boresight direction. In a spherical coordinate system, the angles θ and ϕ define points on the surface of the forward unit hemisphere. The angle θ is measured from boresight and ϕ extends to the plane of scan from the x-axis. The projection of points from the forward unit hemisphere onto the x–y plane yields the coordinate axes labeled as u and v. The coordinates u and v can also be used to describe beam directions and are known as sine-space coordinates.

The relations used to transform angles between the spherical, azimuth/elevation, and sine-space coordinate systems are defined as [31]:

$$u = \sin\theta\cos\phi, \ v = \sin\theta\sin\phi \tag{11.3a}$$

$$u = \cos EL\sin AZ, \ v = \sin EL \tag{11.3b}$$

$$\sin^2\theta = u^2 + v^2, \ \tan\phi = v/u \tag{11.3c}$$

$$\cos\theta = \cos EL\cos AZ, \ \tan\phi = \tan EL/\sin AZ \tag{11.3d}$$

$$\tan AZ = u/\sqrt{1 - u^2 - v^2}, \ \sin EL = v \tag{11.3e}$$

$$\tan AZ = \tan\theta\cos\phi, \ \sin EL = \sin\theta\sin\phi \tag{11.3f}$$

where, as earlier, $0 \le \theta \le \pi$ and $-\pi \le \phi \le \pi$. Note that AZ and EL are not the same as θ and ϕ. They are defined differently, as shown in Figure 11.2(a).

For a perfect planar array in sine space, the scan angle of the beam is determined by the distance of the point (u, v) from the origin. A feature of sine space is that the beam pattern is invariant with respect to scan angle. In other words, as the beam is scanned, every point in the array response is shifted in the same direction and by the same amount as the beam peak. The region inside the unit circle, where $u^2 + v^2 \le 1$, is defined as the visible region (as shown in Figure 11.2(b)). This represents the forward hemisphere from which energy is received by the array. The infinite region outside the unit circle is referred to as imaginary space. Although no power is received from imaginary space, it is useful for describing the motion of grating lobes (see Section 11.2.6) as the beam is scanned.

Figure 11.3(a) illustrates the beamformed output in sine-space coordinates for an LOS signal measured using a 35-by-35-element planar synthetic aperture in a lab setting. Figure 11.3(b) illustrates the same LOS array response in azimuth and elevation coordinates, and Figure 11.3(c) shows an azimuthal cut along the principal plane corresponding to elevation angle equal to 0°.

11.2.2 Beamforming and angular resolution

The far-field response in spherical coordinates (θ, ϕ) for a perfect planar array of $M \times N$ homogeneous elements located entirely in the x–y plane and oriented in the same direction is given by [31]

$$B(\theta,\phi) = E(\theta,\phi)\sum_{m=0}^{M-1}\sum_{n=0}^{N-1} w_{mn}e^{jk(x_m\sin\theta\cos\phi + y_n\sin\theta\sin\phi)} \tag{11.4}$$

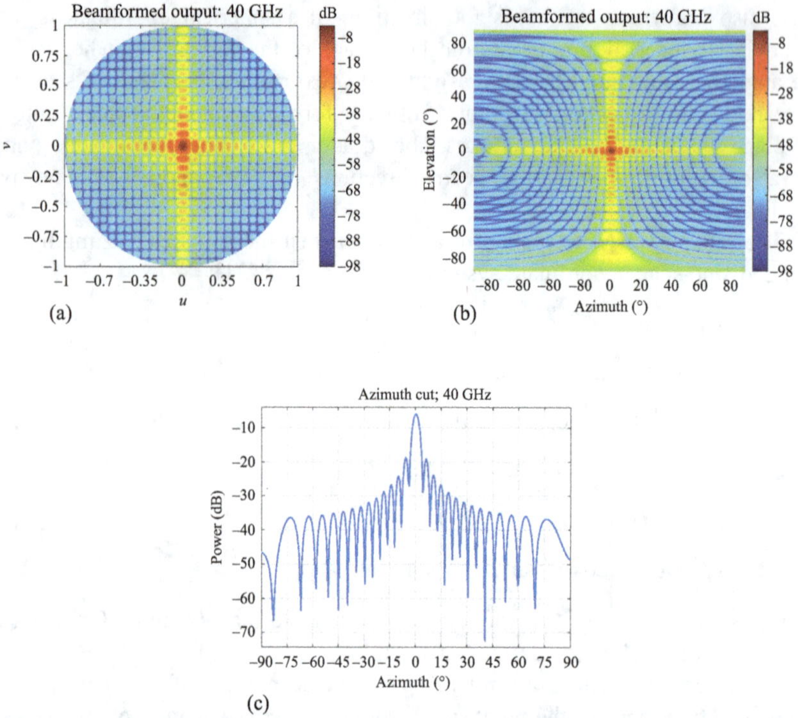

Figure 11.3 Measured array response at 40 GHz (a) sine-space coordinates; (b) azimuth, and elevation angular coordinates; (c) azimuthal cut

where $E(\theta, \phi)$ is the array element pattern, the wavenumber $k = 2\pi/\lambda$, λ is the operating wavelength, and w_{mn} is the array element weighting assigned in the analysis. Weighting is often uniform with $w_{mn}=1$ for all m and n but may be tailored to provide desired beamforming characteristics, such as reduced sidelobe levels discussed in Section 11.2.5. Note that (11.4), while a commonly used approximation, does not rigorously include polarization effects, as described in [33].

If the array elements are uniformly spaced on a rectangular grid, then the element locations are given by $x_m = md_x$ and $y_n = nd_y$, where d_x and d_y denote the spacing between elements in the x and y directions. This equation can be rewritten as a two-dimensional spatial Fourier transform by using sine-space coordinates [31]

$$B(u,\ v) = E(u, v) \sum_{m=0}^{M-1} \sum_{n=0}^{N-1} w_{mn} e^{jk\left(md_x u + nd_y v\right)}. \tag{11.5}$$

Note that the array response defined by (11.4) and (11.5) is symmetric between the forward and backward hemispheres. Theoretically, one could represent the backward hemisphere in sine-space coordinates as well. Since the array beamforming operation coheres all the energy collected at the aperture from all directions, if there are backlobe signals present in the data, the symmetry of the array response will not allow the forward and backward hemisphere signals to be differentiated. If a planar OTA system must satisfy a performance requirement to receive signals arriving from the backward and forward hemispheres simultaneously while also differentiating between them, one must perform measurements at points that do not all lie in the same plane.

Synthetic-aperture sampling lattices that exhibit rotational symmetry, such as cylinders or spheres, are well suited for creating omnidirectional views of the channel. When forming these lattice types, it is equally important to use an antenna element with negligible gain in the backlobes to avoid creating spatial ambiguities in the beamformed array response. The backlobe gain, if not eliminated, may limit the dynamic range of the synthetic-aperture measurement. Note that the array response for cylindrical or spherical scans is not governed by (11.4) and (11.5) and is treated separately.

With hardware planar phased arrays, the radiation pattern of each array element may be slightly different, or the antenna elements may not be perfectly angularly aligned, in which case factoring out the term $E(\theta, \phi)$ in (11.4) is an approximation. With synthetic-aperture systems, however, there is only a single receive antenna element that is common to all spatial sample locations, so (11.4) is an exact description of the array response, provided that the antenna elements all point in the same direction. The directivity of the element pattern is one of the factors that determines the scan angle limits of the main beam. If the element pattern is highly directive, such as with a high-gain horn, then, at large scan angles the element pattern will severely attenuate the array response, resulting in reduced array gain. If the power gain of the receive antenna is less directive, then the array main beam can scan to larger angles with less scan loss.

11.2.3 Beamforming measured S_{21} data

Measured S_{21} data can be beamformed by summing the products of the S_{21} data measured across the aperture and steering-vector weights. An estimate of the angular distribution of signal energy can be formed by beamforming the S_{21} values measured across the synthetic aperture. To beamform the measured data for a single fixed frequency, the complex S_{21} values, y_k, from all MN array elements are stacked into a vector, $\mathbf{y}$, known as the array output vector,

$$\mathbf{y} = \begin{bmatrix} y_1 & \cdots & y_{MN} \end{bmatrix}^T. \tag{11.6}$$

Next, the complex beamformer output $z(\theta, \phi)$ for a specified direction (θ, ϕ) can be formed by projecting the array output vector onto the steering vector $\mathbf{v}(\theta, \phi)$:

$$z(\theta, \phi) = \mathbf{v}(\theta, \phi)^H \mathbf{y} \tag{11.7}$$

where the mnth component of the steering vector is

$$[v(\theta, \phi)]_{mn} = e^{jk(x_m u + y_n v)} \tag{11.8}$$

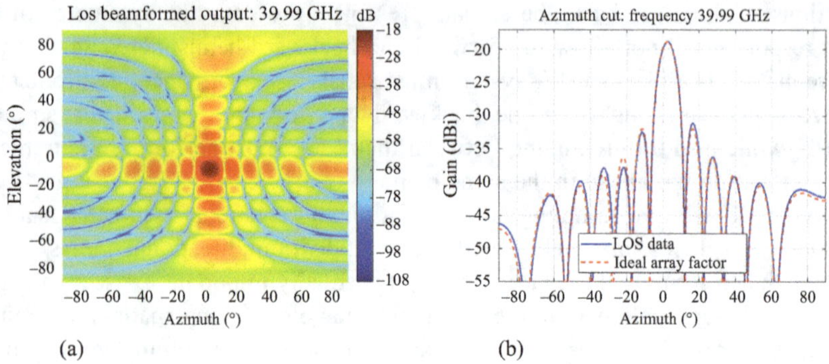

Figure 11.4 Beamformed output for a 15-by-15-element planar array for an LOS signal: (a) power gain of the received signal in azimuth and elevation; (b) azimuthal cut $(EL = 0°)$ *for measured and ideal data*

with $0 \leq m \leq M - 1, 0 \leq n \leq N - 1$, and u, v as defined in Equation (11.3a). This dot product multiplies the data measured at each array element by a steering phase that, after summation, coherently adds the aperture data of a signal incident on the array from the direction (θ, ϕ). This procedure is equivalent to applying a taper with linear phase in the x and y directions across the aperture. Note that with other aperture geometries, such as cylindrical or spherical surfaces, beamforming is still implemented by applying a phase taper across the aperture, although this taper is no longer strictly linear.

Wideband beamforming is described in Section 11.5.3.1, utilizing frequency-dependent steering vectors to realize a time delay between adjacent array elements, instead of a phase shift. This effectively steers the array response toward a given beam direction at all frequencies.

Figure 11.4(a) and (b) show the beamformed output for a 15-by-15-element planar synthetic array that measured an LOS signal from the transmit antenna. The plots compare the beamformed, measured data to an ideal array factor computed using omnidirectional array elements with constant gain in all directions. Since the receive horn used to generate the synthetic aperture had a wide one-sided 3-dB beamwidth (23°) and the LOS signal was near array boresight, there is not much difference between the plots. The measured 3-dB two-sided beamwidth in the boresight direction $(u = 0, v = 0)$ at 39.99 GHz for the 15-by-15-element synthetic planar array is 7.27°.

The plots in Figure 11.5(a)–(d) investigate the ability to resolve two vertical cylinders placed near each other as a function of frequency. These plots show the beam-steered output measured using a synthetic-aperture system implemented at NIST in Boulder, CO. The spatial samples were arranged on a 35-by-35-element planar grid with 3.7-mm spacing (i.e., $\lambda/2$ at 40 GHz) between the sample points. The VNA stepped over 1,350 frequencies between 26.5 and 40 GHz. Two vertically oriented aluminum cylinders with 2.54-cm diameter were placed in the scene such that they were separated by 9.5° in azimuth. The ability of the synthetic-aperture system to resolve the cylinders in angle by beamforming the array output was evaluated as a function of frequency.

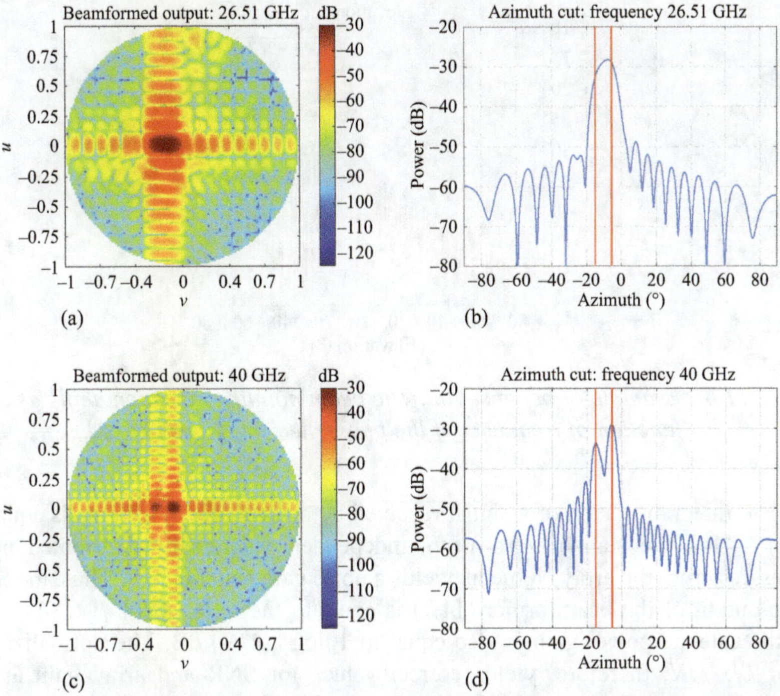

Figure 11.5 Measured 35-by-35-element planar synthetic aperture for 26 GHz [(a) and (b)] and 40 GHz [(c) and (d)]. (a) and (c) Beam-steered array received power in dB; (b) and (d) measured received power: azimuthal cut ($EL = 0°$).

The plots show that as the frequency increases, the width of the main beam decreases because the spatial sample spacing is fixed and the electrical length of the aperture (given by D/λ) is increasing (as discussed in more detail in Section 11.2.6). The variable D corresponds to the dimension of the square aperture in each principal plane. Nominally, the angular resolution of the planar array is taken to be equal to the 3-dB width of the main beam, meaning that two objects within the half-power beamwidth (HPBW) cannot be effectively separated, as shown in Figure 11.5(a) and (b). The approximate location of the cylinders is drawn with red lines in Figure 11.5(b) and (d).

A normalization factor is often applied to (11.4) or (11.5). In situations where an estimate of the signal source power is desired in the pointing direction of the main beam, the factor $1/MN$ is applied to the array response B to normalize the peak of the main beam to unity. In other scenarios, it is desirable to estimate the signal-to-noise ratio (SNR) at the output of the beamformer (11.7) when signal-plus-thermal noise is present at the input, which also corresponds to the gain of the array. In this case, the normalization factor is $1/\sqrt{MN}$, such that the peak value of $|B|^2$ is equal to MN. To understand this choice of normalization factor, note that the

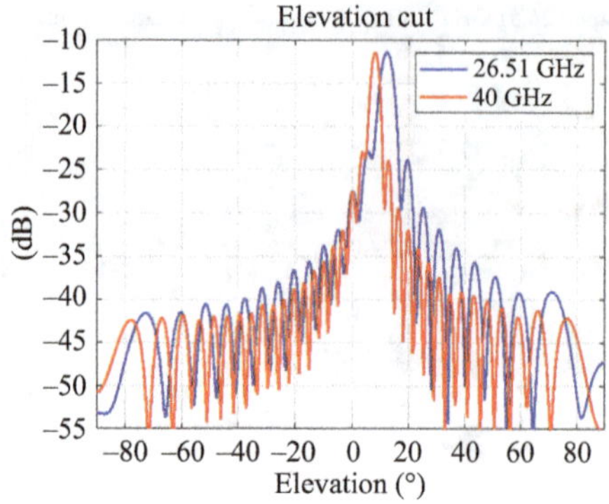

Figure 11.6 Example of beam squint. The beam pointing direction shifts as a function of frequency if the phase steering taper is fixed.

coherent integration gain of a signal received by all *MN* array elements is equal to $(MN)^2$. Also, the sum of zero-mean, independent, identically distributed noise processes across the array elements yields a noise gain equal to *MN*. Thus, the SNR at the output of the beamformer (11.7) is equal to the ratio $(MN)^2/(MN) = MN$ and the antenna power gain is also equal to $10\log_{10}(MN)$ dB. The normalization factor $1/\sqrt{MN}$, therefore, yields correct values for SNR and array gain at the output of the beamformer.

If the signal to be transmitted or received is narrowband, i.e., its bandwidth is much narrower than the carrier frequency, phase-shift steering will point the peak of the main beam in the desired direction. In the case of signals with a wide instantaneous bandwidth, phase-shift steering will "squint" the main beam, meaning that different frequency components will radiate in different directions [34,35].

Beam squint results in a linear distortion of the received signal spectrum for a fixed receiver location because higher frequency components are attenuated as a result of seeing diminishing antenna gain. Furthermore, beam squint causes a deviation or error in the pointing direction of the peak of the main beam compared to the desired direction. Figure 11.6 illustrates the effect of beam squint using measured planar-array data for the case where a phase taper is computed for steering the beam to ($AZ=0°$, $EL=5°$) at 40 GHz and the same phase taper is also applied to steer the beam at 26.51 GHz. As the plot shows, the beam moves away from the desired pointing direction when the frequency changes. An effective wideband beamsteering technique that avoids beam squint is to insert a true time-delay behind each array element as opposed to a phase shift. A frequency domain implementation of true-time-delay beamsteering is described in Section 11.5 of this chapter.

11.2.4 Effect of element pattern on beam steering

Equations (11.4) and (11.5) from Section 11.2.2 described the array response of a 2D planar-array synthetic aperture. The summation term in (11.4) or (11.5) is commonly known as the array factor, $AF(\theta, \phi)$

$$AF(\theta, \phi) = \sum_{m=0}^{M-1}\sum_{n=0}^{N-1} w_{mn}e^{jk(x_m \sin\theta\cos\phi + y_n \sin\theta\sin\phi)}. \tag{11.9}$$

A useful way of describing (11.4) is that the complex synthetic-aperture-array beam pattern equals the array factor times the element voltage pattern, $E(\theta, \phi)$

$$B(\theta, \phi) = E(\theta, \phi)AF(\theta, \phi). \tag{11.10}$$

For the case of a 2D planar synthetic aperture where the array element maintains the same orientation and is pointed in the same direction at each spatial sample, this representation of the array response is exact.

The power gain of the synthetic aperture, $B_G(\theta, \phi)$, is equal to the antenna directivity times an efficiency constant that accounts for losses. This can also be written as

$$B_G(\theta, \phi) = |B(\theta, \phi)|^2. \tag{11.11}$$

The voltage pattern (and corresponding gain characteristic) of the array element is an important consideration in synthetic-aperture design. For example, an array element, such as a horn, with higher directivity and gain may be desirable for increasing the detection range of the synthetic-aperture system. However, the narrower beamwidth of the directional horn will also decrease the accessible scan volume. This reduction in scan volume is a direct consequence of (11.10), which shows that if the element pattern is highly directional, the array response at large scan angles will be attenuated. Horn antennas may be desirable for other attributes, such as bandwidth. Because the impedance of a horn antenna is well matched to free space in the far field, horn antennas can support higher measurement bandwidths. Horn antennas also have lower backlobe levels. Antenna backlobes can create complications in channel measurements by allowing signals that arrive from angles outside the main beamwidth of the antenna to interfere with those within the main beam. Also, conventional beamforming cannot distinguish between signals arriving at the aperture from the backward hemisphere versus the forward hemisphere.

Antennas that exhibit lower directivity, such as an open-ended waveguide (OEWG), have wider beamwidths and are a good choice for measurement scenarios that require large scan angles. The gain pattern of OEWGs is relatively smooth and analytically more tractable than the gain pattern of a horn antenna, which may offer an advantage during the analysis stage.

The sidelobe behavior of an antenna is another important factor that affects channel measurement performance. Antennas with erratic or elevated sidelobes may allow signals to be detected from ambiguous spatial directions. Nominally, the ideal receive-antenna element would be an isotropic antenna with unity gain in all directions for all frequencies. This ideal isotropic antenna has no sidelobes and supports scan angles anywhere in the forward hemisphere.

As described in [31], the power gain of a planar array is decreased according to the cosine of the azimuth and elevation scan angles. This reduction in gain is also

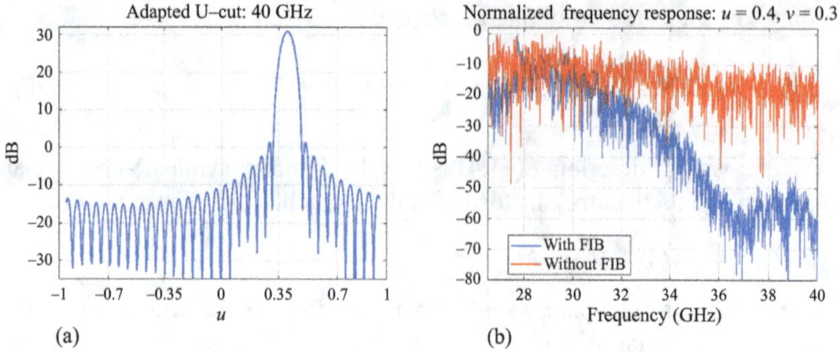

Figure 11.7 (a) U-cut of beam pattern showing reduced sidelobe levels possible with a FIB. (b) Comparison of frequency response with and without the application of a FIB for the sine-space coordinate $u = 0.4$, $v = 0.3$. © 2020 IEEE. Reprinted, with permission, from [32].

called scan loss and is equivalent to a broadening of the array beamwidth since beamwidth is inversely proportional to gain. For example, if the element spacing is $d = \lambda/2$ and the array is scanned to $\theta = 60°$, the power gain B_G of the overall synthetic-aperture array will be $\cos 60° = 0.5$ (corresponding to a 0.707 amplitude gain), relative to the peak value at boresight. Note that in practical settings, the voltage-gain pattern $E(\theta, \phi)$ of many actual antenna elements can be described by a $\cos(\theta)^n$ characteristic as well, where $n = 0.5$, 1, or 2, making $B(\theta, \phi)$ in (11.10) and $B_G(\theta, \phi)$ in (11.11) easy to calculate.

11.2.5 Frequency invariant beamformer with reduced sidelobes

In many cases, it is desirable to steer the main beam to a specified pointing direction (u_0, v_0) while simultaneously minimizing the overall sidelobe level of the adapted antenna pattern over the entire measurement bandwidth. This can be useful when scanning off boresight to form directional power delay profiles (PDPs) or to study the frequency response in an arbitrary direction. Algorithms for implementing this can be found in many references, including [32,36–40]. An example of the application of such a frequency invariant beamformer (FIB) is shown in Figure 11.7 (a) and (b). Note that the principal sidelobes for an unadapted beampattern are 13.2 dB below the main beam peak but the adapted beampattern illustrated in Figure 11.7(a) shows an additional 20-dB reduction in sidelobe levels. This sidelobe reduction is held constant over all measurement frequencies while also simultaneously keeping the beamwidth fixed.

11.2.6 Bandwidth effects

The summation term in (11.4) and (11.5) is known as the array factor. The array factor repeats in the u dimension with period λ/d_x and repeats in the v dimension with period λ/d_y. A single period of the array factor in sine space is equal to the rectangular region $-0.5d_x \le u \le 0.5d_x$ and $-0.5d_y \le v \le 0.5d_y$. This was illustrated in Figure 11.2(b). The visible region of the array factor that exists in physical space corresponds to the interior of the unit circle $u^2 + v^2 \le 1$. The large peaks outside the unit circle are known as grating lobes. In most array designs, grating lobes are always held outside the visible region. If grating lobes are allowed to enter the array visible region, perhaps because the array element spacing is greater than $\lambda/2$, then grating lobes have the same effect as high sidelobes; meaning that they amplify energy received from directions other than the main beam and they create spatial ambiguities.

The Nyquist spatial sampling rate that avoids spatial aliasing or grating lobes is given by $d_x = d_y = \lambda/2$. If the element spacing is greater than $\lambda/2$, the array is undersampled and the grating lobes move closer to the unit circle and may even enter the visible region. If the element spacing is less than $\lambda/2$, the array is oversampled and the grating lobes move farther away from the unit circle. Figure 11.2 (b) illustrates the output spatial Fourier transform (on a decibel scale) and the array factor at 40 GHz for a 35-by-35-element planar aperture. The inter-element spacing in the x and y directions is equal to $\lambda/2$ at 40 GHz or 3.7 mm. If the spacing is held fixed, the array is oversampled at 26.5 GHz so the grating lobes move farther away from the visible region. The peak of the main beam depends only on the number of array elements and is equal to $10\log_{10}(MN)$ or 30.8 dB.

Many properties of Fourier transforms apply to the array factor, in particular, the Fourier shift and the Fourier scaling properties. The Fourier shift property is relevant when the main beam is steered to a direction (u_0, v_0). Beam steering is accomplished by applying the linear phase $e^{-jk(md_xu_0+nd_yv_0)}$ across the aperture. The main beam will shift by an angular distance equal to the slope of the linear phase taper in the u and v directions. Steering the main beam for a single frequency is a linear transformation that does not affect the amplitude of the array factor nor does it affect the shape of the main beam when the main beam is described in sine space. When the beam scans, sidelobes that were originally outside the unit circle in the invisible region of uv space will enter the visible region and, therefore, the sidelobe structure of the beam changes.

The Fourier scaling property states that changing the sample spacing of a 2D discrete sequence $h(m,n)$ expands or contracts the output of the Fourier transform $H(u,v)$ according to

$$H\left(\frac{u}{a}, \frac{v}{b}\right) \Leftrightarrow \frac{1}{|ab|} h(am, bn). \tag{11.12}$$

This property implies that if the physical spacing between array elements is held fixed, then there is an inverse relationship between the operating frequency and the width of every angular lobe in the beam pattern (including main beam and sidelobes). Consequently, a synthetic aperture of fixed dimensions attains higher angular resolution at 40 GHz than at 26.5 GHz due to the narrower width of the main beam (approximately equal to λ/D radians, where D is the largest dimension of the aperture in the principal planes) [31]. Figure 11.8(a) illustrates the Fourier

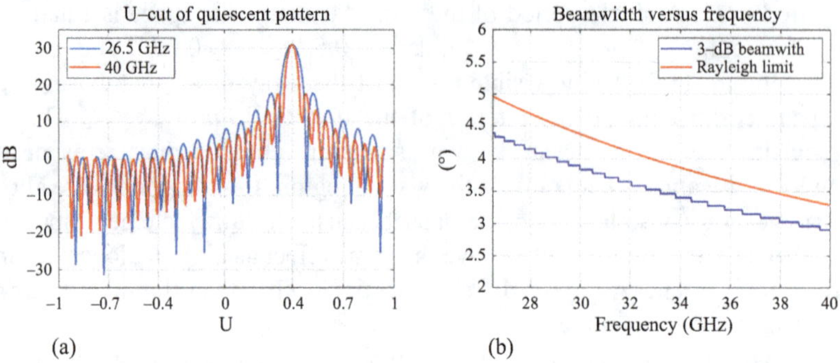

Figure 11.8 (a) Illustration of the Fourier scaling property with the main beam scanned to (u = 0.4, v = 0.3). © 2020 IEEE. Reprinted, with permission, from [32]. (b) 3-dB beamwidth calculated as a function of frequency for a 35-by-35-element planar synthetic-aperture array.

scaling property by comparing a u-dimension cut of the array beam pattern for 26.5 and 40 GHz with the main beam steered to $(u = 0.4, v = 0.3)$.

Figure 11.8(b) plots 3-dB beamwidth versus frequency for a square 35-by-35-element array. Since the array is symmetric, the azimuth and elevation beamwidths are equal. The 3-dB beamwidth was computed by calculating the fraction of the antenna pattern over 180° that exceeds the peak value minus 3 dB. The minimum beamwidth is 2.9° attained at 40 GHz. Also superimposed on the plot is the Rayleigh approximation for the (azimuth or elevation) array beamwidth equal to λ/D. The Rayleigh approximation for the beamwidth at 40 GHz is 3.3°.

11.2.7 Delay resolution

The frequency range measured at each spatial location of the synthetic-aperture antenna element determines the synthesized bandwidth of the channel measurement. The delay resolution of the channel measurement is equal to the reciprocal of the synthesized bandwidth. For a configuration with a 35-by-35-element planar aperture, if the total bandwidth is $F_{TOT} = 13.5$ GHz between $f_{min} = 26.5$ GHz and $f_{max} = 40$ GHz, this would correspond to a delay resolution $\Delta\tau$ of 0.074 ns, which yields a distance resolution of 2.2 cm [29].

The maximum unambiguous delay extent that can be measured by the 35-by-35-element aperture is equal to the increment between frequency samples. If, for example, the frequency step size is $\Delta f = 10$ MHz, this corresponds to a maximum unambiguous delay extent of $T = 100$ ns, or $D = 29.96$ m [41]. This frequency step size yields 1,351 frequency samples within the synthesized 13.5-GHz bandwidth and 1,351 delay samples within the total time duration of the PDPs. Thus, for this case, the temporal sampling rate of each PDP is equal to the measurement bandwidth, or $F_{TOT} = 13.5$ GHz.

Since the ratio of the highest measurement frequency to the total measurement bandwidth is very nearly an integer, $40/13.5 = 2.96 \approx 3$, band-pass sampling considerations for the complex S_{21} parameters suggest that any aliasing due to sampling at a temporal rate equal to F_{TOT} will be small. In general, aliasing is negligible if the complex sampling rate f_{sam} of the PDPs satisfies [42]:

$$f_{\mathrm{sam}} \geq qF_{\mathrm{TOT}}, \text{ where } 1 \leq q \leq \left\lfloor \frac{f_{max}}{F_{\mathrm{TOT}}} \right\rfloor. \tag{11.13}$$

11.2.8 Total received-power delay spread

In channel characterization, it is often of interest to compute the total received power versus delay. This curve is a scaled version of the average PDP. The total (or average) PDP is constructed by summing the magnitude-squared PDP outputs over all beam directions [43].

If the beam directions overlap at the HPBW (3-dB beamwidth), for a planar array, the total-received-power PDP is proportional to an isotropic illumination of the forward hemisphere with uniform gain. If the beam directions do not overlap at the HPBW, then the PDP summation operation imposes an amplitude ripple on the total PDP due to the nonuniform overall gain illuminating the forward hemisphere.

Determining a scan lattice where the beams overlap at the HPBW is nontrivial for the case of synthetic-aperture arrays because the beamwidth increases (or equivalently the antenna gain off boresight decreases) by a factor of $\cos(AZ)\cos(EL)$, where AZ and EL correspond to the beam pointing direction. One solution involves iteratively solving the equations provided in [44].

A simpler approach leverages the fact that the beamwidth is scan angle invariant if the array pattern is described in *uv* sine space. Therefore, the desired scan lattice can be determined by specifying a uniformly spaced grid of beam directions in sine space separated by the HPBW and then converting the *uv* beam coordinates back to azimuth and elevation.

Figure 11.9(a) illustrates a scan lattice where the beams overlap at the HPBW. The resulting synthesized omnidirectional power gain pattern is isotropic and uniform in the forward hemisphere. In Figure 11.9(b), the beams overlap at the 6-dB

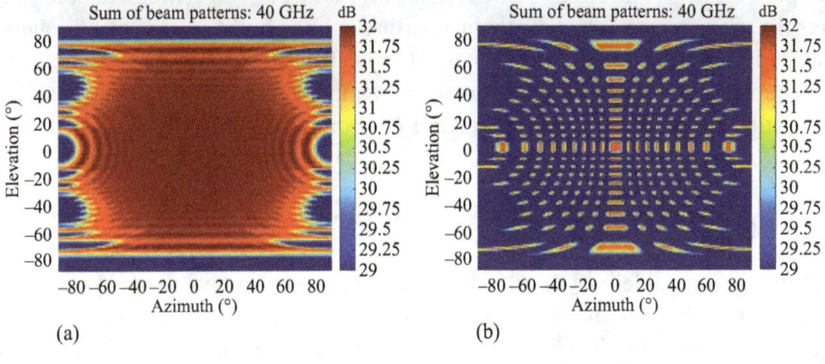

Figure 11.9 Synthetic-aperture scan lattice for a planar array: (a) beams overlap at the 3-dB point and (b) beams overlap at the 6-dB point

beamwidth. For this case, the synthesized omnidirectional power gain pattern exhibits ripples in the forward hemisphere. Any amplitude ripple embedded in the total PDP can affect the relative values of the multipath components and affect the outcome of a thresholding process that identifies signal peaks.

11.2.9 Dynamic range

The instantaneous dynamic range of a measurement at a single spatial sample of the synthetic aperture is determined by the intermediate-frequency (IF) bandwidth and the amount of sweep-to-sweep averaging performed in the VNA. For example, VNA measurements that use a 10-Hz IF bandwidth with no averaging applied to the data will have a maximum specified dynamic range close to 120 dB over the measurement bandwidth [45]. A particular advantage of synthetic-aperture systems is that since there is a dedicated receive chain behind each spatial sample and no beamforming occurs before the signal is digitized, the instantaneous dynamic range of the entire system measured at the output of the post-processor beamforming operation can be very high as compared to other instruments.

Another advantage of synthetic-aperture systems is that the effective aperture area can be made almost arbitrarily large, which allows for maximum detection range and sensitivity to weak signals. In general, for either hardware-based phased arrays or synthetic-aperture systems, the effective aperture area that receives signal energy is proportional to the number of array elements. In a hardware-based phased array, the number of array elements is fixed. However, in a synthetic-aperture system, the number of spatial samples that can be collected is constrained only by the positioner's range of motion and the acquisition time available.

11.2.10 Using system parameters to design a synthetic-aperture system

Given these system parameters, a synthetic-aperture system can be designed to measure features with a known size, distance, and over a specific frequency of operation. After a brief discussion on measurement hardware and uncertainties, two examples follow, respectively, in Sections 11.4 and 11.5. The first example illustrates the use of a cylindrical synthetic aperture applied in the microwave frequency bands to characterize a spatial-channel-emulating test chamber. The second example illustrates the use of an mmWave synthetic-aperture system in the characterization of a high-multipath industrial environment.

11.3 Measurements and uncertainties

With the design guidelines detailed in the previous section, the user may develop a synthetic-aperture system with the physical characteristics necessary to resolve objects of various physical sizes for various frequency ranges. In this section, we discuss this instrumentation and highlight some of the uncertainty elements unique to synthetic-aperture measurements.

11.3.1 Measurement approaches

Most synthetic-aperture measurement systems consist of a few basic components: a positioner for the receive aperture, a set of transmit and receive antennas, and measurement equipment to transmit and receive RF waveforms.

11.3.1.1 Instrumentation for VNA-based synthetic-aperture systems

As mentioned in the Introduction, a VNA is often used to create and sample the received signal for several reasons, including the measurements can have a high dynamic range; VNAs have well-established calibrations that can be used to de-embed cables and connectors; and measurements made with VNAs can be made traceable to fundamental physical quantities, yielding traceable OTA spatial-channel measurements.

The narrower the IF bandwidth in the VNA measurement, the higher the dynamic range, but the longer the measurement time [45]. Consequently, as discussed in Section 11.2.7, the use of VNAs is generally limited to static or slowly varying environments. However, VNAs allow a wide range of measurements involving the characterization of three-dimensional multipath channels in controlled laboratory environments, test chambers, and in-field tests.

Some synthetic-aperture measurement systems are based on large-signal network-analyzers (LSNAs) that extend the range of measurements by providing calibrated magnitude and relative phase between measured frequency components [28,29]. LSNAs are used to determine conventional scattering parameters from the ratios of the forward and backward waves at their measurement ports, as do conventional VNAs, but LSNAs can also measure the amplitudes and cross-frequency phases of those forward and backward waves. Prior to a measurement, the LSNA is calibrated as a standard VNA with a scattering-parameter calibration kit. Additionally, power and phase calibrations are carried out through the use of calibration transfer standards such as power meters and comb generators. All of these calibrations can be made traceable to fundamental physical units after propagating the uncertainties.

These additional calibrations allow the synthetic-aperture system to transmit and receive modulated RF signals with known phase relationships between frequency components. This can aid in assessing distortion from phased-array antennas, including the characterization of down-converting phased arrays, in which the transmitted signal is measured at RF, while the received signal is measured at IF [28].

As an example, the RF–RF channel may be characterized in magnitude and relative phase by the LSNA-based synthetic-aperture system under modulated-signal conditions. The synthetic-aperture system can then be removed and the phased-array under test placed in the same location, where it is exposed to a known channel with a known modulation. Because the distortion introduced by the synthetic-aperture measurement system is quantified by its uncertainty analysis, additional distortion in the magnitude and/or phase components of the received signal would have been introduced by the phased array. For the frequency converting case, the same procedure may be followed, but the channel is characterized in terms of its RF–IF response by the synthetic-aperture system.

11.3.1.2 Antenna-positioning apparatus

Both antenna positioners (e.g., linear translation and rotational stages) and robotic arms are used for synthetic-aperture measurements. Many of these have tight mechanical tolerances that provide repeatable positioning even when the accuracy of the location is not perfect. Because of this, it can be advantageous to monitor the location of the positioner with an external instrument such as an optical camera system or laser tracker.

One important consideration in the choice of translation stages as opposed to robotic arms is that the array element should point along array boresight or the array element's pattern should be known and de-embedded (see Section 11.3.2.2). It is especially important to consider this when lateral positioners are oriented toward a compact transmitting antenna. Reflections from both positioners and robotic arms can be minimized by covering surfaces with RF absorbing material or using positioners made of low-reflectivity materials. Section 11.3.2.4 discusses methods for accounting for residual reflections.

11.3.1.3 Antennas

Many synthetic-aperture measurements are made with directional antennas as the array element. These antennas have the advantage of higher gain on boresight, which can be useful for OTA tests in which weak multipath components may be received. The reduced sidelobe and backlobe levels for a directional antenna can simplify the analysis of the scanned data, depending on the scan type.

The advantage of an omnidirectional array element is that the antenna pattern is easier to de-embed, especially if the main beam is steered off boresight. As shown in Section 11.2, the array sidelobes may take a complicated shape, and a directional array element pattern can complicate this even further.

11.3.2 Uncertainties

In this section, we provide a brief overview of some key components of uncertainty for synthetic-aperture measurements. The list is not exhaustive: for example, temperature effects are not addressed here. However, the components listed later represent significant contributors to uncertainty for synthetic-aperture measurements.

In the approach presented next, the components of uncertainty may be cascaded to allow the propagation of uncertainty from the VNA measurements to the OTA signal measured by the synthetic-aperture system. The uncertainties may then be propagated further, to the post-processed metrics such as AoA, time delay, and even further, to models of the environment.

11.3.2.1 Instrumentation and cabling for VNA-based synthetic-aperture systems

The use of an LSNA or a VNA in taking frequency domain measurements provides traceability to fundamental quantities through dimensional measurements of the calibration standards. Uncertainty in the dimensions of the calibration standards can then be propagated through to the corrected measurements, providing an uncertainty analysis of each individual frequency domain measurement [46]. The uncertainty in the responses of the calibration standards is propagated into the VNA calibration algorithm and then through to the synthetic-aperture measurements as they are corrected by this algorithm.

In addition to errors in the calibration standard models, other uncertainties related to VNA measurements include drift and uncertainty in the cable response due to bending. Drift may be assessed by performing VNA calibrations before and after the synthetic-aperture-system measurement. Because these measurements can take many hours if a low IF bandwidth is used, drift can be a significant contributor to the uncertainty in the estimate of the spatial channel. To minimize the effect of reflections from the VNA on the measurement of the channel, the VNA and other instrumentation are often located away from the main scene. This means that the cables between the VNA and the antenna positioner may be long and, as the positioner moves from one aperture location to another, cable bending may introduce uncertainty as well. A method for assessing this is described in Section 11.4.4.

11.3.2.2 Antenna response

In order to obtain accurate estimates of the OTA signal impinging on the synthetic aperture, the response of the antenna should be de-embedded from the measurements. One method for doing this requires obtaining the far-field antenna pattern across azimuthal angle, elevation angle, frequency, and polarization. An estimate of the on-axis antenna gain is first obtained by applying the techniques in [33]. Next, measurements of the antenna on a grid of azimuthal, elevation, and polarization angles are taken, as in [47], and normalized to the gain. An advantage of this approach is that one can incorporate the VNA calibration uncertainties, cable-bending uncertainties, and other uncertainties.

11.3.2.3 Positioning uncertainties

The physical location of the antenna element in synthetic-aperture measurements enters into the post-processing analysis. Errors in these positions induce uncertainties in estimates of the spatial-channel signal impinging on the aperture. One method for accounting for positioning errors is to explicitly track antenna locations, use these data (rather than ideal antenna locations) in the analysis, and incorporate an estimate of residual antenna-location errors into the uncertainty analysis.

If the positioning system errors were independent and identically distributed, then they would have a negligible effect on the beamformed response, since the resulting phase errors would tend to average out [48]. However, positioners may have correlated positioning errors that can lead to significant changes and uncertainty in beamformed responses.

Methods for tracking positioning errors include the use of an optical camera system or laser tracker to provide accurate position information. The errors in the positions are accounted for in the post-processing algorithms, and the residual errors are accounted for in the uncertainty analysis. A camera-based method is presented in Section 11.5 for the second measurement example.

11.3.2.4 Scattering from antenna positioners and fixturing

Most antenna positioners and fixtures include metallic components that may introduce reflections that change the RF channel. As synthetic-aperture measurements typically contain hundreds of individual measurement positions, each with the antenna positioner in a different location, evaluating the uncertainty due to scattering off of the positioner and related fixturing is nontrivial.

One approach to characterizing this effect is to perform multiple measurements of nominally the same channel but with different scan configurations. As each synthetic-aperture measurement consists of a set of specific positioner locations and implicitly incorporates the scattering associated with this set of positions, the effect of this scattering can be estimated from the spread of the multiple different synthetic-aperture measurements. The spread of channel measurements due to scattering from the positioner and fixturing would then be associated with error in the estimate of the received signal and may be incorporated into the uncertainty analysis.

11.3.2.5 Combining components of uncertainty

Methods for combining components of uncertainty are provided in the Guide to Uncertainty in Measurement [49]. At low frequencies and for narrowband measurements, it is often assumed that the components of uncertainty are independent and may be combined using the common methods outlined in [49]. However, for mmWave measurements, for wide bandwidths, and for measurements where the same VNA and calibration kit have been used to calibrate multiple parts of the measurement setup, correlations may exist between components of uncertainty. An example of the latter is when the same VNA and calibration kit are used both to measure the received signal in the synthetic-aperture-system measurement as well as to characterize the antenna response. Systematic errors in the calibration kit standards will be correlated in both the channel and the antenna measurements. If not accounted for, the calculated uncertainty will typically be too low. Some components of uncertainty may also have a structure, in which correlation occurs between measured samples. An example of this is impedance mismatch, which often exhibits a structure as a function of frequency. In synthetic-aperture measurements, one must also be concerned with the structure of the errors in antenna locations.

To account for potential correlations within and between sources of uncertainty, the Supplement 1 to the Guide to Uncertainty in Measurement [50] allows the use of Monte Carlo methods. In this approach, models of the various components of uncertainty are randomly exercised to cover the expected range of values. For example, if the tolerances on the pin diameter of a calibration standard are 0.5 ± 0.01 mm, the Monte Carlo iteration would select a pin diameter randomly distributed between 0.49 and 0.51 mm and apply this to the VNA calibration, which is then propagated through the synthetic-aperture analysis. The signals incident on the synthetic aperture are then estimated from the resulting distribution. This method allows the propagation of uncertainties from individual uncertainty mechanisms through to each associated system response. An advantage of this method is the preservation of physicality in the uncertainty of the estimate of the received signal due to the underlying sources of uncertainty, in contrast to estimating the uncertainty via the root-sum-of-squares method, which ignores physically important correlations. An implementation of this approach is provided in [51], with an example in Section 11.4.

11.4 Characterization of a microwave-band spatial-channel "hybrid" test chamber with a synthetic-aperture system

In this section, we illustrate the design and characterization of a repeatable, reconfigurable 3D spatial-channel test environment to be used for OTA testing of wireless devices in reference spatial channels. The characterization covered a frequency range of 1.8–2.2 GHz, corresponding to common cellular bands. By the use of synthetic-aperture measurements, the channel was characterized in terms of arrival angle and temporal delay. As discussed earlier, such measurements allow for the determination of uncertainties in the channel properties.

A reverberation chamber serves as the test environment, providing multipath reflections in shielded, repeatable conditions. However, unlike traditional reverberation-chamber applications, the chamber is heavily loaded with RF absorber to limit and control the multipath environment. We call the result a "hybrid chamber," as it is a cross between a reverberation chamber and an anechoic chamber. With shorter propagation distances and a limited number of multipath components, the hybrid chamber illustrated here can readily accommodate device testing that involves environments that are a few meters per side, such as factory work cells or offices. With the addition of external delays, much larger spatial channels can be emulated. This approach could be used, for example, to reproduce 3GPP channel models [52] inside the hybrid chamber.

Note that the hybrid chamber was equipped with metallic paddles for mechanical stirring, but, unlike for standard reverberation-chamber measurements, they were kept in a single stationary position during the synthetic-aperture measurements.

11.4.1 The environment: chamber design and configuration

The design of the spatial channel to be created inside the hybrid chamber is accomplished with a simple model of the chamber and a ray-tracing simulation [53]. Ray tracing was used to identify multipath-reflection locations and departure angles, which are then used to inform the placement of RF absorber in the chamber and orientation of the transmit antenna.

The chamber model includes the interior dimensions of the chamber, which correspond to the boundaries of the plot in Figure 11.10(a), and the locations of the transmit antenna (red dot) and center of the cylindrical synthetic aperture (green dot), also indicated in Figure 11.10(a). A two-path spatial channel constructed using the ray-tracing simulation is shown by the blue and red paths in Figure 11.10 (a). The blue path represents an LOS path, and the red path represents an NLOS

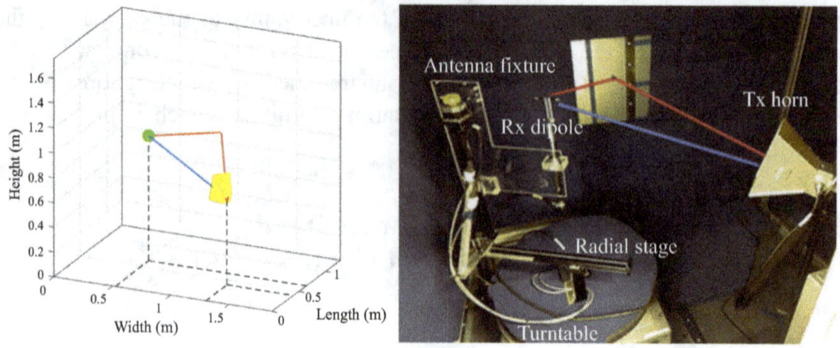

(a) (b)

Figure 11.10 Chamber ray-tracing simulation (with interior dimensions of chamber indicated on axes) (a) and photo of the corresponding chamber setup (b). The blue and red paths are simulated LOS and NLOS paths, respectively. In (b), the exposed metal, which serves as a reflection surface for the NLOS path, can be seen at the center of the photograph toward the rear of the chamber.

reflected path. The simulation provides the locations where NLOS signals reflect from interior chamber surfaces. This information is then used to position the RF absorber inside the chamber, as shown in the photo in Figure 11.10(b). The absorber is magnetically backed and easily mounted to the walls and ceiling of the chamber. For the example presented here, the exposed metal surface area that provided the NLOS path had a width of 39 cm and a height of 41 cm. These dimensions correspond to $\tilde{}2\lambda$ at the lowest measured frequency of 1.8 GHz. Because the antennas were not in the far-field, the ray-tracing model was approximate but, as will be shown, it was adequate for determining the placement of the RF absorber.

11.4.2 Measurement setup

A critical step in the development of the spatial-channel environment for wireless OTA testing purposes is the characterization of the spatial and temporal properties of the environment. Here, three-dimensional synthetic-aperture measurements assess both azimuth and elevation arrival angles of multipath components, as well as arrival times, in the test environment.

The measurement setup is illustrated in Figure 11.10(b), with the chamber having interior dimension of 1.75 m $(h)\times$1.22 m $(w)\times$1.87 m (l). From the ray-tracing model, the LOS path corresponded to an AoA of approximately 73° in azimuth and $-20°$ in elevation, and the NLOS path to an AoA of approximately 153° in azimuth and $-10°$ in elevation. These paths are relative to an arbitrary origin at the transmit antenna.

The synthetic aperture was created by stepping a dipole receive antenna, tuned to 1.8 GHz, through a three-dimensional grid of spatial points. A dipole was used for its azimuthally omnidirectional radiation pattern, which allowed for omnidirectional characterization of the spatial channel.

For measurements in this frequency band, the goal was to create a synthetic-aperture measurement system that covered a relatively large physical space in order to maximize angular resolution while ensuring that the antenna did not travel too close to walls or other metallic structures within the chamber such that its pattern would be significantly distorted. An automated positioning system stepped the receive antenna through a cylindrical grid of points with height and diameter both equal to 30 cm. The corresponding angular resolution of these synthetic-aperture measurements was approximately 43°, as given by the Rayleigh criterion, $\sin\theta = 1.22\lambda/D$ [54].

The antenna positioning system consisted of a vertical translation stage mounted on a radial translation stage, which was mounted on a turntable, as shown in Figure 11.10(b). The repeatability of the positioning system was approximately 50 µm, which was measured in-house with a laser tracking system. The angular resolution of the turntable was 84.5 $\pm$0.1 µrad, and the linear resolution of the translation stages was 6.35 $\pm$0.18 µm, both manufacturer hardware tolerances. The antenna was mounted on the positioning system such that the radiating element was at least 23 cm ($> 1\lambda$) away from metal parts.

For each turntable position, the radial translation stage was stepped through equally spaced positions, starting at the center of the turntable and moving toward the outer diameter. This sequence was repeated for each vertical height until a cylindrical volume with diameter and height equal to 30 cm was traced out. The antenna positioning system was used to create a synthetic-aperture grid consisting of 36 azimuth angles, ϕ, separated by 10°, 7 radial positions, r, separated by 2.5 cm, and 13 vertical heights, z, separated by 2.5 cm. For the center frequency of 2.0 GHz, this grid spacing is ~λ/7. Based on empirical studies, this synthetic-aperture grid was decided upon as a trade-off between the number of spatial samples (3,276) and measurement time (over 10 h). Note that this grid is oversampled with respect to the Nyquist criterion of λ/2 to improve on the resolution of the details of the multipath environment with additional spatial averaging. A complete measurement sequence at the positions described earlier was conducted for three orthogonal dipole polarizations: vertical, radial, and azimuthal.

The directional dual-ridge-guide transmit antenna minimized the transmission of signals in unwanted directions. The transmit antenna was vertically polarized for measurements with a vertically polarized dipole and horizontally polarized for radial and azimuthal dipole polarizations. The transmit and receive antennas were connected to ports 1 and 2, respectively, of a calibrated VNA. At each receive-antenna position, complex scattering parameters were recorded over a bandwidth of 0.4 GHz (1.8–2.2 GHz in 100 kHz steps with 4,001 frequencies). The IF bandwidth was set to 1 kHz and the dwell time to 10 µs.

11.4.3 Measured results

The power-angle profiles for the chamber configuration described in the previous section are shown in Figure 11.11(a) and (b) for one of two orthogonal transmit

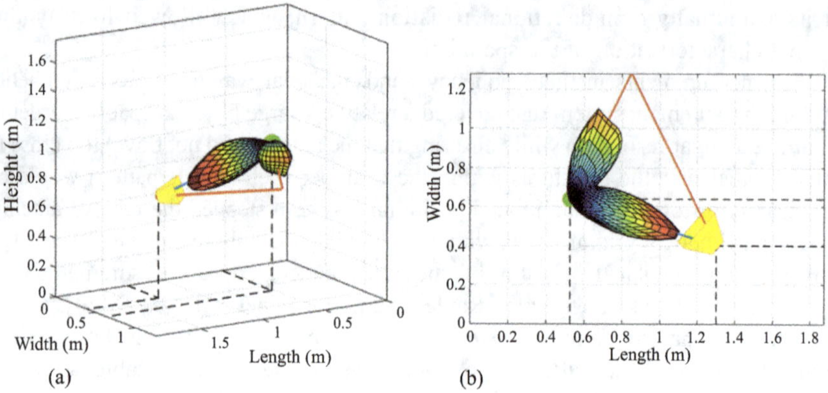

Figure 11.11 Experimentally measured power-angle profiles corresponding to the cylindrical synthetic aperture (shown by the surface plot), compared to the ray-tracing result (shown by the blue and red solid lines): (a) side view and (b) top-down view

antenna polarizations. The results were combined to give a single power-delay-angle profile corresponding to the horizontally polarized transmit antenna.

Parts (a) and (b) of Figure 11.11 compare the ray-tracing model results to the measurement-based calculations for the vertically polarized antenna case. For the LOS path, the errors were approximately 4° in azimuth and 2° in elevation. For the NLOS path, the errors were approximately 1° in azimuth and 1° in elevation. For comparison, the HPBWs of the LOS and NLOS lobes were approximately 19° and 16° in azimuth, and 12° and 18° in elevation.

The PDP corresponding to the simulated azimuth and elevation arrival angles for the LOS and NLOS paths was also extracted from the data. The measurement was conducted twice, with the receive antenna element in the vertical and horizontal polarizations. These were then averaged. The temporal resolution of the PDP was 2.5 ns, which was determined by the 400 MHz bandwidth.

The theoretical delays, based on the propagation of a plane wave traveling at the speed of light from the transmit antenna to the center of the synthetic aperture, were also computed. The delays derived from measurement were ~2 ns longer than the theoretical value. However, given that the temporal resolution of the PDP was 2.5 ns and the electrical delay of the antennas was not accounted for, such an offset is not unexpected.

11.4.4 Measurement uncertainty

The uncertainty analysis presented here, based on [51], includes both systematic uncertainties, corresponding to the calibration of the VNA with imperfect coaxial calibration standards, and an estimation of random errors in our result due to coaxial cable bending. The uncertainty due to cable bending may be crucial because the synthetic-aperture technique involves a significant amount of cable movement. The VNA was calibrated with an electronic calibration module. To estimate the uncertainty in the scattering-parameter measurements due to the VNA calibration, the states of the electronic calibration module were characterized with a precision Type-N mechanical calibration kit. Those errors were propagated from the models for the mechanical Type-N calibration standards to the states of the electronic calibration module using the procedures described in [55]. Finally, those uncertainties were propagated to the synthetic-aperture measurements using the techniques described in Section 11.3.2.5.

To evaluate the effects of cable bending, the cable connecting the reference antenna to the bulkhead inside the chamber was removed from the system to perform bending tests. After calibrating the VNA with mechanical standards, ten different bending measurements were performed, consisting of combinations of looping the cable, bending it, straightening it out, and twisting it. None of these tests placed the cable below the minimum bending radius specified by the manufacturer nor in a configuration that would damage it. For each bending configuration, the scattering parameters of the cable were measured. These bending tests were intended to simulate the possible configurations the cable could experience while moving on the turntable.

Results shown in Figure 11.12(a) and (b) illustrate that the standard uncertainty of the cable frequency response due to cable bending was on the order of 0.01 dB in magnitude and 0.4° in phase. Propagating the uncertainty through the PDP yielded 0.04 and 0.01 dB uncertainty in amplitude for the LOS and NLOS components, respectively. These were somewhat larger than the amplitude uncertainty estimates seen in Figure 11.12(a).

The uncertainties in the time of arrival (not shown explicitly) were small because the phase errors estimated due to the VNA calibration and cable bending were themselves quite small. For example, the roughly 0.4° of phase error in Figure 11.12(b) corresponds to a delay of only 0.55 ps. As a reminder, the temporal resolution is 2.5 ns, which means that the cable bending was a negligible source of

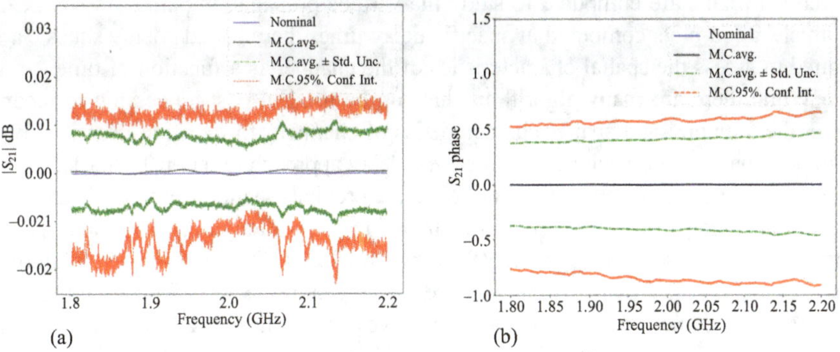

Figure 11.12 Magnitude (a) and phase (b) uncertainty of the cable frequency response due to cable bending derived from 10 measurements of the cable used in the synthetic-aperture system

uncertainty for this configuration. However, this component of uncertainty may increase at mmWave frequencies.

11.4.5 Hybrid-chamber test environment

The hybrid chamber extends the use of reverberation chambers to the emulation of spatial-channel environments that could be useful for testing both beamforming and MIMO antenna-array capabilities. The test environment discussed here is readily extensible to the emulation of three-dimensional channels at mmWave frequencies. The implementation and synthetic-aperture characterization of more complex propagation channels that include diffraction and diffuse multipath components would provide a known reference environment for the verification of other test methods. This, in turn, supports advances in wireless communication technology by providing necessary infrastructure and measurement methods with known uncertainties. The development of such technology for both microwave and mmWave frequency bands is essential for accommodating data-intensive applications of wireless devices, which are expected to increase exponentially in the future.

11.5 Millimeter-wave synthetic-aperture system in a highly reflective industrial environment

The wide measurement bandwidth of many VNA-based synthetic-aperture systems enables high delay resolution in the temporal domain and the ability to accurately separate scatterers that are closely spaced in angle but have different path lengths. Such characteristics are useful in determining the channel characteristics in highly reflective, spatially diverse environments such as factories and industrial plants.

In this section, measurements are presented from an mmWave planar synthetic-aperture system deployed in a highly reflective industrial, utility plant environment. A single transmit/receive configuration is measured (i.e., a single channel), but many directional PDPs are computed to study the various propagation paths. As well, azimuth/elevation plots computed at specific delay times, here called "delay slices," are plotted to assess the spatial characteristics of the channel as a function of time.

While there are many algorithms that can be used in post-processing to reconstruct the temporal and spatial characteristics of channels, this section describes two common analysis approaches. The true-time-delay approach (Section 11.5.3.1), based on the Fourier-space techniques described in the previous sections, provides a detailed electromagnetic picture of the environment and is, thus, often utilized by the research community. The second approach (Section 11.5.3.2) is based on super-resolution analysis techniques in which the response of the channel is described in terms of a summation of plane waves arriving at the receive array, with each a function of power, delay time, and AoA. Super-resolution methods are often used to develop statistical channel models for standards development. A comparison of temporal and spatial characteristics of the utility plant channel derived as computed by these two algorithms concludes this chapter.

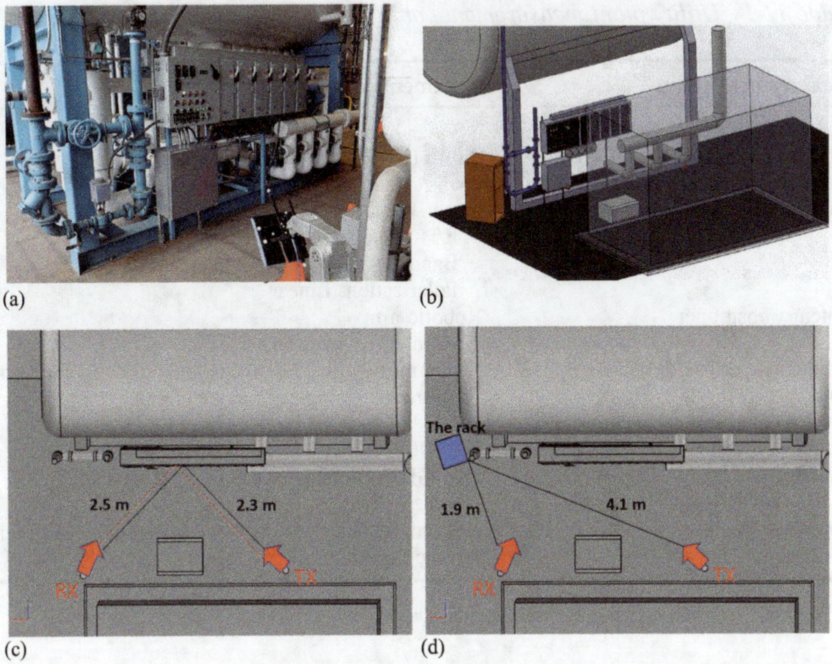

Figure 11.13 (a) Photograph and (b) CAD drawing of the utility plant aisleway. Parts (c) and (d) illustrate possible reflective paths and respective separation distances between the transmit or receive antennas and the object. The red arrows correspond to the boresight of the TX antenna and RX aperture and the blue box in (d) represents a large equipment rack that was placed to increase spatial diversity of the channel.

11.5.1 The environment: industrial utility plant

Measurements were performed in the highly reflective Central Utility Plant at the Department of Commerce Boulder Laboratories in 2019. This environment consists of large boiler tanks, piping, and numerous racks of control hardware, as shown in Figure 11.13(a)–(d). Measurements focused on an aisleway with dimensions approximately $w = 2$ m by $l = 6$ m by $h = 8$ m, with a horizontal overhead tank approximately 4 m above the floor. The dimensions of this aisleway are on the order of a factory "work cell," which is typically less than 10 m per side. In a work cell, one or a few machines complete a specific operation. There is currently great interest in adopting wireless technology in the work cell environment due to the potential for low-latency machine-to-machine communications, increased efficiency, and ability to reuse spectrum [56]. The reflective material, while primarily metal, also included a range of dielectric materials. A complex multipath-channel scenario such as this provides an ideal environment for assessing the performance of models that may rely on simplifying assumptions about the geometry and reflective properties of the scene to make the analysis tractable.

Table 11.1 Utility plant measurement configuration

Measurement setup	Parameter
Aisleway specifications	
Interior dimensions (approx.)	2.0 m (w)×6.0 m (l)×8.0 m (h)
Synthetic-aperture configuration:	
Transmit and receive antennas	Type: WR-28, 17-dBi waveguide horn E/H plane beamwidth: 23°/24° Bandwidth: 26.5–40 GHz Polarization: Linear
Antenna positioner	Robotic arm Number of positions: 1,225
Synthetic-aperture dimensions	Planar array with 12.5 cm width and 12.5 cm height, 3.7-mm spacing ($\lambda/2$ at 40 GHz)
VNA measurement parameters:	
Frequency range	26.5–40 GHz
Bandwidth	13.5 GHz (10-MHz steps, 1,351 frequencies)
IF bandwidth	100 Hz
Reference planes	Coaxial connector at coaxial-to-waveguide adapter
Dwell time	1 msec

11.5.2 Measurement setup

The NIST mmWave, synthetic-aperture measurement system [28,29] was configured to operate over a wide frequency band (26.5–40 GHz) in order to study detailed delay characteristics of the channel. A diagram of the system in a laboratory environment was given in Figure 11.1, and a photograph of it can be seen in the lower right of Figure 11.13(a). The system consists of a fixed, directional transmit antenna, a directional receive-antenna element mounted onto a robotic-arm positioner that forms the synthetic aperture, and a VNA to measure the static channel's complex impulse response. The robotic arm allows a variety of 3D synthetic-aperture scans with different geometries and polarizations, although 2D planar scans were used in the measurements shown here. The arm used provides position accuracy better than 100 μm and repeatability in the tens of μm, as determined by an optical camera system that estimates the positions of the receive antenna as it is scanned. Measurement parameters are given in Table 11.1.

11.5.3 Analysis approaches

As mentioned in the introduction to this section, two analysis approaches are presented, the Fourier-space true-time-delay beamforming technique and a super-resolution technique. Both can be used to accurately extract the received power, delay, and angular characteristics of discrete multipath components, as illustrated in Section 11.5.4. The computational complexity of each approach and the level of detail required in the resulting channel characterization are trade-offs in the choice of method.

11.5.3.1 Approach 1: true-time-delay algorithm

The true-time-delay approach is based on the Fourier-space beamforming techniques described in the previous sections. It consists of a wideband algorithm that jointly estimates delay and AoA for multipath components. This approach provides directional PDPs that include all multipath components at the minimum delay time resolution (Section 11.2.7), including both discrete and diffuse multipath components.

True-time-delay beamforming refers to the practice of inserting a pure time delay instead of a fixed phase shift behind each array element to steer the beam in wideband phased arrays [34]. True-time-delay beamforming can be implemented on wideband synthetic-aperture-system measurements in post-processing to avoid beam squint by applying a steering phase $\psi(m, n)$ that varies linearly with frequency at the *mn*th array element, as in [34]

$$\psi(m,n) = \frac{\omega}{c}(x_m \sin\theta \cos\phi + y_n \sin\theta \sin\phi) \tag{11.14}$$

where $\omega = 2\pi f$ and c is the speed of light.

After summing all the products across every array element of the beamsteering phase and the measured S_{21} parameter, for each frequency, an inverse discrete Fourier transform (IDFT) is computed to yield the beam output received from the direction (θ, ϕ) as a function of delay. This beam output is also known as the power angle delay profile (PADP) for the direction (θ, ϕ) and can be used to locate scatterers in angle and delay. The process may be summarized in the following algorithm:

Algorithm: Wideband Joint Angle-Delay Estimation Algorithm—Computation of Directional PDPs

---

Require: Array output vector $\mathbf{y}(\omega_s)$ at each frequency ω_s for $s = 0, \ldots, S-1$ and a desired beam pointing direction (θ_0, ϕ_0) or (u_0, v_0).
Procedure:

1. Determine the desired beam pointing direction by locating peaks in the spatial Fourier transform output computed at the highest frequency (having the narrowest beamwidth). Compute the phase steering vector for each frequency, $\mathbf{w}(\omega_s; u_0, v_0)$.
2. Beamform the array output vector $\mathbf{y}(\omega_s)$ at each frequency by forming the dot product $b(\omega_s; u_0, v_0) = \mathbf{w}(\omega_s; u_0, v_0)^H \mathbf{y}(\omega_s)$.
3. Compute the length S temporal IDFT to obtain the beam output (directional PDP), $x(\tau_s; u_0, v_0) = \text{IDFT}[b(\omega_s; u_0, v_0)]$.
4. Optional: To increase sampling resolution, compute a low-sidelobe window function $w(s)$ of length S (e.g., a Hamming window). Zero-pad the sequence $w(s)b(\omega_s; u_0, v_0)$ to L times its original length and compute the IDFT.

Computation of delay slices

The wideband true-time-delay algorithm can be leveraged to evaluate delay slices of the four-dimensional channel impulse response by computing directional PDPs at directions (θ_k, ϕ_k) or (u_k, v_k) on a discrete angular grid of $k = 0, \ldots, K-1$ angles that encompass the entire forward hemisphere. If these PDPs are evaluated over all the angles at the fixed delay $\tau = \tau_m$, then $x(\tau_m;u, v)$ is the spatial frequency spectrum of all signal sources impinging on the array and can be used to estimate strong angles of arrival for the delay bin τ_m. The equation for evaluating the IDFT of the beam output $b(f_s;u_k, v_k)$ at the mth delay bin τ_m is

$$x(\tau_m;u_k, v_k) = \frac{1}{S}\sum_{s=0}^{S-1} b(f_s;u_k, v_k)e^{j2\pi ms/S} \tag{11.15}$$

where S is the total number of frequency samples f_s and $0 \leq m \leq S-1$.

This is illustrated in Figure 11.14(a)–(d). In this figure, we see the signal received from the utility plant environment described earlier as a function of azimuth and elevation at four different time delays (note that "delay bin" refers to

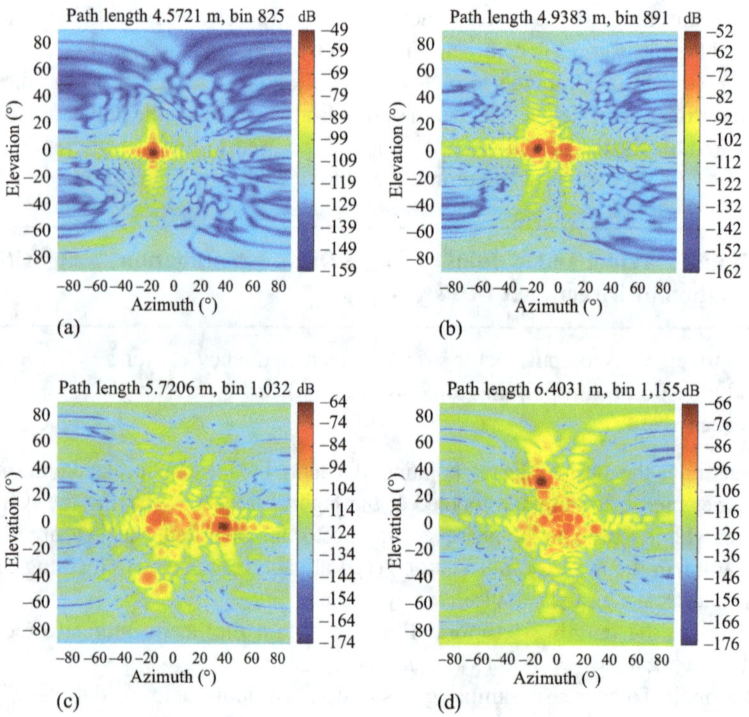

Figure 11.14 Application of true-time-delay beamforming to characterize the received energy density in the utility plant at four different delay times. Strong reflections are evident from the (a) switch bank across from the transmit antenna; (b) the switch bank and other reflectors across the aisle; (c) the rack at 38° azimuth; and (d) the overhead tank at 38° elevation.

the discrete-time index of the delay sample). The relative received power level is given by the color bar. These delay slices show a detailed view of the temporal and angular evolution of the multipath reflections in the utility plant environment.

Practical experience suggests that the true-time-delay and delay slice algorithms are most effective when a candidate set of delay bins to search for multipath components is determined in advance. A particularly useful method for determining candidate delay bins is to integrate (using summation) each available delay slice $x(\tau_m;u,\ v)$ over all angles in order to compute the total energy received versus delay. This procedure yields a PDP $r(\tau_m)$ that describes the total energy impinging on the synthetic aperture from the entire forward hemisphere as a function of delay as

$$r(\tau_m) = \sum_{k=0}^{K-1} |x(\tau_m;u_k, v_k)|^2. \tag{11.16}$$

Note that the implementation presented here follows several standard approaches and omits sensitivity to polarization. An extension of the current algorithm to include polarization effects, as discussed in [33], could improve the characterization.

As described in Section 11.2, the Fourier structure inherent in the beamforming techniques described earlier, including the true-time-delay approach, will generate sidelobes in the spatial and temporal domains. As discussed in Section 11.2.5, sidelobes can be suppressed through windowing techniques (Hamming, Kaiser, etc.) or by the use of more complex optimization techniques [32,36–40]. However, the trade-off between the sidelobe levels and the width of the main beam cannot be escaped in the Fourier-space approaches. One approach for improving detection reliability in the presence of significant components of multipath is the use of a super-resolution technique, described next.

11.5.3.2 Approach 2: super-resolution technique

Super-resolution techniques, such as MUSIC and ESPRIT [57,58], rely on partitioning the data eigenspace into signal and noise subspaces to extract angles of arrival. CLEAN [59] or SAGE [60] are a class of methods that were developed to solve nonlinear optimization problems* [61]. For channel impulse response extraction, CLEAN and SAGE require that the channel is represented as the superposition of a finite sum of plane waves incident on the receive antenna, each having discrete values of angle, delay, and power. Using SAGE, the characteristics of a specific channel are then extracted by maximizing the likelihood ratio of the observed data. Using CLEAN, the parameters of multipath components are extracted by successively subtracting plane wave contributions from the received data.

For example, the CLEAN algorithm iteratively searches for the dominant peak in the beamformed response and subsequently scales (in power) and shifts (in angle and delay) the ideal spatial–temporal response of the system so that it best aligns with the peak. The iterative search is governed by an optimization technique. In the

*Note that CLEAN is not an acronym, rather it was developed to clean up "dirty" astronomical images.

most general case, the algorithm would simultaneously search over all of the variables (delay, azimuth, elevation, and power). The conventional SAGE algorithm attempts to simplify this search by holding three of the four variables constant, while it searches over the other free variables in what is called a "greedy" search. However, other optimization methods may be used as well.

The alignment for the four variables can be finer than the beamwidth and bandwidth resolution of the corresponding Fourier structure because a physical model for the data is assumed, and that is why these algorithms are referred to as super-resolution techniques. The scaled and shifted ideal response is then subtracted from the beamformed response, and the algorithm proceeds to the next iteration and repeats the process until, ideally, the remaining peaks are below the noise floor of the system.

SAGE is similar to CLEAN; however, it considers multiple dominant peaks per iteration jointly. While it can obtain better accuracy, it is more sensitive to any system nonidealities characterized through the uncertainties mentioned earlier.

Super-resolution approaches rely on two important assumptions about the physical scenario under measurement. The first assumption is that the reflection coefficient of the material that causes each multipath component is not frequency-dependent. In other words, the assumption is that multipath components do not result from scattering off of frequency selective material. Reflections from dielectric surfaces and multipath components caused by diffraction may have a frequency dependence that distorts the estimate of the transfer function if not accounted for. The second assumption is that the distance between the transmit and receive antennas and the reflective objects in the channel is sufficiently large to be considered in the far field. This means that the propagation over each path can be represented as a plane wave over the area of interest, typically the size of the antenna-array aperture.

Algorithm: Super-Resolution Technique for Spatial–Temporal Channel Impulse Response

---

The super-resolution algorithms aim to reconstruct the measured received signal $\widehat{Y}(f, \boldsymbol{r})$ by searching for the channel properties $(\alpha_n, \tau_n, \theta_n, \phi_n)$ that minimize the squared error between the measured and modeled received signals.

Require: That the channel frequency response can be modeled as the superposition of N plane waves incident on the receiver:

$$H(f, \mathbf{r}) = \sum_{n=1}^{N} \alpha_n e^{-j2\pi\tau_n f} e^{-j\boldsymbol{k}(\theta_n,\, \phi_n)\cdot\, \boldsymbol{r}}, \tag{11.17}$$

each with its own complex amplitude α_n, delay τ_n, and AoA (θ_n, ϕ_n).

Procedure: Given the ideal response of the channel sounder (i.e., the synthetic-aperture measurement system)

$$X(f, \mathbf{r}) = e^{j2\pi\tau f} e^{j\boldsymbol{k}(\theta,\, \phi)\cdot\, \boldsymbol{r}}, \tag{11.18}$$

the received signal can be modeled as

$$Y(f,\mathbf{r}) = X(f,\mathbf{r})\cdot H(f,\mathbf{r}) = \sum_{n=1}^{N} \alpha_n e^{j2\pi(\tau-\tau_n)f} e^{j\cdot(\boldsymbol{k}(\theta,\,\phi)-\boldsymbol{k}(\theta_n,\,\phi_n))\boldsymbol{r}}. \quad (11.19)$$

The method iteratively estimates multipath-component parameters such as delay, power, AoA over an "M-Step," sequentially estimating multipath components from the measured data [60]. Unlike typical SAGE, the method used here directly extracts the delay and power from measured angle-dependent PDPs without iteration. The directional PDPs are obtained by the use of a wideband beamformer, similar to steps 1–3 of the Wideband Joint Angle-Delay Estimation Algorithm, but with a Kaiser window applied [62]. For AoA estimation, an exhaustive search is performed over the set of peaks in the directional PDPs that exceed a specified threshold. A maximum-likelihood AoA estimation algorithm is applied that uses the relative power/phase between multiple angle-dependent PDPs to obtain high-resolution AoA estimation, which can be at angles finer than the beam-scan grid [63].

In the "E-Step," the peaks corresponding to the ith multipath components are subtracted from the ideal response generated by $(i-1)$th multipath-component parameters as for typical SAGE implementations. Due to system modeling errors, noise, etc., the residual of subtraction could become a false ("ghost") detection. The ghost classifier used here is based on Monte Carlo simulations of the data from the synthetic-aperture measurements to reduce such false detection.

Note that for each separate multipath component, the previous description is for a single polarization direction. For a general description, two orthogonal polarizations (assuming we are in the far-field) should be described.

11.5.4 Measured results

Measurements were performed in the utility plant aisleway described earlier with the transmit antenna and the planar synthetic-aperture array both oriented toward the bank of switches shown in Figure 11.13(a). Orienting the directional antennas away from each other minimized LOS coupling in order to focus on reflected multipath components. To increase the azimuthal diversity, a large metallic equipment rack was placed at the end of the aisleway. In Figure 11.13(b), a CAD drawing based on physical measurements of the environment illustrates the location of the transmit and receive antennas (shown by red dots), the location of the VNA instrumentation (shown by the light gray rectangle located between the transmit and receive antennas), and the equipment rack (shown by the blue box at the end of the aisleway). The switch bank and other reflecting surfaces are shown in simplified form.

Figure 11.13(c) and (d) illustrates some of the anticipated angular reflections. In part (c), two different reflected paths from the central panel are shown by solid black and dashed red lines. In part (d), the primary reflection from the rack is illustrated. The

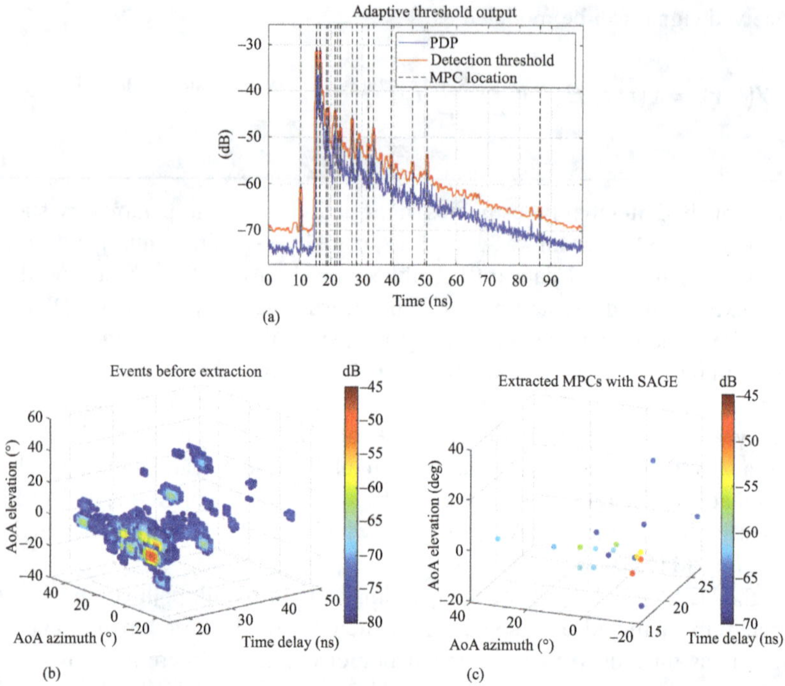

Figure 11.15 Temporal and spatial view of the channel as seen at the synthetic-aperture receive array. In (a), the true-time-delay approach was used to estimate the total received power PDP. In (b), the SAGE algorithm was used to extract several multipath components as a function of angle and delay, with the 19 strongest components in (c). Power is indicated by the color of each sample.

distance between the transmit or receive antennas and the object is estimated from the CAD drawing. The summation of these distances corresponds to the total path length travelled by the electromagnetic wave when received by the antenna.

Figure 11.15(a) illustrates the PDP corresponding to the total received power. This was estimated from the true-time-delay approach by summing the magnitude squared of all directional PDP outputs over all 1,225 measured beam directions. This approach was described in Section 11.2.8. The strongest multipath components are indicated by vertical dashed lines. The directional characteristics of these multipath components at four of these delay times were presented in Figure 11.14(a)–(d).

Figure 11.15(b) and (c) shows the PADP obtained by the use of the super-resolution approach. In Figure 11.15(b), several multipath components are illustrated and reflections corresponding to different features are called out. In Figure 11.15(c), the strongest 19 multipath components were extracted. These multipath components could be used in a channel model to represent the main features of the environment.

Table 11.2 Six strongest multipath components (MPCs) extracted from the utility plant synthetic-aperture measurements for true-time-delay and SAGE analysis approaches

MPC	Path length (m)		Path loss (dB)		Delay (ns)		AoA Az (°)		AoA El (°)	
	Fourier	**SAGE**	**Fourier**	**SAGE**	**Fourier**	**SAGE**	**Fourier**	**SAGE**	**Fourier**	**SAGE**
1. Panel 1	4.57	4.64	−46.1	−47.7	15.3	15.5	−16.5	−16.6	−1.5	−1.7
2. Panel 2	4.94	5.00	−47.7	−49.5	16.5	16.7	−17.0	−17.4	+2.0	+1.7
3. Pipe	5.57	5.64	−60.2	−62.5	18.6	18.8	+17.5	+17.5	−2.5	−2.2
4. Rack	5.72	5.78	−59.3	−62.2	19.1	19.3	+38.5	+38.2	−3.0	−2.9
5. Tank	6.40	6.47	−60.9	−63.7	21.4	21.6	−12.5	−12.5	31.0	31.3
6. 3Refls	8.03	8.11	−60.8	−63.8	26.8	27.0	−18.0	−18.3	−0.5	−0.4

"3Refls" refers to a three-reflection path.

The delay and AoA of six strong, discrete multipath components are presented in Table 11.2, displayed from shortest to longest delay. The source of each multipath component is denoted in the first column. Path-length values were estimated by dividing the delays obtained from the two analysis techniques by the speed of light. The remaining columns in the table compare the delay and AoA extracted with both methods, with the true-time delay method designated "Fourier" and the super-resolution method designated "SAGE." The agreement between the two analysis methods is within 0.2 ns for delay, 0.4° in azimuth, and 0.3° in elevation for the AoA values. Whether these differences are significant depends on whether they fall within the 95% confidence interval, which is the subject of current research.

11.5.5 Summary

This section demonstrated the use of a network-analyzer-based synthetic-aperture measurement system as a channel sounder for field tests in highly reflective, static environments such as a utility plant. Synthetic-aperture systems can help to determine the three-dimensional channel's spatial and temporal characteristics. Once the data have been collected, various analysis approaches can be used. The key to agreement between the approaches is having a synthetic-aperture measurement system that provides temporal and angular resolution that is commensurate with the physical features in the environment.

References

[1] Zhang F and Fan W. Near-Field Ultra-Wideband mmWave Channel Characterization Using Successive Cancellation Beamspace UCA Algorithm. *IEEE Trans. Veh. Technol.* 2019;68(8):7248–7259.

[2] Remley KA, Dortmans J, Weldon C, *et al.* Configuring and Verifying Reverberation Chambers for Testing Cellular Wireless Devices. *IEEE Trans. Electromagn. Compat.* 2016;58(3):661–672.

[3] Kildal PS, Chen X, Orlenius C, *et al.* Characterization of Reverberation Chambers for OTA Measurements of Wireless Devices: Physical Formulations of Channel Matrix and New Uncertainty Formula. *IEEE Trans. Antennas Propag.* 2012;60(8):3875–3891.

[4] Dias MHC and Siqueira GL. Indoor 1.8 GHz AOA-TDOA Measurements Extending the Use of a Wideband Propagation Channel Sounder with the Synthetic Aperture Concept. *European Microwave Conference.* EuMA, Paris, France; October 4–6, 2005.

[5] Marks RB, Formulations of the Basic Vector Network Analyzer Error Model Including Switch Terms. *50th ARFTG Microwave Measurement Conference.* ARFTG, Portland, OR; December 4–5, 1997.

[6] Jargon JA, Williams DF, and Sanders A. Three-Port Vector-Network-Analyzer Calibrations Using the NIST Microwave Uncertainty Framework. *92nd ARFTG Microwave Measurement Conference.* ARFTG, Orlando, FL, January 19–22, 2019.

[7] Hale PD, Remley KA, Williams DF, *et al.* A Compact Millimeter-Wave Comb Generator for Calibrating Broadband Vector Receivers. *85th ARFTG Microwave Measurement Conference*. ARFTG, Phoenix, AZ, May 22, 2015.

[8] Roh W, Seol J-Y, Park JH, *et al.* Millimeter-Wave Beamforming as an Enabling Technology for 5G Cellular Communications: Theoretical Feasibility and Prototype Results. *IEEE Commun. Mag.* 2014;52(2):106–113.

[9] Wei L, Hu RQ, Qi Y, *et al.* Key Elements to Enable Millimeter Wave Communications for 5G Wireless Systems. *IEEE Wirel. Commun.* 2014;21 (6):136–143.

[10] Yu B, Yang K, Sim C-Y-D, *et al.* A Novel 28 GHz Beam Steering Array for 5G Mobile Device With Metallic Casing Application. *IEEE Trans. Antennas Propag.* 2018;66(1):462–466.

[11] Qi Y, Yang G, Liu L, *et al.* 5G Over-the-Air Measurement Challenges: Overview. *IEEE Trans. Electromagn. Compat.* 2017;59(6):1661–1670.

[12] Fan W, Kyösti P, Rumney M, *et al.* Over-the-Air Radiated Testing of Millimeter-Wave Beam-Steerable Devices in a Cost-Effective Measurement Setup. *IEEE Commun. Mag.* 2018;56:64–71.

[13] Ranvier S, Kyro M, Haneda K, *et al.* VNA-Based Wideband 60 GHz MIMO Channel Sounder with 3-D Arrays. *IEEE Radio and Wireless Symposium*. IEEE, San Diego, CA, USA; January 18–22, 2009.

[14] Nguyen H, Mahler K, Peter M, *et al.* Estimation of DoA Based on Large Scale Virtual Array Data. *10th Eur. Conf. Antennas Propag (EuCAP)*. IEEE, Davos, Switzerland; April 10–15, 2016.

[15] Nguyen H, Keusgen W, and Eichler T. Instantaneous Direction of Arrival Measurements in Mobile Radio Channels using Virtual Circular Array Antennas. *IEEE Globecom Workshop*. IEEE, Washington, DC, USA; December 4–8, 2016.

[16] Hayashi H and Ohtsuki T. DOA Estimation in MIMO Radar Using Temporal Spatial Virtual Array with MUSIC Algorithm. *9th International Conference on Signal Processing and Communication Systems (ICSPCS)*. IEEE, Cairns, Australia; December 14–16, 2015.

[17] Mannesson A, Bernhardsson B, Yaqoob M, *et al.* Optimal Virtual Array Length Under Position Imperfections. *IEEE 8th Sensor Array and Multichannel Signal Processing Workshop*. IEEE, A Coruna, Spain; June 22–25, 2014.

[18] Santella G and Restuccia E. Analysis of Frequency Domain Wide-Band Measurements of the Indoor Radio Channel at 1, 5.5, 10 and 18 GHz. *IEEE Globecom Conference*. IEEE, London, UK; November 18–29, 1996.

[19] Howard SJ and Pahlavan K. Measurement and Analysis of the Indoor Radio Channel in the Frequency Domain. *IEEE Trans. Instrum. Meas.* 1990;39 (5):751–755.

[20] Remley KA, Koepke G, Holloway CL, *et al.* Radio-Wave Propagation Into Large Building Structures – Part 2: Characterization of Multipath. *IEEE Trans. Antennas Propag.* 2010;58(4):1290–1301.

[21] Kyro M, Ranvier S, Kolmonen V-M, *et al.* Long Range Wideband Channel Measurements at 81-86 GHz Frequency Range. *4th Eur. Conf. Antennas Propag (EuCAP)*. IEEE, Barcelona, Spain; April 12–16, 2010.

[22] Schulpen R, Konstantinou D, Rommel S, *et al.* Extended Range Ultra-Wideband Millimeter-Wave Channel Sounder With Over-the-Air Calibration. *13th Eur. Conf. Antennas Propag (EuCAP)*. IEEE, Krakow, Poland; March 31–April 5, 2019.

[23] Qiao H, Sarangi P, Alnumay Y, *et al.* Sample Complexity Trade-Offs for Synthetic Aperture Based High Resolution Estimation and Detection. *IEEE 11th Sensor Array and Multichannel Signal Processing Workshop*. IEEE, Hangzhou, China; June 8–11, 2020.

[24] Mbugua AW, Fan W, Ji Y, *et al.* Millimeter Wave Multi-User Performance Evaluation Based on Measured Channels with Virtual Antenna Array Channel Sounder. *IEEE Access*. 2018;6(3):12318–12326.

[25] Chen B, Zhong Z, Ai B, *et al.* Moving Virtual Array Measurement Scheme in High-Speed Railway. *IEEE Antennas Wirel. Propag. Lett.* 2016;15:706–709.

[26] Zhou T, Tao C, Salous S, *et al.* Measurements and Analysis of Angular Characteristics and Spatial Correlation for High-Speed Railway Channels. *IEEE Trans. Intell. Transp. Syst.* 2018;19(2):357–367.

[27] Becker MG, Horansky RD, Senic D, *et al.* Spatial Channels for Wireless Over-the-Air Measurements in Reverberation Chambers. *12th Eur. Conf. Antennas Propag (EuCAP)*. IET, London, UK; April 9–13, 2018.

[28] Weiss AJ, Williams DF, Quimby J, *et al.* Large-Signal Network Analysis for Over-the-Air Test of Up-Converting and Down-Converting Phased Arrays. *IEEE MTT-S Int. Microwave Symp*. IEEE, Boston, MA, USA; June 2–7, 2019.

[29] Weiss AJ, Quimby J, Leonhardt R, *et al.* Setup and Control of a Millimeter-Wave Synthetic Aperture Measurement System with Uncertainties. *95th ARFTG Microwave Meas. Conf.* ARFTG, Los Angeles, CA, USA; August 7, 2020.

[30] Durgin GD. *Space-Time Wireless Channels*. Upper Saddle River, NJ, USA: Prentice-Hall; 2003.

[31] Haupt RL. *Antenna Arrays: A Computational Approach*. Hoboken, NJ, USA: Wiley and Sons; 2010.

[32] Vouras P, Quimby J, Jamroz B, *et al.* Frequency Invariant Beampatterns for Wideband Synthetic Aperture Channel Sounders. *IEEE 21st Int. Workshop on Signal Processing Advances in Wireless Comm. (SPAWC)*. IEEE, Atlanta, GA; May 27–29, 2020.

[33] Newell A, Baird R, and Wacker P. Accurate Measurement of Antenna Gain and Polarization at Reduced Distances by an Extrapolation Technique. *IEEE Trans. Antennas Propag.* 1973;21(4):418–431.

[34] Frigyes I and Seeds AJ. Optically Generated True-Time Delay in Phased-Array Antennas. *IEEE Trans. Microwave Theory Tech.* 1995;43(9): 2378–2386.

[35] Wang B, Jian M, Gao F, *et al.* Beam Squint and Channel Estimation for Wideband mmWave Massive MIMO-OFDM Systems. *IEEE Trans. Signal Process.* 2019;67(23):5893–5908.

[36] Tan CM, Fletcher P, Beach MA, *et al.* On the Application of Circular Arrays in Direction Finding Part I: Investigation into the estimation algorithms. *First COST 273 Workshop*; May 2002.

[37] Do-Hong T, Demmel F, and Russer P. A Method for Wideband Direction-of-Arrival Estimation Using Frequency-Domain Frequency-Invariant Beamformers. *IEEE Antennas and Propagation Society Symp*. IEEE, Columbus, OH, USA; June 22–27, 2003.

[38] Ward DB, Ding Z, and Kennedy RA. Broadband DOA Estimation Using Frequency Invariant Beamforming. *IEEE Trans. Signal Process.* 1998;46 (5):1463–1469.

[39] Yang Y, Wan C, and Sun C. Broadband Beamspace DOA Estimation Algorithms. *IEEE Oceans*. IEEE, San Diego, CA; September 22–26, 2003.

[40] Wax M. Direction Finding of Coherent Signals via Spatial Smoothing for Uniform Circular Arrays. *IEEE Trans. Antennas Propag.* 1994;42(5):613–620.

[41] Dortmans JNH, Quimby JT, Remley KA, *et al.* Design of a Portable Verification Artifact for Millimeter-Wave-Frequency Channel Sounders. *IEEE Trans. Antennas Propag.* 2019;67(9):6149–6158.

[42] Porat B. *A Course in Digital Signal Processing*. Hoboken, NJ, USA: Wiley and Sons;1997.

[43] Sun S, MacCartney Jr. GR, Samimi MK, *et al.* Synthesizing Omnidirectional Antenna Patterns, Received Power and Path Loss from Directional Antennas for 5G Millimeter-Wave Communications. *IEEE Global Comm. Conf. (GLOBECOM)*. IEEE, San Diego, CA, USA; December 6–10, 2015.

[44] Waters WM, Patel DP, and Trunk GV. Optimum Number of Faces of a Volume-Scanning Active Array Radar. *IEEE Trans. Aerosp. Electron. Syst.* 1998;34(3):1032–1037.

[45] Keysight Technologies. Keysight 2-Port and 4-Port PNA-X Network Analyzer: Data Sheet and Technical Specifications; 2018.

[46] Jargon JA, Williams DF, and Hale PD. Developing Models for Type-N Coaxial VNA Calibration Kits Within the NIST Microwave Uncertainty Framework. *87th ARFTG Microwave Meas. Conf.* ARFTG, San Francisco, CA, USA; May 27, 2016.

[47] van den Biggelaar AJ, Jamroz BF, Williams DF, *et al.* Correction of the Measured Phase of the Radiation Pattern of Millimeter-Wave Antennas. Antenna Meas. Techn. Assoc. AMTA, Virtual; November 1–6, 2020.

[48] Denavit J and Hartenberg RS. A Kinematic Notation for Lower-Pair Mechanisms Based on Matrices. *J. Appl. Mech.* 1955;22:215.

[49] Joint Committee for Guides in Metrology. "Evaluation of Measurement Data—Guide to the Expression of Uncertainty in Measurement." International Bureau of Weights and Measures (BIPM), Sèvres, France, September 2008. (Available from: http://www.bipm.org/en/publications/guides/gum.html).

[50] Joint Committee for Guides in Metrology. "Evaluation of measurement data – Supplement 1 to the "Guide to the expression of uncertainty in measurement" – Propagation of distributions using a Monte Carlo method." Intl.

Bureau Weights Measures, Sèvres, France, September 2008. (Available from: http://www.bipm.org/en/publications/guides/gum.html).

[51] Williams DF and Lewandowski A. *NIST Microwave Uncertainty Framework*. National Institute of Standards and Technology, 2011. (http://www.nist.gov/ctl/rf-technology/related-software.cfm:2011).

[52] 3rd Generation Partnership Project. "Technical Specification Group Radio Access Network. Study on Channel Model for Frequencies From 0.5 to 100 GHz." 3GPP TR 38.901V16.1.0, Sophia Antipolis Cedex, France, November 2020. (Available from: https://www.etsi.org/deliver/etsi_tr/138900_138999/138901/16.01.00_60/tr_138901v160100p.pdf).

[53] Li Q, Shirani-Mehr H, Balercia T, *et al.* Validation of a Geometry-Based Statistical mmWave Channel Model Using Ray Tracing Simulation. *IEEE 81st Vehicular Technology Conf.* IEEE, Glasgow, UK; May 11–14, 2015.

[54] Hecht E and Zajac A. *Optics*. Reading, MA, USA: Addison-Wesley Publishing; 1974.

[55] Jargon JA, Williams DF, Wallis TM, *et al.* Establishing Traceability of an Electronic Calibration Unit Using the NIST Microwave Uncertainty Framework. *79th ARFTG Microwave Meas. Conf.* ARFTG, Quebec, QC, Canada; June 22, 2012.

[56] Candell R and Kashef M. Industrial Wireless: Problem Space, Success Considerations, Technologies, and Future Direction. *2017 Resilience Week (RWS)*. IEEE, Wilmington, DE, USA; September 18–22, 2017.

[57] Schmidt RO. Multiple Emitter Location and Signal Parameter Estimation. *IEEE Trans. Antennas Propag.* 1986;34(3):276–280.

[58] Zhang Y, Brown AK, Malik WQ, *et al.* High Resolution 3-D Angle of Arrival Determination for Indoor UWB Multipath Propagation. *IEEE Trans. Wireless Commun.* 2008;7(8):3047–3055.

[59] Högbom J. Aperture Synthesis With a Non-Regular Distribution of Interferometer Baselines. *Astrophys. J. Suppl. Ser.* 1974;15:417–426.

[60] Fessler JA and Hero AO. Space-Alternating Generalized Expectation Maximization Algorithm. *IEEE Trans. Signal Process.* 1994;42(10):2664–2677.

[61] Dempster AP, Laird NM, and Rubin DB. Maximum Likelihood from Incomplete Data via the EM Algorithm. *J. R. Stat. Soc.* 1977;39(1):1–38.

[62] Gentile C, Lopez SM, and Kik A. A Comprehensive Spatial-Temporal Channel Propagation Model for the Ultra-Wideband Spectrum 2-8 GHz. *IEEE Global Comm. Conf. (GLOBECOM)*. IEEE, Honolulu, HI, USA; November 30–December 4, 2009.

[63] Caudill D, Papazian PB, Gentile C, *et al.* Omnidirectional Channel Sounder With Phased-Array Antenna for 5G Mobile Communications. *IEEE Trans. Microwave Theory Tech.* 2019;67(7):2936–2945.

Chapter 12

Using reverberation chambers for mm-wave over-the-air and electromagnetic compatibility measurements

Feng Fang[1], Qian Xu[1], Wei Xue[2], and Xiaoming Chen[2]

As the fifth-generation (5G) wireless communication system is deployed widely, over-the-air (OTA) testing becomes the main method to estimate the radiated frequency (RF) performance of 5G devices. In 5G, there are roughly two wide frequency ranges specified in 3GPP (Third Generation Partnership Project). One is what we usually call sub-6 GHz (FR1) and the other is millimetre-wave (FR2) band. In FR1, we can still adopt the traditional way to test devices, which utilises cables and connectors to connect devices. However, when the frequency comes to FR2, it may not be suitable for conducting a method to measure the multi-antenna devices. If the conducting measurement is adopted, cables and connectors will be needed to connect antennas, which are costly and have not enough space. Besides the previous disadvantages of the traditional method for millimetre-wave (mm-wave) measurements, we need to use the OTA method to characterise the system, e.g., beamforming test. With the development of 5G technology, the OTA test will be used widely to estimate the performance of the wireless devices. Over the past decade, there have been significant industry efforts to move reverberation chambers (RCs) into the mainstream of OTA wireless device testing. This chapter gives an overview of OTA and electromagnetic compatibility (EMC) measurements in RCs. The measurement procedure is introduced in detail. This chapter is organised as follows: the RC background is introduced in Section 12.1, OTA measurements are introduced in Sections 12.2 and 12.3, and EMC measurements are detailed in Sections 12.4–12.6.

[1]College of Electronic and Information Engineering, Nanjing University of Aeronautics and Astronautics, Nanjing, China
[2]School of Information and Communications Engineering, Faculty of Electronic and Information Engineering, Xi'an Jiaotong University, Xi'an, China

12.1 Introduction

RC is an electrically large, over-moded, highly conductive cavity composed of metal walls, where mechanical mode stirrers are utilised to stir the electromagnetic field to create a random-like multipath environment. Different from other reflectionless test environments (such as anechoic chamber (AC), using absorbing materials to build an ideal electromagnetic environment), measurements in RC are concluded by acquiring sufficiently large numbers of independent samples to ensure accurate results from a statistical point of view. Besides using mechanical mode stirrers, there are several other methods to increase the number of independent samples, such as source stirring, multiple probes systems.

For an RC, the most fundamental quantities are field uniformity and chamber quality factor, Q. Field uniformity is an important parameter to characterise the performance of an RC, which is defined to evaluate the statistical uniformity of the environment generated by RC. The boundary condition of the RC is changing when the stirrer is rotating. Electrical field strength is different according to different rotating angles. The maximum value of electrical field strength at each frequency of interest can be calculated when one cycle is finished. If the standard deviation of the maximum values of some positions in RC is less than 3–4 dB, field uniformity meets the requirement of IEC 61000-4-21. The quality factor Q is used to assess the ability of an RC to store energy. It is defined as

$$Q = \frac{\omega U}{P_d} \tag{12.1}$$

where $\omega = 2\pi f$ represents the angular frequency at the operating frequency, f; U represents the energy in the RC; P_d is the power dissipation. Thus, the calculation of the chamber Q requires that all losses are accounted in evaluating P_d in (12.2). A theory has been developed to include four types of loss [1]:

$$P_d = P_{d1} + P_{d2} + P_{d3} + P_{d4} \tag{12.2}$$

where P_{d1} is the power dissipated from the metallic surface (e.g., walls, ceiling, floor, and stirrers), P_{d2} is the power absorbed in loading objects within the chamber, P_{d3} is the power loss from the aperture leakage, and P_{d4} is the power dissipated from receiving antennas.

By applying either mechanical stirring or frequency stirring, the electromagnetic fields within the working volume of an RC become statistically uniform and isotropic. The probability density functions of various electric field magnitudes or squared magnitudes are of chi- or chi-square distributions with an appropriate number of degrees of freedom (DoFs). The magnitude of any of the electric field components, for example, $|E_x|$ is chi-distributed with two DoFs and consequently has a Rayleigh distribution [1]:

$$f(|E_x|) = \frac{|E_x|}{\sigma^2} \exp\left[-\frac{|E_x|^2}{2\sigma^2}\right] \tag{12.3}$$

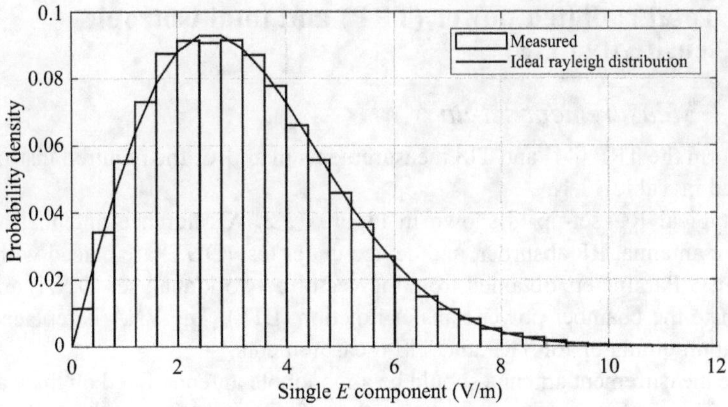

Figure 12.1 Comparison of the measured probability density function of the magnitude of a single rectangular component of the electric field with theory (Rayleigh distribution)

Figure 12.1 shows a comparison of (12.3) with measured data taken at 1 GHz in the NASA Chamber A [2]. The chamber has two stirrers, and the total number of samples (stirrer positions) is 225. The data were taken with a small electric-field probe that was calibrated at NIST [2]. The agreement is about as good as can be expected for 225 samples.

The RC has been widely used in EMC [3] measurements. The use of RC for EMC measurements was first proposed in 1968. It took some time for RC measurements to gain acceptance, while from the 1980s, the use of RCs had been well established in EMC measurements. After many years, the RC has become a common way to estimate the EMC performance of a system. It can be applied in radiated immunity test, radiated emission test, and shielding effectiveness measurement.

Over the past two decades, the multi-antenna communication systems have developed a lot. Antenna measurement has been performed in anechoic environments to determine the performance in an angle-dependent way. This may not be enough for all applications, since multi-antenna terminals would be used in a multipath scattering environment, where the scattering effect could improve the performance of the terminals.

To test a multi-antenna system, a straight-forward approach is to physically emulate a multipath scattering environment in which the device is working. An RC is an advantageous facility to perform these tests, such as antenna efficiency, diversity gain [4–12], total radiated power (TRP), total isotropy sensitivity (TIS) [13–18], and throughput of wireless devices. Unlike the EMC test, active MIMO OTA tests (using either base station emulators or a real base station) usually have specific requirements on the propagation channel (e.g., delay spread, Doppler spread, and *K*-factor) in the RC. The delay spread (or coherence bandwidth) can be readily configured by loading the RC with lossy objects [19]. The Doppler spread (or coherence bandwidth) can be controlled by varying the speed of mode stirrers [20].

12.2 Total radiated power (TRP) and total isotropic sensitivity (TIS)

12.2.1 Measurement set-up

To perform the TRP [44] and TIS measurements in an RC, the required instruments are listed in Table 12.1.

A typical RC set-up is shown in Figure 12.2. A reference antenna, a measurement antenna, RF absorber, and device under test (DUT) are placed within the chamber. *S*-Parameters obtained from the vector network analyser (VNA) are used to estimate the chamber power transfer function (PTF). The VNA is replaced by a base station emulator for TRP and TIS measurements.

The measurement antenna should be a monopole antenna fixed on the wall or a directional antenna pointing away from the reference antenna and the DUT. If a monopole antenna is used, the uncertainty caused by the direct coupling of unstirred energy can be decreased by either cross-polarising it for the reference and DUT antennas or utilising multiple antennas to achieving various polarisations. If a directive antenna is used, it should not point at the reference antenna or DUT, since this will enlarge the direct coupling of unstirred energy and increase uncertainty.

Table 12.1 The instruments needed in OTA test

Instruments	Application
Reference antenna	Receiving antenna
Measurement antenna	Transmission antenna
RF absorber	To change the Q factor
DUT (device under test)	Used for test
VNA (vector network analyser)	To obtain S-parameter

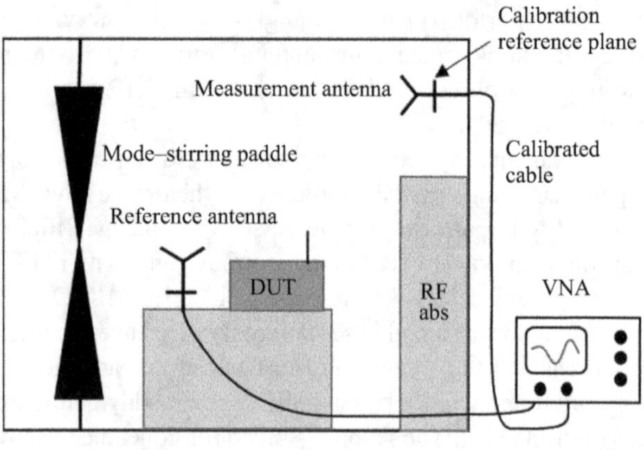

Figure 12.2 Typical set-up for estimating the power transfer function in an RC

The radiation efficiency and free-space refection coefficient of the reference antenna must be known over the operating frequency points. The manufacture may provide efficiency, which can also be tested from direct measurement, such as three-antenna measurement. It is suggested that the reference antenna is mounted on a low-loss dielectric fixture to avoid reflections of the mounting fixture itself.

The VNA must cover the frequency bands for measurement and can be employed to measure the chamber PTF (chamber loss). Calibrate the VNA with a full two-port calibration so that the vector *S*-parameters between the ports of the measurement antenna and the reference antenna can be measured. In Figure 12.2, the VNA's calibration reference plane is located at the antenna ports after calibration. The network analyser is configured to perform a frequency sweep for *N* mode-stirring samples at each frequency point, and an automated test system is highly recommended.

RF absorbing material may be required to broaden the coherence bandwidth of the chamber as specified in Section 12.2.2.2. An RF absorber may be stacked in a single position or distributed inside the chamber. Stacking absorber takes up less physical space in the chamber, but because of the less exposed area for each single block, more stacking absorbers are needed than distributed absorbers inside the chamber. The placement of the antennas and RF absorber within RC has effect on the measurement of the PTF. Each parameter is discussed as follows:

Placement of antennas near the metallic surfaces: In [21], Hill concluded that, for an ideal empty chamber, when the antenna is placed at least $\lambda/2$ from the metallic wall, the average of the measured received power will be approximately uniform (where λ is the free-space wavelength in metres).

Placement of an RF absorber: For OTA measurement, it is expected to obtain a frequency-flat channel for demodulation by loading the chamber with the RF absorbers. The influence of this method is a function of the surface area of RF absorbers in the chamber [22–24]. The more exposed the surface area of the RF absorber, the poorer the spatial uniformity of the statistical electromagnetic fields in the chamber, and the more stirring samples are needed to compensate for uniformity.

Placement of antennas near the absorbing surfaces: If the transmit antennas are placed near absorbing surfaces, energy will be absorbed before undergoing mode-stirring, which reduces the power accepted by the receiving antenna.

12.2.2 Chamber characterisations needed for measurements

12.2.2.1 Calculation of the chamber power transfer function

The PTF, G_{ref}, for a given RC configuration is estimated from *T* sets of the proposed stirring sequence as specified in Section 12.2.6 Step 1, and each of the *T* sets is conducted at spatially uncorrelated reference antenna positions. The stirrer angular correlation is defined in the following equation [25]:

$$\rho_c = \left| \frac{\sum_{t=1}^{T} \left[\left(S_{21,1} - \langle S_{21,1T} \rangle \right) \left(S_{21,2} - \langle S_{21,2T} \rangle \right)^* \right]}{\sqrt{\sum_{t=1}^{T} \left[\left| S_{21,1} - \langle S_{21,1T} \rangle \right|^2 \right]} \sqrt{\sum_{t=1}^{T} \left[\left| S_{21,2} - \langle S_{21,2T} \rangle \right|^2 \right]}} \right| \tag{12.4}$$

For example, if the number of a proposed stirring sequence is 200, the measurement of G_{ref} is performed with T different sets of 200-sample stirring sequence at T uncorrelated reference antenna positions. For large-form-factor devices, it may be necessary to increase T to ensure that the uncertainty in a measurement that uses a given stirring sequence does not exceed the level specified in Table 7-8 of the CTIA (International Association for the Wireless Telecommunication Industry) Test Plan for Over the Air Performance [26]. Uncertainty may also be reduced by increasing the number of mode-stirring samples that define a stirring sequence or modifying the stirring sequence.

At each reference antenna location, the PTF is obtained from the mean value of the measured received power ($|S_{21}|^2$), where the average is calculated over all frequencies and mode-stirring samples. The final PTF is calculated from the mean of the samples from all locations. For each set of stirring sequence, the PTF, $G_{ref,t}$, of the RC should be calculated using the following equation [15,27]:

$$G_{ref,t} = \frac{\frac{1}{NF}\sum_{n=1}^{N}\sum_{f=1}^{F}|S_{21}(f,n)|^2}{e_{mismatch,meas}e_{mismatch,ref}\eta_{meas}\eta_{ref}} \tag{12.5}$$

where $G_{ref,t}$ represents the tth PTF of a given RC configuration; $e_{mismatch,meas}$ and $e_{mismatch,ref}$ are the mismatch factor of the measurement antenna and reference antenna, respectively; η_{ref} and η_{meas} are the radiation efficiency of the reference antenna and measurement antenna, respectively. Note that the value of the mismatch factor and radiation efficiency of the measurement antenna is not required, because they are the same for both the reference and DUT measurements. Therefore, the measurement antenna efficiency will not affect the final results. The term $e_{mismatch}$ can be calculated from $1 - \langle|S_{11}|^2\rangle$ of the antenna in question, where $\langle\bullet\rangle$ denotes the average over N mode-stirring samples and F frequencies across the operating channel bandwidth given in Table 12.1. After finishing the T sets of sequence measurement, the PTF G_{ref} (Figure 12.3) can be calculated from the average over the $G_{ref,t}$:

$$G_{ref} = \frac{1}{T}\sum_{t=1}^{T} G_{ref,t} \tag{12.6}$$

where $T = T_{pre}$ (Section 12.2.6)

12.2.2.2 Calculation of chamber coherence bandwidth

The coherence bandwidth should be larger than the channel over which the DUT will be measured. It is calculated from S-parameters measured from a complete stirring sequence. It depends on the loading effect of the chamber. The coherence bandwidth for DUT tests should not be narrower than the values given in Table 12.1. But with the RF absorbers placed in, the measurement uncertainty will increase; thus it is important to ensure that the uncertainty does not exceed the limit (Table 12.2).

The procedure for evaluating the coherence bandwidth for a particular RC configuration is as follows:

1. Select the adequate stirring sequence for the DUT test, measure the transmission parameter S_{21} for each mode-stirring sample and reference antenna

Table 12.2 Coherence bandwidth for different transmission standards

Transmission standard	Channel bandwidth (MHz)	Coherence bandwidth (MHz)
GSM\GRPS\EGPRS[a]	0.2	0.2
CDMA\1×EVDO\1×RTT[b]	1.23	1.3
UMTS (WCDMA)[c]	3.84	4.0
LTE[d]	Number of resource blocks×180 kHz	4.0[e]

[a]GSM, global system for mobile communications; GPRS, general package radio service; EGPRS, enhanced GPRS.
[b]CDMA, code-division multiple access; 1×EVDO – 1× evolution-data optimised; 1×RTT – single-carrier radio transmission technology.
[c]UMTS, universal mobile telecommunications system; wideband code-division multiple access (WCDMA).
[d]LTE, long-term evolution.
[e]The channel band width (CBW) for LTE is based on the upper limit for loading WCDMA as opposed to the actual channel bandwidth.

position. The minimum value for the reference antenna location should be 1. The VNA frequency step for this test should be less than 100 kHz.

2. Calculate the complex autocorrelation function (ACF) using (12.7) for each mode-stirred sample n_i over a 100-MHz bandwidth (at least 1,001 measured frequency points) surrounding the centre frequency, f_c, of each band to be evaluated, as specified in Chapter 2.12 of the CTIA [26]:

$$R(\Delta f_i, n_i) = \sum_{j=1}^{M} S_{21}(f_j, n_i) S_{21}^*(f_j + \Delta f_i, n_i) \tag{12.7}$$

where $S_{21}(f_i, n_i)$ corresponds to the measured complex, S_{21} at frequency f_j with M frequency points measured in the bandwidth of interest, BW; thus $f_1 = f_c - BW/2, f_M = f_c + BW/2$. The index of n_i is the mode-stirring sample, Δf corresponds to a frequency offset over the bandwidth of interest and the asterisk denotes complex conjugation.

3. Average all ACFs from (12.7) over the N mode-stirring samples at each frequency point and get the expected value of the N channels. Normalise the expected value to a maximum of 1.
4. Determine the coherence bandwidth for the current loading condition from the frequency band that exceeds the threshold value. The threshold value should be 0.5.

12.2.3 Test procedure for TRP measurement

The TRP measurement set-up is similar to the configuration of the reference PTF G_{ref} measurement. The configuration for the TRP test is shown in Figure 12.4. It can be found that the VNA is replaced by the base station emulator and power meter or spectrum analyser. The base station emulator is used to establish and

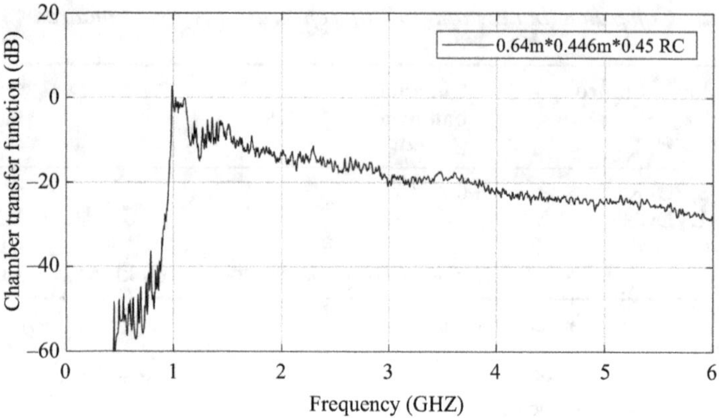

Figure 12.3 Measurement of chamber transfer function in RC

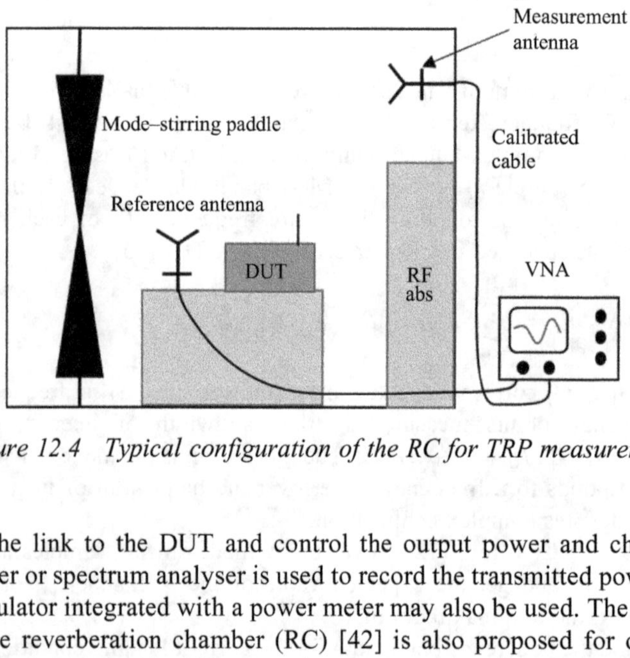

Figure 12.4 Typical configuration of the RC for TRP measurement

maintain the link to the DUT and control the output power and channel. The power meter or spectrum analyser is used to record the transmitted power. A base station emulator integrated with a power meter may also be used. The A compact multi-probe reverberation chamber (RC) [42] is also proposed for over-the-air (OTA) testing. Compared with conventional RCs equipped with limited receiving (Rx) antennas, 16 probe Rx antennas can measure more samples rapidly and enhance the testing efficiency. A vertical stirrer is also used to have a hybrid stirring. Total radiated power (TRP) and pattern correlation measurements have been performed to validate the RC system. The TRP measurement is performed as follows (Figure 12.5):

1. Conduct the reference PTF characterisation procedure as described in the previous section, including the determination of the reference and measurement antennas

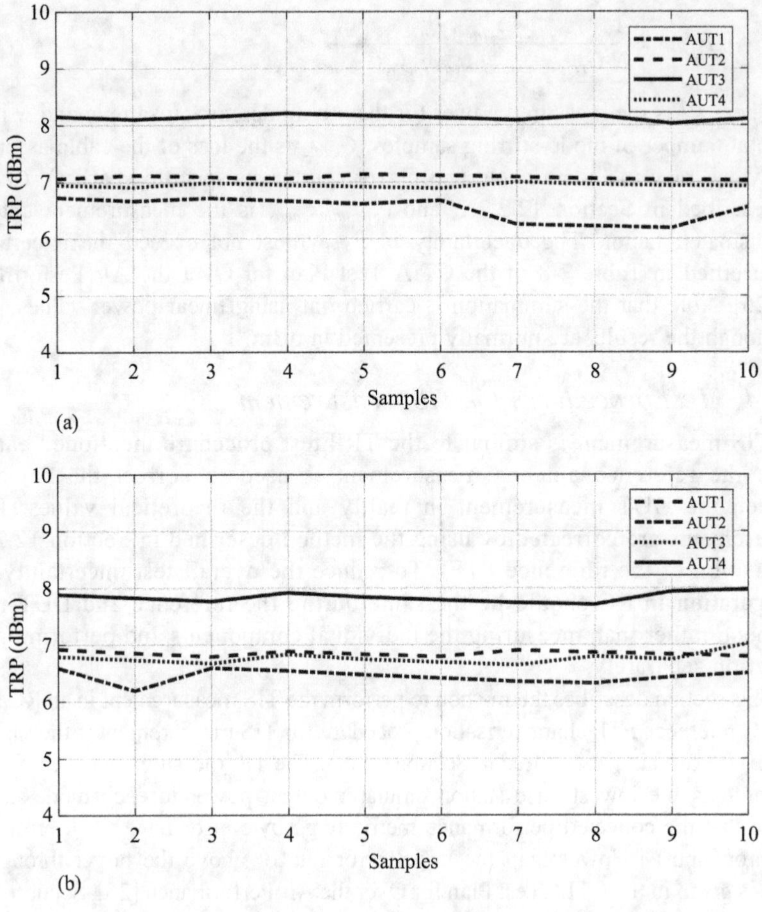

Figure 12.5 TRP for different AUTs calculated from (12.8) (transmitting power = 10 dBm) at (a) 2.4 GHz and (b) 5.18 GHz

mismatch and radiation efficiency, reference PTF, and the measurement of cables-assembly loss. The antenna used in the reference PTF measurement should be terminated with a 50-Ω load and should be still placed within the chamber. The DUT should be located within the valid test volume to ensure that the loading conditions are invariant.

2. Control the base station emulator to establish a link to the DUT and radiate at its maximum output power on the channel to be measured.
3. Using the power meter, spectrum analyser, or base station emulator to measure the power for each mode-stirring sample in the stirring sequence, which is the same sequence determined by the reference PTF characterisation procedure.
4. Calculate the TRP value, P_{TRP}, by taking an average value of all power samples and applying the reference PTF (see [3,15,27,43]):

$$P_{TRP} = \frac{1}{N} \frac{\sum_{n=1}^{N} P_n}{G_{ref} e_{mismatch,meas} \eta_{meas} G_{cable}} \tag{12.8}$$

where P^n is the measured power for the nth mode-stirring sample and N is the total number of mode-stirring samples, G_{cable} is the loss of the cable-assembly (which may include a cable and power splitter, as specified earlier), *Gref* is described in Section 12.2.2.1, and $e_{mismatch,meas}$ is the measurement antenna mismatch factor. The uncertainty in P_{TRP} must not exceed the uncertainty specified in Table 7-8 of the CTIA Test Plan for Over the Air Performance [26]. Note that the summation is carried out using linear power values, even though the results are normally presented in dBm.

12.2.4 Test procedures for TIS measurement

The TIS measurement is similar to the TRP test procedure mentioned earlier, where the reference antenna measurement is used to correct the deviation between the DUT measurement in reality and the theoretical values. DUT measurements are corrected by using the method described in Section 12.2.2.2 (calculation of the reference PTF). To reduce the overall test uncertainty, the configuration in RC should be the same during the reference and DUT measurement, rather than measuring the individual components and performing the correction separately.

This section describes the method to perform the TIS measurement in an RC [28]. The RC reference PTF characterisation procedure for TIS measurement is the same as for the TRP test and is specified in Section 12.2.3. The TIS measurement is carried out by searching the lowest base station emulator output power in each mode-stirring sample that the collected performance metric (e.g., byte error rate, block error rate, and throughput) is lower than the target error rate or above the target throughput which is given in the CTIA Test Plan for Over the Air Performance [26], Section 12.6. The TIS procedure is executed using the following steps:

1. The DUT and reference antenna should remain inside the chamber. The reference antenna should be terminated with a 50-Ω load. Conduct the reference PTF characterisation procedure described in Section 12.2.2.1.
2. Configure the DUT to the channel of interest and place it in the loopback mode to enable a digital error rate or throughput measurement, as applicable to the link technology under test.
3. Set the base station emulator to a specific output power and measure a digital error rate.
4. Increase or decrease the base station output power accordingly, and repeat Step 3 until the lowest output power (power step size should be no more than 0.5 dB when the RF power level is near the final declared sensitivity level) is found, which gives a lower digital error rate than the specified target bit error ratio (BER) as applicable.
5. Repeat Steps 3 and 4 for each of the mode-stirring samples in the measurement sequence.
6. Calculate the TIS value from the following equation [28]:

$$P_{TIS} = G_{ref} e_{mismatch,meas} \eta_{meas} G_{Cable} \left(\frac{1}{N} \sum_{n=1}^{N} \frac{1}{P_{BSS}(n)} \right)^{-1} \tag{12.9}$$

where P_{BSS} is the output power from the base station emulator when it is adjusted to give the specified digital error rate or throughput from the DUT for mode-stirring sample n, N is the total number of mode-stirring positions, G_{cables} is the loss in the cable assembly connecting the base station emulator to the measurement antenna, and G_{ref} is described in Section 12.2.2.1 and $e_{mismatch,meas}$ are the reference PTF and measurement antenna mismatch factor as found in the characterisation steps. The uncertainty in P_{TI} must not exceed the uncertainty specified in Table 7-8 of the CTIA Test Plan for Over the Air Performance [26]. Note that the summation is performed using linear power values, even though the results are normally presented in dBm.

12.2.5 *Analysis of measurement uncertainty in TRP and TIS measurements*

By utilising different stirring techniques (e.g., mechanical stirring, source stirring, and frequency stirring), the electromagnetic modes excited in the RC can be stirred and redistributed. As a result, various electromagnetic modes overlap at specific frequencies. Considering the characteristics of all electromagnetic modes, the averaged electromagnetic fields within the RC can be treated as uniform, isotropic, and homogeneous. Obviously, the uniformity of RC is presented from the viewpoint of statistics, resulting the RC-based measurements necessitate a large number of independent samples. Generally, the RC's stochastic property dominates the measurement uncertainty. In this subsection, we present the statistical analysis of the measurement uncertainty in TRP and TIS measurements.

12.2.5.1 Statistical uncertainty model of TRP

To exhibit the statistical analysis of the measurement uncertainty in TRP measurement in a clear and intuitive way, (12.8) is rewritten as

$$\widehat{P}_{TRP} = \frac{\langle P_n \rangle_N}{\langle |S_{21}(f,n)|^2 \rangle_{NF}} L \tag{12.10}$$

where P^n is the measured power for the nth mode-stirring sample and N is the total number of mode-stirring samples, $\langle \bullet \rangle$ represents the ensemble average of the samples, $L = e_{mismatch,ref} \eta_{ref} / G_{cable}$ is the loss factor and should be a constant for specific measurement configuration. P_{TRP} denotes the true TRP of the DUT, which should be a constant for a specific DUT. By comparing with $\langle |S_{21}(f,n)|^2 \rangle_{NF}$, $\widehat{P}_{TRP}$ denotes the estimated TRP based on measured results. This is a random variable (RV) and follows a certain distribution.

It should be noted that $\langle P_n \rangle / P_{TRP}$ represents the ratio of the averaged power received by the measurement antenna to the power transmitted by the DUT, which is the PTF of the chamber. In order to distinguish from the averaged PTF, $\langle |S_{21}(f,n)|^2 \rangle$, determined by the VNA, $\langle P_n \rangle / P_{TRP}$ is denoted as the equivalent averaged PTF, $\langle |S_{21,TRP}^{equivalent}(n)|^2 \rangle$. Therefore, (12.10) can be transformed into

$$\widehat{P}_{TRP} = P_{TRP} \frac{\langle \left| S_{21,\mathrm{TRP}}^{equivalent}(n) \right|^2 \rangle_N}{\langle |S_{21}(f,n)|^2 \rangle_{NF}} L \tag{12.11}$$

It should be stressed that (12.11) reflects the relationship between the measured and true TRPs, which is fundamental to the statistical analysis of measurement uncertainty. In an RC operating at over-moded state, both PTFs of the chamber, i.e., $X = |S_{21}(f,n)|^2$ and $Y = |S_{21,TRP}^{equivalent}(n)|^2$, follow the exponential distribution [29]:

$$f(t) = \frac{C_{RC}}{Q} \exp\left(-\frac{C_{RC}}{Q} t \right) \tag{12.12}$$

where $C_{RC} = 16\pi^2 V/\lambda^3$ with V being the inner volume of the RC and λ being the wavelength, and Q represents the quality factor of the RC. Suppose that the used N and F samples are obtained through independent mechanical stirring and frequency stirring; thus X_i ($i = 1, 2, \ldots, NF$) and Y_i ($i = 1, 2, \ldots, N$) should be independent and identically distributed (IID) RVs. If we denote $U = \sum_{NF} X_i$ and $V = \sum_N Y_j$, both U and V follow the gamma distribution [30]:

$$f(t) = \frac{A^M}{\Gamma(M)} t^{M-1} \exp(-At) \tag{12.13}$$

where $A = C_{RC}/2Q$, $M = NF$ for U, and $M = N$ for V. Γ represents the Gamma function. When M is an integer, $\Gamma(M) = (M-1)!$ with '!' denoting the factorial operator. By utilising the multivariate transformation, the distribution of RV $Z = V/U$ can be derived:

$$f(z) = \frac{\Gamma(N+NF)}{\Gamma(N)\Gamma(NF)} \frac{z^{N-1}}{(1+z)^{N+NF}} \tag{12.14}$$

Based on the distribution of RV Z, the statistics (i.e., expectation and variance) of RV $R = (NF/N) \times Z$ can be derived straightforwardly:

$$\begin{aligned} \mathrm{E}(R) &= \frac{NF}{NF-1} \\ \mathrm{VAR}(R) &= \frac{NF^2(N+NF-1)}{(NF-2)(NF-1)^2} \end{aligned} \tag{12.15}$$

where E and VAR denote the mathematical expectation and variance operator, respectively. It should be pointed out that $N_S \gg 1$ and $N_F \gg 1$ are satisfied in practical measurements; therefore, the cases of small NF are not considered here. Noting the relationship between $\widehat{P}_{TRP}$ and R, i.e., $\widehat{P}_{TRP} = P_{TRP} \times L \times R$, the expectation and variance of $\widehat{P}_{TRP}$ can be further calculated:

$$\begin{aligned} \mathrm{E}(\widehat{P}_{TRP}) &= \frac{NF}{NF-1} P_{TRP} L \\ \mathrm{VAR}(\widehat{P}_{TRP}) &= \frac{NF^2(N+NF-1)}{(NF-2)(NF-1)^2} (P_{TRP} L)^2 \end{aligned} \tag{12.16}$$

12.2.5.2 Statistical uncertainty model of TIS

Similar to the analysis of statistical uncertainty in TRP measurement, (12.9) is rewritten as

$$\widehat{P}_{TIS} = \frac{1}{L}\frac{\langle |S_{21}(f,n)|^2\rangle_{NF}}{\langle 1/P_{BSS}(n)\rangle_N} \tag{12.17}$$

where $\widehat{P}_{TIS}$ denotes the estimate of the true TIS of the DUT, P_{TIS}. Noting $\langle P_{TIS}/P_{BSS}\rangle$ represents the averaged PTF of the chamber, and following the same thought, we also denote it as the equivalent averaged PTF, $\langle |S_{21,TIS}^{equivalent}(n)|^2\rangle$. By doing so, (12.17) is transformed into

$$\widehat{P}_{TIS} = P_{TIS}\frac{\langle |S_{21}(f,n)|^2\rangle_{NF}}{\langle \left|S_{21,TIS}^{equivalent}(n)\right|^2\rangle_N}\frac{1}{L} \tag{12.18}$$

Note that both $\sum_{NF}|S_{21}(f,n)|^2$ and $\sum_N|S_{21,TIS}^{equivalent}(n)|^2$ follow the gamma distribution; hence, the expectation and variance of $\widehat{P}_{\mathrm{TIS}}$ can be derived in a similar way as the derivation of $\mathrm{E}(\widehat{P}_{TRP})$ and $\mathrm{VAR}(\widehat{P}_{TRP})$:

$$\begin{aligned}\mathrm{E}(\widehat{P}_{TIS}) &= \frac{N}{N-1}P_{TIS}\frac{1}{L}\\ \mathrm{VAR}(\widehat{P}_{TIS}) &= \frac{N(N+NF-1)}{F(N-2)(N-1)^2}\left(P_{TIS}\frac{1}{L}\right)^2\end{aligned} \tag{12.19}$$

It is worth stressing that the measured TRP is the arithmetic mean of the power samples, P_n, while the measured TIS is the harmonic mean of the power samples, P_{BSS}. Intuitively speaking, $\widehat{P}_{TRP}$ and $\widehat{P}_{TIS}$ should possess different distributions. However, by introducing the equivalent averaged PTF, it is interesting to find that $\widehat{P}_{TRP}$ and $\widehat{P}_{TIS}$ have an isomorphic distribution, i.e., a V/U type (where RVs U and V follow the gamma distribution).

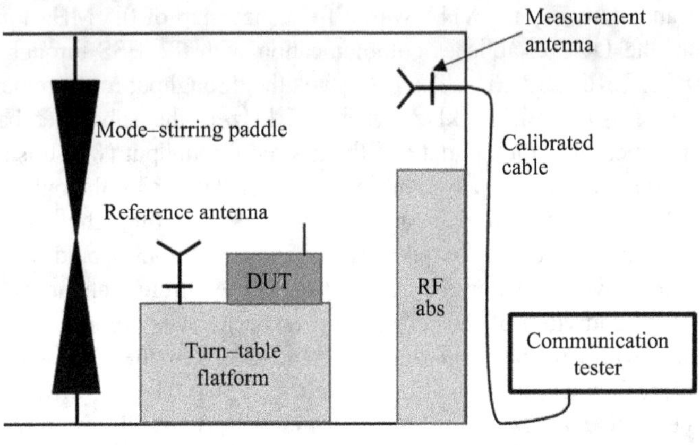

Figure 12.6 Measurement set-up for TRP and TIS measurements

12.2.5.3 Measurements and verifications

In order to validate the proposed statistical uncertainty models of TRP and TIS, extensive measurements are performed in an RC (as shown in Figure 12.6). Based on the measurement results, parameters required in the statistical models are determined, verifications are conducted as well.

Measurement set-up and procedures

The RC has an inner dimension of 1.46 m$\times$1.39 m$\times$0.80 m with a fundamental frequency of about 150 MHz. Two mechanical stirrers (one vertical and one horizontal) and one platform are equipped within the RC. A mobile phone (i.e., iPhone 7) is selected as the DUT. A double-ridged horn antenna and a wideband discone antenna are selected as the measurement antenna and reference antenna, respectively. The reference antenna and the DUT are placed on two trestles, whose height and orientation are adjustable. Both the trestles are located at the edge of the platform to achieve a better platform stirring. The measurement antenna is placed on a trestle located in the corner of the RC and oriented to the vertical stirrer. The same stirring sequence is adopted for both TRP and TIS measurements. Two mechanical stirrers rotate simultaneously stepwise to ten angles evenly distributed over a whole rotation. At each stirrer position, the platform rotates stepwise to ten angles evenly distributed over a whole rotation. In this way, total $10\times10 = 100$ stirring samples are obtained.

In the case of TRP measurements, for the reference measurement, both reference antenna and measurement antennas are connected to the VNA. At each mechanical stirring position, all the *S*-parameters covering the frequency range of 1.95–2 GHz are sampled and stored by the VNA with a frequency step of 0.5 MHz. For the DUT measurement, the BSS (i.e., R&S CMW500) establishes communication with the DUT through uplink channel 18550, i.e., 1.97–1.98 GHz. At each mechanical stirring position, the BSS controls the DUT to operate at its maximum power and records the received power of the measurement antenna.

In the case of TIS measurement, for the reference measurement, both reference antenna and measurement antenna are connected to the VNA. At each mechanical stirring position, all the *S*-parameters covering the frequency range of 2.14–2.19 GHz are sampled and stored by the VNA with a frequency step of 0.5 MHz for the DUT measurement; the DUT establishes communication with the BSS through downlink channel 550, i.e., 2.16–2.17 GHz. The BSS starts the throughput measurement with an output power of -75 dBm/15 kHz (where 15 kHz is the subcarrier bandwidth). Comparing the measured throughput and the desired throughput (which is selected as 95% of the maximum throughput here), increasing or decreasing the output power of the BSS until the minimum output power for the desired throughput is found and recoded. The procedure for all the mechanical stirring positions should be repeated.

In order to provide a better characterisation of the wireless channels within the RC, the testing bandwidth of the reference measurement is commonly larger than the signal bandwidth in the DUT measurement [26]. For the TRP and TIS measurements here, this condition is also satisfied, i.e., 50 MHz $>$ 10 MHz. The validation of the statistical model necessitates independent realisations of TRP and TIS; therefore, the well-known nine-case measurement method [15,26,31] is adopted. Specifically, the DUT measurement is repeated under nine different conditions: adjusting the trestle used to support the DUT to three heights and orienting the DUT to three directions at each height. To ensure the independence of the nine measurements, the distance between two adjacent heights (15 cm) is larger than half-wavelength at the lowest testing frequency and the three directions of the DUT are orthogonal to each other.

Determining the number of independent samples for uncertainty calculation

If we focus back on the derivations of the statistical models, it is found that the measured PTFs are supposed to be IID. Therefore, it is necessary to verify the independence and distribution of the measured PTFs in practice. We start from the determination of the number of independent samples, i.e., N and F. The ACF with adjusted threshold is adopted to determine N [32] as follows:

$$N = \frac{N_{meas}}{\delta(\rho(\delta) = \rho_0)} \tag{12.20}$$

where N_{meas} represents the total number of measured mechanical stirring samples; δ denotes the offset of the shifted samples when the autocorrelation coefficient (ρ) drops below the target threshold (ρ_0). The detailed calculations of mentioned autocorrelation coefficient and target threshold can be found in [31,32].

Considering $N_{meas} = 100$, the autocorrelation coefficient will be 0.2565 when the level of significance is selected as 1%. We applied the ACF with adjusted threshold to the measured data, calculation results shown that all the measured samples are independent, that is, $N = 100$ over the whole bandwidth of interest. Since the measurements at frequencies separated by a certain interval (at least one coherence bandwidth [19]) can be treated as uncorrelated approximately; thus F can be determined as [32]

$$F = \frac{BW^{tot}}{\langle B_C \rangle} \tag{12.21}$$

where BW^{tot} represents the total bandwidth of the reference measurement. $\langle B_C \rangle$ denotes the averaged coherence bandwidth over BW^{tot}, which can be determined following the procedure discussed in Section 12.2.2.2. For both TRP and TIS measurements, calculation results show that $\langle B_C \rangle = 2$ MHz; therefore, $F = 25$.

It is worth pointing out that orthogonal frequency division multiplexing (OFDM) technique provides frequency diversity, and the coherence bandwidth affects the frequency diversity order (i.e., the number of independent OFDM subcarrier channels) [32,33]. Different from the passive measurements, the intrinsic frequency stirring caused by the DUT should be taken into consideration when OFDM technique is used, such as the TIS measurement. The frequency diversity order can be estimated as [32,33]

$$N_I = \frac{BW^{OFDM}}{\langle B_C \rangle} \tag{12.22}$$

where BW^{OFDM} represents the bandwidth of the OFDM signal, $\langle B_C \rangle$ represents the averaged coherence bandwidth over BW^{OFDM}.

In the case of TIS measurement, considering the intrinsic frequency stirring in the DUT measurement, the total number of independent samples N should be modified as $N \times N_I$. In the case of TRP measurement, the single-carrier modulation is adopted to save the battery of the mobile terminals [33]. In other words, the OFDM technique is not utilised in the uplink communication. As a result, the total number of independent samples N should be unchanged. Note that the OFDM technique is not utilised in the passive measurements, such as the reference measurements in both TRP and TIS cases. Therefore, the total number of independent samples NF should also be unchanged.

Verification of field uniformity within testing area

Next, we will verify that whether the distributions of the PTF measured from different spatial positions are the same or not. The distribution parameter (p) of the exponential distribution may be calculated using [29]

$$p = \frac{\langle |S_{21}|^2 \rangle_N}{\eta_{ref}\eta_{meas}e_{ref}e_{meas}} \tag{12.23}$$

According to the CTIA standard for the RC-based OTA testing, the working volume is defined as a cylindrical area if there is a platform within the RC. Obviously, the nine spatial positions adopted here (i.e., 3 heights $\times$ 3 orientations) are within the working volume. Theoretically, the field uniformity within the working volume is perfect, while definitely imperfect in practice. Therefore, it is necessary to verify the uniformity.

For simplicity and without loss of generality, the distributions of three cases are considered, i.e., the distribution of samples from a single TRP measurement at $f = 1.975$ GHz, the distribution of samples from nine TRP measurements at $f = 1.975$ GHz, and the distribution of samples from nine TIS measurements at $f = 2.165$ GHz. Corresponding empirical distributions and exponential distributions with estimated p using (12.23) are shown in Figure 12.7, where p_{max} and p_{min} represent the maximum and minimum estimated p among the nine TRP (or TIS) measurements, respectively. The fitted curves of the empirical distribution are also plotted in the figures to give a clear and intuitive comparison between the empirical and estimated distributions.

The centre frequencies ($f = 1.975$ GHz for TRP and $f = 2.165$ GHz for TIS) are selected from the whole testing bandwidth to validate the distributions. As can be seen from Figure 12.7(a), there exist discrepancies between the empirical and estimated distributions, which is mainly due to the limited number of PTF samples ($N = 100$). However, the consistence of the general trends can still be observed. Note that the estimated distributions with p_{max} and p_{min} are very close; thus the distributions of different spatial positions vary little. Taking all the nine positions into consideration, good agreements between the empirical and estimated distributions can be seen from Figure 12.7(b) and (c). Therefore, it can be concluded that the uniformity within the working volume is reasonably good for both TRP and TIS measurements. Furthermore, the PTFs obtained through nine

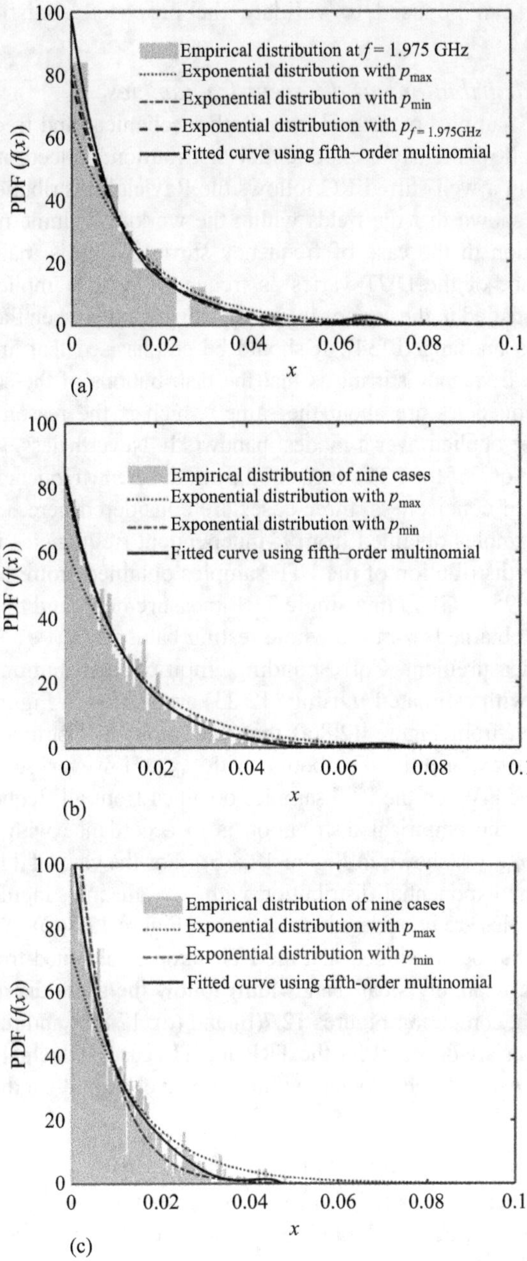

Figure 12.7 Empirical distribution, corresponding fitted curve using the fifth-order multinomial, and exponential distributions with estimated parameter p for the PTF samples obtained from (a) a single TRP measurement at $f = 1.975$ GHz; (b) nine independent TRP measurements at $f = 1.975$ GHz; (c) nine independent TIS measurements at $f = 2.165$ GHz

spatial positions can be used to validate the proposed statistical uncertainty models.

Verification of distributions at different frequencies

Note that the PTF samples measured from both mechanical and frequency stirrings are assumed to follow the same exponential distribution. Indeed, it is well known that the field within a well-stirred RC follows the Rayleigh distribution. On the other hand, it has been shown that the fields within the working volume may not be ideal Rayleigh distribution in the case of frequency stirring. This is mainly because the radiation impedance of the DUT varies as frequency, which implies Rayleigh distribution can be satisfied in the case of frequency stirring only when the radiated power of the DUT keeps unchanged [34]. It should be emphasised that an important prerequisite to utilise frequency stirring is that the distributions of the samples obtained from different frequencies are about the same, which is the reason why frequency stirring can only be applied over a modest bandwidth. Nevertheless, it is important to verify the distributions of the PTFs measured from different frequencies.

For the sake of conciseness, three cases are considered here, i.e., the distribution of the PTF samples obtained from F-independent frequencies in a single TRP measurement, the distribution of the PTF samples obtained from the whole testing bandwidth (i.e., 1.95–2 GHz) in a single TRP measurement, and the distribution of the PTF samples obtained from the whole testing bandwidth (i.e., 2.14–2.19 GHz) in a single TIS measurement. Corresponding empirical distributions and exponential distributions with estimated p using (12.23) are shown in Figure 12.8.

It can be seen from Figure 12.8(a) that the empirical distribution shows good agreement with the exponential distribution with $p_{\min}$. However, some discrepancies can also be observed. When the PTF samples obtained from all frequencies are taken into consideration, the empirical distribution is in accordance with the exponential distribution with $p_{\min}$ (as shown in Figure 12.8(b)). For the case of TIS, the empirical distribution and the exponential distribution with $p_{\min}$ are indistinguishable when all the frequency samples are considered (as shown in Figure 12.8(c)).

Therefore, it can be concluded that the PTF samples obtained from different frequencies (within a relatively small bandwidth) follow the same distribution approximately. In addition, comparing Figures 12.7(b) and (c), 12.8(b) and (c), it can be seen that the distributions are different for the TRP and TIS cases, which also indicates that frequency stirring can only be adopted within a relatively small bandwidth.

12.2.5.4 Evaluation of measurement uncertainty

The mean of the measured samples offers a better estimation than a single sample does; thus the sample mean is adopted to estimate the TRP and TIS. Corresponding standard uncertainty can be calculated as

$$u_S = \sqrt{\frac{\sum_{i=1}^{M}\left(P_i - \langle P_i \rangle_M\right)^2}{M(M-1)}} \tag{12.24}$$

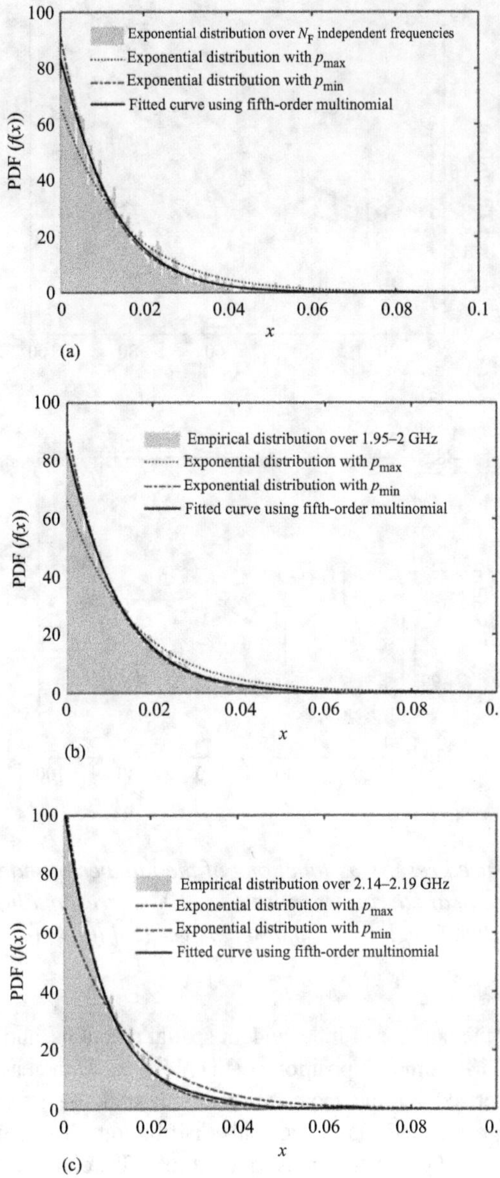

Figure 12.8 Empirical distribution, corresponding fitted curve using the fifth-order multinomial, and exponential distributions with estimated parameter p for the PTF samples obtained from (a) f-independent frequencies in a single TRP measurement; (b) the whole testing bandwidth (i.e., 1.95–2 GHz) in a single TRP measurement; (c) the whole testing bandwidth (i.e., 2.14–2.19 GHz) in a single TIS measurement

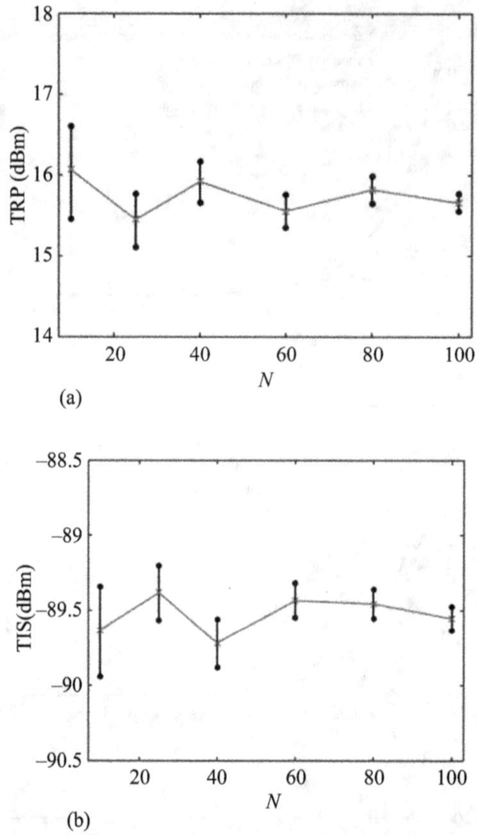

Figure 12.9 Measured results as functions of the number of independent mechanical stirring samples (i.e., N). Corresponding 1 sigma confidence regions are marked as vertical lines: (a) TRP and (b) TIS.

where M represents the number of independent spatial positions, and P_i represents the measured TRP (or TIS) of the ith position ($i \in [1, M]$). As M increases to infinity, the distribution characterised by the expectation, $\mu(x)$, and the standard uncertainty, $u_S(x)$, can be approximated as a normal distribution. Therefore, the interval $[\mu(x) - u_S(x), \mu(x) + u_S(x)]$ corresponds to about 68.27% confidence level [35].

The measured TRP and TIS as functions of N are shown in Figure 12.9(a) and (b), respectively. Corresponding 1 sigma confidence regions are also marked in the figures. As N increases, the standard uncertainties decrease for both TRP and TIS cases. At the same time, the measured TRP and TIS tend to approach a specific value, i.e., the true TRP and TIS.

Note that the statistics of the measured TRP and TIS are functions of the corresponding true values and the loss factor (as shown in (12.16) and (12.19)). In order to validate the statistical uncertainty models, the relative standard deviation

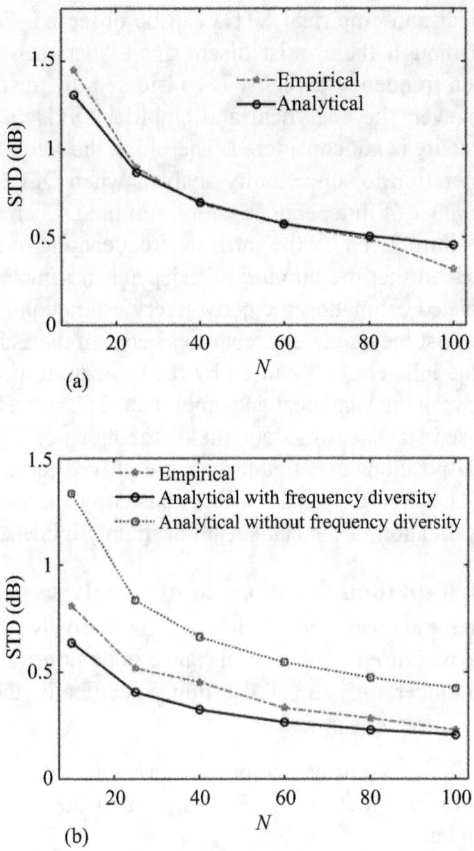

Figure 12.10 Analytical and empirical relative uncertainties as functions of the number of independent mechanical stirring samples (i.e., N) for (a) TRP and (b) TIS

(STD) is introduced and utilised:

$$\sigma_{TRP}^{relative} = \sqrt{\frac{N + NF - 1}{N(NF - 2)}}, \quad \sigma_{TIS}^{relative} = \sqrt{\frac{N + NF - 1}{NF(N - 2)}} \tag{12.25}$$

Corresponding empirical STD can be calculated using

$$\sigma_{emp} = \sqrt{\frac{\sum_{i=1}^{M} (\widetilde{P}_i - \langle \widetilde{P}_i \rangle_M)^2}{M - 1}} \tag{12.26}$$

where $\widetilde{P}_i$ represents the normalised TRP (or TIS) of the ith position ($i \in [1, M]$). Both empirical and analytical STDs of TRP and TIS are shown in Figure 12.10(a) and (b), respectively. The linear uncertainties are transformed into the dB form using $\sigma_{dB} = 10\log_{10}(1 + \sigma)$. For the case of TRP measurement, good agreements between the analytical and empirical STDs can be observed. For the case of TIS measurement, even though there exist discrepancies between the analytical and empirical STDs when frequency diversity is considered, the discrepancy decreases as N increases. However, the analytical and empirical STDs are totally different when frequency diversity is not considered. Therefore, the frequency diversity must be taken into consideration for uncertainty analysis when OFDM technique is utilised. That is, the number of independent samples in the DUT measurement should be modified with N_I introduced by the intrinsic frequency stirring of the DUT.

It should be stressed that the number of independent samples (N, F, and N_I) is determined by associated estimation methods. Every estimation method has its own accuracy; thus there exist inevitable discrepancies between the estimated and the true values. Obviously, the inherent bias caused by the estimation method contributes to the inconformity between the analytical and empirical STDs as well. However, before validating the proposed statistical models, the distributions of the samples obtained from different spatial positions and frequencies are proved to be approximately the same. Therefore, it is believed that the effect caused by the estimation method is insignificant in the validation of the statistical uncertainty models.

12.2.5.5 A unified method for uncertainty analyses

In order to exhibit the statistical method clearly and intuitively, we take the analysis of the uncertainty of measured TRP as an instance to demonstrate the procedure of statistical analysis of uncertainty in OTA testing. Specifically, the whole procedure is composed of the following steps:

1. Transform the single measured nth power parameter, P_n, into the form of the ratio of two power parameters, P_n/P_{TRP}, and treat the ratio as the equivalent PTF of the chamber.
2. Extract the relationship between the measured metric, $\widehat{P}_{TRP}$, and corresponding true value, P_{TRP}.
3. Derive the distribution of the expression obtained in Step 2, and further derive the corresponding statistics based on the distribution.
4. Establish the RC-based measurement set-up, determine the parameters required in the statistical model based on the RC configuration. Verify the distributions of the measured samples that assumed to obey certain rule in the derivations.
5. Validate the derived distribution and associated statistics.

Generally, there exist parameters of measured power in the original estimation model. Obviously, statistical analysis cannot be proceeded with its current form. By introducing the true value of the metric to be determined, the power parameter can be transformed into the ratio of two power parameters. At the same time, the relationship between the measured and true metrics is obtained, which is the basis of statistical uncertainty analysis. Utilising the well-known distribution of the field within the RC and the basis of probability theory, the distribution of the expression obtained in Step 2 can be derived, as well as associated statistics. However, the derivation of distributions and statistics sometimes may be a rather tough task. In these situations, proper approximation methods can be adopted to simplify the original expressions. Combining with the basis of probability theory and mathematics, it is believed that the statistical model can be finally obtained. It should also be stressed that every assumption in the derivation should be well verified in practical measurements, e.g., the IID property of the used samples, which ensures that the measured data are in accordance with the theoretical prerequisite.

Theoretically, the statistical analysis method can be applied to any metrics in OTA testing. With the help of the statistical models, the statistical uncertainty for a given chamber set-up can be easily determined. Moreover, for a desired uncertainty, the minimum number of mechanical stirring samples can also be determined conveniently. In summary, the unified statistical method significantly improves the efficiency, flexibility, and convenience of OTA testing, which is useful and important to the wireless industry.

12.2.6 Uncertainty of spatial uniformity

In an ideal RC, the average transmission between two antennas is independent of the location or orientation of the antennas within the test volume. The uncertainty caused by the lack of spatial uniformity represents how much the actual RC environment deviates from the ideal. Because this uncertainty may need a large effort to estimate, the measurement is performed based on the pre-characterisation test campaign separate from the DUT measurement.

A major factor determining this uncertainty is the loading condition of the chamber. In a pre-characterisation measurement, the RF absorber and antennas should be placed in the chamber where they will be located in the same position as those in DUT TRP/TIS measurement. To simulate an actual DUT, an RF absorber whose dimension is equal or greater than those of the maximum-size DUT should be placed in the location of the DUT. The number and position of all RF absorbers should be recorded to reproduce the condition in the DUT test. Pre-characterisation measurements are made for multiple loading conditions representing the expected range of loading under actual conditions.

The pre-characterisation procedure consists of measuring the PTF, chamber loss, coherence bandwidth, and uncertainty due to a lack of chamber spatial uniformity under various loading conditions. Before moving the reference antenna to the next position, complete all measurements under different loading conditions for each reference antenna position and orientation This operation ensures that the reference antenna position is the same for every loading condition. These data should be stored for the calculation of the RC.

1. Place the reference antenna at one of four uncorrelated locations and each location has three orthogonal orientations. The number of reference antenna measurement samples, T_{pre}, is set as 12. For chambers with turntables, the reference antenna should be placed at $r = R_{\min}$, $r = R_{\max}$, and two heights corresponding to the $z = 0$ and $z = Z_{\max}$ of the test volume. For chambers without turntables, the reference antenna should be placed at the two opposite corners of a cube at the bottom ($z = 0$) of the test volume and the other two corners at the top ($z = R_{\max}$) of the test volume, respectively.

2. The loading placement should meet the requirement for coherence bandwidth. It is general to do this by loading the chamber with the increasing amount of an RF absorber.
3. For a specific loading condition, perform *S*-parameter measurement over the frequency band of interest. The loading conditions relate to the number of blocks of an RF absorber or the exposed area of the RF absorber ranging from zero (or a minimum amount) to the maximum in approximately equal increments. The finer the loading increment is measured, the more accurately uncertainty can be estimated.
4. For each loading condition, calculate the coherence bandwidth as specified in Section 12.2.2.2.
5. For each loading condition i, calculate the reference PTF G^i_{ref}, as reported in Section 12.2.2.1. Using (12.5) with a minimum of $T_{pre} = 12$ reference measurement samples are described in Step 1. This calculation should be operated at the centre frequency for each channel to be tested, averaging the measured frequencies over the channel bandwidth shown in Table 12.1 for each link technology to be supported by the chamber.
6. For each loading condition i, calculate the standard deviation, $\sigma_{G^i_{ref}}$, from the following equation:

$$\sigma_{G^i_{ref}} = \sqrt{\frac{1}{(T_{pre} - 1)} \sum_{t=1}^{T_{pre}} \left(G^i_{ref,t} - G^i_{ref} \right)^2} \tag{12.27}$$

12.3 Pattern correlation

This section will calculate the radiation pattern correlation of two different antennas under test (AUTs), namely, AUT1 and AUT2. The configuration for the pattern correlation set-up is shown in Figure 12.11. A measurement antenna, RF absorber, and AUT are placed within the chamber. *S*-Parameters and the received power are tested using the base station emulator or the power meter. Compared with TRP measurement, the pattern correlation measurement does not need a reference antenna. According to the conclusion given in [25], the angular correlation can be defined over the measured received power, measured complex *S*-parameters. Therefore, the user can calculate the pattern correlation coefficients by the power which the measurement antenna received.

The pattern correlation measurement is performed as follows:

1. Configure the chamber to measure the TRP as specified in Figure 12.11. Place the AUT1, RF abs, and measurement antennas that will be used in the pattern correlation measurement into the RC.
2. Place the AUT1 into the valid test volume of the chamber. Using the power meter, spectrum analyser, or base station emulator, the measured radiated

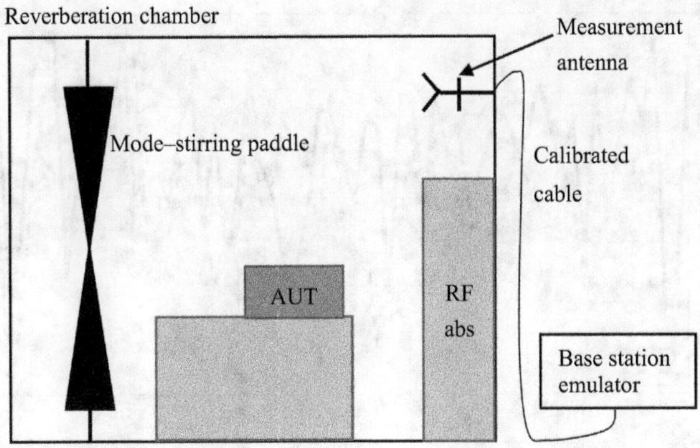

Figure 12.11 Typical configuration of RC for pattern correlation measurement

powers for each mode-stirring sample in stirring sequence can be obtained, which is $P_{11}, P_{12}, \ldots, P_{1n}$ (n represents mode-stirring numbers).

3. Replace the AUT1 with the AUT2 in the same position. Rotate the mode-stirring paddle as the same sequence determined by the AUT1 procedure. The measured radiated powers $P_{21}, P_{22}, \ldots, P_{2n}$ can be received.
4. Calculate the value of radiated pattern correlation R by taking an average of all power samples and formula as in the following equation:

$$R = \frac{\frac{1}{n-1}\sum_i^n (P_{1i} - \langle P_1 \rangle)(P_{2i} - \langle P_2 \rangle)}{\sqrt{\left(\frac{\sum_i^n (P_{1i} - \langle P_1 \rangle)}{n-1}\right)^2}\sqrt{\left(\frac{\sum_i^n (P_{2i} - \langle P_2 \rangle)}{n-1}\right)^2}} \quad (12.28)$$

where P_n is the measured power for the nth mode-stirring sample and n is the total number of mode-stirring samples; $\langle P_1 \rangle$ and $\langle P_2 \rangle$ are the average of the AUT1 and AUT2 total mode-stirring samples, respectively. Note that the results are normally presented in dBm (Figure 12.12).

12.4 Antenna efficiency

The antenna efficiency generally includes the radiation efficiency and the total radiation efficiency. For the quantity of radiation efficiency, the IEEE standard definition of terms can be referred to which describe the quantity [36] as the ratio of the TRP P_{rad} to the net power accepted by the antenna $P_{accepted}$ at its input terminals:

$$\eta_{rad} = \frac{P_{rad}}{P_{accepted}} \quad (12.29)$$

The quantity of radiation efficiency describes the ohmic losses that exist within a given antenna structure; losses that are due to factors such as the material

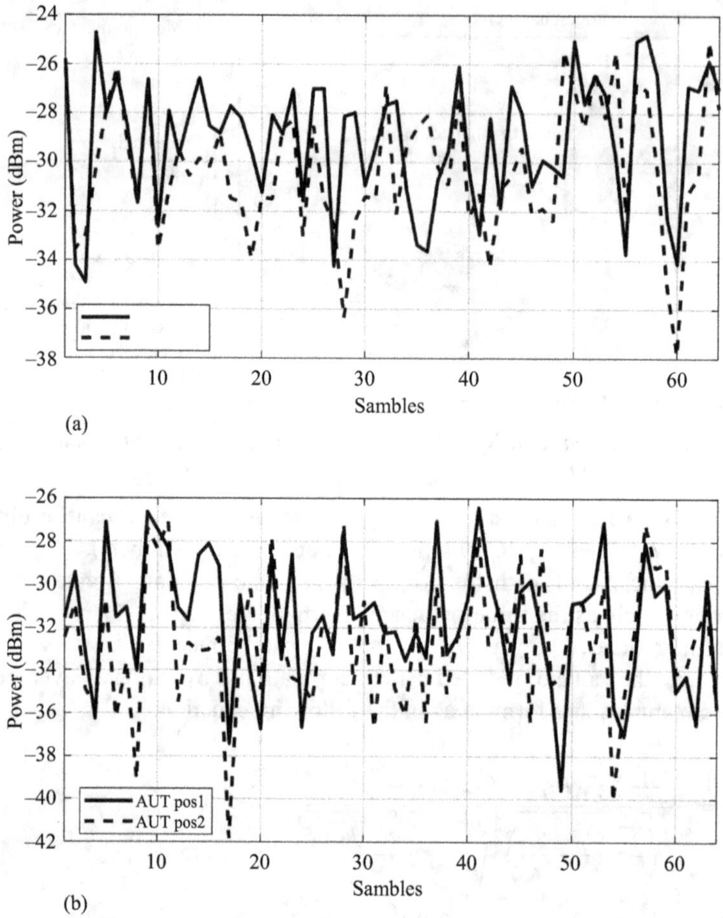

Figure 12.12 Pattern correlation between different positions at (a) 2.4 GHz (R = 0.5082) and (b) 5.18 GHz (R = 0.6786)

characteristics of the antenna. For the quantity of total radiation efficiency, the IEEE standard definition of terms is again referred to the ratio of TRP to the power incident on the antenna port:

$$\eta_{total} = \eta_{rad}(1 - |S_{22}|^2) \tag{12.30}$$

There are many methods to measure antenna radiation efficiency, such as pattern integration [37], Wheeler cap [38], sliding wall cavity [37], ultra-wideband Wheeler cap [39,40], calorimetric [41], and RC methods. This section mainly describes the method to calculate the antenna efficiency using the RC. Standard techniques to measure the efficiency of antennas require a reference antenna which has known performance characteristics. The reference antenna method is described in IEC 6100-4-21 [3]. In recent years, the technique measuring the efficiency without the reference antenna is derived [4], as specified in Section 12.4.2.

12.4.1 Reference antenna method to measure the antenna efficiency

The power transmitted from VNA Port 1 (defined as P_{Source}) will first encounter the reflection by the transmitting antenna (due to impedance mismatch) with the reflection coefficient S_{11} and radiation efficiency η_F. Therefore, the power radiated into the chamber (neglecting the cable and connector loss for the whole system) is

$$P_{Chamber} = P_{Source}\left(1 - |S_{11}|^2\right)\eta_F \tag{12.31}$$

Part of the $P_{Chamber}$ reached the receiving antenna and can be expressed as

$$P_{input} = P_{Chamber} - P_{Loss} \tag{12.32}$$

where P_{Loss} is the loss of power due to the chamber ohmic loss and power spreading.

The received power with the reference antenna used as the transmitting antenna is measured as P_{ref}, and when the reference antenna is replaced by the antenna under test (AUT as shown in Figure 12.13), the measured received power is referred to P_{AUT}:

$$P_{Ref} = P_{in}\eta_{Ref}\left(1 - |S_{22,Ref}|^2\right) \tag{12.33}$$

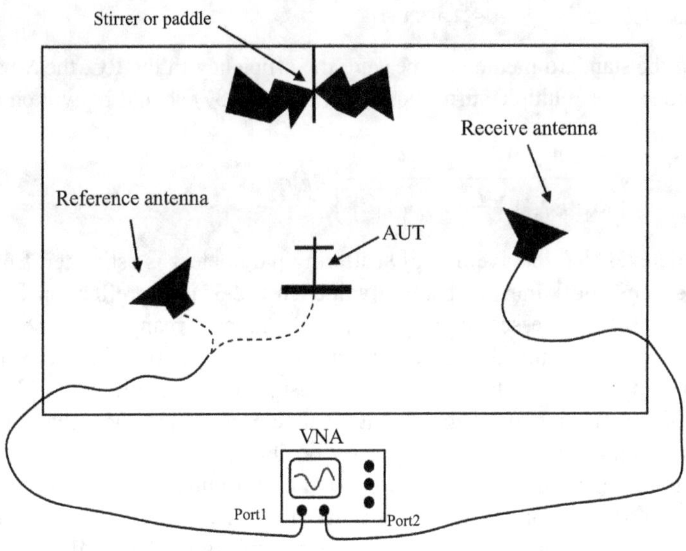

Figure 12.13 RC set-up for the measurement of the efficiency for an AUT using a reference antenna

$$P_{AUT} = P_{in}\eta_{AUT}\left(1 - |S_{22,AUT}|^2\right) \tag{12.34}$$

The received power is assumed to be the same for the power transmitted from either reference antenna or AUT, and it is partial loss due to antenna efficiency and partially reflected because of the mismatch between the antenna and connector. Hence, the power transmitted from the VNA Port 1 to Port 2 is

$$|S_{21,Ref}|^2 = \frac{P_{Ref}}{P_{Source}} = \frac{P_{in}\eta_{Ref}\left(1 - |S_{22,Ref}|^2\right)}{P_{Source}} \tag{12.35}$$

And

$$|S_{21,AUT}|^2 = \frac{P_{AUT}}{P_{Source}} = \frac{P_{in}\eta_{AUT}\left(1 - |S_{22,AUT}|^2\right)}{P_{Source}} \tag{12.36}$$

Dividing (12.35) by (12.36) will get

$$\eta_{AUT} = \left\{\frac{|S_{21,AUT}|^2(1 - |S_{22,Ref}|^2)}{|S_{21,Ref}|^2(1 - |S_{22,AUT}|^2)}\right\}\eta_{Ref} \tag{12.37}$$

The previous derivation means that in addition to the *S*-parameter, the reference antenna that has known performance is needed for the calculation of the radiation efficiency. It can be also found that there is no need for the calibration of the transmitting antenna and the chamber loss will not affect the final result.

For the quantity of the total radiated efficiency, it is a product of the radiation efficiency and mismatch efficiency of an antenna:

$$\eta_{Total} = \eta_{AUT} \times \left(1 - |S_{22,AUT}|^2\right) \tag{12.38}$$

Using the standard method to measure the efficiency in the RC, the *S*-parameter is the average over multiple stirrer positions; thus (12.37) should be written as

$$\eta_{AUT} = \left\{\frac{|S_{21,AUT}|^2\left(1 - |S_{22,Ref}|^2\right)}{|S_{21,Ref}|^2\left(1 - |S_{22,AUT}|^2\right)}\right\}\eta_{Ref} \tag{12.39}$$

where $\langle\rangle$ denotes that the average of scattering parameters is estimated from many stirrer positions and $||$ means absolute value. In (12.37), the reflection coefficient quantities are not expressed as an ensemble average of many stirrer positions. In this form, they are assumed to be acquired in an AC. It is possible to measure the reflection coefficient of antennas accurately using the RC as shown in [38], where the reflection coefficient quantities in (12.39) are preferred to show that they would be formed as an average from many stirrer positions.

As can be seen, this method is direct and easy to understand. But the need for the reference antenna in practice may cause a problem: a reference antenna with known efficiency over the frequency of interest cannot be used or the actual quantity of this efficiency might have changed since the last calibration. Thus, a revised method without a reference antenna is needed.

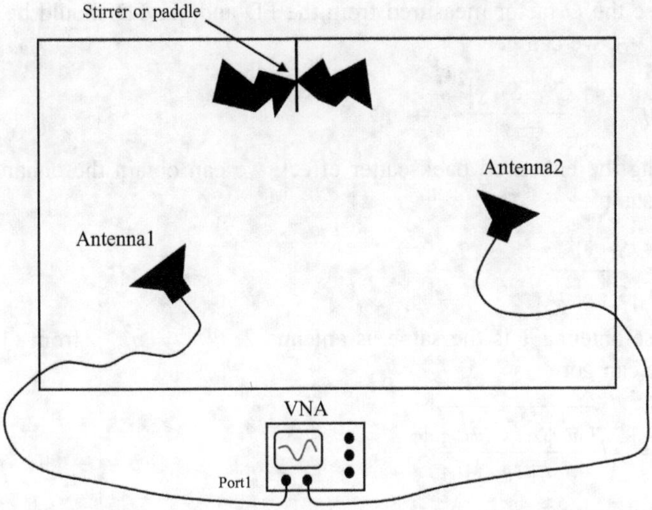

Figure 12.14 Measurement configuration for three-antenna approach within the RC

12.4.2 Revised method to measure efficiency without reference antennas

Holloway *et al.* developed new methods to measure the radiation efficiency of an antenna without using a reference antenna. The RC configuration is shown in Figure 12.14. These methods are based on the Q factor measured from the FD (frequency domain) and the TD (time domain) and utilise the enhanced backscatter effect. The detailed methods consist of three different approaches: one antenna, two antenna, and three antenna.

In the FD, the Q factor can be calculated from Hill's equation [4]:

$$Q_{FD,cor} = \frac{C_{RC}\left|S_{21,S}\right|^2}{\eta_{total}^1 \eta_{total}^2} \tag{12.40}$$

where C_{RC} is a constant:

$$C_{RC} = \frac{16\pi^2 V}{\lambda^3} \tag{12.41}$$

where V is the volume of the RC, λ is the wavelength in free space. $Q_{FD,cor}$ means that the Q factor is corrected given the total radiation efficiency of antennas, and $S_{21,S}$ is stirred part of the S-parameter defined in the following equation:

$$S_{*,s} = S_* - S_* \tag{12.42}$$

The Q factor can be also obtained from the TD [4]:

$$Q_{TD} = \varpi \tau_{RC} \tag{12.43}$$

where τ_{RC} is the chamber decay constant which can be extracted from the TD response of the chamber.

Because the Q factor measured from the FD and the TD should be the same, $Q_{FD,cor} = Q_{TD}$, we can get

$$\eta^1_{total}\eta^2_{total} = \frac{C_{RC}|S_{21,S}|^2}{\varpi\tau_{RC}} \tag{12.44}$$

Utilising the enhanced backscatter effect, we can obtain the enhanced backscatter constant:

$$e_{b1} = \frac{|S_{11,S}|^2}{|S_{21,S}|^2} \tag{12.45}$$

Suppose antenna 1 is the same as antenna 2, $\eta^1_{total} = \eta^2_{total}$, from (12.44) and (12.45), we can get

$$\eta^1_{total} = \sqrt{\frac{C_{RC}|S_{11,S}|^2/e_{b1}}{\varpi\tau_{RC}}} \tag{12.46}$$

For a well-stirred RC, $e_{b1} = 2$, thus

$$\eta^1_{total} = \sqrt{\frac{C_{RC}|S_{11,S}|^2/2}{\varpi\tau_{RC}}} \tag{12.47}$$

And the radiation efficiency of antenna 1 can be calculated from the following equation:

$$\eta^1_{rad} = \frac{\eta^1_{total}}{1-|S_{11}|^2} = \frac{1}{1-|S_{11}|^2}\sqrt{\frac{C_{RC}|S_{11,S}|^2/2}{\varpi\tau_{RC}}} \tag{12.48}$$

From the results, we can find that the use of antenna 1 is enough to finish the measurement, other antennas are not needed, and (12.47) and (12.48) are formulas for the one-antenna approach [4]. If we do not make the assumption that antenna 1 and antenna 2 are identical, the antenna efficiency can be obtained from the two-antenna approach.

The two-antenna approach does not require $e_b = 2$ (for the one-antenna approach), it does require to be identical for antennas placed in different positions within the chamber, $e_{b1} = e_{b2} = e_b$:

$$e_{b1} = \frac{|S_{11,S}|^2}{|S_{21,S}|^2} \tag{12.49}$$

$$e_{b2} = \frac{|S_{22,S}|^2}{|S_{21,S}|^2} \tag{12.50}$$

We can get

$$e_b = \frac{\sqrt{|S_{11,S}|^2|S_{22,S}|^2}}{|S_{21,S}|^2} \tag{12.51}$$

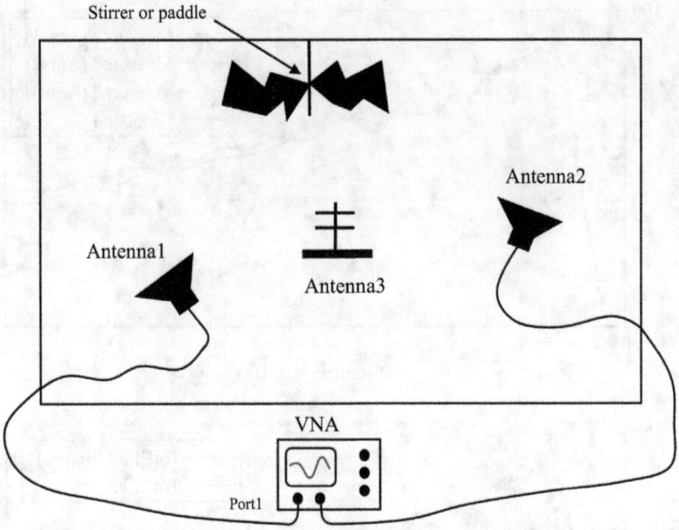

Figure 12.15 Measurement configuration for three-antenna approach within the RC

From (12.51) applied to antenna 1 and antenna 2:

$$\eta_{total}^{1} = \sqrt{\frac{C_{RC}|S_{11,S}|^2/e_{b1}}{\varpi\tau_{RC}}} = \sqrt{\frac{C_{RC}|S_{11,S}|^2/e_b}{\varpi\tau_{RC}}} \tag{12.52}$$

$$\eta_{total}^{2} = \sqrt{\frac{C_{RC}|S_{22,S}|^2/e_{b2}}{\varpi\tau_{RC}}} = \sqrt{\frac{C_{RC}|S_{22,S}|^2/e_b}{\varpi\tau_{RC}}} \tag{12.53}$$

which are the formulae for the two-antenna method [4]. Both the previous approaches make assumptions for enhanced backscatter constant. One-antenna method requires $e_{b1} = 2$, and the two-antenna method requires the enhanced backscatter constant be the same for antennas placed in different positions in RC. By using the three-antenna approach, as shown in Figure 12.15, these assumptions are not needed.

Suppose we have three antennas and no limit applied on e_{b1}, e_{b2}, and e_{b3}. From (12.44), we have a system of equations:

$$\begin{aligned} \eta_{total}^{1}\eta_{total}^{2} &= \frac{C_{RC}|S_{21,S}|^2}{\varpi\tau_{RC}} \\ \eta_{total}^{2}\eta_{total}^{3} &= \frac{C_{RC}|S_{32,S}|^2}{\varpi\tau_{RC}} \\ \eta_{total}^{1}\eta_{total}^{3} &= \frac{C_{RC}|S_{31,S}|^2}{\varpi\tau_{RC}} \end{aligned} \tag{12.54}$$

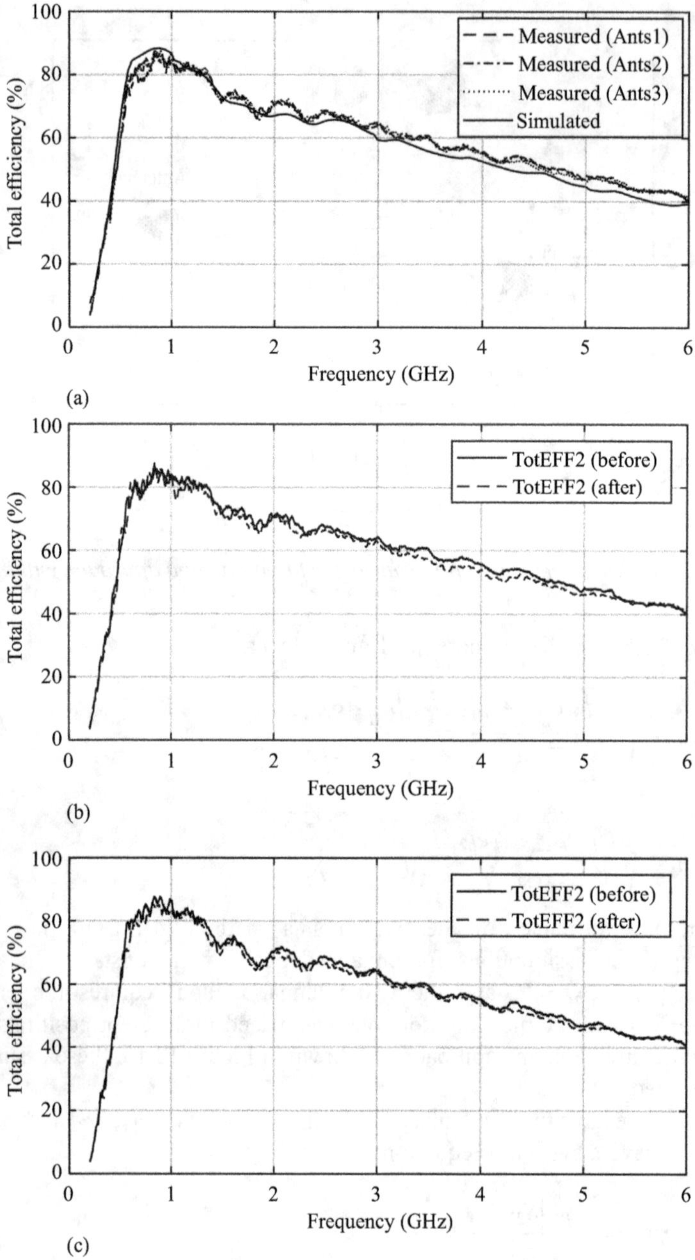

Figure 12.16 (a) Antenna efficiency measured from three-antenna method; (b) comparison between two measurements for antenna 1; (c) comparison between two measurements for antenna 2; (d) comparison between two measurements for antenna 3; (e) absolute error for antennas in measurements using three-antenna method

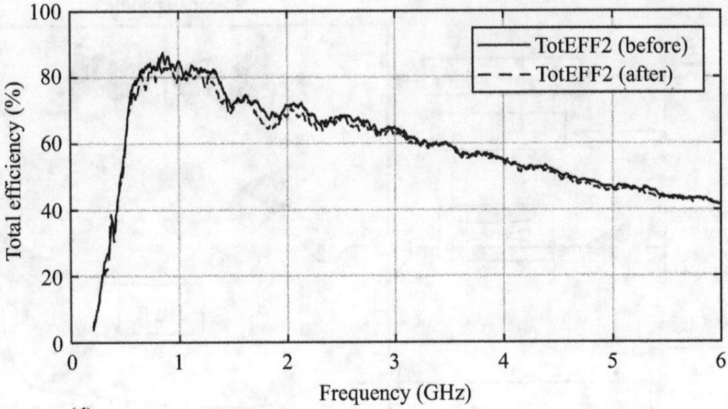

(d)

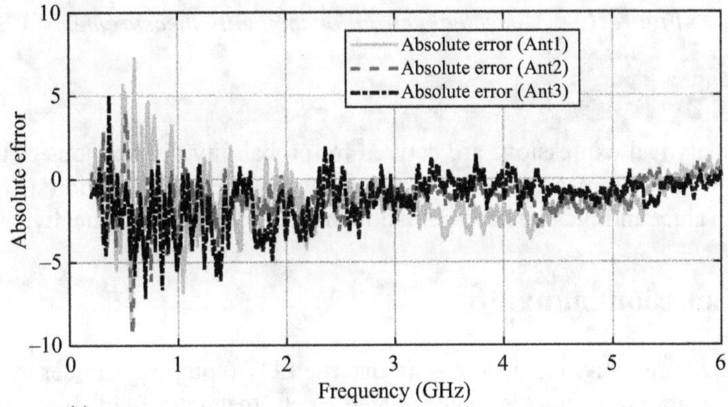

(e)

Figure 12.16 (*Continued*)

The total efficiency can be calculated as follows:

$$\begin{aligned}
\eta_{total}^{1} &= \sqrt{\frac{C_{RC}\left|S_{21,S}\right|^{2}\left|S_{31,S}\right|^{2}}{\varpi\tau_{RC}\left|S_{32,S}\right|^{2}}} \\
\eta_{total}^{2} &= \sqrt{\frac{C_{RC}\left|S_{21,S}\right|^{2}\left|S_{32,S}\right|^{2}}{\varpi\tau_{RC}\left|S_{31,S}\right|^{2}}} \\
\eta_{total}^{3} &= \sqrt{\frac{C_{RC}\left|S_{31,S}\right|^{2}\left|S_{32,S}\right|^{2}}{\varpi\tau_{RC}\left|S_{21,S}\right|^{2}}}
\end{aligned} \tag{12.55}$$

which are the formulae for the three-antenna method (Figure 12.16) [4]. The radiation efficiency of antennas can be obtained in a similar way from (12.48).

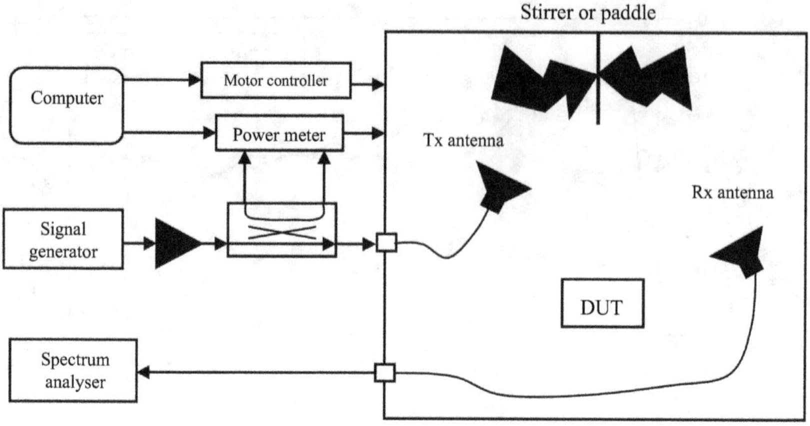

Figure 12.17 Typical set-up for immunity measurement

In [43], analytical expressions are derived for probability distributions of the three antenna efficiency measurement. The probability density function (PDF), the expected value, and the standard deviation are also derived analytically.

12.5 Emission immunity

The radiated immunity test is to ensure that the EUT (equipment under test) works normally when placed in a strong radiated electromagnetic field. Because of the high-quality factor of RC, it is easy to generate high E-field. Thus, it is suitable to estimate radiation immunity in RC. A schematic plot of the radiated immunity test is shown in Figure 12.17. The RC contains a transmitting antenna (Tx antenna), a receiving antenna (Rx antenna), and an EUT.

12.5.1 Immunity test set-up

The EUT should be placed at least $\lambda/4$ from the chamber walls at the lowest usable frequency of the RC. For the EUTs designed for desktop, the distance between the EUT and chamber floor should be at least $\lambda/4$. For the standing EUTs, they should be supported 10 cm above the floor. The support equipment should be non-metallic and low-absorbing.

The chamber should not contain any unnecessary absorbing materials, such as wooden tables, coverings for walls and floor, and ceiling tiles. For a new chamber, an assessment of the chamber is proposed to perform before installing any other support equipment than doors, vents, and access panels. The EUT and all the supporting equipment should not occupy more than 8% of the total chamber volume. The transmit antenna should not point to the EUT or the receive antenna directly. It is recommended to direct the antennas into the corner of RC.

12.5.2 Selecting the frequency sweep/intervals

A computer is used to control the measurement and collect the input power for different stirring positions. To meet the test requirements for the immunity test, we should change the chamber input power P_{input} to obtain the desired electric-field intensity using the following equation:

$$P_{input} = \left[\frac{E_{Test}}{E_{24\ or\ 9}\sqrt{CLF}}\right]^2 \tag{12.56}$$

where P_{input} is the power injected into the chamber to achieve the desired field strength for the immunity test, $E_{24\ \mathrm{or}\ 9}$ is the average of the normalised E-field (average value or peak value) obtained from the empty chamber validation, $CLF(f)$ is the chamber loading factor at frequency f, and E_{Test} is the expected field strength for the immunity test.

Considering the response time of EUT and monitoring equipment, the frequency sweep or step rate should be justified. The IEC standard [3] specifies the number of frequency points needed in the test, which should be no less than 100 per decade. The test frequencies should be logarithmically stepped. A formula which can be used to calculate these frequencies in ascending order is as follows [4]:

$$f_{n+1} = f_n \times 10^{N-1} \tag{12.57}$$

where $n = 1$ to N with f_n is the nth test frequency (f_1 is the start frequency and f_N is the end frequency). The dwell time at each test frequency should be at least 0.5 s to make the EUT exposed to the field for enough time and justified based on EUT and test equipment response time. Because using the sweep frequency test may affect the stationary of the field, it is generally to perform the immunity test in discrete frequency.

12.5.3 Immunity test procedures

Because the RC does not support plane waves at a specified position of the stirrer, the propagation and EUT excitation are different from those in free space. Thus, it is preferred to use the maximum or average of the power density as opposed to field strength to evaluate the physical–statistical nature of the fields inside an RC.

Use either mode-turned or mode-stirrer to perform the test. For the mode-turned method, determine the number of steps and ensure the stirrer rotated in almost spaced steps. If using mode-stirrer operation, it is critical to make sure that the EUT is exposed to enough number of samples and ensure that the EUT is exposed to the field level for the appropriate dwell time.

Monitor and record P_{MaxRec} and P_{AveRec} with the receiving antenna of each frequency band to ensure that the required field strength is being generated. Use P_{AveRec} to ensure that the chamber loading has not changed from the validation. P_{MaxRec} may also be used to estimate the peak E-field generated.

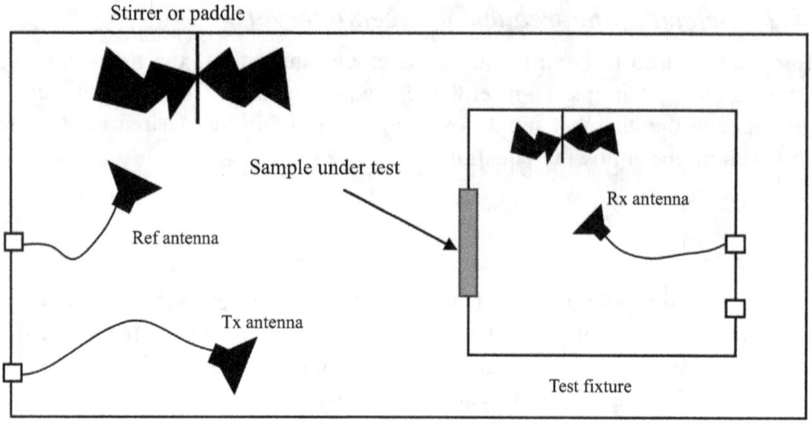

Figure 12.18 Typical material SE measurements in RC

12.6 Shielding effectiveness measurement

There are many test methods to measure the SE (shielding effectiveness) of materials and equipment designed to provide shielding. The typical measurement procedure for SE is to compare the electromagnetic environment inside the enclosure with and without the shielding material.

For many shielding designs, the measurement in RC adopts a 'nested chamber' approach (i.e., an RC within an RC, as shown in Figure 12.18). A receiving antenna and paddle wheel tuner are installed in a test fixture to estimate any RF energy that 'leaks' into the fixture from aperture.

The SE is defined as the ratio of the signal received (from the Tx antenna) without the shield to the signal received inside the shield and can also be considered the insertion loss when the shield is placed between the transmitting antenna and the receiving antenna. The SE can be calculated as

$$SE = 10\log_{10}\left(\frac{P_{Ref}}{P_{Test\ Fixture}}\right) \tag{12.58}$$

where P_{Ref} is the maximum power received by the reference antenna within the chamber, $P_{Test\ Fixture}$ is the maximum power received by the Rx antenna within the test fixture.

Section 12.6.1 will introduce the test procedure for the material SE measurement, and SE measurement for a cavity is specified in Section 12.6.2.

12.6.1 Material shielding effectiveness measurement

The typical SE measurement set-up for the material is shown in Figure 12.18. As can be seen, a nested RC approach is used and a test fixture is constructed to install the material estimated.

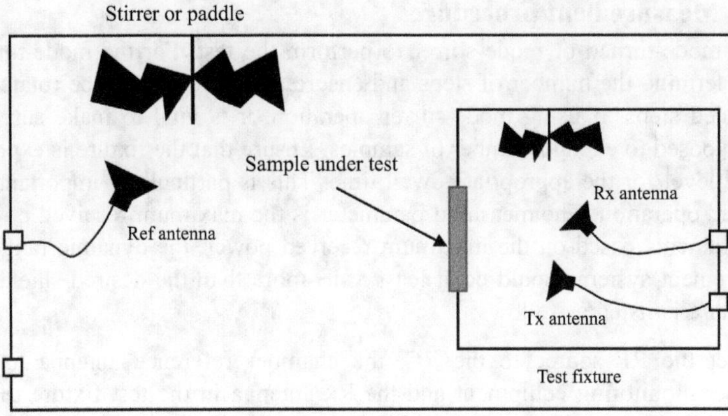

Figure 12.19 Test fixture validation set-up using RC

12.6.1.1 Test fixture validation

The test fixture should be calibrated before the measurement. This calibration should be performed to calculate the quality factor Q of the fixture so that any effect of the Q on the measurement can be explained. The validation steps are as follows:

1. Since the fixture should have at least 60 possible modes ($N = 60$), the lowest usable frequency can be determined for the test fixture by using (12.59) (valid for rectangular enclosures):

$$N = \frac{8\pi}{3} abd \frac{f^3}{c^3} - (a + b + d)\frac{f}{c} + \frac{1}{2} \tag{12.59}$$

   where a, b, and d are dimensions of the test fixture, f is the test frequency, c is the speed of wave propagation within the test fixture.
2. Install the transmitting antenna into the test fixture as shown in Figure 12.19. It is better to direct the Rx antenna into the corner of the RC. Pointing the Tx antenna into the stirrer is recommended. The feed points linking and stirrer power should be shielded well. Shielding of the feed points should be at least 5 dB above the expected SE.
3. Configure the test fixture the same as the test set-up. For each test frequency, input the power to the Tx antenna within the test fixture. Record the maximum receiving power over the one complete cycle of the stirrer inside the test fixture.
4. Taking the ratio between the transmitted power and the maximum received power to get the test fixture validation factor (TFVF):

$$TFVF = \frac{P_{MaxRec}}{P_{Input}} \tag{12.60}$$

12.6.1.2 Measurement procedure

Use either mode-turned or mode-stirrer to perform the test. For the mode-turned method, determine the number of steps and ensure the stirrer should be rotated in almost spaced steps. If using mode-stirrer operation, it is vital to make sure the fixture is exposed to enough number of samples. Ensure that the fixture is exposed to the field level for the appropriate dwell time. This is particularly important for mode-stirred operation. The measured parameter is the maximum received power. For measurements based on the maximum received power, the dynamic range of the measurement system should be at least 5 dB more than the desired shielding. The procedure is listed as follows:

1. Connect the RF source to the RC, the chamber reference antenna to the chamber monitoring equipment and the Rx antenna in the test fixture to the measurement equipment.
2. Install the material under test to the fixture.
3. Rotate the mode-stirring paddles. Start the RF source at a fixed input power.
4. Record the maximum power from the chamber and the EUT monitoring equipment over one complete rotation.
5. Repeat Step 4 across the frequencies of interest.
6. The shielding effectiveness SE (in dB) can be calculated by using the following equation, taking into account the attenuation of the links for both the EUT and reference antenna:

$$SE = 10\log_{10}\left(\frac{P_{Ref}}{P_{Test\ Fixture}}\right) - 10\log_{10}(TFVF) \tag{12.61}$$

7. Use a known SE material to cover the suspected leaking areas. Perform the leakage area test by repeating Steps 4–6. The leakage area test is used to improve the SE for EUT. If the SE measured from Step 6 is 5 dB less than the SE unmodified, the SE can be considered to be the same as before.

12.6.2 Cavity shielding effectiveness measurement

To protect an electronic device from unexpected signals or to reduce the emission of the device itself, an enclosure is normally employed to protect it. The shielding effectiveness is an important figure of merit to characterise the electromagnetic isolation performance of an enclosure.

Generally, the SE can be measured by following the Institute of Electrical and Electronics Engineers (IEEE) standard 299 [2] or using a nested RC, as shown in Figure 12.20. Suppose the radiation efficiency of antennas is high, the SE value can be obtained as

$$SE = 10\log_{10}\left(\frac{P_{Ref}}{P_{Cavity}}\right) \tag{12.62}$$

where P_{Ref} is the maximum power received by the reference antenna within the chamber, P_{cavity} is the maximum power received by the Rx antenna within the cavity.

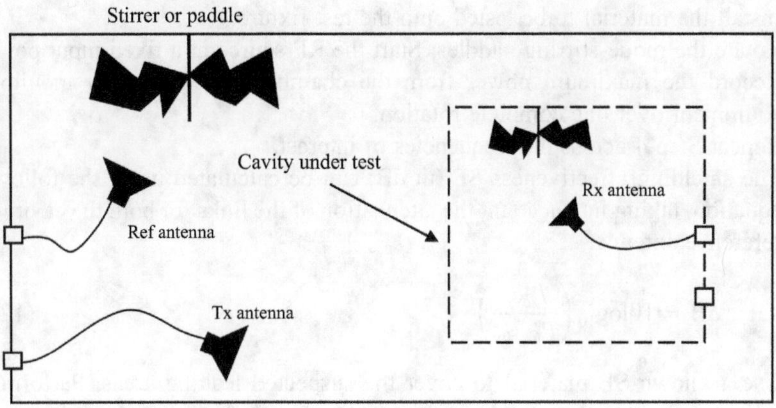

Figure 12.20 Configuration of cavity SE measurement in RC

12.6.2.1 Cavity preparation

Unlike the validation of test fixture for material SE measurement, the test fixture losses should be considered, as the losses caused by the cavity are the part of the shielding that the cavity is expected to have. The range of lowest usable frequency should still be determined, over which the accurate data can be obtained. The preparation procedure is as follows:

1. The cavity should have at least 60 possible modes ($N = 60$), the lowest usable frequency can be determined for the cavity by using (12.59).
2. Fix the transmitting antenna into the cavity as shown in Figure 12.20. It is better to direct the Rx antenna into the corner of the RC. Pointing the Tx antenna into the stirrer is recommended. The feed points linking and stirrer power should be shielded well. Shielding of the feed points should be at least 5 dB above the SE expected.

12.6.2.2 Measurement procedures

Use either mode-turned or mode-stirrer to perform the test. For the mode-turned method, determine the number of steps and ensure the stirrer should be rotated in almost spaced steps. If using mode-stirrer operation, it is essential to make sure the fixture is exposed to enough number of samples. Ensure that the fixture is exposed to the field level for the appropriate dwell time. This is particularly important for mode-stirred operation. The measured parameter is the maximum received power. For measurements based on the maximum received power, the dynamic range of the measurement system should be at least 5 dB more than the desired shielding. The procedure is listed as follows:

1. Connect the RF source to the RC, the chamber reference antenna to the chamber monitoring equipment and the Rx antenna in the test fixture to the measurement equipment.

2. Install the material to be tested onto the test fixture.
3. Rotate the mode-stirring paddles. Start the RF source at a fixed input power.
4. Record the maximum power from the chamber and the cavity monitoring equipment over one complete rotation.
5. Repeat Step 4 across the frequencies of interest.
6. The shielding effectiveness SE (in dB) can be calculated using the following equation, taking into account the attenuation of the links for both the cavity and reference antenna:

$$SE = 10\log_{10}\left(\frac{P_{Ref}}{P_{Cavity}}\right) \tag{12.63}$$

7. Use a known SE material to cover the suspected leaking areas. Perform the leakage area test by repeating Steps 4–6. The leakage area test is used to improve the SE for EUT. If the SE measured from Step 6 is 5 dB less than the SE unmodified, the SE can be considered to be the same as before.

References

[1] Hill D. A., Ma M. T., Ondrejka A. R., Riddle B. F., Crawford M. L., and Johnk R. T. 'Aperture excitation of electrically large, lossy cavities'. *IEEE Transactions on Electromagnetic Compatibility*. 1994, 36(3), pp. 169–178.

[2] Ladbury J., Koepke G., and Camell D. *Evaluation of the NASA Langley Research Center Mode Stirred Chamber Facility*. U.S. Nat. Inst. Stand. Technol. Tech. Note 1508, 1999.

[3] IEC 61000-4-21. *Electromagnetic Compatibility (EMC), Part 4–21: Testing and Measurement Techniques Reverberation Chamber Test Methods*. International Electrotechnical Commission, 2nd ed., 2011.

[4] Holloway C. L., Shah H. A., Pirkl R. J., Young W. F., Hill D. A., and Ladbury J. 'Reverberation chamber techniques for determining the radiation and total efficiency of antennas'. *IEEE Transactions on Antennas and Propagation*. 2012, 60(4), pp. 1758–1770.

[5] Chen X. 'On statistics of the measured antenna efficiency in a reverberation chamber'. *IEEE Transactions on Antennas and Propagation*. 2013, 61(11), pp. 5417–5424.

[6] Fur G. L., Besnier P., and Sharaiha A. 'Time reversal efficiency measurement in reverberation chamber'. *IEEE Transactions on Antennas and Propagation*. 2012, 60(6), pp. 2921–2928.

[7] Hallbjörner P. 'Reflective antenna efficiency measurements in reverberation chambers'. *Microwave and Optical Technology Letters*. 2001, 30(5), pp. 332–335.

[8] Krauthäuser H. G. and Herbrig M. 'Yet another antenna efficiency measurement method in reverberation chambers'. *2010 IEEE International Symposium on Electromagnetic Compatibility*. Fort Lauderdale, FL, USA, July 2010, pp. 536–540.

[9] Li C., Loh T.-H., Tian Z., Xu Q., and Huang Y. 'Evaluation of chamber effects on antenna efficiency measurements using non-reference antenna methods in two reverberation chambers'. *IET Microwaves, Antennas & Propagation*. 2017, 11(11), pp. 1536–1541.

[10] Xu Q., Huang Y., Zhu X., Xing L., Tian Z., and Song C. 'A modified two-antenna method to measure the radiation efficiency of antennas in a reverberation chamber'. *IEEE Antennas and Wireless Propagation Letters*. 2016, 15, pp. 336–339.

[11] Gifuni A., Flintoft I. D., Bale S. J., Melia G. C. R., and Marvin A. C. 'A theory of alternative methods for measurements of absorption cross section and antenna radiation efficiency using nested and contiguous reverberation chambers'. *IEEE Transactions on Electromagnetic Compatibility*. 2016, 58(3), pp. 678–685.

[12] Xu Q., Huang Y., Zhu X., Alja'afreh S. S., and Xing L. 'A new antenna diversity gain measurement method using a reverberation chamber'. *IEEE Antennas and Wireless Propagation Letters*. 2015, 14, pp. 935–938.

[13] Horansky R. D., Meurs T. B., North M. V., Wang C., Becker M. G., and Remley K. A. 'Statistical considerations for total isotropic sensitivity of wireless devices measured in reverberation chambers'. *2018 International Symposium on Electromagnetic Compatibility (EMC EUROPE)*. Amsterdam, 2018, pp. 398–403.

[14] Hussain A., Kildal P., and Glazunov A. A. 'Interpreting the total isotropic sensitivity and diversity gain of LTE-enabled wireless devices from over-the-air throughput measurements in reverberation chambers'. *IEEE Access*. 2015, 3, pp. 131–145.

[15] Remley K. A., Wang C.-M., Pirkl R. J., *et al.* 'A significance test for reverberation-chamber measurement uncertainty in total radiated power of wireless devices'. *Transactions on Electromagnetic Compatibility*. 2015, 58(1), pp. 207–219.

[16] Kildal P.-S, Chen X., Orlenius C., Franzen M., and Lötbäck Patané C. 'Characterisation of reverberation chambers for OTA measurements of wireless devices: physical formulations of channel matrix and new uncertainty formula'. *IEEE Transactions on Antennas and Propagation*. 2012, 60(8), pp. 3875–3891.

[17] Remley K. A., Dortmans J., Weldon C., *et al.* 'Configuring and verifying reverberation chambers for testing cellular wireless devices'. *IEEE Transactions on Electromagnetic Compatibility*. 2016, 58(3), pp. 661–672.

[18] Horansky R. D. and Remley K. A. 'Flexibility in over-the-air testing of receiver sensitivity with reverberation chambers'. *IET Microwaves, Antennas & Propagation*. 2019, 13(15), pp. 2590–2597.

[19] Chen X., Kildal P.-S., Orlenius C., and Carlsson J. 'Channel sounding of loaded reverberation chamber for over-the-air testing of wireless devices: coherence bandwidth versus average mode bandwidth and delay spread'. *IEEE Antennas and Wireless Propagation Letters*. 2009, 8, pp. 678–681.

[20] Karlsson K., Chen X., Carlsson J., and Skarbratt A. 'On OTA test in the presence of Doppler spreads in a reverberation chamber'. *IEEE Antennas and Wireless Propagation Letters*. 2013, 12, pp. 886–889.

[21] Hill D. A. 'Boundary fields in reverberation chambers'. *IEEE Transactions on Electromagnetic Compatibility*. 2005, 47(2), pp. 281–290.

[22] Remley K. A., van de Beek G. S., Holloway C. L., Ladbury J. M., and Leferink F. 'Characterizing large-form-factor devices in a reverberation chamber'. *2013 International Symposium on Electromagnetic Compatibility*. Brugge, Belgium, September 2013, pp. 375–380.

[23] Burger W. T. C., Remley K. A., Holloway C. L., and Ladbury J. M. 'Proximity and antenna orientation effects for large-form-factor devices in a reverberation chamber'. *2013 IEEE International Symposium on Electromagnetic Compatibility*, Denver, CO, USA, 2013, pp. 671–676.

[24] Xue W., Li F., Chen X., Zhu S., Zhang A., and Svensson T. 'A unified approach for uncertainty analyses for total radiated power and total isotropic sensitivity measurements in reverberation chamber'. *IEEE Transactions on Instrumentation and Measurement*. 2021, 70, pp. 1–12.

[25] Xu Q., Xing L., Yan D., Zhao Y., Jia T., and Huang Y. 'Experimental verification of stirrer angular correlation with different definitions in a reverberation chamber'. *The 12th International Symposium on Antennas, Propagation and EM Theory (ISAPE 2018)*, Hangzhou, China, 2018, pp. 1–5.

[26] CTIA. *Test Plan for Wireless Device Over-the-Air Performance; Method of Measurement for Radiated RF Power and Receiver Performance*. Version 3.8.2, April 2019.

[27] Remley C. A., Toorn J. A. D, Holloway C. L., Ladbury, J. M., and Wang C.-M. 'Proximity-effect test for lossy wireless-device measurements in reverberation chambers'. special issue "Recent developments on the use of reverberation chambers for testing wireless systems," *IET Science, Measurement and Technology*. 2015, 9(5), pp. 540–546.

[28] Orlenius, C., Kildal, P.-S., and Poilasne, G. 'Measurements of total isotropic sensitivity and average fading sensitivity of CDMA phones in reverberation chamber'. *2005 IEEE Antennas and Propagation Society International Symposium*. Washington, DC, USA, July 2005, 1A, pp. 409–412.

[29] Hill D. A. *Electromagnetic Fields in Cavities: Deterministic and Statistical Theories*. Wiley-IEEE Press, Piscataway, NJ, USA, 2009.

[30] Grimmett G. and Stirzaker D. *Probability and Random Processes*. Oxford University Press, New York, NY, USA, 3rd ed., 2001.

[31] Chen X. 'Measurement uncertainty of RC and its reduction techniques for OTA tests: a review'. *IET Microwaves, Antennas & Propagation*. 2019, 13(15), pp. 2598–2604.

[32] Chen X. 'Throughput modeling and measurement in an isotropic-scattering reverberation chamber'. *IEEE Transactions on Antennas and Propagation*. 2014, 62(4), pp. 2130–2139.

[33] 3GPP. *LTE; Evolved Universal Terrestrial Radio Access (E-UTRA); Physical Channels and Modulation*. TS 36.211 v11.2. 0, rel. 11 April 2013.
[34] West J. C. and Bunting C. F. 'Effects of frequency stirring on reverberation chamber testing: an analysis as a radiation problem'. *IEEE Transactions on Electromagnetic Compatibility*. 2019, 61(4), pp. 1345–1352.
[35] Taylor B. N. and Kuyatt C. E. *Guidelines for Evaluating and Expressing the Uncertainty of NIST Measurement Results*. National Institute of Standards and Technology, Gaithersburg, MD, USA, 1994.
[36] IEEE. *IEEE Standard Definitions of Terms for Antennas*. IEEE Std 145-1993, 1993.
[37] Johnston R. H. and McRory J. G. 'An improved small antenna radiation-efficiency measurement method'. *IEEE Antennas and Propagation Magazine*. 1998, 40(5), pp. 40–48.
[38] Pozar D. M. and Kaufman B. 'Comparison of three methods for the measurement of printed antenna efficiency'. *IEEE Transactions on Antennas and Propagation*. 1988, 36 (1), pp. 136–139.
[39] Schantz H. G. 'Radiation efficiency of UWB antennas'. *IEEE Conference on Ultra Wideband Systems and Technologies*. Baltimore, MD, USA, 2002, pp. 351–355.
[40] Huang Y. Radiation efficiency measurements of small antennas. *In: Handbook of Antenna Technologies* (ed. Z. Chen). Singapore: Springer, 2015.
[41] Schroeder W. L. and Gapski D. 'Direct measurement of small antenna radiation efficiency by calorimetric method'. *IEEE Transactions on Antennas and Propagation*. 2006, 54(9), pp. 2646–2656.
[42] Qi W., Fang F., and Xia, W., *et al.* 'A compact multi-probe reverberation chamber for over-the-air testing'. *The Applied Computational Electromagnetics Society (ACES) Symposium*, Accepted.
[43] Xu Q., Xing L., Zhao Y., Jia T., and Huang Y. 'Probability distributions of three-antenna efficiency measurement in a reverberation chamber'. *IET Microwaves, Antennas & Propagation,* 2021, 15(12), pp. 1545–1552.
[44] Xu Q., Qi W., and Liu C., *et al.* 'Measuring the total radiated power of wideband signals in a reverberation chamber'. *IEEE Antennas and Wireless Propagation Letters*. 2020, 19(12), pp. 2260–2264.

Chapter 13
Over-the-air testing for autonomous vehicle communications

Erik Kampert[1], Elijah I. Adegoke[1], Reiner Stuhlfauth[2], and Matthew D. Higgins[1]

The future of transportation lies with the development of connected and automated mobility (CAM) solutions, which include intelligent and self-driving vehicles, with a large research and engineering community investigating and overcoming the technical challenges in their development. Although safest CAM solutions are expected to be autonomous and fully self-supporting, the ability to receive and share information with other vehicles or infrastructure holds promise to lead to advanced environment and traffic awareness as well as cooperative behaviour functions. Fast and reliable wireless communication between vehicles and infrastructure will likely rely on 5G NR and beyond technology, with access to already existing cellular communication infrastructure as well as to newly deployed infrastructure for the higher frequency bands that provide access to larger bandwidths. The work presented here evaluates the quality of such over-the-air, vehicle-to-infrastructure signals in a mini-urban campus environment for both low and high carrier frequencies, by comparing the results from detailed channel sounding measurements with signal quality parameters such as the error vector magnitude and the established data throughput. This provides insights into the relationships between fundamental parameters used to establish vehicular communication channel models and the quality of service parameters used in vehicular applications and international standards.

13.1 Introduction

Wireless communication is considered an essential infrastructure component in modern societies. The technology evolution outlines a change from purely circuit switched voice call services to ubiquitous communication networks offering a wide range of data-oriented services within the last decades. The automotive industry is

[1]WMG, University of Warwick, Coventry, United Kingdom
[2]Rohde & Schwarz International GmbH, Munich, Germany

not just service application-based any more, and one may consider it as a vertical with new applications and services driving the technology evolution. Mobility is an essential pillar of modern society, whether it is personal mobility, public transportation of people or the transport and logistics supply chain [1]. The conveyance of goods and passengers is essential for economic growth. With the implementation of communication technologies, the mobility sector is able to introduce sophisticated mechanisms that foster energy preservation and a cleaner environment.

Efficiency in transportation networks can be achieved through conventional road construction; however, this often reaches its limits in terms of land consumption and lack of public acceptance. Engineers and scientists are seeking processes and technologies to enhance traffic management. An intelligent transportation system (ITS) seeks to avoid traffic congestion and increase overall traffic efficiency [2]. End-to-end digitization, starting from a single vehicle to other elements of the transportation system (such as a back-end server), all provide the basis for continuous control and management of traffic flow. Technological progress in the automotive industry towards automated driving and development of advanced driver-assistance systems (ADAS) is propelling a digitally driven transportation system. This has led to the ongoing development of cooperative-ITS (C-ITS), a transportation system in which all road users communicate and cooperate with each other, which promises to increase efficiency even further and additionally reduce the number of road traffic incidents [3]. Reducing the severity of injuries in road traffic accidents is an ambitious intention declared by various government agencies around the globe. Two motivations drive this technology evolution: safety and comfort. Technologies enhancing user comfort will ensure that vehicles offer relaxed spaces, e.g. through sophisticated entertainment services, whereas safety-related applications such as eCall, an in-vehicle-based emergency communication system, will ensure a reduction in the amount of road casualties such as injuries or loss of life.

13.2 Automotive wireless connectivity requirements

In a simplified view, the development of the vehicle of the future follows two major trends: a higher integration level of electronic domain controllers and the support of ADAS leveraging self-driving vehicles. The Society of Automotive Engineers (SAE) defines a scale from 0 (no automation at all) to 5 (full automation) for the levels of automation of an autonomous vehicle [4]. For a level-5 vehicle, it is essential to have independent autonomous driving control systems, because intra-vehicle communications may be interrupted, e.g. the autonomous vehicle cruise control needs to operate as a stand-alone and self-reliant subsystem. Communication-based technologies can play a complementary role supporting these control decisions. Presently, in most vehicles, multiple electronic control units are responsible for diverse tasks and communication is via automotive Ethernet [5].

One of the major requirements for wireless vehicle-based communication technologies is the available frequency spectrum. Historically, governments and industry have pushed for additional or new spectrum allocations. For example, the IEEE 802.11p Wireless Access in Vehicular Environments (WAVE) standardization process originates from the allocation of the dedicated short-range communications (DSRC) spectrum band in the United States in 1999. In 1999, the US Federal Communication Commission (FCC) allocated 75 MHz of spectrum at 5.9 GHz to be used exclusively for vehicle-to-vehicle (V2V) and vehicle-to-infrastructure (V2I) communications. This spectrum is now becoming technology agnostic, and possible implementation of Long-Term Evolution (LTE), LTE vehicle-to-everything communications (LTE-V2X) or 5G NR V2X might take place in the same band. In a timeline approach, it is observed that the primary motivation for wireless vehicular communication technologies is the enhancement of road and vehicle safety. In addition to safety, traffic management is an important use case for ITS and focuses on providing updated local information, maps and other relevant messages limited in space and/or time. It is anticipated that this can, among others, reduce traffic jams, which will result in saved time for drivers and reduced pollution from carbon emissions from idling vehicles.

13.2.1 Legacy and current automotive wireless communication

Cooperative-ITS represents a transportation system in which all road users and fixed infrastructure components, such as road-side units (RSUs), cooperate and communicate with each other. Various terms and projects exist that operate towards this common goal; one may classify these concepts into certain categories. As a first category the region-specific use cases and radio technologies exist, which can be clustered into the geographical regions: North America, Europe and Asia. A second category is the wireless radio technology used for vehicular communication, with the representative technologies being the IEEE-based WLAN technologies, such as IEEE 802.11p, and the 3GPP-based cellular technologies, such as LTE and 5G NR. Furthermore, one may also categorize based on the use cases, i.e. whether they are purely safety-related or also enhancing the driving experience through improved communication services in addition to road safety applications.

In the early 1980s, there had been some trials targeted towards improving road safety, of which one was based on the deployment of a vehicular ad hoc network (VANET). VANETs represent an early attempt to support safety-related applications such as accident warning, crash notification and cooperative cruise control. With VANETs, a vehicle is considered a router- or network-node and neighbouring vehicles are allowed to communicate with each other via V2V communications. Vehicles can also communicate with RSUs via V2I communications to collect road and traffic-related information. The IEEE standardization body describes a methodology to enable vehicular communication within the IEEE 802.11 and IEEE 1609 series of its standards. It can be seen as a complete protocol stack, covering application specific functions, such as basic safety messages (BSMs), up to the radio layer, in which a WLAN device supports IEEE 802.11p radio transmission and reception in direct communication mode.

The major difference between DSRC and LTE/5G NR communication technologies is that DSRC is designed for direct communications only, whereas LTE and 5G NR are cellular technologies that offer both network and direct communication services. With the adoption of 3GPP Release 14, direct communication is possible using sidelink-based communication over the PC5 interface. This PC5 interface is defined as a reference point and combined with the entire protocol stack, and the radio interface is described as the sidelink.

As highlighted earlier in this chapter, two standardization bodies – IEEE and 3GPP have performed a vast amount of work to amend and extend existing wireless communication technologies for the use within automotive scenarios. The radio standards originating from these bodies are IEEE WLAN 802.11p and 3GPP LTE-V2X. With regards to the latter, 5G NR-V2X starts with 3GPP Release 16 and enhances the services offered by LTE-V2X. In the following subsections, a concise summary of these technology variants is discussed.

13.2.1.1 DSRC and ITS-G5 vehicular communication standards

To leverage vehicular communication scenarios, North American standardization body IEEE and the European Telecommunications Standards Institute (ETSI) started technology initiatives to enable direct communication between vehicles, which are also known as DSRC or IEEE Wireless Access in Vehicular Environments (WAVE) [6], and ETSI ITS-G5 [7], respectively. They encompass application layers and radio technologies, with the latter based on IEEE 802.11p; a derivative of the IEEE 802.11a WLAN technology standard which focuses on direct device-to-device (D2D) communication [8]. The IEEE 802.11 standard body has published an amendment of the existing WLAN specification series named IEEE 802.11p with the focus on supporting V2X communication. As mentioned previously, the IEEE 802.11p WAVE standardization process defined for higher layers (with regards to IEEE 1609) originates from the allocation of the DSRC spectrum band in the United States in 1999 and the associated efforts to define the technology that uses that band. The IEEE 802.11p should not be considered a complete and stand-alone standard, and in the overall picture, V2X communication should fulfil safety applications as well as non-safety applications. Technicalities relating to the DSRC channel plan and operational concept are taken care of by the upper layer of the IEEE 1609 standards [6]. The IEEE 1609.4 standard which provides the application layer is placed above IEEE 802.11p and enables the operation of upper layers across multiple channels, without requiring knowledge of the physical (PHY) parameters. These application and service layers on top of 802.11p may vary depending on regional regulations. Nevertheless, in order to guarantee interoperability, the BSM, which is the critical application for V2V communication, remains unchanged.

IEEE 802.11p allows communication via an ad hoc network or via a connection set-up to a basic service set (BSS) infrastructure. This infrastructure approach in WLAN allows a network set-up that is essentially based on a group of nodes anchored to one access point. Most WLAN technologies support setting up an independent BSS between several nodes to form an ad hoc network. The challenge associated with this network is to reduce the complexity and associated overhead, especially in V2X communication. For example, the IEEE 802.11p changes the WLAN concept of frame addressing using the service set identification as it would present excessive signalling overhead. In IEEE 802.11p, a so-called WAVE mode is allowed, permitting a node to transmit and receive data frames with a wildcard ID, independent of any BSS numbering. This represents a major change in the MAC layer between the IEEE 802.11a standard and its derivative – IEEE 802.11p, which is designed for automotive applications.

The WAVE standard introduces a new BSS type – WAVE BSS (WBSS), whereby a node first transmits an on-demand beacon which uses the beacon frame to advertise a WBSS. It contains all necessary information for receiver nodes to understand the services offered within the WBSS. If an entity decides to join the WBSS, it obtains all relevant information to request its membership to this ad hoc WBSS from the advertisement. In other words, a node can decide to join and complete the joining process of a WBSS by only receiving a WAVE advertisement with no further interactions [9]. As highlighted earlier, the IEEE 802.11p standard is derived from the WLAN standard IEEE 802.11a, and by comparing the physical layer parameters, these differences can be observed. With regards to timing, the cyclic prefix (CP), the symbol duration and the guard time in IEEE 802.11p are doubled when compared to IEEE 802.11a. This is to provide robust resilience against increased delay spread as IEEE 802.11p is designed to operate at a higher vehicle velocity and over larger distances. The main system parameters of IEEE 802.11p [IEEE 802.11] are provided in Table 13.1.

In addition to the physical layer amendments, IEEE 802.11p defines some receiver enhancements to cope with cross-channel interference. It also defines an improved transmission mask to reduce channel interference. The IEEE also formed a task group to work on enhancements of the IEEE 802.11p MAC and PHY layers, through the amendment known as IEEE 802.11bd Enhancements for Next Generation

Table 13.1 IEEE 802.11p physical layer parameters

Physical layer parameter	**Values**
System bandwidth	10 MHz
Subcarrier spacing	56.25 kHz
Number of subcarriers	52
Number of data subcarriers	48
Number of pilot subcarriers	4
Symbol duration	8 μs
Guard time	1.6 μs
Modulation schemes	BPSK, QPSK, 16-QAM, 64-QAM
Code rate	1/2, 2/3, 3/4 convolutional code
Frame size	100/1,500 bytes
Peak data rate	7.69 Mb/s (100 bytes/frame)
Peak data rate	23.08 Mb/s (1,500 bytes/frame)

Table 13.2 IEEE 802.11bd physical layer parameters

Physical layer parameter	Values
System bandwidth	10/20 MHz
Subcarrier spacing	156.25 kHz
Number of subcarriers	52/108
Number of data subcarriers	48
Number of pilot subcarriers	4
Symbol duration	8 μs
Guard time	1.6/3.2 μs
Modulation schemes	BPSK, QPSK, 16-QAM, 64-QAM, 256-QAM
Code rate	1/2, 2/3, 3/4, 5/6 LDPC coding
Frame size	100/1,500 bytes
Peak data rate	8 Mb/s (100 bytes/frame)
Peak data rate	30.09 Mb/s (1,500 bytes/frame)

Vehicular Communication. The system parameters of IEEE 802.11bd are summarized in Table 13.2. The main design goals of this standard are as follows [10]:

- Backward compatibility and interoperability with IEEE 802.11p.
- Enhanced throughput measured at the MAC layer.
- Higher reliability resulting from reduced packet collisions and performance improvement under high Doppler shift conditions.
- Support devices with target speed of 250 km/h.
- Double the range with respect to IEEE 802.11p.

Higher data rates can be achieved by adopting existing PHY technologies, such as low-density parity-check (LDPC) coding, multi-antenna technologies, higher order quadrature amplitude modulation (QAM), and a 20 MHz bandwidth support. In order to improve the range, dual-carrier modulation and range extension modes can be adopted from IEEE 802.11ax. The IEEE 802.11p standard is limited by the scope of IEEE 802.11, which is strictly a MAC and PHY level standard that is meant to work within a single logical channel. This is an important difference when compared to cellular V2X (C-V2X) technologies, which specify all protocol layers, even in direct communication mode.

13.2.1.2 Long-term evolution vehicle-to-everything

The 3GPP LTE-V2X specification is strongly motivated by the assumption that V2X applications are at least in part replicated from DSRC, WAVE and ITS-G5. Figure 13.1 represents the technology evolution through the various 3GPP Releases on either network-based communication or direct V2X communication. With 3GPP Release 14, the support of direct V2X communication was introduced into wireless communication standards through the sidelink or PC5 interface for evolved universal terrestrial radio access (E-UTRA), supporting direct communication scenarios such as V2V, V2I and vehicle-to-pedestrian (V2P). This feature sets the starting point for the evolution of applications not previously supported by mobile

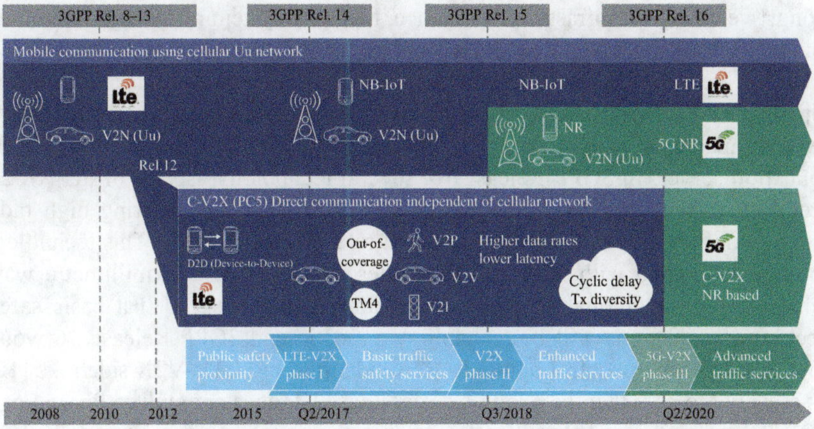

Figure 13.1 3GPP-based cellular mobile communication technology evolution for automotive connectivity, divided into network-based (top) and direct (centre) communication

communications technology and paves the way for ubiquitous and future-proof connectivity in the automotive domain. With focus on road-safety enhancements and interoperability, the application layer basic safety messages (BSM) were aligned between IEEE and 3GPP. Furthermore, it is predicted that non-vehicle devices such as smart phones, backpacks and bicycles will include the technology in order to communicate with vehicles. It is envisaged that these devices would alert drivers to the presence of a cyclist on the road or a pedestrian on a crossing, consequently reducing the number of vulnerable road user casualties.

The evolution of applications and use case scenarios for the PC5 interface of V2X services can be divided into three different phases [11]:

- Phase I, 3GPP Release 14: supports direct V2X communication, distributed random and reservation-based resource usage to exchange basic traffic safety information between vehicles and other devices, e.g. RSU, pedestrians.
- Phase II, 3GPP Release 15: supports low latency, direct V2X communication, high data rate links to exchange data for enhanced automotive applications.
- Phase III, 3GPP Release 16: introduces direct V2X communication using 5G NR technology and supports broadcast, groupcast and unicast communications for advanced automotive applications.

LTE-V2X in Phase I is mainly designed to carry vehicle status and environmental information for safety-related applications. In Phase II, enhanced direct V2X (eV2X) is introduced, intended to support applications such as cooperative perception. Phase III is standardized in Release 16 and introduces the 5G NR concept into direct V2X communication. With Release 14, 3GPP extended the concept of direct communication from D2D proximity services solely for public safety to V2X. The first service use cases involve the exchange of the cooperative awareness message and the decentralized environmental notification message amongst vehicles and infrastructure. Hence, the V2X system reference architecture design refers to the system model specified for 3GPP proximity-based services detailed in Release 12. New standardized interfaces, reference points and functions supplement the previous architecture to offer user equipment (UE) communications services capable of carrying user data to and from the V2X application and the application server. In 3GPP Release 16, several technology aspects of 5G NR are introduced to the direct communication path with NR-V2X, offering high radio access flexibility, low latency and high data rates on the sidelink. This technology evolution is aligned with increased data rates due to the use of millimetre wave (mmWave) and additional spectrum allocations. It is envisaged that basic safety direct communications will be technology agnostic and 3GPP Release 16 would allow the network to control both NR-V2X sidelinks and LTE-V2X sidelinks [12]. It is essential to note that the evolution, e.g. Phase III, does not replace Phase I or II of C-V2X communication, as its primary objective is to extend existing LTE-V2X technology to NR-V2X.

13.2.2 5G automotive wireless communication

NR-V2X or V2X Phase III is expected to address more complex advanced traffic scenarios such as remote driving, high density platooning, cooperative manoeuvres as well as advanced sensor data sharing which require ultra-reliable low-latency communication (URLLC) links. V2X services are realized by various types of V2X applications, i.e. V2V, V2P, V2I and vehicle-to-network (V2N), as shown in Figure 13.2. One of the key technology aspects of NR-V2X is the introduction of flexible 5G NR characteristics, e.g. flexible numerologies, resource scheduling, synchronization and frame structure. 3GPP TS 23.287 [13] defines V2X communication as a communication link that supports V2X services leveraging Uu (V2N) and PC5 reference points. An extended channel structure supports data transmission with greater flexibility. Besides various orthogonal frequency division multiple access (OFDMA) numerologies, NR-V2X offers a dynamic resource allocation scheme, a spatial multiplexing (multiple-input

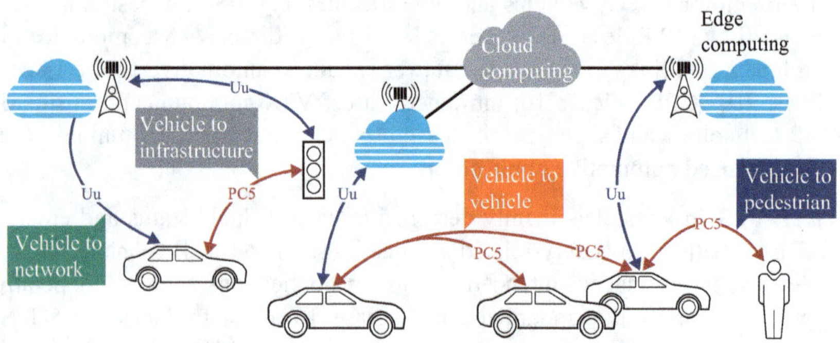

Figure 13.2 Typical examples of vehicle-to-everything communication infrastructure and services leveraging Uu and/or PC5 reference points

multiple-output) scheme (MIMO) and non-constant HARQ timing. As NR-V2X represents a complement of legacy LTE-V2X, the infrastructure needs to be future-proof and backward compatible. In addition to the eNB assigning resources on the LTE-V2X sidelink, 3GPP Release 16 also contains gNB enhancements for the NR Uu to allocate the LTE-V2X sidelink. 5G NR standalone operation is considered as first priority when network-based scheduling is considered, but sidelink operation is also supported with 5G networks operating in NSA mode or NR-DC multi-connectivity [14]. Table 13.3 presents a concise overview of the physical layer differences across

Table 13.3 Physical layer details of sidelink across the 3GPP Releases 12–17

Parameter	LTE D2D (Release 12)	LTE-V2X (Release 14/15)	5G V2X (Release 16/17)
Frequency	All bands possible (e.g. FirstNet 700 MHz)	Target 5.9 GHz	Target 5.9 GHz, FR1 and FR2
Waveform	DFT-s-OFDM	DFT-s-OFDM	CP-OFDM
Subcarrier spacing	15 kHz	15 kHz	NR numerologies, 15/30/60/120 kHz
Cyclic prefix	Normal and extended	Normal	Normal and extended (only 60 kHz)
Modulation	QPSK, 16-QAM	R14: QPSK, 16-QAM R15: + 64-QAM	QPSK, 16-QAM, 64-QAM, 256-QAM
Channel coding	Turbo code	Turbo code	LDPC (data) and polar (signalling) codes
Time scheduling	1 Subframe ≙ 1 ms	1 Subframe ≙ 1 ms	1 Slot flexible slot duration Slot aggregation possible
Number of DMRS symbols per TTI	2 Per subframe	4 Per subframe	2 To 4 per slot
Data/control multiplex	TDM	FDM	TDM and FDM
HARQ			RX UE reports to TX UE TX UE reports to gNB
MIMO	Single layer	R14: single layer R15: TX and RX diversity	Up to 2 layers
Retransmissions	4 (By default)	Up to 2	Up to 32 (configurable and resource reservation)
Communication type	Groupcast, broadcast	Broadcast	Unicast, groupcast, broadcast
Carrier aggregation	No	R14: no R15: up to 8 CCs	No
Peak throughput	≈7 Mb/s	R14: ≈ 32 Mb/s R15: ≈ 72 Mb/s	≈200 Mb/s (256-QAM)

the 3GPP Releases 12-17 for the sidelink [15]. Most notable is the adaptation of flexible numerologies for C-V2X in terms of subcarrier spacing (SCS) and number of slots per subframe from Release 16 onwards, as introduced in the main 3GPP Release 15 and discussed in the following section.

13.2.2.1 5G frequency ranges: FR1 and FR2

3GPP TS 38.101 separates the operating frequency range for 5G NR into two frequency ranges: frequency range 1 (FR1) [16] and frequency range 2 (FR2) [17]. FR1 comprises spectrum from 410 MHz to 7.125 GHz and FR2 from 24.25 to 52.6 GHz. Whereas FR1 will carry most of the traditional cellular mobile communications traffic, the higher frequency bands in FR2 and their larger bandwidths will provide very high data rate capabilities over shorter ranges. In addition to the frequency bands, TS 38.101 also specifies a different transmission bandwidth configuration per band and per subcarrier spacing, which represents the channel bandwidth. For FR1, transmission bandwidth configurations of 5, 10, 15, 20, 25, 30, 40, 50, 60, 70 (gNB only), 80, 90 and 100 MHz are allowed, whereas for FR2, transmission bandwidth configurations of 50, 100, 200 and 400 MHz are specified. 5G is designed to support a variety of different services. Not only have the end-user data rate and system capacity been further increased, also the nature of the services has become much more diverse. In order to address these peak data rate and capacity requirements, 5G NR deploys a much higher bandwidth (up to 400 MHz per carrier) when compared with 4G. Due to spectrum allocation, large bandwidths are only available at the mmWave frequencies in FR2. Consequently, 5G NR is designed for deployment in two main frequency ranges. Flexibility is given to adjust the radio interface by selecting the appropriate numerology/waveform that addresses the respective radio channel characteristics. At lower frequencies in FR1, a narrower subcarrier spacing facilitates equalization and at higher frequencies in FR2 a wider SCS reduces the impact of phase noise. These flexible numerologies enhance overall achievable spectrum efficiency. With regards to the V2X sidelink, if it operates in FR1, 15, 30 and 60 kHz SCS are permitted, and if operating in FR2, only 60 and 120 kHz SCS are used. In order to reduce overall complexity, the extended CP is only allowed with 60 kHz SCS.

13.2.3 Over-the-air testing

OTA or non-conducted connection testing describes a particular test and measurement set-up whereby a device under test (DUT) and the test equipment are connected without a cable-based “direct” connection. Typically, this setup resides in a Faraday-shielded anechoic chamber to avoid reflections and electromagnetic interferences, both with the actual tests and electrical devices that are external to the research lab. Moreover, for outdoor OTA testing, specific licences are required from the national telecommunications regulatory body for most of the 5G spectrum as well as for other commercially used radiofrequency spectrum.

With OTA, it is possible to analyse the signal quality versus the spatial domain, a methodology that is for instance applicable to verify antenna beamforming technologies. A 3D antenna radiation pattern presents the only way to verify whether a smart antenna directs the beam into the desired direction. The most relevant OTA testing parameters for an antenna are the equivalent isotropic radiated power (EIRP), the total radiated power (TRP), the total isotropic sensitivity (TIS), the antenna gain and its directivity. Requirements for OTA test set-ups range from small form factor shielding chambers designed for production environment testing to large chamber sizes for large device sizes, e.g. full vehicle testing. Essential parameters that should be considered in an OTA antenna test set-up are the aperture size D affecting the near-field (NF) and far-field (FF) Fraunhofer distance, the frequency range, shielding effectiveness and the quality of the quiet zone.

In the automotive domain, OTA testing is required in several situations, e.g. for directive antenna testing, situations where access to the DUT via cable is not possible, EMC testing, scenarios where shielding is required, radar testing and full vehicle testing. Today's vehicles support not only simple broadcast radio systems, but also complex connectivity domain controllers. In order to support these connectivity-based functions and services, vehicles require a large number of high-performance antennas integrated at various locations within and outside the vehicle. Full-vehicle antenna testing (FVT) evaluates the entire car as the DUT through the use of a vehicle drive-in full anechoic chamber as an indoor test facility for precise measurement and reproducible testing methodologies, such as shown in Figure 13.3. A key aspect of the requirements for FVT setups is to emulate and test multiple radio technologies in parallel, especially when considering technology coexistence.

In the wider automotive RF domain, real-world outdoor OTA V2X testing is required in order to understand the impact of the typical environments (e.g. urban, rural) of vehicles and traffic scenarios on the performance of their RF communication channels. Ideally this requires access to dedicated test beds, such as the UK's Centre for Connected and Autonomous Vehicles (CCAV) and Zenzic funded Midlands Future Mobility CAV test bed [18] and the Fraunhofer Institute for Integrated Circuits (IIS) funded 5G Bavaria Automotive Test Bed [19]. The work described in Sections 13.3 and 13.4 has been carried out on the former test bed, in particular its deployment on the University of Warwick campus.

13.2.3.1 Crucial parameters to investigate

From an FVT perspective, which includes both vehicular communication with infrastructure and with other vehicles, the following test and measurement aspects need to be considered:

- Frequency range from 400 MHz to 7.125 GHz (FR1) and preferred extension to mmWave (FR2).
- Passive antenna measurements such as directivity, gain, efficiency, radiation patterns and phase centre determination.
- Near field – far field (NF-FF) transformation: magnitude and phase information measured in NF and then converted into FF.
- Active antenna measurements, such as EIRP, TRP, TIS, sensitivity, half-power beam width (HPBW) and adjacent channel leakage ratio [20].

Figure 13.3 WMG's vehicle drive-in anechoic chamber inside the National Automotive Innovation Centre at the University of Warwick, UK

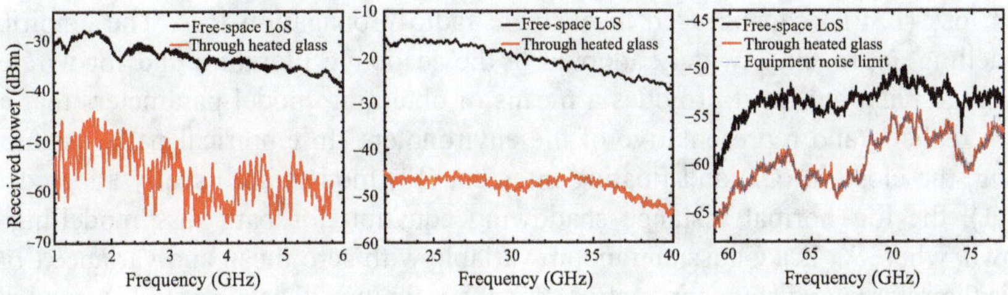

Figure 13.4 RF transparency of automotive electronically heated glass in FR1 (left), the lower end of the FR2 range (middle) and towards the automotive long-range radar frequency range just above FR2 (right)

- Influence of antenna choice (directional versus omnidirectional) on the received signal quality and overall connectivity/coverage.
- Fundamental understanding of the received signals in terms of their reflection- and multi-path-richness, typically summarized in a delay spread value.
- Understanding of the impact on the quality of signal (QoS) and the quality of (communication) experience (QoE), derived from EVM and achieved data throughput.
- Doppler effect due to the relative speed between vehicles or with respect to road-side infrastructure.
- Measurements based on certified test methods, e.g. from the European Conference of Postal and Telecommunications Administrations (CEPT) or the American Cellular Telecommunications Industry Association (CTIA).

As an emphasis of the importance of automotive RF testing, and as an introduction to the following sections, Figure 13.4 displays the results on prototype automotive industrial-grade, electronically heated glass of its transparency for unmodulated signals in both FR1 and FR2, that was tested within WMG's anechoic facilities. As can be seen from comparing the received RF power through the glass with a free-space line-of-sight (LoS) path with the same distance between the transmitter (TX) and receiver (RX) antennas, about 99% (20 dB) of the radiation is attenuated in FR1, and about 99.9% (30 dB) is attenuated in the lower end of FR2. Even in the range towards the automotive long-range radar frequencies, substantial attenuation exists, but its absolute value is unknown as the received power is below the setup's noise level. This attenuation is most likely due to the mesh of metallic wires in this prototype glass that essentially forms a grid that is nearly non-transparent to FR1 and FR2 radiation with wavelengths that are larger than the grid size. Such glass would thus severely impact wireless communication with embedded RF devices within the vehicle's cabin, and with mobile devices from the driver and passengers.

13.2.3.2 Path loss channel modelling

For a given TX–RX separation distance, the path can be modelled using analytical methods or via empirical modelling. In the former, equations such as the free-space path loss (FSPL) can be used to estimate radio propagation loss. The empirical modelling approach is widely adopted in the academic literature and for wireless coverage planning, as it provides a means of obtaining model parameters that are more realistic and representative of the environment. In empirical path loss modelling, the close-in (CI) and floating intercept (FI) models are usually adopted. In (13.1), the log–normal distance shadowing equation for path loss modelling is shown, where X_σ is a Gaussian random variable with zero mean and variance (σ^2), d_o is the reference distance in metres, A is the reference distance path loss, and n is the path loss exponent (PLE):

$$\mathrm{PL} = A + 10n\log_{10}\left(\frac{d}{d_o}\right) + X_\sigma \qquad \text{(dB)} \tag{13.1}$$

Path loss modelling using this approach is referred to as the CI model, and A can be obtained from the FSPL model or from measurements with the antenna gains excluded. In the FI model, the log–normal distance shadow equation is presented in a different manner, as is shown in the following equation (13.1):

$$\mathrm{PL} = \alpha + 10\beta\log_{10}(d) + X_\sigma \qquad \text{(dB)} \tag{13.2}$$

where α is the y-intercept and β is the slope of the regression line. The shadowing variable X_σ can be obtained from the measurements and predicted data using the following equation 13.3 [21]:

$$\sigma = \sqrt{\sum_{i=1}^{N} \frac{\left(P_{meas}(i) - P_{calc}\right)^2}{N-1}} \qquad \text{(dB)} \tag{13.3}$$

where $P_{meas}(i)$ is the ith measured path loss, P_{calc} is the mean predicted path loss and N is the total number of samples.

13.3 CAV V2X measurement set-up and methodology

13.3.1 *Experimental details for OTA V2X measurements in FR1*

The set-up adopted in this work is based on R&S channel sounding and 5G equipment [22]. For path loss and delay spread, a time domain channel sounding technique is used. The channel sounding equipment uses the pulse compression method and for FR1 consists of an R&S SMBV100B Vector Signal Generator (VSG), an R&S BBA150 Broadband Amplifier (BBA), two R&S TSMX-PPS2 GPS modules and an R&S FSVA3007 signal and spectrum analyser (SA). The VSG enables the transmission of frequency band limited signals with a carrier frequeny (CF) up to 6 GHz and an arbitrary modulated waveform with clock frequencies up to 300 MHz. The SA enables the processing of the received signals with a maximum sampling rate of 400 MHz. The resulting I/Q data are forwarded to the R&S Channel Sounding Software, which autocorrelates the received I/Q data with the originally transmitted signal. The equipment is time-synchronized by means of GPS signals using R&S TSMX-PPS2 receiver modules with PPS output and dead reckoning, with a typical root mean square (RMS) jitter of 30 ns. A 300 MHz bandwidth signal is generated and analysed, providing a time resolution of 3.3 ns, equivalent to a spatial resolution of 1 m. A filtered Frank–Zadoff–Chu sequence [23] with a length of 120,000 samples is used as a channel-sounding waveform, resulting in a channel impulse response (CIR) update rate of 400 μs. Per antenna constellation 512 CIR measurements are automatically triggered, and their results are stored and further analysed. From the resulting 512 power–delay profiles (PDPs), the received power values per delay time are then averaged in order to obtain one averaged power delay profile (APDP) for each antenna constellation.

For EVM measurements, the same measurement equipment is used in a different configuration. For M-ary modulations of 16-, 64- and 256-QAM, the bandwidth parts (BWPs) in a single 5G cell are modified using the flexible numerology available in the VSG and SA [24]. Following the downlink channel information in TS 38.101-1 [16], EVM measurements are carried out for three M-ary modulation schemes using a 30 and 60 kHz subcarrier spacing and a channel bandwidth of 100 MHz. The cell deployment range was set to 3 GHz $< f \leq 6$ GHz with a Case B synchronization signal (SS/PBCH). The SS/PBCH blocks use an SCS of 30 kHz and the BWPs use 30 kHz and 60 kHz. From the flexible numerology available within 5G NR, the configuration on the VSG gives rise to three non-overlapping BWPs which are used within the physical downlink shared channel (PDSCH). For ease of configuration, each BWP was configured with a different modulation scheme. Thus, M-ary modulation was allocated to similar resource blocks (RBs) such that the channel bandwidth is split equally. At the SA, the configurations are replicated in the signal description. In addition, phase compensation at 1 GHz was included in the configuration at the VSG and SA. A typical OTA V2X link quality measurement is shown in Figure 13.5.

FR1 measurements presented in this chapter were taken along the pedestrian route from the Professor Lord Bhattacharyya Building (PLBB), home to the National Automotive Innovation Centre (NAIC), to the Nursery on the campus of the University of Warwick, U.K. In order to replicate an urban micro cell (UMi), the transmitter was placed on the third floor outdoor terrace of the PLBB at ≈15 m above street level, as shown in Figure 13.6. The route to the Nursery can be classified as LoS, given that there is a clear path from the terrace towards the Nursery. Sparse foliage exists along the path with low height buildings located on either side of the road. Along the entire route, the average road width is between 4 and 7 m. At RX, measurements were taken at staggered locations spaced 5 m apart with the entire measurement route spanning ≈ 440 m. This route is a cut-down version of a rural area, mapped out for a regional CAM service, also known as MFM Rural. The path loss between the transmitter and receiver can be calculated as

$$\mathrm{PL}(d) = P_T + G_t + G_r - P_r - 2L_{cab} \qquad \text{(dB)} \tag{13.4}$$

where P_T is the RF power supplied to the transmitting antenna in dBm, G_t and G_r are the directional gains of TX, respectively, RX in dBi, L_{cab} is the cable loss and

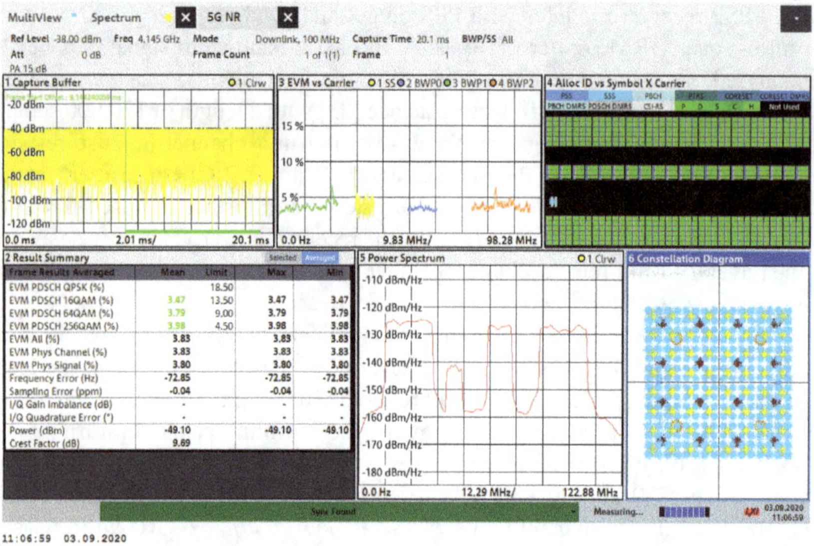

Figure 13.5 *Spectrum analyser screen capture showing the time plan, result summary, power spectrum, constellation and EVM for a 5G NR (FR1) signal for a typical outdoor OTA V2X measurement*

Figure 13.6 *Left: TX boresight view from the PLBB terrace towards the Nursery at the end of an ≈440 m long road. Right: view from the omnidirectional RX at the Nursery. All FR1 measurements were carried out on the pavements along this road, acting as a sample urban environment of the MFM CAV test bed.*

Table 13.4 Link budget and set-up specifications for channel sounding – FR1

Parameter	Symbol	Value (unit)
Directional antenna gain	G_d	11 dBi
BBA amplifier gain	G_{amp}	+10
RF power supplied to TX	P_{tG}	+35.7 dBm
Omnidirectional antenna gain	G_o	3 dBi
Long coaxial cable loss	L_{cab}	8 dB
Maximum measurable path loss		120 dB

P_r is the received power in dBm. When an omnidirectional antenna is used at the receiver, the RX antenna gain is represented as G_o (dBi) and when a directional antenna is used as G_d. In Table 13.4, the link budget of the FR1 channel sounding setup is presented. For this setup, the maximum path loss that can be measured is ≈120 dB. This represents the lower limit of the set-up, as path loss values exceeding this will cause the signal's potential multi-path components (MPC) to be below the system's noise floor. All presented measurements were carried out at off-peak hours or during periods with minimal vehicle and pedestrian traffic; thus the communication channel can be assumed quasi-static.

13.3.2 Experimental details for OTA V2X measurements in FR2

For frequencies ranging from 8 to 85 GHz, the WMG channel sounding equipment consists of an R&S SMW200A Vector Signal Generator, an R&S SMZ90 Frequency Multiplier, an R&S FSW85 signal and Spectrum Analyzer and an R&S RTO2044 Digital Oscilloscope. The former two devices enable the transmission of frequency band limited signals with a CF up to 85 GHz and an arbitrary modulated waveform with clock frequencies up to 2.4 GHz. The latter two devices enable the processing of the received signals with a maximum sampling rate of 2.4 GHz. As for the FR1 setup, the resulting I/Q data are forwarded to the R&S Channel Sounding Software, which autocorrelates the received I/Q data with the originally transmitted signal. The FR2 set-up is also synchronized through the use of two R&S TSMX-PPS2 GPS receiver modules with PPS output. A 2 GHz bandwidth signal is generated and analysed, providing a time resolution of 0.5 ns, equivalent to a spatial resolution of 15 cm. For this bandwidth, the numeric bit resolution of the setup's analog-to-digital converter is 10 bit, resulting in a 60 dBc dynamic range of the spectrum analyser. As for the FR1 setup, but for CFs up to 40 GHz, a filtered Frank–Zadoff–Chu sequence with a length of 120,000 samples is used as a channel-sounding waveform, resulting in a channel impulse response update rate of 50 μs. Per antenna constellation 1024 CIR measurements are automatically triggered, and their results are stored, averaged and further analysed. For CFs between 60 and 90 GHz, a custom chirp sequence with a length of 480,000 samples is used; resulting in a CIR update rate of 200 μs.

The FR2 channel evaluation experiments, which provide both EVM and achievable data throughput measurements, are carried out using the NI mmWave transceiver system (MTS). The MTS is a software defined radio with 2 GHz of real-time bandwidth that can be used to create over-the-air prototypes of 5G NR communications links. A multi-FPGA processing architecture makes it possible to both capture and generate 2 GHz of data and to process the data in real time. The MTS is a modular system and built from a common set of hardware, the PXIe-3610 DAC, the PXIe-3630 ADC, and the PXIe-3620 LO and IF module. Each PXIe-3610 and PXIe-3630 is paired with a PXIe-7902 FPGA. A PXIe-6674T timing and synchronization module provides a trigger that is used to synchronize multiple channels and provides a higher quality 10 MHz clock source for better overall RF performance. External to the PXI chassis, the MTS is connected to mmWave radio heads: an mmRH-3642 transmitter and an mmRH-3652 receiver, optimized for the frequency band from 24.25 to 33.4 GHz, with 2 GHz instantaneous bandwidth and an analogue gain range of at least 50 dB. In addition to the RF hardware in the MTS, four additional FPGAs are added to the system to create a real-time 5G NR physical layer and to allow for real-time encoding and decoding of the communication link's signal, which enables a data throughput calculation. In its current state, the MTS software configuration follows the Verizon 5G specifications, using eight 100 MHz carriers resulting in a total 800 MHz bandwidth, a CP orthogonal frequency-division multiplexing (CP-OFDM) waveform, dynamic time-division duplexing, turbo coding and quadrature phase-shift keying (QPSK), 16-QAM and 64-QAM as possible modulation schemes. Both TX and RX are connected to the channel sounding equipment with 2 m long, low loss, phase stable coaxial cables, which have a typical insertion loss of 4.8 dB each, plus 0.5 dB from a between-series coaxial adapter matching the input connector from the FSW85. Shortest possible, low loss, phase stable coaxial cables are used for the channel evaluation measurements in order to have highest possible TX output and RX input power, thereby the largest possible range over which a high throughput communication link can be established. These cables are 30 cm and 65 cm long and have a typical insertion loss of 1.0 dB, respectively, 1.7 dB. For both the FR2 channel sounding and the FR2 channel evaluation measurements, a directional waveguide horn antenna with a typical gain of 25 dBi and an HPBW of 10° was used as a TX. On the RX-side, either an identical directional antenna was used, or an ultra-wideband omnidirectional antenna with a typical gain of 6 dBi and an HPBW of 20° in its elevation plane.

13.4 CAV V2X OTA results and inferences

13.4.1 FR1 large- and small-scale fading: received power and delay spread

In Figure 13.7(a) and (b), the empirical models obtained from the measurement data are presented for directional and omnidirectional measurements. For the CI model with parameters n and σ in (13.1), A was obtained from measurements at $d_0 = 1$ m and the PLE was obtained by performing a minimum mean square error

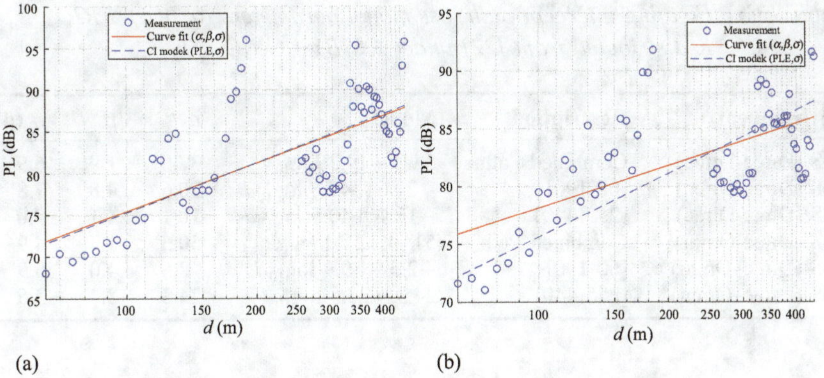

Figure 13.7 LoS path loss model fits for FR1 OTA V2X channel sounding with directional TX and (a) directional or (b) omnidirectional RX. The CI reference distances (d_0 = 1 m) and A are obtained from additional measurements. Redistributed from [22] under CC BY-NC-ND 4.0.

regression to the RX power. The parameters α and β of the FI model were obtained by performing a least squares (LS) regression to the measured path loss. From Figure 13.7(a) and (b), the path loss experienced increases from 150 m to 190 m. This can be associated with obstruction from sparse foliage and the receiver not being boresight aligned with the transmitter. At around 250 m, where the measurements resume after a roundabout, the footpath aligns with the TX boresight; thus, the RX power increases due to reduced obstruction on the LoS path. Thereafter, a prolonged deep fade is experienced along the path, which can be associated with shadow fading from the surrounding buildings. At the end of the route near the Nursery, the footpath bends and causes the path loss to increase as RX drifts into an obstructed LoS. The prolonged shadow fading resulting from the environment along this path is also shown in the delay spread and OTA EVM measurements.

In Table 13.5 and 13.6, a summary of the parameters obtained for the respective FI and CI models is presented. In the model summary tables, the pedestrian path is split into two data sets based on the TX–RX separation distance, as they represent slightly different propagation scenarios. The corresponding models from both data sets show that combining the entire measurement into one data set can skew the model parameters. The PLEs/slopes from Tables 13.5 and 13.6 display a typical PLE of $\approx$2 for both directional and omnidirectional measurements. The obtained PLE corresponds to results from similar measurement campaigns published in the academic literature. The performance of the empirical models is evaluated using the root mean square error (RMSE) as shown in Table 13.7. Given that the TX–RX separation distance in this work is less than 1 km, it is envisaged that different model parameters will be required for distances beyond the

Table 13.5 Floating intercept path loss model parameters (α, β, σ)–FR1. Redistributed from [22] under CC BY-NC-ND 4.0.

Environment	Description	Model	α	β	σ (dB)
LoS (Direc-Direc)	Combined Path	$37.0 + 19.2 \log_{10} d$	37.0	1.9	6.9
LoS (Direc-Direc)	[60 190]	$-20.6 + 47.6 \log_{10} d$	−20.6	4.8	7.8
LoS (Direc-Direc)	[255 440]	$-18.1 + 40.5 \log_{10} d$	−18.1	4.1	5.0
LoS (Direc-Omni)	Combined Path	$51 + 12.7 \log_{10} d$	50.5	1.3	4.9
LoS (Direc–Omni)	[60 190]	$-2.5 + 39.9 \log_{10} d$	−2.5	4.0	5.9
LoS (Direc–Omni)	[255 440]	$24.3 + 23.5 \log_{10} d$	24.3	2.4	3.5

Table 13.6 Close-in path loss model parameters (PLE, σ)−FR1. Redistributed from [22] under CC BY-NC-ND 4.0.

Environment	Description	Model	PLE	σ (dB)
LoS (Direc-Direc)	Combined Path	$35.7 + 19.8 \log_{10} d/d_o$	2.0	6.9
LoS (Direc-Direc)	[60 190]	$35.7 + 20.7 \log_{10} d/d_o$	2.1	7.7
LoS (Direc-Direc)	[255 440]	$35.7 + 19.4 \log_{10} d/d_o$	1.9	5.0
LoS (Direc-Omni)	Combined Path	$39.3 + 20.7 \log_{10} d/d_o$	1.8	4.9
LoS (Direc-Omni)	[60 190]	$39.3 + 19.9 \log_{10} d/d_o$	2.0	5.9
LoS (Direc-Omni)	[255 440]	$39.3 + 17.6 \log_{10} d/d_o$	1.8	3.5

Table 13.7 RMSE for empirical path loss models (combined path). Redistributed from [22] under CC BY-NC-ND 4.0.

Environment	Model	RMSE (dB)
Direc–Direc	FI	5.0
	CI	4.7
Direc–Omni	FI	3.8
	CI	3.2

"breakpoint." This break point is defined for distances where the first Fresnel zone between and around TX and RX includes the ground. Nonetheless, coverage planning is usually required for calculating link budgets. The shadow fading variable provides communication system planners with some flexibility in estimating the path loss from empirical models, as it is expected that in most cases these parameters will vary between environments. In this regard, suitable probability distributions that describe the shadow variable need to be investigated to improve coverage planning from empirical models.

The delay spread at each measurement location was obtained from the respective APDP. In Figure 13.8(a), representative APDPs for directional and omnidirectional measurements are shown for a TX−RX separation of ≈300 m. The

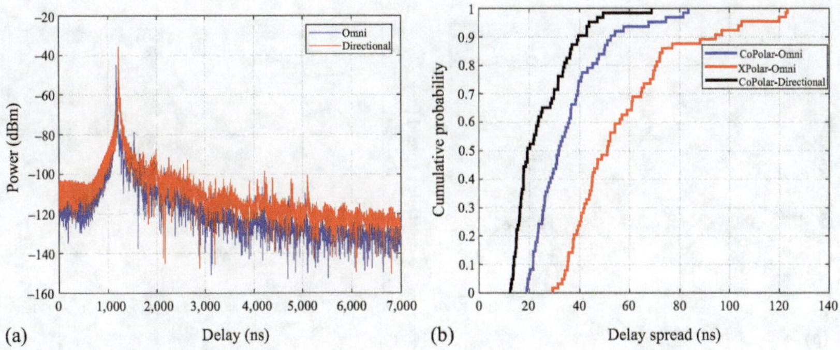

Figure 13.8 (a) Averaged power–delay profile (d ≈ 300 m) and (b) delay spread CDF for omnidirectional (co- and cross-polar) and directional FR1 OTA V2X channel sounding measurements. Redistributed from [22] under CC BY-NC-ND 4.0.

delay spread from the APDPs can be obtained from

$$\tau_{rms} = \sqrt{\frac{\int_0^\infty (\tau - \overline{\tau})^2 \cdot \text{APDP}(\tau) d\tau}{\int_0^\infty \text{APDP}(\tau) d\tau}} \tag{13.5}$$

in which τ is the time delay of each received MPC, and $\overline{\tau}$ is the mean delay of the whole channel. The cumulative distribution function (CDF) of the delay spread from the FR1 measurement campaign is shown in Figure 13.8(b), and the relationship between the delay spread and the TX−RX separation distance is presented in Figure 13.9(a). From both the figures, it can be observed that the *Direc–Direc* setup significantly reduces the time dispersion in the channel. For measurements with the omnidirectional RX, the delay spread generally increases with distance, whereas the measurement results from the directional antenna are bounded for the same antenna locations. Foliage causes the delay spread to increase as the TX−RX separation distance increases. Thereafter, a prolonged deep fade (as observed in the path loss values) can be noticed along the route from 320 m to 370 m.

In Figure 13.9(b), a plot of the path loss versus the RMS delay spread is presented for both an omnidirectional and a directional RX. From the plot, a correlation between path loss and delay spread for both antenna types can be observed. Generally, the results show that as the path loss increases, the delay spread also increases which aligns with the academic literature [25]. With regards to the interdependencies between these parameters, there is a weak correlation between delay spread and path loss for the omnidirectional antenna (low R^2), whereas a strong correlation exists when a directional antenna is used as the receiver ($R^2 = 0.77$).

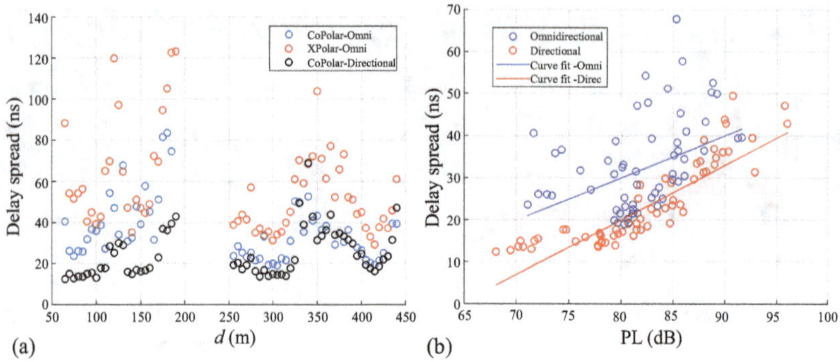

Figure 13.9 *(a) Delay spread versus distance and (b) delay spread versus path loss for omnidirectional and directional FR1 OTA V2X channel sounding measurements, excluding measurement outliers. Redistributed from [22] under CC BY-NC-ND 4.0.*

Table 13.8 *EVM PDSCH limits from 3GPP TS 38.101-1 [16]*

PDSCH modulation	Limit (%)
QPSK	18.50
16-QAM	13.50
64-QAM	9.00
256-QAM	4.50

13.4.2 FR1 error vector magnitude

As an alternative to evaluating the end-to-end (E2E) wireless communication link, EVM has been adopted as a figure of merit for the network. This approach allows wireless network providers to estimate the performance of the link a priori without a complete communications protocol stack. For each of the modulation schemes adopted in FR1 5G NR, the EVM limits as provided by 3GPP are given in Table 13.8. During the measurements, EVM values that exceeded these limits are discarded. In Figure 13.10(a), the OTA EVM versus distance for the measurement route is displayed. From this figure, a similar shadow fading trend can be observed when compared with the channel sounding results presented in Figures 13.7(a) and (b) and 13.9(a). Observing the OTA EVM results, it can be seen that the EVM from the directional RX is below 4% at most measurement locations. When compared with omnidirectional results, it is evident that the directional antenna significantly improves the E2E QoS. To a large extent, this can be explained by the higher received power for directional measurements, because of the additional antenna gain. The OTA EVM for the 5G NR signals have also been compared with the channel sounding results at the same measurement locations. In Figure 13.10(b), the EVM for the *M*-ary 5G NR waveform versus the delay spread is shown, which

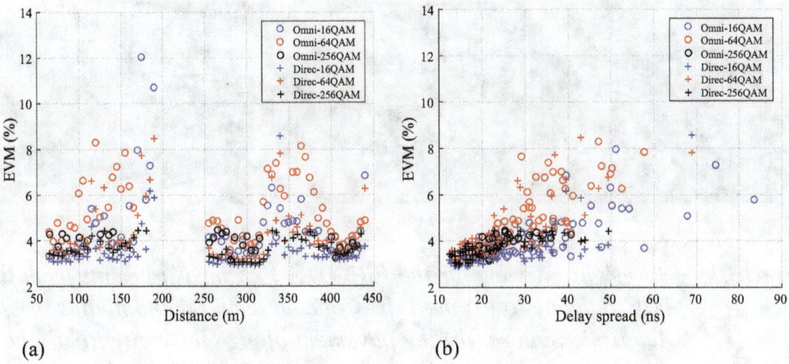

Figure 13.10 EVM for M-ary 5G NR waveforms and directional and omnidirectional FR1 OTA V2X measurements: (a) EVM versus distance and (b) EVM versus delay spread. Redistributed from [22] under CC BY-NC-ND 4.0.

also displays a trend of increasing EVM with delay spread. In a live wireless communication network, it is expected that the receiver communication system adopts techniques, such as dynamic rate switching and error correction schemes, in low signal-to-noise environments. At an additional computational cost, directional antennas that are capable of beam-steering or beam-switching can be adopted at the receiver to further enhance the E2E QoS.

13.4.3 FR2 large- and small-scale fading: received power and delay spread

In the previously published work the impact of a vehicle's bodywork on the absorption and reflection of mmWaves has been explained for various types of vehicles [26,27]. With the acquired knowledge from these fundamental channel evaluation measurements, it is possible to add realism to the tested scenarios by evaluating outdoor V2X performance. The main V2X scenario investigated here is the communication path from the outdoor balcony of the third floor Boardroom of the University of Warwick's International Manufacturing Centre (IMC), to the IMC Concourse, its neighbouring Academic Square and the pavements along Library Road and University Road. Within the MFM CAV testbed, this area contains potential routes on which low-speed last-mile autonomous pods are to be deployed. Hence, combining the outdoor V2X results with the previous channel evaluation inside an anechoic chamber of such full vehicle provides a detailed indication of the potential of FR2 V2X communication for the scenario. The TX location is $\approx$11.7 m above the Concourse, while RX is positioned at a height of 1.2 m, moved to various positions on the Concourse and Academic Square, and manually boresight aligned with TX at every location. In order to enable the largest possible communication distance for the set-up described in Section 13.3.2, two waveguide horn antennas with a typical gain of 25 dBi and an HPBW of 10° are initially used

(a) (b) (c)

Figure 13.11 Boresight TX view for the FR2 OTA V2X measurements from the IMC Balcony to (a) the IMC Concourse and the footpaths of Academic Square, (b) the pavement along University Road and (c) the pavement along Library Road. A waveguide horn antenna with a typical gain of 25 dBi and an HPBW of 10° is used as TX.

Figure 13.12 Photos of the FR2 OTA V2X measurements from the IMC Balcony to the IMC Concourse and the footpaths of Academic Square varying the line-of-sight distance between TX and RX from 32 to 35, 40, 63, 64, 69, 69, 85, 119, 159, 173 and 173 m (RX boresight view, TX encircled). Both a waveguide horn antenna with a typical gain of 25 dBi and HPBW of 10° (see photos); and an ultra-wideband omnidirectional antenna with a typical gain of 6 dBi and HPBW of 20° in the elevation plane are used as RX.

as both TX and RX. For the channel sounding experiments, the FZC-waveform is generated at a fixed output power of 20 dBm, and the excess path loss and RMS delay spread values are determined from the received signals. Dividing the received power by 400 MHz, the largest foreseen channel bandwidth for 5G NR FR2, results in the representative received power spectral density (RPSD). Figures 13.11 and 13.12 display the boresight views of TX and RX, respectively, for various antenna locations on the IMC Concourse and Academic Square.

Figure 13.13 presents the RPSD and the delay spread for various TX–RX separations, for both directional, 10° HPBW TX and RX, as well as an

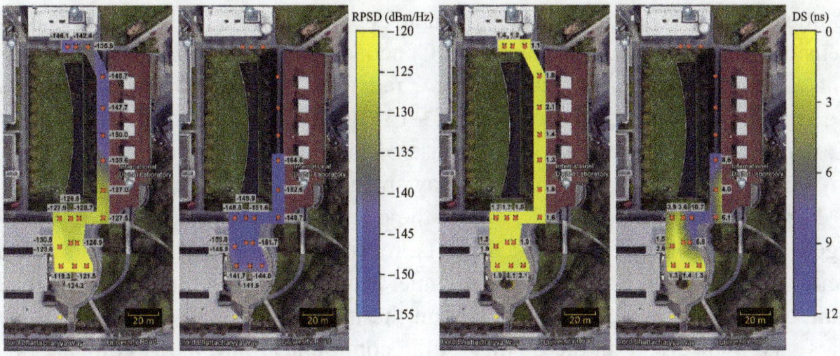

(a) Directional RX (b) Omnidirectional RX (c) Directional RX (d) Omnidirectional RX

Figure 13.13 Received power spectral density (a and b) and RMS delay spread (c and d) at various locations on the IMC Concourse and Academic Square with the transmitter located 11.7 m above the pavement (yellow dot) and output power P_{TX} = 20 dBm. The receiver is positioned 1.2 m above the pavement and its location is step-wise varied (red dots). The measured values (grey boxes) are converted into a contour plot through triangulation, interpolation and smoothing to visualize the highly likely in-between trend. A waveguide horn antenna with a typical gain of 25 dBi and HPBW of 10° is used as TX, and either (a and c) the same directional antenna is used as RX, or (b and d) an ultra-wideband omnidirectional antenna with a typical gain of 6 dBi and HPBW of 20° in the elevation plane is used as RX. Background images were obtained from Google Maps.

omnidirectional 20° HPBW receiver. As will be explained in more detail in Section 13.4.4, an *M*-ary modulation-dependent threshold behaviour exists between RPSD and the possible data throughput, with values below −140 dBm/Hz unable to receive 64-QAM-modulated data. For the directional RX, several RX positions along Academic Square the LoS-path are considerably blocked by tree branches, reducing the RPSD to below −140 dBm/Hz. The RPSD varies significantly due to obscuration by trees and lampposts, as is visualized by the strong colour variation in the contour plots. The lowest RPSD, and thus highest path loss, down to −150 dBm/Hz is recorded in the centre of the Academic Square footpath along the International Digital Laboratory (IDL), at which the branches of the row of trees alongside the path are arranged such that most mmWave radiation is absorbed. Depending on the level of LoS-obscuration due to the leafless branches of trees, the RPSD can vary up to 13 dB for different antenna constellations with very similar TX–RX separation. Surprisingly, for the RX location at the far end of the Academic Square at the other end of the small pond, the RPSD is increased with 5.4 dB if RX is horizontally instead of boresight aligned. Upon boresight alignment, the LoS-path for this antenna constellation is fully blocked by an extension of the IMC building onto the Concourse, but upon horizontal RX alignment, apparent reflections of the mmWave radiation from the pond's surface allow for an indirect LoS-path and an RPSD of −141 dBm/Hz. For the directional TX and RX, very few MPCs are observed because of the directionality of the 10° HPBW horn antennas, resulting in very low delay spread values, smaller than 2 ns.

For the omnidirectional RX, due to the overall gain reduction of ≈19 dB, the range over which mmWave signals can be received is drastically reduced, and the received signal strength for potential channel evaluation measurements is below the required threshold; hence; the focus is on the understanding of MPCs in the scenarios, and their effect on the quality of the communication link. A comparison of Figure 13.13(a) and (b) shows a similar overall RPSD trend for a directional and omnidirectional RX, with decreased signal loss in the centre of the IMC Concourse due to a tree and lamppost, as well as at the start of the Academic Square footpath along the IDL building due to the previously described row of trees alongside the path. A comparison of the delay spread graphs in Figure 13.13(c) and (d) displays that the number of MPCs increases considerably when RX is omnidirectional, resulting in a larger delay spread. Highest delay spread values are determined for RX locations with both IMC Concourse lampposts close to the LoS path and for locations very near the IDL building, but their values are still so small that they will not likely cause intersymbol interference.

The second scenario uses the same TX location on the balcony of the third floor IMC Boardroom, but now boresight aligned to RX locations on pavements along Library Road and University Road. Figure 13.14 displays the boresight views of RX for various antenna locations, with the corresponding boresight views from TX presented in Figure 13.11(b) and (c). Because of live traffic on the road and pedestrians on the pavement, the measurement duration is adapted such that related obstruction effects can be distinguished from obscuring effects related to the stationary communication path, which are of primary interest. Figure 13.15 displays the RPSD and the delay spread for various TX–RX separations, for both a directional and an omnidirectional RX. Similar to the previous directional RX scenario, for different antenna constellations along Library Road with very similar TX–RX separation, the RPSD can vary up to 14 dB. This is again due to the different levels of LoS-path obscuration by a varying number of leafless branches of trees. The particularly low-RPSD region on University Road is due to a huge tree at the corner of the IMC Concourse. Interestingly, for larger TX–RX separations, boresight realignment of the antennas results in paths that are less obscured, probably because the tree's leafless crown has a lower density of branches. As has been mentioned for the previous scenario, very low delay spread values are recorded due to a limited number of observed MPCs because of the directionality of the 10° HPBW horn antennas.

A comparison of the RPSD graphs for directional and omnidirectional RX shows a similar overall behaviour, except for the location nearest to TX. As this location is closest to TX and has a clear LoS, it is best explained as due to imperfect boresight antenna alignment, as the omnidirectional antenna's elevation HPBW is

Figure 13.14 Photos of the FR2 OTA V2X measurements from the IMC Balcony to the pavement along Library Road and University Road varying the line-of-sight distance between TX and RX from 90 to 76, 67, 40, 48, 48, 68, 92, 148, 170, 120 and 90 m (RX boresight view, TX encircled). Both a waveguide horn antenna with a typical gain of 25 dBi and HPBW of 10° (see photos); and an ultra-wideband omnidirectional antenna with a typical gain of 6 dBi and HPBW of 20° in the elevation plane are used as RX.

only 20°. Delay spread values are also very similar for both RX antenna types, except for the location closest to the Library Road entrance of the Humanities Building. There, the combination of several MPCs and a large path loss for the LoS-signal results in an increased delay spread of 10 ns, which is still substantially lower than values that could lead to intersymbol interference.

In Figure 13.16(a) and (b), the empirical models obtained from the combined FR2 measurement data are presented for directional and omnidirectional measurements, respectively. Similar to the analysis of the FR1 measurements in the previous section, the parameters α and β of the FI model were obtained by performing a LS regression to the measured path loss. In Table 13.9, a summary of the parameters obtained for the FI models is presented. The PLEs/slopes are high and between $\approx$3 and 4.5 for the directional and omnidirectional measurements, respectively, corresponding to results from the few similar measurement campaigns that have been published in the academic literature. As for the FR1 results, the large shadow fading variable provides communication system planners with great flexibility in estimating the path loss from empirical models. With σ approaching 9 dB for both antenna types, this means that the actual received power at a location in the presented scenarios can be up to eight times smaller or larger than calculated from the FI model. For this particular environment, it thus highlights the large impact of foliage and street infrastructure that partially obstruct the LoS communication path.

(a) Directional RX (b) Omnidirectional RX

(c) Directional RX (d) Omnidirectional RX

Figure 13.15 *Received power spectral density (a and b) and RMS delay spread (c and d) at various locations on the pavement along Library Road and University Road with the transmitter located 11.7 m above the pavement (yellow dot) and output power $P_{TX} = 20$ dBm. The receiver is positioned 1.2 m above the pavement and its location is step-wise varied (red dots). The measured values (grey boxes) are converted into a contour plot through triangulation, interpolation and smoothing to visualize the highly likely in-between trend. A waveguide horn antenna with a typical gain of 25 dBi and HPBW of 10° is used as TX, and either (a and c) the same directional antenna is used as RX, or (b and d) an ultra-wideband omnidirectional antenna with a typical gain of 6 dBi and HPBW of 20° in the elevation plane is used as RX. Background images were obtained from Google Maps.*

13.4.4 FR2 error vector magnitude and throughput

The FR2 research presented here is aimed to understand how the typical signal quality parameters of a vehicular communication channel depend on its environment. Of particular interest are the EVM and the achieved data throughput, which provide information on the channel's usability for specific OTA V2X information transfer. As the QoS parameter throughput is a performance indicator, its measurement typically requires an E2E data connection. Measuring EVM permits the a priori estimation of the achievable throughput without the need for a full protocol stack. Furthermore, both evaluated parameters can be related to the previously determined RPSD and delay spread values, in order to observe general trends and dependencies.

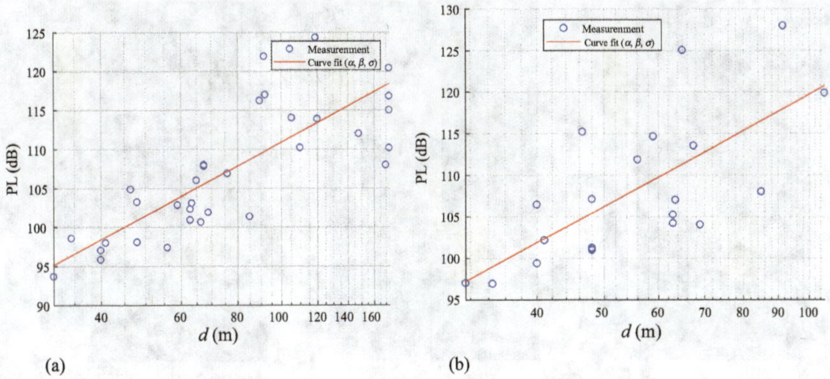

Figure 13.16 FI path loss model fits for omnidirectional and directional FR2 OTA V2X channel sounding measurements: (a) directional RX and (b) omnidirectional RX

Table 13.9 Floating intercept path loss model parameters (α, β, σ)–FR2

Environment	**Model**	***α***	***β***	***σ* (dB)**
LoS (Direc–Direc)	$48.0 + 31.5 \log_{10}d$	48.0	3.2	8.6
LoS (Direc–Omni)	$30.2 + 44.7 \log_{10}d$	30.2	4.5	8.7

In order to enable the largest possible communication distance for the high-order 64-QAM modulation scheme, two waveguide horn antennas with a typical gain of 25 dBi and HPBW of 10° are used as both TX and RX. For all antenna constellations, the TX output power is varied, and EVM and data throughput values are recorded. Because of live traffic on the road and pedestrians on the pavement, the measurement duration is adapted such that related obstruction effects can be distinguished from obscuring effects related to the stationary communication path, which are of primary interest. For both presented FR2 scenarios in Figures 13.17 and 13.18, with TX−RX separations ranging from 32 to 173 m, and TX output power varying from 1 to 24 dBm accordingly, a maximum throughput of 2.8 Gb/s is observed for nearly all positions. Only when the LOS path is obscured by multiple leafless branches of trees alongside University Road, the throughput reduces to 1.7 Gb/s. Depending on the level of LoS-obscuration due to the leafless branches of trees, the TX output power required for maximum data throughput can vary up to 14 dB for different antenna constellations with very similar TX−RX separation, with associated EVM values of ≈–26 dB. For several RX positions along Academic Square the LoS-path is considerably blocked by tree branches, reducing the throughput to zero if the RPSD is reduced below −140 dBm/Hz.

As described in the previous section, a steep RPSD-threshold region exists in which a ≈1 dB change in received power can reduce the maximum throughput to

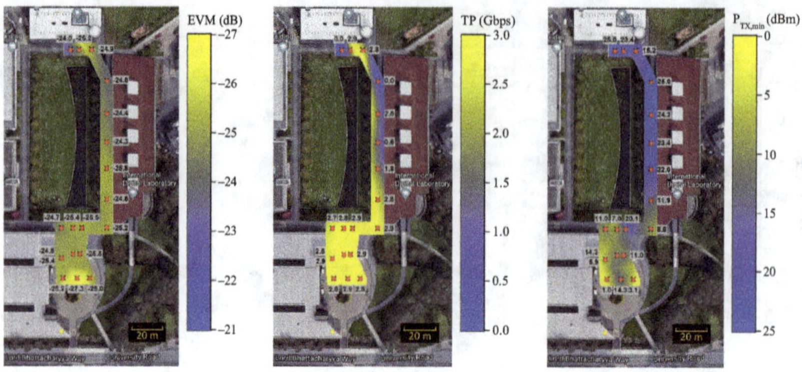

(a) (b) (c)

Figure 13.17 (a) Error vector magnitude corresponding with the maximum achievable throughput in (b), for minimum required P_{TX} in (c), at various locations on the IMC Concourse and Academic Square with the transmitter located 11.7 m above the pavement (yellow dot). The receiver is positioned 1.2 m above the pavement and its location is step-wise varied (red dots). Waveguide horn antennas with a typical gain of 25 dBi and HPBW of 10° are used as both TX and RX. Background images were obtained from Google Maps.

zero. This is represented in Figure 13.19, which displays the modulation scheme-dependent averaged data throughput versus the EVM and RPSD for a LoS communication link. All plots show a clear threshold behaviour that is almost TX–RX separation independent, hence, is 5G NR specific and is hardly influenced by the V2X communication path's environment. The plots display that for full throughput in a 64-QAM, 16-QAM and QPSK modulation scheme, the minimum required EVM is –25 dB, –18 dB and –10 dB, respectively. Moreover, the minimal required RPSD to achieve this is –135 dBm/Hz, –142 dBm/Hz and –148 dBm/Hz, respectively. Further preliminary results show that this behaviour is antenna HPBW-independent. As mentioned in Section 13.4.2, because the modulation scheme determines the maximum possible data throughput, these results can be used to optimize that throughput in a given scenario by dynamic switching between schemes. For instance, the 64-QAM scheme is used for throughputs larger than 2 Gb/s for signals with an EVM lower than –24 dB and an RPSD higher than –137 dBm/Hz; for an EVM between –24 dB and –17 dB and an RPSD between –137 dBm/Hz and –144 dBm/Hz, the 16-QAM modulation scheme delivers throughputs larger than 0.8 Gb/s; and for an EVM lower than –17 dB and an RPSD lower than –144d Bm/Hz, the QPSK modulation scheme still delivers data according to the curves depicted by the triangles in Figure 13.19.

Two open research topics that will be further investigated if time and weather permit are the understanding of the impact of foliage on mmWave communication links, through additional spot measurements with the trees fully leaved in the

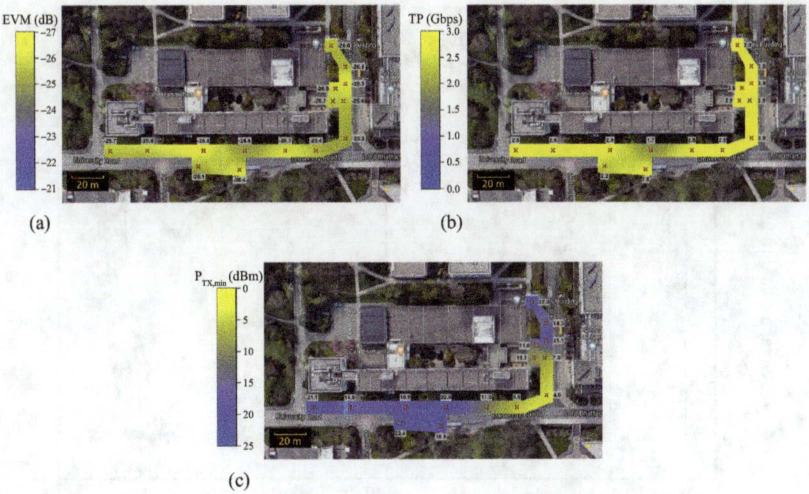

Figure 13.18 *(a) Error vector magnitude corresponding with the maximum achievable throughput in (b), for minimum required P_{TX} in (c), at various locations on the pavement along Library Road and University Road with the transmitter located 11.7 m above the pavement (yellow dot). The receiver is positioned 1.2 m above the pavement and its location is step-wise varied (red dots). Waveguide horn antennas with a typical gain of 25 dBi and HPBW of 10° are used as both TX and RX. Background images were obtained from Google Maps.*

summer, and to understand the HPBW-dependence and constellation scheme dependence of these scenarios, to gain an in-depth understanding of the implications of infrastructure antenna choice and location on typical OTA V2X communication.

13.4.5 Influence of weather, air quality and traffic

Both FR2 scenarios were investigated in good weather conditions (no precipitation) with an outside temperature of 15 °C and relative humidity of 55% (7.0 g/m^3) for the first scenario, and 13 °C and 68% (7.7 g/m^3) for the second scenario. Such temperature differences only affect the results through the temperature deviation from the device calibrations, for which advised corrections were made (<0.3 dB). The water vapour in the air accounted for a negligible calculated attenuation of 0.1 dB/km. The Air Pollution Banding during the measurements was low, with PM2.5 values (atmospheric particulate matter with d < 2.5 μm) from the nearest monitoring station (UKA00592) only ranging from 1.8 to 16.8 μg/m^3. Traffic influences are most difficult to assess, as they critically depend on the vehicular position with respect to the LoS-path. Large vehicles were able to fully block the mmWave signal, or if driving in the opposite lane, reduced the received signal with

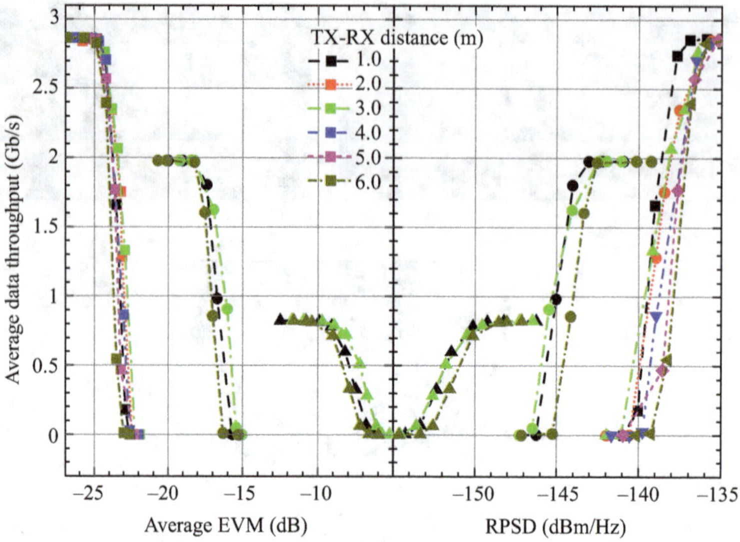

Figure 13.19 Modulation scheme-dependent plots of the measured averaged data throughput versus the EVM and RPSD in a line-of-sight communication link between a 10° HPBW waveguide horn antenna and an ultra-wideband omnidirectional antenna for various antenna separations (see legend). On the left, the throughput is plotted versus the measured averaged EVM, and on the right, the throughput is plotted versus the RPSD. The code rate for 64-QAM (solid squares) and 16-QAM (solid circles) is 7/8; for QPSK (solid triangles) it is 3/4.

13 dB during ≈1.5 s. On the other hand, small vehicles were able to reflect an extra area of the TX-beam towards RX, such that the RPSD increased with ≈1.5 dB and the throughput increased correspondingly.

13.5 Conclusions

This chapter presented a combination of FR1 and FR2 OTA V2X channel sounding and channel evaluation measurements, linking the fundamental properties of such channels with the signal's QoS and QoE properties on which vehicular applications and users rely. The detailed on-campus measurements emphasize the importance of unobstructed wireless communication paths between the transmitter and receiver, and the large path loss due to road-side foliage, which is particularly difficult to overcome by mmWave radiation. The results included large- and small-scale fading effects that have been interpreted in path loss models and through delay spread analysis. The channel quality was evaluated in terms of EVM and throughput, for both a directional and an omnidirectional receiver. The former antenna choice produced a higher signal quality, reduced outages for higher QAM schemes, as well as reduced time dispersion. Real-world vehicular communication applications using directional antennas, however, will require additional overhead for antenna beamsteering or switching to guarantee alignment with the transmitting vehicle, road-side infrastructure or cellular base station. In order to achieve further understanding of the effects of MPCs, and to optimize signal processing overhead versus signal quality, future research work could include investigating the presented scenarios using antennas with other or adaptable HPBWs, as well as for additional modulation and coding schemes.

Acknowledgements

This work was supported in part by the WMG Centre High Value Manufacturing Catapult, University of Warwick, Coventry, UK, in part by the Wireless Infrastructure Group, Bellshill, UK, in part by Midlands Future Mobility, University of Warwick, Coventry, UK and in part by the EPSRC through the UK-RAS Network.

References

[1] Future of Mobility. UK Government Office for Science; 2019. Accessed: September 16, 2021. Available from: https://assets.publishing.service.gov.uk/government/uploads/system/uploads/attachment_data/file/780868/future_of_mobility_final.pdf.

[2] Lu M., editor, *Evaluation of Intelligent Road Transport Systems: Methods and Results*. Stevenage, UK: IET; 2016.

[3] Lu M., editor, *Cooperative Intelligent Transport Systems: Towards High-Level Automated Driving*. Stevenage, UK: IET; 2019.

[4] SAE International. Taxonomy and Definitions for Terms Related to Driving Automation Systems for On-Road Motor Vehicles; 2018. Accessed: September 16, 2021. Available from: https://www.sae.org/standards/content/j3016_201806/.

[5] Matheus K. and Königseder T., *Automotive Ethernet*. 3rd ed. Cambridge, UK: Cambridge University Press; 2021.

[6] IEEE Std 16094-2016. IEEE Standard for Wireless Access in Vehicular Environments (WAVE) – Multi-Channel Operation; 2016. p. 1–94.

[7] ETSI. Intelligent Transport Systems (ITS); Decentralized Congestion Control Mechanisms for Intelligent Transport Systems operating in the 5 GHz Range; Access Layer Part. European Telecommunications Standards Institute; 2018. 102 687. Version 1.2.1.

[8] Arena F, Pau G, and Severino A., A Review on IEEE 802.11p for Intelligent Transportation Systems. *Journal of Sensor and Actuator Networks*. 2020;9(2).

[9] Jiang D and Delgrossi L., IEEE 802.11p: Towards an International Standard for Wireless Access in Vehicular Environments. In: VTC Spring 2008 – IEEE Vehicular Technology Conference; 2008. p. 2036–2040.

[10] Anwar W, Franchi N, and Fettweis G., Physical Layer Evaluation of V2X Communications Technologies: 5G NR-V2X, LTE-V2X, IEEE 802.11bd, and IEEE 802.11p. In: 2019 IEEE 90th Vehicular Technology Conference (VTC2019-Fall); 2019. p. 1–7.

[11] Rosier H., Introduction to the Vehicle-to-Everything Communications Service V2X Feature in 3GPP Release 14. Rohde & Schwarz GmbH & Co; 2018. PD 3607.7970.52. Version 01.00.

[12] Stuhlfauth R., Wireless Communications for Automotive Applications. Rohde & Schwarz GmbH & Co; 2021. PD 3608.6890.5. Version 01.00.

[13] 3GPP. Technical Specification Group Services and System Aspects; Architecture Enhancements for 5G System (5GS) to Support Vehicle-to-Everything (V2X) Services (Release 16). 3rd Generation Partnership Project; 2020. 23.287. Version 16.4.0.

[14] 3GPP. Technical Specification Group Radio Access Network; NR; Study on NR Vehicle-to-Everything (V2X) (Release 16). 3rd Generation Partnership Project; 2019. 38.885. Version 16.0.0.

[15] 3GPP. Technical Specification Group Radio Access Network; Overall description of Radio Access Network (RAN) Aspects for Vehicle-to-Everything (V2X) Based on LTE and NR (Release 16). 3rd Generation Partnership Project; 2020. 37.985. Version 16.0.0.

[16] 3GPP. Technical Specification Group Radio Access Network; NR; User Equipment (UE) Radio Transmission and Reception; Part 1: Range 1 Standalone (Release 15). 3rd Generation Partnership Project; 2020. 38.101-1. Version 16.5.0.

[17] 3GPP. Technical Specification Group Radio Access Network; NR; User Equipment (UE) Radio Transmission and Reception; Part 2: Range 2 Standalone (Release 16). 3rd Generation Partnership Project; 2020. 38.101-2. Version 16.5.0.

[18] Midlands Future Mobility. 5G Research Services – Midlands Future Mobility; 2020. Accessed: September 16, 2021. Available from: https://midlandsfuturemobility.co.uk/services/5g-research-services.

[19] Fraunhofer Institute for Integrated Circuits IIS. 5G Bavaria Automotive Test Bed; 2021. Accessed: September 16, 2021. Available from: https://www.iis.fraunhofer.de/en/ff/kom/mobile-kom/5g-bavaria/5g-testbed-automotive.

[20] Naehring A. Demystifying Over-the-Air (OTA) Testing – Important Antenna Parameters, Test System Setup and Calibration. Rohde & Schwarz GmbH & Co; 2019. PD 5216.3789.52. Version 01.00.

[21] Olasupo TO., Otero CE., Otero LD., *et al.* Path Loss Models for Low-Power, Low-Data Rate Sensor Nodes for Smart Car Parking Systems. *IEEE Transactions on Intelligent Transportation Systems*. 2018; 19(6): 1774–1783.

[22] Adegoke EI., Kampert E., and Higgins MD. Channel Modeling and Over-the-Air Signal Quality at 3.5 GHz for 5G New Radio. IEEE Access. 2021; 9: 11183–11193.

[23] Frank R., Zadoff S., and Heimiller R., Phase Shift Pulse Codes With Good Periodic Correlation Properties (Corresp.). *IRE Transactions on Information Theory*. 1962; 8(6): 381–382.

[24] Naseef M. and Roessler A., 5G NR Signal Generation and Analysis. Rohde & Schwarz GmbH & Co; 2019. GFM322. Version 1e.

[25] Feuerstein MJ., Blackard KL., Rappaport TS., *et al.* Path Loss, Delay Spread, and Outage Models as Functions of Antenna Height for Microcellular System Design. *IEEE Transactions on Vehicular Technology*. 1994;43 (3): 487–498.

[26] Kampert E., Jennings PA., and Higgins MD., Investigating the V2V Millimeter-Wave Channel Near a Vehicular Headlight in an Engine Bay. *IEEE Communications Letters*. 2018; 22(7): 1506–1509.

[27] Kampert E., Schettler C., Woodman R., *et al.* Millimeter-Wave Communication for a Last-Mile Autonomous Transport Vehicle. *IEEE Access*. 2020; 8: 8386–8392.

Chapter 14

Performance evaluation and compliance testing of 5G/6G base stations, user-devices and systems

Lars Jacob Foged[1], Francesco Saccardi[1], Alessandro Scannavini[1] and Nicolas Gross[2]

Antennas and electromagnetic characterizations have been challenging topics for more than 50 years. The measurement of an antenna or device is commonly an evaluation of the radiation properties in the far field (FF) assuming that the origin is positioned at an infinite distance from the observation point. The effort to perform the measurement in sufficiently good approximation to FF condition is a paradox since a major part of modern communications supported by antennas and devices occur at a finite distance and often in near field (NF). The FF condition is also often referred to as plane wave condition since radiation tends to approximate a plane wave at infinity. Far from the antenna or device origin, electromagnetic radiation will have spread out sufficiently that it will appear as uniform amplitude and phase on a plane perpendicular to its direction of travel. Thus, at sufficient distance an antenna will appear to radiate plane waves in all directions. Due to reciprocity, the FF radiation of an antenna can also be measured by exposing the entire antenna locally to an approximate plane wave.

The first successful attempt to measure electromagnetic radiation in free space or antenna patterns as we call them today was made by Heinrich Hertz in 1886 [1]. In his experiment Hertz confirmed Maxwell's theories about the existence of electromagnetic radiation. The unit of frequency, one cycle per second, is named 'hertz', in recognition of this and other pioneering discoveries.

Hertz used a spark transmitter consisting of a dipole antenna with a spark gap powered by high-voltage pulses from a Ruhmkorff coil as his antenna/device under test (AUT/DUT). His receiver was a loop antenna with a micrometre spark gap between the elements. By moving the receiver in 3D space around the transmitter, he was able to determine the radiation intensity with spatial direction from the intensity of the fields (sparks) in the receiver. The technologies employed today in antenna measurements have greatly improved, but even today, the radiation from an antenna or device in each spatial direction is measured as the coupling between

[1]Microwave Vision Italy (MVI), Pomezia, Italy
[2]Microwave Vision Group (MVG), Paris, France

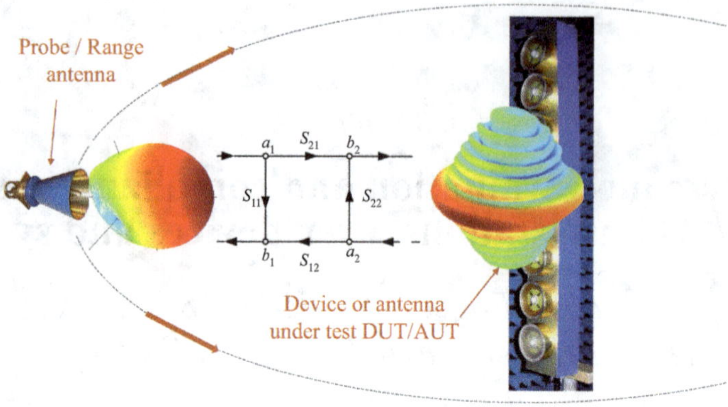

Figure 14.1 Illustration of a typical antenna or device measurements scenario. The coupling between the probe/range antenna and the AUT/DUT is measured in different spatial directions.

the source of the radiation and the probe/range antenna [2–4]. An example of a typical measurement set-up is shown in Figure 14.1. In most measurements, either both the probe/range antenna is fixed and the AUT/DUT is moving or vice versa. As we will see later, in some systems, the discrete positions of the moving probe can be substituted by an array of highly similar probes that can be scanned electronically to achieve fast measurements [5].

A system set-up like the one shown in Figure 14.1 is capable of measuring both passive and active devices. The former are devices which can be connected to an external signal source or receiver (connectorized device) and the latter are devices with an internal signal generator or receiver. For example, antennas of modern wireless devices fall in this second category. Such antennas are in fact strongly integrated with the whole radio frequency (RF) system; hence the users cannot access directly to its outputs/inputs. Measurements of such active devices are typically called over-the-air (OTA) testing and are studied and standardized by different organizations (e.g. CTIA and 3GPP). To measure active and/or passive devices, we must first calibrate the measurement system in gain using a well-known reference antenna.

14.1 Antenna gain and OTA measurements

Classical antenna gain measurements are based on the Friis transmission formula shown in (14.1). Harald Trap Friis was a Danish scientist known for his significant contributions to radio propagation, radio astronomy and radar while working at the US-based Bell Laboratories [6]. His two Friis formulas on noise and transmission remain widely used today. Applied to the scenario in Figure 14.1, the formula links the input power of the transmitting antenna to the output power of the receiving antenna which is proportional to the wavelength square and the gains of the two antennas, and inversely proportional to the square of the probe/DUT separation distance, R. The term $(4\pi R/\lambda)^2$ is also called the *free-space path loss* (FSPL) or *pathloss* and results from the spherical spreading of the radiated energy:

$$S_{21}^{dB} = 10\log\left(\frac{\text{Power}_{\text{OUT}}}{\text{Power}_{\text{IN}}}\right) = G_{\text{Probe}}^{dB} + G_{\text{DUT}}^{dB} - 10\log\left(\frac{4\,\pi\,R}{\lambda}\right)^2 \tag{14.1}$$

The Friis transmission formula is valid if the antennas are perfectly impedance and polarization matched. If the reflection coefficients, Γ, of the probe and/or the DUT are known, a correction factor of the form $1-|\Gamma|^2$ can be inserted in the Friis formula to correct for mismatch for both the probe and DUT. Polarization mismatch can be quantified as well. It is recommended to use well-matched antennas and polarization to align the system when performing such measurements.

The measured S_{21} of the probe and DUT is directly linked to the gain of the two antennas and the distance between them as shown in the previous equation. It is worth noticing that at a separation distance of just one wavelength the path loss is already ~−22 dB. Each time the separation distance increases by a factor of 2, the path loss increases 6 dB.

A widely used gain measurement technique in both FF and NF is the *Gain-Substitution or Gain-Transfer Technique* [3,4]. This technique requires two measurements, involving a second measurement with a calibrated (known gain) reference antenna as shown in Figure 14.2. Any calibrated, thus antenna with known gain characteristics at the desired frequency can be used as reference. Nevertheless, it is a good practice to involve reference antennas as similar as possible to the antenna to be measured (e.g. dipoles for low directive devices and standard gain horn (SGH) for medium/high directive devices).

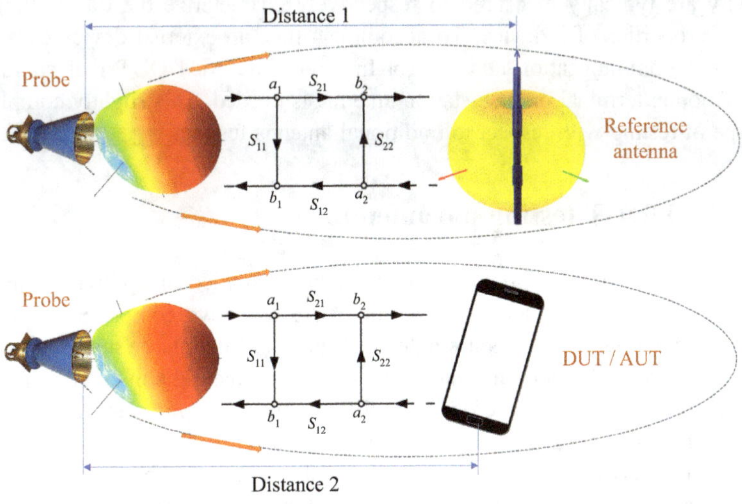

Figure 14.2 System calibration by gain-substitution or gain-transfer technique

The gain of the unknown AUT/DUT is expressed in terms of the known gain of the reference antenna and the measured probe antenna coupling in the two measurements as shown in the following equations:

$$S_{21}^{\mathrm{dB}(1)} = G_{\mathrm{Probe}}^{\mathrm{dB}} + G_{\mathrm{REF}}^{\mathrm{dB}} - \mathrm{FSPL}^{\mathrm{dB}(1)} \quad (14.2)$$

$$S_{21}^{\mathrm{dB}(2)} = G_{\mathrm{Probe}}^{\mathrm{dB}} + G_{\mathrm{DUT}}^{\mathrm{dB}} - \mathrm{FSPL}^{\mathrm{dB}(2)} \quad (14.3)$$

$$G_{\mathrm{DUT}}^{\mathrm{dB}} = S_{21}^{\mathrm{dB}(2)} - S_{21}^{\mathrm{dB}(1)} + G_{\mathrm{REF}}^{\mathrm{dB}} - \Delta\mathrm{FSPL}_{2-1} \quad (14.4)$$

Often the term $\Delta\mathrm{FSPL}_{2-1}$ is considered negligible if the two measurements are performed at the same distance. The term $S_{\mathrm{Cal}}^{\mathrm{dB}}$ is often called the *system calibration* term as it gives the relationship between power in/out and power out/in of the probe and unknown AUT/DUT after system calibration:

$$S_{\mathrm{Cal}}^{\mathrm{dB}} = 10\log\left(\frac{\mathrm{Power}_{\mathrm{Probe}}}{\mathrm{Power}_{\mathrm{DUT}}}\right) = S_{21}^{\mathrm{dB}(1)} - G_{\mathrm{REF}}^{\mathrm{dB}} \quad (14.5)$$

After system calibration, the gain pattern of the unknown AUT can be determined in all spatial directions by adding the system calibration term to the measured coupling values. The *system calibration* is also sometimes called *OTA system* calibration and measurement set-ups such as the one shown in Figure 14.1 are often referred to as a *calibrated OTA set-up*. The purpose of the *OTA system calibration* is to determine the isotropic power 'in-to' or 'out-of' the unknown AUT/DUT antenna, for a given measurement distance. After calibration, the probe becomes a virtual connectorized port of the AUT/DUT. In such calibrated measurement set-up, OTA measurements can now be performed as the pertinent power parameters in-to and out-of the unknown AUT/DUT can be determined from the power parameters measured at the probe input ports. In OTA testing, spatial power quantities related to radiated power and device sensitivity are typically measured to respectively characterize the transmitting and receiving properties of a device. To accomplish this, an external device capable of setting up a communication and control link with the AUT/DUT that could be a wireless/mobile terminal or base-station antenna is needed. This slightly complicates such type of testing with respect to traditional antenna testing.

14.2 AUT/DUT testing parameters

The family of antennas and devices can roughly be divided into active and passive antennas [2]. The passive antenna does not have an active components within its structure. An active antenna is a single- or multi-port antenna system with one or more components integral to the antenna that influence the radiating and/or receiving characteristics. The devices are often amplifiers, but other powered, nonreciprocal, analogue or digital components may also be employed. The term active antenna contracts with passive antenna, whereas the AUT/DUT cannot be both.

Once antennas are integrated and inseparable from an RF system, the testing can no longer be performed at antenna level using traditional antenna performance parameters. Testing must be performed as a system using testing parameters of system performance. Examples of such systems are wireless device and base stations that perform data processing internally and do not allow users to access the outputs/inputs of the antenna directly.

OTA testing of a wireless device involves the determination of transmitter and receiver characteristics such as power, sensitivity and in some cases data throughput. The testing often comprises the effect of antenna, body of the wireless device and other objects often found in the close proximity of the device during use, which can be represented by standardized phantoms: hand, head, torso, etc. The phantoms are representations of average size body parts, with average electrical parameters enabling reproducible and standardized testing of devices.

OTA testing of wireless devices is only slightly more complicated than traditional antenna testing. Such testing requires an external device, capable of setting up a communication, and control link with the devices under test is needed. In the case of wireless device testing such as headsets, this is often a base-station emulator. In the case of base-station testing, it is often a control device on the fibre-optical link with the base station. The external device regulates the communication and maintains the test mode of the device.

By measuring two spatial power quantities, usually in spherical coordinates, such as effective isotropic radiated power (EIRP), and effective isotropic sensitivity (EIS), both transmitter and receiver properties can be characterized respectively. The definitions of these two terms are the following:

1. $\mathrm{EIRP}(\theta, \varphi)$. The equivalent/effective isotropically radiated power is the amount of power that a theoretical isotropic antenna would emit to produce the power density observed in a given direction.
2. $\mathrm{EIS}(\theta, \varphi)$. The EIS of a receiver, including its antenna, expresses the level of minimum received power of an equivalent isotropic antenna to achieve a certain sensitivity threshold according to a pre-established bit error rate (ER).

The definitions of EIRP and EIS are based on radiated power with respect to an isotropic antenna, making the definition very similar to isotropic gain. Like antenna gain, these parameters can be measured directly in a gain calibrated, OTA test set-up as discussed in Section 14.1. The gain calibration can be performed following standard measurement techniques where the gain by substitution method is the most common method to calibrate the OTA system. A reference antenna with a known gain is measured, providing the *OTA system calibration* factor of the system. By virtue of the calibration, the input/output isotropic power of the wireless device, thus EIRP and EIS performance parameters can be measured directly at the input/output port of the test antenna or probe.

The metrics of integrated power pattern or total radiated power (TRP) and total isotropic sensitivity (TIS), sometimes also called total radiated sensitivity or total receiver sensitivity (TRS), are the principal figures of merit in the international standardization groups such as 3GPP and CTIA [7–10]. These terms also represent the average directional line-of-sight performance of the transmitter and receiver, respectively.

TRP is a measure of the radiated power from a wireless device. It is defined as the integral of the power radiates by the wireless device in all directions over the entire sphere for both polarizations:

$$\mathrm{EIRP}(\theta,\varphi) = \mathrm{TRP}\, D_{\mathrm{DUT}}(\theta,\varphi) \tag{14.6}$$

$$\mathrm{TRP} = \frac{1}{4\pi}\iint \mathrm{EIRP}(\theta,\varphi)\,\sin(\theta) d\theta d\varphi \tag{14.7}$$

where $D_{\mathrm{DUT}}(\theta,\varphi)$ is the directivity of the DUT.

If the EIRP is measured in discrete samples on a regular angular grid around the AUT/DUT with N samples in θ and M samples in φ for a total of $N \times M$ measurement points, the TRP can be approximated by the summation:

$$\mathrm{TRP} \approx \frac{\pi}{2NM}\sum_{n=0}^{N-1}\sum_{m=0}^{M-1} \mathrm{EIRP}(\theta_n,\varphi_m)\,\sin(\theta_n) \tag{14.8}$$

TIS or TRS refers to the average receiver sensitivity across the entire radiated sphere in order to achieve a specific ER threshold for both polarizations, which represents the overall receiving capability of the DUT. If the EIS is measured in discrete samples on a regular angular grid around the DUT with N samples in θ and M samples in φ for a total of $N \times M$ measurement points, the TIS or TRS can be approximated by a summation:

$$\text{TRS or TIS} \approx \frac{2NM}{\pi}\,\frac{1}{\sum_{n=0}^{N-1}\sum_{m=0}^{M-1}\frac{1}{EIS(\theta_n,\varphi_m)}\sin(\theta_n)} \tag{14.9}$$

The reported summation formulas for TRP, TIS and TRS are valid for regular angular grids. In case an irregular angular grid is used, a normalization factor on each angular term must be defined before the summation.

The traditional AUT/DUT testing parameters for both active and passive antennas' testing are shown in Table 14.1.

14.3 Near field versus far field

The characterization of most of the radiating performances of an AUT/DUT is strongly influenced by the measurement distance (R), which is the distance between the origin of the DUT coordinate system and the probe (or range antenna).

For example, the measurement of the radiation pattern done at different measurement distances could lead to different results, unless R is chosen to be sufficiently large. The 'proper' choice of the measurement distance depends on the size and the working frequency of the AUT/DUT, or in other words, on its electrical size. Figure 14.3 shows the normalized electric field of an electrically small antenna at different distances from the antenna. As can been seen, the deviation among the different distances is relatively small suggesting that measurement

Table 14.1 Common testing parameters for active and passive antennas

Testing parameters		Passive Any application	Active Space/military	5G and IoT
Pattern	Gain and directivity (φ, θ)	Yes	Yes	Yes
Power	TRP and EIRP (φ, θ)	No	Yes	Yes
Sensitivity	TIS/TRS and EIS (φ, θ)	No	No	Yes
System parameters	Device dependent	Efficiency, polarization, etc.	Gain over temperature and frequency, saturated flux density, group delay, etc.	Blocking, spurious emission, channel leakage, error vector magnitude, throughput and channel emulation, etc.

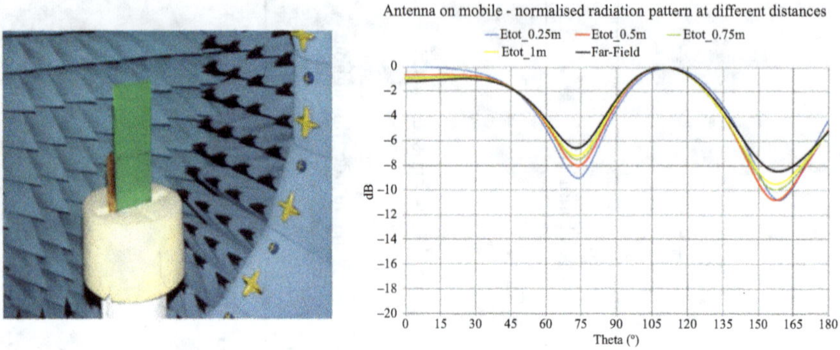

Figure 14.3 *Electrically small device: example of normalized radiation pattern at different distances*

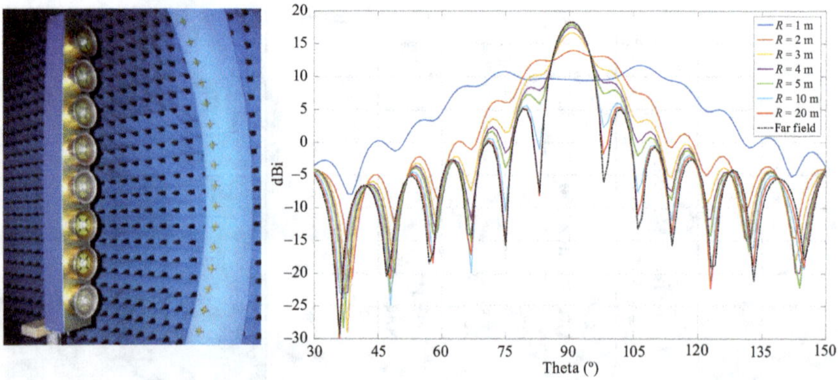

Figure 14.4 *Electrically large device (1.23-m array antenna at 1,970 MHz): example of directivity radiation pattern at different distances*

performed at such distances would lead relative accurate measurements of the radiation pattern.

On the other hand, measuring an electrically large antenna (e.g. like an array) at different distances could lead to significantly different radiation patterns as shown in Figure 14.4.

The main antenna figures of merit (such as gain, directivity, radiation pattern, polarization, etc.) are defined under plane wave or FF conditions. This means that for the receiving parameters, an ideal plane wave (generated by a source placed at infinite distance) impinging on the antenna is assumed. Similarly, the transmission parameters are defined assuming observations points located at infinite distance. The most straightforward way to measure the antenna figure of merits in the 'real world' is to consider a sufficiently large measurement distance so that the plane wave condition is approximated. The larger the measurement distance, the better the plane wave approximation, but unfortunately, such distance in most of the cases cannot be chosen as arbitrarily large because of the limited available spaces and propagation losses which compromise the dynamic range of the measurement. Some guidelines for a good trade-off between measurement accuracy and distance have been suggested [3] and will be described later. Alternatively, proper devices designed to generate plane waves can be considered or data processing can be applied to propagate the acquired data to infinity. These two approaches will be as well discussed in this chapter.

A well-known criterion for a sufficiently large measurement distance to approximate a plane wave condition is described later. Let us assume an AUT/DUT with maximum dimension D_{max} illuminated by a probe placed at distance R. Considering R large enough to avoid the reactive region of the probe [11], the radiation of the probe can be assumed to be spherical. As shown in Figure 14.5, increasing R the spherical wave front tends to approximate a plane wave on the AUT/DUT plane.

The accuracy of this approximation is usually quantified in terms of the phase difference ($\delta_{phase} = 2\pi\delta/\lambda$) between the ideal plane wave and the spherical wave at the edge of the AUT/DUT. The measurement distance is thus usually chosen fixing the maximum acceptable phase difference. A widespread rule of thumb is to set $\delta_{phase} = (\pi/8)$ (or 22.5°). Geometrically we have

$$(R+\delta)^2 = R^2 + \left(\frac{D_{max}}{2}\right)^2 \tag{14.10}$$

which yields

$$\delta = \left(\sqrt{R^2 + \left(\frac{D_{max}}{2}\right)^2} - R\right) \approx \frac{D_{max}{}^2}{8R} \tag{14.11}$$

where the last approximation in (14.11) is obtained with the first-order Taylor expansion of the square root. Imposing a maximum phase difference of $\pi/8$ is

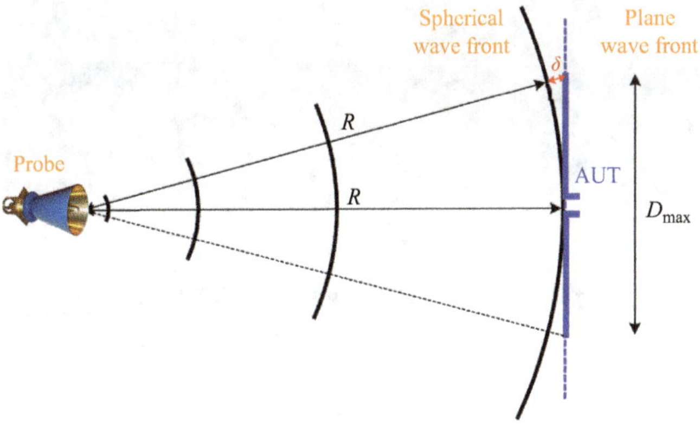

Figure 14.5 Far-field criterion: approximation of spherical wave front to plane wave front

equivalent to consider $= \lambda/16$; hence from (14.11), we obtain

$$R = R_{FF} = \frac{2D_{max}^2}{\lambda} \tag{14.12}$$

where R_{FF} stands for far-field (FF) distance. Larger measurement distances than R_{FF} will lead to phase errors smaller than 22.5° on the edge of the AUT.

As an example, let us consider the array antenna shown in Figure 14.4. Such antenna is 1.23-m long and the shown patterns are at 1,970 MHz ($D_{max} \approx 8\lambda$). Considering (14.12), we obtain $R_{FF} = 20$ m. As expected, the agreement between the pattern measured at 20 m (red trace) and the 'true' FF pattern (black dashed trace) is indeed very good. The minor deviations observable on the nulls depths and on the sidelobes away from the mean beam are due to the plane wave approximation described earlier, which allows 22.5° phase error. Decreasing the measurement distance down to 5 m, the errors on the sidelobes and nulls are further increased but the main beam is still measurable with fairly good accuracy. Instead, considering even more reduced distances would also compromise the measurement of the main beam.

For the same array antenna, the gain error and first sidelobe level (SLL) error varying the measurement distance are shown in Figure 14.6. The gain can either be computed by inverting the Friis equation [see (14.1)] or with the substitution method [3]. The results shown in Figure 14.6 are representative of both methods. In this case the obtained gain error is negligible from distances larger than 10 m, which is half of the FF distance described earlier. The first sidelobe of such antenna is 13 dB below the peak. The SLL error at $R = R_{FF} = 20$ m is approximately 0.2 dB and, at $R = 10$m, is approximately 0.8 dB.

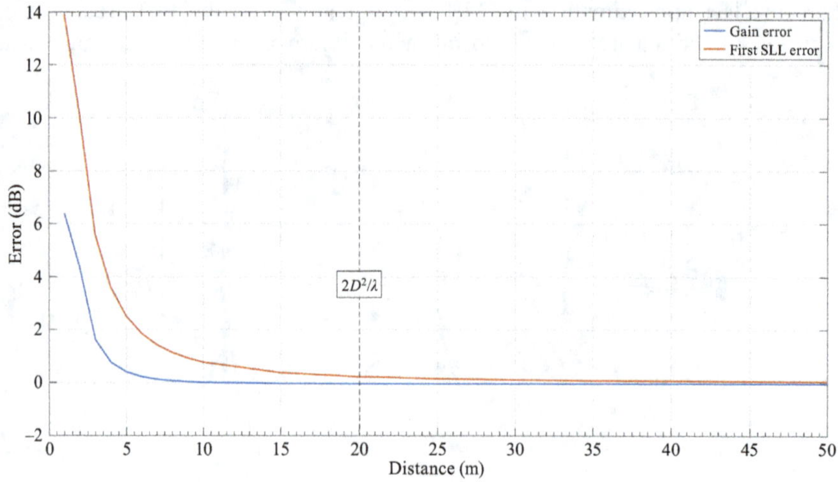

Figure 14.6 Electrically large device (1.23-m array antenna at 1,970 MHz): gain and SLL error varying the measurement distance

			Far-field distance							
			RFIC	Module	Test board	Phone	Tablet	Laptop	BTS	Car
		D (mm)	14.4	30	50	150	300	500	1,000	6,000
Frequency (GHz)	0.07	Far-field	0.1 mm	0.4 mm	1.2 mm	10.5 mm	42 mm	116.7 mm	0.47 m	16.8 m
	0.7		1 mm	4.2 mm	11.7 mm	105 mm	0.42 m	1.2 m	4.7 m	168 m
	6		8.3 mm	36 mm	0.1 m	0.9 m	3.6 m	10 m	40 m	1,440 m
	30		41.5 mm	0.18 m	0.5 m	4.5 m	18 m	50 m	200 m	7,200 m
	60		82.9 mm	0.36 m	1 m	9 m	36 m	100 m	400 m	14,400 m

Figure 14.7 Examples of FF distances for different devices at different frequencies

The previous examples suggest that some antenna figures of merit like the gain and the half-power beamwidth can be measured with good accuracy at distances even smaller than the introduced FF distance R_{FF}. Instead, depending on the required uncertainty, other figure of merits, such as null-depths and side-lobes level, might require measurement distances even larger than R_{FF}. Examples of FF distance for some different and commonly used devices are shown in Figure 14.7 at different frequencies.

An important aspect to consider when choosing the measurement distance is related to the FSPLs defined as

$$\text{FSPL} = \left(\frac{4\pi R}{\lambda}\right)^2 \tag{14.13}$$

While increasing R allows to obtain a better plane wave approximation, the FSPLs are increased at larger measurement distances leading to an attenuated measured signal which will be closer to the noise floor of the system. Instead, measuring at shorter distances would allow for better dynamic ranges.

It is interesting to evaluate the FSPL at the FF distance R_{FF} which can be obtained simply combining (14.13) and (14.12):

$$\text{FSPL}_{\text{FF}} = \left(\frac{8\pi {D_{\max}}^2}{\lambda^2}\right)^2 \tag{14.14}$$

Let us also consider the gain of an antenna G_{AUT} as a function of its effective area A_e [11]:

$$G = \frac{4\pi A_e}{\lambda^2} = \frac{2\pi \eta_a {D_{\max}}^2}{\lambda^2} \tag{14.15}$$

where the last equation is obtained assuming a squared aperture and introducing the term η_a as the aperture efficiency. Expressing ${D_{\max}}^2$ in terms of G and substituting it into (14.14) yields

$$\text{FSPL}_{\text{FF}} = \left(\frac{4G}{\eta_a}\right)^2 \tag{14.16}$$

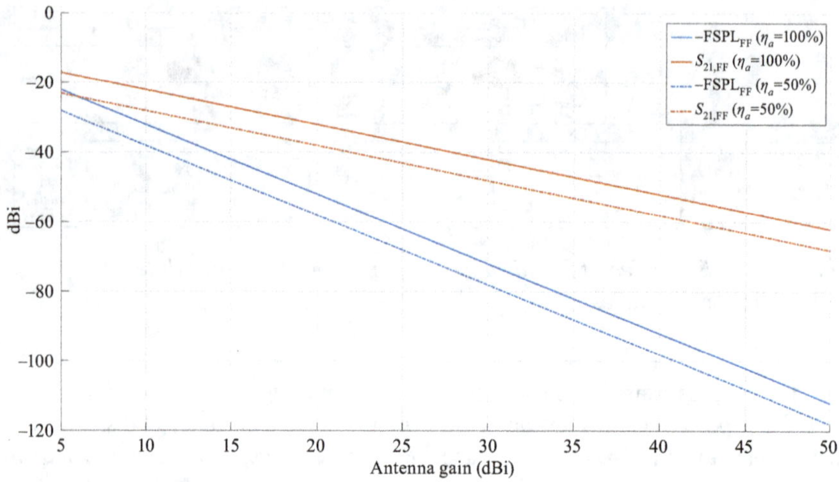

Figure 14.8 Inverse of the FSPL and S_{21} at the FF distance (R_{FF}) varying the gain of the AUT

The FSPL at FF distance increases with the square of the gain of the AUT/DUT. Moreover, the lower the antenna aperture efficiency, the higher the FSPL. Indeed, a highly efficient aperture antenna can reach a certain gain with a smaller physical area (and hence a smaller D_{max}) than a low efficient antenna, leading to a shorter FF distance. Finally, considering the Friis formula, we have

$$S_{21,FF} = \frac{G_{AUT}G_{Probe}}{FSPL_{FF}} = \frac{G_{Probe}\eta_a^2}{16G_{AUT}} \tag{14.17}$$

At the FF distance, the strength of the measured signal is inversely proportional to the gain of the AUT and proportional to the square of the aperture efficiency.

The (inverse of the) FSPL and the S_{21} at FF distance defined by (14.16) and (14.17), respectively, are plotted in Figure 14.8 for different antenna gains. For the S_{21} traces, a probe antenna with 0 dBi is considered. The slope of the FSPL trace is more pronounced than the S_{21} trace, meaning that the gain of the antenna is able to partially compensate the FSPL. However, the S_{21} is not flat and decreases with the gain of the AUT/DUT. This is an important aspect to consider when measuring the FF distance in order to obtain a sufficiently large dynamic range.

In Figure 14.8, it is also highlighted that radiating devices with smaller aperture efficiency ($\eta_a = 50\%$ instead of $\eta_a = 100\%$ in the example) leads to larger FSPL and hence attenuated measurement signals (6-dB attenuation in this case).

In some cases, it is thus convenient to consider the maximum antenna dimension associated with the effective area rather than the physical area ($D_{max,e}$ instead of D_{max}). By doing that, the FF distance can be reduced, but of course this approach requires some a priori information of the AUT/DUT.

Antennas installed on large structures are examples of radiating system with poor efficiency. For example, depending on the operational frequency, antennas mounted on cars might interact with only a small portion of the car structure. The non-interacting part can make the efficiency very small. For example, a 6-GHz antenna installed on a medium size car could lead to aperture efficiency in the range of 1%–10%. Such efficiencies would in turn lead to 40–20 dB extra attenuations of the measured signal if the whole car structure is considered D_{max}. At higher frequencies, since only a reduced portion of the car structure around the antenna (e.g. 10λ) will contribute to the final radiation, it is reasonable and convenient to consider only a representative portion of the car as the maximum dimension of the device, relaxing the FF measurement distance and hence increasing the dynamic range.

Similar examples are 5G NR2 millimetre-wave (mmWave) antennas (e.g. phased arrays at 28 GHz) installed on smartphones or tablets. The area occupied by such antennas is usually in the order of 3×1.5 cm^2 while the sizes of a typical smartphone and tablets are about 15×7.5 cm^2 and 25×15 cm^2, respectively. Also in this case, the efficiency of the whole device is in the order of 1%–10% (the estimated upper bound is obtained considering some possible coupling with the surrounding of the antenna). Considering the effective radiating area instead of the full physical area of the whole device would allow to relax the FF measurement distance and, again, to increase the dynamic range.

14.4 Modern testing solutions

Testing parameters and methods in modern communication testing are strongly influenced by industry needs. Modern testing solutions can very roughly be divided into 'black box' and 'white box' testing:

1. Black box: Testing without a priori knowledge of the test object.
2. White box: Testing in which specific knowledge of the test object is used to facilitate the testing, reduce complexity and time, etc.

The standardization effort often targets specific needs for regulatory purposes such as 'Type Approval' that are common Pass/Fail scenarios. Such needs are in favour of the 'black box' testing. For all other areas of testing, very different parameters of the testing systems are often equally important. Speed, convenience and diagnostics capabilities are important parameters in the development cycle of antennas and antenna systems. In these cases, a 'white box' testing is often preferred.

The testing techniques can roughly be divided into two categories: NF and FF techniques as shown in Figure 14.9. Any system that allows testing at a reduced distance with respect to the classical FF criteria can be considered an NF system. The distinction between the different NF systems is based on how the NF to FF (NFFF) transformation is performed. Classical NF techniques are based on discrete sampling of the radiated NF on a surface around the AUT/DUT and numerical

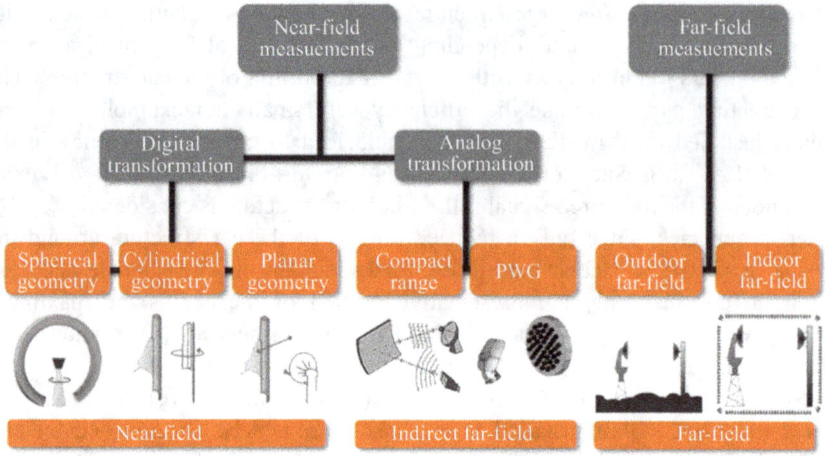

Figure 14.9 Overview of modern AUT/DUT testing techniques

transformation to determine the FF. Reflector or array-based plane wave generators (PWGs) perform an analogue sampling of the NF on the continuous surface of the reflector or the discrete points of the array elements and transform to FF without numerical transformation. In most standardization communities, the reflector and array-based PWGs are considered FF systems and sometimes they are named *indirect FF systems*.

14.5 Near-field measurements with digital NF/FF transformation

Conventional NF techniques are based on the discrete sampling of the radiated field on a surface surrounding the AUT/DUT at relatively close distance to the radiating object. Once the data acquisition is completed, a mathematical NFFF transformation is applied to compute the field at infinite distance.

NFFF transformation techniques are based on the Huygens principle which states that from the knowledge of the radiated field on a closed surface around a generic radiating source (AUT/DUT), it is possible to compute (transform) the field at any point in space outside that closed surface (see Figure 14.10).

The correct application of the NFFF techniques requires both amplitude and phase information of the measured field and proper spatial sampling on the measurement surface. According to the Huygens principle, the knowledge of the field components tangential to the sampling surface is sufficient to correctly perform field transformation. The surface surrounding the AUT/DUT could theoretically be of any shape. However, practical implementations of NF systems use canonical geometries like spherical, cylindrical and planar as shown in Figure 14.9. Such sampling schemes require only two scanning axes and hence their implementation

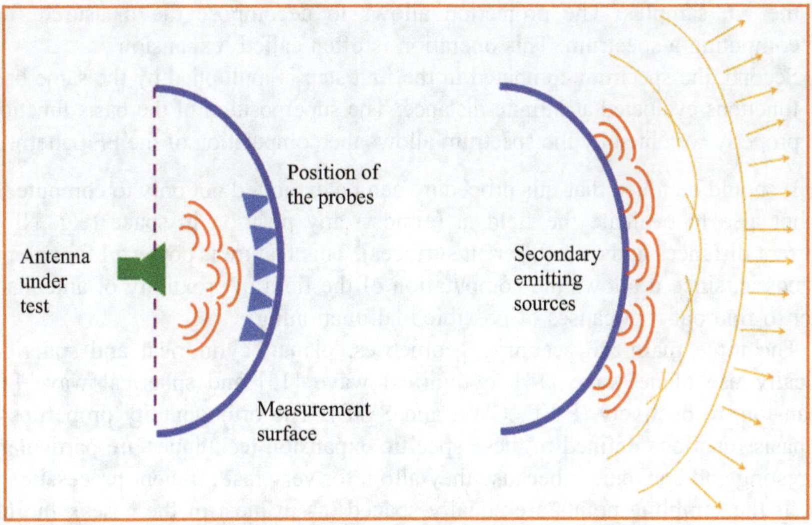

Figure 14.10 Illustration of the Huygens principle

is simplified. Moreover, the data processing is also drastically simplified when such conventional scanning geometries are used.

NF measurement techniques offer several advantages when compared to direct FF measurements done at larger distances. First, the FSPLs are much smaller; hence higher dynamic range can be reached. Second, the reduced distance allows the installation of such systems in smaller, and thus cost-effective, controlled environments like shielded anechoic chambers. Moreover, the application of the NFFF transformation allows the realization of 'virtual' ideal plane wave illumination with a consequent ideal FF condition.

On the other hand, the main disadvantages of the NF techniques are the needs of a sufficiently large number of samples and a closed scanning surface. The sampling requirements are needed to avoid processing errors like the aliasing [12]. Such requirement, in the case of relatively electrically large AUT/DUT, could lead to long measurement times. The closed surface requirement is often not met because of mechanical constraints and/or nature of the scanning geometries, such as planar and cylindrical surfaces. Truncation errors are thus another processing artefact that could occur when partial scanning surfaces are considered. Such errors are more pronounced in the case of low-directive and low-frequency devices than in the case of high-directive and high-frequency devices [13].

Independently of the chosen scanning geometry, the NFFF transformations are based on a two-step procedure:

- First, the measured NF is projected over a set basis functions properly defined according to the nature of the scanning geometry and specific transformation technique to be used. The basis functions are evaluated at the same location of the NF samples. The projection allows to decompose the measured field computing a spectrum. This operation is often called 'expansion'.

- Second, the spectrum computed in the first step is multiplied by the same basis functions evaluated at infinite distance. The superposition of the basis functions properly weighted by the spectrum allows the computation of the FF pattern.

It should be noted that this procedure can be exploited not only to compute the FF but also to evaluate the field at (almost) any position in space (e.g. NF at different distances and over different surfaces). This feature is powerful for antenna diagnostic, since it allows the computation of the field in proximity of antenna in order to find out root causes of possible malfunctioning.

The three main NF scanning geometries, planar, cylindrical and spherical, typically use plane wave [14], cylindrical wave [15] and spherical wave [16] expansion, respectively (PWE, CWE and SWE). The orthogonality properties of the basis functions defined in these specific expansion techniques are particularly interesting and convenient because they allow for very fast/efficient processing. In fact, if the sampling points are equally spaced (as in most of the cases), the fast Fourier transform (FFT) can be exploited within the NFFF transformation algorithms keeping the complexity and the computation cost relatively low.

Spherical NF measurements are considered one of the most accurate measurement techniques [16,17] because they allow the characterization of any AUT/DUT in the whole 3D space around the device. The NFFF transformation is usually performed with the SWE as shown in (14.18) and described in great detail in the following equation [16]:

$$\boldsymbol{E}(r,\theta,\varphi) = \frac{k}{\sqrt{\eta}} \sum_{s=1}^{2} \sum_{n=1}^{N_{\max}} \sum_{m=-n}^{n} Q_{\mathrm{smn}}^{(3)} \boldsymbol{F}_{\mathrm{smn}}^{(3)}(r,\theta,\varphi) \tag{14.18}$$

In (14.18) $\boldsymbol{E}(r,\theta,\varphi)$ is the field measured at the distance r from the origin of the coordinate system and at discrete samples located at (θ,φ). The spherical wave functions $\boldsymbol{F}_{\mathrm{smn}}^{(3)}(r,\theta,\varphi)$ are defined at the same sample location of the measured field. The indexes s, m, n represent the orders of the basis functions. Finally, $Q_{\mathrm{smn}}^{(3)}$ are spherical wave coefficients, or in other words, the spectrum to be evaluated in the first stage of the NFFF transformation. An illustration of some spherical wave functions is illustrated in Figure 14.11. As can be seen, the n-indexes are linked to the field variation along the ϑ-coordinates. The higher the n-order, the more the oscillations along ϑ. Similarly, the m-indexes are linked to the field variation along the φ-coordinates. Higher m-order means more oscillations along φ.

The n-series is truncated at

$$N_{\max} = kR_{\min} + n_{\mathrm{safety}} \tag{14.19}$$

where $k = 2\pi/\lambda$ is the wavenumber and $R_{\min}$ is the minimum sphere cantered in the origin of the coordinate system fully enclosing the AUT/DUT. n_{safety} is a safety margin typically set to 10 in order to ensure a proper convergence of the SWE and thus a good representation of the AUT/DUT [18] (Figure 14.12).

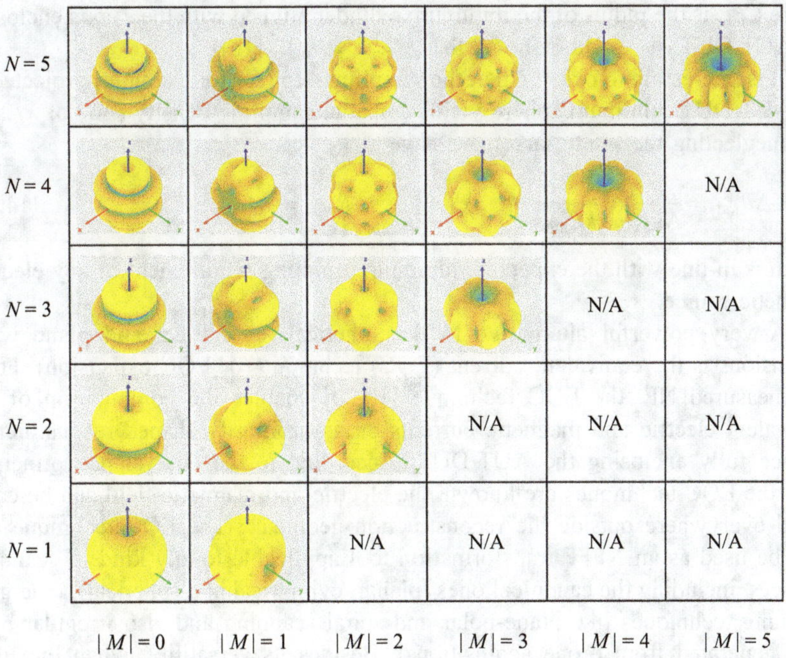

Figure 14.11 Illustration of the spherical basis functions

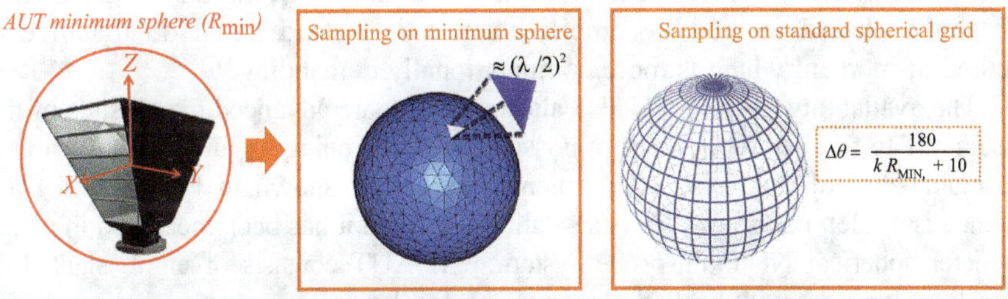

Figure 14.12 Illustration of the AUT/DUT minimum sphere concept and sampling requirement for spherical NF measurements

The sampling requirements in spherical NF measurements are based on the maximum index of the SWE to be computed and hence, according to (14.19), the AUT/DUT minimum sphere and the working frequency. A maximum number of oscillations expected along ϑ are $N_{\max}$. Hence the sampling step should be at least $\Delta\vartheta = 180°/N_{\max}$. Similarly, we also have $\Delta\varphi = 360°/2M_{\max}$, where $2M_{\max}$ is the maximum number of oscillations expected over 360° along φ and is given by

$$M_{\max} = kC_{\min} + n_{\text{safety}} \tag{14.20}$$

where C_{min} is the radius of the minimum cylinder aligned with the *Z*-axis enclosing the AUT/DUT (it should be noted that $M_{max} \leq N_{max}$, [16]).

It is interesting to observe that the sampling step ($\Delta\vartheta$ for example) projected to the AUT/DUT minimum sphere yields a spatial sampling density (ΔS) of $\lambda/2$. In fact, neglecting the safety factor, we have

$$\Delta\vartheta_{rad} = \frac{\pi}{N_{max}} \approx \frac{\pi}{kR_{min}} \xrightarrow{\text{yields}} \Delta S = \frac{\pi}{kR_{min}} R_{min} = \lambda/2 \tag{14.21}$$

which is in-line with the expected minimum sampling requirement of any electromagnetic source.

A very powerful alternative to the spherical, cylindrical and plane wave expansions is the equivalent current (EQC) technique (or EQC expansion). From the measured NF, the EQC technique [19,20] enables the computation of the equivalent electric and magnetic currents on an arbitrarily shaped reconstruction surface fully enclosing the AUT/DUT. According to the equivalence principle, once the EQC techniques are known, the electric and magnetic field can be computed everywhere outside the reconstruction geometry. The EQC technique can thus be used as an NFFF transformation tool applicable to any kind of scanning surfaces, including the canonical ones (planar, cylindrical and spherical), emerging scanning techniques like plane-polar and spiral scanning and also irregular ones (e.g. generated from drone acquisitions). Besides its versatility, when the EQC technique is used as NFFF transformation of partial NF acquisitions, it is also more robust against truncation errors than the other mentioned expansion techniques (see [13] for more details). On the other hand, the main drawback of the EQC technique is the computational cost. Unlike the SWE, CWE and PWE, the EQC expansion is not an FFT-based method. Its implementation is instead based on the inverse method of moment which is more computationally demanding.

The availability of the EQC also allows to evaluate advanced diagnostics of the AUT/DUT to find out root causes of possible malfunctioning. Figure 14.13 illustrates the diagnostics of the same array antenna previously shown in Figure 14.4. The antenna is a reference device for base-station tests, and it has been measured in a 3.2-diameter spherical NF multi-probe system. The AUT consists of eight, slant 45°, linearly polarized cavity-backed elements. As can be seen, reconstruction geometry conformal to the test object has been defined to compute the EQC from the measured field. The co-polarized and cross-polarized electric (J) currents at 1,970 MHz are shown in Figure 14.13. Symmetric element excitation is expected both in amplitude and phase. The power normalized amplitude and phase excitation computed from the EQC is shown on the right side of Figure 14.13. The maximum amplitude unbalance is observed between the first and last elements and is 0.8 dB. The maximum delta between the phases of the element is observed between the second and seventh elements, and it is 14.2°. Such identified errors have then been found to be due to the feeding network of the array and have been corrected accordingly.

Another interesting and powerful feature of the EQC technique is the possibility to use the EQCs, computed from measurements, as source for numerical computations [21]. Antenna placement analysis requires the knowledge of the exact model of the

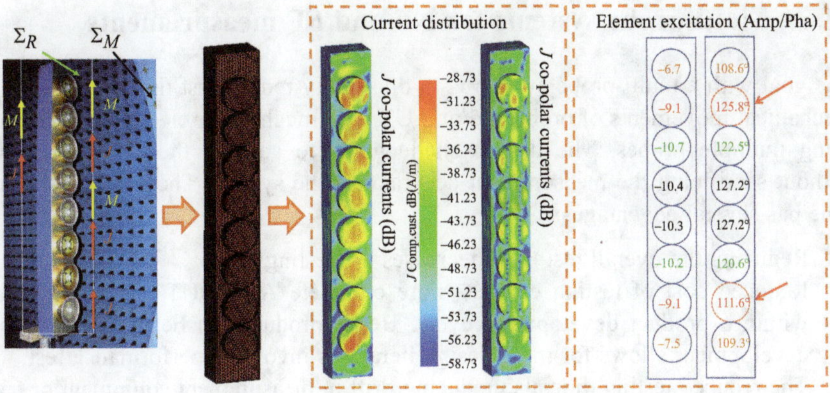

Figure 14.13 Diagnostics of a reference base-station array antenna with the EQC technique

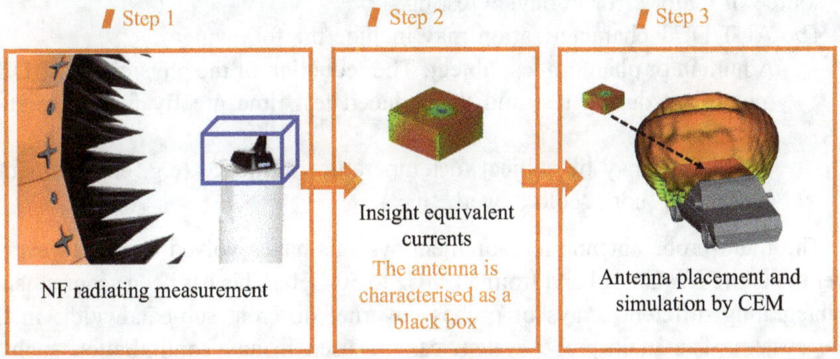

Figure 14.14 Workflow of the measurement-simulation link enabled by the EQC technique

antenna to be used together with the final structure (e.g. vehicles, aircrafts, satellites, ships) in computational electro-magnetic (CEM) solvers. Very often, the antenna comes from third parties that might not be allowed to share details of their device. Moreover, the numerical model of the antenna does not allow to account for possible manufacturing tolerances which might create discrepancies between the modelled and the realized devices. The EQC computed from the measurements of the antenna allows to establish a 'link' between measurements and simulations enabling accurate antenna placement analysis. By doing this, the antenna is seen as a 'black-box' and no additional information are required. The workflow for this type of analysis is schematized in Figure 14.14. The antenna is first measured and then the EQC techniques are computed (usually considering a simple box enclosing the antenna as reconstruction geometry). The EQC techniques are then imported in the CEM solver to be used and used as a source in the final structure to be analysed.

14.6 Multi-probe systems for NF and FF measurements

The goal with a multi-probe system is to drastically reduce test time by replacing mechanical movements of probe and/or AUT with much faster electronic scanning using multiple probes. The design challenge is to achieve the speed increase without sacrificing the measurement accuracy of the system. The reduced testing time has several advantages:

- Reducing the overall test time decreases the testing cost.
- Reduced cost of testing enables more complete AUT/DUT characterization within a product development cycle. Better product can be designed if the development allows for more design iterations involving performance test.
- The reduced testing time also reducing drift in measurement equipment or test environment and improves accuracy. The multi-probe system allows to have easy access to a reference channel that enables calibration for drift.
- Reducing the mechanical handling of the AUT/DUT to one dimension, like a rotation dramatically reduces the AUT/DUT to positioner interaction, a known source of major error in device testing.
- The AUT/DUT characterization may include the following:
  - A human or phantom test object. The reduction of the physical AUT/DUT movement during test and the reduced test time greatly facilitates such investigations.
  - AUT/DUT may be subject to temperature influence (e.g. environmental conditions ‘hot’, ‘cold’, ‘wet’, ‘ice’).

The multi-probe antenna measurement systems have evolved over the years to cover the entire frequency band from 50 MHz to 50 GHz. This has been accomplished by integrating different arrays of probes covering different sub-bandwidths in the same system. The multi-probe system can be used in any configuration such as cylindrical, planar and spherical with the spherical configuration as the most widely used one. In spherical geometry, the AUT/DUT is located in the centre of the system on top of a foam column as shown in Figure 14.15. The full sphere measurement is performed by electronically scanning the probe array in elevation and rotating the AUT/DUT 180° in azimuth. Consequently, a full 3D measurement can be performed very rapidly compared to conventional single-probe systems. Depending on the measurement configuration, a multi-probe system is generally up to ten times faster than a system based on the conventional physical movement of a field probe over the measurement sphere. If the testing diameter of the probe array system is sufficient for the AUT/DUT to be in the FF, the collected data can be used directly as FF performance data. An NFFF transformation of the collected data can also be applied.

The actual radiated field sampling is performed by the wide-band probe array composed of evenly spaced elements along the support structure. A limited mechanical elevation movement of less than a probe spacing can also be added to increase the sampling density. In such configuration the system gives access to unlimited field sampling while taking full advantage of the speed improvement provided by the probe array.

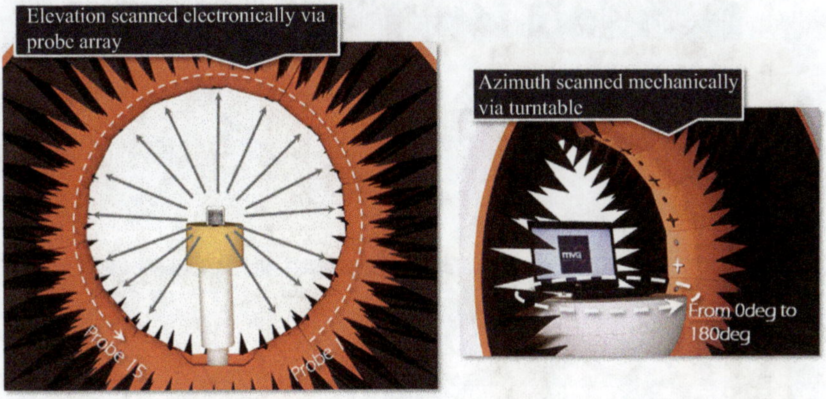

Figure 14.15 Example of multi-probe system. In spherical scan configuration, the probe array is scanned electronically in elevation and AUT/DUT, located on a foam column in the centre of the system, is rotated 180° in azimuth.

The probes are connected to the RF system through a combiner network. The individual probes can interrogate by on/off switching or by advanced modulated scattering technique (A-MST). In an A-MST system, a low-frequency modulation is applied to each probe to perform multiplexing. The measured field information, in terms of amplitude and phase at each probe position, is extracted by coherent detection. For measurements of active systems, the switching network approach is generally preferred.

The multi-probe system can be used in any configuration but the most common are the canonical NF scan configurations such as cylindrical, planar and spherical with the spherical configuration as the most widely used one. Different implementations of the multi-probe system are shown in Figure 14.16. A comparison with typical single probe mechanically scanned NF systems is also shown. Typical multi-probe systems have up to hundreds of elements and are thus often 10–20 times faster than conventional systems.

The probe array elements are wide band printed elements specifically designed and optimized for probe array applications. They are dual polarized and aligned according to vertical and horizontal polarizations. The probes are completely reciprocal and can be used both in receive and transmit modes. Their assembly also houses a circuit board containing the control electronics for the scanning of the probe. The probe array elements are specifically designed to be embedded in multi-layer conformal absorbers. The probe tips protrude through small, crossed slots in the smooth curvature of the absorbers keeping the reflectivity of the probe array at a minimum. The absorbing material also reduces scattering and reflections from the support structure and cabling. Examples of different implementations of multi-probe systems from automotive to telecom antenna testing are shown in Figure 14.17.

Since the array elements have low directivity throughout the operating frequencies and the measurement distance to the AUT/DUT is never too close, the

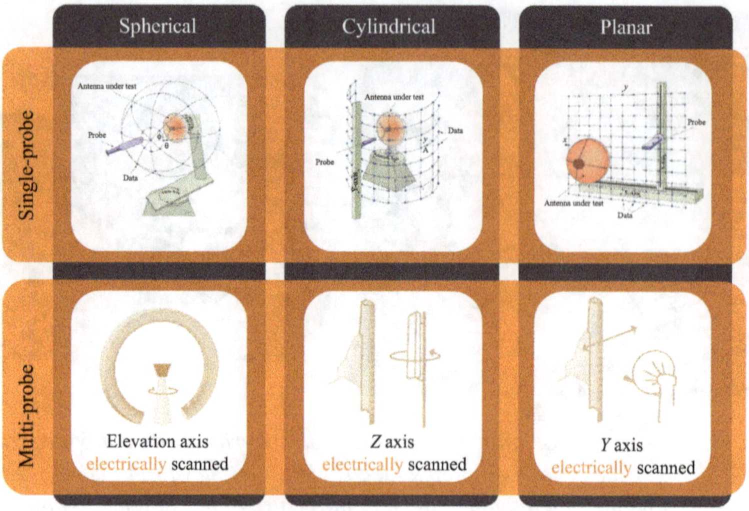

Figure 14.16 Common multi-probe system and comparison to typical mechanically based near field scan configurations such as spherical cylindrical and planar

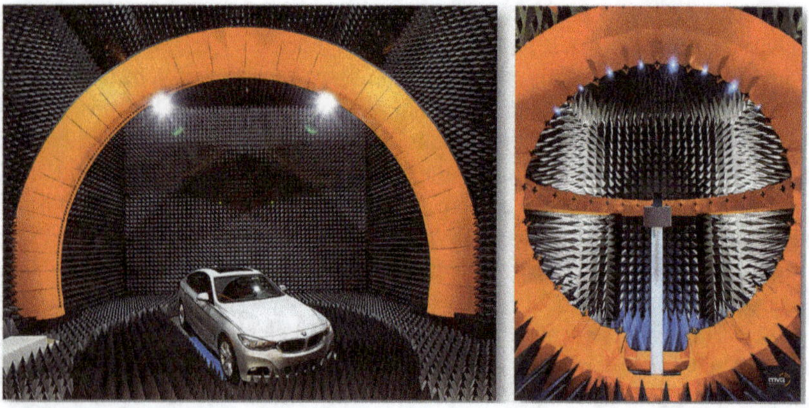

Figure 14.17 Different implementations of multi-probe systems

probes can be assumed to behave like Hertzian dipoles. This implies that the received signal is proportional to the electric field parallel to the polarization of each probe. Traditional probe correction is therefore required in the NFFF transformation. However, since the probes are manufactured using high-volume industrial production techniques, the probe response cannot be assumed to be equal in amplitude, phase and polarization orientation. To correct for these discrepancies a probe array calibration is performed.

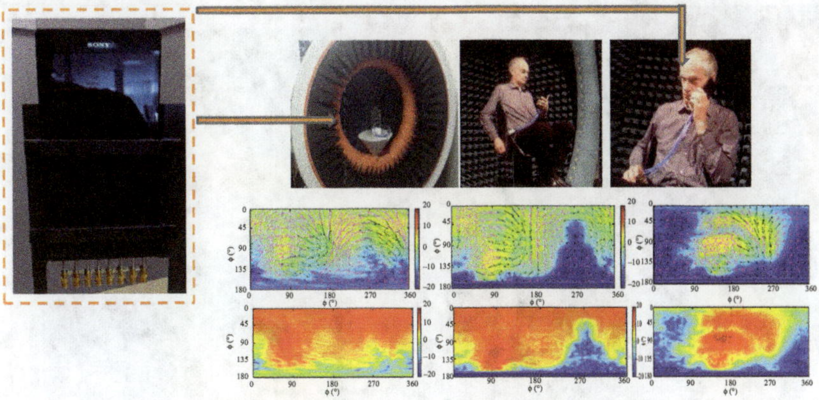

Figure 14.18 Example of full person investigation of mobile device performance using different phased array configurations on the device. The shading effect of head and torso is clearly visible at these frequencies.

Microwave Vision Group (MVG)'s probe arrays are calibrated by pointing a linearly polarized antenna towards each of the probes. Dual polarized, phase and amplitude measurements are taken as the antenna is rotated in polarization in front of each probe. From these data, a set of calibration coefficients are derived for each probe. The calibrated horizontal and vertical components of the electromagnetic field (EMF) are obtained from the measurements. The process is able to compensate for differences among probes, align the polarization axes and reduce the cross-polarization components. After calibration, the typically measured probe uniformity is better than ± 0.1 dB in amplitude and $\pm 1^\circ$ in phase.

At mmWave frequency bands, the impact of the user presence on device performance can be critical. For this reason, most personal devices are equipped with phased arrays in different positions that can be scanned to optimize the system performance. In [22], the body effect is measured with a real user at 15 GHz in MVG multi-probe system to investigate the impact on performance parameters such as total scan pattern and coverage efficiency. The phased array in the study is composed of four notch elements with 9-mm separation at 15 GHz. All elements are linearly located on the top of a phone mock-up, which is shown in Figure 14.18. Two identical sub-arrays with four elements of each are used, in which the digital beam forming progression of the two sub-arrays is independent of each other. Thus, their beams can be steered separately and be feasible for the diversity or MIMO application. In the study, it was assumed that the phased array, controlled by a digital beam shifter, can vary from -180° to 180°.

The TRP of a wireless device can be determined rapidly in a multi-probe system [5,23]. An example of such measurement on a device from SONY Mobile is shown in Figure 14.19. The upper patch array has been measured at 24 GHz in the SL50GHz system. In particular, the EIRP has been measured by first calibrating the

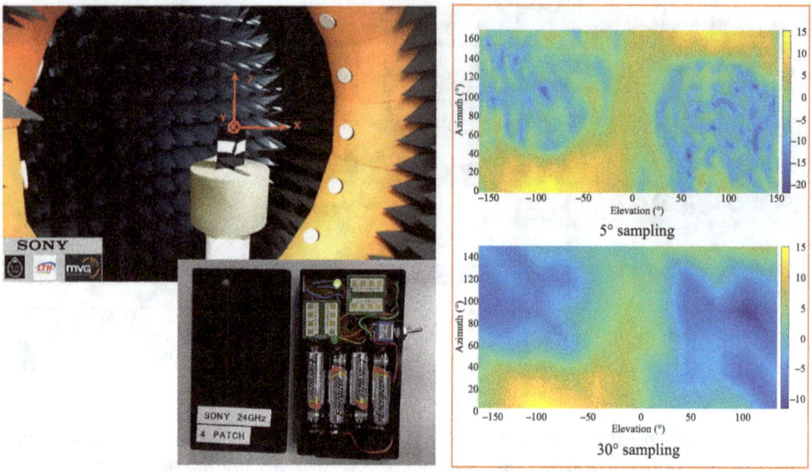

Figure 14.19 Investigation of minimum sampling grid for TRP measurement of wireless device at 24 GHz

Table 14.2 TRP versus sampling grid resolution of SONY wireless device at 24 GHz

$\Delta\theta = \Delta\varphi$ (°)	**TRP (dBm)**
5	4.49
10	4.41
15	4.42
20	4.38
25	3.75
30	4.66

system and then considering different sampling densities $(\Delta\theta, \Delta\varphi)$. Results are reported in Table 14.2. The variation in the TRP is about 0.1 dB when 5°–20° sampling resolutions are considered. Such difference is considered low enough for this type of application; hence, it can be concluded that for such devices 10°–15° sampling resolution is sufficient to measure the TRP.

The fast measurement in multi-probe systems and processing by EQCs can be used to investigate the radiating current distribution of wireless terminals with distributed antennas as shown in Figure 14.20 [24]. The precise knowledge on the distribution of the radiating current on the device allows to investigate the level of interaction between the antennas and the device.

User equipment, such as mobile terminals and tablets, should comply with RF safety guidelines. In mmWave frequencies, human exposure to RF EMFs is evaluated in terms of incident power density, i.e., free-space Poynting vector. The power density is normally measured over rectangular planes at specified distances

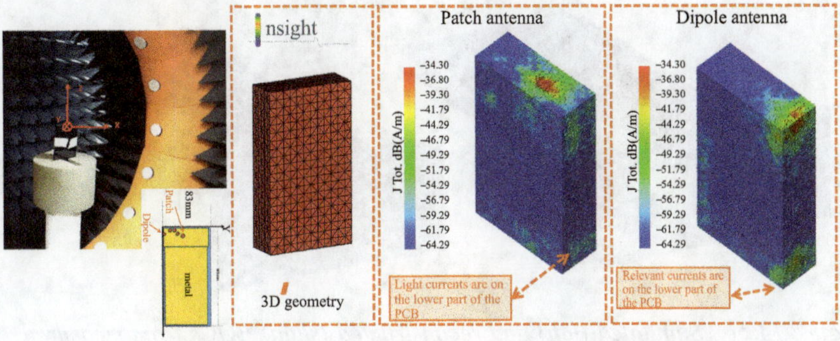

Figure 14.20 Investigation of radiating current distribution at 28 GHz of a wireless device with distributed antennas measured in the 5G StarLab measurement system. The EQCs have been reconstructed on a geometry representing the antenna and the device.

from the DUT using proper scanners and probes with power meter or spectrum analyser. Such measurement procedure can be time-consuming, due to a high number of points to be measured. It is also not very accurate because of the mutual coupling between the DUT and the probe.

An alternative solution is to measure the device in a calibrated system and represent it in terms of EQCs. The EQC techniques are defined on an equivalent geometry fully enclosing DUT. From this representation, the field can be calculated everywhere outside of such geometry. In other words, starting from the NF measured on the scanning surface, an NF to NF transformation on any surface and at any distances can be performed evaluating the power densities according to the safety normative. It should be noted that the EQC technique requires amplitude and phase data; hence, if an OTA set-up is used, a phase recovery technique should also be involved.

In the enclosed example, a 5G mobile terminal mock-up manufactured by Sony Mobile containing a notch element working at 28 GHz is considered. The device has been measured in the multi-probe SL50GHz shown in Figure 14.20. From the measured data, the EQC techniques have reconstructed on box conformal to the DUT. The EQC of the notch element is shown in Figure 14.21, respectively, simulated with a full-wave software (left) and reconstructed from measurements (right). In both cases, the EQC techniques have been used to calculate the NF on planes at distance d from the device also shown in Figure 14.22.

From the computed NF, the incident power density has been evaluated. Figure 14.22 shows the maximum spatial average power density versus the distance d. The discrepancies observed between the measured and the simulated power density with 1 cm^2 are probably due to positioning error of the device. The results suggest that the misalignment could be about 3–4 mm. Such error could be improved by better alignment methods. The discrepancies are instead less pronounced considering the maximum spatial average power density with 20 cm^2 meaning that such metric is more robust against possible measurement error.

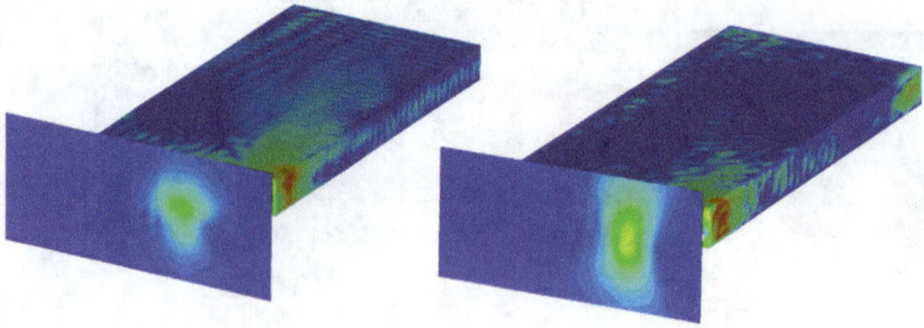

Figure 14.21 Simulated (left) and reconstructed (right) EQCs from measured data of the mobile terminal mock-up

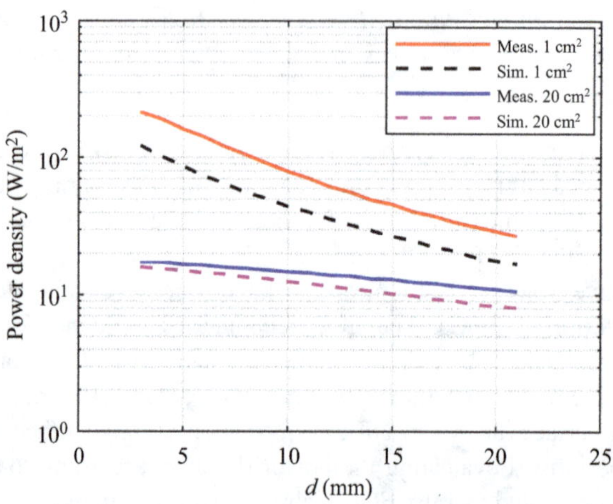

Figure 14.22 Measured (from reconstructed EQCs) and simulated spatial-average power density for the notch element to be compliant with the FCC and ICNIRP power density limits

The testing of active DUTs using NF techniques requires the recovery of the relative phase between the sampled NF data. This can be achieved by a comparison of the measured signal with a stable reference signal. The phase recovery gives access to the features of traditional NFFF transformation techniques.

The application of NFFF in OTA testing of active devices has been validated by experiment. The construction of the validation experiment is shown in Figure 14.23. The pertinent TRP and TRS/TIS performance parameters are determined from conducted experiments on a typical mobile phone with access to an external antenna through a connectorized port. The gain of the reference antenna, an 8 × 1-element base transceiver station (BTS) reference antenna, is measured by conventional measurement techniques. Adding the BTS array as an external antenna to the mobile phone constitutes the reference scenario in which the TRP and TRS/TIS performance parameters are well known.

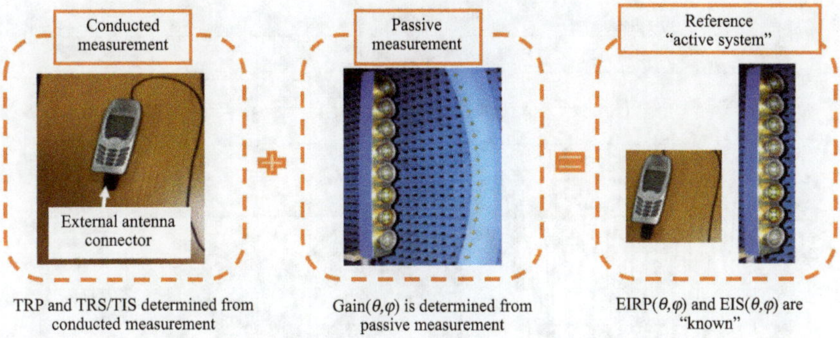

Figure 14.23 Construction of reference scenario for the validation of OTA testing using NFFF and phase recovery

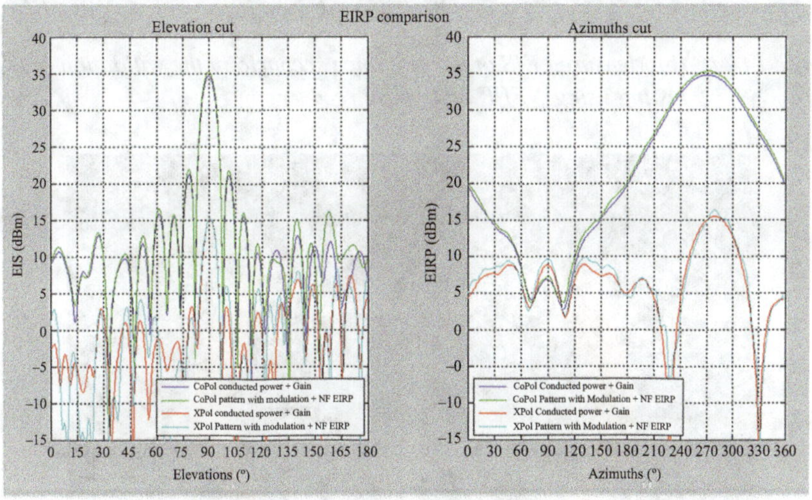

Figure 14.24 (a) Measured EIRP of the reference scenario in the validation of OTA testing using NFFF and phase recovery

The measurement in a multi-probe NF system, with phase recovery, allows to perform NFFF transformation on the tested antenna. This information can be used to determine a correction factor to be applied to data measured directly in NF at a close distance to the AUT/DUT. Power density and sensitivity performance can be determined using this technique. The measured EIRP and EIS of the validation scenario are shown in Figure 14.24(a) and 14.24(b). The differences between the establish reference data from the conducted experiment and the measured OTA performance using phase recovery techniques are within 1 dB. This value is well within the expected uncertainties for both the conducted experiments of the reference case and the NFFF with phase recovery technique.

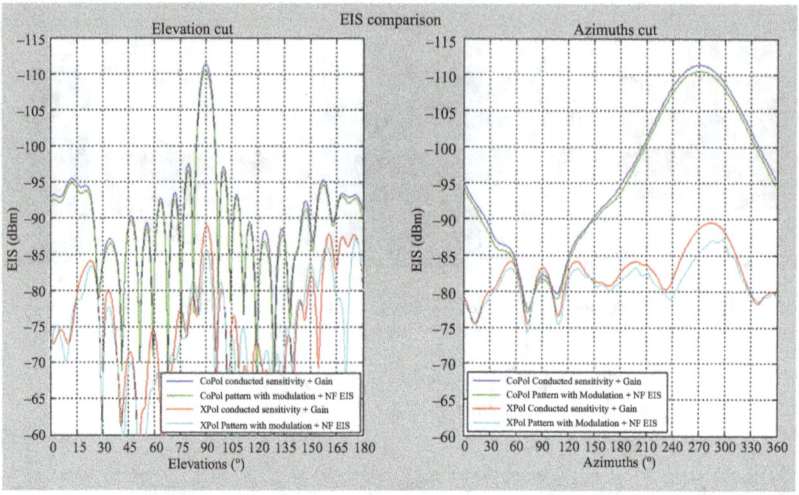

Figure 14.24 (b) Measured EIS of the reference scenario in the validation of OTA testing using NFFF and phase recovery

Figure 14.25 Common multi-probe systems for measurement of massive MIMO base-station antennas at FR1 frequencies using NFFF and phase recovery techniques

A common application of the phase recovery technique and NFFF in a multi-probe system is testing of massive MIMO base-station systems for FR1 applications. Common systems for such measurements are shown in Figure 14.25 [25,26]. An example of measured massive MIMO base-station EIRP pattern at FR1 frequencies using phase recovery techniques is shown in Figure 14.26.

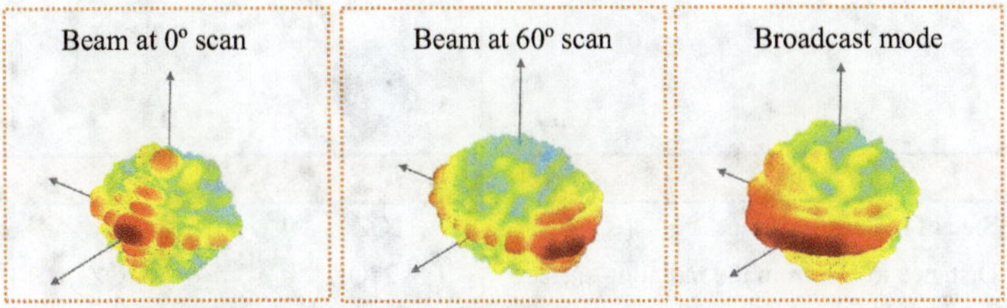

Figure 14.26 Example of measured massive MIMO-base-station EIRP pattern at FR1 frequencies using phase recovery techniques

14.7 Plane wave generators

For some electrically large antennas, its FF distance is too large for a conventional anechoic chamber, and a possible solution is to utilize a PWG. Such system approximates a plane wave, and thus FF condition in the volume occupied by the antenna under test at close distance. This volume is often call quiet zone (QZ). Reciprocity ensures that both transmit and receive testing can be performed with sufficient accuracy. The shorter distance and increased gain reduce the losses due to free-space attenuation and decrease the overall system size. Examples of such solutions are the compact antenna test range (CATR) and array-based PWG.

A PWG has several interesting advantages in comparison to other antenna testing systems. It shares with the CATR the possibility to perform a direct measurement of FF performance in a controlled indoor environment. However, the PWG has the added advantage of having a physical size considerably smaller than the CATR, in particular at lower frequencies. The physical size advantages affect the size and associated cost of the anechoic chamber. This advantage is of particular importance for testing at sub-6-GHz frequencies for OTA testing of 4G, 5G and the upcoming 6G devices. A comparison between the two systems is shown in Figure 14.27.

The CATRs were first described in the 1960s [27]. There are several possible combinations of reflector(s) geometries for CATR's. In the following we will concentrate on the most common single-reflector offset-type configuration. The PWG is less widespread. The original idea was presented in [16,28]. Discussions and implementations of PWG presented in the literature have been limited to theoretical design aspects, or prototypes with narrow bandwidths and single polarization. Recently, solutions with 10:1 bandwidths and dual polarization have been presented, making the PWG a viable complement to the CATR as testing system [29]. In this chapter, we briefly review the PWG design guidelines comprising size and number of elements, including the extension to wideband performance.

14.8 Compact test range

The reflector-based PWG is also known as CATR. The idea is simple. In the CATR, the probe of the FF measurements system shown in Figure 14.1 is replaced

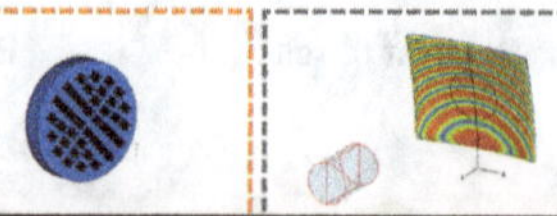

Comparison parameter	PWG		CATR	
Size of radiating surface	1.1*QZ	🙂	1.5*QZ	😐
Distance to QZ from the radiating surface	~1.5–2*QZ	🙂	~ 4–5*QZ	☹
Edge treatment	N/A	🙂	3λ–10λ	😐
Spill–over from feed	N/A	🙂	Yes	😐
Feed radiating into QZ	N/A	🙂	Yes	😐
Feed alignment	N/A	🙂	Yes	😐
QZ uniformity	Yes	🙂	Yes	🙂
Wideband operation	Yes	😐	Yes	🙂
Dual polarization	Yes	🙂	Yes	🙂

Figure 14.27 Indicative comparison of common performance parameters of the CATR and PWG

with a large offset parabolic reflector feed by a range antenna in a conventional anechoic chamber. Thus, a CATR can be considered an optical system with a feed and one or more reflectors which are much larger than the measurement wavelength. The reflector system converts the spherical wave front radiated by the feed into an approximately plane wave over a finite volume in the test area (i.e. QZ).

Common reflector shapes are parabolic or cylindrical. The size of the test area is typically half the size of the reflector and the distance between the reflector, and test area is often three times or more of the size of the reflector. The compact range enables the testing of medium to electrically large antennas at a significantly shorter distance than would be required in a traditional FF test range. An example of a CATR system, indicating the reflector, feed and QZ is shown in Figure 14.28.

The plane wave generated by the CATR is approximately uniform over the test area with some taper towards the edges of the QZ, caused by the amplitude taper of the feed. Edge effects from the reflector are often compensated by adding serrations or edge shaping to direct diffractions away from the QZ. As an optical system, the bandwidth of the CATR is limited only by the finite bandwidth of the feed and the frequency limit of the edge treatment [27,30]. CATRs have been designed to operate at frequencies from as low as 500 MHz to the mmWave and sub-THz range, more details of which can be found in [11,31].

The offset geometry of the parabolic reflector in a single-reflector CATR causes a variation of the polarization tilt angle that induces cross-polar radiation in the QZ even if the reflector is illuminated with a minimized cross-polar feed. The cross-polar levels induced by the reflector are null in the sub-region of the QZ along the plane of symmetry but increase with distance from the symmetry plane. Knowledge of this region can be exploited to perform very accurate polarization measurements if the AUT/DUTs are placed within the region of low cross-polarization. Special feeds and techniques can also be used to improve the cross-polar performance [32].

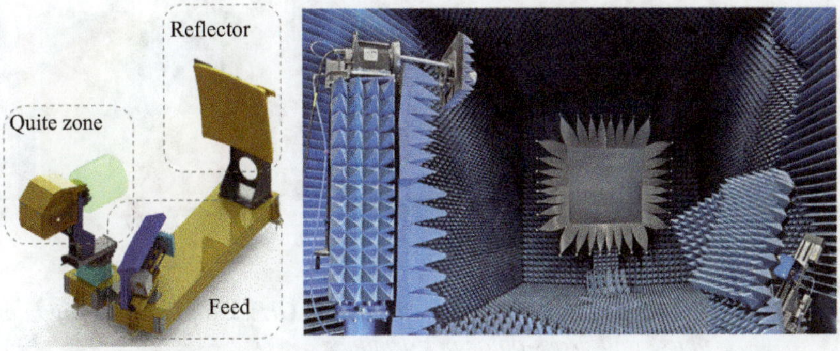

Figure 14.28 Example of CATR system with an indication of feed antenna, reflector and quiet zone

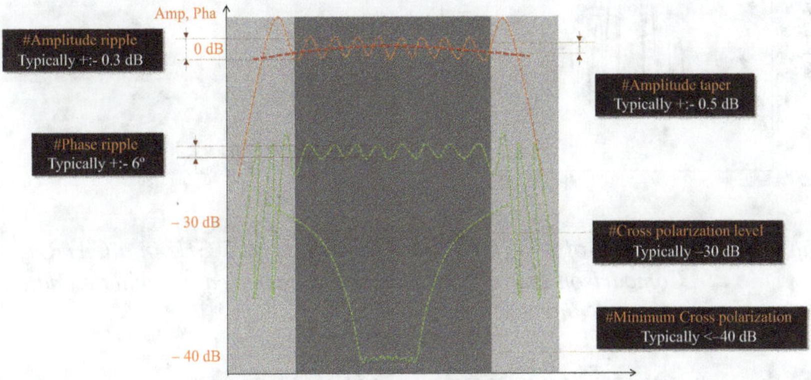

Figure 14.29 Typical co-polar and cross-polar QZ performance characteristics of a single-reflector CATR

The quality of the test volume or QZ is often defined by its co-polarized taper, amplitude and phase ripple. The cross-polarization performance of the QZ is often defined by its on-axis and edge cross-polarization level with respect to co-polar level. Typical QZ performance characteristics for a single-reflector CATR are shown in Figure 14.29.

The QZ evaluation of the CATR is accomplished by directly sampling the amplitude and phase characteristics of the field using an appropriate antenna as a probe. The most common technique for field probing consists of a scan of the QZ on a plane perpendicular to the direction of propagation, where the QZ phase is constant to appreciate both amplitude and phase characteristics. The scan is generally performed over the entire QZ in both orthogonal polarizations. The scan can be accomplished by a two-axis liner scanner or a combination of precision linear and rotation movements. An example of QZ scan results for a 60-GHz compact test range is shown in Figure 14.30 [33].

An example of an AUT/DUT measurement at 29.5 GHz is reported in Figure 14.31. The measurement has been performed on an array antenna with

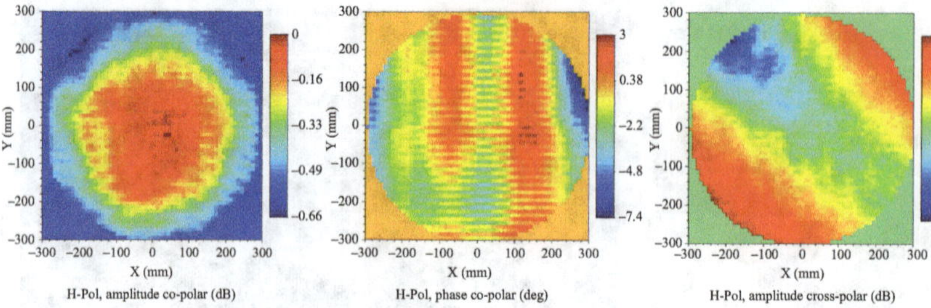

Figure 14.30 Example of QZ scanning results at 60 GHz. Co-polar amplitude and phase and cross-polar performance.

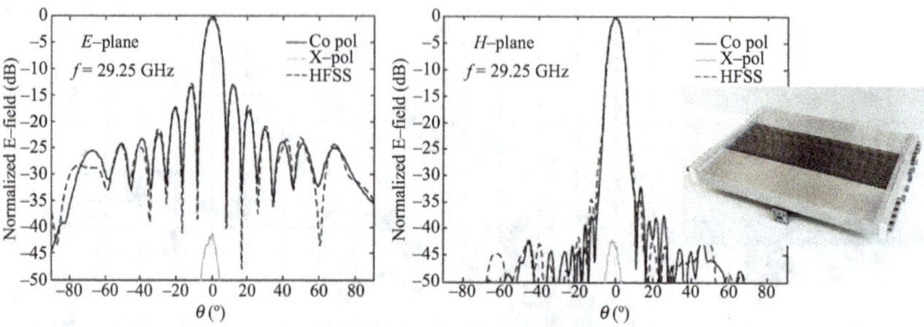

Figure 14.31 Example of AUT/DUT measurement at 29.5 GHz in a CATR. Comparison between measured and numerically simulated data for an array antenna.

comparison with numerical data [33]. As could be expected from the QZ probing of the same system at 60 GHz, the correlation between the numerical simulation and measurement is very good. In general, correlations between state-of-the-art numerical simulations and state-of-the art measurements systems, such as the reported CATR system, are of the order of 45-dB equivalent error level [34].

An important performance parameter of any antenna system AUT/DUT is the polarization purity. Testing in CATR provides convenient FF conditions at a reduced distance but the geometrical optics (GO) effect induced by the offset CATR parabolic single reflector causes a variation of the polarization tilt angle in the QZ as shown in Figure 14.30. The cross-polar fields are very small in the sub-region along the plane of symmetry of the reflector optics but can have a significant impact if the AUT/DUT is placed elsewhere in the QZ. Accurate polarization measurements, in a single-reflector CATR, can only be achieved when the AUT/DUT is positioned at the centre of the QZ and its size is significantly smaller than the CATR reflector.

When testing electrically large antennas and/or fitted to large platforms, the QZ cross-polar performance is often the reason that a more expensive, complex, compensated dual reflector CATR is chosen rather than a single-reflector CATR. This complexity and cost deterrent is why much research has been done on minimizing the QZ cross-polarization of the single-reflector CATR. Solutions such as reflector geometry

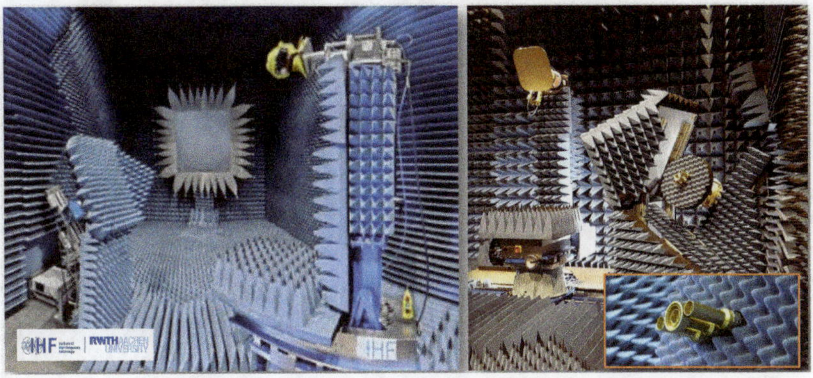

Figure 14.32 CATR system with cross-polar reduction feed (CXR) installation in RWTH Aachen University. Left: view of the system from behind the QZ. Right: view of the system from the reflector showing the positioner with the AUT/DUT and the feed array.

adjustments, other hardware improvements and post-processing techniques have been proposed over the years, but the many drawbacks of these techniques have been a hindrance for their widespread use. Recently, the application of conjugated matched feeds providing wide band dual polarized cross-polar reduction has been presented [32]. The concept provides a cancellation of the GO cross-polar component induced by offset reflector geometry. It consists of a three-element array of wideband dual linearly polarized horns with integrated feeding network: a central corrugated horn, which produces high co-polarization purity illumination of the reflector, such as a standard CATR feed, and two cross-polarized side-elements, creating the conjugate field matching taking an advantage of the array characteristics. Since the GO cross-polar effect is a geometrical effect, the same array can provide the cross-polarization in both polarizations.

An example of such system is shown in Figure 14.32. The cross-polar reduction feed (CXR) is shown on the left. It is rotated 45° to match the edge feed reflector geometry. The CXR array is based on a compact waveguide orthomode junction. A feeding network creates the conjugate field matching taking an advantage of the array characteristics and a suitable excitation of the two cross-polarized side-elements to cancel the GO cross-polar component induced by offset reflector geometry.

The performance improvements in measured cross-polar accuracy are shown in Figure 14.33. A reflector AUT/DUT is measured slightly offset in the QZ and thus into the less favourable cross-polar performance of a standard CATR systems. The CXR feed compensates the GO-induced cross-polar and improves the cross-polar measurement with 10–15 dB. The nominal frequency bandwidth of a CXR feed is 1:1.6. This contrasts with standard CATR feeds that can cover ranges up to 1:2 without measurable QZ performance degradation. The frequency range limitations are imposed by the fixed excitation coefficients of the lateral cross-polar reduction feeds with frequency and the array factor of these elements. The performance is a trade-off between cross-polar reduction level and frequency bandwidth.

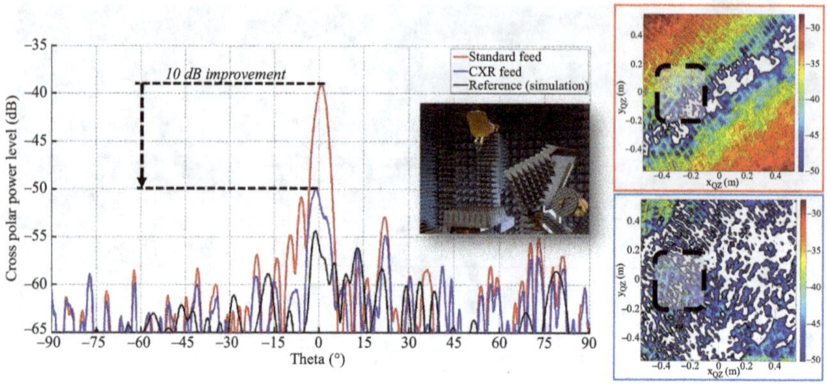

Figure 14.33 Validation of the cross-polar reduction feed (CXR). Comparison of QZ cross-polar and measured cross-polar levels of a reflector antenna with standard feed and CXR feed.

14.9 Array-based plane wave generator

The array-based PWGs consist of a suitable arrangement of radiating elements with optimized excitation. When properly exited, the array generates an approximate plane wave in the close proximity of the array. The PWG can be implemented either as a linear or planar array of elements in 1D or 2D systems for general antenna testing purposes. It consists typically of two main components: the radiating elements arranged in the structure of an array and the feeding system. The feed system may consist of a beam forming network (BFN) or an active system. The PWS can achieve FF testing conditions in a QZ area located in a region close to the radiator, as in the case of a compact range.

The main advantage of the array-based PWG over the reflector-based solution is the reduced physical size for equivalent testing capabilities. At lower frequency, the size advantage is particularly evident as it affects system dimensions and cost. Moreover, the array-based PWG does not require a precisely aligned feed as the CATR and its performance is not affected by the direct leakage from the feed into the QZ. At mmWave frequency bands, the size advantage is less important, but the agility of the system allows to make the PWG movable, while the movements of the AUT/DUT are reduced to increase the overall measurement accuracy of the system. An example of a dual polarized PWG covering 600–6,000 MHz bandwidth is shown in Figure 14.34, [29,35].

A practical and cost-effective solution to make wideband PWGs is to use a combination of passive and digital BFN by dividing the PWG elements into clusters of elements. This way, the elements of the same cluster are always excited with the same amplitude and phase through a passive BFN, while a limited number of parallel transceivers modules can control the relative amplitude and phase weights between the different clusters. With this approach, the large bandwidth capability is obtained by the frequency-dependent excitation, and the overall cost and

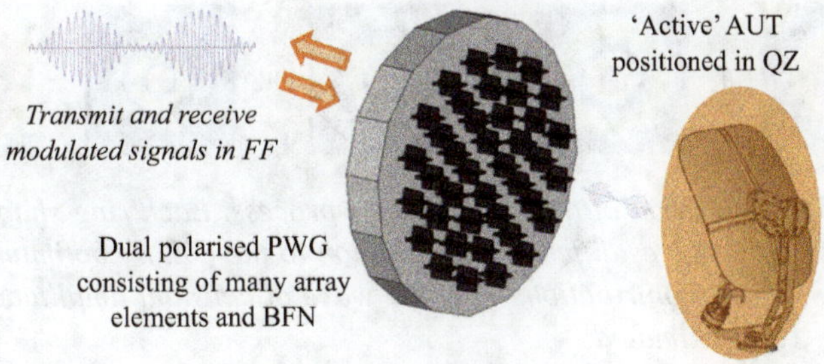

Figure 14.34 PWG example: an approximate plane wave is generated in the QZ by suitably excited array of dual polarized elements

complexity of the PWG is lowered by the reduction in the number of active components. It is worth noting that such PWG is still a reciprocal system and it can be used in transmission or reception, performing active and passive measurements. A further design trade-off for such system is to find the optimum balance between the number of clusters and the performance of the array.

Considering that the desired QZ has circular symmetry, it is convenient to use circular sub-array made of several concentric rings. Each ring array of elements is a cluster with equal amplitude and phase excitation. This can be achieved on a wide bandwidth using high-accuracy passive power dividers. The complex amplitude and phase excitation of each cluster can then be electronically controlled by the transceiver. The array characteristics in terms of the number of rings, the number of uncoupled isotropic elements and ring diameters can be retrieved using the design procedure described in [29], given the PWG diameter, the geometry of the desired QZ and the maximum frequency.

The relative excitation of the ring-shaped sub-arrays is digitally controlled and responsible for the quality of the plane wave approximation in the QZ. The sub-array weights are frequency dependent, although only minimum variation with frequency is observed within the operating frequency range. The optimum coefficients for the PWG can be determined experimentally. The synthesis process is illustrated in Figure 14.35. The different contributions from the sub-arrays are weighted to approximate the plane wave in the QZ. The synthesis domain is represented by a discrete number of points, or stations, fully defining the QZ in 3D space. The excitation synthesis is performed through a non-linear optimization, minimizing an objective function. In this case, the objective function is not only limited to the best approximation of a plane wave by minimizing the field deviation but also to ensure that the total radiated energy is concentrated in the QZ. The advantage of this formulation is to minimize field outside the synthesis domain.

The first step in the synthesis of the PWG excitation consists of the evaluation of the radiated field from each ring-shaped sub-array in the QZ volume. For this purpose, each sub-array is measured individually in an NF system. The NFFF transformation also allows to perform an NFNF transformation where the sub-array radiation can be determined with high accuracy in any point inside the QZ by post-processing [36]. This is done by exploiting the spherical wave expansion theory as shown in Figure 14.36.

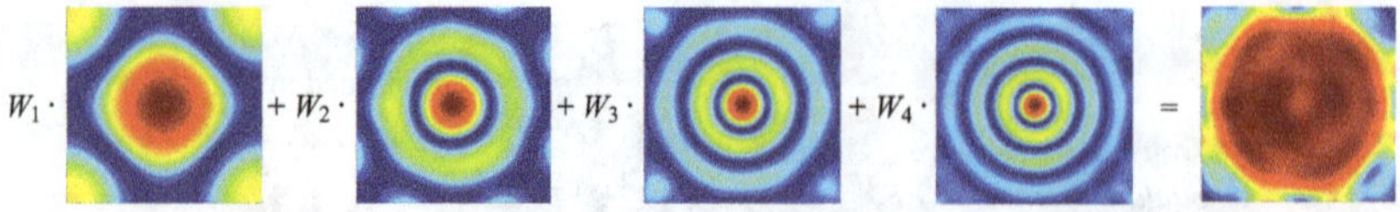

Figure 14.35 Illustration of the QZ synthesis process. Each ring-shaped sub-array gives different contribution to the QZ. By optimum weighting of the contributions, a plane wave of constant amplitude can be approximated.

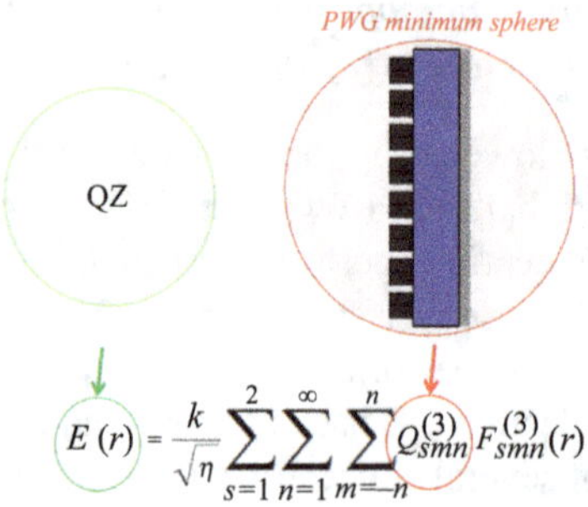

Figure 14.36 Evaluation of the radiated field from each ring-shaped sub-array in the QZ using NF measurements and the spherical wave expansion formula

The PWG can dynamically synthesize different size QZs up to the maximum size depending on the application. This is due to the digital control and measurement-based optimization procedure. A smaller QZ will have less free-space losses due to the gain of the PWG. This can be advantageous in cases where a small device requiring higher signal-to-noise ratios is to be tested.

The measured QZ performance of a 40-element wide band dual polarized PWG covering 600–6,000 MHz is shown in Figure 14.37 [29,37]. An SGH with an aperture dimension of 450 mm × 337.5 mm, thus slightly larger than the QZ, has been measured using this PWG prototype at 4 GHz. The directivity of the SGH is approximately 22 dBi at this frequency constituting a difficult measurement scenario for any measurement system with only 950 mm between the PWG and the QZ centre. At 4 GHz, such distance is too close for the SGH to be FF condition.

The directivity pattern comparison along the *E*-plane of the AUT in the case of the high directivity SGH is shown in Figure 14.38. The proportions of the measurement scenario consisting of the PWG and SGH are also illustrated. In particular, the curves represent the reference pattern (green), the measurement with PWG (cyan) and with the single element (magenta) at 4 GHz. The analysis shows that a single element is unable to generate a decent FF condition to measure accurately. The measured AUT pattern is in fact strongly distorted with a deviation of the boresight directivity of almost 4 dB. Conversely, the PWG generates the proper FF condition in the QZ resulting in an accurate measured pattern whose

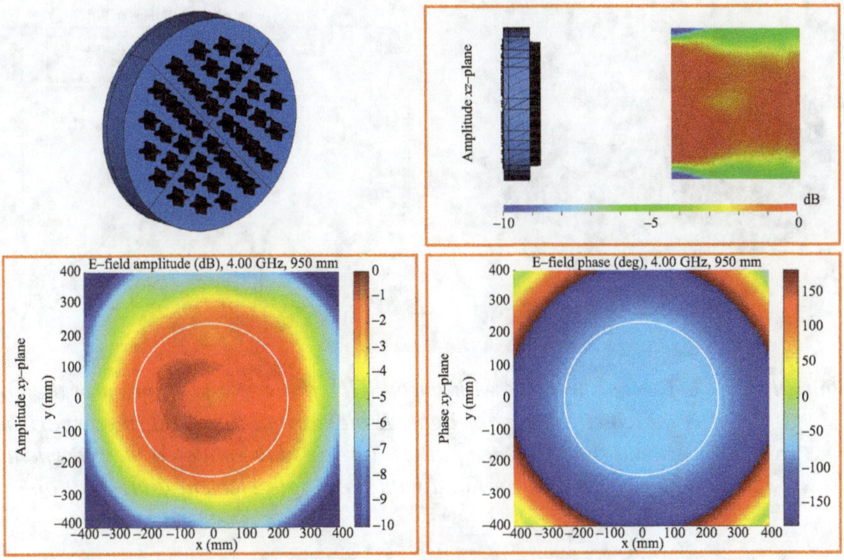

Figure 14.37 Measured QZ amplitude/phase performance at 4 GHz of 40-element 600–6,000 MHz dual-polarized PWG

directivity, SLL and nulls agree very well with the reference. The tests show that in the case of high gain and/or offset antenna, the PWG is cable of testing such challenging antennas in a reduced space with high accuracy.

An mmWave array-based PWG for live person testing is shown in Figure 14.39 [38,39]. In this system, the DUT is rotated only in azimuth while the PWG rotates in elevation. This enables to test devices by themselves, mounted on phantoms emulating a user scenario or even life person testing. Other developments towards mmWave PWG systems are reported in [40,41].

The closed boundary quad-ridge horn QR18000 has been measured in the pilot system as a factory test. The testing scenario is shown in Figure 14.40. The test was performed in non-anechoic conditions to verify the quality of the QZ. The QR18000 was measured in three positions, at 27.5 GHz, with the same orientation but offset 0, −8 and 11 cm in the *Z* directions. This is an effective reduction in PWG to DUT distance. The measured patterns are shown in Figure 14.40 and compared to full-wave simulation. The measurements and simulations are in good agreement. The on-set measurements and the simulations are in very good agreement. Some variations between the offset patterns and the on-set measurement and simulation can be noticed at elevation 45° from boresight around the −20-dB pattern levels. This is believed to be a consequence of the non-anechoic testing conditions in the factory test.

14.10 Conclusion

In FF antenna measurements techniques, the desired performance is determined instantaneously in each single direction. In NF techniques, the measurement

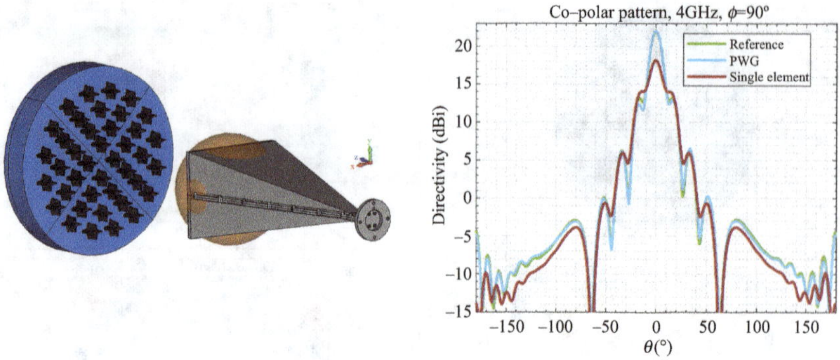

Figure 14.38 Left: measurement with onset SGH, the size and position of the QZ is indicated (orange). Right: E-plane radiation pattern comparison between the reference pattern (green), the emulated measurement with the PWG (cyan) and with the single element (magenta) on the right.

Figure 14.39 Example of mmWave plane wave generator for device and live person testing

requires sampling over a closed surface, but the performance is generally determined over the full 3D pattern after an NFFF transformation step [42]. Other than single-probe systems, typical FF systems are often PWGs; examples of such are CATR and array-based PWG. Multi-probe systems can also be used as FF systems when testing small antennas. The measured results can in some cases be improved by compensating variations in the testing distance.

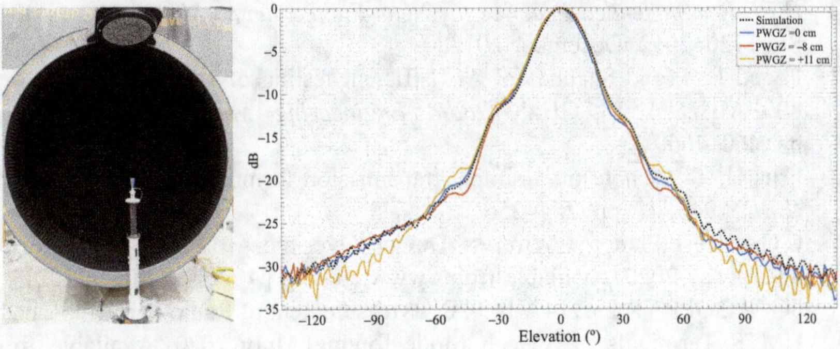

Figure 14.40 Comparison between full-wave simulation and PWG measurements of the QR18000 quad-ridge horn at 27.5 GHz. Measured performed with different offset distances along the Z-axis.

NF antenna measurements are considered the most accurate measurement techniques [43,44]. The NF testing techniques are more compact and versatile than FF testing, when considering performance against overall cost and have become the preferred choice for automotive, satellite antenna and payload testing. In NF antenna measurements the NFFF transformation is typically performed expanding the measured field over a set of basic functions. In the case of orthogonal basis functions such as the generic scan geometries SNF/PNF/CNF, the NFFF is performed by FFT. The NFFF transformation is an expansion of the measured field at a close distance in known (mathematical) functions. Thus, other than being an accurate measurements technique, NF measurements give access to a wide range of diagnostics tools. The measurement post-processing generally gives access to the following:

1. Techniques to understand and debug measurement scenarios. Investigate sources of spurious radiation, coupling and echoes.
2. Understand and give insight into the radiation mechanism of the measured antenna.
3. Provide a link between measurements and numerical simulation tools as measurements, after post-processing can be exported as 'computational or black box' to CEM tools.
4. Determine power density at close proximity to devices for pre-compliance testing.

References

[1] Bryant J. H. *Heinrich Hertz: The Beginning of Microwaves*. Piscataway, NJ: IEEE Service Center. 1988.

[2] *IEEE Standard Definitions of Terms for Antennas*. IEEE Std 145–2013. March 2014.

[3] *IEEE Standard Test Procedures for Antennas*. ANSI/IEEE Std 149–1979. 1979.

[4] *IEEE Recommended Practice for Near-Field Antenna Measurements*. IEEE Std 1720–2012. December 2012.

[5] Foged L. J. and Scannavini A. 'Efficient testing of wireless devices from 800 MHz to 18 GHz'. *Radio Engineering Magazine*, 2009, 18(4), pp. 460–466.

[6] Friis H. T. 'A note on a simple transmission formula'. *Proceedings of the IEEE*, 1946, 41, pp. 254–256.

[7] CTIA. Test Plan for Wireless Device Over-the-Air Performance version 3.9.1. May 2020. Available from: www.ctia.org [Accessed March 2021].

[8] 3GPP. 3GPP TR 25.914 V16.0. 0 – Measurement of Radio Performances for UMTS Terminals in Speech Mode [online]. July 2020. Available from: https://portal.3gpp.org/desktopmodules/Specifications/SpecificationDetails.aspx?specificationId=1343 [Accessed March 2021].

[9] 3GPP. 3GPP TS 34.114 V12.2. 0 – User Equipment (UE)/Mobile Station (MS) Over The Air (OTA) Antenna Performance – Conformance Testing [online]. October 2016. Available from: https://portal.3gpp.org/desktopmodules/Specifications/SpecificationDetails.aspx?specificationId=2361 [Accessed March 2021].

[10] 3GPP. 3GPP TS 37.114 V16.0. 0 – User-Equipment (UE) and Mobile Station (MS) GSM, UTRA, and E-UTRA Over The Air Performance Requirements. July 2020. Available from: https://portal.3gpp.org/desktopmodules/Specifications/SpecificationDetails.aspx?specificationId=3034 [Accessed March 2021].

[11] Balanis C. A, *Antenna Theory: Analysis and Design*, 2nd Ed., John Wiley & Sons, Inc., Hoboken, NJ; 1997.

[12] Newell A. C., 'Error analysis techniques for planar near-field measurements'. *IEEE Transactions on Antennas and Propagation*, 1988, 36, pp. 754–768.

[13] Saccardi F., Rossi F., Scialacqua L., and Foged L. J., 'Truncation Error Mitigation in Free-Space Automotive Partial Spherical Near Field Measurements'. *AMTA 2017*, Atlanta, GA, USA, 15–20 October 2017.

[14] Repjar A. G., Newell A. C., and Francis M. H. 'Accurate determination of planar near-field correction parameters for linearly polarized probes'. *IEEE Transactions on Antennas and Propagation*, 1988, 36(6).

[15] Marshall Leach W. and Paris D. T. 'Probe compensated near-field measurements on a cylinder'. *IEEE Transactions on Antennas and Propagation*, 1973, AP-21(4).

[16] Hansen J. E. Spherical Near Field Antenna Measurements. *IEE Electromagnetic Series*. Peter Peregrinus Ltd., London, UK; 1988.

[17] Breinbjerg O. 'Spherical Near-Field Antenna Measurements – The Most Accurate Antenna Measurement Technique'. *2016 IEEE International Symposium on Antennas and Propagation (APSURSI)*, Fajardo, PR, USA, 2016, pp. 1019–1020, doi: 10.1109/APS.2016.7696217.

[18] Jensen F. and Frandsen A., 'On the Number of Modes in Spherical Wave Expansions'. *AMTA 26th Symposium*, Stone Mountain Park, GA, USA, October 2004.

[19] Araque Quijano J. L. and Vecchi G. 'Improved accuracy source reconstruction on arbitrary 3-D surfaces'. *Antennas and Wireless Propagation Letters, IEEE*, 2009, 8, pp. 1046–1049.

[20] Quijano J. L. A., Vecchi G., Li L., *et al.* '3D Spatial Filtering Applications in Spherical Near Field Antenna Measurements'. *AMTA 32nd Symposium*, Atlanta, GA, USA, October 2010.

[21] Scialacqua L., Foged L. J., Mioc F., and Saccardi F. 'Link Between Measurement and Simulation Applied to Antenna Scattering and Placement Problems'. *EuCAP*, Paris, France, March 2017.

[22] Ying Z., Zhao K., Bolin T., *et al.* 'Study of Phased Array in UE for 5G mm Wave Communication System with Consideration of User Body Effect'. *The 10th European Conference on Antennas and Propagation (EuCAP 2016)*, Davos, Switzerland, April 2016.

[23] 3GPP. R4-1705819 – On TRP Measurements Sampling Grid at >6 GHz. 3GPP TSG-RAN WG4 #83, Hangzhou, CN, 15th–19th May 2017.

[24] Xu B., Ying Z., Scialacqua L., *et al.* 'Radiation performance analysis of 28 GHz antennas integrated in 5G mobile terminal housing'. *IEEE Access*, 2018, 6, pp. 48088–48101.

[25] Huawei. Huawei Launches the World's First SG178 BTS Antenna Testing System [online]. 2016. Available from: https://www.huawei.com/en/news/2016/8/shouge-sg178-jizhan-xitong [Accessed March 2021].

[26] Nokia. Nokia Annual Report on Form 20-F 2020 [online]. 2020. Available from: https://www.nokia.com/about-us/news/releases/2021/03/04/nokia-has-filed-its-annual-report-on-form-20-f-for-2020-and-published-its-nokia-in-2020-annual-report/ [Accessed March 2021].

[27] Johnson R. C., Ecker H. A., and Moore R. A. 'Compact range techniques and measurements'. *IEEE Transactions on Antennas and Propagation*, 1969, 17 (5), pp. 568–576.

[28] Bennet J. C. and Schoessow E. P. 'Antenna near-field/far-field transformation using a plane-wave30 synthesis technique'. *Proceedings of the IEEE*, 1978, 125(3), pp. 179–184.

[29] Scattone F., Sekuljica D., Giacomini A., *et al.* 'Plane Wave Generator for Direct Far-field Over-The-Air Testing of Devices'. *AMTA 40th Symposium*, Williamsburg, VA, USA, November 2018.

[30] Jernberg A., Pinkasy M., Pinchuk G., *et al.* 'Short Focal Length Compact Antenna Test Range for Large L Ku Band Antenna Measurements'. *IEEE Indian Conference on Antennas and Propagation (InCAP 2018)*, December 2018.

[31] Evans G. E. *Antenna Measurement Techniques*. Artech House Inc., New York, NY; 1990.

[32] Giacomini A., Foged L. J., Riccardi A., Pamp J., Cornelius R., and Heberling D. 'Improving the CrossPolar Discrimination of Compact Antenna Test Range using the CXR Feed'. *AMTA 38th Symposium*, Austin, TX, USA, October 2016.

[33] Coq L. Le, Fuchs B., Kozan T., Burgos S., and Iversen Per O. 'IETR Millimeter-Wave Compact Antenna Test Range Implementation and

Validation'. *The 9th European Conference on Antennas and Propagation (EuCap 2015)*, Lisbon, Portugal, April 2015.

[34] Ettorre M., Manzillo F., Casaletti M., Sauleau R., Coq L. Le, and Capet N. 'Continuous transverse stub array for Ka-band applications'. *IEEE Transactions on Antennas and Propagation*, 2015, 63(11), pp. 4792–4800.

[35] Scattone F., Sekuljica D., Giacomini A., *et al.* 'OTA Testing of Antennas & Devices using Plane Wave Generator or Synthesizer'. *The 14th European Conference on Antennas and Propagation (EuCAP 2020)*, Copenhagen, Denmark, March 2020.

[36] Castaner M. S. and Foged L. J. *Post-processing Techniques in Antenna Measurement*. Institution of Engineering and Technology (Electromagnetic Waves series). SciTech Publishing, London, UK; 2019.

[37] Scattone F., Sekuljica D., Giacomini A., *et al.* 'Design of Dual Polarised Wide Band Plane Wave Generator for Direct Far-Field Testing'. *The 13th European Conference on Antennas and Propagation (EuCAP 2019)*, Krakow, Poland, March/April 2019.

[38] Scattone F., Sekuljica D., Giacomini A., *et al.* 'Towards Testing of 5G Millimeter Wave Devices Using Plane Wave Generators'. *The 15th European Conference on Antennas and Propagation (EuCAP 2021)*, Dusseldorf, Germany, March 2021.

[39] Scattone F., Sekuljica D., Giacomini A., *et al.* 'Preliminary Assessment of Millimetre Wave Plane Wave Generator for 5G Device Testing'. *The 15th European Conference on Antennas and Propagation (EuCAP 2021)*, Dusseldorf, Germany, March 2021.

[40] Zhang Y., Wang Z., Sun X., Qiao Z., Fan W., and Miao J. 'Design and implementation of a wideband dual-polarized plane wave generator with tapered feeding nonuniform array'. *IEEE Antennas and Wireless Propagation Letters*, 2020, 19(11), pp. 1988–1992.

[41] Catteau S., Ivashina M., and Rehammar R. 'Design and Simulation of a 28 GHz Plane Wave Generator for NR Measurements'. *The 14th European Conference on Antennas and Propagation (EuCAP 2020)*, Copenhagen, Denmark, March 2020.

[42] Yaghjian A. D. 'An overview of near-field antenna measurements'. *IEEE Transactions on Antennas and Propagation*, 1986, 34(1), pp. 30–45.

[43] Antenna Centre of Excellence. Deliverable A1.2D4 Recommended Practices for Near Field Measurements [online]. 2007. Available from: http://ace1.antennasvce.org/Dissemination/view/download?id_file=48 [Accessed March 2021].

[44] Foged L. J. DeliverableA1.2D2, Recommendations and Comparative Investigations for Near-Field Antenna Measurement Techniques and Procedures. Doc. No. FP6-IST-026957-A1.2D2, SATIMO. 2008.

Part V

Propagation channel metrology

Chapter 15
Metrology for channel sounding

Haowen Wang[1] *and Yunsong Gui*[1]

15.1 Wireless channel measurement requirements of 5G communication

Nowadays, fifth-generation (5G) wireless technology has been applied to various aspects of our daily life, such as public transportation and shopping malls. In addition to these daily applications, compared with the traditional third-generation/fourth-generation (3G/4G) system, 5G new radio (NR) is widely used in other scientific research and technology scenarios because of its ultrahigh traffic volume density, ultrahigh connection density, ultrahigh mobility, and other diversified characteristics. A variety of technologies are used to serve the end users, such as massive multiple-input–multiple-output (mMIMO) channel models [1], millimeter-wave (mm-wave) communications, ultradense networks, and device-to-device (D2D) communications. Therefore, the main requirements of 5G NR channel models could be divided into five areas: (1) higher frequency and larger bandwidth, (2) mMIMO, (3) spatial consistency and deal mobility, (4) wide-range propagation scenarios and diverse network topologies, and (5) high mobility.

15.1.1 Higher communications frequency and larger bandwidth

Considering the growing requirements of available bandwidth and communication capacity, an area that has not yet been ventured for the 5G NR channel is the maximum channel bandwidth. There are ongoing global activities to further explore higher operating bandwidth for 5G system above 400 MHz (a figure that is agreed in the Third Generation Partnership Project (3GPP) Release 15 [2]). Therefore, efforts have been taken to make 5G channel models fulfill this demand. Higher radio frequency (RF) bands such as mm-wave [3] and terahertz frequencies have attracted much attention because of their available huge communication bandwidths. For example, METIS recommends medium- and high-priority frequency bands to be developed, including 10, 28–29, 32–33, 43, 46–50, 56–76, and

[1]Shanghai Institute of Microsystem and Information Technology, Chinese Academy of Sciences, Shanghai, China

81–86 GHz. The frequency band between 6 and 100 GHz has many radio transmission characteristics that are different from those below 6 GHz band, namely, sub-6 GHz as described in the following.

First of all, according to the Friis formula of free-space transmission, shorter wavelength transmission in a higher RF band invites higher path loss. This formula also indicates that antenna gain is proportional to the square of frequency. In certain physical aperture during transmitting power, shorter wavelength transmission requires higher receiving power than low-frequency transmission. Therefore, high-gain directional antennas or beamforming technology should be used in high-frequency bands to ensure the communication for a distance of several hundred meters. Simultaneously, the antenna also needs to be able to adjust the beam pointing toward user terminals (UTs).

Second, due to the inherent nature of short wavelengths, the transmission has quasi-optical characteristics with weak diffraction leading to a lack of abundant scattering for operating frequency bands at 6 GHz and above. In light of sight (LOS), received signals consist of an LOS path and a small number of low-order reflection paths, whereas in non-light of sight (NLOS), signal propagation mainly depends on reflection and diffraction. Therefore, the channel shows the sparsity of time and space. The blocking of the communication link from people or vehicles will greatly weaken received signal level. Since such transmission has quasi-optical characteristics, this enables ray-tracing technology to assist the channel modeling.

Although the propagation characteristics of mm-wave have widely been studied, especially in 60 GHz bands, many important characteristics, such as blockage loss, penetration loss, high-resolution angle characteristics, frequency dependence, and outdoor mobile channel, still need further measurement and exploration. In the aspect of channel models, several higher RF channel models (including 60 GHz) are proposed.

15.1.2 Massive MIMO

It is assumed by the existing channel models that plane wave rebounds from a scatterer (e.g., a scatterer that is located in the far field of transmitting and receiving), and that the size of an antenna array is quite small. At the two ends of an antenna array, the arrival or departure direction of radio wave only causes a phase difference of signals to be very small, while the amplitude of signals remains unchanged, that is, the propagation characteristics of the two ends of an antenna array are similar. When applying larger antenna arrays, including mMIMO and pencil beamforming, the antenna number can reach tens to hundreds, which brings about two impacts, as shown in Figure 15.1.

First, the plane wave assumption is no longer applicable, and the propagation wavefront should be assumed to be spherical. Radio wave propagating in the form of plane wave needs to conform to the far-field assumption, that is, the transmitting/receiving distance between scatterers should be greater than the Fraunhofer distance, $R_f = 2D^2/\lambda$, where D is the aperture size of antenna array, and λ is wavelength. D and the corresponding Fraunhofer distance of array will increase with the

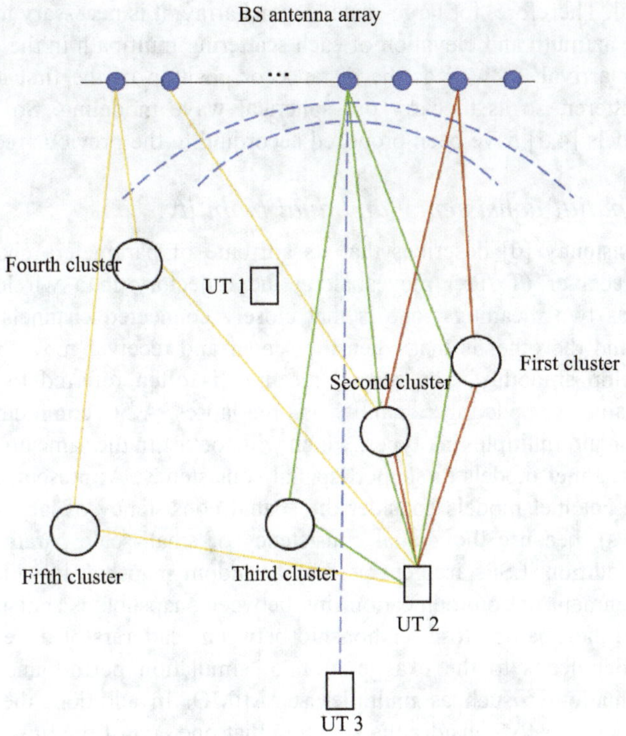

Figure 15.1 Nonstationary and near-field effect in mMIMO [1]

number of antennas, assuming that antenna element spacing and λ remain unchanged, that is, the far field of antenna system expands. The receiver may be located in the near-field region of the transmitter (also known as the Fresnel region), or the scatterer may be located in the near-field region of the transmitter and the receiver. In this case, the wavefront should be simulated as a spherical wave instead of a plane wave.

Second, it is verified by researchers from the Lund University that any scattering cluster in large-scale MIMO system will not affect all antennas in the array. In other words, different parts of an array have different scattering clusters, which affect the signal propagation. In traditional MIMO systems, due to the relative motion among the transmitter, the receiver, and the scatterer, the scattering clusters that affect the signal propagation only change with time. In large-scale MIMO systems, the scattering clusters also change along the axis of the antenna array.

In addition, due to the high spatial resolution of large antenna array, which can distinguish the approaching users in the horizontal and vertical directions, the channel model must provide fine three dimensional (3D), including azimuth and elevation) angular information. The angular resolution should be at least 10° or less and should support a variety of antenna arrays, such as linear, planar, cylindrical, and spherical. Therefore, for large-scale antenna array, it is necessary to accurately simulate the azimuth and elevation of each scattering multipath in the direction of departure or arrival, as well as the distance or position of the first rebound/last rebound scatterer, so as to carry out spherical wave modeling. So far, several channel models [4,5] have been proposed according to the previous requirements.

15.1.3 Spatial consistency and dual mobility

Spatial consistency [6] describes that its attribute of parameters changes when transmitter/receiver (Tx/Rx) moves along the trajectory in a wireless channel model. It has two meanings: one is that closely connected channels are highly correlated, and the other is that when the sender and receiver move, the channel should develop smoothly. The second meaning is often referred to as channel model's parameters updating capability as time lapses. A 5G communication network will contain multiple link types, which will coexist in the same area. All these require 5G channel models to support spatial consistency. At present, most of the widely used channel models consider the spatial consistency of large-scale parameters (LSPs), because the spatial consistency of small-scale parameters is not enough. In addition, LSPs are generated by a random snapshot, in which the scattering environment of communication link between snapshots is statistically independent, and there is no close relationship between scatterers of different mobile devices, which leads to the exaggeration of simulation performance of multi-antenna technologies such as multiple-user MIMO. In addition, the traditional channel models mostly consider the situation that one end of the link is fixed and the other end can move arbitrarily. A new Doppler model should be introduced to support the bidirectional mobility of D2D/vehicle-to-vehicle (V2V).

With the coexistence of links and the increase of density, as well as the application of D2D/V2V, supporting spatial consistency is particularly important for the establishment of wireless channel model. The spatial consistency model can be established by defining the geometric positions of scatterers for the first bounce (emitter-to-satellite) and the last bounce (emitter-to-receiver) of each scattering path. As shown in Figure 15.2, the QuaDriGa channel model library supports the dynamic evolution of the channel caused by the UT movement. The channel coefficient is calculated according to the position of the last scatterer, the distance between the last scatterer and the mobile terminal, the variation of the relative Angle of Arrival (AoA), and the distance between the mobile station and base station (BS).

15.1.4 Wide-range propagation scenarios and diverse network topologies

The vision of 5G is that anyone and anything, no matter when and where, can get information and share data with each other. The China IMT-2020 5G promotion group interprets this as "information at a fingertip, everything is under control." The current cellular mobile communication network serves static and UTs with fixed BS, so the traditional channel model is developed to serve this kind of application. In 5G architecture, the network topology should not only support the cellular network but

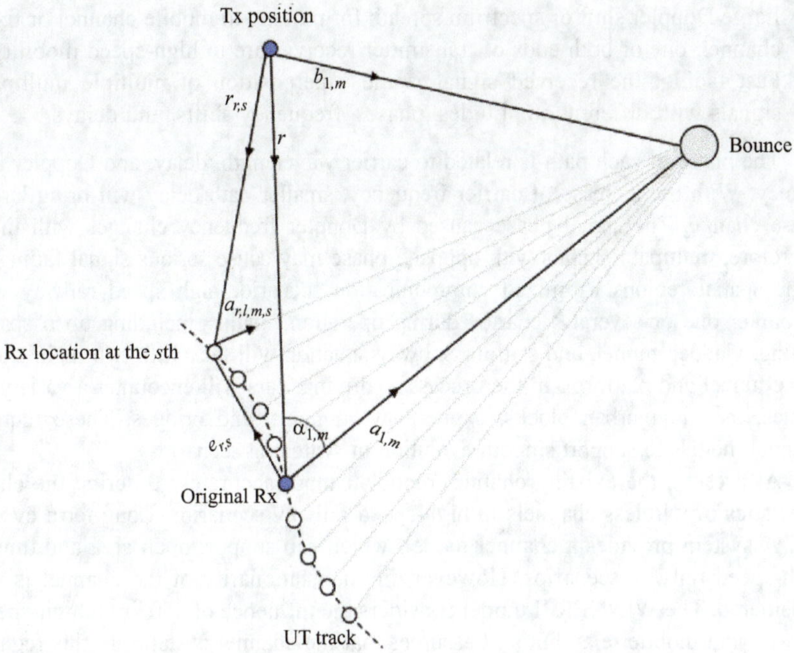

Figure 15.2 Scatterer location and AoA update in dynamic simulation [4]

also support the communication of D2D, machine-to-machine, V2V, and the whole Internet. Therefore, the 5G channel model should support mobile-to-mobile link and network. In D2D/V2V communication, the user antenna is usually as low as 1–2.5 m, which limits its coverage to be smaller than that of ordinary cellular system. In addition, there are a large number of dense scatterers at both transmitter and receiver. Due to the mobility of both ends, the stationary interval of the channel is short. All of these are unique features of D2D/V2V transmission. Many channel measurement activities have been carried out in D2D/V2V, and some preliminary models are given [7–9]. For example, METIS (an EU 5G Research and Development projects) proposed a D2D/V2V model, which contains multiple scenarios, 3GPP proposed a preliminary D2D channel model, and MiWEBA (an EU–Japan joint project) proposed a 60 GHz millimeter band D2D model.

15.1.5 High mobility

The 5G channel model usually needs to support the following scenarios: high-speed rail of more than 500 km/h, ultrahigh user density of six people/m^2 on subway, millisecond end-to-end delay in expressway. Therefore, the channel with high mobility has the following characteristics:

- Fast time varying characteristics: different from low-mobility channel, high-mobility channel has fast time varying and nonstationary characteristics.

- Large Doppler shift or spectrum spread: for moving to mobile channel or fixed channel, one or both ends of transmitter/receiver are in high-speed mobility.
- Fast fading: the received signal is the superposition of multiple multipath signals with different amplitudes, phases, frequency shifts, and delays.

The phase of each path is related to carrier wavelength, delay, and Doppler frequency. With the increase of carrier frequency, smaller path delay will bring larger phase change. The signal phase caused by Doppler frequency changes with time. Therefore, multipath signals with opposite phase may cause serious signal fading in some spatial regions. A mixed communication scenario: high speed railway will encounter one or several scenarios during operation, mainly including open space, cutting, viaduct, tunnel, and station. Subway situation will face with two main situations: tunnel and platform. In the process of driving, cars will encounter a variety of scenes, including urban blocks, expressways, tunnels, and bridges. These require channel models to support smooth evolution in switching scenarios.

At present, the existing channel models cannot accurately describe the characteristics of wireless channels in high-speed railway scenarios. Long-term evolution A system provides a channel model, which can support open area and tunnel high-speed railway scenarios. However, the nonstationarity of the channel is not considered. The WINNER II model considers the influence of 350 km/h high-speed railway and mobile relay but still assumes that the channel is static. In this regard, some measurements have been carried out, such as the high-speed rail/subway measurement of State Key Lab of Rail Traffic Control & Safety in China, the vehicle measurement of Lund University of Sweden, and the high-speed mobile scene measurement of the WINNER II project. As high mobility increases the difficulty and cost of measurement, more and more channel modeling uses ray-tracing technology [10,11].

15.2 Requirements for 5G wireless channel measurement

With the gradual development of wireless communication technology, the study of wireless channel propagation characteristics and channel modeling have become key points to improve wireless system coverage, capacity, and delay performance. Channel sounding systems and methods have received extensive attention as primary tools for studying wireless propagation and modeling.

In general, channel sounders are divided into frequency- and time-domain-based methods according to different detection methods. The method based on the frequency domain is mainly implemented with a vector network analyzer (VNA). A VNA measures the frequency-domain response of the wireless channel based on frequency-domain scanning and then obtains the time delay power profiles (PDP) of the wireless channel through Fourier transform. By setting a narrower IF signal bandwidth, a VNA can achieve a higher measurement dynamic range. Meanwhile, the delay estimation accuracy will be higher as the frequency-domain range of the scan increases. A MIMO channel sounder can also be realized based on a VNA.

However, due to the long measurement time, the VNA-based channel sounder is only suitable for detecting and analyzing static wireless channel scenarios.

The time-domain-based method usually implemented on the basis of the broadband signal transceiver link. The channel sounding can be used to detect dynamic wireless channels based on the design of different transmitted signal waveforms. There are some commonly used channel detection systems for 4G or earlier communication systems for channel detection tasks. These systems include Propsound, RUSK, Aalto University (Aalto U., formerly Helsinki University of Technology, HUT) channel detector, and NTT DoCoMo channel detector. Various channel sounding systems, such as Propsound, Medav RUSK, and Aalto University systems, have been widely used in the development of COST, WINNER, IMT-Advanced, 3GPP-3D, and other channel models. The rest of the section will describe the typical channel sounder in terms of the system architecture, RF antenna array, synchronization system, and sounding signal design.

15.2.1 System architecture

With a widespread use of massive MIMO technology and mm-wave frequency bands, channel detection equipment has gradually evolved toward multichannel, high-frequency band, and high bandwidth. According to the number of channels measured simultaneously, channel measurement methods can be divided into single Tx/Rx channel (SISO) (using time-division multiplexing (TDM)-based MIMO antennas), multiple Tx/Rx channels (MIMO), and compromise single-Tx and multiple-Rx channels (SIMO). The MIMO method is to measure multiple parallel channels at the same time. It has the highest measurement speed and can be used in high-speed moving scenes.

However, the MIMO measurement method requires multiple parallel RF transceiver units. In addition to the complex structure and high cost, the inherent requirements for consistency and synchronization also increase system implementation and calibration difficulty. Besides, multiple simultaneous detection signals will interfere with each other. Even if orthogonal sequences are used as excitation signals, it is difficult to guarantee these sequences' orthogonality with different delays in the multipath channel. Because of the interference, a more complex processing algorithm is needed for separating multiple channels.

The MIMO channel sounder based on SISO mode requires only one RF link at the transmitter and receiver and detects multiple channels iteratively by the TDM method. There is an electronic switch box at each end for synchronous switching of different antenna units connected to the RF chain. The technique is a simple structure and low cost. Since there is only one RF link, it is easier to achieve calibration and consistency between multiple channels. Also, since only one sounding signal sends, orthogonality is not required. The main disadvantage is that when the number of transmitting and receiving antennas increases, its measurement time will significantly increase, limiting its ability to measure dynamic time-varying channels.

SIMO is a compromise between SISO and MIMO. It uses switch signal transmission and parallel signal reception and achieves a certain balance between system complexity, cost, and measurement speed.

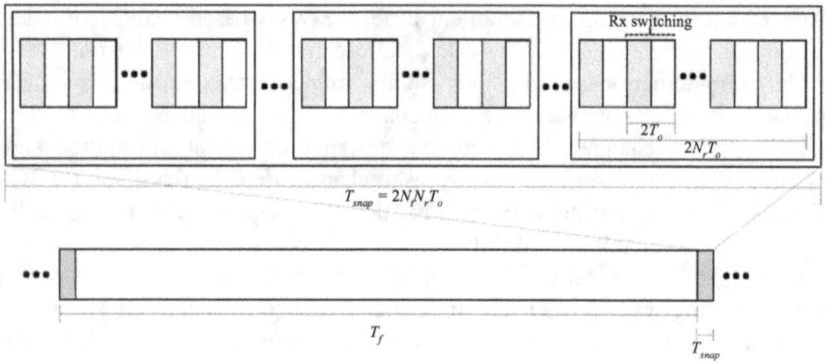

Figure 15.3 The timing sequence in MIMO channel

Figure 15.3 shows MIMO channel detection achieved by sequentially switching the receiver and transmitter antennas using the SISO method. Here is a simple analysis of the dynamic detection capability of the system. N_t represents the number of transmitting antennas, N_r represents the number of receiving antennas, and $\tau_{\max}$ represents the channel's maximum excess delay. Take the pseudo-noise (PN) sequence as the detection sequence, and take the SISO detection as an example. The periodic detection sequence is composed of K chips. The length of the chip is T_p and the length of the sequence is $T_a = KT_p$.

T_a determines that a processing gain can be obtained, but it also limits the measurable Doppler frequency. The period of a detection T_o should be equal to the sum of the length of the detection sequence T_a and $\tau_{\max}$ corresponding to the maximum Doppler spread (DS), $T_o = T_a + \tau_{\max}$. If the antenna switching time T_w is considered, and the measurement time and the switching time are equal, $T_w = T_o$, the measurement time of the subchannel is 2 T_o. The time it takes to iterate all subchannels in the snapshot is $T_{snap} = 2\,T_oN_tN_r$, which is the time to perform a complete channel detection. For time-varying channels, the sounding period is denoted as T_{snap}. If the interval between snapshots is zero, that is, $T_f = T_{snap}$, the maximum Doppler frequency that the system can measure is $f_D = 1/T_{snap}/2 = 1/(4N_tN_r)/(KT_p + \tau_{\max})$.

As shown in Figure 15.4, a unified modular channel sounder is used as a framework to illustrate various methods, techniques, and equipment in channel measurement. We mainly study the measurement system from three aspects: RF unit and antenna system (typically including antenna, and local oscillator (LO) and up/down converter), baseband detection signal, and transceiver synchronization method.

15.2.2 RF and antenna array

The RF unit's main parameters include RF channels, RF range, and RF signal bandwidth. Currently, most models have specified the applicable frequency range or detection bandwidth. For example, 3GPP spatial channel model (SCM) is 1–3 GHz/5 MHz, WINNER II/+ and IMT-A 0.45–6 GHz/100 MHz, 3GPP D2D,

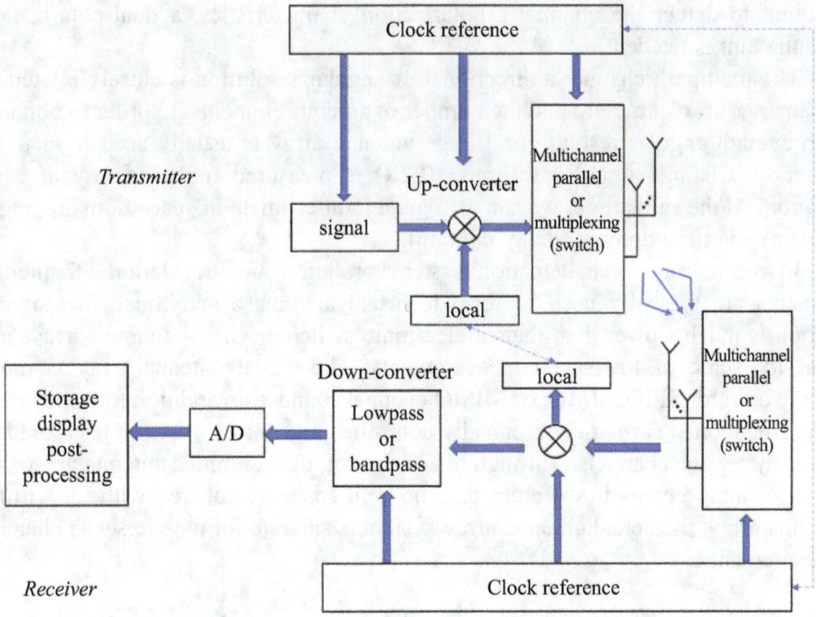

Figure 15.4 Principle sketch of MIMO channel sounder

and 3D 1–4 GHz/100 MHz. For frequency bands below 6 GHz, the existing channel model supporting a signal bandwidth of 100 MHz seems to be sufficient, but the demand for greater bandwidth will continue to emerge. For example, the maximum bandwidth in IEEE 802.11ac is 160 MHz. Therefore, the channel model must support higher bandwidth. In the 60 GHz millimeter band, the bandwidth of the detection system used in the development of IEEE 802.11ad is 800 MHz. In future mm-wave applications, the bandwidth of the signal waveform may exceed 2 GHz, which requires that the RF unit of the mm-wave channel detector should be able to support a bandwidth greater than 2 GHz. A typical RF transmission link consists of a baseband signal generator, a digital-to-analog converter (DAC), a transmitter LO signal generator, up-conversion, amplifiers, filters, etc. A typical RF receiving link includes a low-noise amplifier, filter, down-conversion, a receiver LO signal generator, an analog-to-digital converter (ADC), and a baseband-processing circuit.

There are many types of transmitting and receiving antenna elements for detection. In order to ensure that radio signals are radiated and collected in all spaces and at all angles, omnidirectional or low-gain antennas are used, such as half-wave dipole antennas, patch antennas, cone antennas, double cone antennas, and open waveguides to ensure sufficient signal strength. Use of directional antennas, such as standard gain horn antennas, parabolic antennas, and lens antennas, can only provide signal strength detection at a limit angle range. In order to cover all spatial angles, it is necessary to splice multiple directional antennas in different directions or rotate one directional antenna for scanning. When it is required to detect the channel's polarization characteristics, a dual-polarization antenna unit is needed.

The multipath channel's directional or angular resolution is closely related to the antenna array structure and the number of antenna elements. In order to obtain a high enough angular resolution, a large antenna array is usually used in channel detection. Channel impulse response (CIR) is measured independently at each location. At the same time, we can also get a joint estimate of space domain (angle domain) and time domain (delay domain).

In recent mm-wave detection systems, rotation and translation of antenna elements are typically used to form a virtual antenna array. Since mechanical motion is much slower than that of electronic switches, virtual antenna arrays are ideal for static or low-speed measurements. The actual antenna array is more widely used for SISO, SIMO, or MIMO channel sounder. In addition to the antenna arrays mentioned earlier, electronically controlled antenna arrays have been used to detect mm-wave channels. Through phase control, the combined antenna array can form scanning beams in different directions in space. Compared with the virtual antenna array, the actual antenna array system is suitable for mobile scene channel measurement.

15.2.3 Synchronization system

The channel sounder system could use different reference clocks at the transmitter and receiver ends. The baseband clock, ADC/DAC sampling clock, and LO can be obtained by frequency multiplication and locking based on the reference clock. If they are inconsistent, the difference will be regarded as part of the channel, and the channel parameters will be inaccurate. Therefore, the detection system should reduce the error of the transmitter and receiver reference clock. This requirement can easily be achieved in short-range measurements, such as indoor detection. For a VNA-based broadband sounder, if a VNA is used for sending and receiving simultaneously, the clock sharing between channels should be realized in the VNA.

If it is an outdoor mobile environment, the shared reference clock usually adopts a global positioning system (GPS)-tamed oven-controlled crystal oscillator or a high-precision and high-stability atomic clock (clock or cesium clock), or both. Even so, it still cannot avoid clock skew. The jitter or phase noise of each clock will affect its performance. When using SISO or Simo for MIMO detection, the synchronous switching of the switch array at the transmitter and receiver must be strictly aligned. The clock of the switch control circuit also comes from the same reference clock.

15.2.4 Sounding signal design

According to different parameter extraction algorithms, sounding signals significantly impact the performance of the detection system. How to select the detection signal to obtain the best channel detection performance is a question worth discussing. The set of sounding signal waveforms can be evaluated from the signal duration, signal bandwidth, time bandwidth product, power spectral density, peak-to-average power ratio, etc. The following sections show some commonly used channel detection signals.

15.2.4.1 Periodic pulse signal

Pulse signal with high bandwidth is the most intuitive choice for channel detection. Shortening the pulse width of the sounding signal can improve the resolution of the path delay. However, the sounding signal's very narrow pulse width is resulting in high peak power of the radiosonde signal and severe nonlinear distortion in the circuit. In order to reduce the distortion, the signal power must be reduced, thus limiting the measurement range of the detection system. According to the Shannon–Nyquist sampling theorem, the receiver should have an ADC with an ultrahigh sampling rate, which is very expensive. It is recommended to use compressed sensing technology to reduce the sampling frequency. In the early stages of channel sounder, periodic pulses are more widely used. Later, spreading sequences and flat envelope signals in the frequency domain are widely used.

15.2.4.2 Time-domain flat signal

A flat signal in the time domain is a relatively envelop flat signal in the time domain, which is different from a flat signal in the frequency domain. These signals are sequences with good periodic autocorrelation characteristics, including pseudorandom noise PN sequence, *M* sequence, Chirp sequence, and Zadoff–Chu sequence (also frequency-domain flat signal). The combination of system time-domain signals using this system shows excellent performance in terms of delay resolution and dynamic range, so it has been widely used.

When sending the PN sequence by the transmitter, the receiver can use two different processing methods. One method is to use a correlator or matched filter. At each sampling time, the CIR can be obtained directly, or the signal can be recorded and then processed. Some parameter extraction algorithms with high temporal and spatial resolution, such as maximum likelihood parameter estimation framework (RIMAX) [12] and space-alternating generalized expectation–maximization (SAGE) [13], have been widely used in channel measurement. This processing method requires a high-speed ADC for sampling and analog-to-digital conversion and storing a large amount of data. Another technique is sliding-related technology.

The transmitter and receiver use the same PN sequence, but the chip rate difference is small. Due to the difference in chip rate associated with the clock frequency, the transmitter and receiver's PN sequence will relatively slide. When the two sequence chips are realigned due to the transmission delay, the largest correlation peak will be generated, thereby realizing the spread spectrum signal's despreading. The advantage of this technique is that it does not require high-speed sampling. The single-measurement time is determined by the difference in clock frequency and the maximum DS length of the transmitted and received PN sequences. If the frequency difference is small, the relative "sliding" of the two PN sequences is slow, and it will take longer time to complete the measurement. Therefore, this method is not suitable for high-mobility scene detection.

15.2.4.3 Frequency-domain flat signal

A flat signal in a frequency domain refers to a signal with a relatively flat envelope in the similar domain, which is different from a signal with a flat envelope in a time domain. Early narrowband channel sounder techniques used sinusoidal signals as the detection signal. After the signal passes through the propagation channel, the signal amplitude and phase change is the channel frequency response at that frequency. In a VNA-based broadband detection system, the entire detection frequency band is divided into several frequency points.

At each frequency point, the sine-wave detection signal is still used to scan all frequencies. This method is usually very slow and is not suitable for dynamic scene measurement. At different frequency points, the RF unit of the transmitter and receiver must ensure that the frequency and phase are consistent. In order to overcome the limitations of the frequency sweep method, multi-carrier (OFDM) signals are usually used in wideband frequency-domain measurement to detect the entire frequency band at each measure.

15.3 Channel measurement methods

15.3.1 System architecture

In this section, a code division multiplexing (CDM)-based MIMO channel sounder consists of five subsystems, as shown in Figure 15.5.

For the baseband processing subsystem, orthogonal PN codes and QPSK modulation sequences are generated as baseband signals for different transmission channels on the transmitter side. Upsampling and shaping filters are used to improve these orthogonal signals' frequency-domain spectrum before sending them to the corresponding RF channels. On the receiver side, from the downconverter and high-speed ADC of the parallel RF channel, the baseband measurement signal with MIMO channel information is obtained and stored. These signals contain the complete MIMO channel characteristics of the scenario and frequency band. Therefore, it is critical to keep all valuable raw data in the disk array in real time, but it is also challenging.

For the RF subsystem and antenna arrays, each channel uses an up-converter to perform a single-stage direct conversion to convert baseband I/Q signals to RF signals at a specific LO frequency at the transmitter side. Using a wideband orthogonal correction technique, the method can minimize the residual sideband image and LO power to provide a wide instantaneous RF bandwidth. Eventually, multiple RF signals are sent simultaneously through an antenna array with M elements. In this case, M is equal to 8. At the receiver end, RF signals from multiple paths are captured by an N-element antenna array ($N = 8$) and then converted down to the baseband from a specific LO frequency for digitization, storage, and analysis.

For the clock and synchronization subsystem, to achieve high spatial and temporal resolution in channel parameter estimation, a synchronization accuracy of 70 picoseconds (ps) among MIMO channels must be guaranteed. The two high-precision reference clocks at the Tx and Rx terminals should be precisely

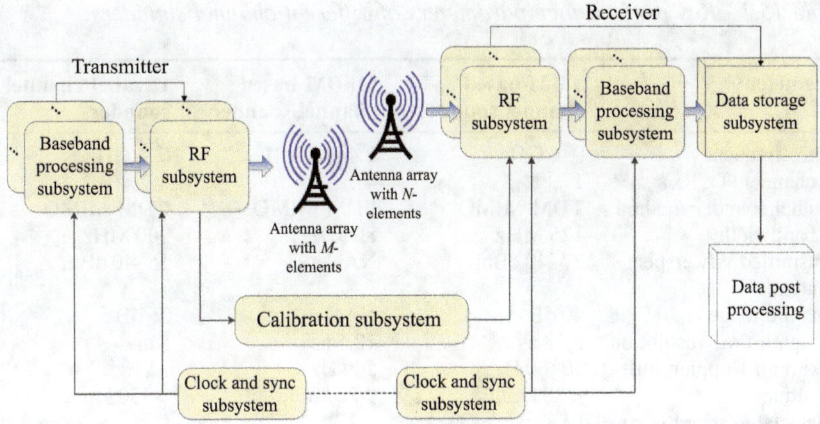

Figure 15.5 System architecture of MIMO parallel channel sounder [14]

synchronized with a constant, coherent phase. All Tx/Rx channels rely on the same LO and trigger mechanisms to generate/sample baseband signals and send/receive phase-coherent RF signals. The GPS provides a stable 1PPS pulse train, an excellent trigger signal for the arbitrary waveform generator to generate the necessary periodic signals to achieve initial rough synchronization between a transmitter and a receiver.

For calibration subsystem, the MIMO parallel channel detector can be considered a multichannel time-varying complex system. In order to ensure the accuracy of channel estimation, a parallel calibration subsystem needs to be designed to carefully compensate for the different nonideal channel responses on each Tx/Rx pair, which are derived from the multichannel clock phase-locked loop (PLL) circuits and RF devices and receivers of the two transmitters. The calibrated emission signal will be stored on the disk and will be used during the post-processing phase of the data.

For the data storage subsystem, raw measurement data from parallel channels are vast in volume and must be stored simultaneously into disk arrays at the receiver. It is essential to ensure continuous data-streaming capacity in the parallel channel sounder to measure channel consistency in time and space domains. In order to capture more detailed time-varying channel characteristics, a real-time high-throughput data-streaming scheme is developed to store large-volume raw measurement data, together with the corresponding GPS and calibration information encapsulated as valuable records. This approach virtually guarantees time-alignment and accuracy of measurement data across parallel channels, thus enabling synchronous multichannel data processing and analysis afterward.

For the data storage subsystem, the MIMO channel's raw measurement data are enormous and must be stored simultaneously in the receiver's disk array. It is essential to ensure the MIMO channel sounder's continuous dataflow capacity to measure channel consistency in both time and space domains. To capture more

Table 15.1 Key performance parameters of different channel sounders

Parameter	**TDM-based channel sounder**	**OFDM-based channel sounder**	**Parallel channel sounder**
Center frequency	0–6 GHz	2,442.5 MHz	0–6 GHz
RF channel/(Tx/Rx);	1	4	8
Channel sounder method	TDM MIMO	FDM MIMO	CDM MIMO
RF bandwidth	125 MHz	80 MHz	200 MHz
Transmitted power per antenna	27–40 dBm	NAN	15–40 dBm
Dynamic range	40 dB	NAN	36 dB
Multipath time resolution	8.33 ns	12.5 ns	5 ns
Maximum Doppler shift	30–60 Hz	500 Hz	2 kHz
Maximum detectable speed of an object	<30 km/h	>80 km/h	>350 km/h
Maximum impulse response	NAN	NAN	20.48 μs
Data stream rate	NAN	NAN	51.2 Gbps
Data stored	CIR	Processing online	Raw data/CIR
PN sequence length	NAN	NAN	4,096

detailed time-varying channel characteristics, a real-time high-throughput data stream transmission scheme was developed to store large amounts of raw measurement data and the corresponding GPS and calibration information encapsulated as valuable records. This method ensures the time alignment and accuracy of the measured data across parallel channels, which can be used for simultaneous multichannel data processing and analysis in the future.

15.3.2 Design challenges and key technologies

As shown in Table 15.1, compared with previous TDM-based and frequency division multiplexing (FDM)-based channel sounders, the CDM-based MIMO channel sounder can estimate higher Doppler shift (up to 2 kHz), at least 350 km/h, by parallel architecture. Simultaneously, our parallel channel sounder can support up to 51.2 Gbps data stream rate, which is necessary to store eight RF channels raw sampling data with near 200 MHz symbol rate. The raw data can provide higher estimation accuracy than CIR. With high-precision calibration and synchronization, our parallel channel sounder can achieve a high dynamic range (around 36 dB) quite enough to do the outdoor measurement, which is hardly achieved in CDM-based architecture. The following subsections show some highlights.

15.3.2.1 An accuracy of 30 ps synchronization across MIMO channels

For a carrier frequency of 3.5G Hz, antenna elements are placed with a distance of 0.5 wavelength, i.e., 4.3 cm. So, the time difference for electromagnetic waves traveling across two adjacent elements is about 140 ps. In order to accurately estimate

Angle of Arrival/Angle of Departure (AoA/AoD) parameters, the synchronization accuracy across all Tx/Rx channels needs to be improved to a level of 70 ps.

In practice, different sampling frequencies may be used in multiple channels. These sampling clocks need to be aligned with a common reference clock. In addition, due to the differences between clock PLL circuits across multiple RF channels, it is crucial to design a robust triggering mechanism for achieving a very accurate synchronization of clock's phases. As detailed in the following, our proposed solution can solve this challenging problem with a synchronization accuracy of 30 ps.

For the 3.5 GHz carrier frequency, the antenna elements are placed at a distance of 0.5 wavelengths, or 4.3 cm. Thus, the time difference between the electromagnetic wave propagating between two adjacent parts is about 140 ps. To accurately estimate the AoA/AoD parameters, synchronization accuracy on all Tx/Rx channels needs to be improved to 70 ps.

Different sampling frequencies can be used across multiple channels in practice, and these sampling clocks need to be aligned with the standard reference clock. In addition, due to the differences between clock PLL circuits across multiple RF channels, it is essential to design reliable triggering mechanisms to achieve an exact synchronization of clock's phases. As described later, the proposed solution can solve this challenging problem with 30 ps synchronization accuracy.

Both Tx and Rx channels share the same 10 MHz reference clock. Considering various sampling frequencies, a common reference clock, named share trigger reference clock (STRC), is set with its period that equals the least common multiple periods of different sampling clocks. The sampling clock for each channel and STRC is phase-locked but not phase-aligned to the 10-MHz reference clock. Another common synchronization signal, named share sync pulse clock (SSPC), is generated by an arbitrary channel. The SSPC is distributed through finely calibrated paths to different channels to ensure accurate phase alignment among these SSPC signals. Finally, channel 1 generates a sync pulse using an external trigger signal and distributes it to all channels. After detecting this sync pulse, every channel measures the delay gap between SSPC and STRC signals' first rising edges.

As shown in Figure 15.6, the delay gap between channel one and channel N is measured as ΔT_2. Every channel adjusts its DAC/ADC output phases according to its delay gap against channel 1. In doing so, the STRC and different sampling clocks of all channels are accurately synchronized and phase aligned.

15.3.2.2 51.2 Gbps parallel data streaming for real-time storage of raw data

Unlike most traditional channel detectors, which provide only a CIR, in this scheme, the MIMO channel sounder will capture and store complete raw measurement data, which is fully post-processed and analyzed using the precise SAGE algorithm. Considering each of the eight Rx channels with a sampling rate of 180 M/s and quantization accuracy of 14 bits, the resultant total speed required for real-time data stream transmission of the sampled raw IQ data is about 40 Gbps (or 180 Mbps$\times 14 \times 2 \times 8$).

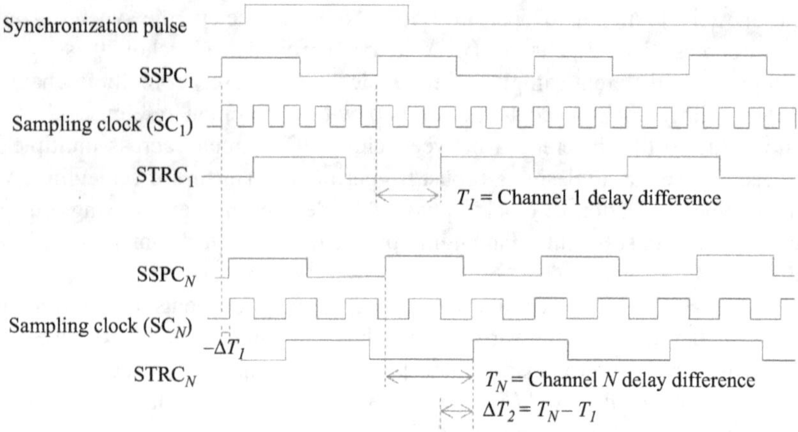

Figure 15.6 Clock phase alignment and synchronization mechanism [14]

To achieve this goal, a direct memory accession (DMA)-first-input–first-output approach is used to take the advantage of the tremendous data transfer rate of 64 Gbps on the backboard bus. Besides, zero-copy technology has been extended to avoid additional copies and state transitions, effectively reducing latency in the operating system kernel.

Specifically, as shown in Figure 15.7, zero-copy technology requires only two DMA copy processes, from the hard drive to the kernel buffer and from the socket buffer to the protocol engine to complete the file transfer. Compared with ordinary file I/O operations, zero-copy technology saves two copying processes, from the kernel buffer to the user buffer and from the user buffer to the socket buffer. It thus significantly reduces the latency and burden of CPU in sizeable data streams and real-time transfers. With the ultrafast, low-latency solution, the parallel channel sounder can achieve a data rate of 51.2 Gbps for real-time storage of raw measurements across all channels.

15.3.2.3 High-precision calibration of parallel RF channels

The hardware and RF devices in multiple Tx and Rx channels are identical. Still, the channel responses' inconsistency among parallel RF channels significantly reduces the accuracy in channel characteristics analysis and cross-channel parameters modeling. It is necessary to develop a high-precision RF channel calibration scheme for the parallel channel sounder. Thus, before the analytical process of channel characterization and modeling, the influence of inconsistent channel responses across multiple channels can be minimized.

In order to capture the calibration data in the parallel RF channel, the back-to-back test is performed using an eight-by-eight (8×8) RF coupler to connect the Tx and Rx terminals before the measurement. The physical response between each input and output port of this RF coupler is premeasured using a VNA. Similar to the standard channel probe process, the orthogonal PN sequences are sent from eight

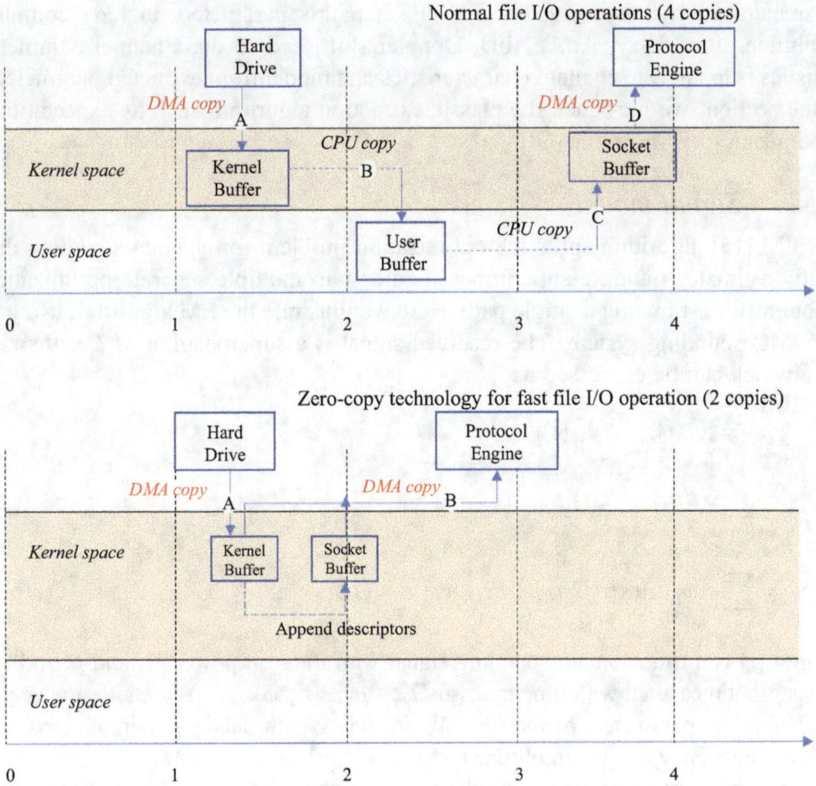

Figure 15.7 Zero-copy technology for fast data storage [14]

Tx channels simultaneously, and the aliased raw IQ data are recorded at the Rx terminal. An eight-by-eight (8×8) matrix for the channel response can be derived by (i) calculating the correlation between the transmitted PN sequence and the received signal on the parallel track, and (ii) eliminating the RF coupler response and output port on each corresponding input pair. Using these 64 Tx–Rx channel responses, a calibrated emission signal can be obtained by performing a series of convolution operations using an orthogonal PN sequence. During the data's post-processing phase, these calibrated emission signals are used to derive the final pure channel response and modeling parameters across parallel wireless channels in the actual measurement environment.

15.4 Channel data processing

In channel measurement and detection, data processing includes two aspects: path parameter extraction and channel parameter statistics. Path parameter extraction uses high-resolution parameter estimation algorithms to extract each path's parameters from the received signal. Each path's parameters include complex amplitude, time delay, AoA, AoD, Doppler shift, and so on. Channel parameter statistics is to analyze channel characteristics and model from extracted parameters. In the section, we introduce the classic extraction algorithm such as expectation–maximization (EM) and SAGE.

15.4.1 Signal model

The EM [15] algorithm splits the optimization problem for channel sounding that jointly estimates multiple superimposed paths into multiple separate optimization problems for estimating a single path. Here we illustrate the EM algorithm using in an SIMO sounding system. The received signal is a superposition of L paths signals, which can be expressed as

$$\begin{aligned} y(t) &\triangleq [y_1(t), \ldots, y_M(t)]^T \\ &= \sum_{l=1}^{L} s(t;\vartheta_l) + N(t) \\ &= \sum_{l=1}^{L} c(\phi_l)\alpha_l \exp\{j2\pi v_l t\} u(t-\tau_l) + N(t) \end{aligned} \tag{15.1}$$

where $u(t)$ is a time-domain sounding signal with transmit power Pu and period T_a. lth snapshot measured with time intervals T_f. $\vartheta_l \triangleq [\tau_l,\ \phi_l,\ v_l,\ \alpha_l]$ denotes the vector containing the parameters of the lth path, including path delay τ_l, azimuth AoA ϕ_l, Doppler frequency v_l, and amplitude $\alpha_l \cdot c(\phi) \triangleq [c1(\phi),\ \ldots,\ cM(\phi)]^T$ denotes the antenna array's steering vector reflecting two aspects of information: the antenna radiation pattern and the array's geometric structure.

$n(t) \triangleq [n_1(t),\ \ldots,\ n_M(t)]^T$ denotes M-dimensional problematic white Gaussian noise.

15.4.2 EM Algorithm

The classical EM algorithm [15] can be described as follows. In step E, the posterior distribution of the complete (unobservable) data are found for the observable (incomplete) data and the assumed (or updated) parameters. It is used to calculate the expected log-likelihood of the complete data. Then, in step M, we search for the parameter with the maximum expected value.

Define complete (unobservable) data sets $x_l(t) \triangleq s(t;\vartheta_l) + \sqrt{N_0/2}n_l(t)$ incomplete (observable) data sets $y(t) = \sum_{l=1}^{L} x_l(t)$. First of all, assuming that $x_l(t)$ is known, the log-likelihood function of complete data $x_l(t)$ in terms of the parameter vector ϑ_l can be expressed as follows:

$$\Lambda(x_l;\vartheta_l) \triangleq \frac{1}{N_0}\left[2\int_{D_0} R\left\{s^T\left(t';\vartheta_l\right)x_l\left(t'\right)\right\}dt' - \int_{D_0} \left\|s\left(t';\vartheta_l\right)\right\|^2 dt'\right] \tag{15.2}$$

The parameters that maximize the ML function are simply written as

$$\left(\widehat{\vartheta}_l\right)_{ML}(x_l) \in \mathrm{argmax}_{\vartheta_l}\{\Lambda(x_l;\vartheta_l)\} \tag{15.3}$$

E Step: In fact $x_l(t)$ is unknown, thus, we need to estimate $x_l(t)$ first based on the incomplete data $y(t)$. One direct way is to use the conditional expectation of $x_l(t)$ with respects to $Y(t)$. Specifically, we can use successive interference cancellation method to obtain $x_l(t)$ assuming that the parameters about other paths are known:

$$\widehat{x}_l\left(t;\widehat{\vartheta}\right) \triangleq y(t) - \sum_{\substack{l'=1 \\ l' \neq l}}^{L} s\left(t;\widehat{\vartheta}'_{l'}\right) \tag{15.4}$$

M Step: The *l*th path parameters ϑ_l can then be reestimated by computing its maximum likelihood estimation based on the estimation of the *l*th signal:

$$\widehat{\vartheta}''_l = \left(\widehat{\vartheta}_l\right)_{ML}\left(\widehat{x}_l\left(t;\widehat{\vartheta}'\right)\right) \tag{15.5}$$

In particular, the ML estimator can be expressed as

$$(\tau_l, \widehat{\varphi_l}, \nu_l) = \mathrm{argmax}_{\tau,\phi,\nu}\{|z(\tau,\phi,\nu;\widehat{x_l})|\} \tag{15.6}$$

And

$$z(\tau,\phi,\nu;x_l) = \sum_{i=1}^{I}\int_{D_i} u^*\left(t'-\tau\right)e^{-j2\pi\nu t'}c^H(\phi)\widehat{x}_l\left(t'\right)dt' \tag{15.7}$$

where D_i is the observation time interval of the *i*th snapshots, $i=1, 2, \ldots, I$. The calculation of *z* can be seen as correlations (match filtering) in both time and spatial domains. Because the EM algorithm needs to update all parameters at the same time in each iteration, the convergence speed is very slow.

15.4.3 SAGE algorithm

The sage algorithm is developed from the classical EM algorithm. However, in step *M*, it divides all parameters of a path into several groups and only updates one group of parameters at a time, while keeping other parameter groups unchanged. This process will be repeated until all parameters are updated. Fleury [13] first applied the SAGE algorithm to the problem of channel parameter extraction. Here we consider an SIMO sounder. The *M* step of the SAGE algorithm can be written as follows:

$$\widehat{\tau}''_1 = \mathrm{argmax}_\tau\left\{\left|z\left(\tau,\widehat{\phi}'_l,\widehat{\nu}'_l;\widehat{x}_l\left(t;\widehat{\vartheta}'\right)\right)\right|\right\} \tag{15.8}$$

$$\widehat{\phi}''_1 = \mathrm{argmax}_\phi\left\{\left|z\left(\widehat{\tau}''_l,\phi,\widehat{\nu}'_l;\widehat{x}_l\left(t;\widehat{\vartheta}'\right)\right)\right|\right\} \tag{15.9}$$

$$\widehat{\nu}_l'' = \mathrm{argmax}_\nu \left\{ \left| z\left(\widehat{\tau}_l'', \widehat{\phi}_l'', \nu; \widehat{x}_l\left(t; \widehat{\vartheta}'\right) \right) \right| \right\} \tag{15.10}$$

$$\widehat{\alpha}_l'' = \frac{1}{I \left\| c\left(\widehat{\phi}_l''\right) \right\| T_a P_u} z\left(\widehat{\tau}_l'', \widehat{\phi}_l'', \widehat{\nu}_l''; \widehat{x}_l\left(t; \widehat{\vartheta}'\right) \right) \tag{15.11}$$

As for the SAGE algorithm's initialization, each path's initial parameters can be set all zeros or obtained using the MUSIC algorithm. In the M step, since no a priori knowledge about the phase of the complex amplitudes, the maximization procedures for τ and ϕ are replaced by two noncoherent estimation programs:

$$\widehat{\tau}_l'' = \mathrm{argmax}_\tau \left\{ \sum_{m=1}^{M} \sum_{i=1}^{I} \left| \int_{D_i} u^*\left(t' - \tau\right) \widehat{x}_{l \cdot m}\left(t'; \widehat{\vartheta}'\right) dt' \right|^2 \right\} \tag{15.12}$$

$$\widehat{\phi}_l'' = \mathrm{argmax}_\phi \left\{ \sum_{i=1}^{I} \left| \int_{D_i} u^*\left(t' - \widehat{\tau}_l''\right) c^H(\phi) \widehat{x}_l\left(t'; \widehat{\vartheta}'\right) dt' \right|^2 \right\} \tag{15.13}$$

where the calculation of Doppler frequency still uses the same procedure as the previous iterative process. Correspondingly, since only the $l-1$ paths are estimated, in the initialization process, the E step of estimating lth path parameters is replaced by

$$\widehat{x}_l\left(t; \widehat{\theta}'\right) = y(t) - \sum_{l'=1}^{l-1} s\left(t; \widehat{\vartheta}_{l'}'\right) \tag{15.14}$$

15.5 Channel measurement

15.5.1 Scenario test planning and measurement

In the planning work of channel scenario test with typical geographical characteristics defined by IMT-2020, Shanghai Research Center for Wireless Communications has completed the following scenario tests to collect massive data of parallel measurement channels:

- anechoic chamber environmental verification,
- office scenarios,
- large-scale indoor scenarios,
- stadium outdoor scenarios,
- rural outdoor scenarios,
- Shanghai Circuit V2V scenarios,
- city hot spot regional scenarios,
- high-mobility scenarios.

15.5.2 Measurement in anechoic chamber (with a reflector)

In order to verify the parallel channel sounder, test is perform in anechoic chamber shown in Figure 15.8.

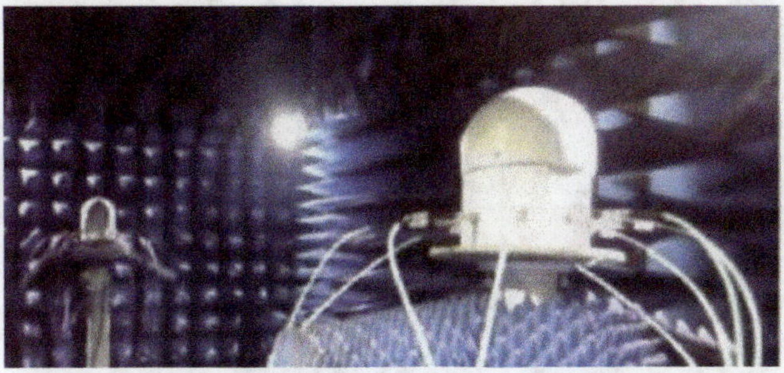

Figure 15.8 *System verification test in microwave anechoic chamber. © 2018 Springer Nature Customer Service Centre GmbH: Springer Nature, 5G Hardware Test Evaluation Platform by Yang et al. Reprinted, with permission, from Figure 6.15 in Reference [16].*

For the static scenario in the chamber, we choose PN sequence with length of 4,096 and an I/Q chip with rate of 100 *M*/s. Each snapshot collected 100 cycles. The total length is the measurement data within $10\,\text{ns} \times 4,096 \times 100 \times 2 = 8.192\,\text{ms}$. We first use the 8-input–8-output connector to make calibration and measurements and then test via air interfaces. The test results are shown in Figure 15.9.

Figure 15.9 is the spectrum of 64 pairs of multipath time delay. They are the air interface data collected from left to right, calibration data, and the results compensated by calibration data. It can be seen from the figure that after using 64 calibration files to calibrate the original signal, the response difference of each transmit and receive channel can be well compensated, and the consistent delay as shown in the right figure can be obtained.

In Figure 15.10, color represents the intensity of spectral components. It can be seen from the figure that the reconstructed signal has a strong similarity with the original signal spectrum. The energy of the reconstructed signal is 37 dB lower than that of the original signal. The verification shows that the wireless channel's space parameters estimated by the post-processing algorithm are very close to the actual situation.

Figure 15.11 is a time-domain view of wireless channel characteristics. The estimated time-delay spectrum is consistent with the original time-delay spectrum. The original signal power can be reduced by more than 10 dB after elimination. It can be seen from the results of time-domain and spatial parameter estimation and reconstruction and elimination that the channel measurement data and estimation results are reliable and can more accurately correspond to the actual cavity environment.

15.5.3 Outdoor measurement

As shown in Figure 15.12, we have measured the suburban wireless channel in China countryside. We use the lift to raise the Rx to a height of about 20 m in blue point, and Tx is in red point beside the plants. By post-processing by SAGE, we

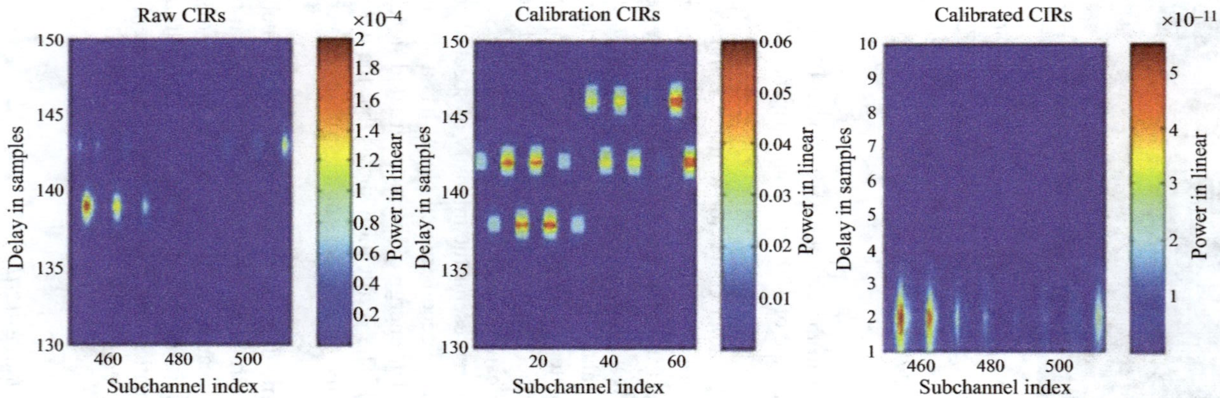

Figure 15.9 *Multipath delay spectrum before and after calibration. © 2018 Springer Nature Customer Service Centre GmbH: Springer Nature, 5G Hardware Test Evaluation Platform by Yang et al. Reprinted, with permission, from Figure 6.16 in Reference [16].*

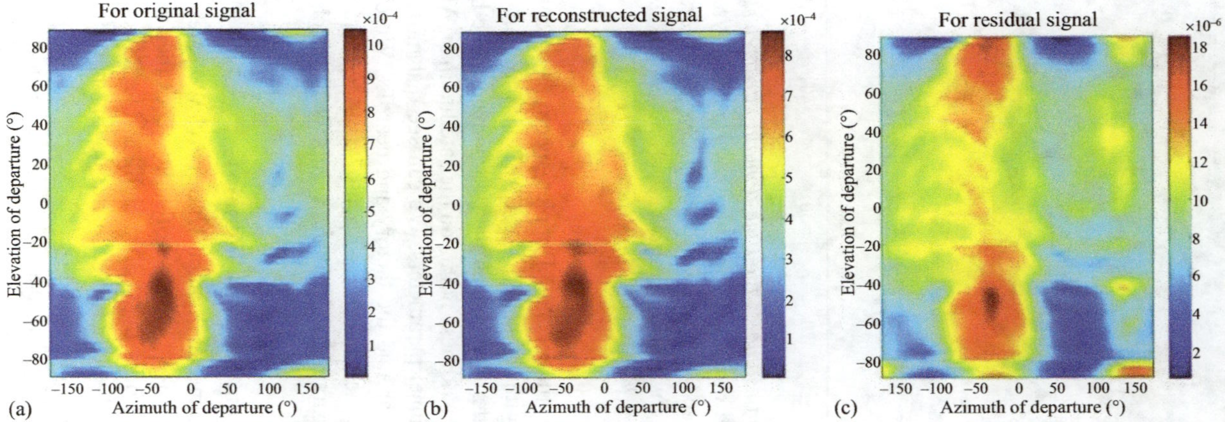

Figure 15.10 PS-SAGE algorithm estimates the spatial parameters and reconstructs the signal spatial spectrum. The figures show (a) the elevation of departure vs azimuth of departure; (b) the power spectra of the reconstructed signal based on parameter estimation; (c) the residual signal after cancelation. © 2018 Springer Nature Customer Service Centre GmbH: Springer Nature, 5G Hardware Test Evaluation Platform by Yang et al. Reprinted, with permission, from Figure 6.17 in Reference [16].

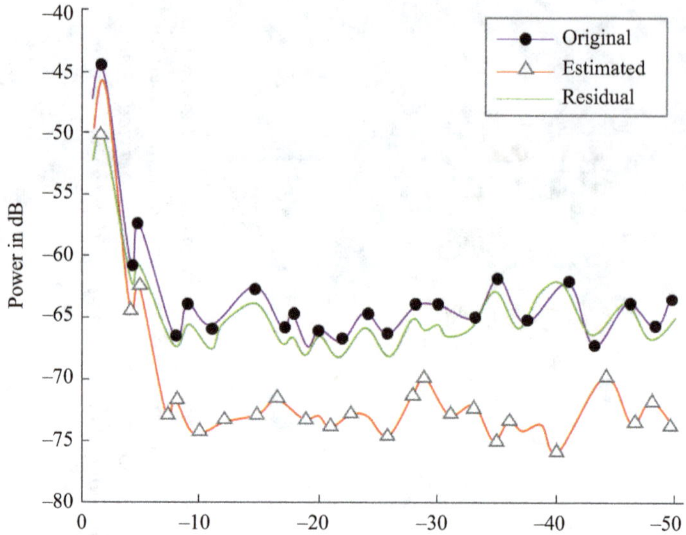

Figure 15.11 SAGE algorithm estimates the time-domain parameters and reconstructs the signal time-domain spectrum. © 2018 Springer Nature Customer Service Centre GmbH: Springer Nature, 5G Hardware Test Evaluation Platform by Yang et al. Reprinted, with permission, from Figure 6.18 in Reference [16].

estimate the radio channel in time delay, Doppler spectrum, and the wireless channel's spatial estimation. With the help of AoA/AoD information, we can find the exact position of each scattering point on the two-domain plane based on the rebound hypothesis. Then we mark these points on the satellite map. All green dots represent the scattering of channel parameters. We can see that most of the scatterers are located in the fence around the factory and near Tx. We can also find scatterers that may exist on the factory roof near point 7 or 14. Generally speaking, the scattering obtained by parameter estimation can well match the actual environment.

15.5.3.1 Outdoor measurement (time domain)

We measured the wireless channel at Xiamen University. We carefully selected several observation points with or without line-of-sight scenes then put Tx on the top of the chemical building (9 floors, about 30 m high), and let Rx (on the ground, 2.1 m high) move around the building. We do this to distinguish the channel characteristics in LOS/NLOS (Figure 15.13).

From the raw data collected by 8×8 parallel Channel Sounder, we have CIR, including 100 snapshots for each observation point. As shown in Figure 15.14, we used SAGE to estimate the time domain's channel parameter for each observation point. Based on the Gaussian or the Laplace distribution hypothesis from the 3GPP SCM channel model, we can deduce the following channel parameters distribution

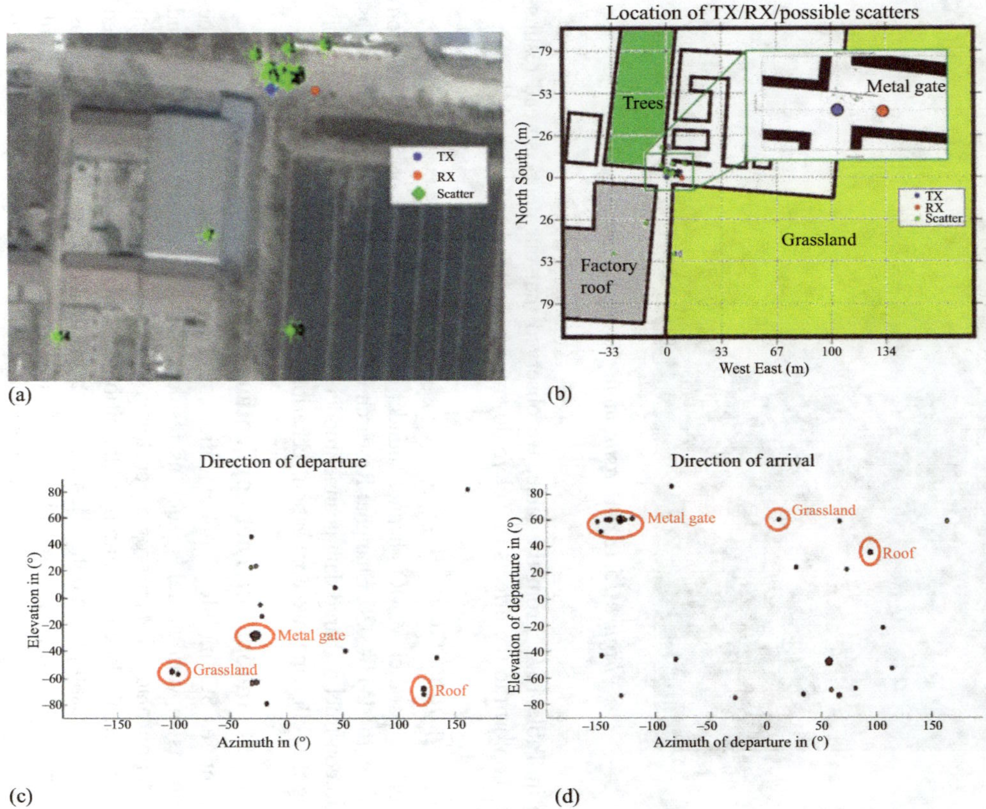

Figure 15.12 Measurement for the suburban scenario: (a) satellite image of the scenario; (b) sketch of the scenario; (c) estimated cluster in the direction of departure; (d) estimated cluster in the direction of arrival

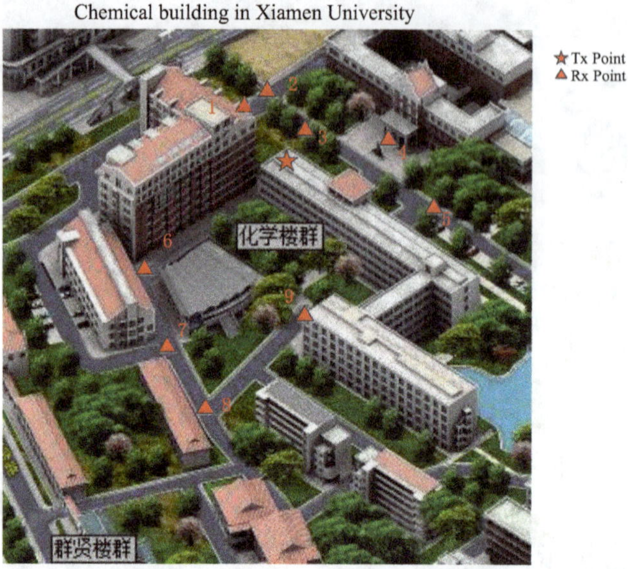

Figure 15.13 Measurement for the campus

shown in Table 15.2. From the table, we observed that our delay spread results roughly conform to the effects of Tokyo downtown crowded area from METIS.

15.5.4 Channel model library

Based on the parallel MIMO channel sounder and the definition of 5G application scenarios in the IMT-2020, Shanghai Research Center for Wireless Communications began to construct a shared data channel model based on real measured data, mainly covering the following two application scenarios.

15.5.4.1 5G mobile Internet application scenarios

In the future, 5G mobile communication system will meet different service requirements in various fields, such as residence, workplace, leisure place, and transportation place. It can provide customers in different scenes with excellent service experience such as ultrahigh definition video, virtual reality, augmented reality, cloud desktop, and online games. Scenarios include dense residential areas, offices, stadiums, open-air assembly places, subway, expressway, high-speed rail, wide area coverage and other application scenarios with ultrahigh mobile dataflow density, ultrahigh mobile connection density, and ultrahigh mobility. The previous application scenarios can be divided into two categories: continuous wide area coverage scenarios and hot high-capacity scenarios.

15.5.4.2 5G Internet of Things (IoT) application scenarios

The future 5G mobile communications system will penetrate to Internet of things (IoT) and various (vertical) industry fields, deeply integrating with industrial

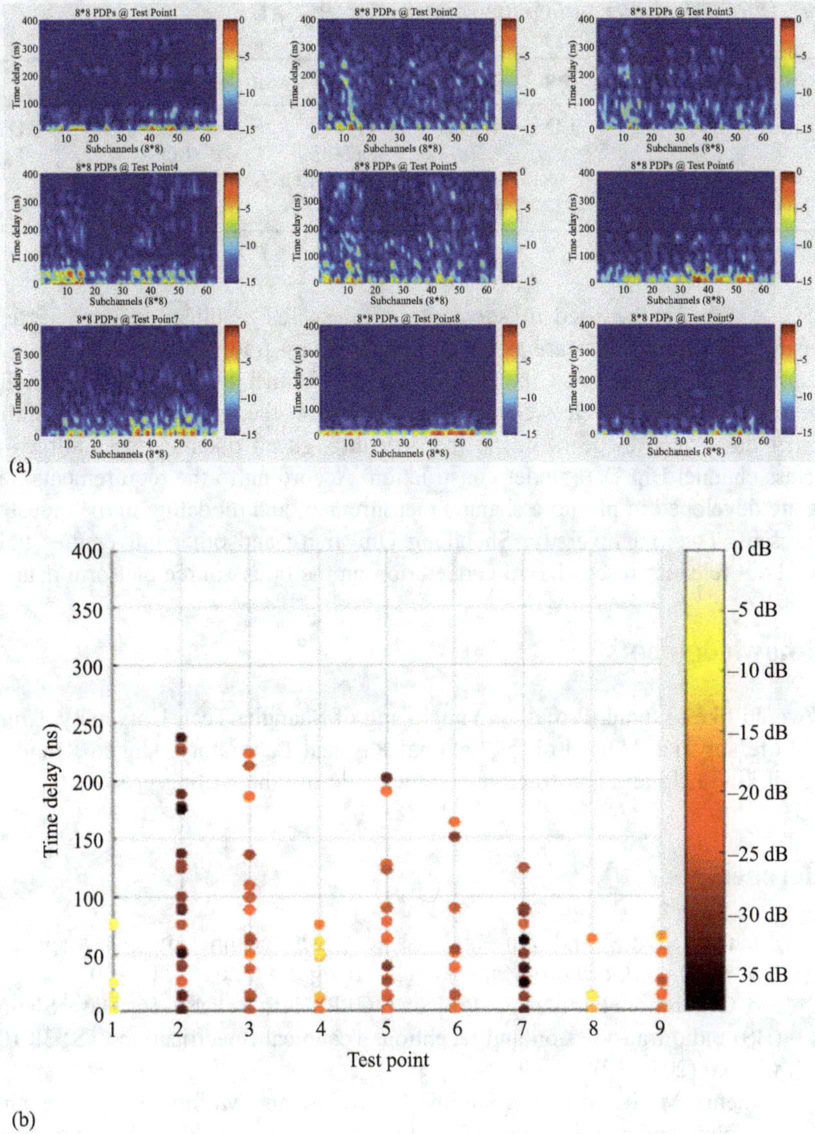

Figure 15.14 (a) Power–delay profile of 9 test points and (b) delay estimation of 9 test points

facilities, medical equipment, vehicles, etc., effectively meet the diversified service needs of industry, medicine, transportation, and other vertical industries and finally realize the real "IoT." The previous application scenarios can be divided into high connection scenarios with low-power consumption and high-reliability scenarios with low latency. We can see that these two applications are the latest exploration

Table 15.2 Parameter estimation results using PS-SAGE

Estimation results using PS-SAGE			**Results from METIS**		
Scenario	LOS	NLOS	ITU-R M2135 UMI (LOS)	Small-cell LOS	D2D LOS
Average delay spread (ns)	32	91	64	38	24

of 5G, which can be used to solve the wireless IoT applications and vertical industry applications that are not well supported by traditional mobile communication systems. Based on the existing scene channel test database, Shanghai Research Center for Wireless Communications shares the data obtained by parallel channel detector with peers in the industry through an open-source platform of wireless channel Big Data under construction. According to the requirements, test data are developed to promote channel measurement and modeling in the industry. At present, Tongji University, Shandong University and other universities have carried out relevant research and cooperation on the open-source platform data.

Acknowledgments

We would like to thank Professor Yang Yang of ShanghaiTech University, China and Professor Tian Hong Loh of National Physical Laboratory, United Kingdom for their kind help and constructive comments during the writing process.

References

[1] Zheng K., Ou S. and Yin X. 'Massive MIMO channel models: a survey'. *International Journal of Antennas & Propagation*. 2014, 11:1–10.

[2] 3rd Generation Partnership Project (3GPP). NR Release 16, Base Station (BS) radio transmission and reception. Technical Specification (TS) 38.104 v16.6.0 (2020-12).

[3] Akdeniz M. R., Liu Y., Samimi M. K., *et al.* 'Millimeterwave channel modeling and cellular capacity evaluation'. *IEEE Journal on Selected Areas in Communications*. 2014, 32(6):1164–1179.

[4] Jaeckel S., Raschkowski L., Börner K. and Thiele L. 'QuaDRiGa: A 3-D multicell channel model with time evolution for enabling virtual field trials'. *IEEE Transactions on Antennas Propagation*. 2014, 62(6):3242–3256.

[5] Rappaport T. S., Maccartney G. R., Samimi M. K. *et al.* 'Wideband millimeter-wave propagation measurements and channel models for future wireless communication system design'. *IEEE Transactions on Communications*. 2015, 63(9):3029–2056.

[6] Zheng K., Ou S. and Yin X. 'Massive MIMO channel models: a survey'. *International Journal of Antennas and Propagation*. 2014, 11:1–10.
[7] METIS D1.4 V1.0. METIS Channel Models [online]. ICT-317669, METIS project, Feb. 2015. https://www.metis2020.com/ [Accessed April 2021].
[8] METIS D5.1. Intermediate Description of the Spectrum Needs and Usage Principles [online]. V1.0, ICT-317669, METIS Project, August 2013. https://www.metis2020.com [Accessed April 2021].
[9] Roh W., Seol J. Y., Park J. *et al.* 'Millimeter-wave beamforming as an enabling technology for 5G cellular communications: theoretical feasibility and prototype results'. *IEEE Communications Magazine*. 2014, 52(2):106–113.
[10] ITU-R M.2135–1. Guidelines for Evaluation of Radio Interface Technologies for IMT-Advanced. International Telecommunication Union (ITU), Geneva, Switzerland, Technical Report, December 2009.
[11] Ghazal A., Wang C.-X., Ai B., Yuan D. and Haas H. 'A non-stationary wideband MIMO channel model for high-mobility intelligent transportation systems'. *IEEE Transactions on Intelligent Transportation Systems*. 2015, 16(2):885–897.
[12] Dupleich D., Fuschini F., Mueller R., *et al.* 'Directional characterization of the 60 GHz indoor-office channel'. *The 31th URSI General Assembly and Scientific Symposium (URSI 2014)*. Beijing, China, 16–23 August, 2014.
[13] Fleury B. H. and Tschudin M. 'Channel parameter estimation in mobile radio environments using the SAGE algorithm'. *IEEE Journal on Selected Areas in Communications*. 1999, 17(3):434–450.
[14] Yang Y., Gui Y., Wang H., *et al.* 'Parallel channel sounder for MIMO channel measurements'. *IEEE Wireless Communications*. 2018, 25(5):16–22.
[15] Moon T. K. 'The expectation-maximization algorithm'. *IEEE Signal Processing Magazine*. 1996, 13(6):47–60.
[16] Yang Y., Xu J., Shi G. and Wang C.-X. *'5G Wireless Systems Simulation and Evaluation Techniques'*, Springer International Publishing, New York, NY; 2018.

Chapter 16

Empirical mmWave channel measurements and modelling in indoor and outdoor complex environments for 5G communications

Sohail Payami[1] *and Mohsen Khalily*[1]

Millimetre-wave (mmWave) communication systems have become one of the key technologies that are used in the fifth-generation (5G) of the mobile networks. The availability of large chunks of lightly licensed/unlicensed spectrum in the mmWave bands provide the opportunity to deliver revolutionary data rates in 5G systems compared to its predecessors. However, the characteristics of radio propagation channels in such frequency bands impose severe constraints on the design and deployment of such systems. For example, the transmitted signals over mmWave channels experience high attenuation, which can significantly limit the communication range and link reliability in practical systems. Such characteristics directly impact the system requirements for mmWave technology in terms of hardware, algorithms and even protocols. In the following, first we present an overview of mmWave technology and the current efforts towards modelling their radio propagation channel. After introducing the necessary theoretical tools and terminology related to channel modelling, we present empirical mmWave channel measurements and modelling in realistic indoor and outdoor environments with the focus on 5G communications. The results are focused on 26, 32 and 39 GHz bands with 2-GHz bandwidths based on several extensive measurement campaigns that were conducted at the University of Surrey, United Kingdom.

16.1 Introduction

The sub-6-GHz spectrum is currently shared between various organisations, including telecommunication operators, satellite systems, military, television and radio broadcasting services. With the significant increase in the demand for wireless connectivity and data transmission over the past decade, the frequency bands below 6 GHz are becoming saturated. To overcome this limitation, 5G systems have also adopted the

[1]Institute for Communication Systems (ICS), Home of the 5G and 6G Innovation Centres (5GIC and 6GIC), University of Surrey, United Kingdom

available spectrum in higher frequencies to deliver higher data rates and connectivity. In particular, there is an abundance of unlicensed, semi-licensed and licensed spectrum in 30–300 GHz, which are referred to as mmWave* frequencies.

In wireless systems, the choice of hardware and transmission methods significantly depends on the characteristics of the radio propagation channels. For example, the amount of the transmit power and antenna gains in the link budget calculations are decided according to the signal attenuation at a desired location. Let P_{Tx}, P_{Rx}, G_{Tx}, G_{Rx} denote the transmitted and received powers, antenna gains at the transmitter (Tx) and receiver (Rx), respectively. In an environment that is empty of all objects except the Tx and Rx antennas, the Friis transmission equation expresses that

$$P_{\mathrm{Rx}} = \frac{G_{\mathrm{Tx}} G_{\mathrm{Rx}} \lambda^2}{(4\pi d)^2} P_{\mathrm{Tx}}, \tag{16.1}$$

where λ and d represent the signal's wavelength and the distance between Tx and Rx antennas. Regardless of the operating frequency, the signal-to-noise ratio (SNR) at the receiver must reach a minimum level for a successful information transfer, depending on the coding and modulation scheme that is employed. Clearly, (16.1) indicates that the received signal levels at higher frequencies can be maintained by increasing the transmit power or using directional antennas with higher gains at the transmitter and receiver. The increased path loss, just by itself, can result in significant design and deployment challenges. Due to the health and safety reasons and also the coexistence of various wireless technologies, there are strict regulations to limit the maximum allowed transmit power. On the other hand, the use of high-gain antennas naturally results in a directional transmission which, in turn, requires more complex signalling for beam alignment between the Tx and Rx antennas. The directional antennas have either a fixed pattern or they can electronically steer the beam towards desired directions by using beamforming.

To apply beamforming techniques at one transmission end, it is required to deploy multiple antennas on that side. By controlling the phase and magnitude of the signals at the antenna elements, beamformers can perform spatial filtering. For example, the signals from specific directions can be enhanced or mitigated [1]. Using more antenna elements, in turn, increases the size, weight, cost, complexity and power consumption of the devices which need to be considered during the design process [2]. The size of each antenna element is inversely related to the operating frequency, and hence, it is possible to place more antennas within the same physical space which alleviates the problems regarding the form factor. However, deploying a large number of antennas with beamforming capabilities can still significantly increase the cost and complexity of the required hardware and the corresponding algorithms [3].

*It should be noted that it is common to also refer to frequencies around 26 GHz, and above, as mmWave frequencies. Although this terminology is not accurate, the radio propagation channels at 26–30 GHz share similar properties with mmWave channels.

If each antenna is directly connected to the baseband processor via a radio frequency (RF) chain, then it is possible to use methods from digital signal processing to effectively control the phase and amplitude of the signal per antenna element. Such flexibility enables the implementation of efficient beamforming algorithms that improve the achievable data rates. However, this increases the power consumption, cost and form factor of the system as more RF chains are used. Another alternative is to connect the antennas to the baseband by using an analogue beamformer. In this approach, a single-RF chain and a network of phase shifters are used to connect the baseband to the antennas. Such systems, however, limit the number of simultaneously transmitted data streams and the achievable data rates. To reach a trade-off between the two structures, a third beamforming approach, which is referred to as hybrid analogue and digital beamforming, has emerged [2,4]. In this method, a small number of RF chains are connected to a larger number of antennas using different types of phase shifter and switch networks [4]. Such systems provide a trade-off between system cost, complexity, power consumption and spectral efficiency [5].

In addition to the signal attenuation in free-space propagation and the frequency-dependency of the antennas, it is also well known that the propagation processes, such as reflection and signal penetration, also vary with frequency and the electrical size of the objects. For example, consider the diffraction phenomena which can enable signal reception even if the direct line-of-sight (LoS) path between the Tx and Rx antennas is blocked by a large object. The diffraction efficiency significantly reduces with the increase in frequency. Hence, there can be significant blockage [6] and shadowing in the mmWave bands. In addition, the signal penetration through foliage [7] and different objects significantly varies depending on the relative electrical size of the signals and the frequency-dependent dielectric properties of the objects [8]. Such effects introduce further challenges to design of mmWave systems compared with the more common wireless systems in lower frequencies. For example, the use of directional antennas and beamforming approaches can impact the frequency response of the effective channel observed in the baseband and the required transmission methods [9,10].

Considering that the propagation channels at mmWave bands are significantly different from the conventional sub-6-GHz channels, it is necessary to gain a deep understanding of their characteristics as they play a crucial role in designing the corresponding circuitry and algorithms to establish reliable wireless links and deliver high data rates. With this motivation, in the next section we review some of the key attempts and projects that aimed towards characterising mmWave channels by conducting measurement campaigns and propose channel models.

16.1.1 Towards the development of mmWave channel models

As the interest in mmWave bands has risen significantly in the past decade, there have been major efforts within academia and industry to characterise such channels in different scenarios. It should be noted that all generations of mobile networks before 5G operate in sub-6-GHz frequency bands. As a result, a vast amount of information about various propagation scenarios in sub-6-GHz channels exists thanks to the experience accumulated over the years. Although there is also a general understanding about the mmWave channels to anticipate their overall behaviour [11], there is much less information about the propagation characteristics of such channels in different scenarios [12]. To deliver 5G's high data rate requirements, there have been significant efforts and investment dedicated towards developing mmWave channel models for 5G applications.

Generally speaking, the channel models can be classified into two categories of *physical* and *analytical* models based on the modelling approach that is used [13]. Physical models accurately describe the parameters of the channel based on electromagnetic wave propagation between the transmitter and receiver antennas in specific scenarios. These models can be further divided into deterministic and stochastic. In deterministic models, the impact of the physical parameters of the environment is directly included in calculating the behaviour of the electromagnetic wave. These approaches, such as ray tracing techniques or recorded measurements, are site-specific and provide high accuracy at the cost of complex computations or resources. Ray tracing tools simulate the features of the environment by taking various parameters into account, including the map and material of the objects, geometry, and provide a deterministic model for the particular channel. One of the advantages of ray tracing tools is that they provide excellent insights about the propagation behaviour of the signals in the scenarios of interest without the need for conducting expensive measurement campaigns.

Stochastic models, on the other hand, use probabilistic approaches to provide spatio-temporal characteristics of the channel. Such models are based on measurement campaigns and require much less complexity compared with deterministic models. They are widely used in system level simulations and performance evaluation based realistic assumptions about the channel. Some of the well-known stochastic models include geometry-based stochastic model (GSCM) and Saleh–Valenzuela, Zwick model.

Unlike the physical models, the analytical models do not account for wave propagation and they are not environment specific. Such models are commonly used for algorithm development and system analysis as they are easier to implement, and they include i.i.d. Rayleigh fading, Kronecker, Weichselberger, finite scatterer, maximum entropy and virtual channel representation models.

Despite the importance of each of these approaches, the stochastic channel models, which are based on measurement campaigns and empirical data, play a critical role in the development and evaluation of wireless communication systems. As the race towards 5G started in around 2010, the interest in developing mmWave systems also rose among the academia and industry. In the following, we briefly review some of the major channel measurement campaigns and models for mmWave systems:

- METIS project [14]: The Mobile and wireless communications Enablers for Twenty–twenty Information Society (METIS) project provided three models, including map-based, stochastic, and hybrid stochastic and map-based models for 5G applications. The map-based model is built upon three-dimensional (3D) ray tracing; however, it should be noted that this model is not a deterministic ray tracing tool [15]. It provides accurate simulations of the environment for frequencies of up to 100 GHz by considering the LoS component, specular reflection, penetration loss, over the roof propagation, diffuse scattering and many other critical factors. The disadvantage of this model is its complexity as it is a ray tracing-based approach. The stochastic model supports up to 70 GHz with 1-GHz bandwidth and is less complex and less accurate compared to the ray tracing model. This project covered a broad range of scenarios: urban macro (UMa), urban micro (UMi), device-to-device (D-2-D), indoor shopping mall and vehicular-to-vehicular.

- MiWEBA project [16]: The mmWave evolution for backhaul and access proposed a 3D channel model for a 60-GHz band, more specifically 57–66 GHz. This model is quasi-deterministic model that combines geometry and stochastic approaches to characterise both strong and weak multi-path components [17]. It includes the impact of transmitter and receiver beamforming and antenna polarisations. The MiWEBA model mainly focuses on backhaul, D-2-D and some specific access scenarios such as street canyon, hotel lobby and university campus.
- NYU Wireless measurements and models [8,18–20]: The New York University (NYU) Wireless conducted extensive measurement campaigns at the 28, 38, 60 and 73 GHz bands. Evolving from narrowband measurements and models, as discussed in [8,18], towards wideband scenarios, NYU Wireless has now built an open-source channel simulators which offer map-based, stochastic channel, and also the hybrid GSCM and ray tracing based models to create impulse responses over a wider range of frequencies [19]. The measurements include both indoor and outdoor scenarios such as UMi, UMa and rural macrocell.
- 3GPP model [21]: The Third-Generation Partnership Project (3GPP) is one of the key players in proposing channel models for mobile networks. 3GPP TR38.901 supports frequencies between 500 MHz and 100 GHz, and it is mainly based on stochastic models. However, 3GPP also offers a hybrid model to better describe channel parameters in various scenarios. 3GPP models are widely used in the literature as a benchmark for various channel models and to evaluate different algorithms.
- QuaDRiGa [22]: The Quasi Deterministic Radio Channel Generator (QuaDRiGa) model is an open-source platform that was initially developed for sub-6-GHz channels, but then it was extended to mmWave channels. It is based on GBSM and can be considered an evolution of WINNER+ channel model. This model supports 3D antenna modelling, massive MIMO and multiusers in multicells.
- mmMAGIC channel model [23]: This project proposed geometry-based stochastic channel models based on a series of measurement campaigns for a broad range of scenarios, including indoor office, airport, UMi street canyon and open square for frequencies between 6 and 100 GHz and a bandwidth of up

to 2 GHz. This model combines the channel modelling methodologies of 3GPP and QuaDRiGa.

- University of Surrey channel measurement [24–28]: The focus of this intensive campaign was to build stochastic models for mmWave frequencies that were considered a 5G band in the United Kingdom [29] and United States [30]. The path loss and root mean square (RMS) delay spread models are proposed for centre frequencies at 26, 32 and 39 GHz with 2-GHz bandwidths. Several indoor and outdoor scenarios, the impact of the antenna height and polarisation have been considered.

For further references about measurement campaigns, the reader can refer to the excellent survey papers such as [31,32]. In the next section, we present the necessary theoretical aspects related to channel modelling and analysis in wireless communication systems, before focusing on the mmWave systems in the subsequent sections.

Summary

Until 5G, all mobile networks operated at sub-6-GHz frequencies. As the demand for more connectivity and higher data rates grew, the availability of large chunks of spectrum in the mmWave bands motivated the telecommunications industry to include mmWave systems in 5G. On the one hand, the use of large bandwidth provides the opportunity to achieve high data rates. On the other hand, the propagation channels at high frequencies impose several challenges for the system designers. To better understand such challenges, intensive measurement campaigns were conducted to propose mmWave channel models focusing on different 5G scenarios. As a result, different map-based, stochastic and hybrid models have been proposed by different projects and organisations such as METIS, MiWEBA, NYU Wireless, 3GPP, QuaDRiGa and the University of Surrey.

16.2 Background theory and terminology

This section presents the background knowledge related to the analysis of wireless channels. Consider a single-input–single-output (SISO) system, shown in Figure 16.1, that operates in a time-invariant channel. Let us assume that the properties of the radio propagation channel[†] does not change over the time. At frequency f, the relationship between the transmitted and received signals, denoted by $X(f)$ and $Y(f)$, respectively, is

$$Y(f) = H(f)X(f) + N(f), \tag{16.2}$$

[†]It should be noted that the discussions in this section are based on the assumption that wide stationary and uncorrelated scattering holds for the channel. For more details about this assumption, please refer to [1].

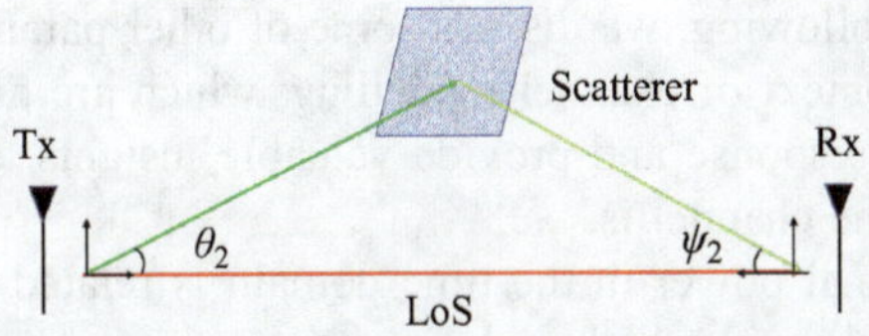

Figure 16.1 SISO system with a double-directional channel in the middle. The red line presents the LoS path and dark and light green indicate the path from the transmitter to the scatterer, and from the scatterer to the receiver respectively.

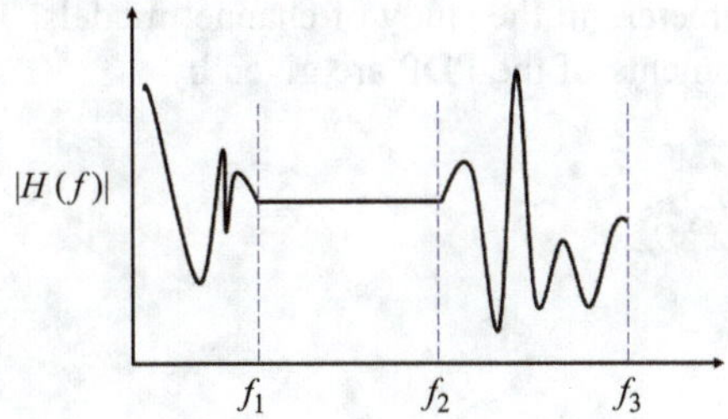

Figure 16.2 The channel response between $[f_1, f_2]$ and (f_2, f_3) is frequency-flat and frequency-selective, respectively

where $H(f)$ and $N(f)$ denote the channel and noise response, respectively. Over the transmission band $f \in [f_1, f_2]$, where f_1 and f_2 denote the start and stop frequencies of the transmit signal, the channel may provide the same or different response to each frequency of the input-signal $X(f)$. If the $H(f)$ provides the same response over the frequency band of interest, then it is referred to as frequency-flat channel; otherwise, it is frequency-selective channel, as demonstrated in the example shown in Figure 16.2.

The impulse response $h(\tau)$ of the channel, given by inverse Fourier transform of $H(f)$, can be used to extract its time-domain characteristics. To build an intuition about $h(\tau)$, consider the environment in Figure 16.1. In this case, there are only two paths, namely the LoS and the non-LoS (NLoS) path, for the signal to reach from the transmitter to the receiver. Since the length of LoS and NLoS paths are different, the signals will arrive with different delays τ_1 and τ_2. The channel impulse response for this scenario can be expressed as

$$h(\tau) = h(\tau_1)\delta(\tau - \tau_1) + h(\tau_2)\delta(\tau - \tau_2), \tag{16.3}$$

where $\delta(\tau)$ is the Dirac delta function. From Fourier analysis, it can be easily verified that if $h(\tau_2) = 0$, i.e., the channel only consists of LoS, then the propagation channel becomes frequency-flat where $|H(f)| = |h(\tau_1)|$. On the other hand, the frequency-selectivity of the channel can significantly increase if $h(\tau) \neq 0$, for $\tau \neq \tau_1$. Consequently, inter-symbol interference occurs and equalisation techniques are required. In the following, we discuss some of other parameters that are commonly used in the context of channel modelling, which are directly derived from the channel impulse response and provide valuable insights, e.g., indicating how frequency-selective the channel is.

The received signal power in the time domain is related to the power–delay profile (PDP) of the channel which is defined as

$$P(\tau) = |h(\tau)|^2. \tag{16.4}$$

More precisely, the PDP contains information about the received signal power from transmitting a delta pulse with unit energy. The PDP information is typically gathered through channel measurements, which enables the derivation of some of the most popular parameters in the study of channel models. The 0th and the normalised first-order moments of the PDP are given by

$$P_{\mathrm{m}_0} = \int_{-\infty}^{\infty} P(\tau) d\tau, \tag{16.5}$$

and

$$\bar{\tau} = \frac{\int_{-\infty}^{\infty} P(\tau)\tau d\tau}{\int_{-\infty}^{\infty} P(\tau) d\tau}, \tag{16.6}$$

which provide insights about the total received power and the mean excess delay, respectively. One of the metrics that provides valuable insights about the frequency-selectivity of the channel is its RMS delay spread. The RMS delay spread is given by

$$\tau_{\mathrm{rms}} = \sqrt{\frac{\int_{-\infty}^{\infty} P(\tau)\tau^2 d\tau}{P_{\mathrm{m}_0}} - \bar{\tau}^2}. \tag{16.7}$$

The larger the RMS delay spread gets, the more frequency-selective the channel becomes. To accurately quantify the frequency selectivity of the channel, the channel frequency response can be used to calculate the coherence bandwidth. Intuitively, this parameter provides an indication about the frequency range that the channel can be considered to be approximately flat. More accurately, the channel coherence bandwidth is given by

$$B_{\mathrm{C}} = \frac{1}{2}\left[\underset{\Delta f > 0}{\operatorname{argmax}}\left(\frac{R(\Delta f)}{R(0)} = 0.5\right) - \underset{\Delta f < 0}{\operatorname{argmin}}\left(\frac{R(\Delta f)}{R(0)} = 0.5\right)\right], \tag{16.8}$$

where $R(\Delta f) = \mathrm{E}[H^*(f)H(f + \Delta f)]$. It should be noted that the reason that $R(\Delta f)$ is only dependent on Δf, not on frequency f, is because of the wide-sense stationary and uncorrelated assumption that we have made throughout this chapter.

Other important parameters that are required for developing channel models are small- and large-scale fading. As shown in Figure 16.3, small-scale fading

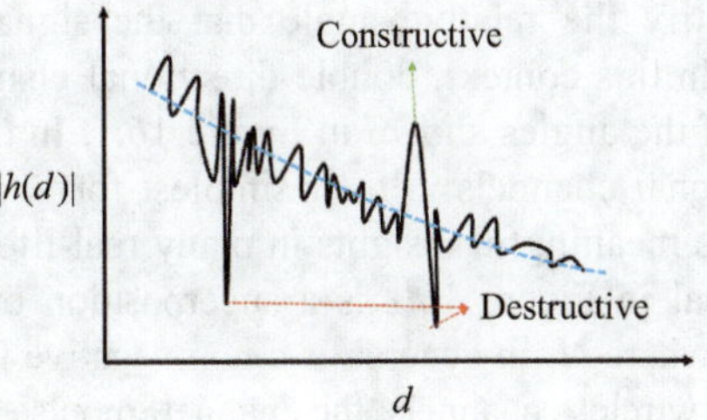

Figure 16.3 Small-scale fading with constructive and destructive interference vs large-scale fading

describes the rapid signal fluctuations over small distances in the order of a few wavelengths. For example, consider a scenario that the transmitter and the propagation environment are fixed. When the receiver moves within a few wavelengths, the signals from multi-path components can be added up in a destructive or constructive manner, which can result in an increase or a decrease in the SNR. Consequently, the received signal levels can experience significant variations, e.g., over 40 dB in some scenarios.

On the other hand, large-scale fading describes the signal variations when the receiver moves over larger distances, e.g., more than ten wavelength, the signal levels gradually change. Such slow variations are due to shadowing and also path loss effects. Shadowing effect is caused by large objects, with respect to the signals wavelength, in the environment. For example, large buildings, trees or even human body in the case of mmWave systems can significantly reduce the received signal strength. Increasing the distance between the transmitter and receiver in communication systems degrades the signal levels as the path loss increases. In addition to distance, path loss also depends on propagation environment, the height of the transmitter and receiver antennas, the electromagnetic characteristics of the propagation medium, and the operating frequency. The most basic approach to estimate the path loss levels is to calculate the free-space path loss for a pair of transmitter and receiver antennas. Using (16.1), the free-space path loss is given by

$$L_{\mathrm{FS}} = \left(\frac{\lambda}{4\pi d}\right)^2. \tag{16.9}$$

Equation (16.9) indicates that the path loss increases with d^{α}, where $\alpha = 2$ is the path loss exponent. In many practical scenarios, α varies between two and four. However, smaller or larger values also exist in specific settings where wave-guiding effect takes place in the environment.

The travelling radio signals experience various propagation mechanisms such as reflection, as elaborated in Figure 16.1. In addition to the shape and material of the scatterers, their impact on the propagation channel also depends on the relative location of the transmitter, receiver and the scatterers. This is of particular interest in mmWave systems as there are only a few scatterers which have a significant contribution to the propagation properties of the channel. Hence, it is necessary in such systems to identify the relative angle that the signals can be effectively transmitted/received. In this context, double-directional channel impulse response includes the impact of the angles shown in Figure 16.1. In the following, we only discuss double directional channels in their simplest form. However, even such a simple model provides meaningful insights in many real-life scenarios.

The received signal in Figure 16.1 is a superposition of the signals from the $N = 2$ different paths where N, in general, is a non-negative integer number. Due to the linear nature of the wireless channels, the channel impulse response of the overall channel $h(\tau)$ can be written as the sum of the channel impulse response $h_n(\tau)$ that includes the channel related to the corresponding scatterer. More precisely,

$$\begin{aligned} h(\tau, \psi, \theta) &= \sum_{n=1}^{N} h_n(\tau)\delta(\tau - \tau_n) \\ &= \sum_{n=1}^{N} h_n(\tau_n, \psi_n, \theta_n)\delta(\tau - \tau_n)\delta(\psi - \psi_n)\delta(\theta - \theta_n), \end{aligned} \tag{16.10}$$

where τ_n, θ_n, ψ_n denote the delay, angle of arrival and angle of departure associated with the nth path. The received signal power over each path is characterised by double-directional delay power spectrum:

$$P(\tau, \psi, \theta) = \sum_{n=1}^{N} |h_n(\tau_n, \psi_n, \theta_n)|^2 \delta(\tau - \tau_n)\delta(\psi - \psi_n)\delta(\theta - \theta_n). \tag{16.11}$$

Similar to the omnidirectional channel impulse response, the properties of signal over the nth path can be separately studied in the delay domain. Such information can be very useful in the choice of waveforms, precoding and equalisation techniques when directional antennas are used [9]. In an extreme scenario, if both transmitter and receiver employ directional antennas, there may be only a single path to communicate which significantly reduces the RMS delay spread. As a result, the channel can be viewed as frequency-flat channel and there may be no requirements for complex equalisation techniques as there is no intersymbol interference caused by the delays between different paths. Finally, the analysis of the propagation channel in the angular domain can be useful in deciding the directions that the receiver and transmitter establish a link. On the receiver side, the angular power spectrum of the channel is defined as

$$P_\Theta(\theta) = \int_{-\pi}^{\pi} \int_{0}^{\infty} P(\tau, \psi, \theta) d\tau d\psi. \tag{16.12}$$

Similar to mean and RMS delay spread, statistical properties of the angular power spectrum can be studied as it will be discussed later.

Now that we have introduced the required terminology and the background information about channel modelling, in the next section we use these tools to present empirical channel measurements and modelling for several 5G-related scenarios.

Summary

This section presented an overview of the key concepts and definitions in the analysis of the channel models. The channel impulse response characterises the behaviour of the propagation channel in the delay domain. Depending on this parameter, the channel can be frequency-flat or frequency-selective which, in turn, impacts the choice of the equalisation and precoding techniques. Other parameters that are closely related are RMS delay spread and coherence bandwidth. The RMS delay spread is calculated based on power–delay profile and can be used to measure how time-dispersive the channel is. Coherence bandwidth, on the other hand, expresses the frequency range that the channel can be assumed to be nearly flat. Double-directional channel models are used to capture the directional nature of the channel and analyse the properties of the signals in the environment towards/from different angles.

16.2.1 Empirical channel measurements and modelling

In this section, we present the measurement results and analysis for several indoor and outdoor scenarios at 26, 32 and 39 GHz based on the wideband channel measurement campaigns that were conducted at the University of Surrey [24–28]. Section 16.2.2 introduces the measurement set-up and the required equipment to collect data. Then, Sections 16.2.3 and 16.2.4 show how the theoretical knowledge of Section 16.2 can be applied in practice and present the measurement methodology and results in realistic 5G scenarios.

16.2.2 The measurement equipment and set-up

The diagram in Figure 16.4 demonstrates the measurement equipment and set-up. A Rohde & Schwarz signal generator R&S®SMW200A transmits a continuous wave signal with 17-dBm signal strength. The sounding signal is based on Zadoff–Chu sequence because of the its autocorrelation properties. The receiver consists of a Rohde & Schwarz signal analyser R&S®FSW67 in combination with an R&S®RTO1044 oscilloscope to receive and digitise complex-valued signals with a 2-GHz bandwidth. There is also a direct connection between SMW and FSW using two cables to synchronise the reference frequency and send a trigger signal to start the measurements. Finally, a computer carries out the hardware control and data processing using the related software tools, including R&S®TS-5GCS to analyse the measured signals.

For the measurements at 26 GHz, the transmitter has quasi-directional pyramid horn antenna with 61° half power beam width (HPBW) and 6.8-dB gain. To capture the directional nature of the channel, the receiver has a wideband lens antenna with HPBM of 5° and 24-dBi gain in both E- and H-planes. For the measurements at 32 and 39 GHz, the transmitter is equipped with a quasi-directional horn antenna with HPBM and gain of 54° and 10 dBi, respectively. The gain of the receiver antenna at 32 and 39 GHz is 23 and 25 dBi, respectively. Table 16.1 presents the parameters and specifications of the measurements and antennas. As indicated in Figure 16.4,

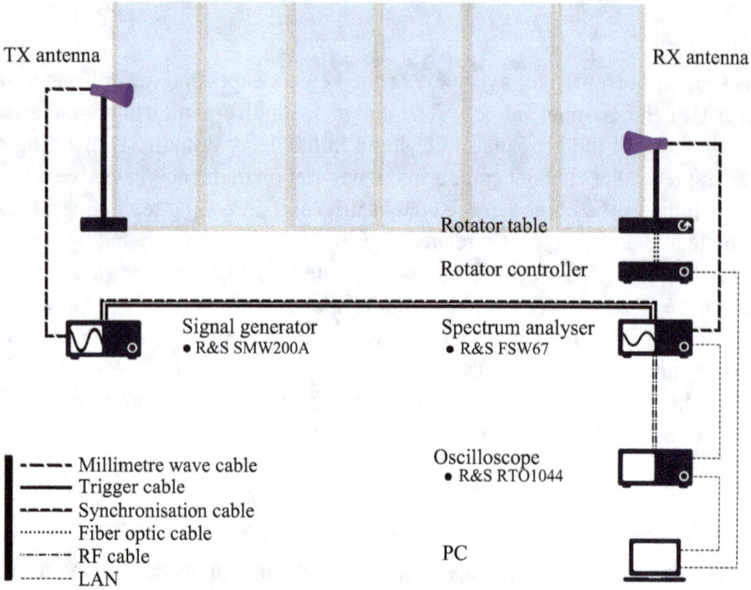

Figure 16.4 Block diagram of the system set-up [25]

Table 16.1 Wideband channel sounding set-up specifications for thee 26, 32 and 39 GHz

Carrier frequency	**26 GHz**	**32 GHz**	**39 GHz**
Sounding waveform	Frank–Zadoff–Chu 65535		
RF bandwidth	2 GHz		
Transmit power	17 dBm		
Delay resolution	0.5 ns		
TX polarisation	Vertical		
RX polarisation	Vertical		
TX E-plane HPBW	78°	54°	
TX H-plane HPBW	61°	54°	
RX E-plane HPBW	5°	8°	
RX H-plane HPBW	5°	8°	
TX antenna gain	6.8 dBi	10 dBi	
RX antenna gain	24 dBi	23 dBi	25 dBi

the receiver antenna is placed on rotating table which is controlled by the same computer connected to the SMW and FSW. It should be noted that all the measured data are calibrated based on the system's response when the transmitter and receiver are directly connected.

The next subsection describes the details of the outdoor measurement campaigns and presents the corresponding analysis and results.

Figure 16.5 Measurement sites: (a) Route 1 environment, (b) Route 2 environment, (c) aerial picture of Route 1, (d) aerial picture of Route 2, (e) aerial picture of both Routes 1 and 2 inside the campus of the University of Surrey [26,27]

16.2.3 Outdoor measurements

The outdoor measurements have been carried out in two different sites inside the campus of the University of Surrey. Figure 16.5 presents the details of these sites, and the approximate locations of the transmitter (Tx) and receiver (Rx). During the measurements, the location and direction of the Tx antenna remains fixed while the Rx moves across the routes shown in Figure 16.5 and scans the propagation channel over different angles. In the following, Route 1 refers to the pavement next to the Institute for Communication Systems (ICS) at University of Surrey, shown in Figure 16.5(a) and (c). The building is located next to an open field and a wide street, and its exterior contains large pieces of glass and metal surfaces. In this route, the heights of the transmitter and receiver antennas are always set as $h_{\text{Tx}} = 5$

and h_{Rx}=1.7, respectively. Route 2 is in a canyon-type street where the surface of the surrounding buildings contains large areas of brick, glass and metal, as shown in Figure 16.5(b) and (d). To investigate the impact of the antenna height, two sets of measurements are carried out for $h_{\mathrm{Tx}} = 5$ and $h_{\mathrm{Tx}} = 8$ m during the measurements in Route 2. In all scenarios, i.e., Routes 1 and 2, the Rx is initially placed at 5-m horizontal distance from the Tx. Then, $I = 18$ measurements are performed by increasing the distance by 5 m until the distance between Tx and Rx reaches 90 m. For each location, the channel impulse response is measured over $K = 200$ snapshots. The direction of the Rx antenna horizontally varies in the range of $[0, 360)$ with 5° steps for 26 GHz and 8° steps for 32 and 39 GHz scenarios. As a result of the difference in HPBW at different frequency bands, $J = 72$ measurements were required to scan 360° for the channels at 26 GHz, while $J = 45$ measurements were conducted for 32 and 39 GHz channels. In the following, the notations V–V, V–H, H–V, H–H represent the measurement results for different combinations of vertical (denoted by 'V') and horizontal (denoted by 'H') polarisations where the first and last symbols refer to the Tx and Rx antenna polarisation, respectively. In the following, we present the methodology to analyse the data and present the main results of the measurements.

Let $h_{i,j,k}^{\mathrm{D},p}(n)$, for $p \in \{\mathrm{V-V}, \mathrm{V-H}\}$, denote the measured directional channel impulse response where $i \in \{1, \ldots, I\}, j \in \{1, \ldots, J\}$, $k \in \{1, \ldots, K\}$ and n denote the location, angle, snapshot and delay bin indexes, respectively. It should be noted that the delay resolution between two delay bins is $1/B = 0.5$ ns where B is the measurement bandwidth. Then, the instantaneous and average power–delay profiles of the directional channels are given by

$$P_{i,j,k}^{\mathrm{D},p}(n) = |h_{i,j,k}^{\mathrm{D},p}(n)|^2, \tag{16.13}$$

and

$$P_{i,j,\mathrm{ave.}}^{\mathrm{D},p}(n) = 1/K \sum_{k=1}^{K} P_{i,j,k}^{\mathrm{D},p}(n), \tag{16.14}$$

respectively. Based on the theory discussed in the previous section, (16.10) can be used to capture the omnidirectional behaviour of the propagation channel based on directional measurements. In particular, an omnidirectional channel impulse response $h_{j,k}^{\mathrm{O},p}(n)$ can be created by superimposing the directional channel impulse responses over all angles per location, i.e.,

$$h_{i,k}^{\mathrm{O},p}(n) = \sum_{j=1}^{J} h_{i,j,k}^{\mathrm{D},p}(n). \tag{16.15}$$

Similar to the notation for directional channel impulse responses, let us define $P_{i,k}^{\mathrm{O},p}(n)$ and $P_{i,\mathrm{ave.}}^{\mathrm{O},p}(n) = 1/K \sum_{k=1}^{K} P_{i,k}^{\mathrm{O},p}(n)$, $\tau_{i,k}^{\mathrm{O},p}$ as the PDP of the omnidirectional channel at location i and snapshot k, its average over K snapshots, and the RMS delay spread of the omnidirectional channel at location i and snapshot k, respectively.

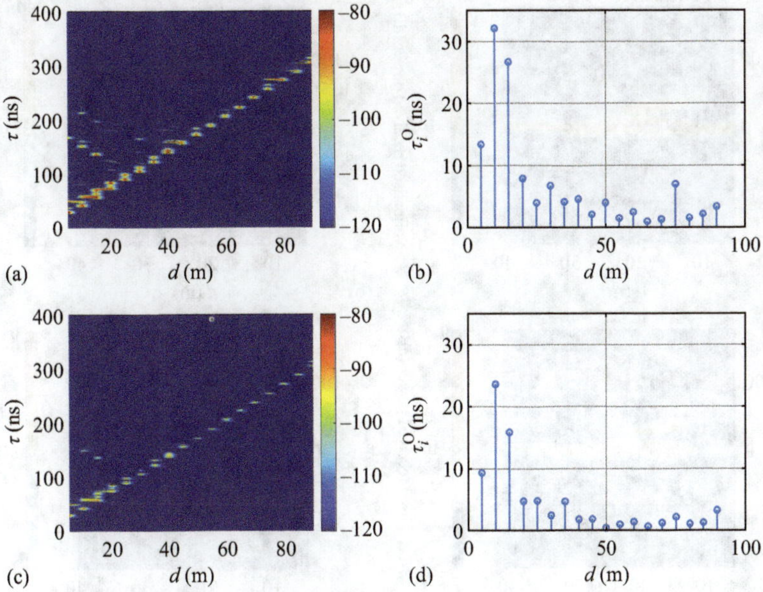

Figure 16.6 Measurement results for 26 GHz band in Route 2: (a) average PDP for V–V, (b) average RMS delay spread for V–V, (c) average PDP for V–H, (d) average RMS delay spread for V–H [24]

Before using the equations in Section 16.2 to extract the channel parameters from the measurements, it must be noted that the measurement data include the impact of noise. In this section, we present two approaches that are commonly used to preprocess the data. In the first approach, the data that are S dB stronger than the noise floor are considered desired signal. In the second approach, the data that are S dB below the maximum received signal at a location are considered noise. Generally speaking, the choice of the threshold level S and the suitable approach depends on the specific measurement and equipment. Hence, it is recommended to experimentally investigate the impact of the threshold level and noise removal approach based on the measured data and the corresponding results. Once a suitable threshold level is chosen, the choice of the approach has less impact on the results.

Figure 16.6 presents $P_{i,\text{ave.}}^{\text{O},p}(n)$ and $\tau_i^{\text{O},p} = \sum_k \tau_{i,k}^{\text{O},p}/200$ for V–V and V–H polarisations and different distances d between Tx–Rx antennas at 26 GHz in Route 2 where 200 is the number of snapshots. Figure 16.6(a) shows the presence of two scatterers within the range of 20 m in addition to the LoS component. It is evident from Figure 16.6(a) and (c) that the V-V channel provides higher gains and consequently the NLoS components are detectable in this setting compared to V–H. However, Figure 16.6(b) and (d) indicates that only locations 2 and 3 experience higher RMS delay spread. This translates into the fact that the scatterers in the environment are more significant for these two locations. As stated earlier and

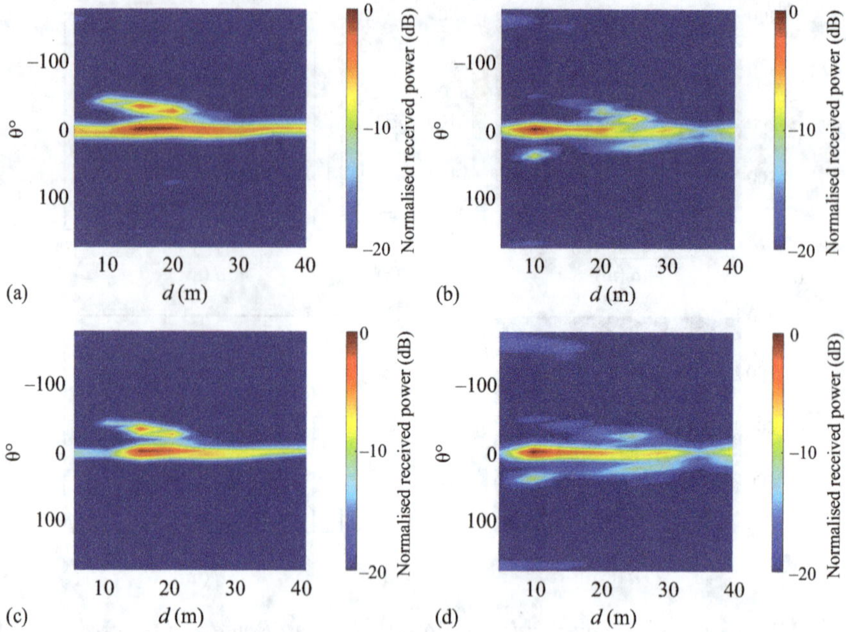

Figure 16.7 Normalised omnidirectional power–delay profile at (a) 32 GHz (Route 1), (b) 32 GHz (Route 2), (c) 39 GHz (Route 1), (d) 39 GHz (Route 2), for H–H polarisations [26]

evident from Figure 16.5(b), the receiver is surrounded by buildings with glass and metal surfaces which evidently contribute to larger RMS delay spreads. Figure 16.6 also shows that the average RMS delay spread for the V–V channel is higher than V–H due to the number of observable multi-path components in the channel.

Similar to the discussions for the 26-GHz channels, the omnidirectional PDPs for the 32 and 39 GHz measurements are created by superimposing the corresponding directional PDPs. Figure 16.7 presents the normalised received power from different angles, for the measurements at 32 and 39 GHz in Routes 1 and 2. In all scenarios, it is evident that the LoS provides the dominant path. However, Figure 16.7(a) and (c) indicates that there is also limited reflection and scattering effects in Route 16 at around 15 and 20 m. Similarly, the LoS is the dominant link between Tx and Rx although limited energy is received from the scatterers in the environment. According to Figure 16.7(b) and (d), the measured channel has significantly attenuated at $d = 35$ m. This is due to some antenna misalignment in elevation between the transmitter and receiver as the slope of the pavement suddenly changes at this point.

Figure 16.8 presents the cumulative distribution function (CDF) of the RMS delay spreads for all combinations of polarisations for 32 and 39 GHz measurements over Routes 1 and 2. As discussed earlier, it is necessary to remove the noise effects from the measured PDPs before calculating the RMS delay spreads. In

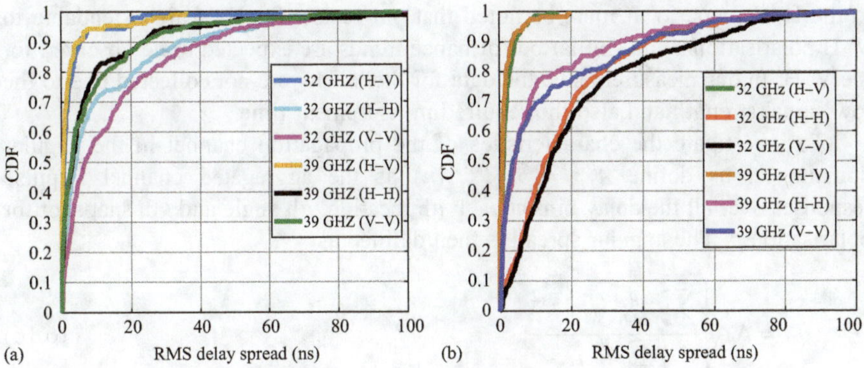

Figure 16.8 CDF of RMS delay spreads of omnidirectional PDPs at 32 and 39 GHz: (a) Route 1 and (b) Route 2 [26]

Figure 16.8, this procedure is performed by setting the threshold level such that all the points below 35 dB of the maximum received signal are set to zero for V–V and H–H measurements. However, the threshold value is set to 20 dB for V–H and H–V as the dynamic range of the PDPs is much smaller due to the opposite polarisation of the Tx and Rx antennas. In this manner, the PDPs have a high dynamic range, which also enable a more fair comparison between the measurements at 32 and 39 GHz as they have different noise floor values.

According to Figure 16.8(a), the RMS delay spreads for the H–H and V–V channels at 32 GHz in Route 1 are 33.3 and 40 ns at 90th percentile, respectively. These values, however, are 22.6 and 23.6 ns at 39 GHz which indicate that there is less energy received from the scatterers in the channel, compared with 32-GHz measurements. As expected, the RMS delay spreads for the PDPs with opposite polarisations, i.e., V–H and H–V, are significantly smaller than V–V and H–H scenarios. In other words, the LoS path is the only link which can provide a connection between Tx and Rx with opposite antenna polarisations. Figure 16.8(b) presents a similar trend in the behaviour of the RMS delay spread over Route 2. In this scenario, 90% of the measured values for the H–H and V–V channels are less than 41.35 and 56 ns, which indicates that the channel in Route 2 has stronger scatterers compared to Route 1. Figure 16.8 also shows H–H has smaller RMS delay spreads than V–V polarisation in nearly all scenarios. The majority of the scatters, e.g., walls and windows, are vertically positioned in the environment. Depending on the polarisation of the transmit signal, the vertical/horizontal orientation of the objects in the environment impacts the strength of the scattered signals form the surface of the scatterers. In the V–V transmission, the transmit signal with vertical polarisation impinges on a vertical scatterer. Consequently, the reflected wave nearly retains its polarisation. However, a horizontally polarised signal experiences depolarisation effects when it bounces a vertical scatterer. In short, the vertical scatterers can better preserve the signal strength of vertically polarised signals in the V–V transmission which increases the RMS delay spread compared to the H–H scenario. It must be noted that our results for H–V are extendable to V–H polarisations; and similar performance trends are expected to be observed for the V–H. In our measurements, the data for the V–H were not collected due to the low signal strength and also the required measurement time.

To investigate the characteristics of the propagation channel in the angular domain, let us define $A^p_{i,j,k} = \sum_n P^{\mathrm{D},p}_{i,j,k}(n)$ as the aggregated channel impulse responses over all the delay bins at each ith location, jth angle and kth snapshot for polarisation p. The angular spread is then defined as

$$\sigma_\theta = \Delta_\theta \sqrt{\frac{\sum_j A^p_{i,j,k}(j-\bar{j})^2}{\sum_j A^p_{i,j,k}}}, \tag{16.16}$$

where $\bar{j} = \sum_j jA^p_{i,j,k} / \sum_j A^p_{i,j,k}$, and the angular step $\Delta_\theta = 5^\circ$ for the measurements at 26 GHz. Figure 16.9 presents the received powers and the RMS angular spreads averaged over 200 snapshots for the measurements at 26 GHz. For the distances within 50 m in the V–V measurements, there are received signals from almost all directions which result in angular spreads of as high as around 100°, shown in Figure 16.9(a) and (b). However, the LoS component becomes the major path for receiving signals after 50 m. The discussion becomes more clear by inspecting the

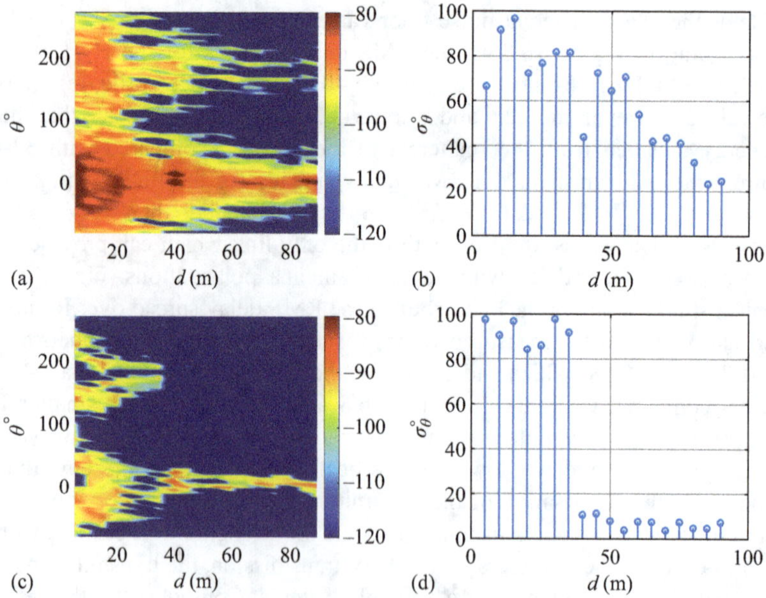

Figure 16.9 Angular properties of the channels at 26 GHz: (a) average received signal per angle-location for V–V, (b) average angular spread per location for V–V, (c) average received signal per angle-location for V–H, (d) average angular spread per location for V–H [24]

average received signal per angle for each location in the V–H scenario, shown in Figure 16.9(c) and (d). This effect is expected to be due to the fact that the surrounding of the receiver changes at such distances, as shown by Figure 16.5(c), and the receiver is placed close to brick walls. For the V–V scenario, it is observed that after 35 m, the only detectable path is the LoS component which results in much lower angular spreads. It is noted that the presented results in Figure 16.9 heavily rely on the chosen threshold level for noise; and the observable power levels from different angles will be significantly reduced if one increases the threshold levels to, e.g., −90 dBm. After careful examination of data, we have intentionally chosen −100 dBm to provide an insight about the propagation channel at 26 GHz against the conventional belief that at such high frequencies the channel is highly directional. In other words, our data indicate that for certain locations, NLoS signals may be received from various directions although they are weaker compared to LoS.

Figure 16.10 presents the CDF of a coherence bandwidth for directional and omnidirectional channels at 26 GHz with different polarisations. The measurement results provide an indication about the impact of using directional antennas in the measurements. For example, the 50th and 90th percentiles of the coherence bandwidth for omnidirectional V–V data are around 280 and 404 MHz compared to 1.3 and 2 GHz for the directional channel, respectively. As is evident by Figure 16.10, the V–H channels result in a significantly flatter frequency response compared to V–V; however, despite this fact, the PDPs in Figure 16.6 indicate that the received signal levels for the V–H scenario are significantly weaker compared to V–V.

As stated earlier, other key parameters in the study of propagation channels are path loss and shadowing. Depending on the measurement scenario and the curve-fitting techniques, the path loss models can vary. Hence, there are various path loss models to estimate the path loss at a desired setting in the literature. Among them, close-in (CI) reference is one of the common approaches which we also use in the

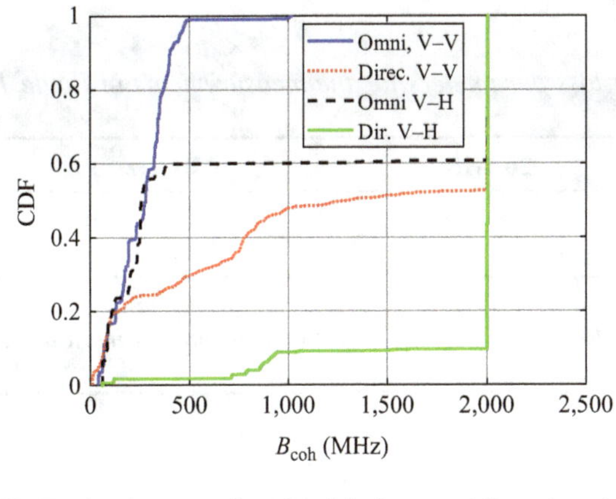

Figure 16.10 CDF of coherence bandwidth for omnidirectional and directional V–V and V–H channels at 26 GHz [24]

following. In the CI model, the path loss $L_{CI}(f,d)$, in dB scale, at frequency f and distance d is given by

$$L_{CI}(f,d) = L_{FS}(f,d_0) + 10n\log_{10}(d/d_0) + X_\sigma^{CI}, \tag{16.17}$$

where $L_{FS}(f,d_0)$ is the free-space path loss in dB scale at the physical reference distance $d_0 = 1$ m, n denotes the path loss exponent, and X_σ^{CI} is a zero-mean Gaussian random variable with σ being its standard deviation which is also expressed in dB scale in (16.17). In the CI model, X_σ^{CI} can be interpreted as the large-scale signal fluctuations around the mean path loss at distance d. The minimum mean square error (MMSE) criteria and least squares approach are then used to find the best linear fit to the data such that it minimises σ at d_0. The path loss exponent and the corresponding σ for all scenarios are presented in Tables 16.2 and 16.3.

As a benchmark for our path loss model, we use 3GPP channel model. Let C, d_{2D} and d_{3D} denote the speed of light, the 2D and 3D distances between Tx and Rx antennas, respectively. According to 3GPP report in [33], the path loss model for LoS scenarios in the Urban Micro Street Canyon is

$$L_{Los}^{3GPP} = \begin{cases} L_1, & 10\text{m} \leq d_{2D} \leq d'_{BP}, \\ L_2, & d'_{BP} \leq d_{2D} \leq 5 \text{ km}, \end{cases} \tag{16.18}$$

where

$$L_1 = 32.4 + 21\text{1log}_{10}(d_{3D}) + 20\log_{10}(f_c), \tag{16.19}$$

$$L_2 = 32.4 + 40\log_{10}(d_{3D}) + 20\log_{10}(f_c) - 9.5\log_{10}((d'_{BP})^2 + (h_{Tx} - h_{Rx})^2), \tag{16.20}$$

$$d'_{BP} = \frac{4(h_{Tx}-1)(h_{Rx}-1)f_c \times 10^9}{C}. \tag{16.21}$$

Table 16.2 Path loss parameters for the measurements in Route 1 [27]

f_c	**26 GHz**	**32 GHz**	**39 GHz**
n	1.8	2.1	1.95
σ (dB)	2.03	2.85	2.24

Table 16.3 Path loss parameters for the measurements in Route 2 [27]

Carrier frequency	**26 GHz**	**32 GHz**		**39 GHz**	
h_{Tx} (m)	8	5	8	5	8
n	2.04	1.81	1.94	1.8	1.91
σ (dB)	2.2	3.77	3.38	3.61	3.16

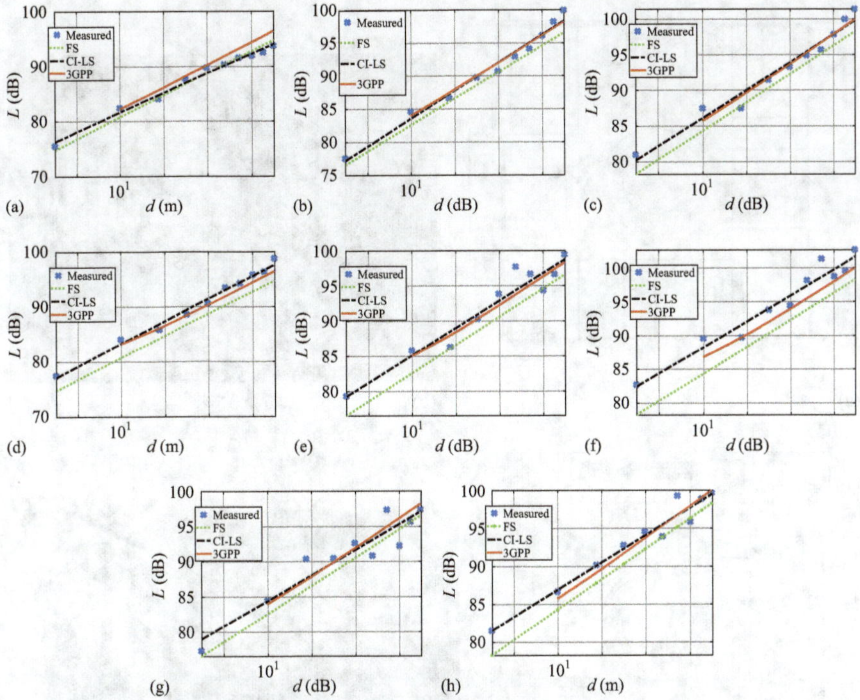

Figure 16.11 *Measured directional, free-space (FS), close-in model with least squares (CI-LS) and 3GPP path loss models for (a) 26 GHz, Route 1, $h_{Tx} = 5$ m; (b) 32 GHz, Route 1, $h_{Tx} = 5$ m; (c) 39 GHz, Route 1, $h_{Tx} = 5$ m; (d) 26 GHz, Route 2, $h_{Tx} = 8$ m; (e) 32 GHz, Route 2, $h_{Tx} = 8$ m; (f) 39 GHz, Route 2, $h_{Tx} = 8$ m; (g) 32 GHz, Route 2, $h_{Tx} = 5$ m; (h) 39 GHz, Route 2, $h_{Tx} = 5$ m [27]*

Figure 16.11 shows the path loss models based on CI with LS approach for curve fitting and 3GPP are in agreement when $h_{Tx} = 5$-m scenario. However, the 3GPP model results in slightly lower path loss than the LS fitting for $h_{Tx} = 8$ m.

16.2.4 Indoor measurements

In the following, we present the results of two sets of indoor measurements conducted inside ICS at the University of Surrey to evaluate the propagation properties of 26, 32 and 39 GHz channels. The first set of measurements took place on the ground floor of ICS, shown in Figure 16.12(a) and (b). The second set of the measurements only focus on 26-GHz channels and the campaign takes place on the second floor of ICS, shown in Figure 16.12(c) and (d). Both floors are a typical open-plan modern office environment and include desks, chairs, elevator, drywall and large pieces of glass. However, there are more desks and chairs in the second floor, as it is evident in Figure 16.12. Except the antenna heights, the rest of the

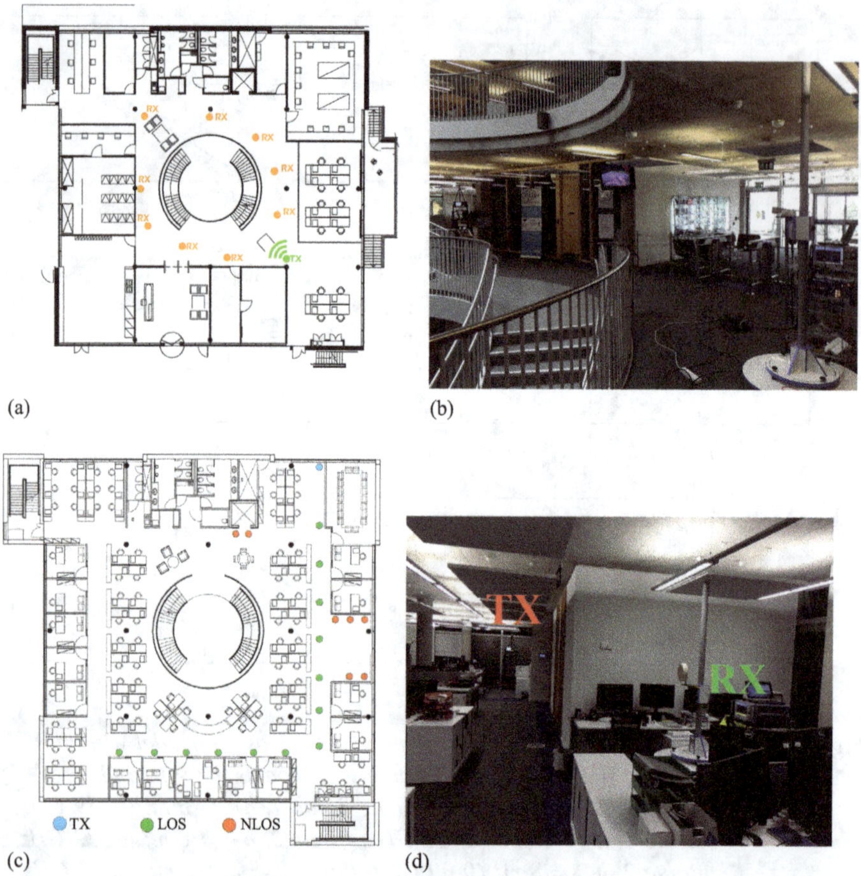

Figure 16.12 (a) Map and (b) the measurement environment of the ground floor [25]; (c) map and (d) the measurement environment of the second floor [28] of ICS, University of Surrey

equipment set-up and measurement parameters are the same as in Table 16.1. In particular, $h_{\mathrm{Tx}} = 2.7$ whereas $h_{\mathrm{Rx}} = 1.6$ m and $h_{\mathrm{Rx}} = 1.5$ m in the first and second measurements, respectively. In both scenarios, the Tx antenna is fixed and the Rx was moved on the path shown in the maps in Figure 16.12. The ground floor measurements only include LoS points where the adjacent distance between points is 2 m, and the distance between Tx and Rx varied from 4 to 20 m. In the second floor, however, both LoS and NLoS scenarios are considered evident in the map in Figure 16.12. The distance between any two green dots on the L-shaped path in the second floor is 2 m where the shortest distance between the Tx and Rx antennas is 4 m. In the following, let us define $d_{\mathrm{C}} = 4 + (j - 1)$ and $d_{\mathrm{L}} = 4 + 2(i - 1)$ where the indexes $j = 1, 2, \ldots, 9$ and $i = 1, 2, \ldots, 12$ are integer numbers to label the LoS measurement points in the ground and second floors, respectively.

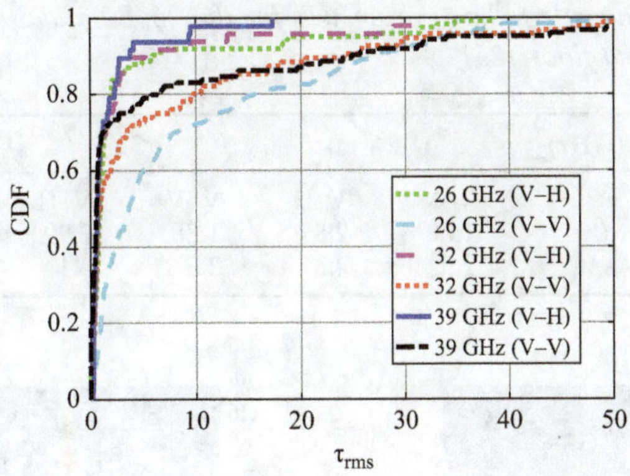

Figure 16.13 The CDF of the omnidirectional channels at 26 GHz, 32 GHz, and 39 GHz with different antenna polarisations in the ground floor [25]

Table 16.4 The average RMS delay spreads at 26, 32 and 39 GHz in the ground floor measurements [25]

Carrier frequency (GHz)	**26**		**32**		**39**	
Polarisation	V–V	V–H	V–V	V–H	V–V	V–H
RMS delay spread (ns)	26.78	5.88	24.27	3.57	20.48	2.72

Figure 16.13 presents the CDF of the RMS delay spreads of the measured channels in the ground floor. Similar to our discussion for the outdoor scenario, a noise threshold level must be applied before calculating the RMS delay spreads. In the following, the elements of the channel impulse response that are more than 35 and 20 dB below the maximum for the V–V and V–H measurements are set to zero, respectively. For the V–V scenario, Figure 16.13 shows that 90% of the RMS delay spreads are less than 26.78, 24.27 and 20.48 ns at 26, 32, 39 GHz, respectively. However, this parameter becomes significantly smaller for the V–H scenario. More specifically, the 90th percentile RMS delay spreads are 5.88, 3.57 and 2.72 ns for the 26, 32 and 39 GHz channels, respectively. Similar to the outdoor scenario, Table 16.4 indicates that the RMS delay spreads reduce as the frequency increases.

As discussed in Section 16.2.3, the CI provides one of the most popular path loss models. As discussed in (16.17), it characterises the path loss as a function of frequency of f and distance d between Tx and Rx antennas. To determine the path loss exponent n, MMSE criteria are used and σ at a physical distance $d_0 = 1$ m is minimised. Table 16.5 shows the path loss exponent n and standard deviation of shadow fading parameter σ based on least squares fitting approach. For the V–V scenario, the path loss exponents are 1.68, 1.88 and 1.81 at 26, 32 and 39 GHz, respectively. It should be noted that all n values are smaller than the free-space path loss exponent ($n = 2$). This indicates that the scattering environment acts as a wave

Table 16.5 Large-scale fading parameters for the 26, 32 and 39 GHz in the ground floor [25]

Carrier frequency (GHz)	**26**		**32**		**39**	
Polarisation	V–V	V–H	V–V	V–H	V–V	V–H
Path loss component (n)	1.68	2.87	1.88	2.99	1.81	2.93
σ	2.9	4.3	2.5	4.1	2.2	3.8

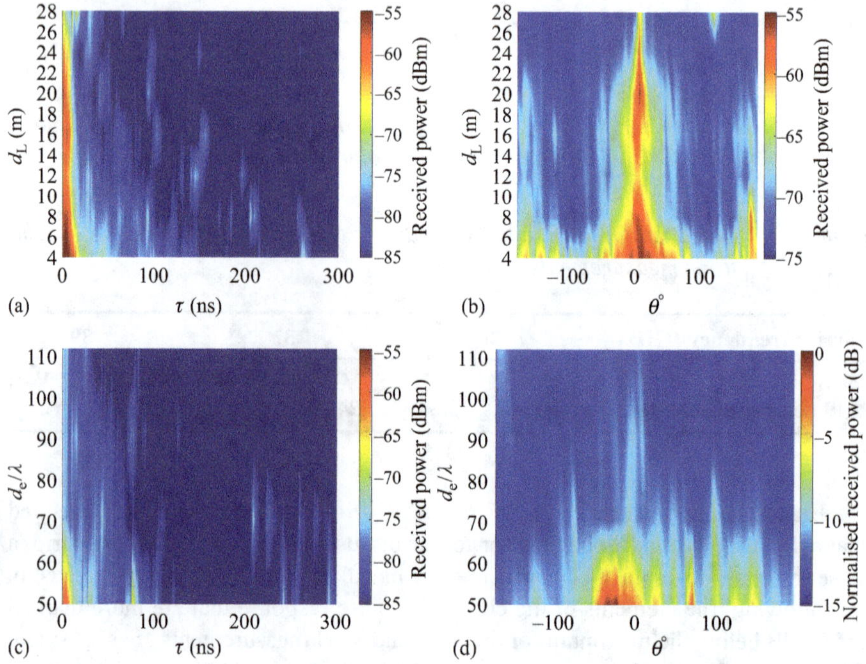

Figure 16.14 PDPs and angular power spectra of 26-GHz channels in the second floor [28]: (a) Los PDP, (b) LoS angular power spectrum, (c) NLoS PDP, (d) NLoS angular power spectrum

guide and the multi-path components are added up constructively. As expected, the path loss exponents for the V–H scenarios are much higher than V–V propagation.

Figure 16.14(a) presents the omnidirectional PDPs at each measurement point to analyse the behaviour of the channel in the delay domain regardless of the angle of arrival of the signal. Similar to our approach in analysing the outdoor channels, the omnidirectional PDPs are obtained by adding the channel impulse responses from the all azimuth angles. Figure 16.14(b) presents the angular power spectrum for the LoS points over d_L for V–V polarisation in the second floor. It is observed that although the LoS component at $\theta = 0°$ provides the strongest signal, the multi-path components from scatterers also have significant contribution. In other words, the impact of the clusters is noticeable when the Rx antenna rotates towards their direction. When the receiver is near the transmitter, both parts (a) and (b) of Figure 16.14 indicate that the NLoS components of the channel can have a major impact.

The presence of strong NLoS components in the measurements motivates further study of NLoS channels. As shown in Figure 16.14(c), the receiver antenna was placed in seven NLoS points at integer multiples of the signal wavelength. In this setting, the Rx antenna is placed at distance d_e from the edge of the walls. It should be noted that in the analysis of channel parameters, such as small scale fading, reflection and scattering, the ratio between physical distance and the signal wavelength is a key factor rather than each of the parameters by their own. To further emphasise on this point, Figure 16.14(c) and (d) presents the normalised angular power spectrum and omnidirectional PDPs over the NLoS channels, respectively. Both figures indicate that the received signals are stronger when the Rx antenna is closer to edges of the walls.

Summary

This section presented empirical mmWave channel measurements and analysis in several indoor and outdoor environments for 5G applications. It also showed how the theoretical discussions of Section 2 can be applied to analyse the characteristics of propagation channels using the data from intensive polarimetric measurement campaigns that were conducted at the University of Surrey. The investigated parameters include power–delay profiles, angular power spectrum, coherence bandwidth, path loss and shadowing effects for systems that operate at 26, 32 and 39 GHz centre frequencies with 2-GHz bandwidths. The field of channel measurements and analysis for mmWave and higher frequencies is still an active research area in which measurement campaigns still play a critical role in their development.

References

[1] Molisch AF. *Wireless Communications*. Chichester, UK: Wiley-IEEE Press, 2005.

[2] Payami S, Khalily M, Araghi A, *et al.* Developing the First mmWave Fully-Connected Hybrid Beamformer With a Large Antenna Array. *IEEE Access*. 2020; 8: 141282−141291.

[3] Zhang X, Molisch AF, and Kung S-Y. Variable-Phase-Shift-Based RF-Baseband Codesign for MIMO Antenna Selection. *IEEE Transactions on Signal Processing*. 2005; 53(11): 4091−4103.

[4] Payami S, Ghoraishi M, Dianati M, *et al.* Hybrid Beamforming With a Reduced Number of Phase Shifters for Massive MIMO Systems. *IEEE Transactions on Vehicular Technology*. 2018; 67(6): 4843−4851.

[5] Payami S, Mysore Balasubramanya N, Masouros C, *et al.* Phase Shifters Versus Switches: An Energy Efficiency Perspective on Hybrid Beamforming. *IEEE Wireless Communications Letters*. 2019; 8(1): 13−16.

[6] Jain IK, Kumar R, and Panwar SS., The Impact of Mobile Blockers on Millimeter Wave Cellular Systems. *IEEE Journal on Selected Areas in Communications*. 2019; 37(4): 854–868.

[7] Nashashibi AY, Sarabandi K, Oveisgharan S, *et al.* Millimeter-Wave Measurements of Foliage Attenuation and Ground Reflectivity of Tree Stands at Nadir Incidence. *IEEE Transactions on Antennas and Propagation*. 2004; 52(5): 1211–1222.

[8] Rappaport TS, Sun S, Mayzus R, *et al.* Millimeter Wave Mobile Communications for 5G Cellular: It Will Work! *IEEE Access*. 2013; 1: 335–349.

[9] Payami S, Sellathurai M, and Nikitopoulos K. Low-Complexity Hybrid Beamforming for Massive MIMO Systems in Frequency-Selective Channels. *IEEE Access*. 2019; 7: 36195–36206.

[10] Payami S, Khalily M, Loh TH, *et al.* Hybrid Beamforming With Switches and Phase Shifters Over Frequency-Selective Channels. *IEEE Wireless Communications Letters*. 2020; 9(8): 1305–1308.

[11] Rappaport TS. *Wireless Communications: Principles and Practice*. 2nd ed. Upper Saddle River, NJ: Prentice Hall PTR, 2001.

[12] Agiwal M, Roy A, and Saxena N. Next Generation 5G Wireless Networks: A Comprehensive Survey. *IEEE Communications Surveys Tutorials*. 2016; third quarter; 18(3): 1617–1655.

[13] Almers P, Bonek E, Burr A, *et al.* Survey of Channel and Radio Propagation Models for Wireless MIMO Systems. *EURASIP Journal on Wireless Communications and Networking*. 2007; 2007(1): 56. Available from: https://doi.org/10.1155/2007/19070.

[14] METIS. Deliverable D1.4 METIS channel models; Feb. 2015. Deliverable ICT-317669-METIS/D1.4. Available from: www.metis2020. com/documents/ deliverables.

[15] Carton I, Fan W, Kyösti P, *et al. Validation of 5G METIS map-based channel model at mmwave bands in indoor scenarios*. In: 2016 10th European Conference on Antennas and Propagation (EuCAP); 2016. p. 1–5.

[16] MiWEBA. D1.1: Definition of scenarios and use cases; Dec. 2013. Deliverable FP7-ICT 608636/D1.1. Available from: www.miweba.eu/#Publications.

[17] Weiler R, Peter M, Keusgen W, *et al.* Quasi-Deterministic Millimeter-Wave Channel Models in MiWEBA. *EURASIP Journal on Wireless Communications and Networking*. 2016; 2016.

[18] Akdeniz MR, Liu Y, Samimi MK, *et al.* Millimeter Wave Channel Modeling and Cellular Capacity Evaluation. *IEEE Journal on Selected Areas in Communications*. 2014; 32(6): 1164–1179.

[19] Sun S, MacCartney GR, and Rappaport TS., *A novel millimeter-wave channel simulator and applications for 5G wireless communications*. In: 2017 IEEE International Conference on Communications (ICC); 2017. p. 1–7.

[20] Rappaport TS, MacCartney GR, Samimi MK, *et al.* Wideband Millimeter-Wave Propagation Measurements and Channel Models for Future Wireless Communication System Design. *IEEE Transactions on Communications*. 2015; 63(9): 3029–3056.

[21] 3GPP. Study on channel model for frequencies from 0.5 to 100 GHz; Dec. 2017. 3rd Gener. Partnership Project, Tech. Rep. 38.901 V14.3.0.

[22] Jaeckel S, Raschkowski L, Börner K, *et al.* QuaDRiGa: A 3-D Multi-Cell Channel Model With Time Evolution for Enabling Virtual Field Trials. *IEEE Transactions on Antennas and Propagation*. 2014; 62(6): 3242−3256.

[23] mmMagic. 6-100 GHz channel modelling for 5G: Measurement and modelling plans in mmMAGIC; Feb. 2016. Available from: bscw.5g-mmmagic.eu/pub/bscw.cgi/d76988/mmMAGIC_WhitePaper-W2.1.pdf.

[24] Payami S, Khalily M, Taheri S, *et al.* Channel measurement and analysis for polarimetric wideband outdoor scenarios at 26 GHz: Directional vs omnidirectional. In: 2020 14th European Conference on Antennas and Propagation (EuCAP); 2020. p. 1−5.

[25] Khalily M, Taheri S, Payami S, *et al.* Indoor Wideband Directional Millimeter Wave Channel Measurements and Analysis at 26 GHz, 32 GHz, and 39 GHz. *Transactions on Emerging Telecommunications Technologies*. 2018; 29(10): e3311. Available from: https://onlinelibrary.wiley.com/doi/abs/10.1002/ett.3311.

[26] Khalily M, Ghoraishi M, Taheri S, *et al.* Polarimetric wideband directional channel measurement and analysis for outdoor small cell scenarios at 32 GHz and 39 GHz. In: 12th European Conference on Antennas and Propagation (EuCAP 2018); 2018. p. 1−5.

[27] Khalily M, Ghoraishi M, Taheri S, *et al.* Millimeter-wave directional path loss models in the 26 GHz, 32 GHz, and 39 GHz bands for small cell 5G cellular system. In: 12th European Conference on Antennas and Propagation (EuCAP 2018); 2018. p. 1−5.

[28] Khalily M, Taheri S, Xiao P, *et al.* 26 GHz indoor wideband directional channel measurement and analysis in LoS and NLoS scenarios. In: 12th European Conference on Antennas and Propagation (EuCAP 2018); 2018. p. 1−5.

[29] Ofcom. Ofcom Notes: Update on 5G spectrum in the UK; 8 Feb. 2017. Available from: https://www.ofcom.org.uk/__data/assets/pdf_file/0021/97023/5G-update-08022017.pdf.

[30] Federal Communications Commission. FCC takes steps to facilitate mobile broadband and next generation wireless technologies in spectrum above 24 GHz; Jul. 2016. Available from: https://www.fcc.gov/document/fcc-adopts-rules-facilitate-next-generation-wireless-technologies.

[31] Wang C, Bian J, Sun J, *et al.* A Survey of 5G Channel Measurements and Models. *IEEE Communications Surveys Tutorials*. 2018; 20(4): 3142−3168.

[32] Hemadeh IA, Satyanarayana K, El-Hajjar M, *et al.* Millimeter-Wave Communications: Physical Channel Models, Design Considerations, Antenna Constructions, and Link-Budget. IEEE Communications Surveys Tutorials. 2018; 20(2): 870−913.

[33] 3GPP. TR 38.900: Technical Specification Group Radio Access Network; Study on channel model for frequency spectrum above 6 GHz; Sep. 2016. 3GPP TR 38.900 version 14.1.0 release 14.

Chapter 17

Massive MIMO: real-world trials and practical solutions for 5G and beyond

Wael Boukley Hasan[1], Paul Harris[2] and Mark Beach[1]

In 2010, Thomas Marzetta published the seminal paper on massive multiple-input, multiple-output (M-MIMO) which laid the theoretical foundations for a plethora of new research in spatial domain signal processing for wireless connectivity [1]. This work and much that initially followed was predominately based upon an assumption that each of the antenna elements exposed a channel with independent and identically distributed (IID) Rayleigh fading. Whilst this provided a mathematical insight to the M-MIMO concept, many in the wireless research community were also eager to see how this would translate to real-world propagation environments using practical hardware.

This chapter describes the world's first real-time 128-antenna M-MIMO testbed implementation and how it was used to evaluate the technology in real-world trials. Subsequently, these trials revealed insights on the structure of signal to interference plus noise ratio (SINR) estimation and a mechanism of predicting error vector magnitude (EVM) for connected terminals in a multi-user configuration.

17.1 A real-time 128-antenna massive MIMO testbed

In 2014 Lund University, the University of Bristol and National Instruments joined forces to undertake a real-world evaluation of M-MIMO. Through this collaboration the world's first 128-antenna real-time M-MIMO testbed [2] was designed and deployed. The system was designed to realise M-MIMO in-line with Marzetta's original definition: time division duplex (TDD) operation, channel state information (CSI) via reciprocity and fully digital beamforming (BF). An overview of the testbed specification is provided in Table 17.1 and pictured in Figure 17.1.

The base station (BS) architecture consists of 64 dual-channel software-defined radios (SDRs)with built-in field-programmable gate arrays (FPGAs) networked across peripheral component interconnect express (PCIe) fabric to a further four MIMO processor FPGAs. For those more familiar with the typical terminology used

[1]Communication Systems and Networks (CSN), University of Bristol, Bristol, United Kingdom
[2]Vodafone, Newbury, United Kingdom

Table 17.1 Bristol testbed specification

No. of antennas	128
No. of layers	12
Carrier frequency	3.51 GHz
Bandwidth	20 MHz
Sampling rate	30.72 MS/s
Waveform	LTE OFDM
Modulation	QPSK, 16-QAM, 64-QAM, 256-QAM
Duplexing	TDD
MIMO detection/precoding	MMSE, ZF, MF (reciprocity based)

Figure 17.1 A 128-antenna M-MIMO testbed at the University of Bristol

in commercial network deployment, one can think of the SDRs as the remote radio heads (RRHs) and the central MIMO processor FPGA as the base band unit (BBU). The (UE) implementation consists of a single SDR connected to a laptop computer.

Designing and implementing a M-MIMO system of this nature brings many challenges with it, but one of the biggest is data shuffling. This arises from the need to have all of the antenna data in one place in order to perform the matrix inversions required for interference-suppressing BF techniques such as zero-forcing (ZF). A 20-MHz bandwidth sampled at 30.72 MS/s over 128 antennas results in an aggregate rate of almost 12 GB/s, which far exceeded the PCIe gen 2 interconnect capacity available on the hardware at the time. Thus, by implementing the orthogonal frequency division multiplexing (OFDM) cyclic prefix and fast Fourier transform processing within the RRHs and only transferring the used subcarriers, the aggregate rate to and from the BBU could be reduced to more manageable 6.5 GB/s. To make the BBU processing of this amount of data possible in real-time with a feasible FPGA clock rate, the processing was split by bandwidth across four FPGAs. These FPGAs each implemented matrix inversions using partially parallel QR decomposition to meet the overall processing rate of 1.4 million 128 × 12 matrices/s. In terms of precoding latency, the time taken from receiving an uplink (UL) pilot symbol to having precoded downlink (DL) samples ready for transmission was 135 μs, and over 70% of this latency is attributed to the OFDM modulation and demodulation. Regardless, this latency was sufficient to meet a channel coherence time requirement of 500 μs [3], permitting UL data to immediately follow a front-loaded pilot and the second half of the 7-symbol slot to contain at least two precoded DL symbols.

In addition to managing and processing the data, a fundamental requirement of any MIMO system is the need to maintain phase coherence and synchronously sample all antenna elements. To achieve this, a network of 8-channel clock distribution modules were used in combination with an oven-controlled crystal oscillator. Reciprocity calibration was then possible by leveraging the mutual coupling between antenna elements and applying a generalised least squares estimation algorithm [4]. Since hardware drifts with temperature and time, a periodic re-calibration is also required to maintain DL performance.

17.2 Evaluating capacity gains from massive MIMO in real-time

17.2.1 An evaluation of indoor LOS massive MIMO performance

The first measurement campaigns using the testbed introduced in Section 17.1 were conducted within the Merchant Venturers Building atrium at the University of Bristol. Initial results were published in both the 2016 IEEE Global Communications Conference [5] and the 2016 IEEE International Workshop on signal-processing systems [6], and world record figures for spectral efficiency were obtained. The latter received coverage from the IEEE Spectrum [7] and led to multiple international awards from National Instruments, Xilinx and Hewlett Packard Enterprise [8]. This section covers some of the key results and insights obtained from one of the earliest campaigns.

17.2.1.1 Experiment configuration

Figure 17.2 (left) shows the measurement environment as viewed from the BS. This figure also pictures the UE placement nearly 25 m away with all antennas vertically polarised. Twelve dual-channel SDRs operating as UEs were placed directly next to one another in a straight line with a minimum inter-element spacing of 10 cm. There was line-of-sight (LOS) between most of the BS array and the central portion of user antennas, but the staircase provides a partial LOS obstruction of the patch array for some users. When the system was configured to use 256 quadrature amplitude modulation (QAM), the inter-device spacing was increased to 50 cm and the antenna polarisation was also alternated for additional diversity. Measurements took place within normal university hours and there were moderate dynamics in the environment due to students walking around.

17.2.1.2 Channel analysis

The results presented here were obtained using UL channel estimates that were simultaneously captured across all 128 elements for 24 single-antenna UEs. Whilst

Figure 17.2 Indoor measurement trial with the UE 24.8 m away (Figure 1 in [9])

the system was only capable of 12 spatial layers of data throughput in real-time, it was still possible to capture UL pilots for a larger number of UEs by reducing the subcarrier density (still remaining well within the coherence bandwidth expected for the scenario). For all the analysis that follows, K refers to the number of single-antenna UEs and M refers to the number of BS antennas.

In M-MIMO, how the pairwise orthogonality increases with the number of BS antennas is of particular interest as this effectively determines how much spatial isolation can be achieved prior to an interference-suppressing BF technique such as ZF. If the orthogonality is great enough to achieve an acceptable SINR for all UEs, then simple matched filtering (MF) BF could be used, negating the need for matrix inversions and some of the data aggregation challenges described in Section 17.1. Pairwise orthogonality can be observed in the user-side Gram matrix [10], which is calculated for a single frequency domain resource block (RB) and time instance as

$$J = HH^H \tag{17.1}$$

where H is the $K \times M$ frequency-domain channel matrix in the UL for one 24-subcarrier RB. This calculation was repeated for all RBs and over multiple OFDM symbols to form an average, and the value of each complex element in the resulting Gram matrix is taken as its absolute $|j_{i,j}|$ for all i and j. The diagonal elements of the Gram matrix represent the UE channel magnitudes and the off-diagonals for a given UE represent the level of correlation with other UEs. By looking at the off-diagonals relative to the diagonals, an approximation to the SINR can be obtained.

Figure 17.3 shows the Gramian intensity plot results for 32, 64 and 128 elements of the 4×32 patch array normalised to 0 dB on the diagonal. For the 32-antenna case, the results were obtained using the second 4×8 panel from the left when facing the array, and for 64 antennas, the two centre panels were used. By increasing the number of antennas at the BS, the off-diagonal elements of the Gram matrix become increasingly weaker compared to the diagonal elements. This phenomenon is called channel hardening and becomes more prominent as the number

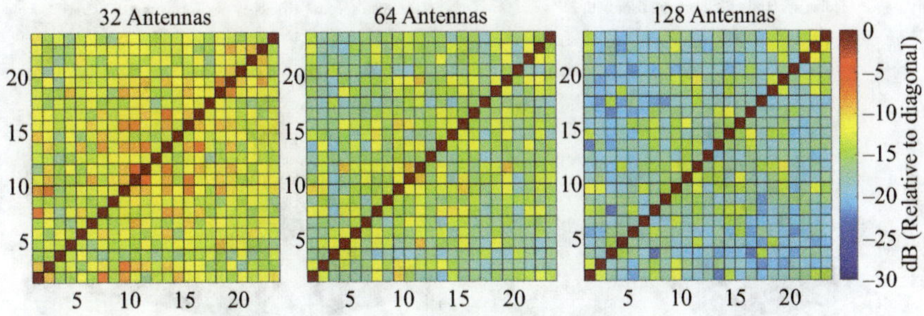

Figure 17.3 User-side Gramian intensity plots for 24 UEs

of antennas is increased [10]. The 128-antenna results show that an SINR of 15 dB and beyond can be achieved for nearly all user pairs, with the most highly correlated neighbouring antenna pairs still experiencing an SINR of over 10 dB. On the other hand, for 32 antennas, it can be seen that the SINR is in the 5–7 dB range for quite a number of UEs, clearly demonstrating a significant growth in pairwise orthogonality as M increases. As the UEs were only spaced 10–50 cm apart and few of them had clear LOS to all of the BS array elements, multipath scattering within the atrium will have likely made a greater contribution to the overall level of spatial isolation.

Inspecting the channel-vector orthogonality between UEs provides insight into the fundamental impact of increasing M, but it is also helpful to understand how this translates into the achievable spectrum efficiency (SE). The aggregate achievable UL rate for all UEs for a given frequency-domain RB b, time instance t and noise N can be found with

$$C_{\Delta}(b,t) = \sum_{i=1}^{K} \log_2 \left(1 + \frac{\left| h_{i,b}(t)\ g_{i,b}^{T}(t) \right|^2}{\sum_{j \neq i} \left| h_{i,b}(t)\ g_{j,b}^{T}(t) \right|^2 + N} \right) \tag{17.2}$$

where $h_{i,b}$ and $g_{i,b}$ represent the channel estimate and linear decoding vectors, respectively, for user i and frequency RB b. $g_{i,b}$ is derived using either ZF or MF. For ZF, $G_b = \left(H_b H_b^H\right)^{-1} H_b^H$, and for MF, $G_b = M^{-1} H_b^H$, where the latter division by the antenna dimension ensures that the decoded MF power for a single user (no interference) is equal to 1 and matches that of ZF. N is then chosen to provide the desired average signal-to-noise ratio (SNR) to all users relative to 1. To avoid noise power correlation, the decoding matrix G_b is derived using the CSI estimate obtained one capture prior to the CSI estimate it operates on, and the period between UL pilot OFDM symbol captures was 200 ms. Figure 17.4 shows the median achievable per user UL SE results for 128 BS antennas, up to 24 active users and an SNR range of 0–50 dB. The achievable sum SE is also shown in Figure 17.5. For both of these results, users were introduced in order of their average contribution to inter-user

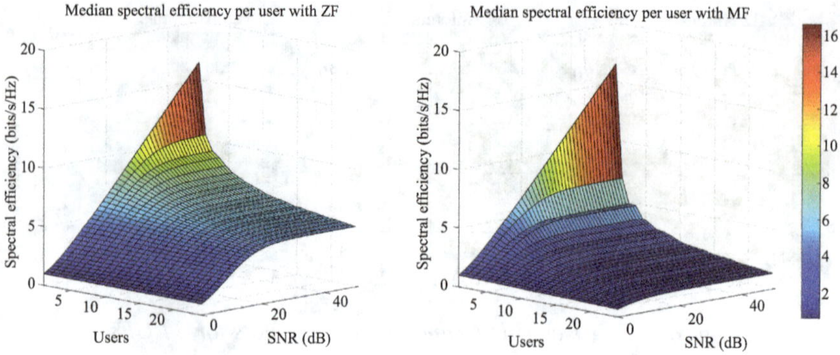

Figure 17.4 Median per-user spectral efficiency with 128 BS antennas. Active users introduced in order of their average contribution to IUI

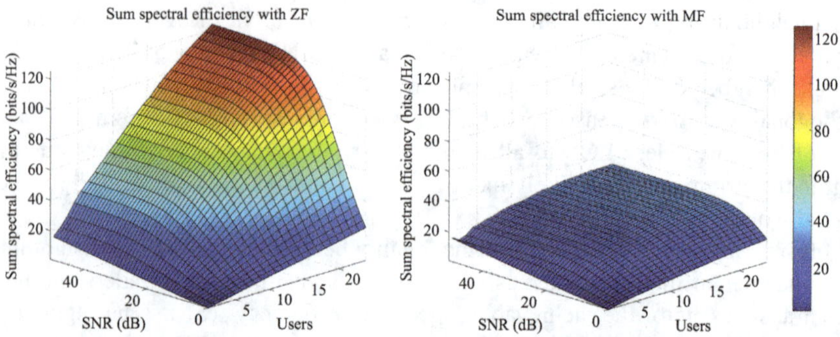

Figure 17.5 Median sum spectral efficiency with 128 BS antennas for 1–24 active users. Active users introduced in order of their average contribution to IUI

interference (IUI). When applying the MF decoder, it can be seen that the achievable per-user SE is halved after the second UE is introduced, with the median per-user SE dropping below 5 bits/s/Hz once five UEs are active. For ZF decoding, the initial drop in median UL user SE is not as sharp and a more graceful decline is seen thereafter till it plateaus at 5 bits/s/Hz. The sum SE when using ZF also grows linearly to exceed 120 bits/s/Hz at 24 users with an SNR of 20 dB. These results indicate that an interference-cancelling decoder such as ZF will still be required to obtain high UE data rates with a practical number of BS antennas, but MF could be well suited to a scenario where many low data-rate devices need to be scheduled together on the same time–frequency resource.

17.2.1.3 Real-time performance

It was possible to connect up to 22 users to the BS with 256-QAM modulation and obtain the decoded UL constellations shown in Figure 17.6. Each constellation

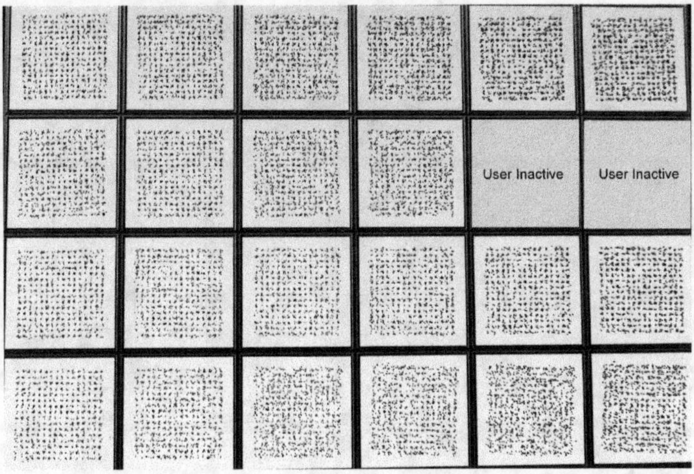

Figure 17.6 256-QAM UL constellations from 22 UEs. BS configured with 128 antennas (Figure 2 in [9]).

represents one single-antenna UE transmitting on the same time–frequency resource. As the FPGA implementation was not capable of processing more than 12 spatial streams in real-time, one 10 ms frame was retrieved from the FPGA every 100 ms and processed using 128-bit complex floating point processing to provide a quasi-real-time result of the decoded constellations. Coupled with the captured channel estimates, it was possible to deduce the overall system performance for this scenario. Whilst it was not possible to measure the real-time throughput at the BS for this number of users, the captured channel was used in an offline emulation that performs UL 256-QAM 128 × 22 MIMO transmission using ZF decoding with hard decisions. In order to consider the effect of CSI estimation error in the emulation, system noise and CSI estimate variance were both measured under lab conditions with a cabled channel. By isolating the contribution of system noise, the magnitude of additional contributions to the CSI estimate variance could then be found. Within the emulation, a normally distributed error is added to the channel matrix used to form the ZF decoder with variance equal to this additional magnitude. Figure 17.8 shows the error rate (BER) result across a 0–40 dB average SNR range. At 35 dB average SNR, an uncoded BER of 3×10^{-4} is achieved which equates to a UL sum rate of over 2.9 Gbps and a SE of 145.6 bits/s/Hz. This result indicates that the spatial separation achievable with 128 antennas is great enough to allow a ZF decoder to obtain a mean 256-QAM BER performance across 22 UEs that is only 2–3 dB worse than an additive white Gaussian noise channel. This is a record-reported result for spectral efficiency [11] and truly demonstrates the tremendous potential of M-MIMO technology for improving system throughput without bandwidth expansion [12].

Connecting a 23rd user caused significant degradation in the constellation quality and the physical location of this additional user within the line of 24 UEs (see Figure 17.7) had no impact. This is because all users were clustered as shown

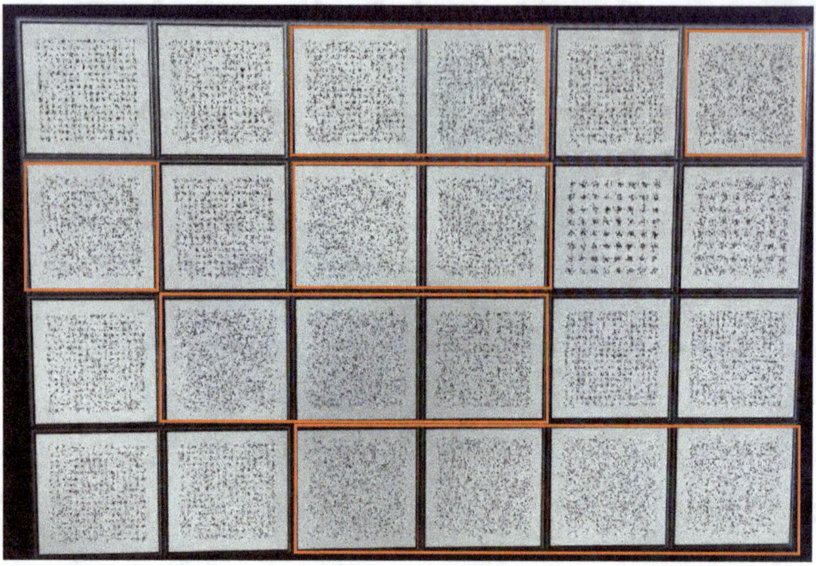

Figure 17.7 256-QAM UL constellations from 24 UEs with 50 cm client separation. High EVM values can be observed from the constellations in the orange collared rectangular compared to the one in Figure 17.6.

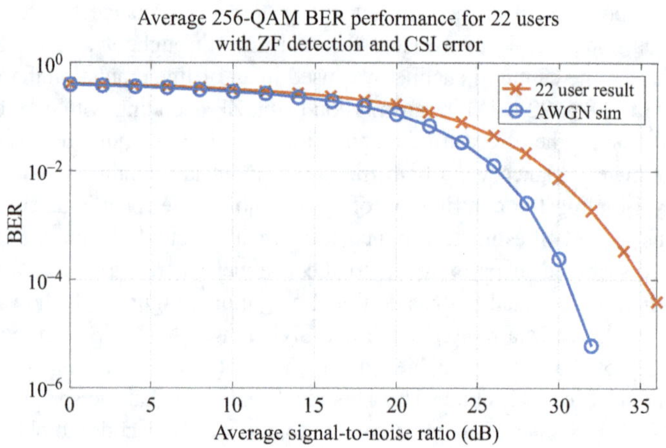

Figure 17.8 Average 256-QAM UL BER performance

in Figure 17.2 and have the same SNR value. Therefore, it is believed that this could be related to the CSI accuracy introduced in [13]. Further analysis through trials and simulations is now provided in this chapter to investigate the impact of inaccurate CSI and the number of UEs on M-MIMO performance.

17.3 Impact of inaccurate CSI in real-time

17.3.1 System model

A single-cell M-MIMO architecture with imperfect CSI is considered in this section. Each UE transmits a frequency-orthogonal UL pilot for the channel estimation. The actual UL channel matrix during the uplink data transmission is denoted by $H_{ul} \in \mathbb{C}^{M\times K}$, which is given by

$$H_{ul} = \widehat{H}_{ul} + E_{ul} \tag{17.3}$$

where $E_{ul} \in \mathbb{C}^{M\times K}$ is the difference between the estimated channel and the actual channel during the uplink data transmission. This error could be caused by hardware impairments, interpolation across frequency, large-scale channel attenuation and any other potential sources of interference. In this chapter, the error in the UL channel estimation between the UE and the BS is modelled as a complex Gaussian distribution $\sim \mathscr{CN}(0, \sigma_e^2 I_M)$, where σ_e^2 is the error variance [14]. Whilst this model may not be so accurate in reality, it serves to illustrate the potential effects of UL CSI inaccuracies in M-MIMO. The equalised UL signal $\widehat{x} \in \mathbb{C}^{\mathbb{K}}$ can be expressed as

$$\widehat{x} = W\left(\sqrt{\rho_{ul}} H_{ul} x + n_{BS}\right) \tag{17.4}$$

where ρ_{ul} is the corresponding UL transmit power. Simple UL power control is assumed in this chapter. The UL transmit power is adjusted so the received SNR from all users is the same. The additive noise vector in UL is denoted by n_{BS}. The noise variance from the antennas at the BS is modelled as $\sim \mathscr{CN}(0, \sigma_{n_{BS}}^2 I_M)$, where $\sigma_{n_{BS}}^2$ is the noise variance.

While it is generally agreed that the propagation channel is reciprocal in time [15], the transceiver radio frequency chains at both ends of the link are generally not [16]. Therefore, the actual DL channel matrix during the DL data transmission is denoted by $H_{dl} \in \mathbb{C}^{K\times M}$. The estimated DL channel matrix $H_{dl} \in \mathbb{C}^{K\times M}$ and how it is affected by H_{ul} are covered in Section 17.3.3. The received downlink signal $\widehat{\mathbf{s}} \in \mathbb{C}^k$ can be written as

$$\widehat{s} = \sqrt{\rho_{dl}} H_{dl} P s + n_{UE} \tag{17.5}$$

where s represents the transmitted symbol vector from the BS to users in the same cell, normalised as $E\left\{|s_k|^2\right\} = 1$. The corresponding DL transmit power is denoted by ρ_{dl} . $P \in \mathbb{C}^{M\times K}$ is the linear precoder matrix, formed by using minimum mean square error (MMSE), ZF or MF. n_{UE} is the additive noise vector in DL. The noise variance from the users' antenna is modelled as $\sim \mathscr{CN}(0, \sigma_{n_{UE}}^2 I_K)$, where $\sigma_{n_{UE}}^2$ is the noise variance.

17.3.2 Accumulation of error amplification in UL

With imperfect channel estimation, the error in (17.3) will always be added to the estimated channel. From (17.4), the equalised signal vector can be written as follows:

$$\widehat{x} = \sqrt{\rho_{ul}} W E_{ul} x + \sqrt{\rho_{ul}} W \widehat{H}_{ul} x + Wn \tag{17.6}$$

A number of users affect the actual equalised signal for the target user k. This can be shown by rearranging (17.6) as follows:

$$\widehat{x}_k = \underbrace{\sqrt{\rho_{ul_k}} \sum_{i=1}^{K} (W E_{ul})_{k,i} x_i}_{\text{cumulative error}} + \overbrace{\sqrt{\rho_{ul_k}} \sum_{i=1}^{K} \left(W \widehat{H}_{ul} \right)_{k,i} x_i}^{\text{desired signal+interference}} + z_k \tag{17.7}$$

where z_k is the amplified noise for user k caused by the decoder. The 'cumulative error' part represents the interference introduced by CSI estimation inaccuracies. The 'desired signal + interference' part consists of the UL transmitted symbol from user k and the interference caused by the inter-user spatial correlation.

17.3.2.1 Match filter

To understand the impact of inaccurate CSI on the MF decoder $\left(W = \widehat{H}_{ul}^{H} \right)$, the equalised received signal from user k can be rewritten as follows:

$$\widehat{x}_k = \underbrace{\sqrt{\rho_{ul}} \sum_{i=1}^{K} \left(\widehat{H}_{ul}^{H} E_{ul} \right)_{k,i} x_i}_{\text{cumulative error}} + \overbrace{\sqrt{\rho_{ul}} \sum_{i=1}^{K} \left(H_{ul}^{H} \widehat{H}_{ul} \right)_{k,i} x_i}^{\text{desired signal+interference}} + z_k \tag{17.8}$$

By increasing the number of antennas at the BS, the ratio between diagonal elements and non-diagonal elements of the Gram matrix $G = \widehat{H}_{ul}^{H} \widehat{H}_{ul}$ will also increase. This is known as the channel-hardening effect as previously explained. The 'desired signal + interference' part is extremely sensitive upon the number of antennas at the BS. The 'cumulative error' part is less impacted by the number of antennas since the matrix elements resulting from the matrix multiplication of $\left(\widehat{H}_{ul}^{H} E_{ul} \right)$ are far smaller than the channel Gram matrix when UL power control is applied. The interference from both parts is increased by increasing the number of users.

17.3.2.2 Zero forcing

To address the impact of inaccurate CSI on the ZF decoder $\left(W = \left(\widehat{H}_{ul}^{H} \widehat{H}_{ul} \right)^{-1} \widehat{H}_{ul}^{H} \right)$, the equalised received signal from user k can be rewritten as follows:

$$\widehat{x}_k = \underbrace{\sqrt{\rho_{ul_k}} \sum_{i=1}^{K} \left(\left(\left(\widehat{H}_{ul}^{H} \widehat{H}_{ul} \right)^{-1} \widehat{H}_{ul}^{H} E_{ul} \right)_{k,i} x_i \right)}_{\text{cumulative error}} + \sqrt{\rho_{ul_k}} x_i + z_k \tag{17.9}$$

Unlike MF, the interference is only caused from the 'cumulative error' part which is introduced by CSI estimation inaccuracies. This interference is increased by increasing number of users. Similar to MF, the channel Gram matrix impacts the interference value. Although ZF suppresses the interference from the estimated channel between users, it amplifies the interference introduced by the inaccurate CSI. This interference is caused by the inverse of the channel Gram matrix in the 'cumulative error' part. Although the value of this part in ZF is larger than the one in MF, the overall interference in MF is still higher than ZF.

17.3.2.3 Minimum mean square error

To cover the impact of inaccurate CSI on the MMSE decoder $\left(\left(\widehat{H}_{ul}^{H}\widehat{H}_{ul}+\frac{1}{\gamma}\mathrm{I}\right)^{-1}\widehat{H}_{ul}^{H}\right)$, the equalised received signal from user k can be rewritten as follows:

$$\widehat{x}_k = \overbrace{\sqrt{\rho_{ul_k}}\sum_{i=1}^{K}\left(\left(\left(\widehat{H}_{ul}^{H}\widehat{H}_{ul}+\frac{1}{\gamma}I\right)^{-1}\widehat{H}_{ul}^{H}E_{ul}\right)_{k,i}x_i\right)}^{\text{cumulative error}}$$
$$\underbrace{+\sqrt{\rho_{ul}}\sum_{i=1}^{K}\left(\left(\widehat{H}_{ul}^{H}\widehat{H}_{ul}+\frac{1}{\gamma}I\right)^{-1}\widehat{H}_{ul}^{H}\widehat{H}_{ul}\right)_{k,i}x_i}_{\text{desired signal+interference}}+z_k$$

The interference and the cumulative error parts are highly affected by the SNR value and the number of users. With a low SNR value, MMSE is less affected by the CSI accuracy as it operates similar to MF, while with a high SNR value, MMSE becomes more sensitive to CSI accuracy as it operates similar to ZF.

17.3.3 Imperfect channel reciprocity and accumulation of error amplification in DL

Channel reciprocity is essential for DL data transmission in a M-MIMO TDD system. Not only propagation conditions determine the radio channel, but also the transceiver front-ends at both sides of the radio link. Since different transceiver chains are used at the BS and the UE, the actual downlink channel $h_{dl} \in \mathbb{C}^{1\times \mathrm{M}}$ between the BS and one user can be written as follows:

$$h_{dl} = h_{ul}^{T}D_b \tag{17.10}$$

where $D_b = diag(cb_1,\ldots,cb_M)$ is the calibration matrix and b_i is the calibration coefficient [17]. The estimated calibrated version of the downlink channel $\widehat{h}_{dl} \in \mathbb{C}^{1\times M}$ between one user and the BS can be written as follows:

$$\widehat{h}_{dl} = \widehat{h}_{ul}^{T}\widehat{D}_b \tag{17.11}$$

where $\widehat{D}_b = diag\left(\widehat{b}_1, \ldots, \widehat{b}_M\right)$ is the estimated calibration matrix. $\widehat{b}_i = c(b_i + e_i)$ is the estimated calibration coefficient where e_i is an IID random process representing the calibration error and c is the unknown common scaling factor. From equations (17.3), (17.10) and (17.11), the actual downlink channel h_{dl} between the BS and one user can be written as follows:

$$
\begin{aligned}
h_{dl} - \widehat{h}_{dl} &= h_{ul}^T D_b - \widehat{h}_{ul}^T \widehat{D}_b \\
h_{dl} &= \widehat{h}_{dl} + \left(\widehat{h}_{ul}^T + e_{ul}^T\right) D_b - \widehat{h}_{ul}^T \widehat{D}_b \\
h_{dl} &= \widehat{h}_{dl} + \widehat{h}_{ul}^T \left(D_b - \widehat{D}_b\right) + e_{ul}^T D_b \\
h_{dl} &= \widehat{h}_{dl} + \widehat{h}_{ul}^T E_b + e_{ul}^T D_b \\
h_{dl} &= \widehat{h}_{dl} + e_{dl}
\end{aligned}
\tag{17.12}
$$

where

$$
e_{dl} = \widehat{h}_{ul}^T E_b + e_{ul}^T D_b \tag{17.13}
$$

$E_b = diag(-ce_1, \ldots, -ce_M)$ is the calibration error matrix. From (17.12), the downlink channel matrix $H_{dl} \in \mathbb{C}^{K \times M}$ between the BS and the UEs can be decomposed as follows:

$$
H_{dl} = \widehat{H}_{dl} + E_{dl} \tag{17.14}
$$

where $\widehat{H}_{dl} \in \mathbb{C}^{K \times M}$ is the estimated calibrated version of the downlink channel matrix between the UEs and the BS, while $E_{dl} \in \mathbb{C}^{K \times M}$ is the difference between the estimated calibrated channel and the actual channel during the downlink transmission.

The UL CSI accuracy has greater impact on increasing the difference between the estimated calibrated channel and the actual channel during the downlink transmission than it has on increasing the difference between the estimated channel and the actual channel during the uplink data transmission. This can be seen in (17.13) where the error introduced by inaccurate UL CSI (e_{ul}) is multiplied by the calibration matrix (D_b). That means the difference between the estimated calibrated channel and the actual channel during the downlink transmission could vary with different M-MIMO hardware using the same reciprocity calibration algorithm and the same radio prorogation environment due to varying the hardware impairment. This was experimentally validated in Section 17.3.4. The hardware impairment error value in DL data transmission is greater than that in UL data transmission. This can be seen in (17.13) where the calibrated DL CSI error is composed from multiplying the UL CSI error to the calibration matrix then adding the multiplication result of the estimated UL CSI to the calibration error matrix. The received signal in DL can be written from (17.5) and (17.14) as follows:

$$
\widehat{s} = \sqrt{\rho_{dl}} E_{dl} P s + \sqrt{\rho_{dl}} \widehat{H}_{dl} P s + n_{UE} \tag{17.15}
$$

By rearranging the previous equation, the received signal for user k can be written as follows:

$$\widehat{s}_k = \underbrace{\sqrt{\rho_{dl_k}}\sum_{i=1}^{K}(E_{dl}P)_{k,i}s_i}_{\text{cumulative error}} + \overbrace{\sqrt{\rho_{dl_k}}\sum_{i=1}^{K}\left(\widehat{H}_{dl}P\right)_{k,i}s_i}^{\text{desired signal+interference}} + n_{UE_k} \tag{17.16}$$

where $$P = \begin{cases} \widehat{H}_{dl} & \text{MF} \\ \left(\widehat{H}_{dl}\widehat{H}_{dl}^{H}\right)^{-1}\widehat{H}_{dl} & \text{ZF} \\ \left(\widehat{H}_{dl}\widehat{H}_{dl}^{H} + \frac{1}{\gamma}I\right)^{-1}\widehat{H}_{dl} & \text{MMSE} \end{cases}$$

Since $E_{dl} > E_{ul}$, the interference caused by the 'cumulative error' part in DL data transmission from (17.15) is greater than the one caused in UL data transmission from (17.6). Therefore, adding more users increases the interference between users in DL data transmission more than for UL data transmission. So the sum SE achieved in UL data transmission is greater than that in DL data transmission. This was experimentally validated in Section 17.3.4.

17.3.4 Real-time evaluation using massive MIMO testbed

In this section, real-time results are presented using the M-MIMO testbed in Section 17.1. These results demonstrate the impact of inaccurate CSI on the SE and show that increasing number of antennas at the BS can decrease this impact. MF, ZF and MMSE were used for UL and DL data transmission by using the exact same static environment but with two different UL CSI accuracy values.

17.3.4.1 Low CSI accuracy

For low CSI accuracy, over-the-air synchronisation was used based on Zadoff–Chu sequence. The phase of the sync signal was generated randomly from each antenna at the BS which created nulls in different locations. This caused sync losses for the users based on their locations and the sync signal phases from the antennas at the BS. The synchronisation error causes inter-carrier interference (ICI), inter-symbol interference (ISI) and interference between data streams due to the inaccurate CSI. The BS and the users save the sequence number of each transmitted radio frame. When the user does not detect the sync signal, it saves the radio frame number. The sum SE calculation does not consider the data transmission between the user who lost the sync signal and the BS. This can be done by removing the data transmitted during the saved radio frame number in both UL and DL for that user. By doing that, the impact of ICI and ISI was removed before calculating the sum SE while the interference caused by the inaccurate CSI is only considered in this scenario.

17.3.4.2 Measurement environment

The Communication Systems & Networks (CSN) lab was used for LOS measurements between the BS and 12 UEs from 6 universal software radio peripherals.

Figure 17.9 Measurement environment in the CSN Research Laboratory (Figure 14 in [9])

Measurements were performed outside of university hours and the M-MIMO testbed was controlled remotely to ensure that the same static environment was applied in all scenarios. An overview of the setup can be seen in Figure 17.9. At the BS side, a 64-element array was used providing half-wavelength spacing at 3.5 GHz. A floor plan of the laboratory is shown in Figure 17.10 with the UE locations.

17.3.4.3 Validation results

Tables 17.2 and 17.3 show the maximum uncoded sum SE results for the UL scenario with high and low CSI accuracy, respectively. Maximum sum SE of 56.59 bits/s/Hz is achieved by using ZF or MMSE with $M = 64$ and high CSI accuracy serving all the 12 UEs with 64-QAM. By reducing the number of antennas to 32, the maximum sum SE is decreased by 6.39 bits/s/Hz due to the higher interference level although the same number of UEs and modulation and coding schemes (MCS) are used. The interference caused by 'cumulative error' part in (17.7) when $M = 64$ is less than the one when $M = 32$ due to the channel hardening effect. With low CSI accuracy, the maximum sum SE with ZF or MMSE is decreased by 7.49 bits/s/Hz and 13.2 bits/s/Hz for $M = 64$ and $M = 32$, respectively. Decreasing the CSI accuracy has greater impact when $M = 32$ with ZF and MMSE. These results validate the theoretical work on M-MIMO showing that by increasing number of antennas at the BS, the interference caused by inaccurate CSI with ZF and MMSE will decrease. With MF, the CSI accuracy has no impact on the maximum sum SE. Only a number of antennas affect the maximum sum SE achieved by MF in the UL scenario. This can be explained from (17.7) where the interference caused by the 'desired signal + interference' part, which is sensitive upon the number of antennas, is very large compared to the one caused by the 'cumulative error' part which is affected by the CSI accuracy.

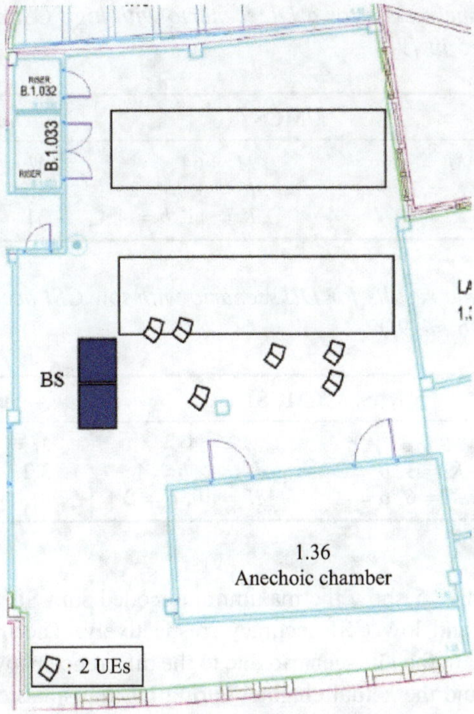

Figure 17.10 CSN Lab floor plan showing the BS and UEs locations (Figure 15 in [9])

Table 17.2 Real-time results for UL scenario with high CSI accuracy (Table 3 in [9])

	***K*/MCS**		**Sum SE (bit/s/Hz)**	
	$M = 32$	$M = 64$	$M = 32$	$M = 64$
MF	$K = 3, b = 2$	$K = 2, b = 4$	4.4	6.2
ZF and MMSE	$K = 12, b = 6$	$K = 12, b = 6$	50.2	56.59

Table 17.3 Real-time results for a UL scenario with low CSI accuracy (Table 4 in [9])

	***K*/MCS**		**Sum SE (bit/s/Hz)**	
	$M = 32$	$M = 64$	$M = 32$	$M = 64$
MF	$K = 3, b = 2$	$K = 2, b = 4$	4.4	6.2
ZF and MMSE	$K = 9, b = 6$	$K = 11, b = 6$	37	49.1

Table 17.4 Real-time results for a DL scenario with high CSI accuracy (Table 5 in [9])

	K/MCS		**Sum SE (bit/s/Hz)**	
	$M = 32$	$M = 64$	$M = 32$	$M = 64$
MF	$K = 3, b = 2$	$K = 3, b = 2$	3.8	4.5
ZF and MMSE	$K = 10, b = 4$	$K = 11, b = 4$	31	34.4

Table 17.5 Real-time results for DL scenario with low CSI accuracy (Table 6 in [9])

	K/MCS		**Sum SE (bit/s/Hz)**	
	$M = 32$	$M = 64$	$M = 32$	$M = 64$
MF	$K = 3, b = 2$	$K = 2, b = 4$	3.2	4.2
ZF and MMSE	$K = 6, b = 2$	$K = 10, b = 2$	8	15

Tables 17.4 and 17.5 show the maximum uncoded sum SE results for the DL scenario with high and low CSI accuracy, respectively. The maximum sum SE results are decreased in the DL scenario due to the difference between the estimated calibrated channel and the actual channel during DL transmission. With high CSI accuracy and ZF or MMSE precoders, the maximum sum SE is decreased by 22.19 and 19.2 bits/s/Hz with $M = 64$ and $M = 32$, respectively, compared to the UL scenario with high CSI accuracy. While by applying the same comparison for the low CSI accuracy, the maximum sum SE is decreased by 34.1 and 29 bits/s/Hz with $M = 64$ and $M = 32$, respectively. By comparing the SE degradation between UL and DL scenarios with high and low CSI accuracy, the maximum sum SE with low CSI accuracy was shown to be affected more. The additional maximum sum SE degradation due to low CSI accuracy was 11.91 and 9.81 bits/s/Hz with $M = 64$ and $M = 32$, respectively, despite using the same reciprocity calibration algorithm. This can be explained in (17.13) where the error introduced by inaccurate UL CSI (e_{ul}) is multiplied by the calibration matrix (D_b). These results validate that UL CSI accuracy has greater impact on the DL data transmission than it has on the UL data transmission.

17.4 SINR estimation and EVM prediction

A common measurement of signal quality used in Third-Generation Partnership Project (3GPP) long-term evolution (LTE) standards is the EVM [17]. This is a comprehensive metric because it considers all impairments upon the transmitted symbol as seen at the receiver. Therefore, using the EVM metric covers the interference caused by inaccurate CSI. The work in [18,19] introduce a relationship between EVM and SNR performance metrics. This relationship holds for a single-input–single-output

(SISO) link or a single-user (SU)-MIMO system. The interference between users in multi-user (MU)-MIMO and M-MIMO affects the EVM performance but will not affect the SNR value. The work in [20] replaced the SNR by the SINR to introduce a relationship between SINR and EVM for M-MIMO with a ZF decoder. This relationship could be used in real-time to estimate the SINR value from the EVM performance. Measuring the EVM value after removing any user requires all the remaining users to retransmit data symbols. Hence relying on the measured EVM to estimate the SINR for a subset of the available users is not applicable. Many user grouping and power control algorithms in M-MIMO are based on selecting a subset of the available users and maximising the SINR value. These algorithms cannot be applied in real M-MIMO system just by relying on the EVM performance.

This section introduces an estimation method with high accuracy associating EVM to SINR values for M-MIMO with ZF and MMSE. It also introduces a novel EVM prediction method for a subset of users taken from the original set of simultaneous users in a single cell M-MIMO. Two different scenarios were considered to evaluate the accuracy of the SINR estimation and the EVM prediction. An IID Rayleigh channel was used in the first scenario with perfect and imperfect CSI, where the number of antennas (M)=128 and $K \in \{2, 12, 22\}$. In the second scenario, the real channels captured from the trial described in Section 17.2.1.1 were used. These scenarios were run through an uplink M-MIMO simulator developed at the University of Bristol. In addition to randomly generated channels, new vectors can be transmitted through channels previously captured by the physical system for more extensive analysis. For imperfect CSI in both scenarios, the error in the channel estimation between the UEs and the BS is modelled as a complex Gaussian distribution $\sim \mathcal{CN}(0, \sigma_e^2 I_M)$ [14] where an error variance of 0.01 (1%) was used for the simulations shown. Whilst this value may be higher or lower and the model may not be so accurate in reality, it serves to illustrate the potential effects of CSI inaccuracies in M-MIMO. Real-world experimental datasets with real-time results are carried out to validate the EVM prediction method using SDR M-MIMO testbed.

From [18,21], the EVM can be given as

$$\mathrm{EVM}_{RMS} = \sqrt{\frac{\frac{1}{N}\sum_{n=1}^{N}|S_r(n) - S_t(n)|^2}{\frac{1}{N}\sum_{n=1}^{N}|S_t(n)|^2}} \tag{17.17}$$

where N is the number of symbols the EVM was measured over. $S_r(n)$ is the nth normalised received symbol and $S_t(n)$ is the ideal value of the nth symbol. The relation between the EVM and the SNR in a single cell was addressed in many studies such as [18,19], which was estimated as follows:

$$\mathrm{SNR} \approx \frac{1}{\mathrm{EVM}_{\mathrm{RMS}}^2} \tag{17.18}$$

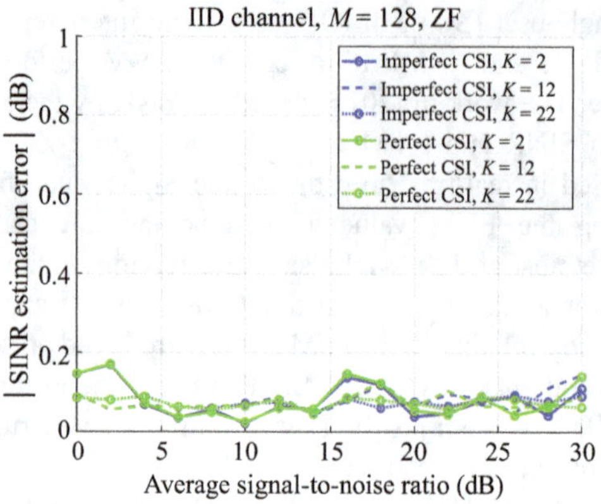

Figure 17.11 SINR estimation error using ZF (IID Rayleigh channels). The SINR estimation values are obtained from (17.19) (Figure 1 in [23]).

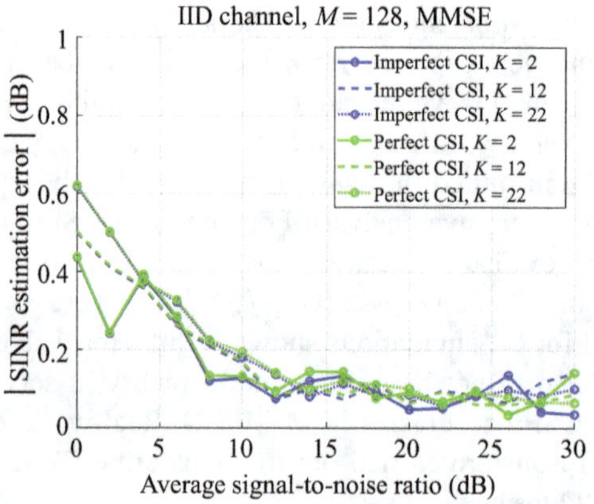

Figure 17.12 SINR estimation error using MMSE (IID Rayleigh channels). The SINR estimation values are obtained from (17.19) (Figure 2 in [23]).

The previous approximation can be used in SISO and SU-MIMO. While in MU-MIMO and M-MIMO, the interference between users highly impacts the accuracy of this approximation. In order to increase the estimation accuracy, the SINR can be used instead of the SNR as follows:

$$\mathrm{SINR} \approx \frac{1}{\mathrm{EVM}_{\mathrm{RMS}}^2} \tag{17.19}$$

The accuracy of the previous equation can be seen in Figures 17.11 and 17.12 for the first scenario and Figures 17.13 and 17.14 for the second scenario. The same approximation method was also used in [20]. The UL SINR in (17.6) can be

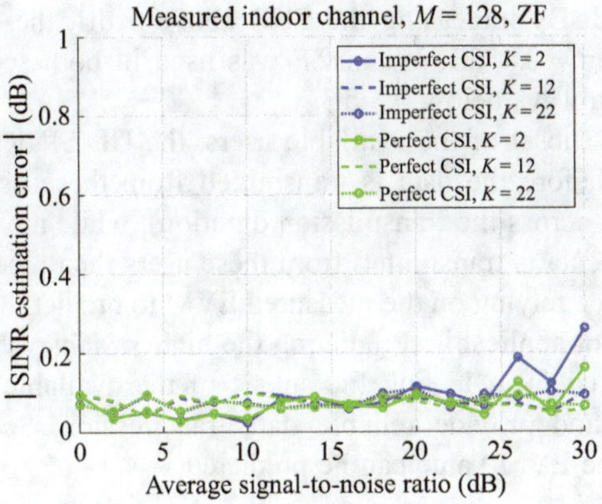

Figure 17.13 SINR estimation error using ZF (measured indoor channels). The SINR estimation values are obtained from (17.19) (Figure 3 in [23]).

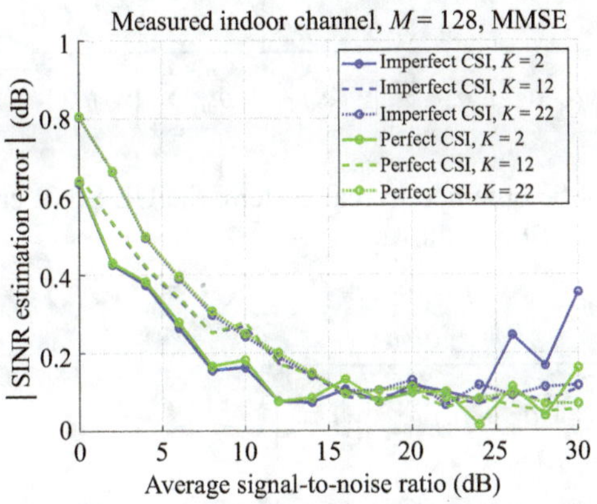

Figure 17.14 SINR estimation error using MMSE (measured indoor channels). The SINR estimation values are obtained from (17.19) (Figure 4 in [23]).

calculated as follows [22]:

$$\mathrm{SINR}_{ul_k} = \frac{\rho_{ul_k} |w_k h_{ul_k}|^2}{\|w_k\|^2 \sigma_n^2 + \rho_{ul_k} \sum_{i \neq k}^{K} |w_k h_{ul_i}|^2} \tag{17.20}$$

The errors in the SINR estimation based on the EVM value from (17.19) were plotted in Figures 17.11 and 17.13 using ZF while MMSE was used in Figures 17.12 and 17.14. When MMSE was used, the SINR estimation error is always below 1 dB. By increasing the SNR above 5 dB, the error in the SINR estimation falls below 0.4 dB. When ZF was used in both scenarios, the SINR estimation error is always below 0.3 dB.

By selecting a subset of the available users, the UL SINR can be calculated based on the CSI before the data is transmitted from these users (assuming the channel is coherent across the transmission duration), while the EVM can only be measured after the data is transmitted from these users then received and decoded at the BS side. Hence relying on the measured EVM to predict the MCS order for a subset of users is not applicable. Predicting the highest achievable MCS order for each user is essential for selecting the subset of the available users in order to maximise the SE and provide reliable data transmission. Sections 17.4.1 and 17.4.2 show how the EVM value can be predicted.

17.4.1 EVM prediction based on SINR calculation

The UL SINR in (17.20) is calculated based on perfect CSI. Whilst with imperfect CSI, the equation can be modified as follows:

$$\mathrm{SINR}_{ul_k} = \frac{\rho_{ul_k}\left|w_k\widehat{h}_{ul_k}\right|^2}{\|w_k\|^2\sigma_n^2 + \rho_{ul_k}\sum_{i=1}^{K}|w_k e_{ul_i}|^2 + \rho_{ul_k}\sum_{i\neq k}^{K}\left|w_k\widehat{h}_{ul_i}\right|^2} \tag{17.21}$$

Since e_{ul} cannot be obtained in real system, the UL SINR can be estimated by ignoring the error in CSI as follows:

$$\widehat{\mathrm{SINR}}_{ul_{\tilde{k}}} = \frac{\rho_{ul_{\tilde{k}}}\left|w_{\tilde{k}}\widehat{h}_{ul_{\tilde{k}}}\right|^2}{\|w_{\tilde{k}}\|^2\sigma_n^2 + \rho_{ul_{\tilde{k}}}\sum_{i\neq\tilde{k}}^{\tilde{K}}\left|w_{\tilde{k}}\widehat{h}_{ul_i}\right|^2} \tag{17.22}$$

where $\tilde{K}$ is the total number of users in the set that the SINR_{ul_k} is being predicted for $\left(\tilde{K} \leq K\right)$. From (17.19), the EVM value for user k can be predicted as follows:

$$\widetilde{\widetilde{\mathrm{EVM}}}^{\,2}_{RMS_{\tilde{k}}} = \frac{1}{\widehat{\mathrm{SINR}}_{ul_{\tilde{k}}}} \tag{17.23}$$

The accuracy of the predicted EVM $(\widetilde{\widetilde{\mathrm{EVM}}})$ square from the previous equation is high with perfect CSI. This can be seen in Figures 17.15 and 17.16 for the first scenario and Figures 17.17 and 17.18 for the second scenario. The green curves show the error between the predicted EVM from (17.23) before data is transmitted and the actual EVM from (17.17) after the data is received and decoded. When ZF or MMSE is used, the EVM% prediction error is below 1% at 12 dB SNR ($\mathrm{EVM}\% = \mathrm{EVM}_{RMS} \times 100$). By increasing the SNR, the EVM% prediction error is reduced until $\widetilde{\widetilde{\mathrm{EVM}}}_{\tilde{k}} = \mathrm{EVM}_k$ at 20 dB SNR.

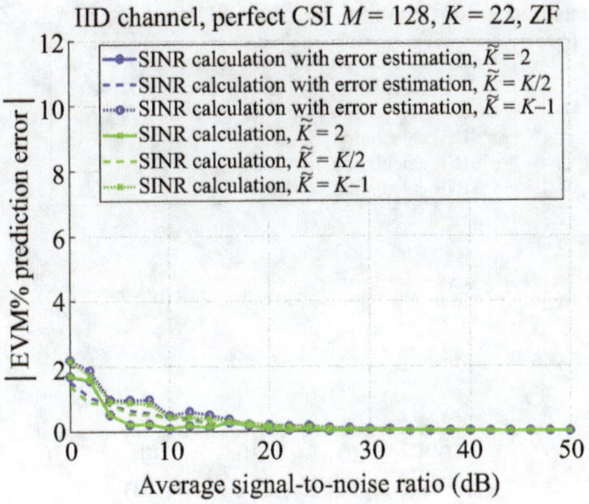

Figure 17.15 EVM prediction error with perfect CSI using ZF (IID Rayleigh channels) (Figure 5 in [23])

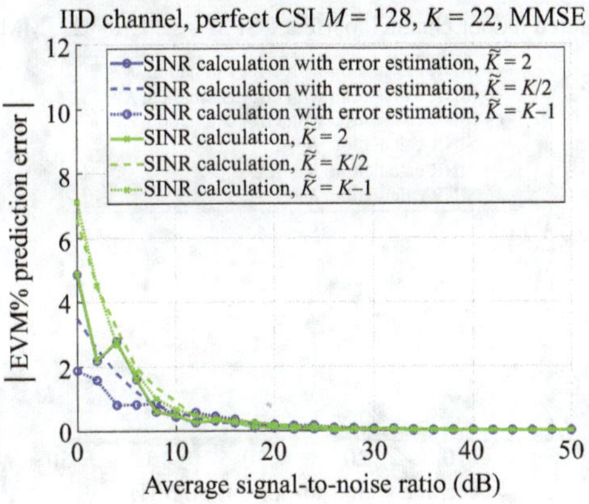

Figure 17.16 EVM prediction error with perfect CSI using MMSE (IID Rayleigh channels) (Figure 6 in [23])

The accuracy of the CSI in (17.22) and the number of users can be seen to significantly impact the accuracy of the predicted EVM in (17.23). This can be seen in Figures 17.19 and 17.20 for the first scenario and Figures 17.21 and 17.22 for the second scenario. When the SNR is increased above 9 dB and $\tilde{K} = 2$, the EVM% prediction error is always below 1% with ZF and MMSE in both scenarios. When increasing $\tilde{K}$ to 11 $\left(\tilde{K} = K/2\right)$, the EVM% prediction error begins to increase at 20 dB SNR in the first scenario and 18 dB SNR in the second scenario. When the SNR value is increased further, the EVM% prediction error is seen to increase as well. The reason for the increment in the EVM% prediction error is the CSI accuracy. The estimated SINR in (17.22) does not consider the interference caused by inaccurate CSI as the interference term embeds the CSI error. The CSI accuracy has a greater

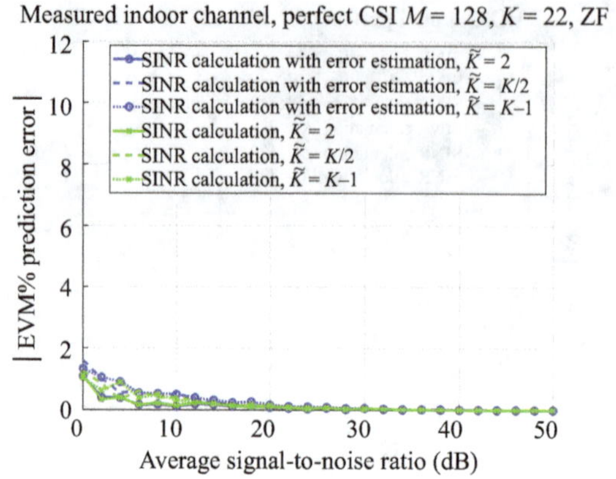

Figure 17.17 EVM prediction error with perfect CSI using ZF (measured indoor channels) (Figure 7 in [23])

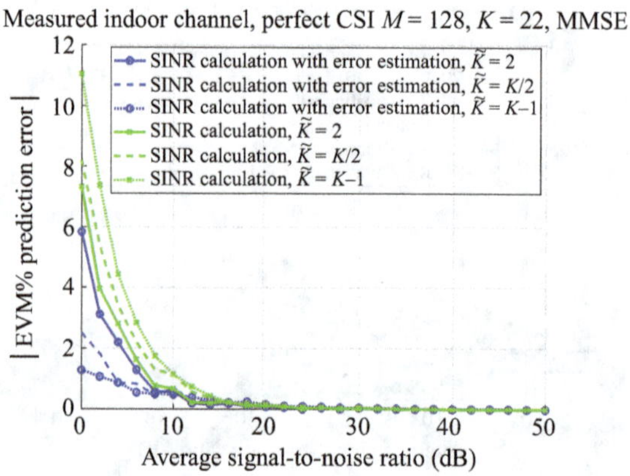

Figure 17.18 EVM prediction error with perfect CSI using MMSE (measured indoor channels) (Figure 8 in [23])

impact on the EVM% prediction error when the number of users is increased. This can be seen when $\tilde{K}$ is increased to 21 $(\tilde{K} = K/2)$. Thus, the CSI accuracy plays a major role in the EVM prediction based on the SINR calculation method.

17.4.2 *EVM prediction based on SINR calculation with error estimation*

The impact of inaccurate CSI is considered in this EVM prediction method. The UL transmit power is adjusted so the received SNRs from all users are the same. The EVM value in (17.17) covers the interference caused by inaccurate CSI, whilst the estimated SINR in (17.22) does not. So the error caused by inaccurate CSI can

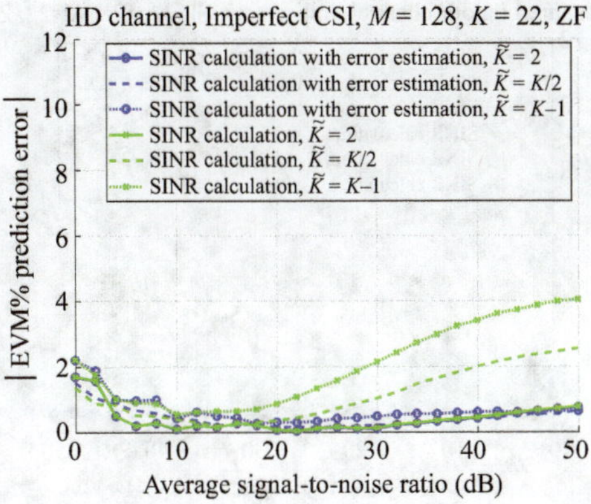

Figure 17.19 EVM prediction error with inaccurate CSI using ZF (IID Rayleigh channels) (Figure 9 in [23])

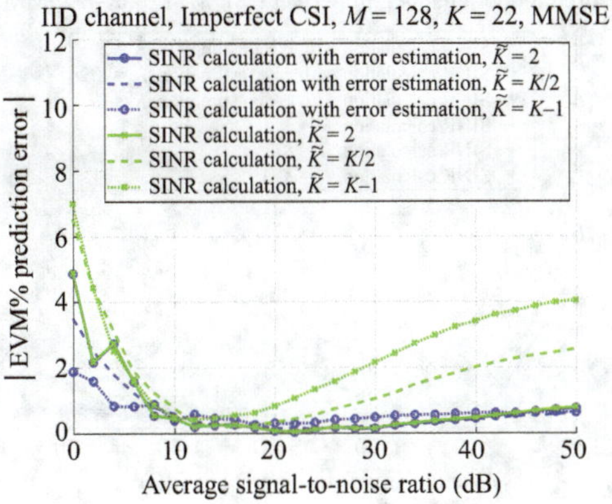

Figure 17.20 EVM prediction error with inaccurate CSI using MMSE (IID Rayleigh channels) (Figure 10 in [23])

be estimated by calculating the SINR value in (17.22) for user k $\left(\text{when } \tilde{K} = K\right)$ and comparing it with the actual EVM value in (17.17) as follows:

$$\zeta_k = \left| EVM^2_{\mathrm{RMS}_k} - \frac{1}{\widehat{\mathrm{SINR}}_{ul_k}} \right| \tag{17.24}$$

The previous error is mainly caused by the inaccurate CSI from all the simultaneous users (the error between SINR and EVM estimation is very small as shown earlier in this section). Since the spatial correlation between users has a significant impact on the interference value, the estimated error in the EVM is

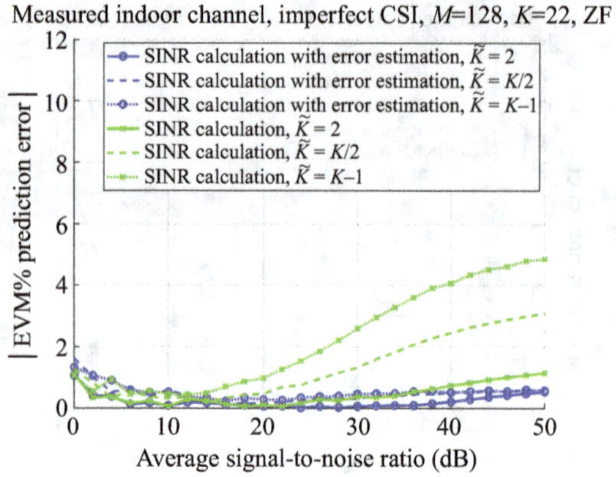

Figure 17.21 EVM prediction error with inaccurate CSI using ZF (Measured indoor channels) (Figure 11 in [23])

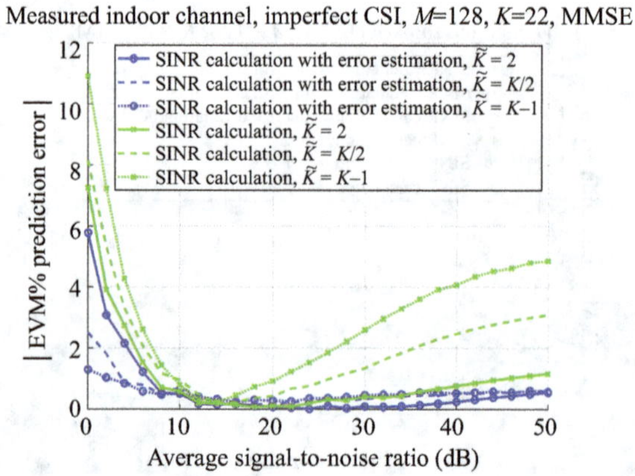

Figure 17.22 EVM prediction error with inaccurate CSI using MMSE (measured indoor channels) (Figure 12 in [23])

divided between users based on the spatial correlation ratio in this method. This can be represented in $\Upsilon \in \mathbb{C}^{K\times K}$ where $\Upsilon_{k,l}$ is the partial EVM^2 error estimation value for user k which is caused only by user l :

$$\Upsilon_{k,l} = \begin{cases} \zeta_k \dfrac{\left|\hat{h}_{ul_k}^H \hat{h}_{ul_l}\right|^2}{\sum_{i\neq k}^{K} \left|\hat{h}_{ul_k}^H \hat{h}_{ul_i}\right|^2} & \text{if } k \neq l \\ 0 & \text{otherwise} \end{cases} \tag{17.25}$$

After calculating all the partial EVM^2 error estimation values for all the users, the predicted EVM^2 error value for user $\tilde{k}$ (in the subset of $\tilde{K}$ users) can be written as follows:

$$\tilde{\tilde{\xi}}_{\tilde{k}} = \sum_{i=1}^{\tilde{K}} \Upsilon_{\tilde{k},i} \tag{17.26}$$

By taking into account the previous EVM^2 error prediction, (17.23) can be modified as follows:

$$\widetilde{\widetilde{\mathrm{EVM}}}^2_{RMS_{\tilde{k}}} = \frac{1}{\widehat{\mathrm{SINR}}_{ul_{\tilde{k}}}} + \tilde{\tilde{\xi}}_{\tilde{k}} \tag{17.27}$$

The accuracy of the $(\widetilde{\widetilde{\mathrm{EVM}}})$ from the previous equation is evaluated with perfect CSI and inaccurate CSI. With perfect CSI, this method provides the same accuracy as the one in Section 17.4.1 with ZF as shown in Figures 17.15 and 17.17. When MMSE is used, the proposed method in this section provides higher accuracy compared to the one in Section 17.4.1. The reason behind this is the low accuracy of the approximation in (17.19) when MMSE is used with low SNR as shown in Figures 17.16 and 17.18. Meanwhile, the proposed method in this section has taken into account the difference between the estimated EVM based on the approximation in (17.19) and the actual EVM from (17.17).

The main advantage of the proposed EVM prediction method in this section is its high accuracy even with inaccurate CSI and a large number of users. This can be seen clearly in Figures 17.19 and 17.20 for the first scenario and Figures 17.21 and 17.22 for the second scenario. Unlike the EVM prediction method proposed in Section 17.4.1, this method maintains the EVM% prediction error below 1% when SNR is greater than 8 dB for both scenarios, with ZF and MMSE, even when $\tilde{K}$ is increased from 2 to $K-1$.

17.4.3 Experimental campaign

The experimental evaluation was a part of the UK 5G Testbeds and Trials Programme [24]. Funded by the UK Government's Department of Digital, Culture, Media and Sport, this event was the world's first urban 5G showcase*. In all the considered scenarios, UL power control was used to adjust the UL transmit power so the received SNRs from all users were the same.

17.4.3.1 Experimental environment

The 5G showcase took place at Millennium Square in Bristol. The balcony of 'We The Curious' building was used to deploy the BS to serve 12 user clients placed 37.5 m away down in the square. The M-MIMO system operated at a 3.51 GHz carrier frequency sharing a common 20 MHz radio channel. The patch panel

*Further details about the world's first urban 5G showcase is available at https://www.bristol.ac.uk/engineering/research/smart/5g-demonstrations/massive-mimo/.

Figures 17.23 Outdoor trial environment with the 12 UEs 36.5 m away

antenna array was set up in a 4×32 configuration with vertical polarisations for all 128 antennas. User clients' antennas were in LOS and were mounted outside along the top edge of the marquee in a straight line with 432 mm spacing as shown in Figure 17.23.

17.4.3.2 Real-time results

During the 5G showcase, a total of 24 high-definition (HD) videos were streamed in real-time in both UL and DL. In order to establish reliable live HD video streams, the adaptive user grouping algorithm in [25] was used. The user grouping algorithm jointly relies on the channel correlation between users and the EVM performance which are the key elements behind ensuring reliable data transmission and increasing the system throughput. The channel correlation provides the interference level before equalising the signal assuming perfect CSI. While the EVM performance considers the interference from the channel and the interference caused by the inaccurate CSI after equalising the signal. Relying on the measured EVM values for a set of users to predict the MCS order for a subset of users is the key element behind the adaptive user grouping algorithm in [25]. This is achieved by extracting the EVM based on EVM prediction method proposed in Section 17.4.2. During the real-time data transmission, EVM% prediction error was always less than 1% by using the MMSE decoder. The highest EVM% prediction error achieved by any user was 0.81%, while the lowest EVM% prediction error was 0.08%. By using ZF decoder, the highest EVM% prediction error was 0.71%. When EVM% was calculated based on perfect CSI, the highest EVM% prediction error was 4.3% and 4.2% for MMSE and ZF, respectively.

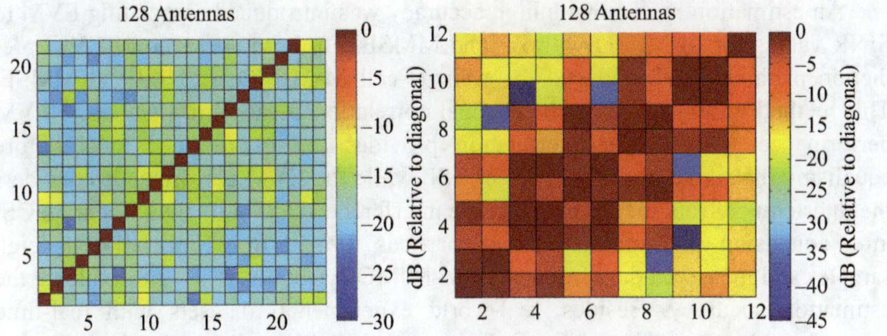

Figure 17.24 User-side Gramian intensity plots. Left: indoor trial with 22 UEs; right: outdoor trial with 12 UEs.

By comparing the results obtained in this outdoor LOS real-time trial to the one in Section 17.2.1.1, performance degradation can be observed. The maximum uncoded throughput and the maximum SE were decreased by 1.664 Gbps and 133.3 bits/s/Hz, respectively. Two main reasons are behind this performance degradation. First, the rich multipath scattering in an indoor environment creates a greater level of pairwise orthogonality between the user channel vectors and thus increases the achievable capacity. While in an outdoor environment where the LOS is the dominant signal, lower pairwise orthogonality is achieved between the user channel vectors. This can be seen in Figure 17.24, where the user-side Gramian matrix for the LOS indoor environment in Section 17.2.1.1 is on the left and the LOS outdoor environment from the 5G showcase is on the right. The high spatial correlation level between users in the outdoor environment increases the interference between users. This will either decrease the number of users that are using the same radio resources or the MCS order. The second main reason behind the performance degradation is the accuracy of the CSI. The distance between the users and the BS in the indoor environment trial was 24.8 m. While in the outdoor trial the distance was increased by 11.7 m. The maximum mean power level used in both trials for UL transmission was −16 dBm. By increasing the distance between the users and the BS, the CSI accuracy was decreased because of the noise floor effect on the pilots during the channel estimation process. This increased the interference level between users which decreased the uncoded throughput and SE.

17.5 Conclusions

In this chapter, the highest SE of 145.6 bits/s/Hz achieved for any wireless system to date and its limiting factors were presented. Common linear decoders/precoders, MF, ZF and MMSE, were used in UL and DL data transmission cases. It was shown and experimentally validated that the inaccurate UL CSI has greater impact on the DL data transmission than it is on the UL data transmission due to the reciprocity calibration error amplification.

An estimation method with high accuracy was introduced associating EVM to SINR values for M-MIMO with ZF and MMSE. Furthermore, a novel EVM prediction method was introduced for a single-cell M-MIMO with ZF and MMSE. This method jointly relies on the channel correlation between users and the EVM performance. The channel correlation provides the interference level before equalising the signal assuming perfect CSI, while the EVM performance considers the interference from the channel and the interference caused by the inaccurate CSI after equalising the signal. Channel matrices formed from both IID Rayleigh samples and measured data from a real M-MIMO system were used to evaluate the estimation accuracy. Besides, real-world experimental datasets with real-time results have been used to validated the proposed EVM prediction method. Results from the M-MIMO simulator show that the EVM% prediction error is less than 1% when SNR is above 8 dB. This was experimentally validated by using SDR M-MIMO testbed.

References

[1] Marzetta TL. Noncooperative Cellular Wireless with Unlimited Numbers of Base Station Antennas. *IEEE Transactions on Wireless Communications*. 2010; 9(11): 3590–3600.

[2] Malkowsky S, Vieira J, Liu L, *et al.* The World's First Real-Time Testbed for Massive MIMO: Design, Implementation, and Validation. *IEEE Access*. 2017; 5: 9073–9088.

[3] National Instruments. R1-162238 – A Dynamic TDD Radio Interface; 2016. 3GPP TSG RAN WG1 Meeting #84bis. http://www.3gpp.org/ftp/tsg_ran/WG1_RL1/TSGR1_84b/Docs/.

[4] Vieira J, Rusek F, and Tufvesson F., *Reciprocity Calibration Methods for Massive MIMO Based on Antenna Coupling*. In: 2014 IEEE Global Communications Conference; 2014. p. 3708–3712.

[5] Harris P, Zhang S, Beach M, *et al.* LOS Throughput Measurements in Real-Time with a 128-Antenna Massive MIMO Testbed. In: 2016 IEEE Global Communications Conference (GLOBECOM); 2016. p. 1–7.

[6] Harris P, Hasan WB, Malkowsky S, *et al.* Serving 22 Users in Real-Time with a 128-Antenna Massive MIMO Testbed. In: Proceedings of 2016 IEEE International Workshop on Signal Processing Systems (SiPS). Dallas; 2016. p. 266–272.

[7] Harris P, Malkowsky S, Liu L, *et al. Building Massive Antenna Arrays to Handle the Explosion of Wireless Signals*. *IEEE Spectrum*. 2017.

[8] University of Bristol. Bristol and Lund Set a New World Record in 5G Wireless Spectrum Efficiency; 2016. http://www.bristol.ac.uk/news/2016/march/massive-mimo.html.

[9] Hasan WB, Harris P, Doufexi A, and Beach M. Real-Time Maximum Spectral Efficiency for Massive MIMO and its Limits. *IEEE Access*. 2018; 6: 46122–46133.

[10] Narasimhan TL and Chockalingam A. Channel Hardening-Exploiting Message Passing (CHEMP) Receiver in Large-Scale MIMO Systems. *IEEE Journal of Selected Topics in Signal Processing*. 2014; 8(5): 847–860.

[11] IEEE Spectrum, 5G Researchers Set New World Record For Spectrum Efficiency; 2016, https://spectrum.ieee.org/5g-researchers-achieve-new-spectrum-efficiency-record#toggle-gdpr.

[12] Harris P, Malkowsky S, Vieira J, *et al. Communications Technology Changing the World*. Honorary Certificate of Appreciation. IEEE Communications Society; 2016.

[13] Marzetta TL, Larsson EG, Yang H, and Ngo HQ. *Fundamentals of Massive MIMO*. Cambridge, UK: Cambridge University Press; 2016.

[14] Wagner S, Couillet R, Debbah M, and Slock DTM. Large System Analysis of Linear Precoding in Correlated MISO Broadcast Channels Under Limited Feedback. *IEEE Transactions on Information Theory*. 2012; 58(7): 4509–4537.

[15] Balanis CA. *Antenna Theory: Analysis and Design*. Hoboken, NJ: Wiley-Interscience; 2005.

[16] Kaltenberger F, Jiang H, Guillaud M, *et al. Relative Channel Reciprocity Calibration in MIMO/TDD Systems*. In: 2010 Future Network Mobile Summit; 2010. p. 1–10.

[17] Björnson E, Hoydis J, Kountouris M, and Debbah M. Massive MIMO Systems With Non-Ideal Hardware: Energy Efficiency, Estimation, and Capacity Limits. *IEEE Transactions on Information Theory*. 2014; 60(11): 7112–7139.

[18] Mahmoud HA and Arslan H. Error vector magnitude to SNR conversion for nondata-aided receivers. *IEEE Transactions on Wireless Communications*. 2009; 8(5): 2694–2704.

[19] Shafik RA, Rahman MS, and Islam AR. On the Extended Relationships Among EVM, BER and SNR as Performance Metrics. In: *2006 International Conference on Electrical and Computer Engineering*; 2006. p. 408–411.

[20] Brown TWC, Humphreys DA, Hudlicka M, *et al.* Prediction of SINR Using BER and EVM for Massive MIMO Applications. In: *12th European Conference on Antennas and Propagation*; 2018.

[21] 3GPP TS 36.141, LTE; Evolved Universal Terrestrial Radio Access (E-UTRA); Base Station (BS) conformance testing, version 10.1.0 Release 10, 2011.

[22] Cho YS, Kim J, Yang WY, *et al. MIMO OFDM Wireless Communication with Matlab*. Singapore: John Wiley & Sons (Asia) Pte Ltd; 2010.

[23] Hasan WB, Doufexi A, Oikonomou G, and Beach M. EVM Prediction for Massive MIMO. In: *2019 IEEE 30th Annual International Symposium on Personal*, Indoor and Mobile Radio Communications (PIMRC); 2019.

[24] UoB, Layered Realities Weekend - A 5G Showcase, 2018, https://www.bristol.ac.uk/engineering/research/smart/events/layered-realities-weekend/.

[25] Hasan WB, Harris P, Bromell H, Doufexi A, and Beach M. Adaptive User Grouping Based on EVM Prediction for Efficient Robust Massive MIMO in TDD. *IEEE Access*. 2019; 7: 162683–162696.

Chapter 18

Millimetre-wave radio propagation measurements towards 5G NR standardizations

Peize Zhang[1], *Cheng Yi*[1] *and Haiming Wang*[1,2]

18.1 Introduction

This chapter deals with the topic related to millimetre-wave (mmWave) radio propagation measurements and indoor channel characteristics from 5G mmWave new radio (NR) standardization aspect. The mmWave wireless channel, in some way as the traditional microwave band channel, is characterized by series of relevant spatial-temporal propagation parameters as a function of traversed distance, carrier frequency and application scenarios; however, mmWave signals experience severe path loss and are more sensitive to physical blockages due to the relatively shorter wavelength [1]. Accordingly, it is hoped that this will allow readers to become acquainted with the mmWave channel sounding methodologies and propagation characteristics across multiple frequency bands in indoor environments.

The mobile communication standard organizations (e.g., International Telecommunication Union-Radiocommunication and Telecommunication Standardization Sectors (ITU-R/T), the 3rd Generation Partnership Project (3GPP), Institute of Electrical and Electronics Engineers (IEEE)) primarily concentrate on the modelling of the wireless channel, which can further support network planning and transmission design based on the observation of large-scale and small-scale channel characteristics, respectively. To fully exploit the potential of mmWave communications, the cluster delay line channel model is widely adopted which characterizes both time-varying and three-dimensional (3D) spatial propagation properties. The statistics of delay and angular spreads for both composite and cluster-level channels are naturally extracted from the field sounding data, while the frequency-dependent behaviour of these site-specific channel parameters is more obvious in mmWave band. Moreover, with the number of antenna elements increasing, the interaction between antennas and propagation channel needs to be investigated to cater to the requirements of future mmWave massive MIMO systems. Consequently, the custom-designed channel

[1]State Key Laboratory of Millimeter Waves, Southeast University, Nanjing, China
[2]Purple Mountain Laboratories, Nanjing, China

sounders need to support flexible configurations for multiband double-directional channel measurements.

In this chapter, we first overview the global update on spectrum for 5G mmWave identified by multiple standards bodies, as well as the scenarios of interest. From channel measurements and modelling perspectives, frequency- and time-domain channel sounding methods and their corresponding implementations are then introduced and compared. Based on the extensive channel measurement campaigns conducted in multiple indoor environments and mmWave bands, both large-scale and small-scale channel characteristics are analysed, including path loss, shadow fading, delay spread and angular spread. Moreover, the empirical-based analyses of the electromagnetic characterization of different building materials are promoted.

18.1.1 5G spectrum overview

With commercial 5G networks in full rollout mode, the potential and success of 5G services are heavily reliant on national governments and regulators for the assignment of the right amount and type of affordable spectrum. To meet existing and future demands for higher data rates and more capacities, global 5G communities have agreement on allocating new spectrum within the mmWave bands for 5G which can provide around 1 GHz per operator. Therefore, 5G will operate on three different spectrum bands across low, mid and high spectrum ranges.

At the World Radiocommunication Conference in November 2019 (WRC-19), the member states of the ITU have agreed on the global IMT identification of 24.25–27.5, 37.0–43.5 (expected portions allocated per region) and 66.0–71.0 GHz for 5G [2]. Also, 45.5–47 and 47.2–48 GHz are partly identified for 5G in 60 countries around the globe and up to 100 countries (including Region 2, Africa, Middle East and some in Asia and Europe), respectively. Specifically, 3GPP has developed an NR access technology for 5G, called 5G NR, using two frequency ranges (FRs) where FR1 corresponds to 410–7,125 MHz and FR2 corresponds to 24.25–52.6 GHz [3]. The available FR2 operating bands used for NR are detailed in Table 18.1. Note that 5G will first adopt TDD for cellular networks, which means transceivers use the same channel for uplink and downlink at different times. Moreover, the IEEE 802.11 wireless networking standard also has developed two

Table 18.1 NR operating bands in FR2 defined in 3GPP TR38.101-2

Operating band	Uplink and downlink operating band (GHz)	Duplex mode
n257	26.50–29.50	TDD[a]
n258	24.25–27.50	TDD
n259	39.50–43.50	TDD
n260	37.00–40.00	TDD
n261	27.50–28.35	TDD

Source: 3GPP TS 38.101-2 V16.4.0 (2020-06) [3].
[a]TDD: time division duplex.

amendments IEEE 802.11aj and IEEE 802.11ad operating in 45 and 60 GHz, respectively.

As previously discussed, the high bands expected to be deployed early on for 5G include 26 and 39 GHz bands, as well as 45 and 60 GHz bands for wireless local area network. Channel measurement campaigns, in turn, are generally preferred to be conducted in these mmWave bands.

18.1.2 Reference scenarios for mmWave communications

Wireless signals in mmWave band suffer from extremely high path loss, meaning that the coverage of mmWave communication systems is limited in a small range. However, thanks to the employment of large-scale antenna arrays, pencil beam of which can be steered towards expected directions, the coverage could significantly be enhanced in both line-of-sight (LOS) and non-LOS (NLOS) scenarios. According to the description in 3GPP and ITU-R channel models, several outdoor and indoor physical operating environments are illustrated, such as urban microcell (UMi), urban macrocell (UMa), suburban macrocell (SMa), rural macrocell (RMa), indoor hotspot (InH) and other specific environments (e.g., warehouse, stadium and gym). The key differences among these scenarios of interest include the height of base station, speed of user equipment and surrounding physical objects (e.g., human bodies, buildings, foliage and vehicles). Additionally, with the complexity of wireless network architecture increasing, there are more application scenarios of 5G mmWave NR, such as outdoor backhaul and data centre links.

Extensive cellular-type channel measurement campaigns have revealed that the cell radius of mmWave networks in urban NLOS scenario can be up to 200 m due to rich reflections from buildings. As for LOS scenario, coverage performance of mmWave networks from channel sounding aspect mainly depends on the equivalent isotopically radiated power of transmitter and the receiver sensitivity level. Accordingly, mmWave channel sounders need to support long-range multi-scenario measurements, and empirical analysis of scenario-specific propagation characteristics can be subsequently delivered for channel modelling.

18.1.3 Standardized mmWave channel models

In this section, we will overview the state-of-the-art channel models proposed by standard organizations and research projects, which can support a wide FR (including mmWave band).

The latest 3GPP TR 38.901 channel model proposes the stochastic model and map-based hybrid model [4]. As for the stochastic channel model, higher delay and angular resolutions are achieved due to larger bandwidth and antenna arrays, meaning that each multipath component (MPC) within a cluster has different propagation delay, angle of arrival (AoA) and angle of departure (AoD). Note that both AoA and AoD include azimuth and elevation angles for 3D spatial channel modelling. The statistics of inter- and intra-cluster parameters, such as cluster and subpath number, Ricean K factor, delay spread, angular spread and their correlations, are detailed for several scenarios. Apart from the small-scale multipath fading, the distance-dependent path loss models are given considering blockage effect and oxygen absorption in mmWave band. As for the map-based hybrid channel model, the stochastic part is developed following the above-mentioned method, and the deterministic part is obtained via ray-tracing simulation combining with a digital map to precisely rebuild the propagation environments. However, the default values of these channel parameters are extracted from the field sounding data collected in different environments and frequency bands using different sounder, or directly substituted by the results for sub-6 GHz channel. Therefore, the frequency-dependent channel parameters given in 3GPP model are a bit unreliable for mmWave channel characterization.

The ITU has developed the IMT-2020 channel model which can support a wide FR up to 100 GHz [5]. Apart from the basic modelling concept, the IMT-2020 model considers the foliage effect in terms of scattering by leaves and diffraction over the canopy of trees. Moreover, the default values of large-scale and small-scale parameters are more practical in comparison with the values in the 3GPP channel model. For example, the mean value of cluster number ranges from 3 to 10 which is significantly smaller than the values in the 3GPP model and is in line with the recent mmWave channel measurement results. On the other hand, ITU-R has released a series of recommendations on indoor and outdoor radio propagation, as well as some specific propagation characteristics. However, ITU-R mainly concentrates on physical propagation mechanism (e.g., rain attenuation, building entry loss (BEL), blockage effect in vegetation) and omnidirectional channel characteristics for terrestrial wireless communication systems.

The IEEE 802.11ad channel model is the extension of Saleh–Valenzuela channel model considering both temporal and spatial cluster characteristics only in indoor environments. Note that transmitter and receiver are only configured with single antenna [6]. Consequently, IEEE 802.11aj extends the IEEE 802.11ad SISO channel model to MIMO channel model operating at 45 GHz, which also focuses on the same indoor semi-closed reference environments, such as conference room, cubicle office and living room [7]. However, these models cannot support massive MIMO and dual-mobility considering the practical in-building user habits.

Apart from the above-mentioned channel models developed by several standard organizations, some research projects (e.g., MiWEBA, QuaDRiGa, METIS, 5GCMSIG and mmMAGIC) also proposed the channel models which can support mmWave band channel generation based on the stochastic or semi-deterministic modelling methods [8]. The key differences among these models lie in the characterization of various channel features with the increase of frequency and antenna array, corresponding to a trade-off between accuracy and computational complexity. An accurate mmWave channel model strongly depends on field channel measurement results, including sounder design, system calibration, data preprocessing, channel parameter estimation and propagation characteristic analysis. The following sections, in turn, will detail the mmWave channel sounding methods and their implementations, as well as indoor measurement results across multiple frequency bands and scenarios.

18.2 Measurement methodologies

In the earlier section, we discussed the basic requirements and challenges of mmWave channel sounding from 5G NR standardization aspect. It emerges that there are certain criteria which provide an adequate description of mmWave propagation channel modelling and it remains now on the channel sounder design to collect sufficient sounding data for the exploitation of new channel characteristics.

There are various channel sounding approaches using different instruments, transmit (Tx) and receive (Rx) antennas and miscellaneous, to collect data for the detailed analysis of space-time channel properties in typical environments and network configurations. However, these measurement approaches adopted by different companies, institutes or universities mainly belong to frequency- and time-domain measurement approaches. There are significant differences between these two methods in terms of the measurement platform and data preprocessing methods [9]. In general, the selection of a channel sounding method depends on the application environments, such as outdoor, indoor, and outdoor-to-indoor (O2I) scenarios. Thus, Sections 18.2.1 and 18.2.2 would detail the principles of two kinds of sounding methods, including the typical implementations, calibration methods and data preprocessing methods.

18.2.1 VNA-based frequency-domain channel sounding method

In what follows we will introduce the frequency-domain broadband channel measurement approaches using a vector network analyser (VNA). The usage of the VNA enables the measurement of both amplitude and phase of the received signal. As a result, the channel impulse response (CIR) can be obtained via inverse Fourier transform based on channel transfer function (CTF). The frequency-domain sounding is a coherent measurement technique that can be performed with a frequency-sweep stimulus or a frequency-stepping stimulus [10].

According to Fourier transform, complex exponentials are the eigenfunctions of LTI systems where impulse response in the time domain can be obtained via measurements in the frequency domain. The response of a linear system with an impulse response $h(t)$ to a complex exponential $e^{j\omega_n t}$ is $H(j\omega_n)e^{j\omega_n t}$, where

$$H(j\omega_n) = \int_{-\infty}^{\infty} h(t)e^{-j\omega t}dt \tag{18.1}$$

The magnitude-phase representation of the continuous-time Fourier transform $H(j\omega_n)$ is

$$H(j\omega_n) = |H(j\omega_n)|e^{j\angle H(j\omega_n)} \tag{18.2}$$

As for ultra wideband signal which can be decomposed into a linear combination of harmonically related sinusoidal signal with different centre frequency. Therefore, the measurement system could use VNA to capture the CTF of the radio

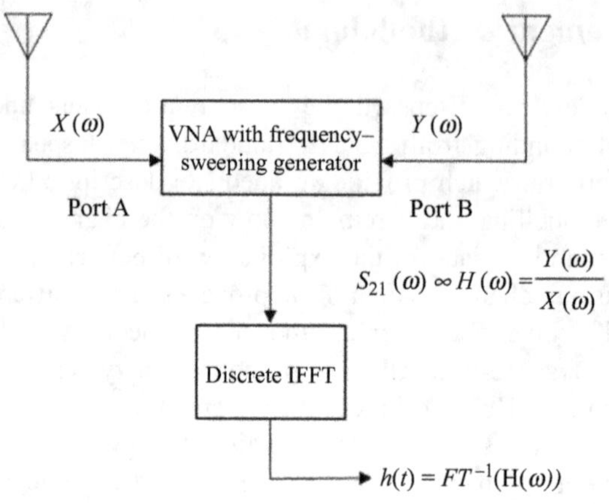

Figure 18.1 Schematic diagram of the VNA-based frequency-domain channel measurement system

channel. By using Inverse Continuous-Time Fourier Transform, CIR can be computed as

$$h(t) = I\text{CTFT}^{-1}[H(j\omega)] \tag{18.3}$$

Figure 18.1 shows the diagram of a VNA-based sounding system. It scans the discrete frequency points of given frequency band with an equal step, which means that VNA could measure the transmission parameter

$$S_{21}(\omega) \propto H(j\omega) = \frac{Y(j\omega)}{X(j\omega)} \tag{18.4}$$

Consequently, VNA could use it to estimate $H(j\omega)$ in the form of

$$h(n) = \frac{1}{N}\sum_{k=0}^{N-1} \exp\left(j\frac{2\pi}{N}nk\right)H(k) \tag{18.5}$$

$$H(k) = \sum_{n=0}^{N-1} \exp\left(-j\frac{2\pi}{N}nk\right)h(n) \tag{18.6}$$

where $\Omega_0 = 2\pi/N$ is the fundamental frequency of $h[n]$ and the frequency of the kth sinusoid in the superposition is $k\Omega_0$.

Dynamic range is the difference between the Rx's maximum input power and the minimum measurable power (i.e., a threshold above the noise floor). In VNA-based measurement, dynamic range can be increased by raising transmitting power or reducing the receiver noise floor. The maximum output power of VNA is limited; however, we can add a power amplifier or additional RF signal generator which will significantly enlarge the range of measurable scenarios. Using sweep averaging and lower IF bandwidth will reduce and smooth the noise floor. The achievable time-domain resolution and the longest recognizable time delay depend only on the channel bandwidth and FR. In fact, this method requires strictly and accurately synchronization between the transceiver and both Tx and Rx are integrated by VNA, which means the measurable distance is subject to the cable length. Moreover, the frequency scanning time is also limited, because this method would fail if frequency response changes quickly, indicating that frequency points should also be limited. A back-to-back calibration should be performed before the measurements to remove the effect of the measurement system from the measurement results, including cables, up and down converters, waveguides and antennas.

The setup for the frequency-domain mmWave channel measurements depends on the performance of VNA and measurement requirements. Thus, there are various typical VNA-based sounders with different external devices.

Figure 18.2 illustrates a platform used for wideband propagation measurement at the 15-, 28- and 60-GHz frequency bands [11]. The sounders are based on a VNA that allows phase-synchronized measurements of scattering parameters with the radio channels being device-under-test. This type of VNA-based channel sounder uses optical fibre cables to replace lossy coaxial cables and mixes VNA signals with LO signals in up- and downconverters, which can support up to 140-GHz measurements with 200-m T–R separation distance.

Figure 18.3 represents an automatic channel sounding measurement system suitable for indoor mmWave wireless communications [12]. The measurement system mainly is composed of a signal generator and a VNA, both of which can generate or analysis signals up to 67 GHz. The generator here is accessory to VNA which can increase its output power, as well as the dynamic range. A coupler is used to separate the signal generated by the generator into two parts: a main part to the transmitting antenna and a weaker part to the input port of the VNA as a

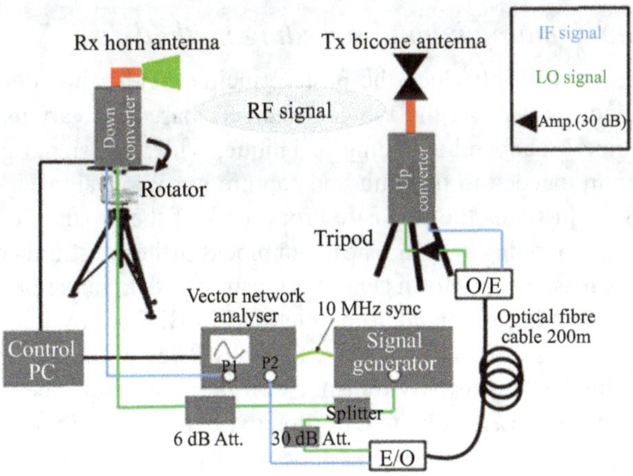

Figure 18.2 Schematic of 28- and 60-GHz channel sounder; the 15-GHz measurements are performed without using up- and down-converters [11]

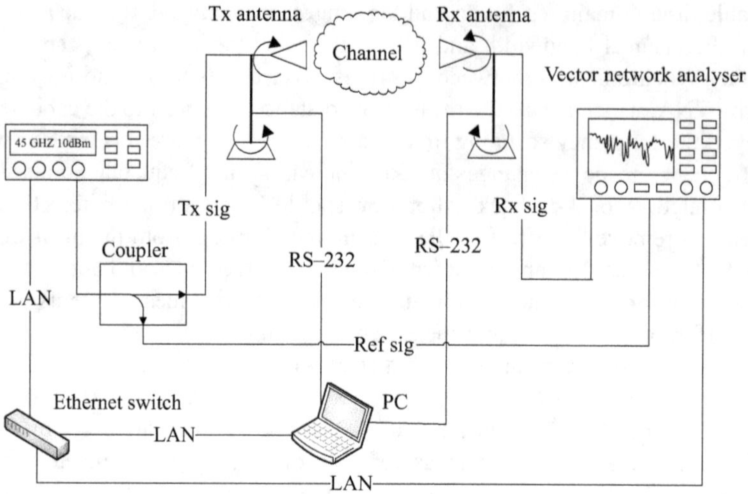

Figure 18.3 Block diagram of frequency-domain channel sounder using high-performance VNA [12]

reference. The control centre of this measurement system is a laptop. The 3D motion of transmitting and receiving antennas are controlled by roll-over-azimuth positioning system through RS-232 serial port line connected to the laptop. By using this auto-controlling system, indoor path loss and space-time characteristics can be tracked and measured.

18.2.2 Time-domain channel sounding method

In this section, we will introduce the basic principle of the time-domain channel sounding method, as well as the typical sounders based on various devices. To employ a genuine broadband sounding technique, wideband signal generator and signal analyser are needed to transmit and capture mmWave signals.

Figure 18.4 illustrates the requisite procedures of the separate transceiver for wideband channel sounding system. The major blocks in the transmitter are the digital source and transmission waveform shaper. Transmitted data in the practical system are always random and confidential to complete as low BER as possible; however, for channel sounding, it is essential to acquire the knowledge of the multipath channel, thus measurement systems usually transmit well-designed sequence with excellent autocorrelation properties. To obtain CIR directly, two frames of the transceiver are illustrated in Figure 18.5, in which we assume the channel is an LTI system.

As shown in Figure 18.5(a), the transmitted signal is ideal impulse function, $\delta(t)$, but do not exist with extremely short duration and infinite amplitude. Consequently, an improved transceiver structure is shown in Figure 18.5(b) and satisfies

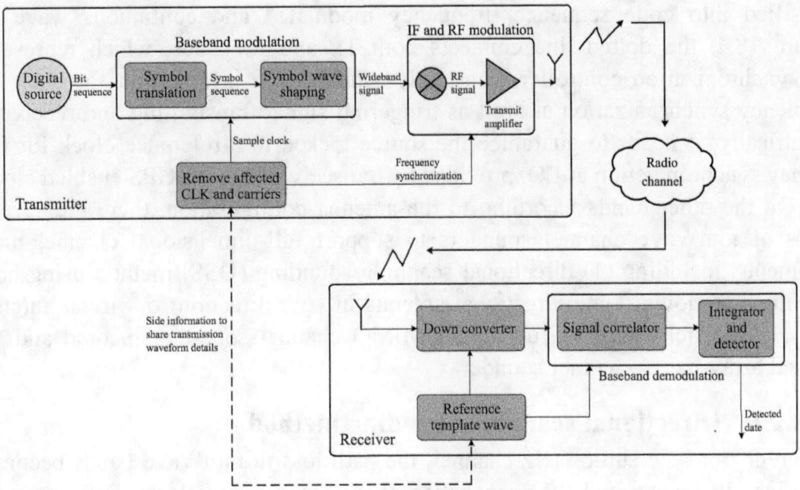

Figure 18.4 System model of time-domain wideband channel sounding with the separate transceiver

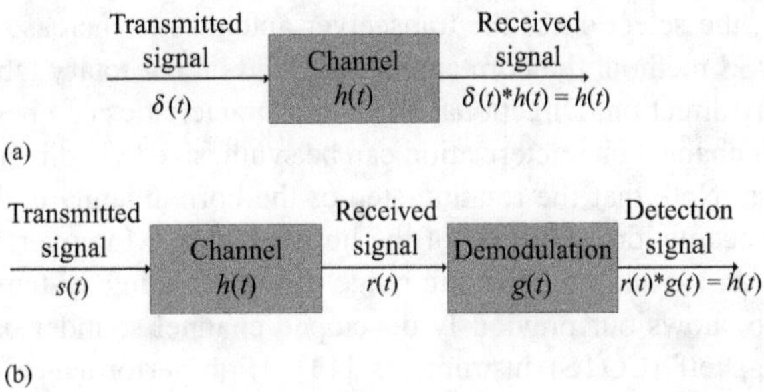

Figure 18.5 Two approaches to obtain channel impulse response with different transmitted signals: (a) unit impulse function and (b) Pseudo-Noise sequence

$$s(t) * g(t) = \delta(t) \tag{18.7}$$

where $s(t)$ is a determined signal and * represents the linear convolution. Therefore, it is crucial to design broadband signal or code sequence which is significantly contrary to simple waveform used in the frequency-domain method. Equation (18.7) just reflects the relationship between $s(t)$ and $g(t)$, where one part corresponding to the impulse response of the matched filter $g(t)$ with uncertain time delay and the others due to the choice of $s(t)$ which has ideal autocorrelation function. For the first issue, it can be solved by accurate synchronization and system calibration to obtain accurate absolute propagation delay. As for the second one, the waveforms (used in Figure 18.5(b) usually called pulse compression) are classified into code sequence, frequency modulated and continuous wave. In Figure 18.4, the dotted line connects both Tx and Rx sides, which represents the synchronization controller, constantly sending side information for time and frequency synchronization as well as triggering signal transmitting and receiving. Specifically, it needs to guarantee the source locked to a reference clock for frequency synchronization at LO, provided by transceiver itself or GPS enabled clock.

On the other hand, according to the antenna configuration, there are several kinds of mmWave channel sounders to support full-dimensional channel measurements, including (1) directional scanning sounding (DSS) method using horn antenna, (2) moving omnidirectional antenna in *x*/*y*/*z* direction for virtual antenna array, (3) switch-based or full-digital MIMO channel sounding method and (4) phased array-based channel sounder.

18.2.2.1 Directional scanning sounding method

Different from the sub-6 GHz channel, the path loss in mmWave bands becomes significantly severe, and the propagation characteristics are closely related to the environments, which is vulnerable to the weather and blockage such as vegetation and buildings. The spatial propagation characteristic is essential for the design of mmWave massive MIMO systems. Therefore, the mmWave channel measurement method must consider the accuracy of the measurement system in the angular domain, that is, the selection of the transceiver antennas to increase angular resolution. In the DSS method, the horn antenna is fixed on the rotary table, which can be steered to any direction. Directional channel characteristics can be captured, and omnidirectional channel characterization can be synthesized via directional channel sounding results. Note that the rotating step of the horn antenna is determined by the half-power beamwidth (HPBW) of the horn antenna. Moreover, the high-gain horn antennas can expand the dynamic range of the sounding system.

Figure 18.6 shows our previously developed channel sounder based on commercial off-the-shelf (COTS) instruments [13]. High-performance vector signal generator (VSG) and vector signal analyser (VSA) are, respectively, employed in Tx and Rx sides, which can support measurements in the FR of 10–40 GHz without external up- and down-converters. The maximum measurable FR of the system depends on the performance of devices, but it can be extended with a mixer. We add an arbitrary baseband generator for using a maximal length 4096 Golay complementary codes to replace the pseudorandom noise sequence. It has been proven this sequence performs well as for periodic auto-correlation property and increase the accuracy of CIR data when doing the relevant operation with raw data stored from a spectrum analyser [13]. Furthermore, programmable controlled multi-dimension positioners which can remote in arbitrary azimuth and elevation directions are used in both Tx and Rx sides for installing high gain horn antennas with small HPBW.

To guarantee a quasi-static environment, we use two wireless routers to connect all measurement instruments in a local area network, which can, in turn, ensure auto-measurement. Additionally, we use two cables to provide precise reference clock signals to keep radio frequency signal generators synchronized with the

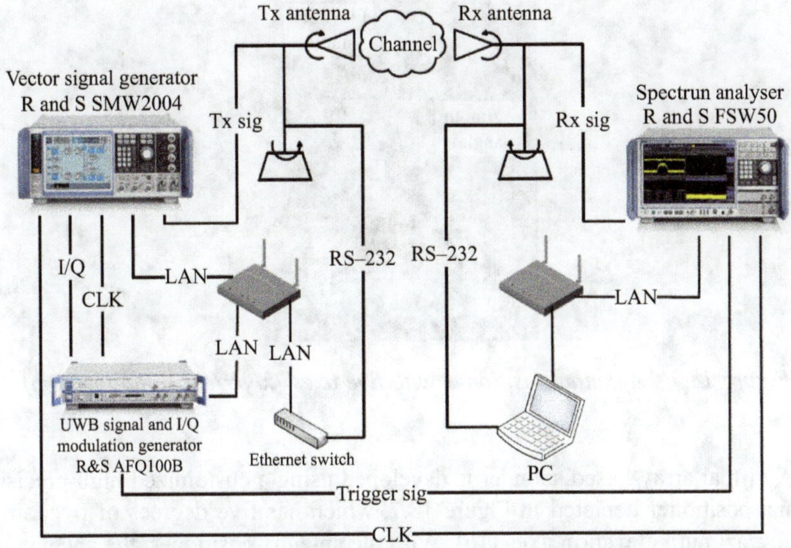

Figure 18.6 Block diagram of mmWave channel sounding system using COTS instruments [14]

baseband and receiver in IF and LO frequencies. Due to the sounding signal constantly transmitted, an external trigger is added to capture data from the RF path after rotator moving to a specific direction. Recently, to separate Tx and Rx with flexible configurations, two separate GPS rubidium (Rb) standard references are connected to the VSG and the VSA, respectively, for frequency synchronization with stable 10-MHz outputs and data reception with periodical 1 pulse-per-second (1 PPS) trigger signals. Thanks to the employment of Rb standard references, absolute propagation delay can be obtained following the calibration method developed by MacCartney [15].

18.2.2.2 Uniform virtual array-based method

In the actual measurement, virtual MIMO antenna array is used to replace the real MIMO antenna array to obtain the mmWave MIMO channel characteristics, which can greatly reduce the complexity and cost of the measurement system. This method uses omnidirectional antennas moving in *x*/*y*/*z* direction to form a virtual array, which overcomes the disadvantage of the directional scanning method in embedding the antenna pattern during data preprocessing. The angle information of the MPCs can be obtained via high-resolution algorithms. Note that the measured CIR is essentially only one snapshot of an environment for static indoor measurements so that the classical beamforming and high-resolution algorithms like MUSIC cannot be directly applied due to the limited temporal samples. Different algorithms have been proposed to address this issue, such as the spatial smoothing technique and the iterative 2D Unitary ESPRIT method.

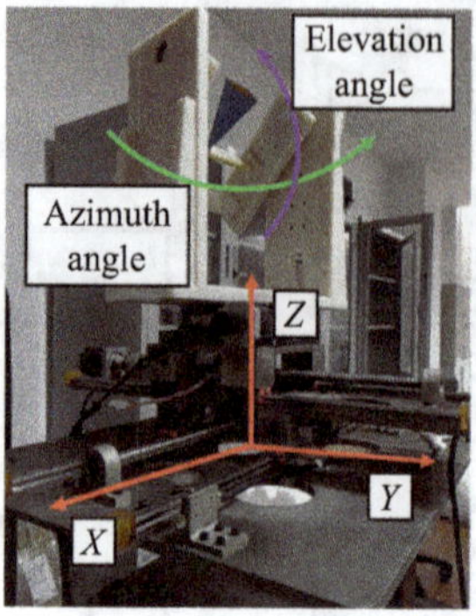

Figure 18.7 Antenna positioner with five axes (x/y/z/azimuth/elevation)

A virtual array-based sounder is developed using a customized high-precision antenna positioner depicted in Figure 18.7, which has five degrees of freedom in the *x/y/z*/azimuth/elevation axes [16]. With the antenna positioner, the antenna can scan in azimuth and elevation 3D angular domains and scan in *x/y/z* axes to form a 3D virtual antenna array. The spacing between adjacent positions in each surface is related to the candidate frequency. With this large 3D virtual antenna array, the precise and high-resolution angle parameters can be estimated using high-resolution parameter estimation algorithms.

18.2.2.3 Switch-based channel sounding method

The DSS method and virtual antenna-array-based method both need to move antennas in 3D space, indicating that the measurement duration of these sounding systems is very slow, and only static channels or channels with very low Doppler frequencies can be measured. Additionally, the directional horn antenna used in these sounding systems is much larger than the wavelength in mmWave frequency band, which does not meet the requirement that the interval of the antenna elements should be half of the wavelength. To address these challenges, the switch-based channel sounding method has been developed. Figure 18.8 shows an example of a hybrid mmWave channel sounder system [17]. The sounder system supports both MIMO and SISO channel measurements depending on the concerned scenarios. Thanks to using COTS instruments, the sounder has fewer difficulties in hardware debugging and software programming with standardized hardware and software interfaces. At Tx side, the RF signal output from VSG will be fed to either multiple elements antenna array (for MIMO) or single horn antenna (for SISO). The solid-state RF switch and corresponding switching controller are used to perform the sequential time-division transmission of the sounding stimulus signal from different Tx antenna elements. Antenna calibration is done by importing the measured antenna pattern data and antenna gain at the measured frequency into the channel parameter estimation algorithm during the data preprocessing. The other system impairments can be calibrated using a two-step back-to-back calibration approach.

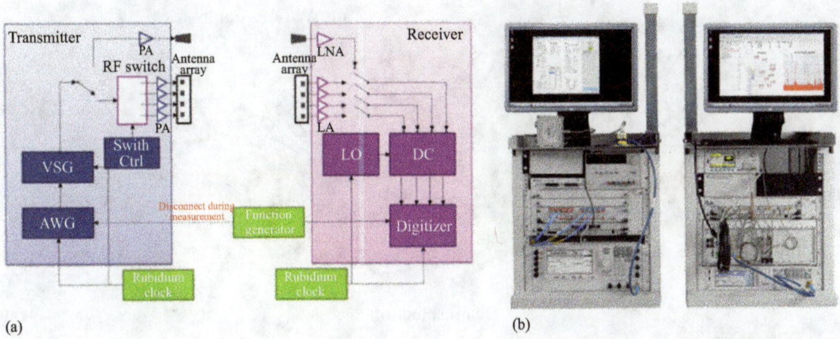

Figure 18.8 *(a) mmWave channel sounder system architecture and (b) an example of mmWave channel sounder system [17]*

18.2.2.4 Phased-array-based channel sounder

To meld the speed of the switched array system with the precision of gridded elements, the real phased-array antennas have been implemented in phased-array-based channel sounder. In conventional MIMO approaches, optimal performance requires prohibitively high transceiver complexity. By combining the concept of beamspace MIMO communication with a hybrid analogue–digital transceiver, continuous aperture phased (CAP) MIMO achieves near-optimal performance, which can dramatically reduce the system complexity. Figure 18.9 presents a lens array-based multi-beam CAP-MIMO prototype, which can simultaneously measure 4 spatial channels by electronically selecting 4 out of 16 beams [18].

The CAP-MIMO transceiver architecture is designed on the basis of the concept of beamspace MIMO communication, which can realize efficient access to the p communication modes of an n-dimensional mmWave link. A hybrid analogue–digital front end is used for analogue spatial beamforming. Thanks to the high-resolution discrete lens array (DLA), the CAP-MIMO can achieve near-optimal performance with dramatically lower transceiver complexity, compared to the order n complexity of conventional MIMO. The DLA acts like a convex lens and maps the signals in different directions to different locations on the focal surface. The high-resolution, low-loss DLA used in the CAP-MIMO is composed of sub-wavelength, nonresonant phase-shifting elements that can be distributed on a planar surface and act as bandpass filters. The DLA was chosen over alternative analogue beamformers because of its low loss and relative ease of construction for antennas with a high-dimensional spatial signal space.

Figure 18.10 shows another type of phased-array-based MIMO channel sounder for 28-GHz channel measurement [19]. The sounder is capable of performing horizontal and vertical beam steering with the help of phased arrays. The sounder can perform measurements that are directionally resolved both at the Tx and Rx due to the fast beam-switching capability. It is worth noting that the high phase stability limits the phase drift between Tx and Rx during the short measurement time, which enables phase-coherent sounding of all beam pairs even when Tx and

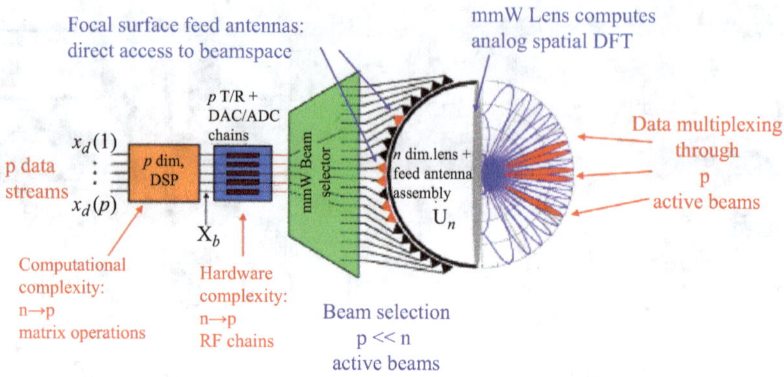

Figure 18.9 DLA-based analogue beamforming in a CAP-MIMO transceiver [18]

Rx are physically separated and have no cabled connection for synchronization. The channel sounder is a beam-switched multicarrier setup with 400-MHz instantaneous bandwidth.

18.2.3 Comparison and discussion

The two mmWave radio channel measurement methods are briefly introduced in this section. For the VNA-based frequency-domain channel sounding method, the frequency response of the radio channel is measured, and then the impulse response in the time domain can be obtained by using the discrete inverse Fourier transform. The VNA-based frequency-domain channel sounding method can obtain the channel CIR with high precision in the time domain by adjusting the sweep range and sweep step, but the measurement method has some limitations. However, the measurement method of frequency sweep itself is time-consuming, so that man-made interference cannot be avoided due to the change of workers' activity. Therefore, this method is more suitable for quasi-static radio channel measurements than for fast time-varying scenarios. Besides, as the two ports of VNA are connected to both ends of the transceiver through cables, the maximum measurement range is limited by the cable length. The severe transmission loss in the cable at high frequency also limits the dynamic range of the channel measurement system. Therefore, the VNA-based frequency-domain sounding method is mostly used for channel measurements in indoor and outdoor short-distance environments.

As for the time-domain channel sounding method, the CIR of the radio channel can be obtained by correlation operation between the receiving sequences and the sounding sequences. In this method, the Tx and Rx can be separated from each other, which eliminates the limitation on the measurement range of the cables connected to the Rx in the VNA-based frequency-domain sounding method. Moreover, the time-domain method can effectively save the measurement time because of no need for frequency sweep. Therefore, the time-domain method can

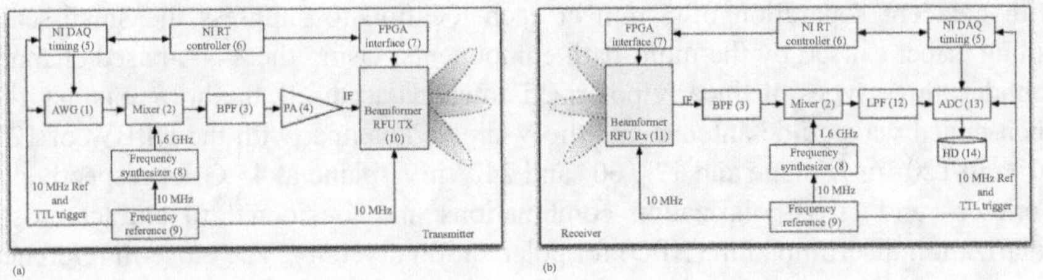

Figure 18.10 Real-time beamformer-based mmWave MIMO channel sounder: (a) Tx block diagram and (b) Rx block diagram [19]

be applied to outdoor long-distance measurements and is more widely used than the frequency-domain method. However, the measurement bandwidth of this method is limited due to hardware limitations. In terms of antenna configurations, both mmWave virtual and real MIMO channel sounders are overviewed, and there is a trade-off among complexity, efficiency and cost.

18.3 Measurement results

Using the VNA-based or COTS instrument-based channel sounder, measurement campaigns have been conducted across multiple 5G identified bands in different typical indoor scenarios, including semi-closed scenario, indoor transitional or blocked scenarios and O2I scenario. In semi-closed rooms, LOS propagation is dominantly combining with the reflected paths from surrounding walls and furniture. In the transitional and blocked scenarios, only NLOS paths exist due to the diffraction over building corners and pillars, as well as the penetration through in-building materials. Additionally, measurement also has been conducted in the O2I scenarios to evaluate the performance of continuous coverage. On the other hand, the accuracy of the ray-tracing simulation strongly depends on the structures and electric parameters of building materials. Thus, site-specific analysis of reflection loss is used to extract reflection coefficients varying with frequency at the end of this section.

18.3.1 Indoor semi-closed scenarios

The semi-closed rooms with the dimensions of a few tens of metres are the most common deployment scenarios for mmWave wireless networks, which can provide ultra-high throughput for 5G and beyond systems.

Following the description in IEEE 802.11aj, three environments are considered, conference room, cubicle office and living room, as shown in Figure 18.11. For each scenario, the detailed layout and the measuring arrangement are also illustrated. In the LOS scenario, Tx and Rx antennas are rotated to be aligned with each other; and in the NLOS scenario, a double-layer aluminium sheet with the size of 2×1 m^2 is vertically placed between Tx and Rx to obstruct the direct path. It is worthy to mention that at each Rx location, 40 local measurements are performed with adjacent separation of 5 mm at each location to suppress the small-scale fading effect caused by the multi-path components. Using the VNA-based channel sounder, three types of linearly polarized antennas are used, the horn antenna, the open-ended waveguide antenna and the V-dipole antenna, with the HPBW of 12°, 60° and 120° in *E*-plane and 12°, 60° and 240° in *H*-plane at 45 GHz, respectively. Both co- and cross-polarization combinations are considered to extract cross-polarization discrimination (XPD) for polarization diversity. The close-in reference distance path loss model is adopted to fit with measured data, expressed as

$$L(d) = \bar{L}(d_0) + 10n\,\lg\left(\frac{d}{d_0}\right) + X_\sigma(dB) \tag{18.8}$$

where $\bar{L}(d_0)$ represents the free-space path loss at the reference distance d_0 (usually set as 1 m), n is path loss exponent (PLE) obtained via minimum mean square error fitting, and X_σ is a zero-mean Gaussian random variable with standard deviation σ in dB. Figure 18.12 depicts the path loss fitting results in the conference room under different configurations. In the comparison of the co-polarization measurement results in LOS scenario with different antenna beamwidth and gain, it can be observed that wider beamwidth and lower radiation gain reduce the power contrast of the LOS component and rich reflected components, corresponding to a random constructive or destructive superposition of received signals. As for cross-polarization channels, the XPD decreases as the beamwidth increases for both LOS and NLOS scenarios. Consequently, the fluctuation of received power around the fitted path loss law is described as a variable following log-normal distribution (see Figure 18.13) with the validation of the Kolmogorov–Smirnov goodness-of-fit test.

Additional multiband channel measurement campaign has been conducted in the same conference room (see Figure 18.11(a)) using the DSS method to analyse indoor space-time propagation characteristics. The 3D power angular profile (PAP) of arrival is obtained via double-directional CIRs $h(\cdot)$ as

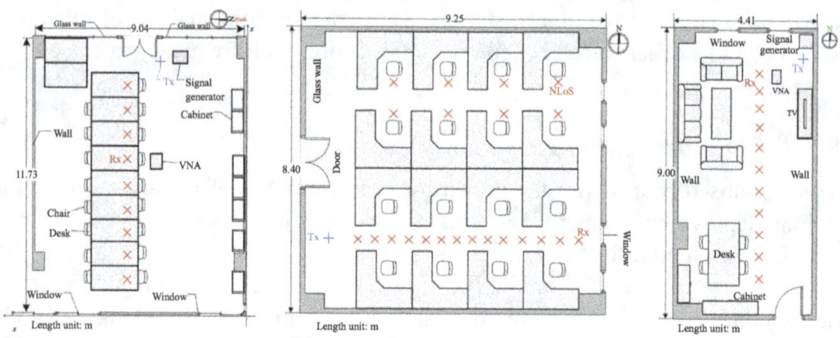

(a) (b) (c)

Figure 18.11 The three typical indoor reference scenarios defined in IEEE 802.11ad/aj [12]: (a) conference room, (b) cubicle office and (c) living room

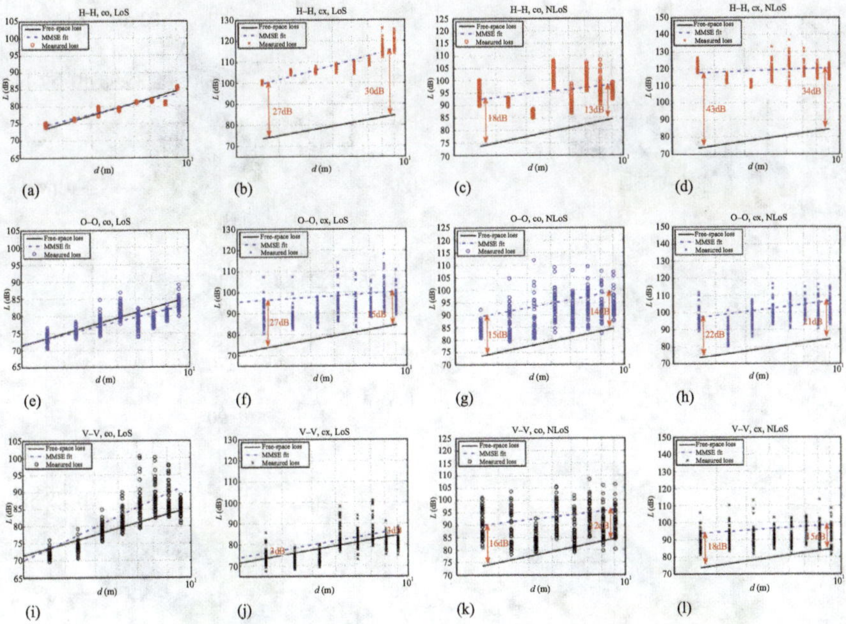

Figure 18.12 Fitting results of measured path loss using the CI model at 45 GHz in the conference room, where the H–H, O–O, and V–V in the subcaptions of (a) to (l) respectively represent horn-to-horn, open ended waveguide (OEW)-to-OEW, and V-dipole antenna (VDA)-to-VDA, and co and cx respectively represent co- and cross-polarization combinations [12]

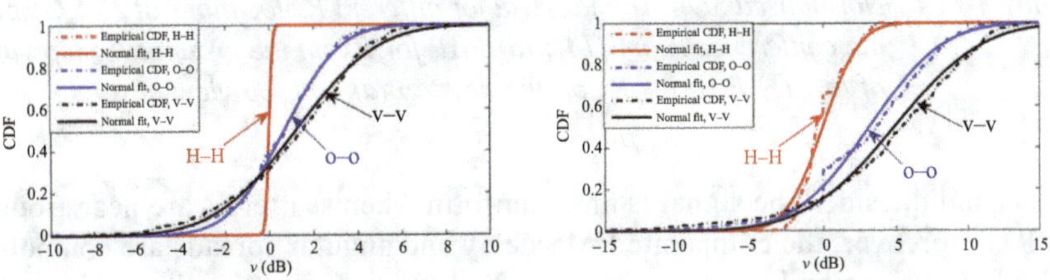

Figure 18.13 Empirical and log-normal fitting CDFs of the received power fluctuation in the conference room [12]: (a) LOS scenario and (b) NLOS scenario

$$P(\theta_R, \phi_R) = \sum_{\tau} \sum_{\phi_T} \sum_{\theta_T} |h(\tau, \theta_T, \phi_T, \theta_R, \phi_R)|^2 \tag{18.9}$$

where ϕ and θ denote azimuth and elevation angles. Figure 18.14 shows normalized 3D PAP examples for the RXs on the ground and desktop, indicating that multipath arrivals are more concentrated in several specific azimuth angles of

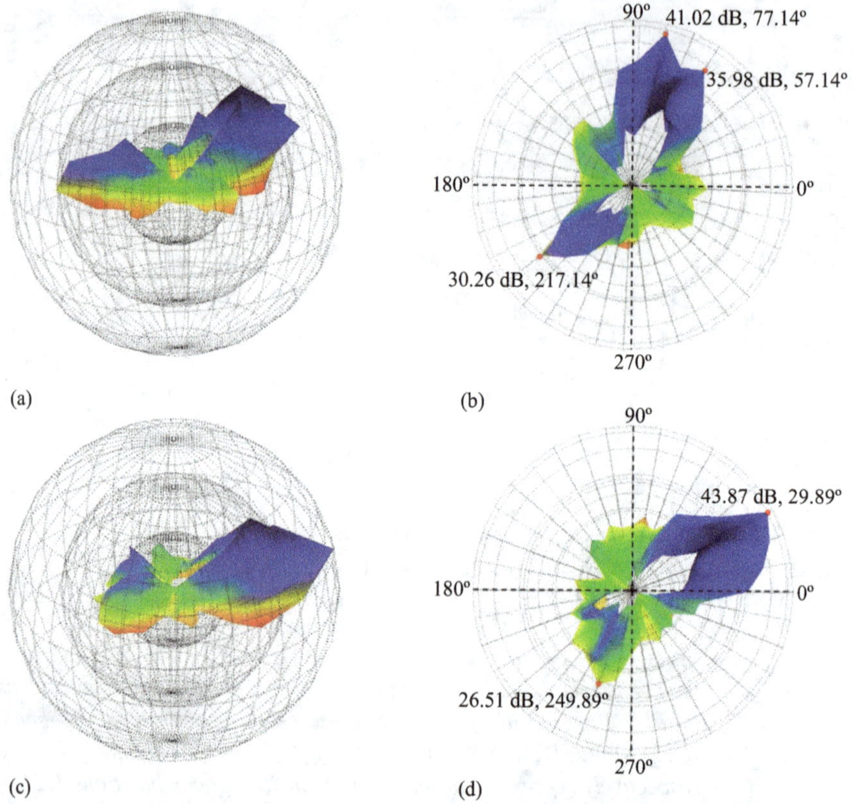

Figure 18.14 Normalized 3D PAP of arrival for different Rx locations at 25.5 GHz in the conference room [14]: (a) PAP for Rx on the ground; (b) top view of (a); (c) PAP for Rx on the desktop (d) and top view of (c)

arrival and the side lobe signal is more uniform when scatterers are near around the Rx. Moreover, the composite RMS delay and angular spreads are computed following the method developed by Fleury [20] and, in turn, the statistics of these channel parameters are modelled as a function of the carrier frequency f_c in GHz

$$\Lambda(f_c) = \alpha \lg(1 + f_c) + \beta \tag{18.10}$$

where α and β are the fitting parameters using the least square method. The frequency-dependent fitting results show that delay and angular spreads in general decrease with frequency increasing. Comparing with the mmWave channel measurement results in the waveguide-like environment, fewer MPCs in the semi-closed environment lead to smaller mean values of delay and azimuth angular spreads, attributed to the collection of multiple high-loss reflection signals from surrounding furniture.

18.3.2 Indoor transitional and blocked scenarios

In the semi-closed indoor environments, the coverage performance of mmWave networks will be significantly improved due to the existing LOS path along with ultra-dense deployment of access points (APs). Accordingly, cost-efficient network planning is urgently needed to reduce the density of APs while maintaining the coverage probability. From channel measurement and modelling perspectives, propagation characteristics in indoor NLOS scenario are of essential importance for the transmission scheme design. Figure 18.15 depicts the layout of indoor NLOS double-directional channel measurement, where Tx is placed in the hall and RXs are in the corridor, conference room and cubicle office. For the RXs in the corridor (*case 1*), diffraction over the building corner and reflection due to waveguide-like effect are dominant. On the other hand, RXs are separated from Tx by a glass wall and composite door (*case 2*), indicating that MPCs are blocked by obstacles.

The coverage performance of wireless communication systems mainly depends on the large-scale path loss model. The CI model provides a fair comparison of the path loss measurement results in terms of the PLE across different configurations such as frequency, bandwidth, and transceiver siting. However, despite the single-slope model that can well characterize the path loss varying with the T–R separation distance in simple environments, it ignores the blockage effect when the Rx locations are isolated via environmental obstacles. Thus, a dual-slope path loss model with a break-point (BP) distance of d_{BP} is developed [21] and written as

$$L(d) = \begin{cases} \bar{L}(d_0) + 10\alpha_0 \lg(d/d_0) + X_{\sigma_0}, & d \le d_{BP} \\ L(d_{BP}) + \beta_1 + 10\alpha_1 \lg(d/d_{BP}) + X_{\sigma_1}, & d > d_{BP} \end{cases} \tag{18.11}$$

where α_0 is PLE, α_1 and β_1 are optimized slope and intercept, and σ_0 and σ_1 in dB are the standard deviations of the zero-mean-Gaussian-distributed shadow fading components. Note that d denotes the propagation distance rather than the Euclidean distance between Tx and Rx, which ensures a precise path loss estimation or link budget in the NLOS setting. The path loss model in (18.11) becomes a CI model for $d \le d_{BP}$, where the set (α_0, σ_0) is obtained through a minimum mean square error fit to the measurement data in the lit region (i.e., LOS scenario) or shadow region

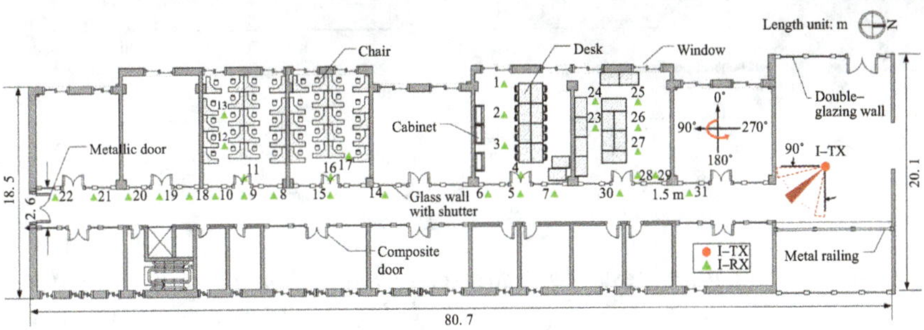

Figure 18.15 The layout of indoor transitional propagation measurement, including two kinds of NLOS scenarios [21]

(i.e., deep fading due to reflection and diffraction). Specifically, they are fitted based on the data measured with RXs in the corridor. For the post-BP segment, the floating intercept model is fit to the difference of $L(d) - L(d_{BP})$, where the path loss at the BP distance can be computed as

$$L(d_{BP}) = \bar{L}(d_0) + 10\alpha_0 \lg\left(\frac{d_{BP}}{d_0}\right) \tag{18.12}$$

Figure 18.16 shows the proposed dual-slop path loss model fitting results for the omnidirectional and directional channel. Comparing the fitting results with different models in the blocked region, it can be observed that the proposed model

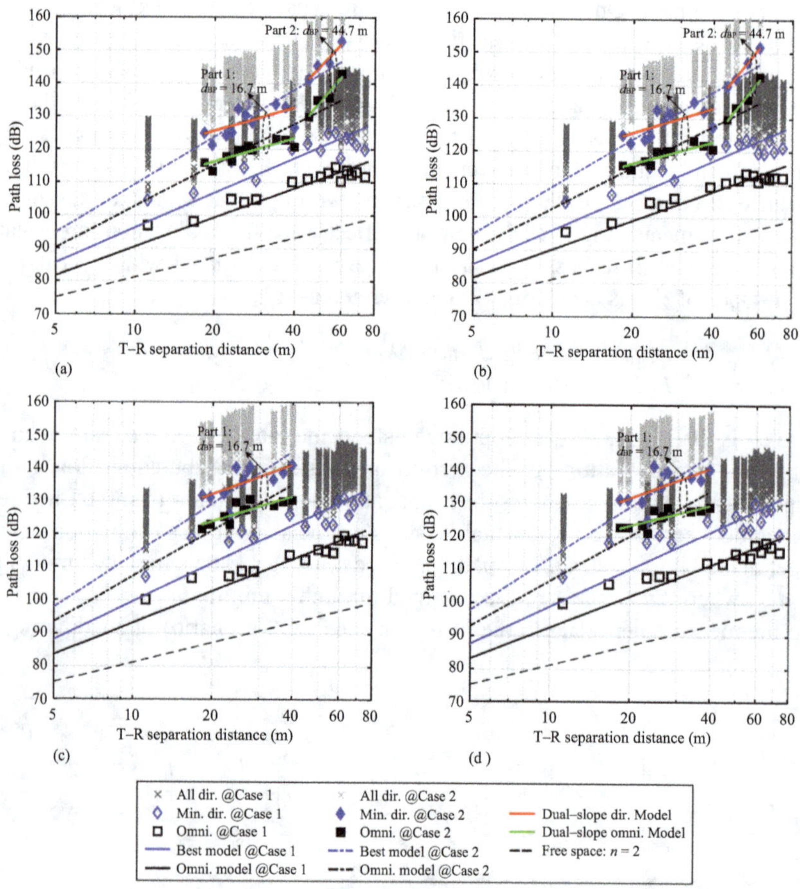

Figure 18.16 The fitting results of measured omnidirectional and directional path loss with single-slope and dual-slope path loss models, where (a) and (b) are obtained using the data in 27.5 GHz band with sounding bandwidths of 200 and 600 MHz, respectively, as well as (c) and (d) in 39.5 GHz band [21]

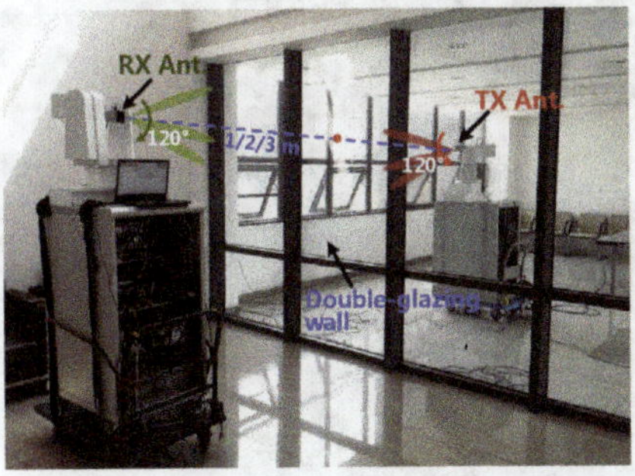

(a)

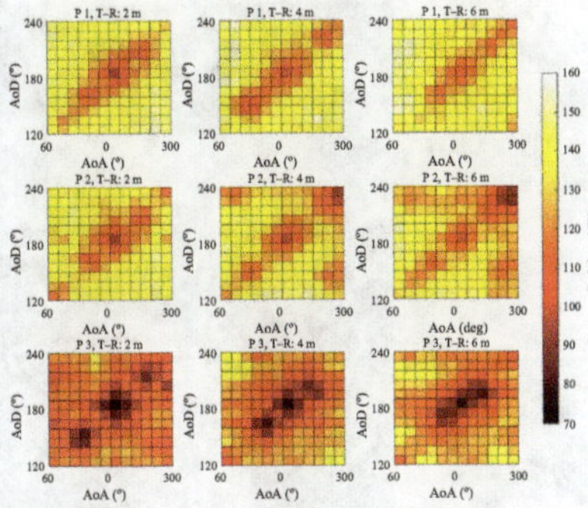

(b)

Figure 18.17 The site-specific penetration loss measurements at 27.5 GHz for various in-building materials with double-directional angles of arrival and departure combinations [22]: (a) measurement environment and (b) double-directional path loss

is more suitable when existing physical blockages. Therefore, it is reasonable to assume that β_1 denotes the indoor-to-indoor excess loss relative to $L(d_{BP})$ due to physical obstruction by building materials. Consequently, this dual-slope path loss model is verified by site-specific penetration loss measurements as shown in Figure 18.17 using the same sounding configurations. At P1 and P2, Tx and Rx are separated by a 25-mm double-glazing wall with a 15-mm air layer; and at P3, reference measurements are conducted without any obstacles to calculate free-space path loss. The penetration loss of the glass wall is computed as the difference between the directional transmission loss at P1 (or P2) and P3. It can be observed that the penetration loss of the double-glazing wall is 22.34 dB for the perpendicular incident, which is slightly larger than the β_1 (i.e., 20.4 dB at 27.5 GHz with 600-MHz bandwidth) due to lacking the combination of multipath signals from different directions.

18.3.3 Outdoor-to-indoor scenarios

Building a real-world 5G wireless network is a balance between capacity and coverage. 5G NR mmWave deployments can provide extremely high data rate in the indoor environment and outdoor cellular with the inter-site distance of 200 m but suffer in other areas such as coverage over distance and building penetration. Furthermore, in-building continuous coverage of mmWave wireless networks not only depends on the channel state information in indoor distributed or collocated antenna systems but also relates to the performance of outdoor BSs in the O2I scenario.

To help assess this deployment challenge for 5G in-building NR mmWave, O2I channel measurements have been conducted to obtain BEL following the guideline in ITU-R recommendation P. 2040. Figure 18.18 depicts the layout of Tx and RXs locations, where Tx is on the rooftop of the building in front of Building A5 with a height of 15 m above the ground, and indoor RXs are in the cubicle office and

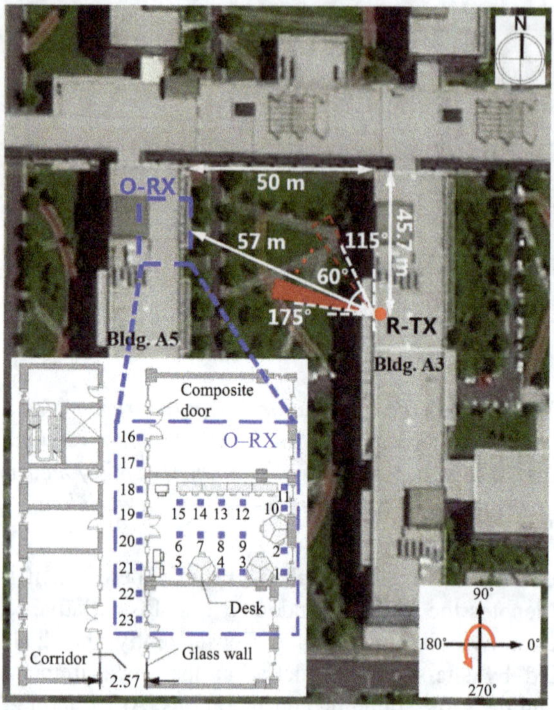

Figure 18.18 The general plan view of O2I scenario [21]

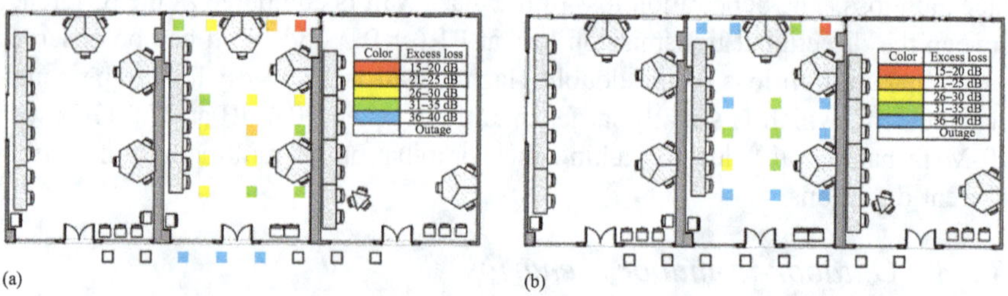

Figure 18.19 The composite BEL versus indoor Rx locations, where the colour of the square indicates the BEL at each Rx location [21]: (a) 27.5 GHz and (b) 39.5 GHz

corridor. The external building wall is composed of brick and concrete with standard double-pane windows, and the window-to-wall ratio is approximately 2:1. There are no obstacles between Tx and the facade of Building A5. The spatial density of points has been carefully considered to ensure that these RXs are uniformly distributed and able to collect sufficient channel data from all the unique aspects of the building.

Figure 18.19 shows the composite BEL at each Rx position measured at 27.5 GHz with sounding bandwidth of 600 MHz. The BEL is calculated as the excess

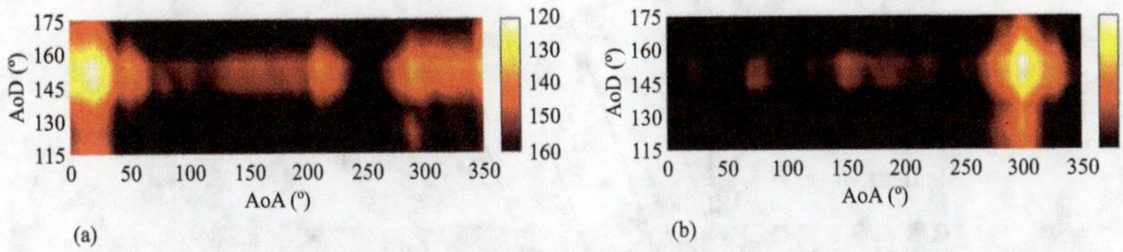

Figure 18.20 The double-directional path loss distributions in the angular domain at 27.5 GHz with different Rx [21]: (a) Rx1 near the window and (b) Rx blocked by the wall

propagation loss relative to the free-space path loss. It is clear that the BEL for the Rx nearest to the external wall and window is smaller than that for the Rx on the side of the office further from the Tx, albeit with some exceptions. For example, the BEL at Rx 14 is lower than the results at surrounding RXs because its location is in line with the incident direction and the penetration loss for double-pane glass windows is significantly smaller than that for brick walls. The maximum BELs are 32.1 and 38.9 dB at 27.5 and 39.5 GHz, respectively; and the differences between maximum and minimum BELs are 17.0 and 19.6 dB. These observations indicate that mmWave signals penetrating through the building blockages undergo substantial attenuation and BELs strongly depend on Rx locations and building structures. Figure 18.20 depicts the distribution of the directional path loss in the angular domain at two specific Rx positions. It can be observed that received signals mainly concentrate on the small angular region, and most signals will be blocked by external walls if wide-beam antennas are employed for sector coverage. Therefore, to realize energy-efficient deployment of 5G NR mmWave networks in the O2I scenario, it is necessary to establish stable communication links with very narrow beams.

18.3.4 Indoor building material reflection measurements

The cellular-type indoor channel measurements introduced before mainly focus on the extraction of large-scale and small-scale channel parameters, as well as their statistics. The composite and cluster-level characterizations can help to describe physical mmWave propagation channels. The multipath effect is caused by the reflection and diffraction over surrounding scatterers where first- and second-order reflections are dominant to enhance the coverage of mmWave networks. On the other hand, the accuracy of deterministic channel models, such as ray-tracing simulation and point cloud modelling, strongly depends on the reconstruction of propagation environments described by their physical dimension and electrical parameters.

Without loss of generality, only specular reflection is considered which means the incident angle equals the reflection angles exactly each time the antennas are moved. In this case, the reflection coefficient depends on the complex permittivity (permittivity and loss tangent) of the material, its thickness and surface roughness, the polarization and the incident angle. If the material is thick or lossy enough, successive internal reflections inside the material become insignificant. In this case,

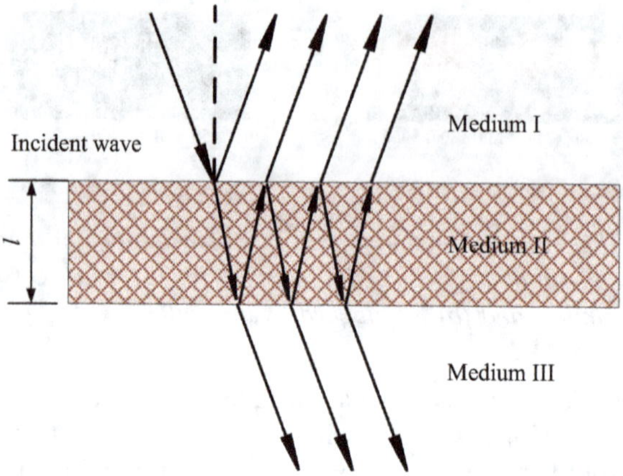

Figure 18.21 Reflection and transmission model of thin material

the polarization and angle dependence of the reflection coefficients can be well described as the Fresnel reflection formulas. Consider a slab of thickness l and real relative permittivity, ε_r, as is shown in Figure 18.21. The reflection and transmission coefficients of this slab are given by

$$R = \Gamma - \frac{\Gamma(1-\Gamma^2)\exp(jk_0 d \sin\,\theta)\exp(-j2ks)\exp(-2\alpha s)}{1-\Gamma^2\exp(jk_0 d \sin\,\theta)\exp(-j2ks)\exp(-2\alpha s)} \tag{18.13}$$

$$T = \frac{(1-\Gamma^2)\exp(-jks)\exp(-\alpha s)}{1-\Gamma^2\exp(jk_0 d \sin\,\theta)\exp(-j2ks)\exp(-2\alpha s)} \tag{18.14}$$

where θ is the angle of incident, $k_0 = 2\pi/\lambda$ is the phase constant in free space, $k = 2\pi\sqrt{\varepsilon_r}/\lambda$ is phase constant in the medium, $\alpha = \omega\sqrt{\mu_0\varepsilon_0\varepsilon_r}\tan\delta/2$ is attenuation constant in the medium and $\tan\delta$ is loss tangent, $s = l/\sqrt{1-\sin^2\theta/\varepsilon_r}$ is path length inside the slab, $d = 2l/\sqrt{\varepsilon_r/\sin^2\theta - 1}$ is path length difference, and Γ for perpendicular polarization is expressed as

$$\Gamma_\perp = \frac{E_\perp^r}{E_\perp^i} = \frac{\sqrt{\mu_2/\varepsilon_2}\cos\,\theta_i - \sqrt{\mu_1/\varepsilon_1}\cos\,\theta_t}{\sqrt{\mu_2/\varepsilon_2}\cos\,\theta_i + \sqrt{\mu_1/\varepsilon_1}\cos\,\theta_t} \tag{18.15}$$

and for parallel polarization is expressed as

$$\Gamma_\parallel = \frac{E_\parallel^{ri}}{E_\parallel} = \frac{-\sqrt{\mu_1/\varepsilon_1}\cos\,\theta_i + \sqrt{\mu_2/\varepsilon_2}\cos\,\theta_t}{\sqrt{\mu_1/\varepsilon_1}\cos\,\theta_i + \sqrt{\mu_2/\varepsilon_2}\cos\,\theta_t} \tag{18.16}$$

As illustrated in Figure 18.21, (1.6) and (1.7) suppose that the multiple bounces of the wave within the slab finally contribute to the total reflected wave and transmitted wave.

If the surface of the material is not smooth enough, an additional scattering reduction factor has to be employed, which is defined by

$$\rho_s = \exp\left[-8\left(\frac{\pi\sigma_h \sin\theta_i}{\lambda}\right)^2\right] \tag{18.17}$$

where σ_h is the standard deviation of the surface height about the surface's average height. The reflection coefficients listed in previous parts should be multiplied by a scattering loss factor ρ_s.

As shown in Figure 18.22, reflection measurements are conducted using the VNA-based channel sounder combining with the custom-designed arch support. The arch support has two racks for the antennas which can move along the arc to make the incident angle change from 0° to 75° with an interval of 5°. For each incident angular, the reflection coefficient is measured at 101 different frequency points from 40 to 67 GHz with an interval of 270 MHz to extract materials' characteristics varying with frequency. Also, the antenna polarization mode is changed from parallel polarization to perpendicular polarization during the measurement. When the material, angular and polarization mode are confirmed, the S_{21} parameters at each single frequency point can be obtained directly from the VNA, with both the amplitude and phase. However, when the incident angular is 0°, the S_{11} parameter is recorded using only one antenna on account of mechanical constraints. Before the practical measurements, reference measurements are needed to normalize the data due to free-space propagation. In general, a smooth metal plate can be used where the metal can be regarded as a perfect reflector in the mmWave band. The other method is to align Tx and Rx horn antennas directly with a distance twice as the distance between antennas and the material under test. The flowchart of mmWave reflection measurements of different kinds of building materials is illustrated in Figure 18.23.

Here, the reflection coefficient of the outer wall, which is made of concrete with plasterboard on its surface, is obtained as a function of the incident angle and frequency from 40 to 67 GHz. Figure 18.24 shows the reflection coefficient of the outer wall, where bold lines show the theoretical simulation results and the measured data fits well with these lines. Consequently, the complex permittivity of the outer wall can be derived. Moreover, the angular dependence of the reflection coefficient is similar to the frequency changes. As shown in Figure 18.25, the reflection coefficient cyclically fluctuates with the frequency when the electric field is parallel polarized; however, it can be modelled as a linear function of the frequency when the electric field is perpendicular polarized. Based on the measured reflection coefficient, the dielectric constants at different frequency points can be obtained, where the red and blue dotted lines in Figure 18.26 represent the results extracted via perpendicular and parallel polarization measurement data, respectively. Accordingly, the dielectric constant can be modelled as a linear function of the frequency written as $\varepsilon_r = -0.0371f + 5.1601$ with centre frequency f in GHz.

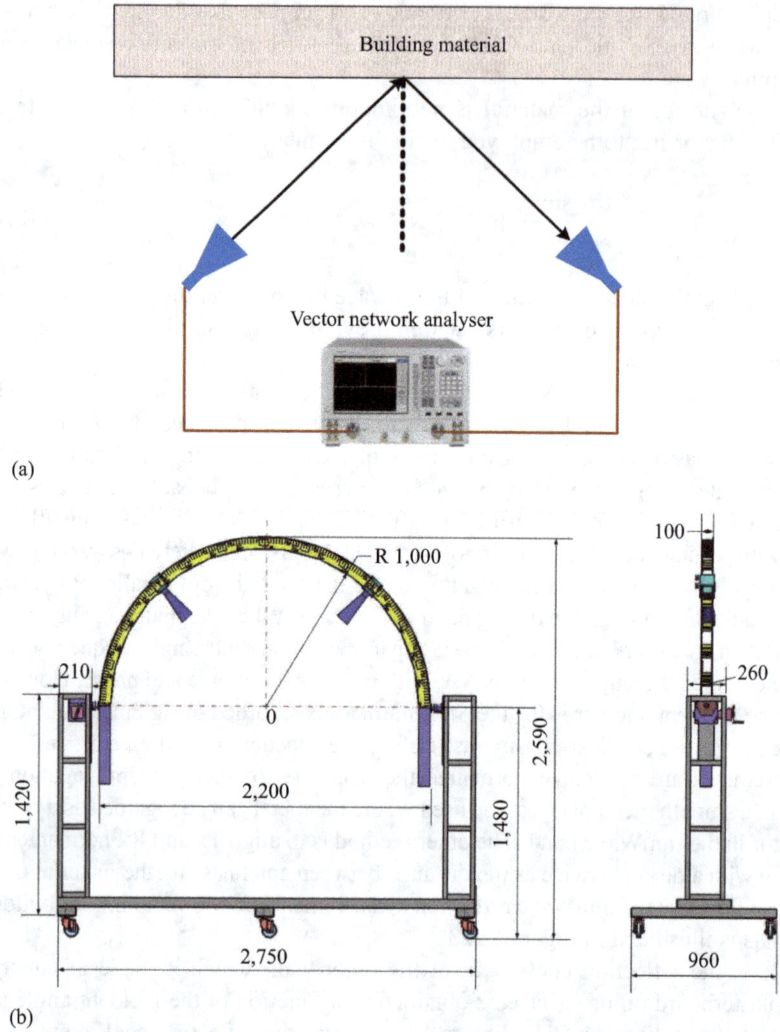

Figure 18.22 Illustration of reflection measurements for practical building materials: (a) the sketch map of the measurement setup and (b) design drawing of the arch supporter

18.4 Summary

In this chapter, we have discussed basic radio channel sounding requirements, challenges and methodologies towards commercial deployment of 5G NR mmWave networks. The standard organizations mainly focus on the channel characterization across multiple scenarios and frequency bands, which can be

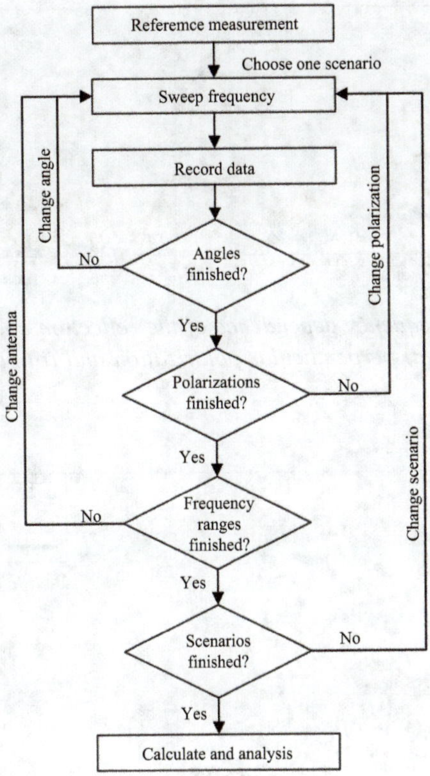

Figure 18.23 The flowchart of mmWave reflection measurement

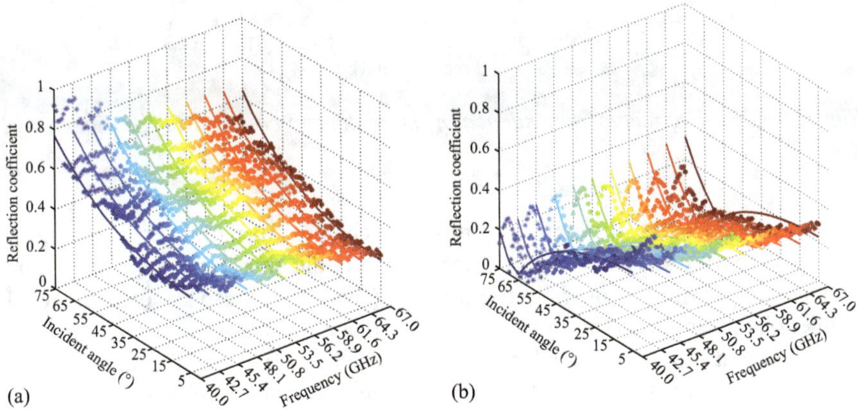

Figure 18.24 Comparison of measured and calculated data of the outer wall: (a) perpendicular polarization and (b) parallel polarization

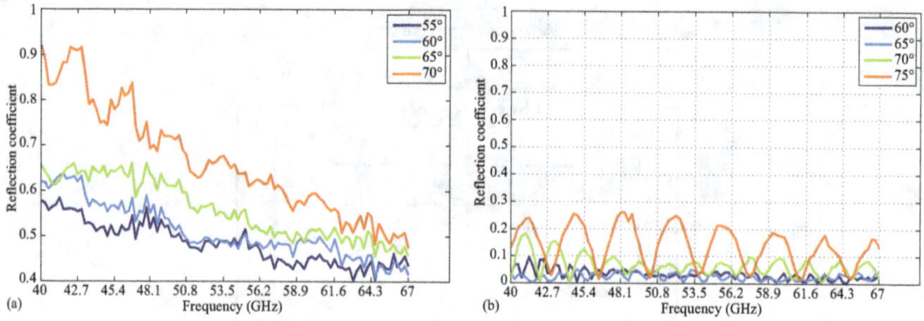

Figure 18.25 The frequency dependence of the reflection coefficient of the outer wall: (a) perpendicular polarization and (b) parallel polarization

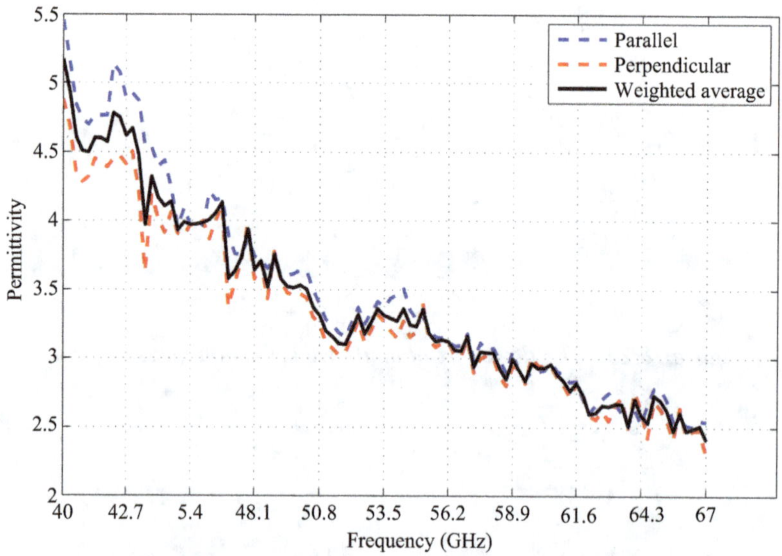

Figure 18.26 Frequency dependence of the dielectric constant of the outer wall

flexibly applied under different transmission schemes. In particular, severe path loss in the mmWave band can be compensated via the employment of large-scale antenna arrays with significant beamforming gains. Beamspace or directional channel characterization, in turn, is a natural choice for mmWave channel modelling. To obtain full-dimensional space-time mmWave propagation characteristics, both frequency-domain and time-domain channels using the different antenna and RF frontend configurations are overviewed, including DSS method for omnidirectional synthesis, virtual antenna array with single antenna moving in *x/y/z* directions, high-speed electronic-switch-based MIMO measurements and phased-array-based real

MIMO measurements. All these sounding methods can be implemented to support long-range mmWave measurements with high delay and angular resolutions combining with elaborate channel parameter estimation algorithms. Consequently, extensive indoor channel measurement campaigns have been conducted in various indoor LOS and NLOS scenarios. The empirical-based analyses of mmWave large-scale and small-scale channel characteristics are promoted, as well as the site-specific analysis of in-building reflection. A fully understanding of mmWave propagation characteristics in different scenarios and network architectures would accelerate the commercial deployment of 5G NR mmWave wireless networks.

References

[1] Rangan S, Rappaport TS, and Erkip E. 'Millimeter-Wave Cellular Wireless Networks: Potentials and Challenges'. *Proceedings of the IEEE*. 2014;102 (3):366–85.

[2] Marcus MJ. 'ITU WRC-19 Spectrum Policy Results'. *IEEE Wireless Communications*. 2019;26(6):4–5.

[3] 3GPP. *3GPP TS 38.101-2 V16.6. 0 (2021-01): NR; User Equipment (UE) Radio Transmission and Reception; Part 2: Range 2 Standalone* [online]. 2021. Available from https://www.3gpp.org/DynaReport/38101-2.htm [Accessed 14 Mar 2021].

[4] 3GPP. *3GPP TS 38.901 V16.1. 0 (2019-12): Study on Channel Model for Frequencies From 0.5 to 100 GHz* [online]. 2020. Available from https://www.3gpp.org/DynaReport/38901.htm [Accessed 14 Mar 2020].

[5] ITU-R M. *ITU-R IMT-2020 Channel Model Platform* [online]. 2018. Available from http://www.zjhlab.net/publications/imt-2020_cm_bupt/ [Accessed 14 Mar 2020].

[6] IEEE. *IEEE Std 802.11ad Channel Model* [online]. 2012. Available from https://ieeexplore.ieee.org/stamp/stamp.jsp?tp=&arnumber=6392842 [Accessed 14 Mar 2020].

[7] Haiming W, Wei H, Jixin C, Bo S, and Xiaoming P. 'IEEE 802.11aj (45GHz): A New Very High Throughput Millimeter-Wave WLAN System'. *China Communications*. 2014;11(6):51–62.

[8] Wang C, Bian J, Sun J, Zhang W, and Zhang M. 'A Survey of 5G Channel Measurements and Models'. *IEEE Communications Surveys & Tutorials*. 2018;20(4):3142–68.

[9] Parsons JD, Demery DA, and Turkmani AMD. 'Sounding Techniques for Wideband Mobile Radio Channels: A Review'. *IEE Proceedings I – Communications, Speech, and Vision*. 1991;138(5):437–46.

[10] Hammoudeh A, Scammell DA, and Sanchez MG. 'Measurements and Analysis of the Indoor Wideband Millimeter Wave Wireless Radio Channel and Frequency Diversity Characterization'. *IEEE Transactions on Antennas and Propagation*. 2003;51(10):2974–86.

[11] Haneda K, Nguyen SLH, Järveläinen J, and Putkonen J, editors. 'Estimating the omni-directional pathloss from directional channel sounding'. *The 10th*

European Conference on Antennas and Propagation (EuCAP); 10–15 April 2016.

[12] Zhu J, Wang H, and Hong W. 'Large-Scale Fading Characteristics of Indoor Channel at 45-GHz Band'. *IEEE Antennas and Wireless Propagation Letters*. 2015;14:735–8.

[13] Wu T, Wang H, Yu C, and Hong W. 'Wideband channel fading characteristics in corridor environment at millimeter-wave bands'. *The 11th European Conference on Antennas and Propagation (EUCAP)*; 19–24 March 2017.

[14] Zhang P, Li J, Wang H, Wang H, and Hong W. 'Indoor Small-Scale Spatiotemporal Propagation Characteristics at Multiple Millimeter-Wave Bands'. *IEEE Antennas and Wireless Propagation Letters*. 2018;17 (12):2250–4.

[15] MacCartney GR, and Rappaport TS. 'A Flexible Millimeter-Wave Channel Sounder With Absolute Timing'. *IEEE Journal on Selected Areas in Communications*. 2017;35(6):1402–18.

[16] Wu X, Wang C, Sun J, *et al.* '60-GHz Millimeter-Wave Channel Measurements and Modeling for Indoor Office Environments'. *IEEE Transactions on Antennas and Propagation*. 2017;65(4):1912–24.

[17] Keysight. *5G Channel Sounding, Reference Solution* [online]. 2019. Available from https://www.keysight.com/be/en/assets/7018-04917/brochures/5992-0983.pdf [Accessed 14 Mar 2020].

[18] Sayeed A. *MmW Multi-beam CAP-MIMO Testbed* [online]. 2017. Available from https://dune.ece.wisc.edu/the-cap-mimo-project/695-2/ [Accessed 14 Mar 2020].

[19] Bas CU, Wang R, Sangodoyin S, *et al.* 'Real-Time Millimeter-Wave MIMO Channel Sounder for Dynamic Directional Measurements'. *IEEE Transactions on Vehicular Technology*. 2019;68(9):8775–89.

[20] Fleury BH. 'First- and Second-Order Characterization of Direction Dispersion and Space Selectivity in the Radio Channel'. *IEEE Transactions on Information Theory*. 2000;46(6):2027–44.

[21] Zhang P, Yi C, Yang B, Wang C-X, Wang H, and You X. 'In-Building Coverage Of Millimeter-Wave Wireless Networks From Channel Measurement and Modeling Perspectives'. *Science China Information Sciences*. 2020;63(8):180301.

[22] Zhang P, Wang H, and Hong W. 'Empirical Analysis of millimeter-wave propagation in indoor transitional environments'. *Proceedings of the IEEE International Symposium on Antennas and Propagation and North American Radio Science Meeting 2020*; July 2020.

Chapter 19

Terahertz communications on various beyond 5G real-world use cases for IoT, railway, train, drone communications—propagation channel characterization and challenges

Ke Guan[1], Haofan Yi[1], Danping He[1] and Zhengrong Lai[2]

19.1 Introduction

With the full deployment of the fifth-generation wireless communication system (5G), research on beyond 5G or the future sixth-generation wireless communication system (6G) is foreseen to utilize upper millimeter wave (mmWave) (100–300 GHz) and terahertz (THz) band (0.1–10 THz) frequencies. As a new generation of mobile communication systems appears about every 10 years, the industry and academia expect the new communication system to be commercially deployed before 2030. It usually takes more than 10 years for new technology from birth to commercial use. Therefore, there is an urgent necessity to carry out the research on new beyond 5G and/or 6G technologies.

The rapid development of smart terminal applications requires future communication systems to achieve ubiquitous ultrahigh data-rate access in various complex environments. To this end, one solution is to improve spectrum efficiency through advanced modulation schemes and signal-processing techniques. However, due to the basic limitation of the current operating frequency band (narrow bandwidth), it is difficult to reach 100 Gbps or even a few Tbps. Another solution is to use a higher carrier frequency to increase the channel bandwidth to provide sufficient transmission capacity.

The upper mmWave and THz frequencies have been identified as the candidate frequency bands to address the problems faced by current and future wireless communication systems [1]. Compared with the mmWave and the optical communications, THz communications have attracted great attention in recent years owing to its unique advantages in terms of [2]:

[1]State Key Laboratory of Rail Traffic Control and Safety, Beijing Jiaotong University, Beijing, China
[2]Guangdong Communications & Networks Institute, Guangzhou, China

1. Tbps level data transmission rate:
   The effective bandwidth of the THz band is much wider than an mmWave band. It can provide a wireless transmission link at the Tbit/s level. However, the mmWave and optical communications can only provide a data rate of 10 Gbit/s.
2. Little impact of meteorological factors [3]:
   The optical communication systems can be applied to satellite-to-ground and inter-satellite communication links. However, the performance of the optical communication system is influenced seriously by unpredictable environmental meteorological factors, such as clouds, fog, rain, and so on. The short THz wavelength is relatively not easily scattered. With sufficient link budget established for both optical and THz communications, when encountering weather such as fog, dust, and turbulence, THz communication performance is relatively stable compared with optical communications.
3. Realization of multiusers communications:
   Compared with THz communications, optical communications have higher directivity. However, this requires extremely high directivity at the transceiver end, so only point-to-point communication can be achieved for IR and visible light frequency band. On the contrary, mmWave and THz systems might realize multiusers to single-user communications due to non-line-of-sight (NLOS) propagation paths thanks to the fast growing research effort on designing reconfigurable intelligent surfaces (RISs) [4]. To break the barriers set by blockage, recently, some researches explain how RIS can be used to mitigate the impact of blockage and introduce the idea of reflected links [5,6]. Without the help of RIS, the NLOS path with mmWave relies on a bounce from scatterers whereby it is envisaged tougher challenges for THz waves reflected and scattered from the rough surfaces.

By considering the shortcomings of current communication systems and unique advantages of the THz frequency band, THz communication technology has attracted widespread attention from academia and industry. It is considered to be a key wireless technology that meets the real-time traffic demand of mobile heterogeneous network systems and realize ultrahigh data-rate wireless communications. The ultrawide bandwidth of the THz band and the supportable ultrahigh transmission rate will enable a large number of new applications and services to become reality. According to the channel modeling document of IEEE 802.15-TG3d [7], close proximity peer-to-peer (P2P) applications, intra-device communications, backhaul/fronthaul, wireless local area networks (WLAN)/wireless personal area networks (WPAN), and data center network will be realized and bring a better user experience than wired communication systems. For the close proximity P2P applications, the kiosk downloading scenario, which offers ultrahigh downloads of multimedia content, is uniformed to line-of-sight (LOS) channel with a transmission distance of a quite short range. The intra-device communications build wireless links inside computers or any other electronic devices. It is of high relevance because wired connectors and micro strip lines on printed circuit boards potentially become a bottleneck of upcoming bus systems and inter chip connections.

WLAN/WPAN offers ultrahigh-speed ad hoc connections in indoor environments. The deployment of wireless fronthaul/backhaul links will be outdoors with transmission ranges of several 100 m up to a few kilometers. Very directive antennas will be static deployed, allowing antenna alignment during the installation process by radio engineers. The wireless data center network uses wireless data links to replace/complement the traditional cable connections, which brings various advantages, e.g., high flexibility and low maintenance cost. The high data rate requirement makes the THz technology a competitive candidate because of its high available bandwidth up to 50 GHz.

The interaction of theory, simulation, and measurement is the core of various channel modeling approaches, of which the essential difference is the trade-off in terms of simulation complexity and measurement difficulty. In sub-6-GHz bands, there was no big challenge in conducting extensive channel measurements. Along with the increase of carrier frequency, it is much more challenging to conduct the wideband channel measurements at the THz band. In order to compensate for the high path loss, the highly directive antennas are used in the channel measurement. In addition, in order to catch multipaths from all directions, THz channel sounders often resort to rotating high-gain steerable antennas. This measurement method is relatively high time costing (manually or mechanically and therefore cannot capture the dynamically changing channel effects) [8]. Moreover, the acquisition time of many THz channel sounders exceeds the coherence time of nonstationary channels because of using the stepped-frequency systems [9]. Thus, very few systems to date can measure both the delay and angular properties of a nonstationary wideband THz channel. Compared with the lower frequency band, besides the difficulty in channel measurements, the surface roughness has a clear effect on the diffuse scattering energy when THz waves propagate. The behavior of scattering is obvious [10]: the rougher the surface is, the higher is the diffuse scattering contribution, i.e., the contribution coming outside the specular reflected component increases as the surface roughness increases. By considering the roughness factor into the specular reflection, the modified Fresnel reflection equations [11] can sufficiently describe the reflected paths. Therefore, for the THz channel modeling, it is not only an increase in frequency, it also shows a clear difference from lower frequency channels.

Hence, in order to effectively support the design, evaluation, and development of the THz communication of different applications and scenarios, the fundamental knowledge to address specific challenges in metrology for THz propagation channel characteristics and channel modeling is of paramount significance.

19.2 THz channel-modeling approach

The channel model that captures the propagation characteristics of the THz band has far-reaching significance for the practical applications and a specific deployment of THz communications. To design a THz wireless communication system, we need first to perform channel measurements, study the characteristics of THz wave propagation, and model the THz channel.

The THz frequency band has the characteristics of high frequency and short wavelength, which makes the THz waves experience great free path loss during wireless transmission. It is precise because of the short wavelength that a large number of antennas can be placed in a very small space [12]. How to make a full use of the superiority of the THz frequency band and establish effective and accurate THz channel models is the key problem. Generally, the methods for channel modeling can be categorized as stochastic approach [13], deterministic approach [14], and hybrid approach [15]. These three methods for THz channel are reviewed as follows.

19.2.1 Stochastic channel modeling

Stochastic channel modeling approach is usually based on the empirical channel measurements. Compared to the lower frequency communication system, the major difference in THz wireless system is the ultrawide bandwidth, which results in the resolvable multipath components (MPCs) [16]. Thus, a stochastic channel impulse response (CIR) model should be developed in the THz band, rather than reusing narrowband fading models like Rayleigh or Rician. The stochastic standard channel models, for instance, the whole cluster-based standard channel model families [17]—WINNER-style (e.g., 3GPP spatial channel model (SCM), extended SCM (SCME) [18], WINNER (WIM1), WINNER II (WIM2) [19], WINNER+ (WIM+) [20], three-dimensional (3D) multiple input and multiple output channel models [21]) and COST2100 [22]—were majorly parameterized by channel sounding data. These models are named measurement-based pseudo-geometric models (MBPGMs) as well. After parameterized by measurement data, the MBPGM families can generate channels in certain types of scenarios for link level and system level simulations. Another typical stochastic model family is called geometry-based stochastic models (GBSMs) [23,24]. General speaking, GBSMs can realize CIRs for performance analysis from the assumed scattering geometry without any measurement [25–27]. Nevertheless, when MBPGMs and GBSMs are utilized for a specific scenarios, it is still necessary to determine the values of probability distribution parameters according to measurements [28]. The CIR h of cluster-based stochastic standard channel models can be characterized as follows [29]:

$$\begin{aligned} h(t) = &\sum_{l=0}^{L}\sum_{k=0}^{K_l}\beta_{k,l}e^{j\chi_{k,l}}\delta(t - T_l - \tau_{k,l}) \\ &\times \delta(\theta_{\mathrm{AoA}} - \Omega_l - \omega_{k,l})\delta(\theta_{\mathrm{EoA}} - \Psi_l - \psi_{k,l}), \end{aligned} \tag{19.1}$$

where T_l, Ω_l, and Ψ_l represent delay, azimuth angle of arrival (AAoA), and elevation angle of arrival (EAoA) of the lth cluster, respectively. $\tau_{k,l}$, $\omega_{k,l}$, and $\psi_{k,l}$ denote the delay, AAoA, and EAoA of the kth path in the lth cluster, respectively. $\beta_{k,l}$ denotes the amplitude of the kth path in the lth cluster and $\chi_{k,l}$ is used to describe the statistically independent random variables, which are normally distributed within $[0, 2\pi)$. A stochastic channel modeling at 300 GHz for the indoor scenario is well studied in [13]. By a clustering observation of the MPCs, the cluster-based model provides the specific parameters, including the complex amplitude, the delay, and the angles of arrival (AoAs) and angles of departure (AoDs) in the indoor scenario. In [30], the famous Saleh–Valenzuela model is proposed that the delays of the clusters and the intra-cluster paths are following the Poisson processes.

However, along with the increase of carrier frequency, it is much more challenging to conduct the wideband channel sounding at a THz band. In order to compensate high path loss and catch multipaths from all directions, channel sounders often resort to rotate high-gain steerable horn antennas, which is either relatively slow (manually or mechanically and therefore cannot capture the dynamically changing channel effects) or very costly using phase shifters (PSs). Thus, very few systems to date can measure both the delay and angular properties of the THz channel. Therefore, the stochastic channel based on the empirical channel measurements is relatively hard to be applied in the THz band.

19.2.2 Deterministic channel modeling

Deterministic channel models can accurately describe the wave propagation based on electromagnetic (EM) theory. The deterministic channel models are usually site-specific channel models, which require the detailed geometry of the propagation environment, the information and EM properties of the materials, and the locations of the transceivers. Generally, the deterministic approach can provide good consistency between the simulations and measurements. Moreover, this approach can be performed with higher degrees of freedom at the THz frequency band.

As a typical representative of the deterministic model, ray-tracing (RT) is gradually spreading in 5G and beyond wireless communications. RT based on optical theory is more relevant to the physical properties of THz waves. The results of RT simulations give the full information of the propagation rays through a series of geometric and EM calculations, such as the ray type (LOS, reflection, scattering, penetration, diffraction, as shown in Figure 19.1), the complex amplitude, the azimuth angles and elevation AoA and departure. These ray properties from RT simulations can accurately describe the determined channel characteristics. Besides, RT is suitable for the prediction and modeling of the time-varying and multi-antenna channels in different frequency bands. Since in the wireless communication system the bandwidth is limited, the rays calculated by RT cannot be completely distinguished in the corresponding delay resolution. Therefore, in order to obtain the CIR that conforms to the system settings in the real communication system, the sub-band division method is adopted to obtain the channel transfer function (CTF) of the entire bandwidth. Figure 19.2 shows the workflow of RT procedure for obtaining the CIR of the target frequency bandwidth. At first, the whole bandwidth is divided into N sub-bands. The division of sub-bands requires the frequency independence of the materials in this sub-band in order to meet the channel stationary in the frequency domain. Then, the center frequency point of each sub-band is as the frequency point to calculate the multipath delay distribution in each sub-band. Due to the material frequency dependence, even if the same

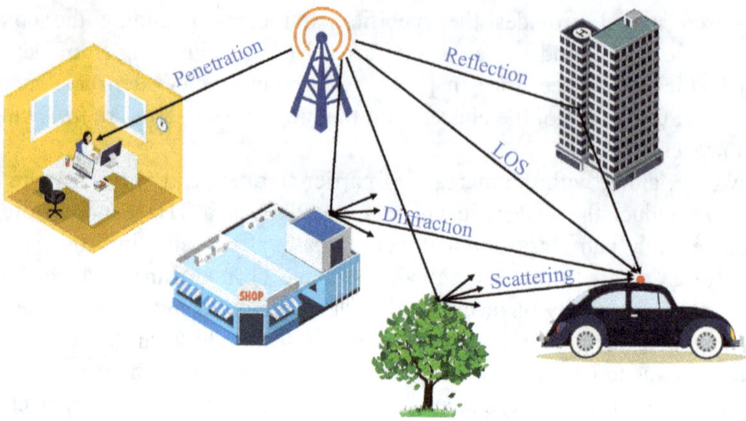

Figure 19.1 The propagation mechanisms of ray-tracing method

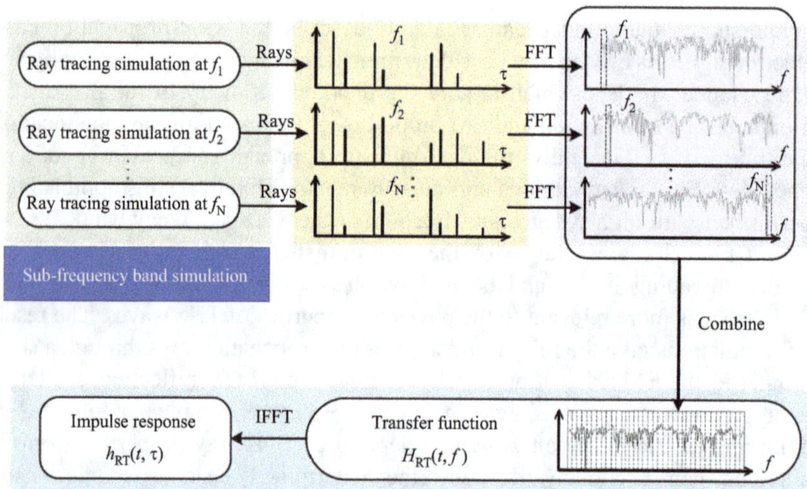

Figure 19.2 The workflow of RT procedure for obtaining CIRs. © 2019 IEEE. Originally from Figure 3 of [31].

multipath components in each sub-band are consistent in time delay, they may cause different distribution lines in each sub-band due to inconsistent amplitudes. After that, through fast Fourier transform (FFT), each sub-band is transformed from the time domain to the frequency domain. These CIRs of these different frequency bands are accurate only near their respective center frequency points. At last, the CTF of the whole frequency band is combined by each sub-band, and thus, the CIR in the time domain is transformed by inverse FFT (IFFT).

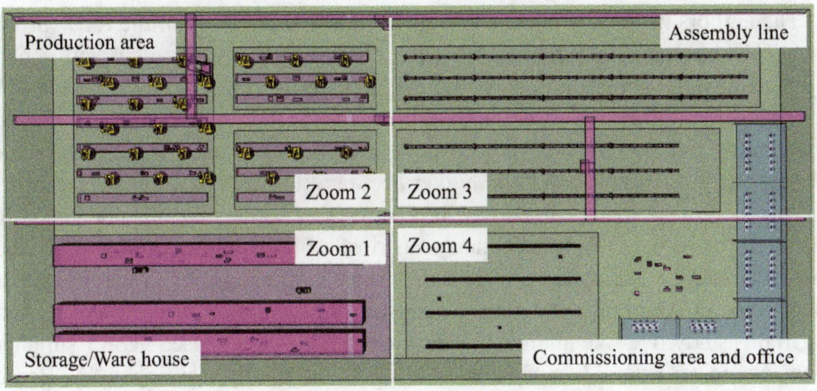

Figure 19.3 Panorama of smart industries and automation

Therefore, the CTF H_{RT} of each ray that can be defined as a function of delay, frequency, and angle, given by [32]

$$\begin{aligned} H_{\mathrm{RT},i} &= a_i e^{j\phi_i} D_i(f) e^{-j2\pi\tau_i(f-f_0)} \\ &\times \delta(\varphi_{\mathrm{AoD}} - \varphi_{\mathrm{AoD},i})\delta(\theta_{\mathrm{AoD}} - \theta_{\mathrm{AoD},i}) \\ &\times \delta(\varphi_{\mathrm{AoA}} - \varphi_{\mathrm{AoA},i})\delta(\theta_{\mathrm{AoA}} - \theta_{\mathrm{AoA},i}), \end{aligned} \tag{19.2}$$

where a_i denotes the amplitude of ray i at reference frequency f_0, ϕ_i denotes its phase, τ_i is the time of arrival (TOA), φ_{AoD} denotes the azimuth angle of departure (AAoD), θ_{AoD} denotes the elevation angle of departure (EAoD), φ_{AoA} denotes the AAoA, and θ_{AoA} denotes EAoA. Moreover, $e^{-j2\pi\tau_i(f-f_0)}$ accounts for the frequency-dependent phase rotation on the ray, and $D_i(f)$ denotes a frequency dispersion function. Note that H_i, a_{i,f_0}, ϕ_i, τ_i, and all the angles are time-varying. In order to have a concise expression, the time variable t is omitted in (19.2). The multipath amplitudes in a_i can further be decomposed into their co- and cross-polarized components, becoming a fully polarimetric channel matrix as

$$a_i = \begin{pmatrix} a_{i,vv} & a_{i,vh} \\ a_{i,hv} & a_{i,hh} \end{pmatrix}, \tag{19.3}$$

where $a_{i,vv}$ and $a_{i,vh}$ denote the amplitude of vertical–vertical and vertical–horizontal polarizations of ray i, respectively; $a_{i,hh}$ and $a_{i,hv}$ denote the amplitude of horizontal–horizontal and horizontal–vertical polarizations of ray i, respectively. For the LOS ray, $a_{i,vh}$ and $a_{i,hv}$ are both zero.

In addition, in order to obtain the CTF of the entire channel, the path-specific CTFs of all the rays in (19.2) need to be summarized and weighted with the polarimetric Tx and Rx antenna radiation patterns G_{tx} and G_{rx}:

$$H = \sum_{i=1}^{N_{\mathrm{Rays}}} G'_{tx} H_{\mathrm{RT},i} G_{rx}, \tag{19.4}$$

where N_{Rays} denotes the number of rays.

The authors in [33] point out the significance and accuracy of the RT approach in the THz frequency band. In [14], the authors develop a complete and generic multipath channel model based on the RT method for the entire THz band. The open problems and future research directions for the RT approach are to develop the advanced acceleration techniques for the higher efficiency and the veracious EM properties of materials for the higher accuracy in the THz frequency band.

Since computational complexity is always the bottleneck of conducting RT simulations in realistic communication scenarios, a promising solution is to develop the RT that can be deployed on high performance computing (HPC) platforms. This will effectively speed up the compute-intensive tasks via parallel hardware technology. In this study, we take an in-house-developed HPC cloud-based 3D RT simulator as an example. This RT simulator—CloudRT—is jointly developed by Beijing Jiaotong University and Technische Universität Braunschweig. More information on CloudRT can be found in tutorial [34] and http://www.raytracer.cloud. Later in this paper, CloudRT will be used in some of the case studies.

19.2.3 Hybrid channel modeling

The traditional empirical measurement-based statistical channel modeling approach gives the high efficiency to characterize the MPCs but has more challenging to conduct the full-dimensional of the propagation channels at the THz frequency band. Thus, the accuracy of the channel models is weak. The deterministic channel model is site specific and with more computational complexity. Therefore, a statistical and RT hybrid method should be mentioned at the THz frequency band.

In the standardization organization, such as 3GPP [35] and ITU-R WP-5D [36], they use map-based hybrid channel modeling method to evaluate 5G communication technology, and RT is a major step of this approach. Besides, in the IEEE standard working group 802.15 3d-2017 [37], RT has been successfully applied to assist in the establishment of statistical channel models for various scenarios in mmWave and THz frequency bands. The main benefits of the hybrid modeling approach are the high accuracy and low complexity. The reason is that the main critical channel parameters are computed deterministically and that rich MPCs are included by the statistical modeling.

19.3 Real-world use case for Internet of things

To realize the Internet of everything, the 6G Internet of things (IoT) network needs to provide wireless access with low power, massive connectivity, and wide coverage. According to the prediction of Ericsson Mobility Report [38], there will be 29 billion IoT devices worldwide by 2022. From the fact that most IoT devices are connected through advanced wireless technologies, their coexistence issues are very important [39]. Thus, first of all, the characteristics of the propagation scenarios and the channel models should be studied.

The IoT communication includes vehicular communications in intelligent transportation systems, communication between intelligent robots and control centers in Industry 4.0, a large number of sensor devices in smart cities uploading data to base stations, and so on. Unlike traditional mobile phone users, IoT devices do not require high transmission rates. On the contrary, it has high requirements on energy consumption, time delay, reliability, and the number of connected devices. For example, robots in smart factories often perform dangerous operations. Therefore, the communication network in the industrial Internet of things (IIoT) must reliably transmit control commands to the robots within a very low time delay. Based on the new features of IoT communication, commercial 5G cellular communication technology lists large-scale massive machine-type communication and ultrareliable and low latency as two major application scenarios. Beyond 5G/6G communication technologies are given high hopes for a better realization of the two major scenarios as well.

Figure 19.3 shows a typical smart factor as an IIoT scenario with multi-functional subareas [40]. According to the different densities, the factor can be divided into four different zooms. In IoT environments, various robots, sensors, and mechanical devices need massive connections in a robust and efficient manner [41]. The IoT channel meets many new challenges to characterize the channel, such as the varied path loss due to the different densities of a scenario, a large numbers of scatterers, blockages, and multi-mobility. Only a few channel measurements have been conducted for the IIoT scenario. Most of these are at the sub-6-GHz frequency band [42–44], which is in-line with current IoT standard. However, the frequency at mmWave and THz bands will surely promise the IIoT scenarios with more massive connections with higher transmission data rates.

19.4 Real-world use case for railway and train

As a vital vertical industry empowered by the reliable high data rate transmissions of 6G [45], "smart rail mobility" is expected to become reality from a concept [46,47]. To meet the goals with respect to efficiency, safety, and convenience, the new era of "smart rail mobility" requires that the infrastructure, trains, travelers, and goods will be increasingly interconnected. For smart rail mobility, railway communications are required to evolve from handling only the critical signaling applications to various high-data-rate applications: on-board and wayside high-definition (HD) video surveillance, onboard real-time high-data-rate connectivity, train operation information, real-time train dispatching HD video, and journey information [48]. These applications can be realized in various rail scenarios. Table 19.1 summarizes the relevant scenarios and high-data-rate applications therein [48]. Figure 19.4 illustrates the panorama of all the smart rail mobility scenarios, requiring the seamless wireless connectivity with the data rate on the order of 100 Gbps or even more, forming a strong driving force for exploiting the THz band, i.e., beyond 300 GHz, where there are huge continuous available spectrum resources [49,50]. Generally, current channel models are mainly for standard indoor and outdoor environments. Usually, railway appears as a small use case of moving hotspot in standard documents. Thus, how to include the railway channel features in channel models is still an open challenge. In

Table 19.1 Smart rail mobility scenarios and high-data-rate applications therein

Smart rail mobility scenarios	On-board and wayside HD video surveillance	On-board real- time high-data rate connectivity	Train operation information	Real-time train dispatching HD video	Multimedia journey information
Train-to-infrastructure (T2I) [54]	*	*	*	*	*
Inside station [55]	*	*		*	*
Train-to-train (T2T) [56]		*	*		*
Infrastructure-to-infrastructure (I2I) [57]	*		*	*	*
Intra-wagon [58]		*	*		*

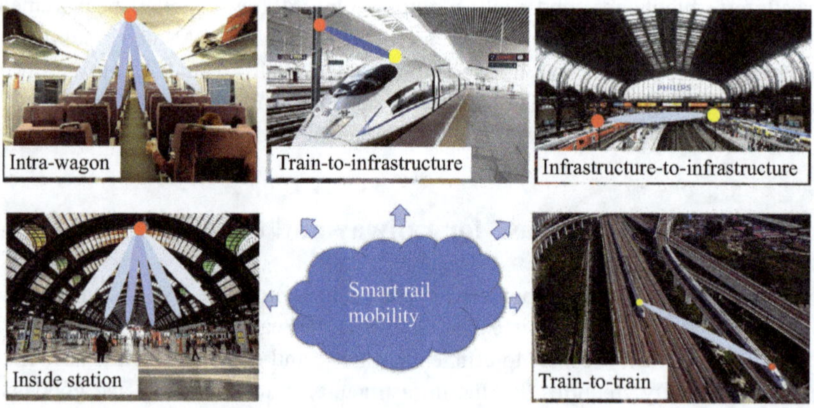

Figure 19.4 Panorama of smart rail mobility scenarios

terms of the modeling approach, for high-speed environments, geometry-based models are generally better suited. Another challenge is channel dynamics. Most of the existing measurements for the railway were done for static channels. In 2014, Samsung presented the worlds first demonstration for 1.2-Gbps transmission at 28 GHz at 110 km/h cruising speed using dual beamforming [51]. This evidence shows the strong potential of THz bands for mobile applications. This should be theoretically and experimentally studied in five rail traffic scenarios for future work. Without any doubt, to fully understand the propagation channel in these scenarios is of importance [52,53].

To carry out the channel modeling for the smart railway mobility scenario, a novel M-sequence correlation-based UWB channel sounder [59] is utilized in a

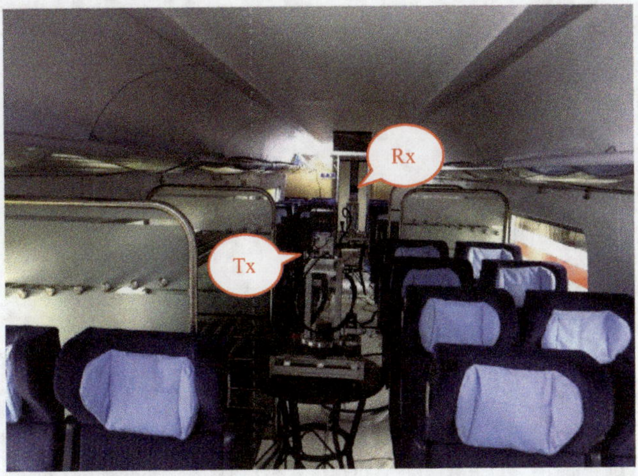

Figure 19.5 Measurement campaign for intra-wagon channel. © 2021 IEEE. Originally from Figure 2 of [60].

measurement campaign in an HST test center inside station scenario [55] and Intra-wagon scenario [58,60]. This channel sounder is composed of a controlling laptop, a base unit, sensor nodes, frequency extensions, and multifunctional cables. Different from the VNA-based channel measurement system, a channel sounder transmits a broadband maximum length sequence (M-sequence) as a reference signal, which has a Dirac-impulse-shaped autocorrelation function in time domain. Thus, the broadband CIR can be obtained through correlation between the transmitted M-sequence and the received signal. Between 300 and 308 GHz with an 8-GHz bandwidth, this channel sounder can measure 17,590 CIRs/s with a dynamic range over 30 dB. As stated in our recent study [61], in which it can be seen that the M-sequence is generated by the clock frequency of 9.22 GHz, which has a length of 4,095 chips and duration of 444.14 ns, allowing a maximum path length of around 133 m. Correspondingly, the delay resolution is the chip duration of 108.5 ps that corresponds to a length of 3.25 cm. Since the high-speed analog-to-digital converters are expensive and have a worse performance in terms of dynamic range and noise, a subsampling technique is used in this channel sounder with a fixed subsampling factor of 128 and the fixed 12th-order M-sequence. Thus, the recording of one CIR takes 56.9 μs ($= 128\times 4,095 \times (1/9.22)$ GHz). Correspondingly, the measurement rate is 17,590 CIR/s. As an example, Figures 19.5 and 19.6 show measurement environment and the synthesized omnidirectional power delay profile (PDP) measured in the Intra-wagon scenario, respectively, which is obtained by combining the measured horizontal-directional-scan-sounding CIRs and is detailed in [58,62].

CloudRT is employed to do RT simulations in frequency domain, generating the CIR for one snapshot by a number of sub-bands at multiple center frequency points [63,64]. For the channel with an 8-GHz bandwidth, we set 3,200 sub-bands to resolve the multipath with the maximum excess delay of 400 ns, which is similar to the length of the M-sequence of the channel sounder (444.14 ns). The LOS ray and reflected rays

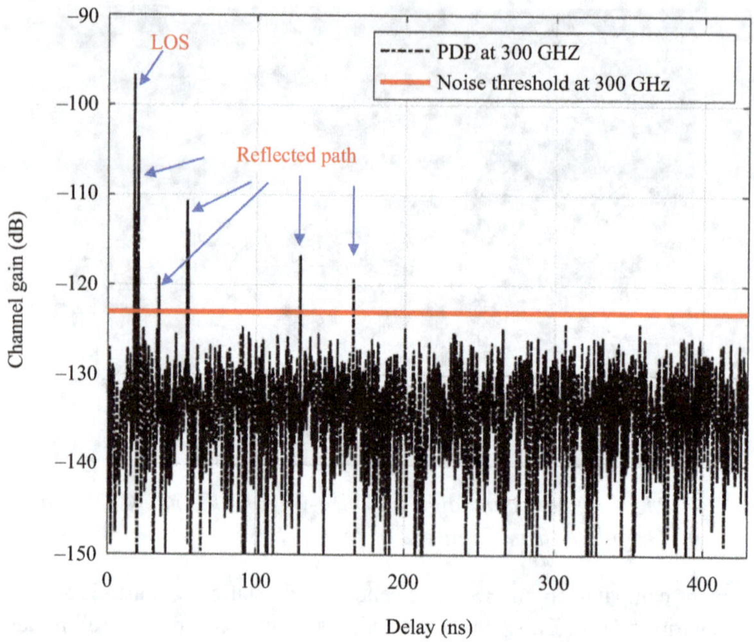

Figure 19.6 The synthesized PDP of the Intra-wagon channel. © 2021 IEEE. Originally from Figure 8 of [60].

Table 19.2 Calibrated EM property of materials at 300 GHz

Material (objects)	**Metal (train, track, (train, track, lamppost)**	**Glass (window)**	**Brick (ground)**	**Rubber (junction of wagons)**
ε_r'	1.000	4.200	3.026 [11,77]	1.021
ε_r''	10^7	0.342	0.159 [11,77]	1.491

(up to the fourth order for inside station and up to second order for Intra-wagon scenario) are traced in simulations for the trade-off between simulation accuracy and computational complexity. All the other setups are in-line with the measurement campaign. The real and imaginary parts of the relative permittivity (ε_r' and ε_r'') of relevant materials are calibrated and summarized in Table 19.2.

The comparison of PDP between the measured and simulated PDP of the Intra-wagon scenario is shown in Figure 19.7, where the good agreements of the significant paths are achieved both in terms of power and delay. The mean absolute error of the gain and delay of all the significant paths is 1.13 dB and 0.56 ns, respectively. This means that the RT simulator together with the 3D model are validated and, therefore, can effectively reproduce the channel sounding measurement results.

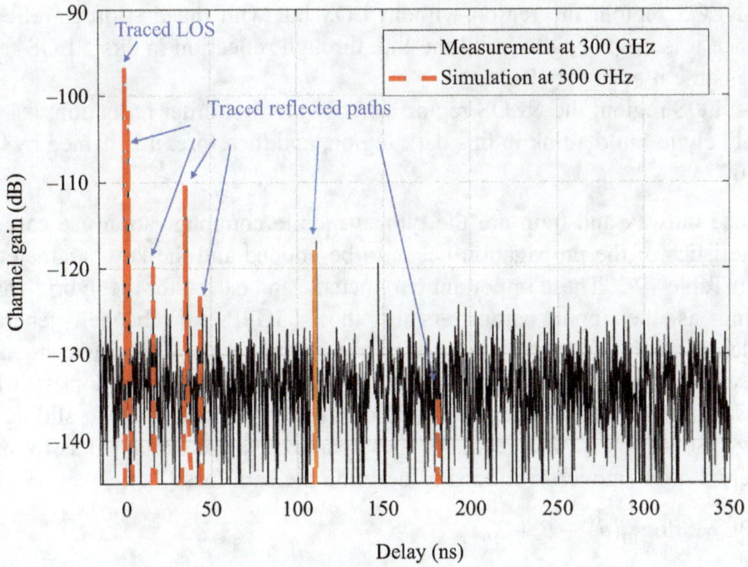

Figure 19.7 Comparison of PDP between measurement and RT simulation. © 2021 IEEE. Originally from Figure 16 of [60].

After calibration, this measurement-validated RT simulator can be utilized to generate more realistic channel data with various setups for comprehensive characterization in order to model the channel by the hybrid channel modeling. For the Inside station channel, totally four cases are simulated:

- P-NT: Tx is deployed above the platform and there is no other train on the adjacent track.
- P-T: Tx is deployed above the platform and there is another train on the adjacent track.
- C-NT: Tx is deployed on the catenary mast between two tracks and there is no other train on the adjacent track.
- C-T: Tx is deployed on the catenary mast between two tracks and there is another train on the adjacent track.

Detailed simulation setups and results can be found in our previous study [55].

For the Intra-wagon channel, extensive RT simulations are done with two Tx deployment strategies—Tx-center and Tx-side. Details can be found in our recent study [60]. Because of the high diffraction loss and transmission loss at THz band, the first-order reflection becomes almost the only chance to build the link if the LOS is blocked (which is very possible in Intra-wagon scenario). Thus, we further classify the NLOS region into L-NLOS and D-NLOS regions:

- LOS region: the region with the LOS signal.

- L-NLOS region: the region without LOS but with the first-order reflection. Since it is promising to build the link through reflection in this NLOS region, we name it as "light (L)."
- D-NLOS region: the NLOS region without the first-order reflection. It is very unlikely to build a link in this dark region, and therefore, it is named by "deep (D)."

Since railway and train are the typical mobile communication use cases, the characteristics of the propagation needs to be noticed and the key parameters are listed in Table 19.3. These important parameters can be used for the hybrid channel modeling as the input parameters of those 3GPP-like channel generators. Considering all the snapshots and passing the Kolmogorov—Smirnov test, each of the previous channel parameters is fitted as a log–normal distribution with mean value, standard deviation, and correlated distance (the threshold: 0.7, sliding window size: 80 samples). The details of these parameters are explained as follows.

Here, "A–B" model is utilized to describe the path loss (PL):

$$\mathrm{PL} = A\log_{10}(d) + B + X_\sigma \tag{19.5}$$

where d denotes the distance between the transceivers; A denotes the slope; B denotes the intercept; and X_σ denotes the shadow fading (SF), which can be expressed as a Gaussian variable with zero mean value and a standard deviation of σ_{SF}.

The Rician K-factor (KF) is defined as the ratio of the signal power of the LOS path to the signal power of other multipaths components, which can be expressed as

$$\mathrm{KF} = \frac{P_{\mathrm{LOS}}}{\sum P(i) - P_{\mathrm{LOS}}} \tag{19.6}$$

where $P(i)$ is the energy of all the paths of the signal during transmission, and P_{LOS} is the energy of the LOS path. μ_{KF} and σ_{KF} in Table 19.3 denote the mean value and the standard deviation of KF.

The RMS delay spread (DS) for the delay time extent is one of the primary parameters in a multipath radio channel. It is defined in [65], namely, the square root of the second central moment of the PDP by the following equation:

$$\sigma_\tau = \sqrt{\frac{\sum_{n=1}^{N} \tau_n{}^2 \cdot P_n}{\sum_{n=1}^{N} P_n} - \left(\frac{\sum_{n=1}^{N} \tau_n \cdot P_n}{\sum_{n=1}^{N} P_n}\right)^2} \tag{19.7}$$

where σ_τ denotes the RMS DS, P_n and τ_n denote the power and the excess delay of the nth multipath, respectively. μ_{DS} and σ_{DS} in Table 19.3 denote the mean value and the standard deviation of DS.

The angular spread (AS) defines the distribution of the arrival and departure angles of each ray in 3D environments for the Tx and Rx, respectively. Each ray will be assigned an azimuth angle in the horizontal plane and an elevation angle in the vertical plane. Thus, four values of ASs, namely, azimuth angular spread of

Table 19.3 Channel parameters for inside station and intra-wagon scenarios of smart rail mobility at 300-GHz band

Scenario	**Inside station**				**Intra-wagon**					
Case	**P-NT**	**P-T**	**C-NT**	**C-T**	**Tx-center**			**Tx-side**		
LOS condition	LOS	LOS	LOS	LOS	LOS	L-NLOS	D-NLOS	LOS	L-NLOS	D-NLOS
A	18.65	18.41	18.34	17.25	21.66	22.12	9.48	20.68	25.81	34.05
B	81.35	81.75	82.24	81.80	79.77	82.16	101.42	79.95	78.96	76.04
σ_{SF} [dB]	5.22	5.37	5.85	5.47	5.59	5.54	8.23	5.61	5.14	7.61
λ_{SF} [m]	0.02	0.02	0.02	0.02	0.03	0.02	0.02	0.04	0.02	0.02
μ_{KF} [dB]	7.30	10.37	3.04	−1.01	6.90	0.06	−13.60	4.27	0.22	−13.57
σ_{KF} [dB]	10.63	9.43	4.95	8.21	8.84	5.11	15.35	7.98	5.26	12.18
λ_{KF} [m]	0.25	0.25	0.25	0.25	0.25	0.25	0.11	0.24	0.25	0.15
μ_{DS} [$\log_{10}$[s]]	−7.99	−7.98	−8.18	−8.26	−8.59	−8.81	−8.94	−8.71	−8.73	−8.88
σ_{DS} [$\log_{10}$[s]]	0.27	0.27	0.20	0.16	0.36	0.27	0.35	0.42	0.24	0.68
λ_{DS} [m]	0.23	0.23	0.24	0.25	0.24	0.25	0.15	0.18	0.23	0.16
μ_{ASD} [$\log_{10}$[°]]	1.12	1.13	0.33	1.70	1.15	1.35	0.33	1.14	1.19	0.97
σ_{ASD} [$\log_{10}$[°]]	0.32	0.32	0.69	0.43	0.56	0.53	0.88	0.30	0.50	0.81
λ_{ASD} [m]	0.25	0.25	0.25	0.25	0.25	0.26	0.14	0.25	0.25	0.22
μ_{ESD} [$\log_{10}$[°]]	0.73	0.74	0.18	0.27	1.37	0.38	−0.05	1.18	0.55	0.59
σ_{ESD} [$\log_{10}$[°]]	0.09	0.09	0.38	0.29	0.31	0.25	0.43	0.35	0.45	0.64
λ_{ESD} [m]	0.25	0.25	0.25	0.25	0.25	0.25	0.14	0.25	0.25	0.25
μ_{ASA} [$\log_{10}$[°]]	1.83	1.86	1.79	1.82	1.30	1.8	1.38	1.24	1.73	1.67
σ_{ASA} [$\log_{10}$[°]]	0.35	0.32	0.43	0.37	0.67	0.30	0.54	0.70	0.34	0.37
λ_{ASA} [m]	0.25	0.25	0.25	0.25	0.25	0.25	0.10	0.26	0.25	0.18
μ_{ESA} [$\log_{10}$[°]]	0.96	1.02	0.88	0.80	0.62	0.97	0.78	0.52	0.90	1.00
σ_{ESA} [$\log_{10}$[°]]	0.13	0.14	0.20	0.23	0.52	0.33	0.56	0.61	0.26	0.35
λ_{ESA} [m]	0.25	0.25	0.25	0.25	0.25	0.25	0.18	0.25	0.25	0.16
μ_{XPR} [dB]	3.05	3.10	5.90	8.53	21.91	2.05	−0.06	12.99	2.62	1.92
σ_{XPR} [dB]	1.89	1.86	1.81	2.32	5.16	2.62	1.91	4.83	2.99	2.98

arrival (ASA), azimuth angular spread of departure (ASD), elevation angular spread of arrival (ESA), and elevation spread of departure (ESD), can be calculated as follows:

$$\sigma_{AS} = \sqrt{\frac{\sum_{n=1}^{N} (\theta_{n,\mu})^2 \cdot P_n}{\sum_{n=1}^{N} P_n}} \tag{19.8}$$

where σ_{AS} denotes the angular spread (AS), P_n denotes the power of the nth multipath, and $\theta_{n,\mu}$ is defined by

$$\theta_{n,\mu} = mod(\theta_n - \mu_\theta + \pi, 2\pi) - \pi \tag{19.9}$$

where θ_n is the AAoA/AAoD/EAoA/EAoD of the nth ray. μ_n is

$$\mu_\theta = \frac{\sum_{n=1}^{N} \theta_n \cdot P_n}{\sum_{n=1}^{N} P_n} \tag{19.10}$$

In Table 19.3, μ_{ASA}, μ_{ASD}, μ_{ESA}, and μ_{ESD} are the mean values of ASA, ASD, ESA, and ESD, respectively; σ_{ASA}, σ_{ASD}, σ_{ESA}, and σ_{ESD} are the standard deviations of ASA, ASD, ESA, and ESD, respectively.

The cross-polarization ratio (XPR) is to describe the variety of the polarization for the multipaths components and is calculated by its definition as follows:

$$XPR = 10\log_{10}\left(\frac{P_{co}}{P_{cross}}\right) \tag{19.11}$$

where P_{co} is co-polarization defined by the received power when Tx and Rx antenna polarize in the same manner. Conversely, P_{cross} is the power received in vertical (horizontal) polarization and transmitted in horizontal (vertical) polarization. In Table 19.3, μ_{DS} and σ_{DS} denote the mean value and the standard deviation of XPR.

The autocorrelation function of a parameter X is to describe how fast a certain variable changes with the distance interval Δd (or time delay τ), which is calculated as follows:

$$R(\Delta d) = \frac{E[(X_d - \mu)(X_{d+\Delta d} - \mu)]}{\sigma_X^2} \tag{19.12}$$

where E is the expected value operator. Correspondingly, the decorrelation length in Table 19.3 signed as λ is expressed as

$$\lambda = \underset{\Delta d > 0}{\text{argmax}} \left(|R(\Delta d)| < T\right) \tag{19.13}$$

where T denotes the threshold with the value of 0.7.

All mentioned these parameters provide valuable insights into the system design and evaluation for the use case of smart railway and train communication systems.

19.5 Real-world use case for drone

With the gradual maturity of small drone technology, unmanned aerial vehicles (UAVs) have been widely used in military and civilian fields in recent years due to their high maneuverability and flexibility. Among them, the use of drones to carry mobile communication base stations to provide network coverage for surface users has become an outstanding representative of the next generation of emergency communication technology. Disasters, such as earthquakes and floods, will cause terrestrial base stations, cables, and signal towers to be damaged and difficult to recover immediately, which will interrupt communications and affect further disaster relief organizations and rescue activities. Existing emergency satellite communication systems, such as International Maritime Satellite, are not suitable for use due to the complex terrain and large numbers of people at the rescue site. Besides, the most commonly used emergency communication vehicles are often unable to be used at the rescue site due to ground collapse. Therefore, the portable drone flight base station, that is, the aerial emergency communication based on the drone, has become an important emergency communication solution. The ultralow delay, high data transmission rate, and large throughput of beyond 5G/6G THz communication systems can be used as the information transmission carrier needed by UAV base stations and even large-scale UAV base station arrays in the future [66].

Combining the advantages of drone and THz technologies, the UAV THz communication system can provide high-speed data transmission and wide network coverage capability, which leads to a broad application prospect in military communication systems. The accurate channel model is vital for the UAV THz communication system design and performance evaluation. Different from the conventional mobile communication scenarios, the UAV THz communication scenario has obvious 3D propagation characteristics. The key factors, namely, 3D propagation environment, 3D flight trajectory, and 3D antenna beam pattern, should be taken into account in the channel modeling.

- 3D propagation scenario

For terrestrial mobile communication scenarios, although the scatterers around the transmitting and receiving antennas are in a 3D environment, the Tx and Rx are usually relatively close; the vertical position changes within a short period of time are relatively small. Thus, the elevation angle has a relatively small impact on the propagation. Therefore, early channel models often ignore the influence of height parameters on channel characteristics and approximate a two-dimensional (2D) channel model description. With the deepening of 5G and beyond system research, researchers have upgraded the traditional 2D model, taking into account the height

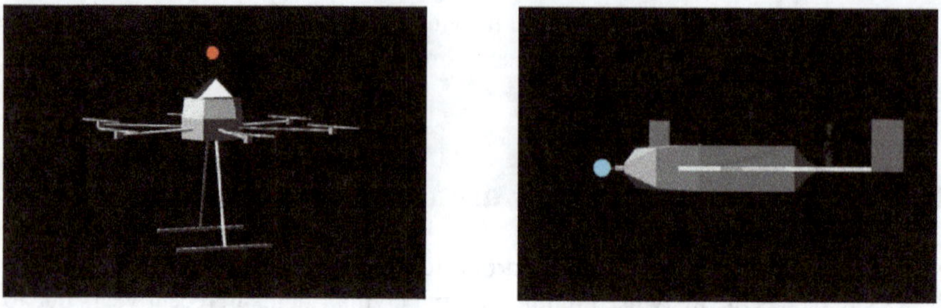

(a) (b)

Figure 19.8 Two different types of UAVs: (a) multi rotor UAV and (b) fixed wing UAV

(a) (b)

Figure 19.9 The 3D propagation environment for drone communication systems: (a) real propagation environment and (b) reconstructed 3D environment

parameters of Tx and Rx, realizing the support for the 3D scenario, and improving the accuracy and scientificity of channel parameter characterization. The 3D scenario should consist of the UAV model (as shown in Figure 19.8) and the propagation environment model (as shown in Figure 19.9).

The 3D channel model uses vectors to describe the instantaneous geographic location of the Rx and the moving scatterers. The characteristics of the departure direction and arrival direction of the MPCs are further described in detail as the azimuth angle and the EAoA and departure. In particular, UAV THz communication can make a full use of 3D large-scale scenarios. UAV performs large-scale flight in the 3D environment, and the vertical position and elevation angle have very obvious changes during the entire communication process. Besides, the channel measured results show that the UAV speed, flight angle, and position altitude have a significant impact on the statistical characteristics of the channel model [67].

- 3D velocity and trajectory

In terrestrial communication systems represented by cellular mobile and vehicular communications, the mobile positions of the transceivers in a short time can be approximately limited to a 2D plane. However, the UAV's moving velocity has obvious 3D characteristics, and the flight trajectory is also distributed in the entire 3D environment. The drone channel model must support the 3D moving velocity.

The authors in [68] analyze the velocity parameters of the Tx and Rx and concluded that the velocity has a crucial influence on the space–time correlation of the mobile communication channel. Based on this conclusion, the V2V channel models successively proposed in [69] extend the moving speed of the mobile terminal to a 3D vector representation. In [70], the authors propose a UAV channel model with arbitrary 3D velocity and flight trajectory for UAV communication scenarios. It needs to be emphasized that the speed of terrestrial mobile terminals such as pedestrians and vehicles only has 2D directions, which can be completely described with only a single plane of angle parameters. However, during UAV flight, there are aircraft-specific behaviors such as tilting and turning of the fuselage, and more angle parameters are needed to describe the 3D attitude information of the fuselage [71]. Therefore, the 3D flight velocity and trajectory should be fully incorporated into the drone channel model.

- 3D antenna beam pattern

Large-scale antenna array technology can make a full use of multipath propagation characteristics and double the channel capacity without increasing the spectrum resource and the transmission power. For THz communications, the antenna array can form strong directional beams to improve the quality of the communication link, thereby making up for the shortcomings of large THz wave propagation loss. As shown in Figure 19.10, the first antenna array employed 4 ×4 antenna elements can form a beam with a gain of 20 dBi; the second antenna array employed 16 ×16 antenna elements can form a beam with a gain of 32 dBi. In

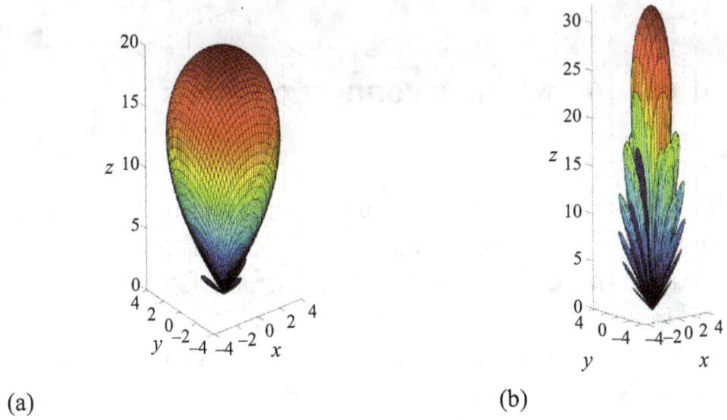

Figure 19.10 The 3D antenna array beam pattern: (a) 4 ×4 antenna array beam pattern and (b) 16 ×16 antenna array beam pattern

addition, the optimized design of the beam shape can also enhance the main lobe gain and suppress the sidelobe energy, thereby further achieving the purpose of effective coverage and interference suppression.

The terrestrial base station often uses a fixed pitch antenna to adjust the 2D beam according to the horizontal dimension information for beam alignment. UAV is located in a large-scale 3D environment and can flexibly use the vertical direction to obtain 3D beams, thereby increasing antenna array gain and effective communication distance. Meanwhile, it can reduce interference between different links and the probability of signal interception. In addition, during UAV flight, it is difficult to ensure a constant elevation angle between Tx and Rx. Thus, dynamic beam searching, tracking, and alignment must be performed. This feature will also have an important impact on the construction of the UAV channel model.

- Challenge of drone channel modeling

For drone THz communication scenarios, the wave propagation may encounter scatterers such as the ground, buildings, mountains, and even water surfaces, resulting in a large amount of reflected, diffracted, and scattered paths. Among them, the paths with very close delays can be approximated as a cluster, and the received CIRs are composed of space–time clusters arriving at different times and directions. With the change of the UAV's position, velocity, and flight height, the movement of the ground, or the change of the propagation environment, some scatterers will disappear and the corresponding clusters will disappear as well. Meanwhile, new scatterers will appear to produce new space–time clusters. Compared with terrestrial THz communication scenarios, UAV communication scenarios have more frequent birth and death of space–time clusters. The propagation is more complex and changeable. These new features due to high-frequency bands and ultra-wideband should be all included in the consideration of model construction. In addition, the new model not only needs to accurately characterize the channel parameters of the drone communication system but also should consider the compatibility of the traditional scenarios and the channel characteristics at lower frequency band.

19.6 Real-world use for wireless connections on a desktop

Since it is not feasible to perform extensive channel sounding measurements with high degrees of freedom at THz frequency band, the empirical or stochastic modeling approaches based on the channel parameters extracted from measurement data no longer stand. In order to break through the bottleneck of scarce full-dimensional channel, a viable paradigm for realizing the real-world THz channel is proposed. The paradigm proposed consists of six steps, including the limited channel sounding, calibration of RT simulator, extensive RT simulations, ray-based novel stochastic channel model, cluster-based standard channel model, and realistic THz channel realization [72].

At the beginning of the proposed paradigm, a typical LOS channel on an optical table is measured for the module of "limited channel soundings." A VNA with mmWave extenders is applied to measure the channel at 330–365 GHz with 5001 frequency point. The VNA measures the narrowband CTFs sequentially in frequency domain. For each narrowband measurement, the CTF is described by a vector and the broadband CTF is obtained by aggregating all narrowband CTFs. The corresponding broadband CIR is the IFFT of the CTF. The corresponding sampling interval, namely, the narrowband Δf is 7 MHz. With such a measurement configuration, the corresponding delay resolution is 0.029 ns, which is short enough to distinguish the MPCs. Besides, the corresponding maximum excess delay is 143 ns, and therefore, it is long enough to capture MPCs in the desktop channel. The VNA-based approach has the advantages of narrowband channel measurements such as low measurement noise and individual calibration for each frequency point as the broadband CTF is composed of many narrowband measurements. On the other hand, this frequency-domain feature makes the measurement time long and therefore, the channel variations caused by, e.g., the Doppler effect and the dynamic blockage cannot be captured. The utilized antennas are with the gain of approximately 24 dBi and with the HPBW of theoretically 10°. The full two-port calibration approach–short, open, load, and Thru–is applied for the calibration of the entire measurement system, including the waveguides, mmWave extenders, cables, and surely the VNA itself. As shown in Figure 19.11, a desktop THz channel with approximately 1-m length is measured in a typical laboratory scenario.

Next, the module of "calibration of RT simulator" is started through RT simulations. As a representing deterministic channel modeling approach, RT becomes increasingly prominent in 5G, beyond 5G, and 6G communications. The precondition of employing RT simulators by CloudRT to do extensive simulations is to calibrate it with the limited channel sounding data. According to the well-known Fresnel formula as described in [73], the reflection coefficient is determined not only by geometry but also the real and imaginary parts of the relative

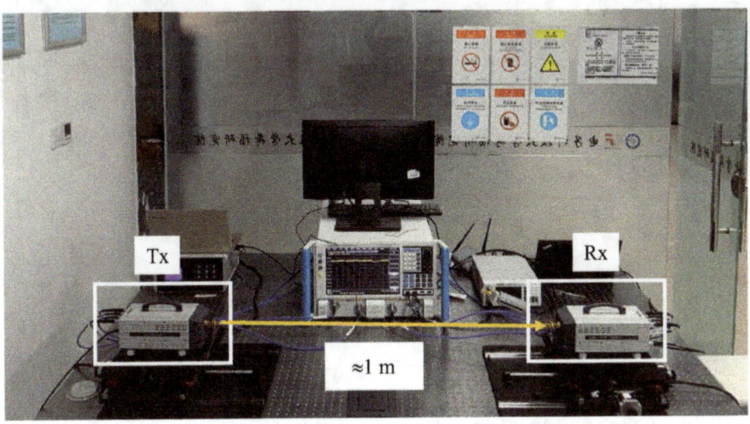

Figure 19.11 Measurement setup for the desktop THz channel

permittivity (ε_r' and ε_r'' [74]). Thus, these two parameters of all the relevant materials need to be calibrated through minimizing the error of the power of the significant paths (reflected rays) between measurements and RT simulations. The initial values of EM property can be obtained from open literature (for instance, metal and rubber [75], glass [76], and brick [11,77]). Then, by using simulated annealing algorithm [78], the corresponding values of relevant materials can be calibrated. Apart from this, the geometry of the 3D environment model needs to be calibrated as well; otherwise, the TOA of multipaths will not agree between measurements and simulations. After calibration, the RT simulator can be utilized for extensive simulations. The simulation configuration remains unchanged with the measurement configuration. At the beginning of the simulation, the LOS ray and up to the sixth order of reflected rays are simulated in the desktop THz channel with 35-GHz bandwidth. The metallic optical table is assumed as a perfect electric conductor and meanwhile, the painted metallic plate of the extenders, which can be treated as the hardware design of the handheld devices and laptops, needs to be calibrated. The EM properties of the painted metallic plate after calibration are with the values of $\varepsilon' = 1.05$ and $\varepsilon'' = 12.075$. As shown in Figure 19.12, the errors of each significant traced rays in power domain are smaller than 1 dB. Meanwhile, the errors are smaller than 0.1 ns in time domain. The traced rays are mainly reflected from the surfaces of devices. In good agreement with the simulated and measured results, the calibrated RT simulator can correctly repeat the desktop channel measurement.

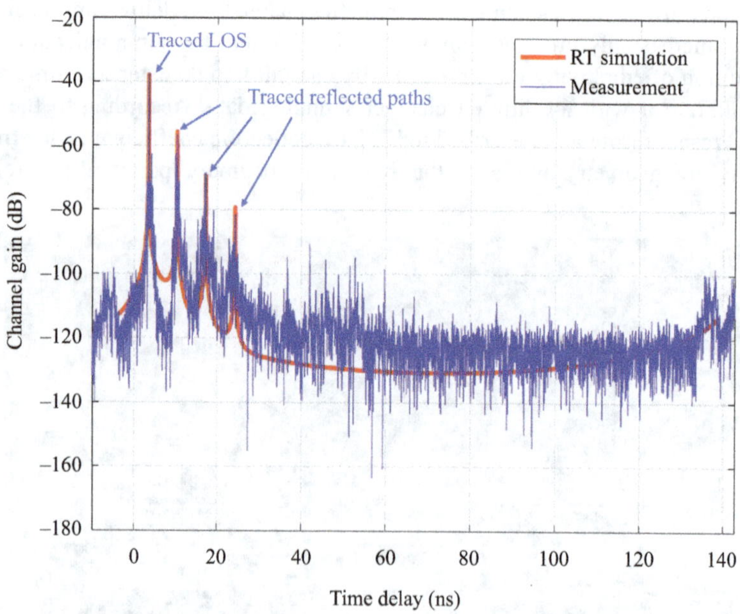

Figure 19.12 Comparison of the channel gain between measurement and calibrated simulation for the desktop channel

As already discussed, the THz broadband channel measurement is usually realized by 2D scanning mechanisms using directional antennas. Thus, the channel needs to be characterized more comprehensively by using the measurement-validated RT simulator, which can effectively extend the scenario setup and channel degrees of freedom in terms of various positions of Tx and Rx and the full polarization combinations. After extensive RT simulations conducted with various configurations in the previous aspects, a panorama of full-dimensional THz channel for specific 6G use cases can be formed. Thus, for the wireless connections on a desktop, in practice, users have more freedom as a matter of fact and might place the devices on the desktop in different locations. However, due to the limitation of the measurement equipment, the comprehensive measurements are too difficult to be conducted and the possible alternative way is with the aid of RT simulator to generate the channel data, mainly the abundant CIRs. Therefore, to carry on the module of "extensive RT simulations," we place the fixed Tx on one side of the table. To find the correlation between the snapshots, the sampling interval of Rx is set to be 0.8 mm, which is approximately equal to the wavelength of the center frequency. The simulated communication distance range is between 0.2 and 1.5 m for a typical desktop channel.

The cluster-based stochastic channel modeling approach has been widely utilized by 3GPP standardization community, spawned entire standard channel model families. With the required channel characteristics such as path loss, SF, Rician KF, from RT simulation results, the cluster-based standard channel models can generate corresponding channel data in the target scenario without running RT simulations or doing measurements. Therefore, in order to generate realistic THz channels by "cluster-based standard channel model families" for the desktop channel, the key parameters should be extracted from the RT simulation results. The key parameters illustrated in Table 19.4 are in log-normal distribution, which is in accord with the distribution of 3GPP-like channel model families. In Table 19.4, μ_X and σ_X denote the mean value and the standard deviation of the X parameter (X can be DS, KF, SF, ASD, ASA, ESD, and ESA), respectively. The path loss is fitted by "A–B" model. The fitted values of A and B are 19.21 and 80.03, respectively. Due to the sparsity of THz channel and with the aid of the power k-means clustering algorithm, the per-cluster parameters for three clusters of the desktop channel are determined and listed in Table 19.5. As a consequence of the THz short

Table 19.4 Key parameters of desktop THz channel

	SF [dB]	**KF [dB]**	**DS [Log$_{10}$[s]]**	**XPR [dB]**
μ_X	–	2.40	−9.13	7.11
σ_X	4.16	5.09	0.15	4.45
	ASD [log$_{10}$[°]]	ASA [log$_{10}$[°]]	ESD [log$_{10}$[°]]	ESA [log$_{10}$[°]]
μ_X	1.88	1.88	1.06	1.06
σ_X	0.002	0.002	0.14	0.14

Table 19.5 Per-cluster parameters for desktop THz channel

Parameter	**SF**	**ASD**	**ESD**	**ASA**	**ESA**
Value	4.29 dB	29.31°	7.31°	0.01°	0.02°

Table 19.6 Cross-correlations for desktop THz channel

	DS	**KF**	**SF**	**ASD**	**ASA**	**ESD**	**ESA**
DS	1	−0.86	0.00	0.99	0.99	0.97	0.96
KF		1	−0.01	−0.88	−0.87	−0.89	−0.89
SF			1	0.00	0.00	0.00	0.00
ASD				1	0.99	0.98	0.98
ASA					1	0.96	0.96
ESD						1	1.00
ESA							1

wavelength, the large-scale parameters are very sensitive to the environment when Rx moves through the scenario, and therefore, the decorrelation distances of KF, DS, and four ASs are 0.0216 m, respectively. The decorrelation distance of SF is 0.0024 m. The fact indicates a much shorter correlation length in the propagation scenario compared to the sub-6-GHz band and mmWave band, which have a major influence on the link- and system-level design. In addition, the cross-correlations of the key parameters are recorded in Table 19.6. After feeding all these parameters into the quasi-deterministic radio channel generator (QuaDRiGa) [79], which is a 3GPP reference channel generator following the geometry-based cluster stochastic channel modeling, the realistic THz channel on the desktop can be realized in a much more straightforward way without intensive measurement data.

Surely, compared to RT simulations or channel sounding, the stochastic models based can achieve a higher generality and can generate the channels much faster. Thus, a huge mass of realistic THz channel data can be realized for link-level or system-level simulations. Last but not least, since the stochastic models are the simplification of RT, it is necessary to verify the channel characteristics generated by the model with RT simulations. By running QuaDRiGa with the same deployment of the transceivers and the moving trajectory in the RT simulation, the last module "realistic THz channel realization" is trigged and validated. As shown in Figure 19.13, the comparisons between QuaDRiGa and RT in terms of RMS DS and ESA are as the examples. The differences between QuaDRiGa and RT of these two parameters are 0.14 ns and 0.33°, respectively. The good agreement in the delay-power domain and angular domain indicates that the proposed paradigm can reproduce the channel data generated by the calibrated RT simulator. Therefore, the extracted parameters can be used for producing more channel data without the complicated and time-consuming computation of RT simulator.

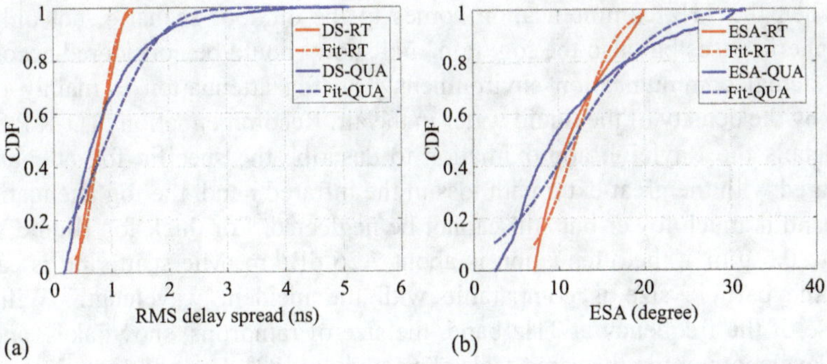

Figure 19.13 Comparison between QuaDRiGa and RT result for desktop THz channel: (a) RMS delay spread and (b) ESA

19.7 New concerns of THz channel modeling

Despite the pleasing vision of THz communications and the feasible THz channel modeling approaches, when it comes to the THz band, new concerns will be faced because of the distinguishing features of the THz wave. Besides the high propagation and transmission loss compared to the lower frequencies, the impact of meteorological factors, the beam split effect related to the antenna array, and multifold noises should be noticed. In this part, we discuss new concerns of THz channel modeling regarding propagation, antenna array, and device aspects.

19.7.1 Propagation aspect—meteorological impact

Compared to the wireless channels at lower frequencies, channels at THz bands are more vulnerable to different meteorological factors, including atmospheric gases, fog, rain, and snow. These meteorological impacts cannot be neglected in channel modeling.

Atmospheric attenuation is due to the energy absorption by the atmospheric molecules. Since some excited molecules vibrate with the certain frequency, a portion of the EM energy is transformed into kinetic energy [80]. The major gaseous molecules at THz band are the oxygen and the water vapor. The intensity of the atmospheric attenuation is determined by several items, including the temperature, pressure, altitude, and the wave frequency as well. Recommendation ITU-R P.676-11 [81] gives the prediction method to evaluate atmospheric attenuation. From Recommendation ITU-R P.676-11, the atmospheric attenuation can define several significant transmission windows, the position and the width of which are mainly determined by the carrier frequency and the gas composition in the propagation environment, mostly the water vapor. Because of this feature, a new communication scheme, that can dynamically conform the waveform according to the carrier frequency and propagation distance, can be explored to match the window shape [1] and avoid the absorption narrow peaks.

When the THz communication comes to the outdoor scenario, not only the atmospheric gases but also the fog, rain, and snow should be considered according to the actual communication environment. The fog attenuation is mainly determined by the density of the liquid water in the air. Recommendation ITU-R P.840-7 [82] details the Rayleigh approximation to describe the specific fog attenuation. Compared with the great extra path loss in the infrared band, the fog attenuation at THz band is much lower but still cannot be neglected. For thick fog (liquid water density: 0.5 g/m^3), the attenuation is about 7.26 dB/km. Mie scattering is caused when the particle size is comparable with the incident wavelength. With the increase of the frequency at THz band, the size of raindrops, snowflakes, and the THz wavelength are in the same scale. The study in [83] draws the conclusion that the rain attenuation increases with the density of the rainfall, whereas the temperature has little influence. The authors in [84] evaluate the rain attenuation prediction approach of ITU-R P.838-3 [85] by comparing with the experiment results at 300 GHz. They find that not only the rainfall rate but also the drop size distribution need to be modeled to predict the rain attenuation. The snow attenuation is forecasted to be more serious, since the structure of snowflakes is fluffier and is primarily made by air, liquid water, and ice. Because the unfriendly weather conditions can easily harm the sensitive and expensive measurement equipment, few studies on snow attenuation have been reported. The prediction method from ITU-R for snow attenuation is absent as well. The valuable measurement results in [86] show that the wet snow might cause larger attenuation in comparison with the dry snow. Besides, the wet snow may cause three times larger attenuation compared to the rain attenuation at the same falling rate. In [87], an 8-m LOS path measurement was conducted in a significant snowstorm day. The measurement result implies that the snow attenuation is very significant with the value of 2.8 dB/8 m.

To the best of the authors' knowledge, until now, the various channel modeling approaches have not considered the impact of the different meteorological factors. Since the high spreading loss and considerable meteorological attenuation due to atmospheric gases, fog, rain, and snow, they should be a new concern in THz channel modeling compared to the channel modeling at lower frequencies.

19.7.2 Antenna array aspect—beam split

For the compensation of the large propagation loss and meteorological attenuation, very large antenna arrays with PSs are widely considered in THz communications. Generally, the phase shift is determined by the carrier frequency and the corresponding wavelength. Since the PS is frequency independent, the directional beams are squinted to different directions at different subcarrier frequencies. This effect, which causes the loss of antenna array gain, is called beam squint in wideband mmWave systems. In THz band, the available bandwidth becomes larger, and meanwhile, the beam generated by the traditional frequency-independent PS becomes narrower. Thus, the beam squint effect will be significantly aggravated, where the beams at different subcarrier frequencies might split far away from the ideal direction. This misalignment effect, which is called beam split, will cause

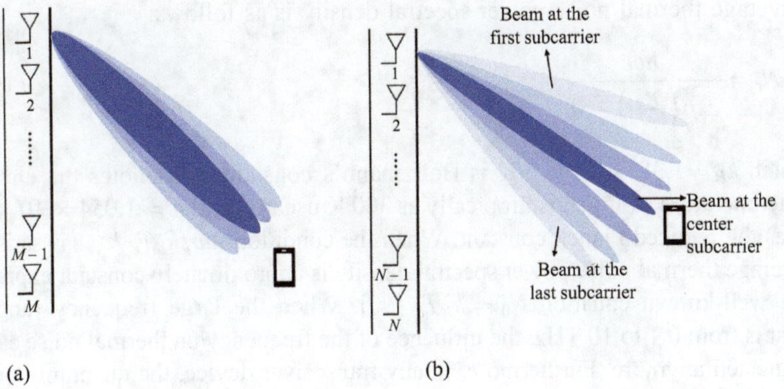

Figure 19.14 (a) Beam squint effect in mmWave and (b) beam split effect at THz band

more obvious array gain loss at THz band and therefore result in the undesirable deterioration of the data rate performance [88]. The comparison between the beam squint effect at mmWave band and the beam split effect at THz band is shown in Figure 19.14.

The theoretical reason for these two effects is mainly because the antenna array with N elements introduces the array response vector as follows:

$$g(\theta_F, \theta, f) = \frac{1}{\sqrt{N}} \sum_{N}^{n=1} e^{j\left[2\pi c^{-1}(n-1)d(f\sin\theta - f_c\sin\theta_F)\right]} \tag{19.14}$$

where θ denotes either AoA for receiving or AoD for transmission, θ_F is ideal beam focus angle, c denotes the light speed, and d is the distance of fixed adjacent antenna elements with $d = c/2f_c$. The antenna array response varies with frequency across the ultra-wideband systems. The authors in [89] propose a delay-phase precoding method to fulfill frequency-dependent PSs through time delay. In the beginning, the PSs engender frequency-independent beams. Then, the time-delay elements provide the beam compensation based on the user physical direction and the subcarrier frequencies. With accurate channel modeling and the fabrication of very large THz array with novel hybrid precoding methods, it is possible to achieve good performance and low power consumption in THz communication systems.

19.7.3 Device aspect—noise impact

There are several noise sources that can affect the performance of THz communications. One major noise source is the phase noise, which can randomly lead to fast fluctuations in the phase of transmitting signals. The receiver will face a great challenge on processing synchronization and demodulation. Moreover, the thermal noise, which is caused by the irregular vibration of free electrons in conductors and semiconductors of the transceivers, needs to be taken into consideration.

The average thermal noise power spectral density is as follows:

$$N_t = \frac{\hbar\omega}{exp\left(\frac{\hbar\omega}{k_B T_e}\right) - 1} \tag{19.15}$$

in which $k_B = 1.38 \times 10^{-23}$ J/K is Boltzmann's constant, T_e denotes the environment temperature assumed isotropically as 300 K usually, and $\hbar = 1.054 \times 10^{-34}$ J s denotes the reduced Planck constant. When the condition $\hbar\omega / k_B T_e \ll 1$ is satisfied, the average thermal noise power spectral density is approximately constant expressed as the well-known equation $N_0 = k_B T_e$ [90]. When the large frequency range of interest is from 0.1 to 10 THz, the influence of the frequency on thermal noise cannot be neglected anymore. Furthermore, in any transceiver device, the quantum noise is unavoidable and is proportional to frequency. Thus, the total noise power spectral density of an ideal linear amplifier or phase detector is the sum of the thermal noise and quantum noise expressed as

$$N = \frac{\hbar\omega}{\exp\left(\frac{\hbar\omega}{k_B T_e}\right) - 1} + \frac{\hbar\omega}{2} \tag{19.16}$$

When the carrier frequency is larger than 7 THz, the quantum noise exceeds the thermal noise and dominant the total noise. Despite the frequency proportionality of the ideal total noise, it is proportional to the bandwidth as well, particularly in the ultrawide bandwidth THz communication systems. Thus, the effect of multifold noises should be considered when the model generates realistic channels for THz communications.

19.8 Conclusion

In recent decades, the demand for higher data rates has almost doubled year by year. Very large swaths of the spectrum (tens to hundreds of gigahertz) are still available in the THz band, which is unused for communications so far. In this chapter, a comprehensive introduction of various real-world use cases at THz band for 5G and beyond wireless communications as well as their propagation channel characterizations and challenges are presented. We provide a brief survey and most relevant scenarios for THz-band communications, including IoT, railway, train, drone, and wireless connections on a desktop. Last but not least, we highlight new concerns in THz communication systems from the meteorological impact, beam split effect of large antenna arrays, and multifold noise sources from devices themselves. Thus, this chapter schemes the foundation for future work that aims to streamline the design, simulation, and development of 5G and beyond THz communication systems.

Acknowledgments

This work was supported by Key-Area Research and Development Program of Guangdong Province, China (2019B010157001), the ZTE Corporation, and the

State Key Laboratory of Rail Traffic Control and Safety (Contract No. RCS2020ZZ005), Beijing Jiaotong University. The authors would like to thank Dr. Jiaowu Dou from the ZTE corporation for his discussion and constructive comments, which greatly helped improve this chapter.

References

[1] Akyildiz I, Jornet J, and Han C. TeraNets: Ultra-broadband communication networks in the terahertz band. *IEEE Wireless Communications*. 2014; 21 (4): 130–135.

[2] Chen Z, Ma X, Zhang B, *et al.* A survey on terahertz communications. *China Communications*. 2019; 16(2): 1–35.

[3] Lai Z, Yi H, Guan K, *et al.* Impact of meteorological attenuation on channel characterization at 300 GHz. *Electronics*. 2020; 9(7): 1115.

[4] Boulogeorgos AAA and Alexiou A. Coverage analysis of reconfigurable intelligent surface assisted THz wireless systems. *IEEE Open Journal of Vehicular Technology*. 2021; 2: 94–110.

[5] Hum SV and Perruisseau-Carrier J. Reconfigurable reflectarrays and array lenses for dynamic antenna beam control: A review. *IEEE Transactions on Antennas and Propagation*. 2013; 62(1): 183–198.

[6] Di Renzo M, Ntontin K, Song J, *et al.* Reconfigurable intelligent surfaces vs. relaying: Differences, similarities, and performance comparison. *IEEE Open Journal of the Communications Society*. 2020; 1: 798–807.

[7] Seto I, Hiraga K, Kürner T, *et al.* Channel Modelling Document (CMD). In: IEEE 802.15 Plenary Meeting. Macau, March 2016.

[8] Priebe S, Kannicht M, Jacob M, *et al.* Ultra broadband indoor channel measurements and calibrated ray tracing propagation modeling at THz frequencies. *Journal of Communications and Networks*. 2013; 15(6): 547–558.

[9] Mobile and Wireless Communications Enablers for Twenty-Twenty Information Society (METIS). METIS Project; 2014. ICT-317669-METIS/D2.1.

[10] Kokkoniemi J, Lehtomäki J, and Juntti M. Measurements on rough surface scattering in terahertz band. In: *2016 10th European Conference on Antennas and Propagation (EuCAP)*. IEEE; 2016. p. 1–5.

[11] Piesiewicz R, Jansen C, Mittleman D, *et al.* Scattering analysis for the modeling of THz communication systems. *IEEE Transactions on Antennas and Propagation*. 2007; 55(11): 3002–3009.

[12] Jornet JM and Akyildiz IF. Channel modeling and capacity analysis for electromagnetic wireless nanonetworks in the terahertz band. *IEEE Transactions on Wireless Communications*. 2011; 10(10): 3211–3221.

[13] Priebe S and Kürner T. Stochastic modeling of THz indoor radio channels. *IEEE Transactions on Wireless Communications*. 2013; 12(9): 4445–4455.

[14] Han C, Bicen AO, and Akyildiz IF. Multi-ray channel modeling and wideband characterization for wireless communications in the terahertz band. *IEEE Transactions on Wireless Communications*. 2014; 14(5): 2402–2412.

[15] Thiel M and Sarabandi K. A hybrid method for indoor wave propagation modeling. *IEEE Transactions on Antennas and Propagation*. 2008; 56(8): 2703–2709.

[16] Han C and Chen Y. Propagation modeling for wireless communications in the terahertz band. *IEEE Communications Magazine*. 2018; 56(6): 96–101.

[17] Cheng X, Yu B, Yang L, *et al.* Communicating in the real world: 3D MIMO. *IEEE Wireless Communications*. 2014; 21(4): 136–144.

[18] Baum DS, Hansen J, and Salo J., *An interim channel model for beyond-3G systems: Extending the 3GPP spatial channel model (SCM)*. In: 2005 IEEE 61st Vehicular Technology Conference. vol. 5; 2005. p. 3132–3136.

[19] Kyösti P, Meinilä J, Hentilä L, *et al.* WINNER II Channel Models; 2007.

[20] Bian J, Sun J, Wang CX, *et al.* A WINNER+ based 3D non-stationary wideband MIMO channel model. *IEEE Transactions on Wireless Communications*. 2018; 17(3): 1755–1767.

[21] Zhang J, Zhang Y, Yu Y, *et al.* 3D MIMO: How much does it meet our expectations observed from channel measurements? *IEEE Journal on Selected Areas in Communications*. 2017; 35(8): 1887–1903.

[22] Zhu M, Eriksson G, and Tufvesson F. The COST 2100 channel model: Parameterization and validation based on outdoor MIMO measurements at 300 MHz. *IEEE Transactions on Wireless Communications*. 2013; 12(2): 888–897.

[23] Cheng X, Yao Q, Wen M, *et al.* Wideband channel modeling and ICI cancellation for vehicle-to-vehicle communication systems. *IEEE Journal on Selected Areas in Communications*. 2013; 31(9): 434–448.

[24] Ghazal A, Wang CX, Ai B, *et al.* A nonstationary wideband MIMO channel model for high-mobility intelligent transportation systems. *IEEE Transactions on Intelligent Transportation Systems*. 2015; 16(2): 885–897.

[25] Yuan Y, Wang CX, He Y, *et al.* 3D wideband non-stationary geometry-based stochastic models for non-isotropic MIMO vehicle-to-vehicle channels. *IEEE Transactions on Wireless Communications*. 2015; 14(12): 6883–6895.

[26] Cheng X, Wang CX, Laurenson D, *et al.* New deterministic and stochastic simulation models for non-isotropic scattering mobile-to-mobile Rayleigh fading channels. *Wireless Communications and Mobile Computing*. 2011; 11 (7): 829–842.

[27] Kwon SC and Stuber GL. Wireless polarized-channel simulator. In: 2014 IEEE 80th Vehicular Technology Conference (VTC2014-Fall); 2014. p. 1–5.

[28] Cheng X, Wang CX, Laurenson D, *et al*, An adaptive geometry-based stochastic model for non-isotropic MIMO mobile-to-mobile channels. *IEEE Transactions on Wireless Communications*. 2009; 8(9): 4824–4835.

[29] Yang M, Ai B, He R, *et al.* A cluster-based three-dimensional channel model for vehicle-to-vehicle communications. *IEEE Transactions on Vehicular Technology*. 2019; 68(6): 5208–5220.

[30] Akdeniz MR, Liu Y, Samimi MK, *et al.* Millimeter wave channel modeling and cellular capacity evaluation. *IEEE Journal on Selected Areas in Communications*. 2014; 32(6): 1164–1179.

[31] Yang J, Ai B, You I, *et al.* Ultra-reliable communications for industrial Internet of things: Design considerations and channel modeling. *IEEE Network*. 2019; 33(4): 104–111.

[32] Guan K, Ai B, Peng B, *et al.* Towards realistic high-speed train channels at 5G millimeter-wave band – Part II: Case Study for Paradigm Implementation. *IEEE Transactions on Vehicular Technology*. 2018; 67(10): 9129–9144.

[33] Kürner T and Priebe S., Towards THz communications – status in research, standardization and regulation. *Journal of Infrared, Millimeter, and Terahertz Waves*. 2013; p. electronic publication (10 pages), http://dx.doi.org/10.1007/s10762-013-0014-3.

[34] He D, Ai B, Guan K, *et al.* The design and applications of high-performance ray-tracing simulation platform for 5G and beyond wireless communications: A tutorial. *IEEE Communications Surveys Tutorials*. 2019; 21(1): 10–27.

[35] Series P., Propagation data and prediction methods for the planning of short-range outdoor radiocommunication systems and radio local area networks in the frequency range 300 MHz to 100 GHz. tech rep, ITU, Tech Rep ITU-R; 2015.

[36] "Study on channel model for frequencies from 0.5 to 100 GHz (Release 14)", 3GPP Technical Specification Group Radio Access Networks Tech. Rep. TR 38.901 V14.0.0, March 2017.

[37] Seto I, Hiraga K, Kürner T, *et al.* Channel modelling document (CMD). In: DCN: 15–14-0310-19-003d, IEEE 802.15 Plenary Meeting Macau; 2016.

[38] Collela P., Ushering in a Better Connected Future. 2017. Available from: https://www.ericsson.com/en/about-us/company-facts/ericsson-worldwide/india/authored-articles/ushering-in-a-better-connected-future.

[39] Sohn I, Yoon SW, and Lee SH. Distributed scheduling using belief propagation for Internet-of-things (IoT) networks. *Peer-to-Peer Networking and Applications*. 2018; 11(1): 152–161.

[40] "RMS delay spread model in IIOT scenario", 3GPP TSG RAN WG1 #97, Reno, USA, May 2019.

[41] Wang W, Capitaneanu SL, Marinca D, *et al.* Comparative analysis of channel models for industrial IoT wireless communication. *IEEE Access*. 2019; 7: 91627–91640.

[42] Liu L, Zhang K, Tao C, *et al.* Channel measurements and characterizations for automobile factory environments. In: *2018 20th International Conference on Advanced Communication Technology (ICACT)*. IEEE; 2018. p. 234–238.

[43] Wassie DA, Rodriguez I, Berardinelli G, *et al.* Radio propagation analysis of industrial scenarios within the context of ultra-reliable communication. In: 2018 *IEEE 87th Vehicular Technology Conference (VTC Spring)*. IEEE; 2018. p. 1–6.

[44] Hanssens B, Kshetri S, Tanghe E, *et al.* Measurement-based analysis of dense multipath components in a large industrial warehouse. In: 12th European Conference on Antennas and Propagation (EuCAP 2018), April 2018.

[45] Ai B, Molisch AF, Rupp M, *et al.* 5G key technologies for smart railways. *Proceedings of the IEEE*. 2020; 108(6): 856–893.

[46] Moreno J, Riera JM, De Haro L, *et al.* A survey on future railway radio communications services: Challenges and opportunities. *IEEE Communications Magazine*. 2015; 53(10): 62–68.

[47] Ai B, Guan K, Rupp M, *et al.* Future railway services-oriented mobile communications network. *IEEE Communications Magazine*. 2015; 53(10): 78–85.

[48] Guan K, Li G, Kürner T, *et al.* On millimeter wave and THz mobile radio channel for smart rail mobility. *IEEE Transactions on Vehicular Technology*. 2016; 66(7): 5658–5674.

[49] Rappaport TS, Xing Y, Kanhere O, *et al.* Wireless communications and applications above 100 GHz: Opportunities and challenges for 6G and beyond. *IEEE Access*. 2019; 7: 78729–78757.

[50] Yu Z, Chen Y, Wang G, *et al.* Wideband channel measurements and temporal-spatial analysis for terahertz indoor communications. In: 2020 IEEE International Conference on Communications Workshops (ICC Workshops); 2020. p. 1–6.

[51] Samsung Electronics Sets 5G, Speed Record at 7.5 Gbps, Over 30 Times Faster Than 4G LTE. October 2018. [Online]. Available: http://global.samsungtomorrow.com/?p=43349.

[52] Peng B, Guan K, Kuter A, *et al.* Channel modeling and system concepts for future terahertz communications: Getting ready for advances beyond 5G. *IEEE Vehicular Technology Magazine*. 2020; 15(2): 136–143.

[53] Chen Y and Han C. *Time-varying channel modeling for low-terahertz urban vehicle-to-infrastructure communications*. In: 2019 IEEE Global Communications Conference (GLOBECOM); 2019. p. 1–6.

[54] IMT Vision – Framework and overall objectives of the future development of IMT for 2020 and beyond, Tech. Rep. ITU-R M.2083-0; 2015.

[55] Guan K, Peng B, He D, *et al.* Measurement, simulation, and characterization of train-to-infrastructure inside-station channel at the terahertz band. *IEEE Transactions on Terahertz Science and Technology*. 2019; 9(3): 291–306.

[56] Goikoetxea J. *Roadmap Towards the Wireless Virtual Coupling of Trains*. Mendizabal J, Berbineau M, Vinel A, *et al.* editors. Cham: Springer International Publishing; 2016.

[57] Popovski P, BraunV Mayer H, *et al.* “EU FP7 INFSO-ICT-317669 METIS, D1. 1 scenarios, requirements and KPIs for 5G mobile and wireless system,” document ICT-317669-METIS/D1.1, METIS, 2013. [Online]. Available: http://publications.lib.chalmers.se/records/fulltext/213055/local_213055.pdf.

[58] Guan K, Peng B, He D, *et al.* Channel characterization for intra-wagon communication at 60 and 300 GHz bands. *IEEE Transactions on Vehicular Technology*. 2019; 68(6): 5193–5207.

[59] Rey S, Eckhardt JM, Peng B, *et al.* Channel sounding techniques for applications in THz communications: A first correlation based channel sounder for ultra-wideband dynamic channel measurements at 300 GHz. In: 2017

9th International Congress on Ultra Modern Telecommunications and Control Systems and Workshops (ICUMT); 2017. p. 449–453.

[60] Guan K, Peng B, He D, *et al.* Channel sounding and ray tracing for intra-wagon scenario at mmWave and sub-mmWave bands. *IEEE Transactions on Antennas and Propagation*. 2021; 69(2): 1007–1019.

[61] Guan K, Ai B, He D, *et al.* Channel sounding and ray tracing for THz channel characterization (invited paper). In: The 13th UK-Europe-China Workshop on Millimetre Waves and Terahertz Technologies (UCMMT 2020); 2020. p. 1–3.

[62] Liang J, Lee J, Kim M, *et al.* Synthesis techniques of narrow beam-width directional antenna measurements for millimeter-wave channel characterization. In: 2015 International Conference on Information and Communication Technology Convergence (ICTC); 2015. p. 689–693.

[63] Priebe S. *Towards THz communications: Propagation studies, indoor channel modelling and interference investigations. PhD Dissertation*, Technische Universität Braunschweig; 2013.

[64] Zhao Y, Hao Y, Alomainy A, *et al.* UWB on-body radio channel modeling using ray theory and subband FDTD method. *IEEE Transactions on Microwave Theory and Techniques*. 2006; 54(4): 1827–1835.

[65] Rappaport T. *Wireless Communications: Principles and Practice*. 22nd ed. Upper Saddle River, NJ, USA: Prentice-Hall; 2002.

[66] Chiaraviglio L, Blefari-Melazzi N, Liu W, *et al.* Bringing 5G into rural and low-income areas: Is it feasible? *IEEE Communications Standards Magazine*. 2017; 1(3): 50–57.

[67] Matolak DW and Sun R., Air-ground channel characterization for unmanned aircraft systems—Part I: Methods, measurements, and models for over-water settings. *IEEE Transactions on Vehicular Technology*. 2017; 66(1): 26–44.

[68] Zhu Q, Li W, Wang C, *et al.* Temporal correlations for a non-stationary vehicle-to-vehicle channel model allowing velocity variations. *IEEE Communications Letters*. 2019; 23(7): 1280–1284.

[69] Jiang H, Zhang Z, Dang J, *et al.* A novel 3-D massive MIMO channel model for vehicle-to-vehicle communication environments. *IEEE Transactions on Communications*. 2018; 66(1): 79–90.

[70] Zhu Q, Yang Y, Wang C, *et al.* Spatial correlations of a 3-D non-stationary MIMO channel model with 3-D antenna arrays and 3-D arbitrary trajectories. *IEEE Wireless Communications Letters*. 2019; 8(2): 512–515.

[71] Lakew DS, Masood A, and Cho S. 3D UAV placement and trajectory optimization in UAV assisted wireless networks. In: 2020 International Conference on Information Networking (ICOIN); 2020. p. 80–82.

[72] Guan K, Yi H, He D, *et al.* Towards 6G: Paradigm of realistic terahertz channel modeling (invited paper). *China Communications*. 2021. Accepted.

[73] Vaughan R and Andersen JB. *Channels, Propagation and Antennas for Mobile Communications*. Electromagnetics and Radar Series. Institution of Engineering and Technology; 2003. Available from: https://books.google.de/books?id=9GO0ZKgKoIYC.

[74] Chen LF, Ong CK, Neo CP, *et al. Microwave Electronics: Measurement and Materials Characterization*. Chichester, UK: John Wiley & Sons, Ltd; 2005. Eq. (1.13).

[75] ITU-R P. 1238-7 Propagation data and prediction methods for the planning of indoor radio communication systems and the radio local area networks in the frequency range 900 MHz to 100 GHz, Geneva: ITU-R Recommendations, 2015.

[76] Piesiewicz R, Kleine-Ostmann T, Krumbholz N, *et al.* Terahertz characterisation of building materials. *Electronics Letters*. 2005; 41(18): 1002–1004.

[77] Piesiewicz R, Jansen C, Wietzke S, *et al.* Properties of building and plastic materials in the THz range. *International Journal of Infrared and Millimeter Waves*. 2007; 28(5): 363–371.

[78] Corana A, Marchesi M, Martini C, *et al.* Minimizing multimodal functions of continuous variables with the simulated annealing algorithm Corrigenda for this article is available here. *ACM Transactions on Mathematical Software (TOMS)*. 1987; 13(3): 262–280.

[79] Jaeckel S, Raschkowski L, Börner K, *et al.* QuaDRiGa-Quasi deterministic radio channel generator, user manual and documentation. *Fraunhofer Heinrich Hertz Institute*, Tech Rep v2; 2019.

[80] Tamosiunaite M, Tamosiunas S, Zilinskas M, *et al.* Atmospheric attenuation of the terahertz wireless networks. In: *Broadband Communications Networks-Recent Advances and Lessons from Practice*. InTech; 2017.

[81] ITU-R. P.676-11: Attenuation by atmospheric gases. ITU Recommendations; 2005.

[82] ITU-R. P 840-7: Attenuation due to clouds and fog. ITU Recommendations; 2017.

[83] Luo Y, Huang W, and Luo Zy. Attenuation of terahertz transmission through rain. *Optoelectronics Letters*. 2012; 8(4): 310–313.

[84] Norouzian F, Marchetti E, Gashinova M, *et al.* Rain attenuation at millimeter wave and low-THz frequencies. *IEEE Transactions on Antennas and Propagation*. 2019; 68(1): 421–431.

[85] ITU-R. P.838-3: Specific attenuation model for rain for use in prediction methods. ITU Recommendations; 2005.

[86] Norouziari F, Marchetti E, Hoare E, *et al.* Low-THz wave snow attenuation. In: *2018 International Conference on Radar (RADAR)*. IEEE; 2018. p. 1–4.

[87] Ma J, Adelberg J, Shrestha R, *et al.* The effect of snow on a terahertz wireless data link. *Journal of Infrared, Millimeter, and Terahertz Waves*. 2018; 39(6): 505–508.

[88] Tan J and Dai L. THz precoding for 6G: Applications, challenges, solutions, and opportunities. arXiv preprint arXiv:200510752; 2020.

[89] Tan J and Dai L. Delay-phase precoding for THz massive MIMO with beam split. In: *2019 IEEE Global Communications Conference (GLOBECOM)*. IEEE; 2019. p. 1–6.

[90] Molisch AF. *Wireless Communications*. vol. 34. Chichester, UK: John Wiley & Sons; 2012.

Part VI

RF exposure metrology

Chapter 20

Measurement of 5G new radio-base stations

Frederic Pythoud[1]

20.1 Introduction

The ability to measure the exposure of new radio (NR)-base stations is an important requirement towards the deployment of 5G in our countries. According to manufacturers, the 5G technology allows larger transmission capacity as compared to the previous generations. Different techniques are used to increase the transmission capacity, one of them being beamforming. The 5G technology is thus very attractive, but at the same time significantly more complex than the previous generations of mobile communication technologies.

The measurements of the actual exposure by using broadband electric field probe are insufficient to provide meaningful information on the conformity of a base station with any regulation. The reason for it is that the broadband measurements provide an instantaneous value of the electric field. This instantaneous value is changing rapidly in time and in space, and it does not provide any value on the maximum exposure of the base station, nor on any kind of expected time-average of exposure (Figure 20.1). Predictive measurement of the exposure of an NR-base station, in terms of maximum or expected mean, is therefore a difficult challenge that many regulation bodies are facing worldwide. This contribution provides a comprehensive review of measurements strategies for the exposure of NR-base stations below 6 GHz.

20.2 The resource grid

The 5G technology, also called NR, has been standardised by the Third-Generation Partnership Project (3GPP) grouping seven telecommunications standard development organisations: the Association of Radio Industries and Businesses, Japan; the Alliance for Telecommunications Industry Solutions; the China Communications Standards Association; the European Telecommunications Standards Institute (ETSI); the Telecommunications Standards Development Society, India; the Telecommunications Technology Association, Korea and the Telecommunication

[1]Federal Institute of Metrology METAS, Bern-Wabern, Switzerland

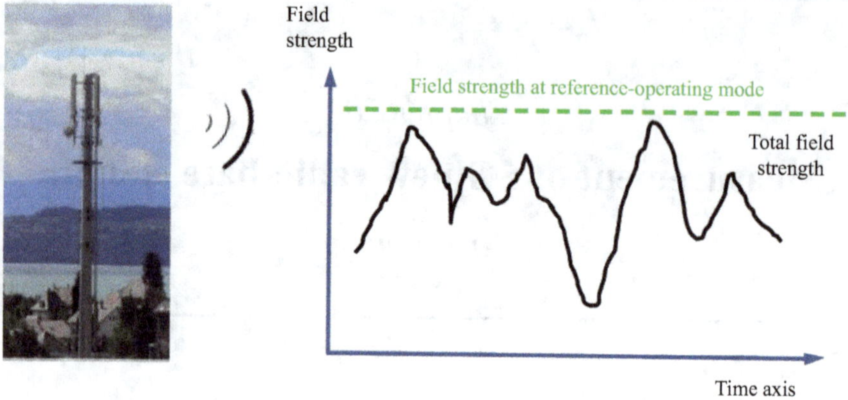

Figure 20.1 *Representation of the exposure of a base station as measured by a broadband electric field probe. Ideally, one would like to measure the field of a well-defined reference-operating mode.*

Technology Committee, Japan. The 3GPP has also standardised in the past the 2G, 3G and 4G. The standards are available as a series of ETSI standards [1–4] on Internet. The 5G technology claims transmission capacity up to 10 Gb/s, which is about 100 times larger than the 4G in its initial phase. The theoretical latency of 5G is 1 ms, whereas for 4G it is 50 ms. The high transmission capacity is achieved using by larger transmission bandwidth, and also using the beamforming technology, which can simultaneously transmit different information in different directions, known as massive multiple-input–multiple-output.

According to release 15 of the 5G standard [1], the NR technology covers two frequency ranges: the first frequency range (FR1) from 410 MHz to 7.125 GHz and the second frequency range (FR2) from 24.5 to 52.6 GHz (extension up to 71 GHz is even under study). The transmission bandwidth ranges from 5 to 100 MHz in the frequency range FR1 and from 100 to 400 MHz in the frequency range FR2. In comparison, 4G has a maximum transmission bandwidth of 20 MHz.

In order to understand how to measure the exposure of an NR-base station, it is important to understand how the physical signal is generated and transmitted by the transmit antenna. The modulation scheme for 5G is orthogonal frequency division multiplexing (OFDM), the same as for 4G. The basic principle of the modulation can be presented in two steps: first, the digital information (bits), for example '10010000 ....', is encoded into complex-valued symbols using a modulation scheme like quadrature phase-shift keying, quadrature amplitude modulation (QAM) with 16 states (16QAM), with 64 states (64QAM) or with 256 states (256QAM). For example a 16QAM constellation defines 16 possible complex values. Hence, 4 bits are needed to identify a symbol within this constellation in a unique way as shown in Figure 20.2.

The constellations are normalised (in the case of Figure 20.2, the divisor $\sqrt{10}$) in such a way that the quadratic average of the modulus of all symbols in the

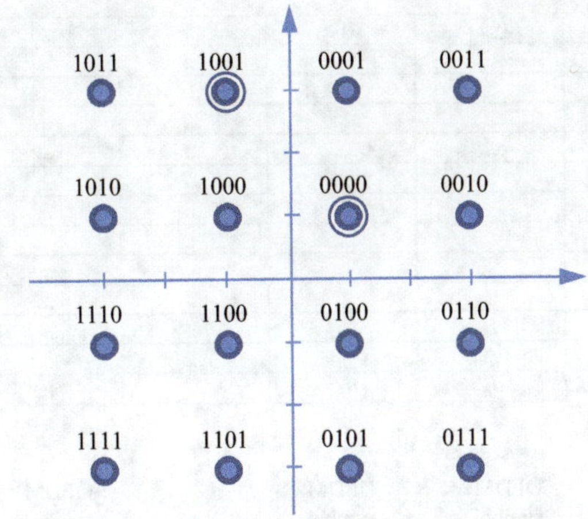

Figure 20.2 Representation of the 16QAM constellation according to the 5G standard. The series of four bits '1001' will be encoded into the complex valued symbol $d_1 = (-1 + 3i)/\sqrt{10}$*, whereas the series of four bits '0000' will be encoded into* $d_2 = (1 + i)/\sqrt{10}$*.*

constellation is one. This is important since it means that the average 'power' per symbol is always the same, independently of the modulation scheme.

In a second step, the streams of symbols are grouped together to build packets of a given size n, each packet representing one OFDM symbol. The OFDM symbols are directly modulated into an electrical signal in terms of a Fourier transform, for example, the first OFDM symbol according to our example will be encoded as

$$s(t) = \sum_{k=0}^{n-1} d_k e^{i\,2\pi\,k \cdot t \cdot \Delta f}, \tag{20.1}$$

where t is the time and Δf the subcarrier spacing, and d_k is a constellation symbol. The signal $s(t)$ is a complex-valued representation of the baseband transmit signal that is then transmitted over the air as $\mathrm{Re}\left(s(t)e^{i2\pi f_c \cdot t}\right)$, f_c being the carrier frequency, where Re(.) returns the real part of a signal. This is of course a schematic description of the OFDM transmission as applied in the 5G standard, since in reality strategies are used to increase the robustness of the signal transmitted, like the use of a cyclic prefix before each OFDM symbol. It consists of repeating the end of the OFDM signal before it, thus slightly increasing the duration of the OFDM symbol. It may be interesting to decrease the effect of discontinuities between the signals of different OFDM symbols, because discontinuities have the effect of generating out-of-band noise, which may be filtered as it is the case for LTE. In the case of 5G, it seems that no specific filtering is foreseen [2]. It is probably motivated by the fact that the OFDM symbols have already a large bandwidth and that a small discontinuity would be imperceptible.

Since the time signal transmitted to the antenna is generated based on the Fourier transform, one can simply say that each symbol d_i corresponds to a

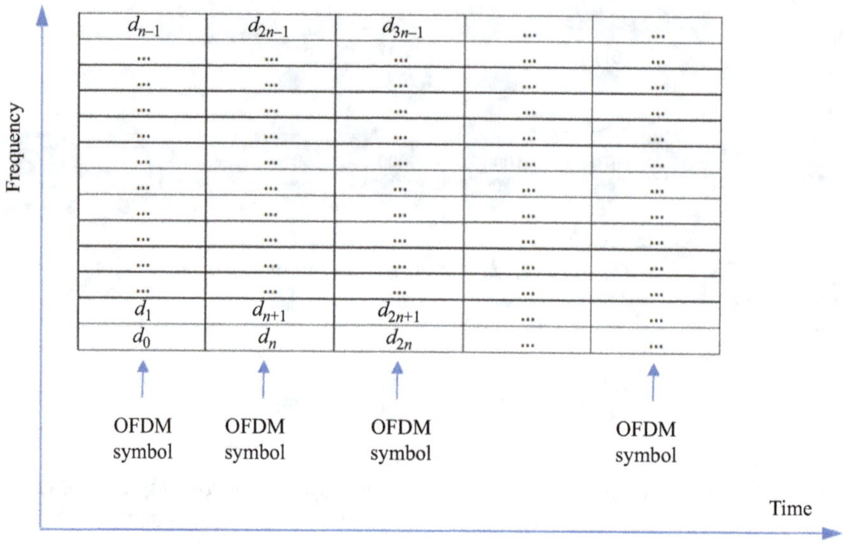

Figure 20.3 Schematic representation of the resource grid. Each element in the resource grid is called a resource element, and it has the information of one complex symbol d_k.

subcarrier frequency. The OFDM modulation scheme is therefore often represented as a resource grid, the resource grid being a time frequency grid as shown in Figure 20.3. According to the standard [2], the resource grid has a duration of one subframe, meaning 1 ms, whereas the concept of resource grid may also be used to refer more generally to the time frequency grid without particular time limitations.

The subcarrier spacing Δf foreseen by the 5G standard [1] may have different values depending on the numerology. It can be 15, 30 and 60 kHz for carrier frequencies up to 7.125 GHz, and 120 and 240 kHz for carrier frequencies above 24 GHz. Different numerologies (subcarrier spacing) may be multiplexed within the same OFDM symbol as mentioned in the same reference. The OFDM symbols are organised in frames of constant duration of 10 ms, independently of the numerology, each frame being divided in ten subframes of 1-ms duration. The number of OFDM symbols within a subframe depends on the numerology, typically from 14 symbols for 15-kHz subcarrier frequency to 224 symbols for 240-kHz subcarrier frequency. In all numerologies, the product of the subcarrier spacing with the duration of the OFDM symbol without cycling prefix is constant: 1 sHz = 1, which is quite easy to remember. As an example, for a subcarrier spacing $\Delta f = 15$ kHz, and a subframe duration of 1 ms, one would expect 15 OFDM symbols. Considering the fact that the symbols are emitted with a cyclic prefix, the effective number of symbols within a frame is thus 14 and not 15. The understanding of the resource grid is very important for the measurements of the exposure of an NR-base station, since it is a schematic representation of the electric signal that is transmitted over the air.

20.3 The principles of the 5G transmission

Several resource grids, generated as explained in Section 20.2, are transmitted simultaneously to the logical antenna ports (Figure 20.4). The definition of a logical antenna port is, according to [2],

> An antenna port is defined such that the channel over which a symbol on the antenna port is conveyed can be inferred from the channel over which another symbol on the same antenna port is conveyed.

This strange definition is not really new. It is more or less the same as in the 4G standard [5], where a very similar wording has been used. It can be understood in the sense that, to each antenna port, there is an associated physical channel. In our case, we interpret this physical channel as being one beam of an adaptive antenna for example. The beam is characterised by its radiation pattern, but also by other features like polarisation. It is produced by adding a phase-shift to the different dipoles of the adaptive antenna. All possible combinations are possible, i.e. by using the same dipoles for different ports simultaneously, each with its phase shift, or using dedicated dipoles for each beam like in hybrid antennas. The way in which adaptive antennas are designed is not essential in this discussion, but the fact that each resource grid is transmitted via its own beam is important.

In the downlink transmission [2], the ports starting with number 1000 (1000 to 1011) are used for the Physical Downlink Shared Channel (PDSCH) (see Figure 20.4). The PDSCH is typically used for payload data. The ports starting with

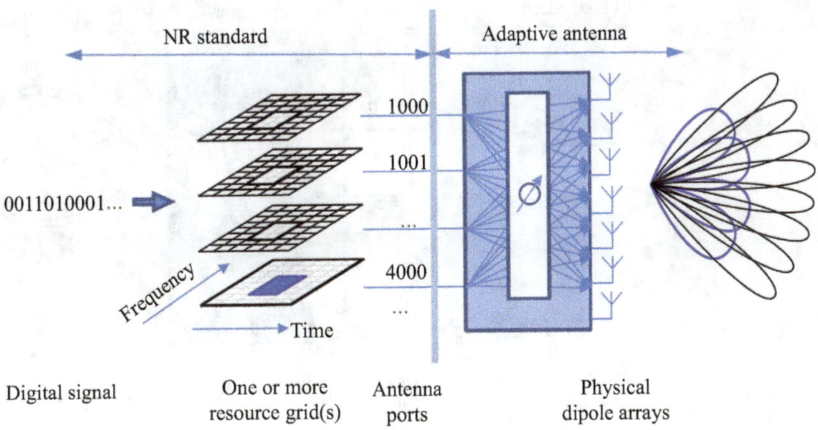

Figure 20.4 Schematic representation of the downlink transmission according to the 5G standard. The PDSCH resource grids are transmitted via the ports ranging from 1000 to 1011, to the physical PDSCH beams (thin black beams on the right of the picture). The SS/PBCH blocks (a block is represented as the blue rectangle in the resource grid of port 4000) are transmitted as the thick blue beams.

the number 2000 are used for the Physical Downlink Control Channel (PDCCH), the ports starting with 3000 for channel-state information reference signals, and the antenna ports starting with 4000 for the Synchronization Signal of the Physical Broadcast Channel (SS/PBCH) block transmission. The signal power of the PDSCH beams depends on the PDSCH traffic: the more traffic, the more resource elements (REs) will be filled up with information, and the more power will be transmitted over the PDSCH beams.

On the other hand, the SS/PBCH blocks (up to 8 blocks for frequencies in the FR1, and up to 64 blocks in the FR2 according to [3]) are transmitted in a constant manner, independently of the traffic load. They use a different antenna port than the PDSCH beams, and consequently, they will be radiated with a different radiation pattern than the PDSCH beams. As far as we know, the NR standard covers the transmission right up to the antenna ports. The 5G-transmission protocol does not have any direct view and any control in the configuration of the adaptive antenna. It is not able to adapt the structure and number of beams. The 5G transmission protocol simply sees different ports and it has to select by itself between the ports in order to optimise the transmission quality. This choice is performed by the evaluation of the physical transmission properties of the different physical channels using physical layer measurements.

At this point, it can be interesting to take a step backward and to remember how this was done in 4G. The schematic representation of the downlink transmission for 4G, derived from [5], is available in Figure 20.5. At first glance, both 4G

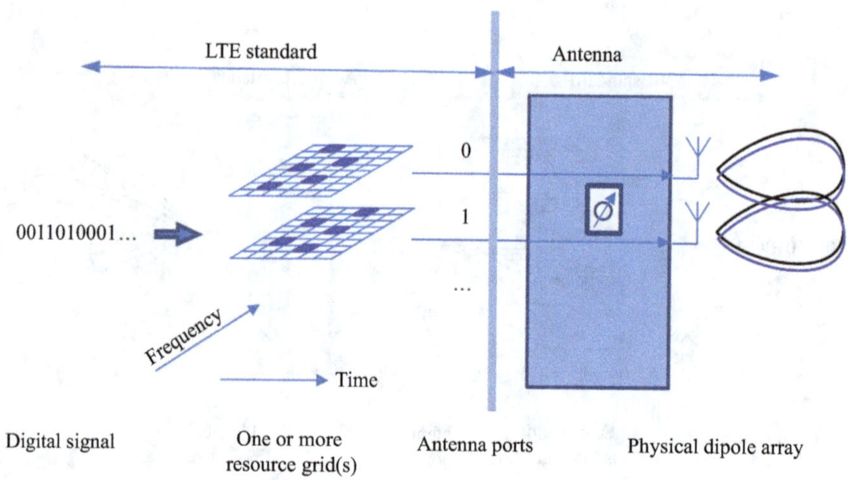

Figure 20.5 Schematic representation of the downlink transmission according to the 4G standard. Each resource grid has is cell-specific reference signals (blue squares). The radiation patterns of the cell-specific reference signals are identical to the radiation patterns of the corresponding resource grid. The difference between the beams of port 0 and port 1 are usually the polarisation, e.g. +45° and −45°.

and 5G look very similar in many points. There are however few important differences: first, the antenna ports for 4G traffic transmission are numbered from 0 to 4 (port numbers 5 or higher are available for special applications), in many situations two of them being only used. Antennas are significantly less complex than the typical adaptive antennas as foreseen for 5G, and the logical antenna ports are usually mapped directly to the physical ports of a dipole array. Moreover, each resource grid transmits its own cell-specific reference signals, independently of the traffic load. Thus, the radiation pattern of the cell-specific reference signals is the same as the corresponding resource grid.

Figure 20.6 presents a typical use of 5G technology, and it illustrates the potential of beamforming. In order to increase the data transmission capacity, different beams can transmit signals in parallel. According to our knowledge, there is no downlink power management for the beams: the average power needed to transmit an RE is the same for all REs of the same antenna port. The average total transmit power of the beam depends on the number of REs used. The 5G technology also integrates the MIMO feature as shown in Figure 20.6: using the concept of codebooks, the symbols of two ports (represented in Figure 20.6 as the direct and the indirect reflected beam to the house) can be correlated in order to produce constructive interference and thus to increase the quality of transmission. It is also possible to use two beams to transmit different information to the same user, for example with a direct beam and a reflected beam in order to increase the transmission capacity.

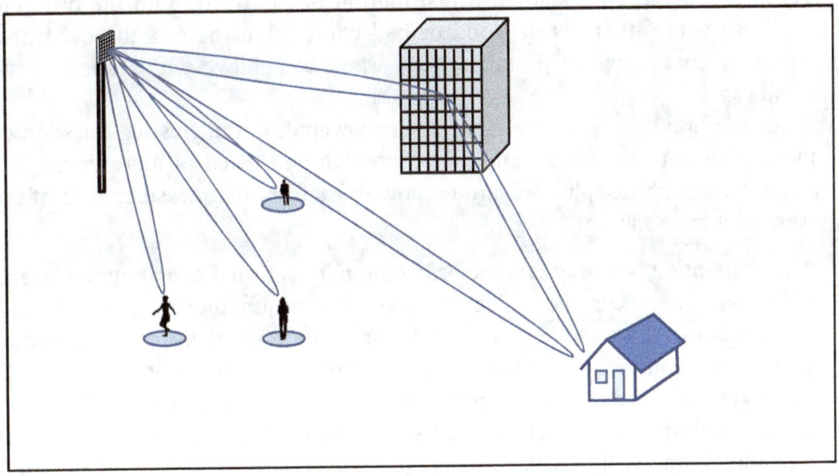

Figure 20.6 Illustration of beamforming: an NR-base station is capable of transmitting different signals to different users, and even to transmit the same signal (phase shifted) in different directions in order to create constructive interference via reflections, as shown by the beams directed to the house

20.4 Strategies to measure the exposure of an NR-base station

Before quantifying the exposure of a base station, it is important to define its measure or in other words the quantity to be measured. In this respect, there are two important questions that need to be answered:

1. At which precise point/points the exposure has to be measured?
2. What is the reference-operating mode of the base station?

The answer to the previous questions does not rely on scientific or technical arguments, but purely on arbitrary decisions. Let us first take the question of the exact measurement location in a given room or place. Knowing that the field is not homogeneously distributed within the room, one might want to define an a priori location, or to define some spatial average on a certain number of pre-defined points [6]. Another possibility is to find the local maximum of the electric field strength and to provide the exposure measurement at this point. None of the previous options is correct or wrong, and each scenario has its own pros and cons. The choice of one or the other option is just a matter of decision (e.g. by the authorities).

The same discussion can be held concerning the operating mode of the base station that is chosen as reference. One might choose to define the reference-operating mode as the mode with maximum traffic at maximum power (as required by Swiss regulations [7] for example), or one may choose to define some typical traffic situations, for example 50% of the maximum traffic at maximum power, and to agree on a statistical distribution of the traffic into the different beams. Here again, there is no good or bad choice, but there is the need of a consensus on the degree of protection one wants to achieve with respect to the limit values.

Once the previous questions have been answered, the interesting question of the measurement of the base station exposure can be treated in more detail. We identify different philosophies regarding how to perform the measurements of the exposure of a base station:

1. The first option is to measure the base station in its actual operating mode, and to correct the measured values to meet the requirements of the desired reference-operating mode. This correction is performed in terms of one or more correction factors, taking into account the actual parameters of the base station as well as the parameters in the reference-operating mode.
2. The second option is to let the base station transmit in the desired reference-operating mode, and to make measurements in this test-mode. In this case, it is most probable that the traffic of the base station needs to be interrupted and that a special 'test mode' has to be played by the base station.
3. The third option is to let the base station transmit at a constant data rate in the direction of the customer device. This transmission is triggered by the download of a large file transfer [8].

4. The fourth option is to perform long-term measurements of the exposure of the base station, for example with a frequency-selective field probe (adjusted to the frequency of the base station), without any further corrections.

All options have their pros and cons. In the case of option (a), the traffic does not need to be interrupted, but on the other hand, complex correction factors have to be determined. Moreover, the operator and the measuring laboratory must trust each other very well, since the correction factors depend on the actual and on the reference-operating mode, part of this information comes directly from the operator. In the case of option (b), the traffic has to be interrupted and the base station must be operated in a 'test mode'. This requires that the test mode is applied correctly during the measurements, and that no intentional manipulation of the 'test mode' is performed to mislead the measurements. Assuming that it is the case, the measurements can then be performed according to the defined requirements. Option (c) requires that the base station has practically no users connected to it and that the full capacity of the base station is available for the requested data transfer. Option (d) is not considered here any further since it suffers from reproducibility. Indeed the exposure of the base station strongly depends on the traffic that varies during 1 day. Such measurements would only make sense if long-term measurements over several days or weeks could be collected. It would have the great advantage of providing the real exposure of the base station, but on the other hand, it would be extremely time-consuming and would not have a general significance.

Here, we focus on option (a), which is the measurement of the base station in its actual operating mode, with extrapolation to the reference-operating mode. This option was chosen because it provides the maximum flexibility corresponds to the actual measurement methods [7] for NR-base stations in Switzerland, with the maximum worst case being chosen as reference-operating mode.

20.5 Measurement of the synchronisation signals

The measurement of the base station in its actual operating mode requires the measurement of time constant signals. The synchronisation signals are good candidates for the NR-base stations, since these are the only constant signals transmitted by the base station. In comparison to 4G, the cell-specific reference signals are transmitted by the base station in a constant manner in addition to the synchronisation signals.

The synchronisation signals are transmitted via the port 4000 and may be distributed over up to eight different beams in FR1. The SS/PBCH blocks for each beam are transmitted sequentially according to Figures 20.7 and 20.8. According to [3], all SS/PBCH blocks are transmitted within the same half frame, and one may assume [3] that this half frame is transmitted with a periodicity of two frames, meaning 20 ms. The SS/PBCH block has a bandwidth of $240\Delta f$ where Δf is the subcarrier spacing of the PBCH block. Despite the fact that according to NR numerology, the subcarrier spacing can be 15, 30 and 60 kHz for carrier frequencies up to 7.125 GHz, the possible subcarrier spacings Δf for the PBCH are only 15 and 30 kHz according to [3] (60 kHz is not used for PBCH). Therefore, only both options of 15 and 30 kHz are

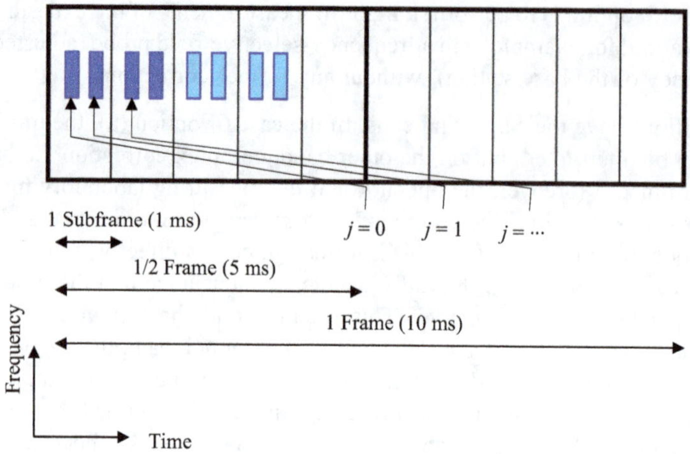

Figure 20.7 Position of the SS/PBCH blocks in the NR-signal according to [3] for an SS/PBCH subcarrier spacing of 15 kHz. Up to four blocks (dark blue) are used for $L_{\max} = 4$ beams. Up to eight blocks are used for $L_{\max} = 8$ beams.

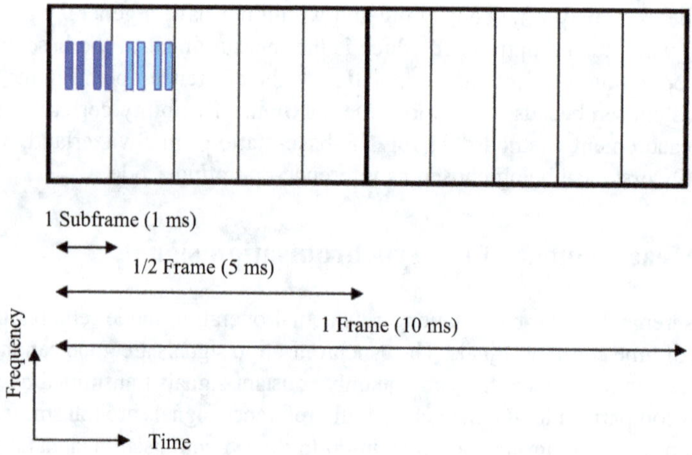

Figure 20.8 Position of the SS/PBCH blocks in the NR signal according to [3] for an SS/PBCH subcarrier spacing of 30 kHz. Up to four blocks (dark blue) are used for $L_{\max} = 4$ beams. Up to eight blocks are used for $L_{\max} = 8$ beams.

represented in Figures 20.7 and 20.8. This corresponds to a bandwidth of 3.6 and 7.2 MHz, respectively. The SS/PBCH block is not necessarily centred with the downlink carrier frequency. Each SS/PBCH block has its own beam identity j, also called SS/PBCH block index. The SS/PBCH block index is encoded in the demodulation reference signal present in the SS/PBCH block. The measurements of the SS/PBCH blocks require a code-selective measurement receiver capable of identifying the SS/PBCH blocks and their block index.

Each SS/PBCH block occupies a set of four consecutive OFDM symbols as shown in Figure 20.9. Some of the REs of the SS/PBCH block are filled up with zeros. These REs can therefore not be used to quantify the exposure of the base station. On the other hand, the secondary synchronisation signal (SSS) REs are stable. The SSS occupies a bandwidth of 127 REs, meaning 1.905 or 3.810 MHz for subcarrier spacing of 15 and 30 kHz, respectively. In other words, each SS/PBCH block has 127 SSS REs. This is interesting, since from the metrological point of view one can build an average of the signal strength all 127 REs in order to obtain a more robust estimate of the exposure. There is in fact no special preference to choose between the primary synchronisation signal (PSS) and the SSS. One motivation to choose the SSS is that by demodulating it, one gets the full identity of the base station cell, which is only partially the case with the PSS. Therefore, the SSS is a signal that allows distinguishing between different cells of the base station. The concept of cell refers here to the fact that a base station mast contains three or four antennas to cover different sectors of typically 90° to 120°. In terms of NR protocol, each sector is covered by a cell having its own identification number i.

The present method therefore assumes that the field strength $E_{i,j}^{\mathrm{SSS(RE)}}$ of the SSS of each SS/PBCH block j of each cell i can be measured. The field strength $E_{i,j}^{\mathrm{SSS(RE)}}$ is the field strength per RE of the SS/PBCH block j of the cell i, defined as the root mean square (or quadratic average) of the field strength of all SSS RE within the same SS/PBCH block. This requires that the antenna factor of the measuring antenna has been taken into account.

In the example of Figure 20.10, we assume that there are no reflections of the radiation against neighbouring buildings. One may therefore conclude that a significant contribution from the SS/PBCH beams with block indexes $j = 2$ and $j = 3$ is present at the measurement location indicated in Figure 20.10, but practically no contributions from the beams with block indexes $j = 0$ and $j = 1$, because the corresponding beams radiate in other directions.

20.6 Summing the contributions of all SS/PBCH block indexes

The next step is to summarise the SSS measurement values of all SS/PBCH block indexes j into one value $E_i^{\mathrm{SSS(RE)}}$. This value is obtained as the quadratic average of the SSS electric-field strength per RE $E_{i,j}^{\mathrm{SSS(RE)}}$ of all SS/PBCH blocks of cell i:

$$E_i^{\mathrm{SSS(RE)}} = \sqrt{\sum\nolimits_j \left(E_{i,j}^{\mathrm{SSS(RE)}}\right)^2}. \tag{20.2}$$

The sum is performed on all available SS/PBCH blocks indexes j located within the same half frame. In the example of Figure 20.10, the contributions of the SS/PBCH beams with index $j = 2$ and $j = 3$ are summed. Instead of summing the

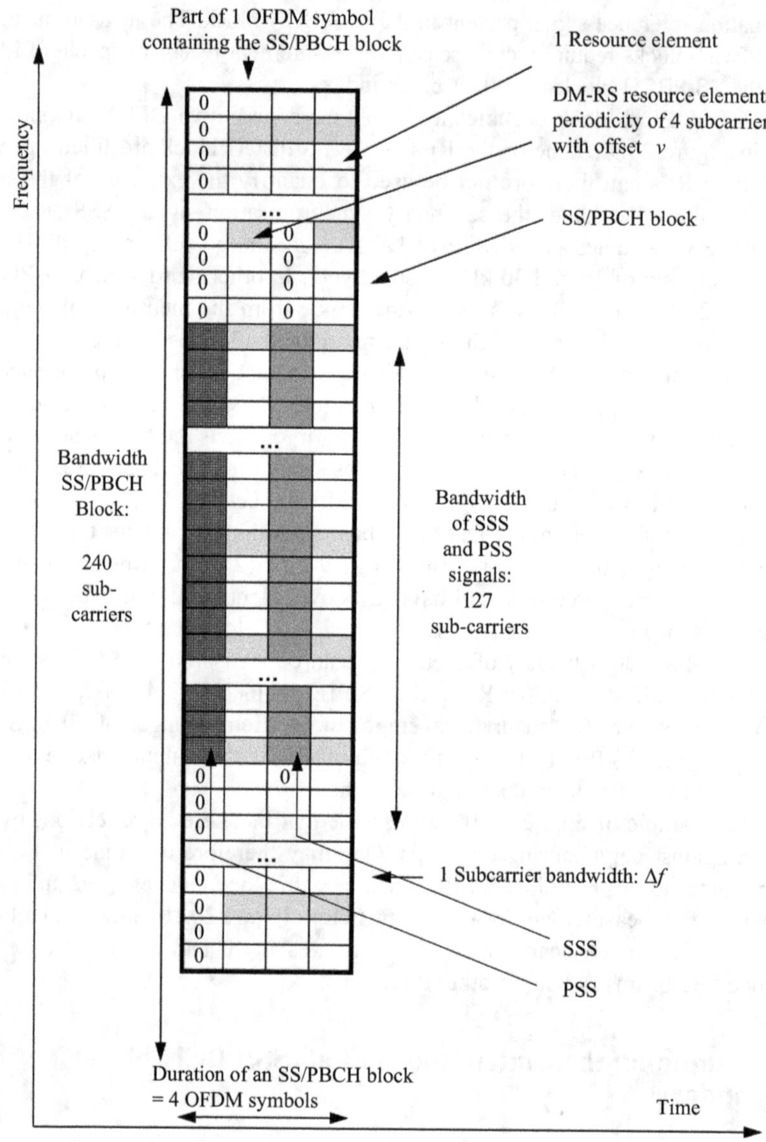

Figure 20.9 NR downlink SS/PBCH block (reconstructed from [9])

contributions of each SS/PBCH signal, other authors have proposed to take the maximum of the SSS over all PBCH beams [9]. There are two motivations to build the quadratic sum of the different beams contributions.

The first motivation is to simplify the calculation of the extrapolation factor. As shown in Figure 20.11, the SSS field strength of the radiation measured in location P1

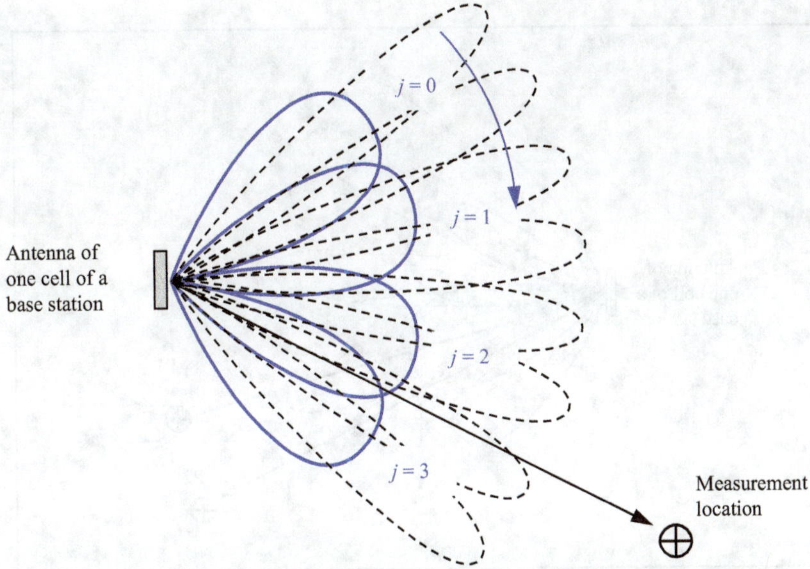

Figure 20.10 Schematic representation of the radiation of the cell of an NR-base station. The SS/PBCH beams are represented by thick blue lines, and the PDSCH beams are represented by dashed black lines. The SS/PBCH blocks are transmitted sequentially in the same frame, starting with the SS/PBCH block index $j = 0$.

consists primarily of the contribution of the SSS of the SS/PBCH block with index $j = 2$. In the measurement location P2, two contributions of the SS/PBCH blocks index $j = 2$ and $j = 3$ are measured. Each contribution is smaller than the contribution of the SS/PBCH block $j = 2$ at the measurement location P2, but the sum of both contributions should be roughly equal to it. Therefore, by summing the contributions of the SSS of different SS/PBCH blocks, one can virtually determine the radiation of the sum of all SS/PBCH beams, even if they are not transmitted simultaneously. One expects the summed SSS radiation pattern to be 'smoother' than the envelope of all SS/PBCH beams. The angle dependence of the measured values should thus be decreased, compared to the measurements of the envelope of all SS/PBCH.

There is a second motivation to build the sum of the SSS contribution of different SS/PBCH beams. It is to take into account the reflections as illustrated by Figure 20.12. In this figure, the reflection against the roof of another building increases the exposure as measured in the apartment facing the base station main beam. An adequate measurement method shall of course be able to evaluate this situation faithfully, and therefore it is important to consider the sum of all radiations in a given measurement location, and not simply the worst case. The situation in Figure 20.12 represents the interference of two radiations in the vertical plane. This addition principle also applies to reflections happening in the horizontal plane (not represented here).

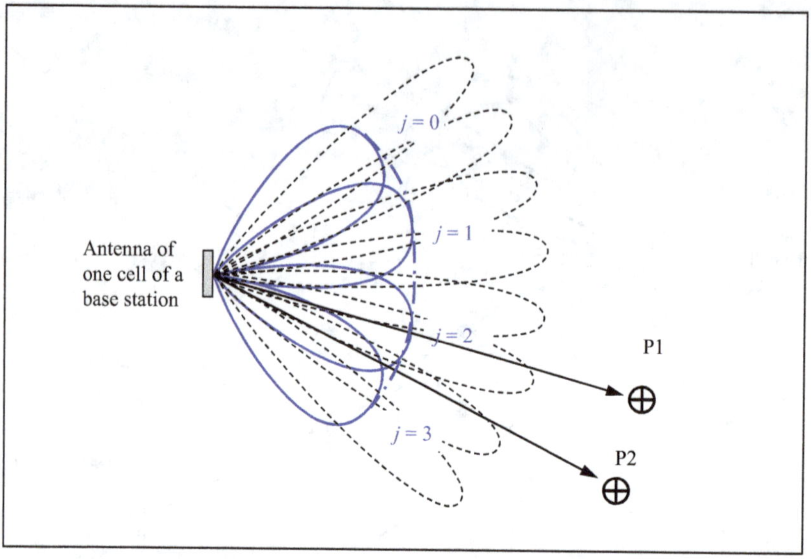

Figure 20.11 *Schematic representation of the radiation of the cell of an NR-base station. The SS/PBCH beams are represented by thick blue lines, and the PDSCH beams are represented by dashed black lines. The summed SSS radiation pattern (blue dashed line) is obtained by summing the radiated power per resource element of all SS/PBCH beams. Two measurement locations P1 and P2 are considered.*

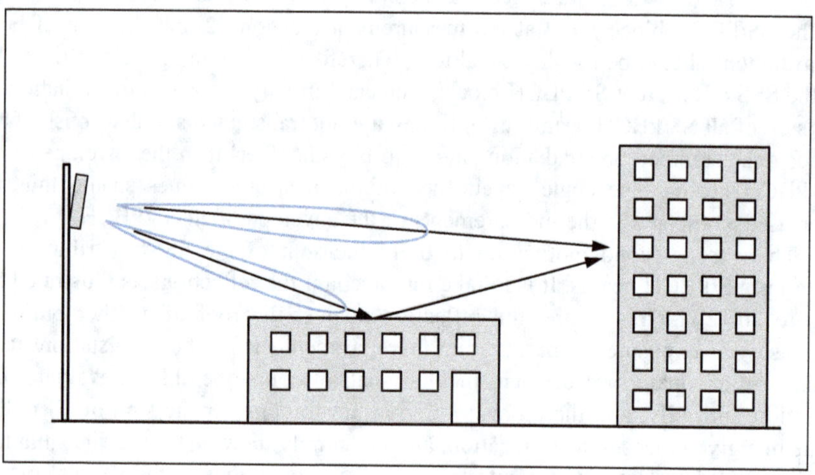

Figure 20.12 *Schematic representation of a situation where reflections increase the exposure in an apartment*

20.7 Extrapolation

At this stage, we assume that the measured value of $E_i^{\mathrm{SSS(RE)}}$ has been determined experimentally at the measurement location. Depending on the choice about the exact measurement point discussed in Section 20.4, the value of $E_i^{\mathrm{SSS(RE)}}$ may represent different quantities. It might be the value of the SSS at one given point, the quadratic average of the SSS measured at well-defined points within a given grid, or the maximum of all SSS values obtained by sampling the measurement volume. As mentioned in Section 20.4, this is only a matter of preference that has no influence on the extrapolation technique. If a spatial maximum scheme is chosen, the following aspects should be considered:

- In order to take into account the standing waves in the measurement volume, the spatial resolution during scanning shall be significantly smaller than the wavelength of the incident radiation.
- In the case of measurements with a linearly polarised antenna, the polarisation of the measuring antenna (receive antenna), as well as its orientation (azimuth and elevation), shall be scanned in all directions during the maximum search of the SSS signal.
- In the case of measurements with an isotropic probe, the reflections on the person or the robot performing the scanning shall be minimised.
- In order to get an acceptable reproducibility, minimum requirements on the measurement volume shall apply. The minimum distance to walls, floor, ceiling, furniture and windows is typically 50 cm. The height above the floor is typically between 0.5 and 1.75 m.

The extrapolation of $E_i^{\mathrm{SSS(RE)}}$ is

$$E_i = E_i^{\mathrm{SSS(RE)}} \cdot K_i(\varphi_i,\ \theta_i). \tag{20.3}$$

The factor $K_i(\varphi_i, \theta_i)$ is the global extrapolation factor for cell i. It depends on the azimuth φ_i and on the elevation θ_i of the measurement location with respect to the main beam direction of the transmit antenna of cell i. The azimuth φ_i and the elevation θ_i are the horizontal and the vertical angle in a spherical coordinate system as shown in Figure 20.13.

The global factor $K_i(\varphi_i, \theta_i)$ consists of the following four terms:

$$K_i(\varphi_i,\ \theta_i) = K_i^{\mathrm{SSS(RE)}} \cdot K_i^{\mathrm{antenna}}(\varphi_i,\ \theta_i) \cdot K_i^{\mathrm{stat}} \cdot K^{\mathrm{duplex}}, \tag{20.4}$$

where

$K_i^{\mathrm{SSS(RE)}}$	the SSS extrapolation factor for cell i. This is the main extrapolation factor and it takes into account the difference between the gain of the PDSCH beams and the gain of the SS/PBCH beams in the reference-operating mode of the base station, as well as the scaling from the power of one

(Continues)

(Continued)

	resource element for SSS to the full bandwidth of the NR downlink transmission
$K_i^{\text{antenna}}(\varphi_i, \theta_i)$	the antenna correction factor takes into account the difference between the antenna diagram of the SS/PBCH signal of cell i and the antenna diagram of the total signal in the reference-operating mode. The antenna correction factor depends on the azimuth φ_i and on the elevation θ_i. The factor $K_i^{\text{antenna}}(\varphi_i, \theta_i)$ is a correction factor and we expect it to be significantly smaller than the SSS extrapolation factor
K_i^{stat}	the beam statistic factor for cell i. It is a reduction factor that the user can freely choose. It needs to take into account the fact that for adaptive antennas, the probability that the power of the base station is focussed into only one beam is quite low. Like for the definition of the reference-operating mode, the definition of this factor has to be defined by the authority
K^{duplex}	the duplex factor. It is also a reduction factor that takes into account the fact that the transmit time in downlink is not 100% of the time in a time-domain duplex (TDD) transmit mode

The extrapolation principle is formulated in general terms as illustrated in Figure 20.14 and it may be potentially adapted to any kind of definition of the reference-operating mode of the base station.

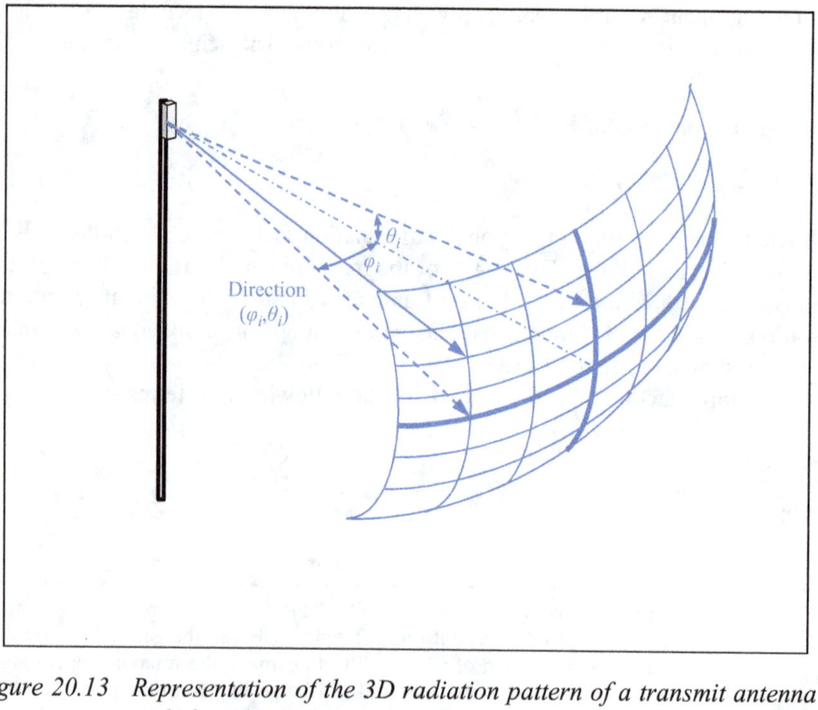

Figure 20.13 Representation of the 3D radiation pattern of a transmit antenna with the spherical coordinates of the azimuth φ_i and the elevation θ_i

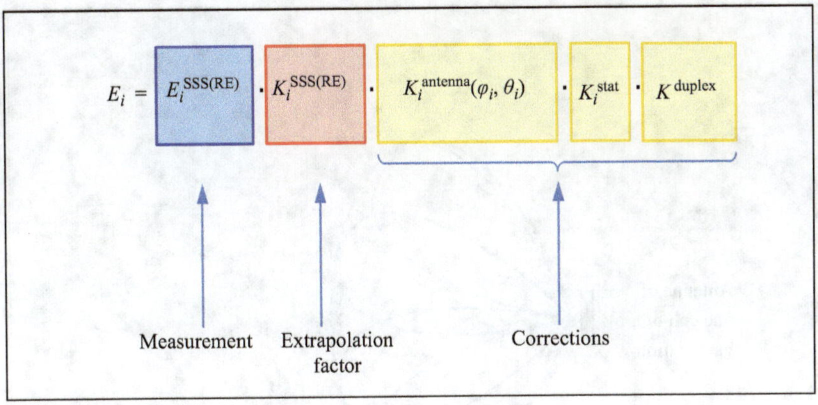

Figure 20.14 Schematic representation of the extrapolation process

20.8 Extrapolation factor for the SSS

For each cell i of the base station, an extrapolation factor $K_i^{\text{SSS(RE)}}$ is defined as

$$K_i^{\text{SSS(RE)}} = \sqrt{\frac{P_{i,\text{ref}}}{P_i^{\text{SSS(RE)}}}}, \tag{20.5}$$

where $P_i^{\text{SSS(RE)}}$ is the actual effective isotropic-radiated power (EIRP) per RE of the SSS of the SS/PBCH block of cell i in W. The variable $P_{i,\text{ref}}$ is the EIRP in W of the reference-operating mode. In the example of an operating mode defined as the maximum data traffic at maximum power, it simply corresponds to the permitted EIRP. The EIRP $P_{i,\text{ref}}$ takes into account the signals of all antenna ports of the cell i: PDSCH, PBCH and PDCCH. The EIRP $P_i^{\text{SSS(RE)}}$ corresponds to the maximum in all directions of the 'summed EIRP SSS radiation power' $P_i^{\text{SSS(RE)}}(\varphi_i, \theta_i)$, and it is given by

$$P_i^{\text{SSS(RE)}} = \max_{\varphi_i,\, \theta_i} P_i^{\text{SSS(RE)}}(\varphi_i,\, \theta_i). \tag{20.6}$$

The definition of the main extrapolation factor requires the concept of the 'summed SSS radiation pattern'. This radiation pattern corresponds to a radiation pattern that cannot be directly measured, but that is obtained by summing the EIRP-radiated power per RE of all SS/PBCH beams as defined by the following equation:

$$P_i^{\text{SSS(RE)}}(\varphi_i,\, \theta_i) = \sum_j P_{i,j}^{\text{SSS(RE)}}(\varphi_i,\, \theta_i). \tag{20.7}$$

The variable $P_{i,j}^{\text{SSS(RE)}}(\varphi_i,\, \theta_i)$ is the actual 'effective isotropic-radiated power' per RE of the SSS of the SS/PBCH block of cell i and index j in the direction given by the azimuth φ_i and elevation θ_i (Figure 20.15).

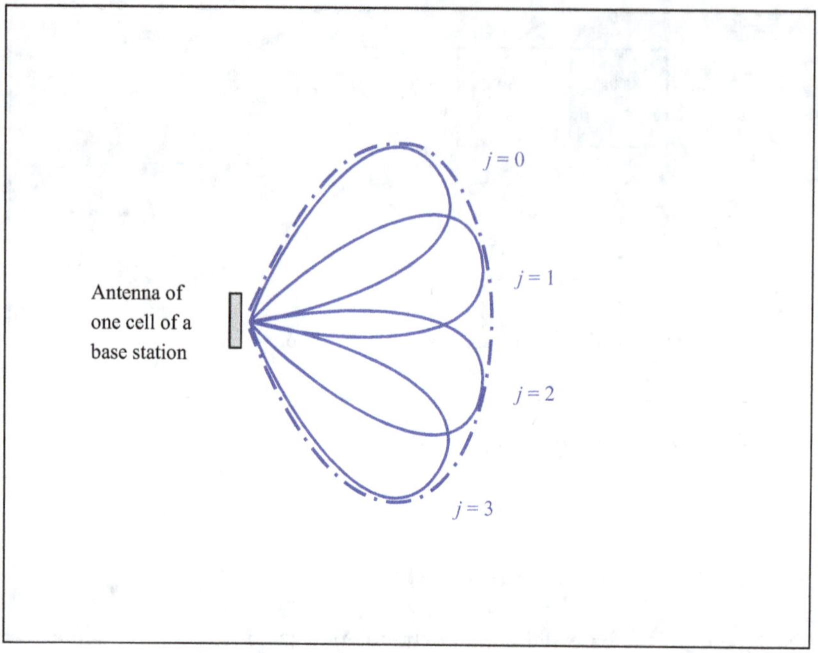

Figure 20.15 Schematic representation of the radiation of the cell of an NR-base station (seen from above). The SS/PBCH beams are represented by thick blue lines, and the summed SSS radiation pattern (dashed blue line) is obtained by summing the radiated power per RE of all SS/PBCH beams.

To determine the 'summed SSS radiation pattern', one needs the radiation pattern of each single SS/PBCH beam, the gain of the corresponding antenna array arrangement and finally the power per RE transmitted to the corresponding logical antenna port. It may well be that one does not have any access to the radiation pattern of each SS/PBCH beam individually. In this case, indirect calculation and/or experimental methods should be used to estimate the 'summed SSS radiation pattern' as well as the EIRP per RE $P_i^{\mathrm{SSS(RE)}}$.

In the report [7], the concept of effective radiated power (ERP) is used instead of EIRP. This is due to historical reasons without any impact on the extrapolation factors since the ratio of EIRP to ERP is per definition 1.64.

20.9 The antenna correction factor

The antenna correction factor $K_i^{\mathrm{antenna}}(\varphi_i, \theta_i)$ takes into account the difference between the antenna diagram of the 'summed SSS radiation pattern' and the antenna diagram of the signal in the reference-operating mode. In the special case where both radiation patterns have the same shape, we simply get $K_i^{\mathrm{antenna}}(\varphi_i, \theta_i) = 1$, i.e. no correction is made. In the general case, the antenna correction factor $K_i^{\text{antenna}}(\varphi_i, \theta_i)$ is defined for each azimuth φ_i and elevation θ_i, as

$$K_i^{\text{antenna}}(\varphi_i, \theta_i) = \begin{cases} 1 & \text{if } A_i^{\text{SSS(RE)}}(\varphi_i, \theta_i) < 10 \\ & \text{and } A_i^{\text{SSS(RE)}}(\varphi_i, \theta_i) \leq A_i^{\text{ref}}(\varphi_i, \theta_i) \\ A_i^{\text{SSS(RE)}}(\varphi_i, \theta_i)/A_i^{\text{ref}}(\varphi_i, \theta_i) & \text{if } A_i^{\text{SSS(RE)}}(\varphi_i, \theta_i) < 10 \\ & \text{and } A_i^{\text{SSS(RE)}}(\varphi_i, \theta_i) > A_i^{\text{ref}}(\varphi_i, \theta_i) \\ K_{i,\max}^{\text{antenna}} & \text{if } A_i^{\text{SSS(RE)}}(\varphi_i, \theta_i) \geq 10 \end{cases} \tag{20.8}$$

with

$$A_i^{\text{SSS(RE)}}(\varphi_i, \theta_i) = \sqrt{\frac{P_i^{\text{SSS(RE)}}}{P_i^{\text{SSS(RE)}}(\varphi_i, \theta_i)}} \tag{20.9}$$

$$K_{i,\max}^{\text{antenna}} = \max_{\{\varphi_i, \theta_i | A_i^{\text{SSS(RE)}}(\varphi_i, \theta_i) < 10\}} A_i^{\text{SSS(RE)}}(\varphi_i, \theta_i)/A_i^{\text{ref}}(\varphi_i, \theta_i). \tag{20.10}$$

The previous formulas involve the attenuation of the radiation diagram in a direction given by its azimuth φ_i and elevation θ_i. The attenuation radiation diagram is determined independently for the 'summed SSS EIRP radiation pattern', called $A_i^{\text{SSS(RE)}}(\varphi_i, \theta_i)$ and for the reference-operating mode diagram $A_i^{\text{ref}}(\varphi_i, \theta_i)$, including the contribution of all signals. In the example of an operating mode defined as the maximum data traffic at maximum power, it simply corresponds to the worst case antenna diagram. An illustration of the concept of attenuation is given in Figure 20.16. The formulas rely on the assumption that the attenuation is provided in terms of 'pure voltage ratio', and not in dB. If the attenuation A_{dB} is given in dB, one needs to determine the absolute voltage ratio as $10^{A_{\text{dB}}/20}$.

The variable $K_{i,\max}^{\text{antenna}}$ is defined as the maximum value of the ratio $A_i^{\text{SSS(RE)}}(\varphi_i, \theta_i)/A_i^{\text{ref}}(\varphi_i, \theta_i)$, where the maximum is taken on all directions for which the attenuation $A_i^{\text{SSS(RE)}}(\varphi_i, \theta_i)$ of the SS/PBCH beam is less than 10 (corresponds to 20 dB).

Equation (20.8) can be explained using Figure 20.17:

- In direction 1, we have approximately $A_i^{\text{SSS(RE)}}(\varphi_i, \theta_i) \cong 1$ (0 dB) and $A_i^{\text{ref}}(\varphi_i, \theta_i) \cong 1$ (0 dB). In this case, the first part of (20.8) applies: $K_i^{\text{antenna}}(\varphi_i, \theta_i) = 1$.
- In direction 2, let us assume that $A_i^{\text{SSS(RE)}}(\varphi_i, \theta_i) = 1$ (0 dB) and $A_i^{\text{ref}}(\varphi_i, \theta_i) = 1.1$ (0.83 dB). The first part of (20.8) applies: $K_i^{\text{antenna}}(\varphi_i, \theta_i) = 1$. This means that no reduction factor is applied despite the fact that the total radiated beam in direction 2 is more attenuated than the SS/PBCH beam in this direction. The first part of (20.8) prevents a reduction of the measured values, especially if the cases where the reference radiation pattern is not smooth.
- In direction 3, let us assume that $A_i^{\text{SSS(RE)}}(\varphi_i, \theta_i) = 1.25$ (1.94 dB) and $A_i^{\text{ref}}(\varphi_i, \theta_i) = 1.1$ (0.83 dB). The second part of (20.8) applies: K_i^{antenna}

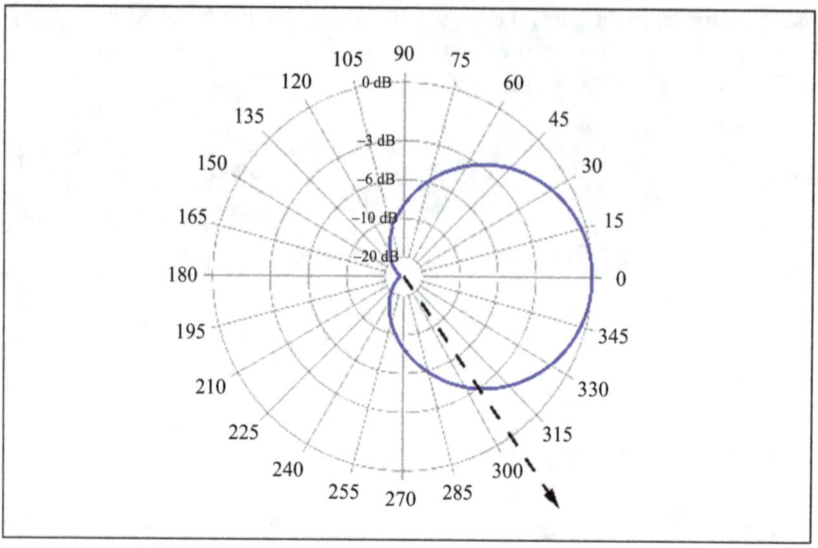

Figure 20.16 *Representation of an antenna diagram. In the direction of the arrow, the attenuation of the radiation is about 3.0 dB. In an absolute voltage ratio, this means that the attenuation is* $A = 1.41$.

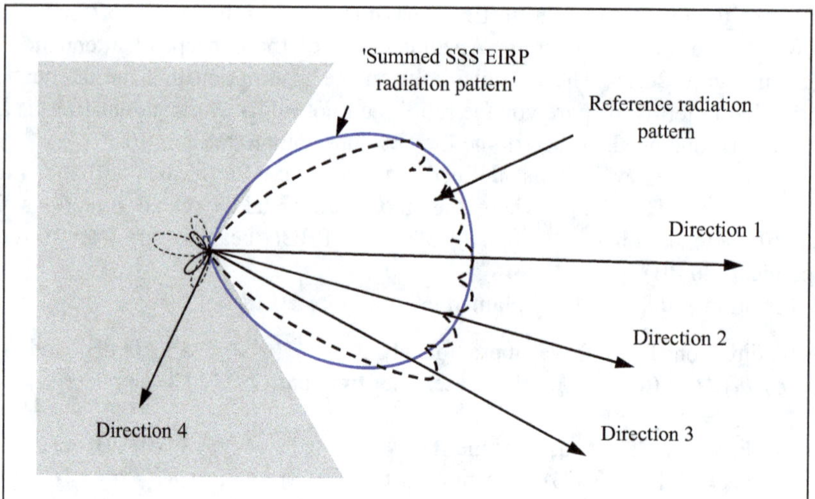

Figure 20.17 *Schematic representation (seen from the previous text) of the horizontal radiation pattern of an NR-base station cell. Both radiation patterns have been normalised in such a way that the attenuation is 1 (0 dBi) in the main direction (direction '1').*

$(\varphi_i, \theta_i) = 1.14$. This means that an extrapolation factor is applied to take into account the fact that the SS/PBCH beam in this direction is more attenuated than the total radiated beam.

- In direction 4, we are behind the transmit antenna. The radiation pattern does not totally vanish, but the radiation is small compared to the radiation in the front direction. Let us assume that $A_i^{\mathrm{SSS(RE)}}(\varphi_i, \theta_i) = 25$ (27.96 dB) and $A_i^{\mathrm{ref}}(\varphi_i, \theta_i) = 5.0$ (13.98 dB). In this case, the third part of (20.8) applies: $K_i^{\mathrm{antenna}}(\varphi_i, \theta_i) = K_{i,\max}^{\mathrm{antenna}}$, meaning the maximum value of the ratio $A_i^{\mathrm{SSS(RE)}}(\varphi_i, \theta_i)/A_i^{\mathrm{ref}}(\varphi_i, \theta_i)$, where the maximum is taken in all directions for which the attenuation $A_i^{\mathrm{SSS(RE)}}(\varphi_i, \theta_i)$ of the SS/PBCH beam is less than 10 (20 dB). This solid angle is represented in white in Figure 20.17, whereas the region where this condition is not fulfilled is represented in grey. Since the worst case antenna correction factor is approximately given by direction 3, we have $K_i^{\mathrm{antenna}}(\varphi_i, \theta_i) \cong 1.14$.

These examples are didactic illustrations of (20.8) for a horizontal cut of the antenna diagrams as represented in Figure 20.17. However, (20.8) is more general and takes into account the elevation θ_i as represented in Figure 20.13. The antenna correction factor $K_i^{\mathrm{antenna}}(\varphi_i, \theta_i)$ depends on the type and orientation of the antenna. This factor may be available, for example in a database or from the antenna manufacturer. In some situations, one may simplify the correction factor. The following situations are considered:

Both the 'summed SSS EIRP radiation pattern' and the reference radiation pattern are identical or very similar in shape. In this case,

$$K_i^{\mathrm{antenna}}(\varphi_i, \theta_i) \cong 1. \tag{20.11}$$

Only the horizontal diagrams of the 'summed SSS EIRP radiation pattern' and of the reference radiation pattern are identical or very similar in shape. In this case,

$$K_i^{\mathrm{antenna}}(\varphi_i, \theta_i) \cong K_i^{\mathrm{antenna}}(\theta_i), \tag{20.12}$$

which means that the correction factor only depends on the vertical angle.

The deviation between the 'summed SSS EIRP radiation pattern' and the reference radiation pattern is limited to a maximum value $K_{i,\max}^{\mathrm{antenna}}$, one could take the following worst case simplification:

$$K_i^{\mathrm{antenna}}(\varphi_i, \theta_i) \cong K_{i,\max}^{\mathrm{antenna}}. \tag{20.13}$$

20.10 Beam statistic factor

Based on the observation that the full power of the NR-base station is not always used for the same traffic beam, a statistical factor K_i^{stat} may be defined (Figure 20.18). This factor should have a value less than 1 (reduction factor) and should be defined by the authority. The definition of a specific value may take into account measurement campaigns [8,10–16]. Since this definition of a precise value for this factor is out of

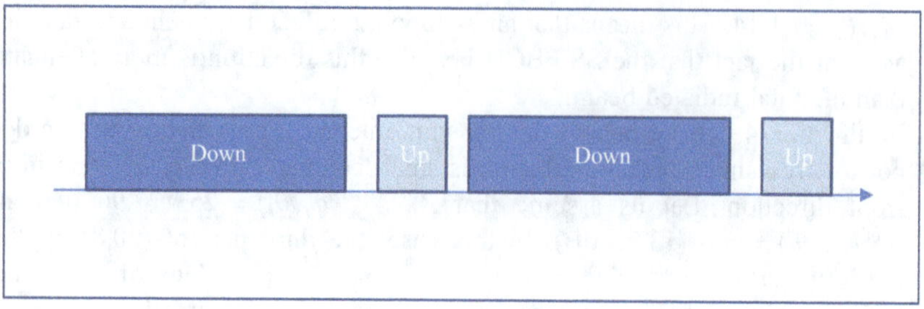

Figure 20.18 Representation of the transmit time of the downlink and of the uplink in a time-domain duplex (TDD) transmission scheme

the scope of this contribution, we simply propose the conservative value that can be freely adapted according to the need of the reader:

$$K_i^{\text{stat}} = 1. \tag{20.14}$$

20.11 Duplex factor

The duplex factor K^{duplex} takes into account the fact that the transmit time in downlink is not 100% of the time in a TDD transmit mode:

$$K^{\text{duplex}} = \begin{cases} \sqrt{r_{\text{DL}}} & \text{for} \quad \text{TDD} \\ 1 & \text{for} \quad \text{TDD with unknown } r_{DL} \\ 1 & \text{for} \quad \text{FDD} \end{cases}, \tag{20.15}$$

where r_{DL} denotes the maximum ratio of the downlink transmission time in a given time interval (duty cycle). This choice is determined by the interpretation of the E-field limits as a quadratic time average of the electric field strength according to the International Commission on Non-Ionizing Radiation Protection (ICNIRP).

20.12 Compliance assessment of base stations

In order to decide on the compliance of a base station, it is necessary to sum the electric-field strengths, E_i, determined by (20.3), of every cell i of the base stations, including NR-cells operating at different frequencies, as well as the cells of other technologies. According the ICNIRP [17], the simultaneous exposure to multiple frequencies with limits $E_{L,i}$ can be unequivocally evaluated as

$$\sqrt{\sum_{i=1}^{n} \left(\frac{E_i}{E_{L,i}}\right)^2} \leq 1. \tag{20.16}$$

In the case where the limit E_L is the same for all measured cells i, the equation reduces to

$$E_{\text{total}} = \sqrt{\sum\nolimits_{i=1}^{n} E_i^2} \leq E_L. \qquad (20.17)$$

According to the conformity assessment principles [16], the measurement uncertainty can be considered or not, depending on the decision rule that species the role of measurement uncertainty in formulating the acceptance criteria. An expanded uncertainty with a coverage factor of $k = 2$ may be used to increase the confidence to 95%, whereas (20.16) or (20.17) when applied without any uncertainty considerations would correspond to a shared risk decision rule. In this case, a maximum uncertainty must imperatively be specified. This means that in any cases, one should be aware that to assess the conformity of base stations, the measurement uncertainty has always to be estimated. The measurement uncertainty shall take into account measurement instruments uncertainty as well as the sampling uncertainty, describing the reproducibility of the measurements. Due to the nature of electromagnetic fields, the experience shows that sampling uncertainties in the range of 10%–15% are realistic to take into account the interference patterns produced by the radiation in a given location.

20.13 Example

For didactic purpose, we present a theoretical example that is inspired from a real case. Consider an installation with three cells as shown in Figure 20.19. All three cells operate in the 3,500-MHz band. The main beam directions of the antennas are 120° from each other.

The technical data of the installation is listed in Table 20.1.

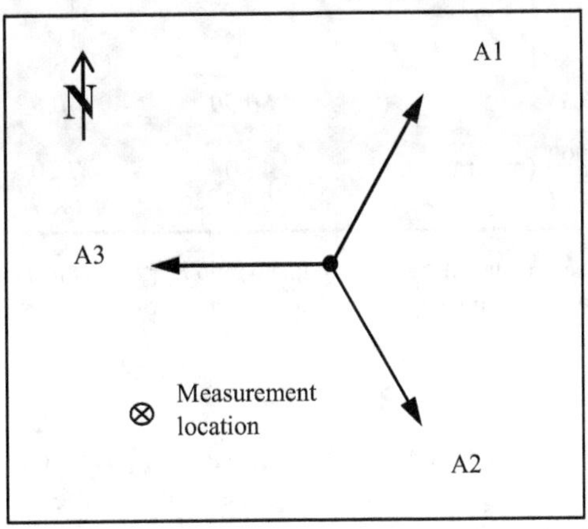

Figure 20.19 Representation of an installation with the three antennas (three cells) and the measurement location

Table 20.1 Technical data sheet of the installation

Cell ID	**214**	**215**	**216**
Antenna	A1	A2	A3
Main beam direction (azimuth)	30°	150°	270°
Main beam direction (elevation)	−10°	−12°	−12°
Number of PBCH beams	1	4	4
Service	NR-3500		
Centre frequency (MHz)	3,515		
Centre frequency of the PBCH (MHz)	3,509		
Bandwidth (MHz)	30		
Numerology (kHz)	30		
Actual EIRP of the SSS per resource element $P_i^{\mathrm{SSS(RE)}}$ (mW)	200	120	120
Reference EIRP $P_{i,\mathrm{ref}}$ (W)	400	200	200

Table 20.2 Example of an extrapolation process

Cell ID	**214**	**215**	**216**
Extrapolation factor for the SSS $K_i^{\mathrm{SSS(RE)}}$	*44.72*	*40.82*	*40.82*
Measurement-location-specific correction			
Horizontal angle of the measurement location with respect to the main beam	−160°	80°	−40°
Vertical angle of the measurement location with respect to the main beam	−15°	−13°	−14°
Attenuation of the SS/PBCH beam in measurement location direction $A_i^{\mathrm{SSS(RE)}}(\varphi_i, \theta_i)$	14.13 (23 dB)	7.94 (18 dB)	1.78 (5 dB)
Attenuation of the reference diagram in the measurement location direction $A_i^{\mathrm{ref}}(\varphi_i, \theta_i)$	31.62 (30 dB)	12.59 (22 dB)	1.41 (3 dB)
Antenna correction factor $K_i^{\mathrm{antenna}}(\varphi_i, \theta_i)$	*1.80*	*1.00*	*1.26*
Other corrections			
Statistical factor K_i^{stat}	1	1	1
Duplex factor K^{duplex}	1	1	1
Global factor			
Global factor $K_i(\varphi_i, \theta_i)$	*80.50*	*40.82*	*51.40*
Measurements			
Measured value $E_{i,\max}^{\mathrm{SSS(RE)}}$ (mV/m)	4.30	7.20	88.00
Extrapolated value $E_{i,h}$ (V/m)	*0.35*	*0.29*	*4.52*

For this calculation, we have assumed that the maximum ratio $K_{i,\max}^{\mathrm{antenna}}$ defined by (20.10) was 1.8. Cells with italic characters can be determined by calculation from other cells values.

We assume that measurements of the SSS signals have been performed using a code-selective measurement equipment. We assume that the electric field strength $E_i^{\mathrm{SSS(RE)}}$ of the SSS of each cell within the measurement volume has been determined (see the second-to-last row of Table 20.2). The extrapolation process is represented in Table 20.2. The extrapolation factor $K_i^{\mathrm{SSS(RE)}}$ is directly obtained by applying (20.5). The next six lines are documented to determine the antenna correction factor. We assume that the horizontal and the vertical attenuations of the beam in the direction of the measurement location are obtained from a database. The antenna correction factor $K_i^{\text{antenna}}(\varphi_i, \theta_i)$ is then fully determined from the previous lines of the table, with the exception of the correction factor for antenna A1, where we assume a worst case value of 1.8. The statistical and duplex correction factors are set to 1.0 in this example. The global extrapolation factor for each cell can thus be determined independently by multiplying the extrapolation and correction factors of each cell. Finally, the measured values are multiplied with the global extrapolation factor according to (20.3).

If the only service installed on this mast is the NR-service measured here, the total value of the electric field strength extrapolated to the reference-operating mode is

$$E_{\text{total}} = \sqrt{\sum_i E_i^2} = \sqrt{0.35^2 + 0.29^2 + 4.52^2} = 4.54\ \text{V/m}.$$

20.14 Summary

This chapter presents a measurement method to quantify the exposure of NR-base stations up to 7.125 GHz. As mentioned in Section 20.4, two strategies to measure the exposure of an NR-base station are available. For the first strategy, the code-selective measurements of the synchronisation signals followed by an extrapolation are needed. According to the second strategy, a well-defined operating test-mode is required, eventually also followed by an extrapolation. Both strategies have their advantages and disadvantages.

We have chosen here the first strategy and we present the method based on the measurement of the synchronisation signals that are transmitted in a constant manner, independently of the traffic load. This method requires a code-selective measurement instrument able to determine precisely the field strength of each SS/PBCH beam and to add them together. This approach offers the advantage that it does not require any interruption of the base station traffic. On the other hand, diverse extrapolations are required in order to provide estimations of the NR-base station exposure in the reference-operating mode.

The extrapolation is performed by splitting the main extrapolation factor into four terms, each related to a different effect (see Figure 20.14): the first term, also referred to as the main extrapolation factor, simply relates the ratio of the EIRP power of the total signal in the reference-operating mode to the EIRP power of the synchronisation signal. The other contributions to the extrapolation are considered corrections. The first correction is related to the difference between the antenna diagrams of the total signal compared to the antenna diagram of the synchronisation signal. This antenna correction factor depends on the location of the measurement location (azimuth and elevation from the main beam direction). The second correction takes into account the statistics of the traffic beam occupation. Finally, the duplex correction factor takes into account the duty cycle of TDD transmission.

The extrapolation method is presented in a very general frame. It provides flexibility concerning the definition of the reference-operating mode of the base station, the sampling method, as well as the evaluation of the statistical nature of the traffic beam occupation. This means concretely that this generic method can be adapted to every particular need of the partners (authorities, operators, measurement laboratories) involved in establishing the conformity of base stations. The decision taken by the relevant authority will directly impact the numerical value of some of the extrapolation factors.

At a first glance, the measurement procedure might look quite complex due to the number and complexity of the correction factors. However, it is important to remember that the 5G technology itself is complex, significantly more complex than the previous generations. From this point of view, it is clear that the measurement procedure must be able to take into account this complexity, and therefore the measurement complexity is in-line with the complexity of the NR-protocol itself.

20.15 Outlook and future work

The method to assess the exposure of base stations requires measurement and extrapolations based on assumptions. Despite the fact that the extrapolation factors are determined as precisely as possible, there will always remain questions about whether the estimation of the exposure is in good agreement with the assumptions. In other words, there is the need for some kind of verification that the full exposure assessment process has been performed correctly. Therefore, it would be interesting to compare two evaluation methods that entirely differ in their principles.

The second strategy mentioned in Section 20.4 could be a candidate for a cross-verification of the measurement method. It consists of operating the base station in a test mode implementing the reference-operating mode. This could be an interesting possibility to validate the extrapolation method presented in this contribution. This kind of validation would not need to be performed systematically on every base station, but only on representative samples of NR-base stations.

We would like to mention here a third possibility consisting of measuring the full time-frequency resource grid (see Figure 20.8) at the measurement location. From the measurements of the full resource grid, it would be possible to measure the power of every RE within the resource grid. One could thus identify the empty REs and measure the field strength of the occupied REs, assuming that a minimum data traffic is available on this base station. From this knowledge, it should be possible to estimate the exposure of the base station in the reference-operating mode, as well as to experimentally determine the extrapolation factors and to compare them with the factors obtained numerically. In this way, we would benchmark the method, and if necessary, provide additional empirical correction if needed. This strategy has not been implemented yet, but we plan to do it in the future.

20.16 Conclusion

The 5G technology is significantly more complex than the previous generations of mobile communication technologies. At a first glance, one may be under the impression that it is impossible to measure any exposure reliably. On the other hand, the fully digital modulation (OFDM) offers numerous possibilities, if one is willing to leave the broadband or the frequency-selective measurement systems and to start performing code-selective measurements. From a metrological standpoint, it is a very challenging task because it requires a new approach with considerable pre-processing to decode the signal without losing the traceability to the International System of units. On the other hand, it opens new horizons with the possibility to quantify thousands of different sub-signals within the same NR signal and to perform some customised statistical analysis of them. The expectations are even that it will be possible to gain not only more precise evaluations of the exposure, but also more information on the current operating-mode and more accurate extrapolations for the reference-operating mode.

We are therefore very confident that in spite of the increasing complexity of systems in the future, we will be able to evaluate them correctly and even more precisely than traditional techniques by implementing digital signal metrology.

References

[1] ETSI TS 138 104, “5G; NR; Base station (BS) radio transmission and reception (3GPP TS 38.104 version 15.3.0 Release15)”, October 2018.

[2] ETSI TS 138 211, “5G; NR; Physical channels and modulation (3GPP TS 38.211 version 15.2.0 Release 15)”, July 2018.

[3] ETSI TS 138 213, “5G; NR; Physical layer procedures for control (3GPP TS 38.213 version 15.6.0 Release 15)”, July 2019.

[4] ETSI TS 138 214, “5G; NR; Physical layer procedures for data (3GPP TS 38.214 version 15.3.0 Release 15)”, October 2018.

[5] ETSI TS 136 211, “LTE; Evolved universal terrestrial radio access (E-UTRA); Physical channels and modulation (3GPP TS 36.211 version 14.2.0 Release14)”, April 2014.

[6] IEC 62232, “Determination of RF field strength, power density and SAR in the vicinity of radiocommunication base stations for the purpose of evaluating human exposure”, 2017.

[7] METAS-Report 154.1-2020-5218-1016, “Technical report: measurement method for 5G NR base stations up to 6 GHz”, 18 February 2020, available at www.metas.ch.

[8] ANFR, “Evaluation de l’exposition du public aux ondes électromagnétiques 5G – Volet 2: premiers résultats de mesures sur les pilotes 5G dans la bande 3400–3800 MHz”, Agence Nationale des Fréquences ANFR, 34p., 2020.

[9] H. Keller (2019): On the assessment of human exposure to electromagnetic fields transmitted by 5G NR base stations, *Health Physics* 117, 541–545.

[10] B. Thors, A. Furuskär, D. Colombi, and C. Törnevik. (2017): Time-averaged realistic maximum power levels for the assessment of radio frequency exposure for 5G radio base stations using massive MIMO, *IEEE Access* 5, 19711–19719.

[11] P. Baracca, A. Weber, and C. Grangeat. "A Statistical Approach for RF Exposure Compliance Boundary Assessment in Massive MIMO Systems", International Workshop on Smart Antennas (WSA), Bochum (Germany), 2018.

[12] M. Matalatala, M. Deruyck, S. Shikhantsov *et al.* (2019): Multi-objective optimization of massive MIMO 5G wireless networks towards power consumption, uplink and downlink exposure, *Applied Sciences* 9, 4974.

[13] S. Shikhantsov, A. Thielens, S. Aerts *et al.* (2020): Ray-tracing-based numerical assessment of the spatiotemporal duty cycle of 5G massive MIMO in an outdoor urban environment, *Applied Sciences* 2020, 10, 7631; doi:10.3390/app10217631.

[14] R. Werner, P. Knipe, and S. Iskra. (2019): A comparison between measured and computed assessments of the RF exposure compliance boundary of an in-situ radio base station massive MIMO antenna, *IEEE Access* 7, 170682–170689.

[15] D. Colombi, P. Joshi, B. Xu, F. Ghasemifard, V. Narasaraju, and C. Törnevik. (2020): Analysis of the actual power and EMF exposure from base stations in a commercial 5G network, *Applied Sciences* 2020, 10, 5280.

[16] JCGM 106, "Evaluation of measurement data – the role of measurement uncertainty in conformity assessment", 2009.

[17] ICNIRP Commission (1998): Guidelines for limiting exposure to time-varying electric, magnetic, and electromagnetic fields (up to 300 GHz), *Health Physics* 74, 4, 494–522.

Chapter 21
Metrology for RF-exposure from massive MIMO system

Fabien Héliot[1], David Cheadle[2], Yunsong Gui[2] and Tian Hong Loh[2]

Massive multiple-input–multiple-output (MIMO) antenna systems are one of the key enablers for the fifth generation (5G) of cellular systems and are so far one of the main differences between 5G and the previous generations at the physical layer. The idea behind massive MIMO (mMIMO) is to equip base stations (BSs) with a very large number of antenna elements to create narrow directional beams for transmitting information to different users on the same frequency/time resource and, hence, greatly improve the spectral efficiency of the system. New technology comes with new challenges, the high directionality of the transmission when using mMIMO, compares to existing cellular technologies, raises some questions about the level of radio frequency (RF)-exposure such system generates on the general population and how to evaluate this exposure. Indeed, traditional BS technology radiates RF power more or less uniformly in all directions, while BSs relying on mMIMO focus their power on the user equipment (UE) devices, such that the distribution of power, which becomes scenario related (e.g. position of the UE, duration of usage), is not necessarily uniform anymore. Hence, it is expected that the RF-exposure would be quite different when mMIMO technology is used at the BS instead of the traditional sectorised BS. In turn, it creates a need for the current exposure regulation, which lacks of harmonisation and is highly conservative in some parts of the world, to be updated in order to fully reflect the way mMIMO operates and generates RF-exposure on the general population.

This chapter first makes a case about the importance of the metrology for mMIMO when it comes to RF-exposure in order to provide meaningful data for the regulatory bodies so they can, in turn, make informed and sensible decisions on how to set exposure limits for 5G BS. It then describes some of the existing works on the topic and focus in more detail on experimental work. It then explains the way forward that is currently taken to define how to measure and regulate the RF-exposure

[1]Institute for Communication Systems, 5G/6GIC, University of Surrey, Guildford, United Kingdom
[2]National Physical Laboratory, 5G and Future Communications Technology Group, Teddington, United Kingdom

of 5G BS, by mainly using statistical approaches to better reflect the beamforming nature of the mMIMO transmission. Statistical models based on simple theory and system level simulations have first been proposed, before data from real 5G BS have been acquired to confirm the results of these models. The main finding is that using statistical approaches for defining the exclusion zone of 5G BS seems reasonable in comparison with the traditional method that creates unrealistically large exclusion zones. However, more works need to be done to fully understand how the RF-exposure generated by mMIMO BS fluctuates over time and space; a metrology method based on a fully reconfigurable 5G mMIMO testbed is fully detailed to do so. The main findings of this method are: (1) the RF exposure decays in a quadratic manner as a function of the distance, even when there is no main beam pointing directly into this direction; (2) the peak of the RF exposure at a particular point in space is reached as an active beam is directly steered towards it; (3) the overall RF-exposure at a given point increases with the number of users that are served simultaneously by the mMIMO BS and this effect is more significant close to the mMIMO BS than further away; and (4) the RF-exposure of a mMIMO BS varies significantly as a function of the amount of data it transmits.

21.1 Motivation for 5G RF-exposure new metrology and guidelines

The new generation of mobile networks, which promises among other things higher data rate, lower latency and higher reliability than the previous generation, is being rolled out (in 2020) throughout the world. The roll-out is driven by the demand for higher speed communication for a range of diverse applications for the general public and industries. As it is the case with previous mobile technology, 5G uses non-ionising RF to operate, and the public concerns over the potential health risks from RF-exposure generated by mobile network equipment, such as BSs, have led in the past to very strict RF-exposure compliance regulation in some parts of the world. In addition, the level of RF-exposure experienced by the general public is foreseen to steadily increase in the decades to come [1], while the impact of non-ionised electromagnetic field (EMF) radiations on the human body is still a debated issue in the scientific community [2], such that concerns about RF-exposure in the general public are currently being reignited by the roll-out of the 5G technology, despite the strict regulation.

At the international level, the International Commission on Non-Ionizing Radiation Protection (ICNIRP) [3] provides guidelines on RF-exposure for both mobile network equipment, e.g. BS, and mobile network devices, e.g. UE. These guidelines have been used to form regulations in many countries and regions in the world (including Europe), however, in an un-harmonised and often stringent manner to deal with the concerns of the population in countries, regions and even cities. For instance, in Europe, at the country level, Switzerland and Italy have different RF-exposure limits at 4 and 6 V/m, respectively, while similarly at the city level, Paris and Brussels have different RF-exposure indoor limits, i.e. 5 and 6 V/m, respectively. All these limits go well beyond the guidelines of 61 V/m set by

ICNIRP for the general public [3]. This un-harmonised and stringent regulatory framework obviously creates challenges for the deployment of new BS technology like mMIMO in 5G. It is expected to affect the design and deployment of effective 5G networks [4,5] – meaning that 5G communication systems may not work in some geographical areas and that the seamless connectivity promised by 5G turns out to be impossible to implement. It is also likely to impact the coverage and quality of service provided to the consumers as it has already been the case to some extent in the fourth generation (4G) of mobile networks [4,6].

The traditional approach utilised by operators for meeting the RF-exposure regulations in the third-generation (3G) and the 4G of mobile networks, in a particular geographical area, is to rely on the concept of exclusion zone, also known as compliance boundary [7]. An exclusion zone is an area around a BS where the general public is not allowed access to. It is traditionally defined, in a conservative way, based on the theoretical maximum power that can be transmitted in all possible directions for a defined time-period of usually several minutes [8]. In effect, it assumes uniform power/energy distribution around the BS. This might be a correct assumption for sectorised 4G BS, but not anymore for 5G BS due to the usage of the mMIMO technology, such that the current way of defining an exclusion zone is deemed to not be fit for purpose anymore [9]. Indeed, the way mMIMO BSs operate is quite different from previous generations of BSs; 3G or 4G BSs were designed to radiate power in an omnidirectional (over 360°) or in sectorised manner (over typically 120°), respectively, to ensure a more or less uniform coverage of their transmit signals within a geographical area, while mMIMO BSs are designed to transmit signals in a very directive manner (e.g. by using 5° narrow beam) with higher antenna gain, i.e. by relying on beamforming. Hence, applying the current approach, i.e. by considering the theoretical maximum transmit power in all the directions, for defining the exclusion zone of mMIMO BS would result in unrealistic large exclusion zone areas, which would prevent operators from deploying 5G BSs at sites with pre-existing 3G and 4G BSs [10]. Another key difference with traditional BSs is that beamforming update and UE scheduling can be done in a matter of a few milliseconds, i.e. a BS can transmit a beam in one direction for a few milliseconds and then change the direction of this beam for the next few millisecond, such that computing the RF-exposure of a transmission based on an average over several minutes does not make that much sense anymore. Another important aspect to take into account is that an mMIMO BS might not be fully loaded every millisecond and when multiple users are served simultaneously on different beams (in different directions), the transmit power is split amongst these beams.

21.2 Measuring 5G RF-exposure

In the light of the obvious shortcomings of the current RF-exposure metrology and its implications on the ability of 5G mMIMO BS to meet the current regulations, which are directly linked to this metrology, there is a clear need to design and adopt a new metrology and update/reshape the regulations accordingly. In this regard, it is envisaged that a robust RF-exposure metrology for 5G mMIMO system should rely on suitable statistical approaches. For instance, approaches based on rigorous scientific evidences that ensure a good balance between user experience and public safety, such that user services requiring high-power beams are only utilised on a need-to-basis [4,9]. A few years ago, the International Electrotechnical Commission (IEC), which is in charge of standardising the method for setting up BS exclusion zone, introduced a statistical approach to do so [8]. The approach evaluates in a more accurate manner the actual transmit power that is radiated in realistic scenarios, such that this power is defined as the 95th percentile of the measured values instead of the theoretical maximum value. Interestingly, it turns out that the exclusion for mMIMO BS is smaller when using this statistical method contrary to the traditional method. Similar conclusions have then been reached in [9,11], where the benefits of using a statistical approach for assessing the exclusion zone around the BS have been studied. The work in [9] proposed a statistical model of the time-averaged maximum power (i.e. defined as in [8]) when considering BS utilisation, mMIMO technology, time-division duplexing (TDD) and spatial distribution of the UEs. Their results show that the exclusion zone can be reduced by 15% as compared to the traditional method based on maximum transmit power. Meanwhile, the work in [11] has evaluated by means of system level simulations how the power is focused when an mMIMO BS is used and realistic assumptions regarding the UE distribution and traffic models are taken into account. Based on the cumulative distribution function of the mMIMO BS transmit power, the exclusion zone has been calculated and it turns out to be half the size of the one obtained by the traditional method.

Meanwhile, measurement campaigns of operating 5G BSs have been performed in countries where 5G is already deployed, e.g. the United Kingdom [12], France [13], Australia [14], to better understand the RF-exposure generated by mMIMO BSs in real environments. In [12], the RF-exposure level of 5G BSs in different test sites throughout the United Kingdom has been evaluated by using a field strength analyser attached to an isotropic field probe and following a procedure based on the in situ RF-exposure measurement method set out by IEC. It was found that the RF-exposure level from a typical 5G BS is only a small fraction of the ICNIRP guideline value, with the highest level recorded being approximately only 1.5% of this value. In [13], similar to [12], the RF-exposure level of 5G BSs in different test sites has been evaluated and a new wave exposure measurement indicator considering the specific characteristics of 5G networks has been proposed based on these evaluations. Accordingly, it is foreseen in [13] that the RF-exposure of a 5G BS should be at worst similar to that a current 4G BS and at best 35% lower to it. In [14], experimental data on the spatial distribution of 5G BS transmit power have been collected over a 24-h period. These data have revealed that assuming constant peak power transmission in a fixed beam-direction leads to an unrealistic EMF exposure assessment. They also indicate that the maximum time-averaged power per beam direction is well below the theoretical maximum, i.e. in-line with the prediction of the statistical model of [11], since they lead to an exclusion zone half the size of the one calculated by using the theoretical maximum power.

Despite the aforementioned works on defining measurement methods and exclusion zone for 5G BS, the definition of robust and effective model and/or experimental-based methods is still being an open problem under evaluation by international organisations. The IEC Technical Committee 106 and IEEE Technical Committee 34 have established IEEE/IEC joint working groups that develop standards in the area of EMF compliance assessment for 5G technology. In addition, the internal telecommunication union-T Study Group 5 is also closely collaborating with these technical committees. More works need to be done to fully understand how the RF-exposure generated by mMIMO BS fluctuates as a function of the environment, i.e. spatial and temporal variations (note that most measurement campaigns using real 5G BS are based on outdoor data), user traffic profile and number as well as positions of the users by using more traceable and reproducible methods in a control environment. An example of such method is detailed in the following, where a 5G mMIMO testbed, based on software-defined radio (SDR) and hence fully reconfigurable, has been set up and utilised to build an experimental-based statistical model of the RF-exposure of 5G BS [15,16].

21.3 Experimental assessment of the RF-exposure of massive MIMO base station via a reconfigurable testbed

In order to define a new metrology for the RF-exposure from mMIMO BS, a fully configurable mMIMO testbed can be used to first gather experimental evidences of 5G BS RF-EMF exposure within a real-world environment and then build a robust statistical model based on the spatial variations of the RF-EMF exposure surrounding the mMIMO testbed when taking into account a large variety of different settings, e.g. number of UEs, position of UEs, number of antenna per UEs, traffic profile. Such a process is fully detailed in the following where an mMIMO testbed with an mMIMO BS capable of mimicking realistic 5G BS new radio functionalities like beamforming has been utilised, as is shown in Figure 21.1, to map the 5G RF-exposure within a real-world indoor environment.

21.3.1 mMIMO testbed set-up

The mMIMO BS of Figure 21.1 comprises four main elements: (1) a BEE7 control and synchronisation/trigger generator (see Figure 21.1(b)); (2) MegaBEE transceiver modules, each module has two times four input/output RF ports and hence could support up to eight IQ channels (see Figure 21.1(b)); (3) a White Rabbit time distribution system (clocking network); (4) an mMIMO transmit antenna (Tx) array with 128 patch antenna elements (see Figure 21.1(c)), with its geometry further detailed in Table 21.1; a network router to connect all the units together in order to programme them through a PC.

In addition to the mMIMO BS of Figure 21.1, the testbed uses several receivers, as is shown in Figure 21.2. The receivers, which mimic MIMO UEs, comprise up to eight vertically polarised dipole antenna elements mounted on a mobile trolley and a MegaBEE transceiver module for performing the MIMO baseband processing. The

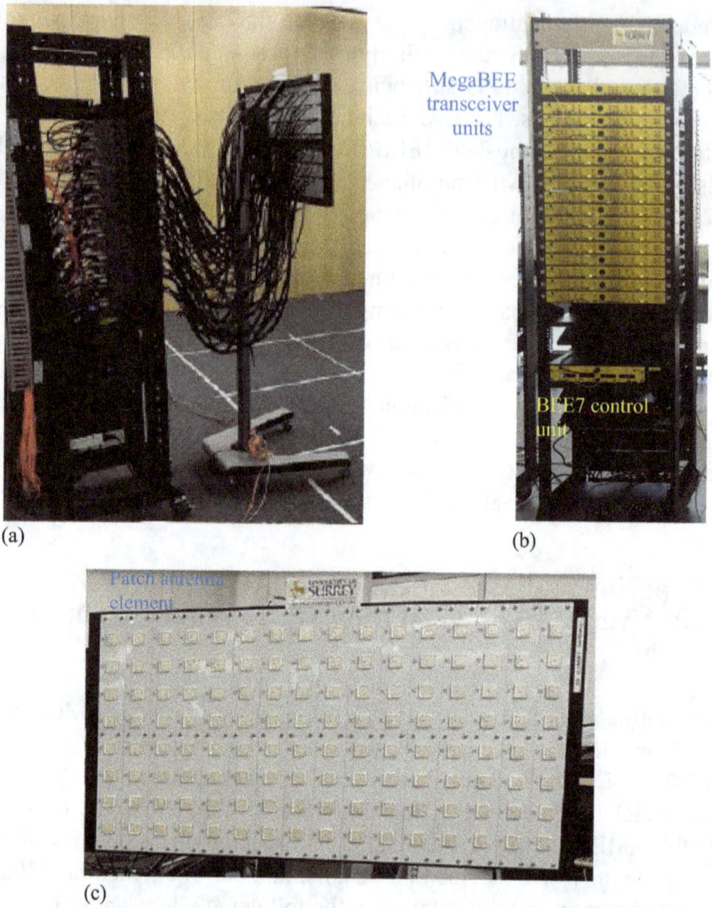

Figure 21.1 (a) mMIMO BS and its main components; (b) MegaBEE transceiver and BEE7 control units; (c) array of 128 patch antenna elements

Table 21.1 Geometry structure of the 128-element antenna array

Number of antenna elements	128
Number of rows	8
Number of columns	16
Distance between two adjacent columns	$\lambda/2$
Distance between two adjacent rows	$\lambda/2$

BEE7 unit is in charge of controlling all the MegaBEE transceiver units and ensure that they are all synchronised with each other before transmission/reception of data. The White Rabbit time distribution helps with the synchronisation by providing sub-nano second time synchronisation between the MegaBEE transceiver modules

Figure 21.2 mMIMO testbed receivers

through the use of a reference clock signal that is delivered over an optical fibre link, which uses SFP+ (small form-factor pluggable plus) network adaptors. The optical fibre link can also be used for transporting data, e.g. channel state information (CSI) feedback.

The mMIMO testbed of Figure 21.1 has up to 128 downlink channels with user-programmable SDR capability. In turn, it is useful for implementing and evaluating new communication algorithms and protocols as well as enabling the evaluation of the relevant over-the-air (OTA) link performance. For instance, by connecting the receivers to the mMIMO BS through the optical fibre link, closed loop MIMO technique like pre-coding/beamforming can be utilised to mimic the 5G BS transmission. At the receiver, up to 32 channels can be used, e.g., up to 8 UEs with 4 antenna elements each or 4 UEs with 8 antenna elements each.

21.3.2 Calibration of the mMIMO testbed equipment

21.3.2.1 mMIMO BS calibration

Prior to the start of the traceable measurement campaigns, the 128-element antenna array of the mMIMO BS, 4-element mMIMO testbed receiver antennas, RF power for the modulated signals of the mMIMO testbed and RF-exposure acquisition system needs to be calibrated.

The multi-antenna calibration was performed in situ (in the same place that the experiments took place) by relying on an OTA-based multi-channel transmitter calibration method. Given the inherent uncertainty of phase and delay caused by multiple RF channels, multi-channel calibration is a crucial factor that affects beamforming performance for a typical mMIMO antenna array. The traditional multi-channel calibration uses cable-based calibration methods or self-calibration by using a dedicated feedback circuit. The cable-based calibration method tends to require a long calibration period whenever a single-port receiver or a complex and costly multi-port receiver set-up is used. This can cause a problem over non-synchronisation calibration on the multi-channel measurements, which becomes more severe as the number of antenna elements increases. On the other hand, the self-calibration method, which relies on dedicated hardware circuit, increases the testbed design complexity and, in turn, this makes it harder for the testbed to achieve its required performance.

Contrary to the aforementioned methods, the OTA-based transmitter multi-channel calibration method does not require extra hardware circuits to obtain the RF calibration factors of multiple channels simultaneously. This method achieves the best performance in an anechoic chamber, but it could still work effectively in a multi-path environment when verified by experimental measurements. More details about the RF multi-channel non-ideal factors of the mMIMO testbed, our multi-channel OTA calibration method, the verification of the calibration, and the beamforming performance after calibration within the real-world environment are provided in the following.

The impairments (i.e. delay, amplitude, phase) due to the transmission through multiple RF chains for a typical mMIMO antenna array can be modelled as

$$x_{i,j}(t) = s(t - \tau_{i,j})\alpha_{i,j}\exp(j\phi_{i,j}), \tag{21.1}$$

where $x_{i,j}(t)$ is the actual waveform transmitted by the ith row jth column element of the array of the 128-element antenna array of Figure 21.1(c) when the RF channel delay, $\tau_{i,j}$, amplitude, $\alpha_{i,j}$, and phase, $\phi_{i,j}$, errors are accounted for, and $s(t)$ is the ideal transmit waveform without impairments. The signal $x_{i,j}(t)$ (or the part of it in the case of multi-path channel) that travelled over the line-of-sight (LOS) path to the receiver can be expressed as

$$y_{i,j}^{LOS}(t) = x_{i,j}(t)\beta_{i,j}\exp(\mathrm{j}\theta_{i,j}), \tag{21.2}$$

where $y_{i,j}^{LOS}$ is the received signal corresponding to the LOS path of the transmit antenna element (i,j), $\beta_{i,j}$ is the amplitude of this LOS path, and $\theta_{i,j}$ is the phase of this LOS path.

A typical OTA-based calibration method uses a multi-antenna pilot scheme based on frequency division multiplexing (FDM) that simultaneously transmits ten antenna pilots for each orthogonal FDM (OFDM) symbol by inserting equally spaced comb pilots in the frequency domain. By using enough OFDM symbol resources, the scheme can realise the pilot pattern for very large numbers of active transmit antennas. At the receiver, the channel for each transmission antenna is estimated based on the pilot signal waveform.

According to the estimated frequency-domain channel response, the time-domain channel impulse response (CIR) can be obtained. In an ideal environment such as a microwave anechoic chamber, estimations of the time delay, $\tau_{i,j}$, the combined amplitude, $\alpha_{i,j}\beta_{i,j}$, and the aggregated phase, $\phi_{i,j} + \theta_{i,j}$, of the direct LOS path can be easily obtained based on the CIR. However, in the multi-path scenario, it is necessary to consider the impacts of the parameter estimation accuracy such that a signal separation algorithm for multi-paths overlapped CIRs based on the maximum likelihood criterion needs to be utilised to estimate these parameters. Once $\tau_{i,j}$ is estimated, it can then be easily compensated for, i.e. its effect can be removed, in $x_{i,j}(t)$. The other important parameter that needs to be compensated for to ensure that the array steers its beam in the desired direction is $\phi_{i,j}$.

In order to understand how $\phi_{i,j}$ can be removed from $x_{i,j}(t)$, it is important to first understand how to extract $\phi_{i,j}$ from $\phi_{i,j} + \theta_{i,j}$. The phase of the LOS path for the *i*th row *j*th column antenna element depends on the relative position between the transmit array and the receiver such that $\theta_{i,j}$ can be expressed as

$$\theta_{i,j} = \theta_{ref} + \Delta\boldsymbol{\theta}_{i,j}^{(a,b)}, \tag{21.3}$$

where θ_{ref} is the phase of the LOS path between one of the antenna elements, chosen as a reference element among all the antenna elements of the array, and the receiver. In addition, $\Delta\boldsymbol{\theta}_{i,j}^{(a,b)}$ is the *i*th row *j*th column component of the steering matrix $\Delta\boldsymbol{\theta}^{(a,b)}$ that assumes the *a*th row *b*th column element of the array, i.e. antenna element (*a*, *b*), to be the reference element. The steering matrix $\Delta\boldsymbol{\theta}^{(a,b)}$ includes all the phase differences of any antenna element of the array relative to the phase of the reference element (*a*, *b*), such that

$$\Delta\boldsymbol{\theta}^{(a,b)} = \begin{bmatrix} (1-b)\Delta_1 + (1-a)\Delta_2 & \cdots & (16-b)\Delta_1 + (1-a)\Delta_2 \\ \vdots & \ddots & \vdots \\ (1-b)\Delta_1 + (8-a)\Delta_2 & \cdots & (16-b)\Delta_1 + (8-a)\Delta_2 \end{bmatrix}. \tag{21.4}$$

These phase differences account for the fact that each antenna element has a slightly different position relative to the receiver, with the position of element (*a*, *b*) as a reference position. In $\Delta\boldsymbol{\theta}^{(a,b)}$, Δ_1 and Δ_2 are the steering factors; they can be expressed, when considering that the distance between the elements is lambda/2 in both horizontal and vertical dimensions, as

$$\Delta_1 = j\pi\sin(\varphi), \tag{21.5a}$$

$$\Delta_2 = j\pi\sin(\vartheta), \tag{21.5b}$$

with φ and ϑ being the horizontal and vertical boresight angles of the transmission antenna array, respectively, and $j = \sqrt{-1}$. For instance, if the antenna element (1,7) is selected as a reference, as is shown in Figure 21.3, then

$$\Delta\boldsymbol{\theta}^{(1,7)} = \begin{bmatrix} -6\Delta_1 & \cdots & -\Delta_1 & 0 & \Delta_1 & \cdots & 9\Delta_1 \\ -6\Delta_1 + \Delta_2 & \cdots & -\Delta_1 + \Delta_2 & \Delta_2 & \Delta_1 + \Delta_2 & \cdots & 9\Delta_1 + \Delta_2 \\ \vdots & \ddots & \vdots & \vdots & \vdots & \ddots & \vdots \\ -6\Delta_1 + 7\Delta_2 & \cdots & -\Delta_1 + 7\Delta_2 & 7\Delta_2 & \Delta_1 + 7\Delta_2 & \cdots & 9\Delta_1 + 7\Delta_2 \end{bmatrix}, \tag{21.6}$$

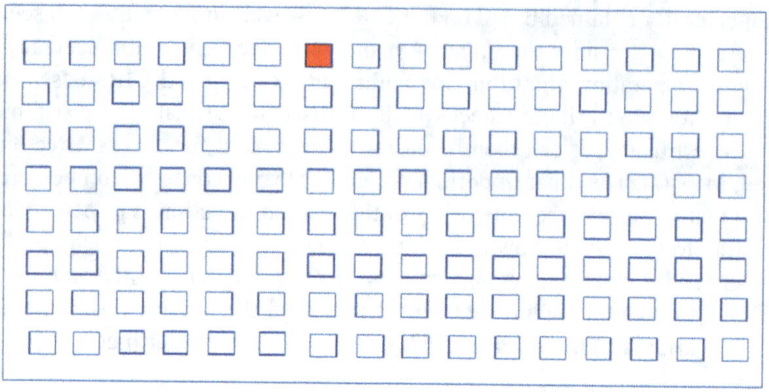

Figure 21.3 Example of reference element for a 128-element antenna array; the antenna element (1,7) is chosen as a reference for the OTA calibration procedure

By knowing $\Delta\boldsymbol{\theta}^{(a,b)}$, its effects can be removed for each antenna element from $\phi_{i,j} + \theta_{i,j}$ such that $\phi_{i,j} + \theta_{i,j} - \Delta\boldsymbol{\theta}_{i,j}^{(a,b)} = \phi_{i,j} + \theta_{ref}$. Then, by subtracting the residual phase of the *i*th row *j*th column antenna element, i.e. $\phi_{i,j} + \theta_{ref}$, from other element residual phases, it becomes possible to easily obtain the relative phase between all the elements of the array and to compensate for any phase misalignments. Thus, the effects of the phase and delay caused by multiple RF channels can be removed from the transmission and beams can be steered in the desired direction. The different steps of the OTA-based calibration procedure are summarised in Figure 21.4.

The calibration of the mMIMO BS and its antenna array was performed in the same indoor set-up that the one used for the actual EMF experiments; more details about the set-up of the experiment are provided in Section 21.3.3. A receiver was placed 6 m away from the mMIMO BS, with the height of both equipment being similar. Several calibration tests were performed with each test producing independent time delay and phase estimations according to the procedure mentioned in Figure 21.4. The median of the estimated phase and delay in several tests was calculated based on the measurement results. Figure 21.5 shows the timing and phase difference across multiple channels before the calibration. From the results, it can clearly be seen that every four channels/antennas form a group in the sense that every channel/antenna within one of these groups has roughly the same timing. However, jitters can be seen between the different four-channel groups. This was somehow expected given the multiple subsystem's architecture of the mMIMO BS. Indeed, each MegaBEE transceiver module is composed of two sub-modules of four input/output RF ports, where each sub-module is internally synchronised with itself, while each sub-module is externally synchronised with the other modules via the White Rabbit time distribution system. Meanwhile, there are significant phase differences between the channels regardless of whether they belong to the same sub-module.

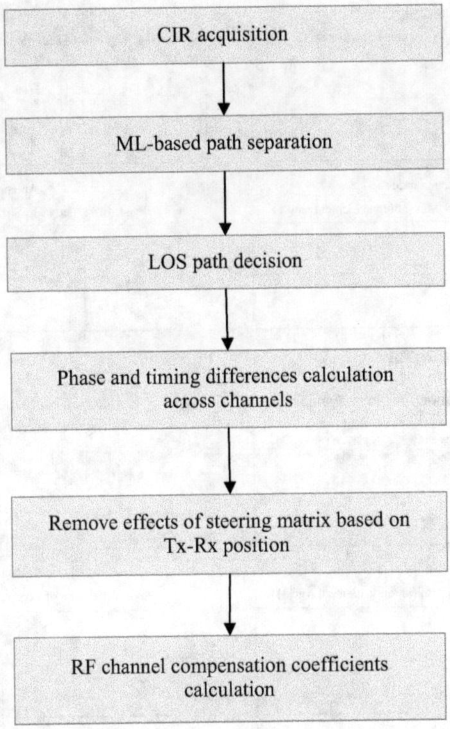

Figure 21.4 Calibration OTA-based scheme: estimation and compensation procedure

Figure 21.6 shows the timing and phase differences across multiple channels after the calibration. These results clearly demonstrate the effectiveness of the OTA-based calibration procedure. After applying it to the multiple RF channels, the phase and timing differences across most of them are adequately compensated, i.e. most of the delays are ten symbols long and most of the relative phase values are very close to 0 rad.

In order to validate the OTA-based calibration process, two beamforming tests were performed:

- In the first test, a beam was generated from the mMIMO BS at an angle of 9.46° from the boresight, i.e. pointing towards the point B, as it is depicted in Figure 21.7(a). At the same time, one receiver was moved over a 2-m distance, i.e. from point A to point C in Figure 21.7(a) (dark blue arrow), by steps of 25 cm. The results of the test in Figure 21.7(b) show that the received power gradually increases from point A to B, with the maximum reached at point B, and then gradually decreases afterward from B to C. This is expected given that the beam points directly to B.

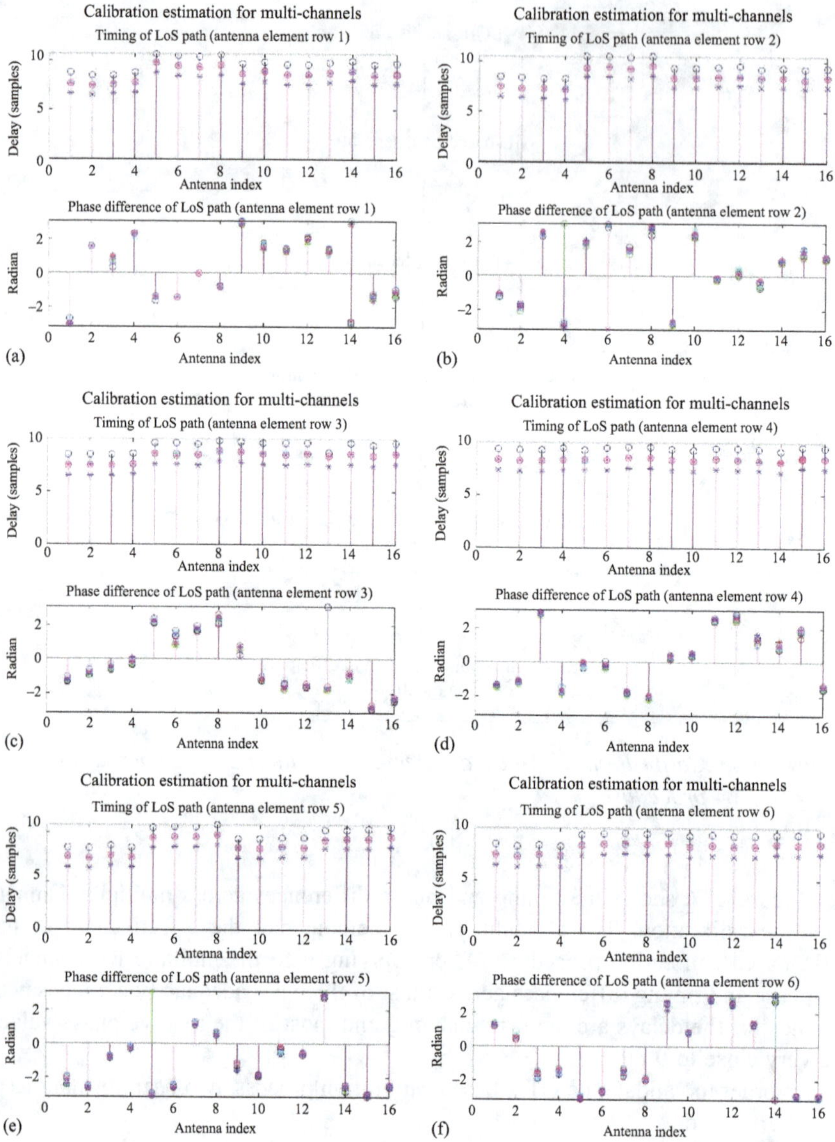

Figure 21.5 Phase and timing differences across the multiple channels prior to OTA-based calibration for the first six row (i.e. (a)–(f)) of the 128-element mMIMO BS array

- In the second test, the receiver was placed at point B, while the beam was steered over 14° (following the orange arrow in in Figure 21.7(a)), i.e. from −1.46° to −15.46° of the boresight direction, by steps of 1°. Similar to first test, the results of the second test in Figure 21.7(c) show that the received power gradually increases when the beam is steered from −1.46° to −9.46° (i.e. the direction of

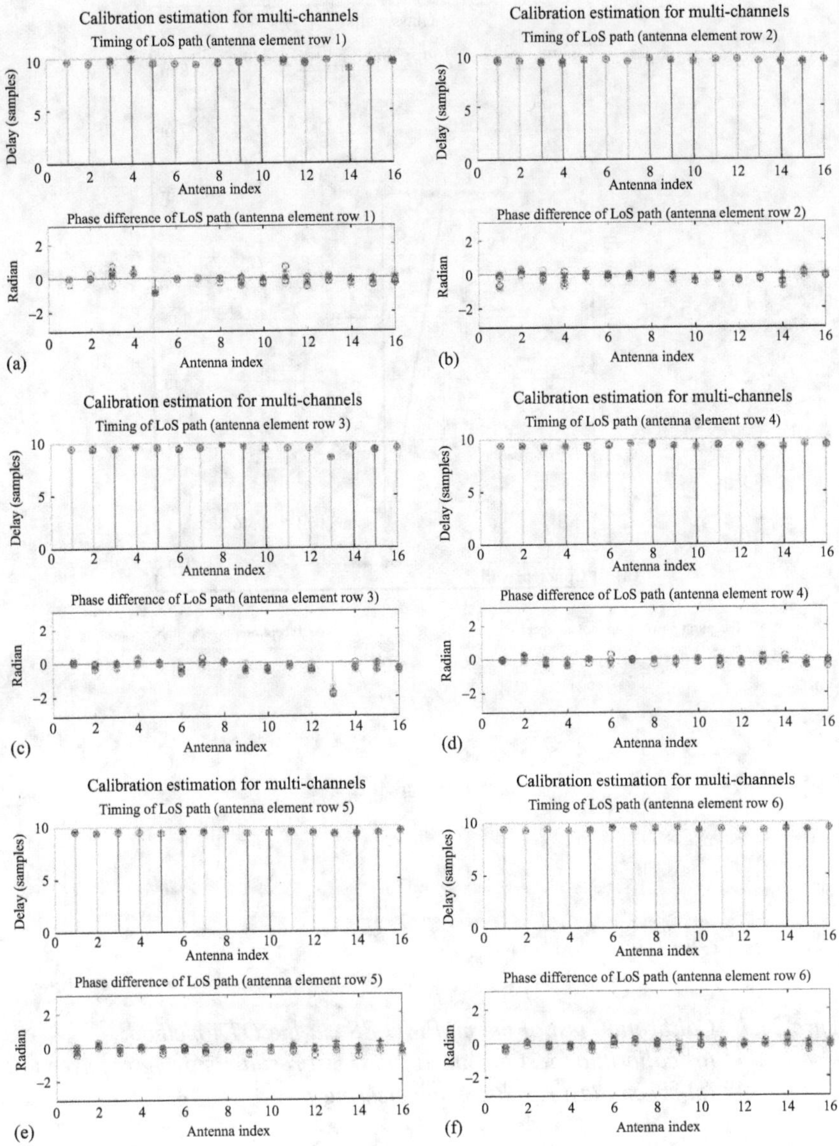

Figure 21.6 Phase and timing differences across the multiple channels after OTA-based calibration for the first six row (i.e. (a)–(f)) of the 128-element mMIMO BS array

point B – corresponds to 0° in Figure 21.7(c) since B is used as a reference point), with the maximum reached for a −9.46° angle, and then gradually decreases afterwards when the beam is steered from −9.46° to −15.46°, with a similar rate of increase/decrease on both sides of point B direction. This further validates the effectiveness of the OTA method for calibrating the mMIMO BS.

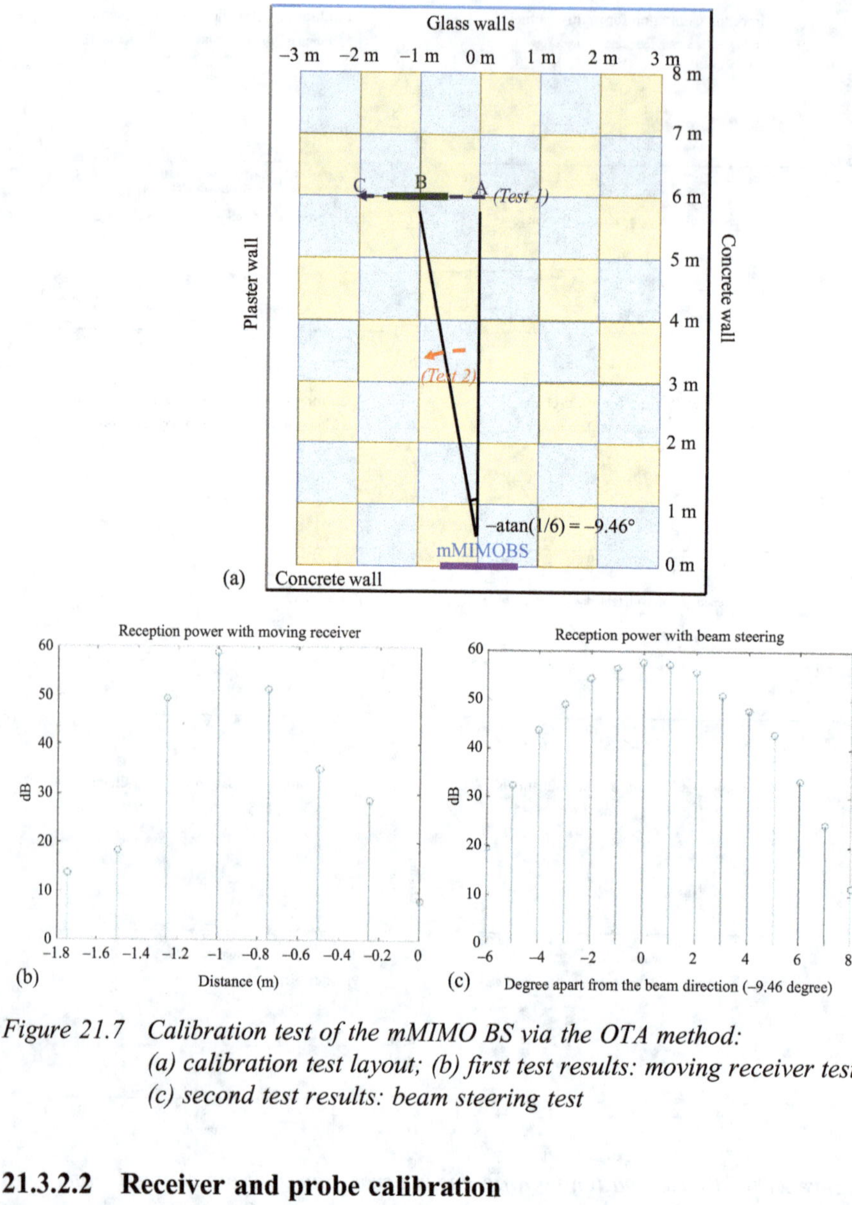

Figure 21.7 Calibration test of the mMIMO BS via the OTA method: (a) calibration test layout; (b) first test results: moving receiver test; (c) second test results: beam steering test

21.3.2.2 Receiver and probe calibration

The E-field measurements were made with two different receivers systems and both were calibrated at the UK National Physical Laboratory (NPL). The first receiver was an RF-exposure acquisition system formed of a Keysight FieldFox handheld spectrum analyser and three-axis probe. This system was calibrated at NPL in its Power Flux Density laboratory against known generated E-field by using established techniques, the measurement of which is depicted in Figure 21.8(a). The second receiver system was the aforementioned receivers from the Surrey mMIMO system. These were

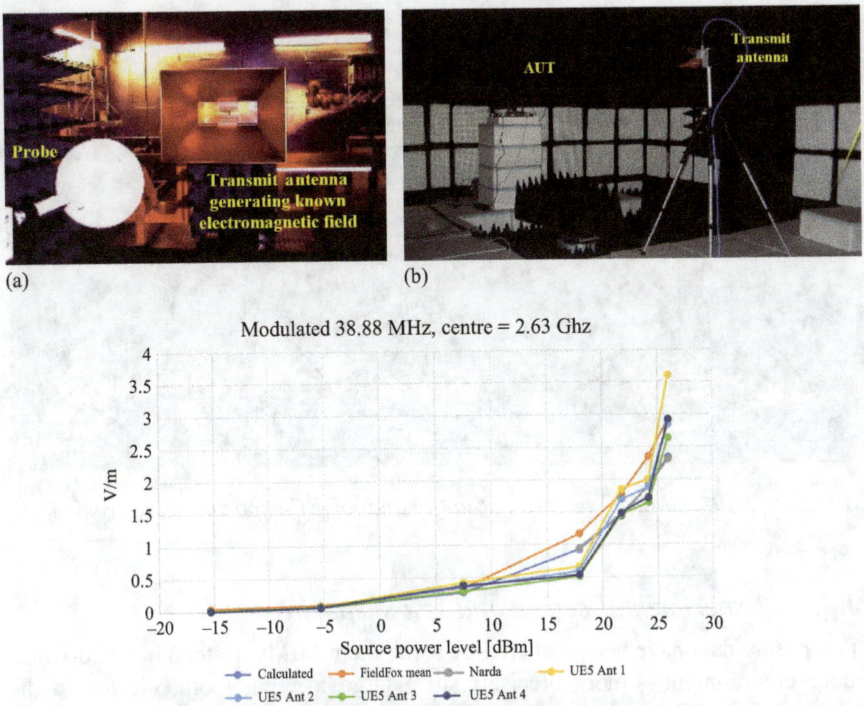

Figure 21.8 Measurement set-up for (a) field probe calibration in NPL's Power Flux Density laboratory; (b) antenna calibration in NPL's anechoic chamber, and (c) calibrated traceable E-field results acquired by different receivers for a 2.63-GHz modulated signal with a 38.88-MHz data bandwidth

calibrated component-wise with separate measurements made on the cables, MegaBEE receivers and antennas. The antennas were calibrated within the fully anechoic chamber facility at NPL, which has the following dimension of 7 m×6.2 m×6.2 m [17]. The four-element receiver antenna arrays were calibrated by using the three-antenna method in which the relevant losses are measured by using a Keysight PNA-X model N5242A vector network analyser. The MegaBEE receivers were calibrated for sensitivity to RF power received by using an NI-5681 power sensor.

As depicted in Figure 21.8(b), measurements were made within the NPL screen fully anechoic chamber facility to confirm the calibration process by comparing the performance of the system in measuring E-field results for a modulated signal centred at a carrier frequency of 2.63 GHz with a 38.88 MHz instantaneous data bandwidth; the measurements are shown in Figure 21.8(c). The results of the various receive system and antenna elements match well with each other and they are more or less in-line with the theoretically calculated results and probe results especially at low transmit power. The theoretical calculation and a calibrated commercial Narda probe were used to further validate the obtained calibrated results.

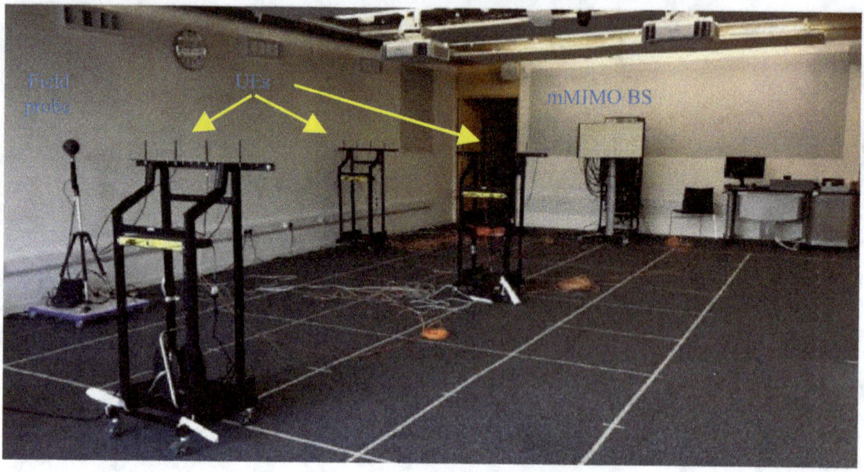

Figure 21.9 mMIMO testbed set-up for indoor RF-exposure measurement

21.3.3 Experimental set-up and test scenarios

RF-exposure data have been gathered by using the mMIMO testbed in a real-world indoor environment – more precisely, in a large meeting room, located in the basement of the 6G Innovation Centre at the University of Surrey. The room, which is 15-m long, 7.5-m wide, and 3-m high, is surrounded by glass, brick and plasterboard walls. The floor, which defines the height reference of Z_0=0 m, is carpeted and made of concrete. On the other hand, its ceiling is also made of concrete with suspended panels, hanging lighting and a projector equipment. All the measurement instruments, i.e. mMIMO BS, receivers, probe, were located inside the room, while the room typical office furniture, e.g. desk and chairs, were placed aside during the measurements. All the details aforementioned about the room set-up can be seen in Figure 21.9. The measurements are performed within an 8 m×6 m area, which is divided into a grid of 48 square of 1 m for the ease of measurement, as is depicted in Figure 2.10.

21.3.4 First experiment

In the first experiment, the mMIMO BS operates with 64 active transmitting antennas, while each receiver/UE operates with four dipole receiving antenna. The mMIMO BS is configured to perform downlink zero-forcing (ZF) pre-coding based on the CSI feedback from the receivers/UEs for different beamforming scenarios. Eight different single-user (SU) and multi-user (MU) downlink communication scenarios with beamforming are considered for different receiver/UE locations based on the position grid of Figure 2.10(a), as is summarised in Table 21.2. Figure 21.10 illustrates scenario (8), where Figure 2.10(a) shows the position of the various equipment on the 2D grid formed on the floor of the room for this scenario,

Table 21.2 Coordinate of UEs in the grid of Figure 2.10(a)

Scenarios	UE1[a]	UE2[a]	UE3[a]
(1)	(0, 8)	N/A	N/A
(2)	(−3, 4)	N/A	N/A
(3)	(3, 2)	N/A	N/A
(4)	(−3, 4)	(3, 2)	N/A
(5)	(0, 8)	(0, 4)	N/A
(6)	(0, 8)	(−3, 4)	N/A
(7)	(0, 8)	(3, 2)	N/A
(8)	(0, 8)	(−3, 4)	(3, 2)

[a]Coordinate units in meter.

and Figure 2.10(b) complements Figure 2.10(a) by depicting the real-life deployment of the equipment.

In all these scenarios, the mMIMO BS transmits OFDM frames of 65,536 samples (i.e. 2^{16}) at 61.44 Msps (mega-samples per second) at a carrier frequency of 2.63 GHz over a 40-MHz wide bandwidth in TDD mode. In addition, the OFDM subcarrier spacing is set to 15 kHz and a 64 quadrature amplitude modulation (64 QAM) is used. The same transmission and measurement process is followed in all the scenarios: (1) first a pilot frame is sent from the mMIMO BS to each receiver/UE to estimate the channel link quality/CSI; the latter is then fed back to the BS via the SFP+ cable; (2) the BS then estimates the CSI and uses it to pre-code the data on each subcarrier, based on the ZF pre-coding scheme; (3) the BS finally transmits the ZF pre-coded data frames in a continuous manner while the field probe is used to measure the RF-exposure at 56 positions within the grid of Figure 21.10(a) (the 56 intersection points between the horizontal and vertical lines in Figure 21.10(a)). A consistent physical probe orientation is ensured for all the measured grid points.

In order to ensure that the ZF pre-coding process works properly and, hence, the mMIMO BS beamforms in the correct direction, the un-coded bit-error-rate (BER) of each link transmission between the BS and any UE is calculated based on the first received data frame. The results of Table 21.3, which report the BER value for each link for each scenario, clearly indicate that the BS beamforms in the expected direction in all scenarios since an un-coded BER of around 10^{-3} is achieved in the worst case scenario. They also indicate that the quality of the beamforming degrades as the number of users/channels increases, which can be expected with ZF. The results are completed by Figure 21.11, which depicts a received constellation in scenario (1) and confirms the high link quality.

The calibrated RF-exposure results for the eight beamforming scenarios as well as their average, based on the field probe measurements of the received channel power at each grid point, are presented in Figure 21.12 by means of a heat map. The intensity of the RF-exposure is expressed in V/m and is colour-coded to illustrate its variation at different points of the room. In each subplot, the circle(s) represent the position of the UE(s).

(a)

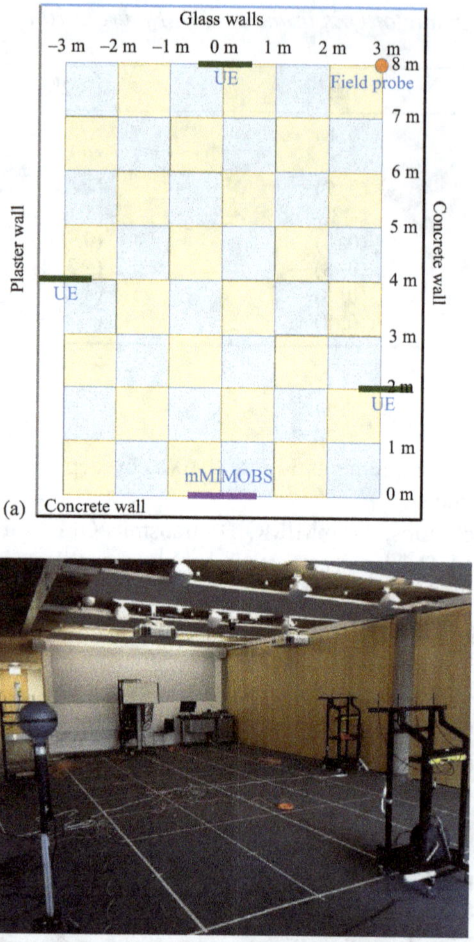

(b)

Figure 21.10 Illustration of three active UEs at (0, 8 m), (−3, 4 m) and (3, 2 m) scenario: (a) position of the equipment on the 2D grid within the room; (b) corresponding picture

The results show that an mMIMO BS does not generate a uniform RF-exposure as a traditional BS will do, but the exposure pattern follows more or less the envisaged 'beam profile'. However, as for a traditional BS, the higher exposure levels are concentrated in the vicinity of the BS, i.e. the highest level of RF-exposure generated by the 64-Tx mMIMO BS is localised at (0, 0 m) in most cases, where the mMIMO BS is located.

In order to get some statistical insights from the RF-exposure results, the results for an offset 0 m and a distance varying from 0 to 8 m, i.e. along the red arrow of Figure 21.12(i), are plotted for all the scenarios and average in Figure 21.13. The general trend, in all cases, is that the RF-exposure is inversely

Table 21.3 Uncoded BER of each BS to UE link for the eight scenarios

Scenario	UE1	UE2	UE3
(1)	1.35×10^{-6}	N/A	N/A
(2)	4.61×10^{-6}	N/A	N/A
(3)	5.12×10^{-6}	N/A	N/A
(4)	2.35×10^{-6}	4.82×10^{-5}	N/A
(5)	8.38×10^{-5}	9.22×10^{-5}	N/A
(6)	3.68×10^{-5}	4.31×10^{-5}	N/A
(7)	4.37×10^{-5}	3.99×10^{-5}	N/A
(8)	8.04×10^{-4}	2.68×10^{-3}	5.11×10^{-4}

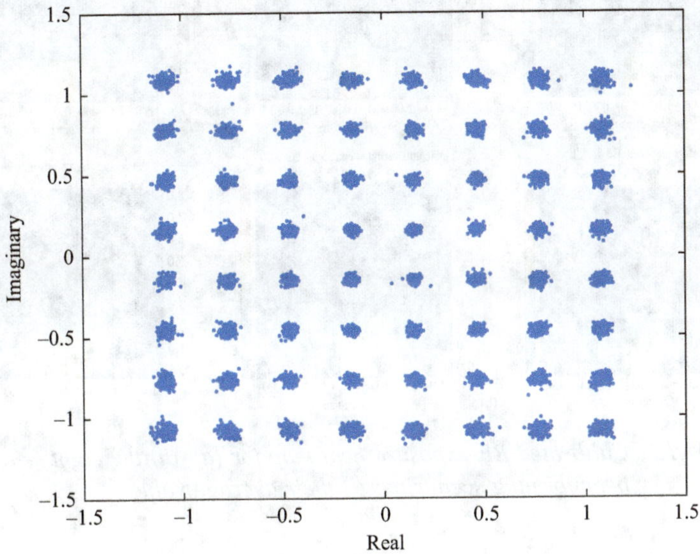

Figure 21.11 Received 64-QAM constellation in scenario (1)

proportional to the square of the transmission distance, even when there is no main beam pointing directly into this direction, i.e. scenario (4). It is also interesting to note that the scenario generating the most RF-exposure among all the scenarios is scenario (1), where the BS transmits only to one UE in the main direction studied in Figure 21.13. On the other hand, in scenario (5) with two UEs in the main direction or scenarios (6), (7) and (8) with at least one UE in the main direction, the mMIMO BS generates less RF-exposure than scenario (1). This tends to imply that the more UEs are served, the more the RF-exposure could be spread and hence be less intense. It can also be noted that the exposure varies between 1.37 and 3.09 V/m in these indoor scenarios.

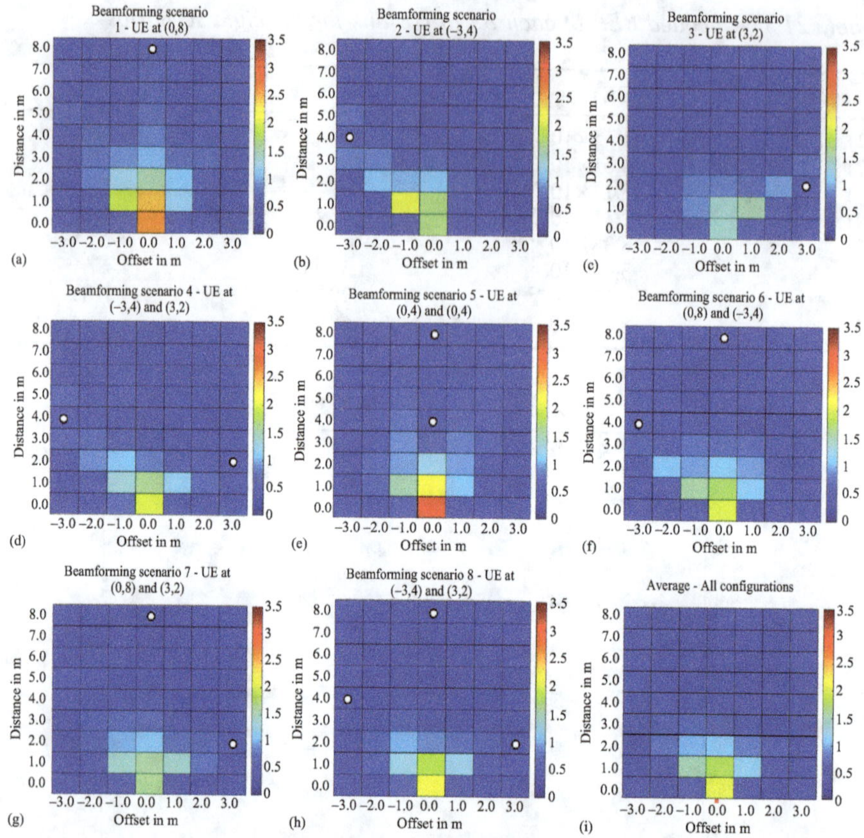

Figure 21.12 Calibrated RF-exposure heat map for (a)–(h) the eight beamforming scenarios; (i) their average value

21.3.5 Second experiment

Contrary to the first experiment, the second experiment is more comprehensive and study, for instance, the impact of the number of beams or the user traffic load with non-ideal mMIMO beamforming operation on the RF-exposure based on a very large amount of acquired data [16]. The other differences are fully detailed in the following.

In the second experiment, the mMIMO BS operates with 96 active transmitting antennas, while each receiver operates with four dipole receiving antennas. Contrary to first experiment where the receivers were considered to be UEs, i.e. receiving pilot signals and feed backing the CSI to the BS such that it can steer beams towards them, the receivers are here used as extra probes to measure the RF-exposure at the location they are placed whilst the mMIMO BS is not necessarily transmitting/steering a beam towards them; this allows to acquire more measurements and get a better statistical view of the RF-exposure. During the measurement

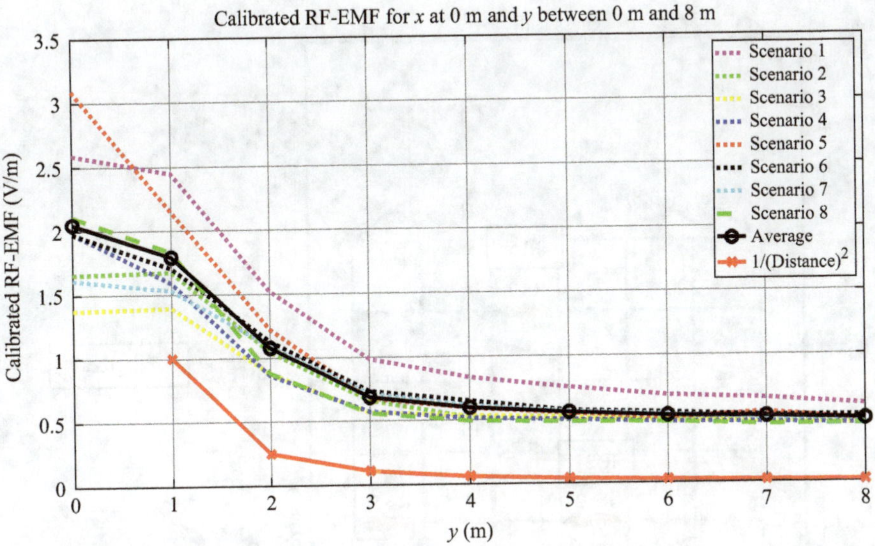

Figure 21.13 Calibrated RF-exposure of all scenarios for an offset of 0 m and a distance varying from 0 to 8 m

campaign, various SU and MU MIMO downlink communication beamforming scenarios were considered, with different combinations of active beams and data traffic loading to mimic the performance of a realistic 5G BS. The RF-exposure was measured by using five receivers and the probe placed at different locations within the grid, as it is depicted in Figure 21.14. In Figure 2.14, Rx # indicates the receiver number with its corresponding IP address, while P1 represents the probe.

In the second experiment, fully compliant 5G new radio (NR) waveforms are used to transmit data. The 5G NR beamforming baseband signals are generated by using Keysight PathWave System Design platform (also known as SystemVue). This software can be used to generate up to 19 potential beam directions equally spaced every 5° between ±45° in azimuth and elevation=0°. Each generated beam contains a physical downlink shared channel with a configurable data payload that is transmitted within a 1-ms period. The data rate of each beam payload can be controlled by adjusting the modulation coding scheme, i.e. 4, 16, 64 QAM, as well as the number of allocated resource blocks (NRBs) used in the transmission. In the experiment, up to 216 NRBs are used per beams and up to 4 out of the 19 beams are active simultaneously to mimic an mMIMO BS beamforming data transmission towards up to 4 active users at the same time. Note that up to 864 total NRBs are used for 4 simultaneously active beams. As in the first experiment, the carrier frequency of the downlink transmission is 2.63 GHz with a bandwidth of 38.88 MHz, and a subcarrier spacing of 15 kHz.

One envisages that the variations of the beam profiles and data rates are useful for assessing the spatial variation of RF-exposure surrounding the mMIMO testbed. In order to get insights into the effect on the stochastic nature of RF-exposure

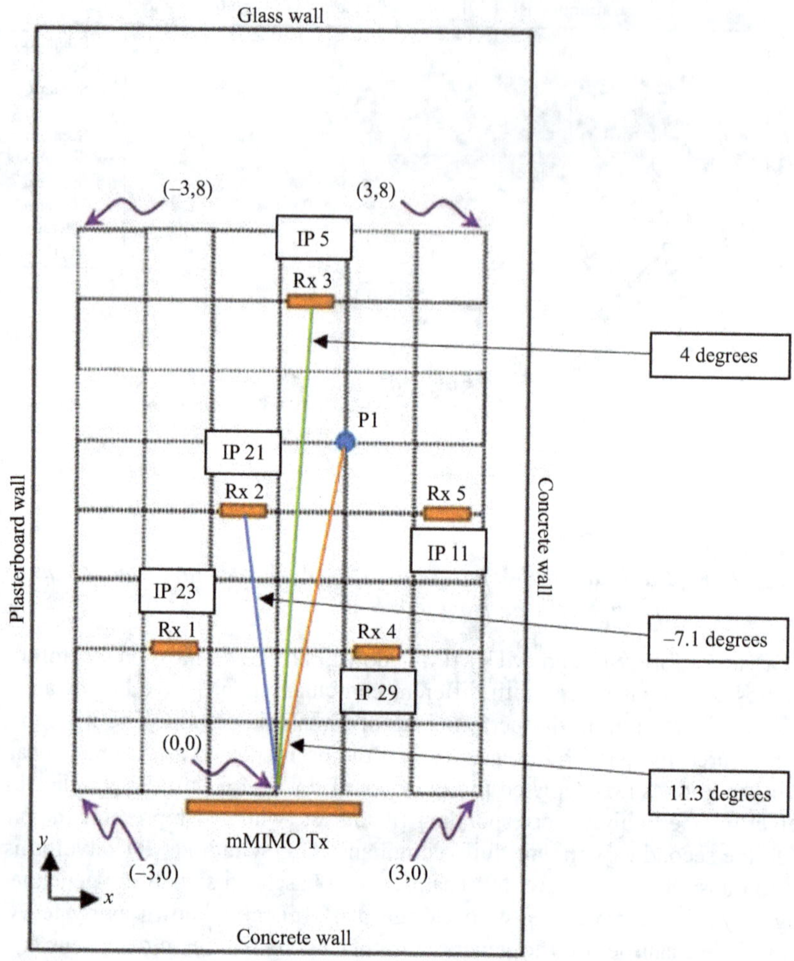

Figure 21.14 Measurement set-up for 96 active mMIMO Tx at (0, 0 m), the five receivers and the probe, in the second experiment

generated by the non-ideal mMIMO beamforming operation due to potential hardware impairments and/or other factors (i.e. the beams may not be well defined and may not steer in the expected direction) in the studied environment, tens of thousands of E-field measurements were acquired, after the multi-channel OTA calibration had been performed, with random active beams and payload data rate.

To understand the statistical variation of the RF-exposure (in V/m) as a function of the number of users in the system, by using the number of beams as proxy (i.e. assuming that each beam is steered towards a given user), Figure 12.15 depicts the empirical calibrated RF-exposure and their cumulative distribution function (CDF) based on the FieldFox measurements. More precisely, Figure 12.15(a) depicts the

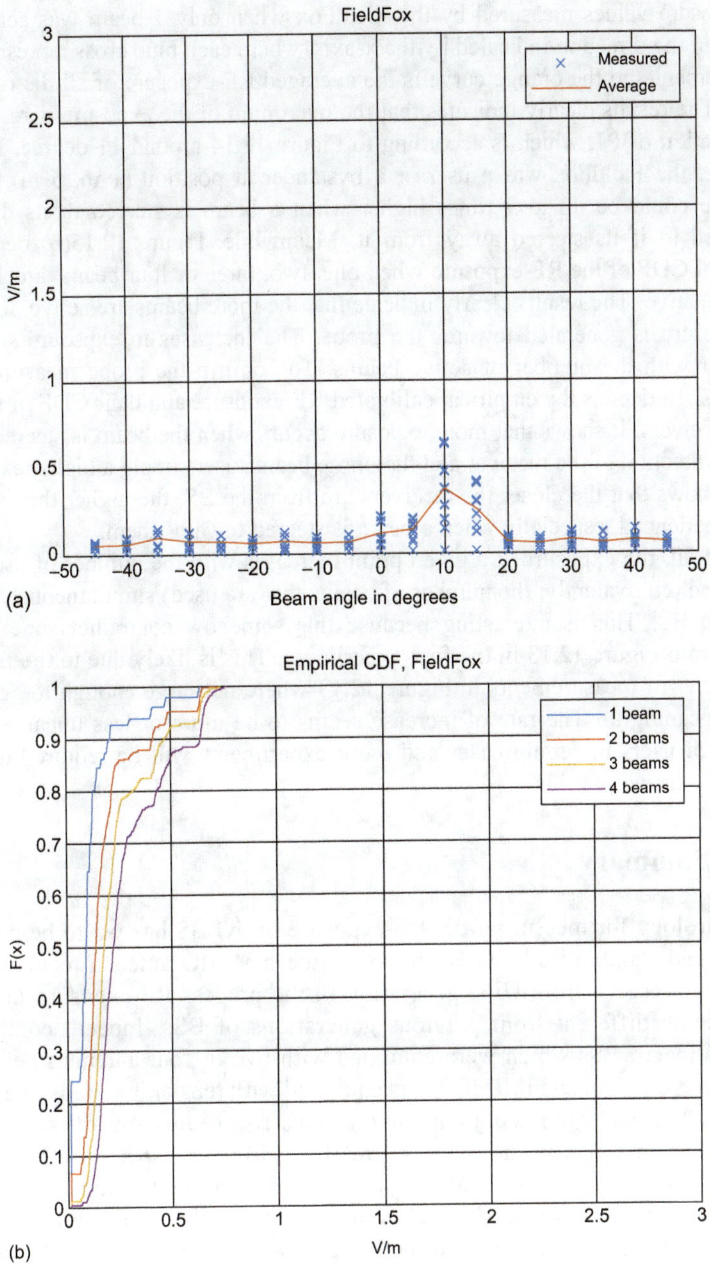

Figure 21.15 The empirical calibrated RF-exposure results of the FieldFox probe (P1) (a) for 1 active beam as a function of the beam angle and (b) empirical CDF of the RF-exposure for 1 to 4 active beams

RF-exposure values measured by the FieldFox when only 1 beam was active and steered in the direction indicated by the x-axis, where each blue cross represents one measurement and the orange curve is the averaged RF-exposure of all the measurements. The results clearly indicates that the maximum of the exposure is reached in the direction of P1, which is according to Figure 12.14 around 11 degree. In other words, if the FieldFox was a user or a bystander at position (1 m, 5 m) then its exposure could be up to 7 times higher when a beam is steered in its direction compared to if its steered away from it. Meanwhile, Figure 12.15(b) depict the empirical CDF of the RF-exposure when one, two, three or four beams are simultaneously active. The results clearly indicate that the more beams are active, the more RF-exposure is generated towards the probe. The increases in expsoure seems to be linear with the number of active beams. To confirm the probe measurements, Figure 12.16 depicts the empirical calibrated RF-exposure and their CDF of the five other receivers. It shows that more exposure occurs when the beam is steered in the general direction of the receiver and that more beams/users imply a higher exposure. It also shows that the closer the receivers are from the BS, the higher the exposure they experienced, especially when a beam is steered towards them.

Overall, the exposure at a given point increases with the number of users that are served (equivalently, the number of beams that are used) simultaneously by the mMIMO BS. This is interesting because this somehow contradicts one insight gained from Figure 12.13 in the first experiment. This is likely due to the fact that the number of measurements in Figure 12.13 where not large enough to be statistically meaningful. The rate of increase seems to be more or less linear with the number of users in certain cases and more experiments will be required to get a definitive answer on this.

21.4 Summary

The metrology for measuring the RF-exposure of 5G BS has yet to be properly defined and standardised. 5G BS relies on the mMIMO antenna technology to transmit to/receive from UEs, generating in the process RF-exposure in a way that is quite different from previous generations of BSs. Indeed, contrary to previous macro BSs which were equipped with two or four antennas per sector (over three sectors), an mMIMO BS is equipped with tens or hundreds of antenna elements that can be used conjointly to create narrow beams towards several UEs at the same time and, in the process, greatly improve the spectral efficiency of mobile systems. Given that mMIMO BS operates differently than previous BS generation, the RF-exposure regulations, e.g. exclusion zone, design for the previous generation, cannot be directly applicable to this type of BS without considerably hampering their deployment. Hence, there is need for a new regulation based on a new bespoke metrology, which is itself grounded in strong and accurate scientific evidences.

Given the statistical nature of the mMIMO transmission, where the spatial pattern of transmission (i.e. number of beam used, direction of the beams) can change every

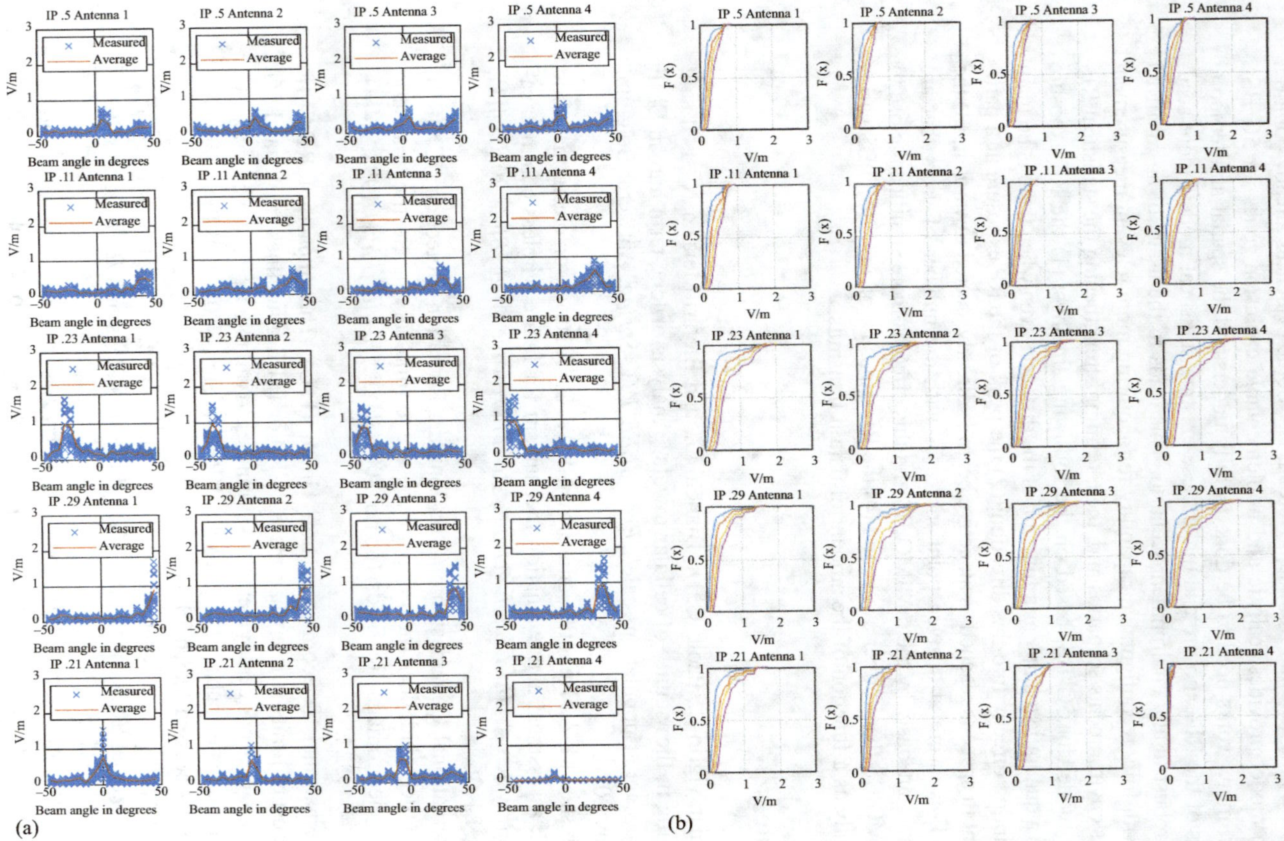

Figure 21.16 *The empirical calibrated RF-exposure results of the four antennas of the five receivers (Rx3, Rx5, Rx1, Rx4, Rx2) (a) for 1 active beam as a function of the beam angle and (b) empirical CDF of the RF-exposure for 1 to 4 active beams (where blue, red, orange and purple colour plots refer to '1 beam', '2 beams', '3 beams', and '4 beams', respectively)*

few milliseconds, it makes some sense that the new metrology for 5G BS is based on statistical RF-exposure models, either based on system level simulations or real measurement data acquired from deployed 5G BS. Some evidences already point out that such model provides realistic and implementable exclusion zones for 5G BSs. However, more works need to be done to fully understand the spatial and temporal variations of the RF-exposure and make these existing statistical models more accurate and robust in any environment/scenario. This leads to the usage of an mMIMO testbed with fully reconfigurable capability for acquiring more relevant measurement in various environments/scenarios and, then, refining/strengthening the statistical RF-exposure model for 5G BS, based on the following findings: (1) the RF exposure decays in a quadratic manner as a function of the distance, even when there is no main beam pointing directly into this direction; (2) the peak of the RF exposure at a particular point in space is reached as an active beam is directly steered towards it; (3) the overall RF-exposure at a given point increases with the number of users that are served simultaneously by the mMIMO BS and this effect is more significant close to the mMIMO BS than further away; and (4) the RF-exposure of a mMIMO BS varies significantly as a function of the amount of data it transmits.

References

[1] Chiaraviglio L., Paolo C. D., Bianchi G., and Blefari-Melazzi N. 'Is it safe living in the vicinity of cellular towers? Analysis of long-term human EMF exposure at population scale'. *Proc. IEEE VTC-Spring*, Antwerp, Belgium, May 2020.

[2] Jamshed M. A., Héliot F., and Brown T. W. C. 'A survey on electromagnetic risk assessment and evaluation mechanism for future wireless communication systems'. *IEEE J. Electrom. RF Microwave Med. Biol.* 2020, 4(1), pp. 24–36.

[3] International Commission on Non-Ionizing Radiation Protection (ICNIRP), 'ICNIRP guidelines for limiting exposure to time-varying electric, magnetic and electromagnetic fields (up to 300 GHz)'. *Health Phys.* 1998, 74(4), pp. 494–522.

[4] International Telecommunication Union (ITU), The impact of RF-EMF exposure limits stricter than the ICNIRP or IEEE guidelines on 4G and 5G mobile network deployment, 2018 [online]. Available from: https://www.itu.int/rec/dologin_pub.asp?lang=e&id=T-REC-K.Sup14-201805-I!!PDF-E&type=items [Accessed March 2021].

[5] Chiaraviglio L., Cacciapuoti A.S., Di Martino G., *et al.* 'Planning 5G networks under EMF constraints: state of the art and vision'. *IEEE Access* 2018, 6, pp. 51021–51037.

[6] Global System for Mobile Communication Association (GSMA), Arbitrary radio frequency exposure limits: impact on 4G network deployment, 2014 [online]. Available from: https://www.gsma.com/publicpolicy/wp-content/

uploads/2014/03/Arbitrary-Radio-Frequencyexposure-limits_Impact-on-4G-networks-deployment_WEB.pdf [Accessed March 2021].

[7] Blanch S., Romeu J., and Cardama A. 'Near field in the vicinity of wireless base-station antennas: an exposure compliance approach'. *IEEE Trans. Antennas Propag.* 2002, 50(5), pp. 685–692.

[8] International Electrotechnical Commission (IEC), IEC 62232:2017, "Determination of RF field strength, power density and SAR in the vicinity of radiocommunication base stations for the purpose of evaluating human exposure," August 2017.

[9] Thors B., Furuskär A., Colombi D., and Törnevik C. 'Time-averaged realistic maximum power levels for the assessment of radio frequency exposure for 5G radio base stations using massive MIMO'. *IEEE Access* 2017, 5, pp. 19711–19719.

[10] Degirmenci E., Thors B., and Törnevik C. 'Assessment of compliance with RF EMF exposure limits: approximate methods for radio base station products utilizing array antennas with beam-forming capabilities'. *IEEE Trans. Electromagn. Compat.* 2016, 58(4), pp. 1110–1117.

[11] Baracca P., Weber A., Wild T., and Grangeat C. 'A statistical approach for RF exposure compliance boundary assessment in massive MIMO systems'. *Proc. 22nd Int. ITG Workshop Smart Antennas (WSA)*, Bochum, Germany, March 2018, pp. 1–6.

[12] The Office of Communications (OfCom), Electromagnetic Field (EMF) measurements near 5G mobile phone base stations, technical report, April 2020 [online]. Available from: https://www.ofcom.org.uk/__data/assets/pdf_file/0015/190005/emf-test-summary.pdf [Accessed March 2021].

[13] Agence Nationale des Fréquences (ANFR), Assessment of the exposure of the general public to 5G electromagnetic waves, technical report, April 2020 [online]. Available from: https://www.anfr.fr/fileadmin/mediatheque/documents/5G/20200410-ANFR-rapport-mesures-pilotes-5G-EN.pdf [Accessed March 2021].

[14] Colombi D., Joshi P., Xu B., Ghasemifard F., Narasaraju V., Törnevik C. 'Analysis of the actual power and EMF exposure from base stations in a commercial 5G network'. *Appl. Sci.* 2020, 10(15), 5280.

[15] Loh T. H., Héliot F., Cheadle D., and Fielder T. 'An assessment of the radio frequency electromagnetic field exposure from a massive MIMO 5G testbed'. *Proc. EuCAP 2020*, Copenhagen, Denmark, March 2020.

[16] Loh T. H., Cheadle D., Heliot F., Sunday A., and Dieudonne M. 'A study of experiment-based radio frequency electromagnetic field exposure evidence on stochastic nature of a massive MIMO system'. *Proc. EuCAP 2021*, Düsseldorf, Germany, March 2021.

[17] National Physical Laboratory, Small antenna radiated test range. 2010 [online]. Available from: https://www.npl.co.uk/products-services/radio-frequency/smart-chamber [Accessed March 2021].

Chapter 22
Conclusions and future perspectives

Tian Hong Loh[1]

22.1 Conclusion

This book presents a comprehensive overview of the state-of-the-art measurement capabilities, testbeds and relevant research and development (R&D) activities for fifth generation (5G) and emerging wireless technologies at a wide range of frequencies up to THz frequency bands. Several real-world field trials and use cases are also presented, for example, massive multiple-input–multiple-output (mMIMO), autonomous vehicles, drones, railways and Internet of Things (IoT). The focus has been on R&D of measurement techniques and metrology for 5G and beyond that underpin all aspects, from new radio (NR) signals, devices, antennas, systems and propagation environments to RF exposure. The presented material will support readers from universities, industry, standard bodies and national measurement institutes to enable further metrological R&D activities by describing advances in the triad of measurement system design, measurement techniques and underpinning metrology required to cover many communication aspects from the network-edge to the network-core. In next section, the open challenges that seed the future research directions on metrology for 5G and emerging wireless technologies are discussed.

22.2 Future perspectives

The rapid growth of the wireless communication market, with increasing requirements for functionality, energy efficiency and robustness of communications, has resulted in huge demands in the development of new and advanced emerging wireless technologies to improve performance and usage of the available spectrum in the most efficient way. These include, for example, light-fidelity (Li-Fi), orbital angular momentum (OAM), smart antennas, reconfigurable intelligent surfaces (RISs), antenna-in-package (AiP), nanoelectronics. Moreover, the growing demands on various applications of these technologies has driven new requirements

[1]5G and Future Communications Technology Group, Electromagnetic and Electrochemical Technologies Department, National Physical Laboratory (NPL), Teddington, United Kingdom

of multi-parameter and multi-physics measurements in a range of situations both in the laboratory and on-site. The range of complex measurements expected in both the frequency-domain and time-domain will continue to grow. In all cases, the propagation of uncertainties between the domains continues to be of importance and must be addressed, with specific challenges for communication parameters. Metrology must support these developments by adequate measurement methods, calibration capabilities, and traceability to International System of Units (SI).

Operating 5G and emerging wireless technologies at higher frequencies such as mm-wave, THz and optical bands is a route to enable higher data bandwidth. Moreover, the recent advancements on graphene-based intelligent beamforming antennas with frequency agility ability has driven its application to THz bands. The high density of users in 5G and future wireless networks means a critical consideration is the interference from nearby users. Furthermore, the link performance of wireless communication systems may degrade in real-world operation due to, for example, the proximity of lossy objects (such as human hands and heads), multipath fading, unwanted interference, weather conditions. The traceability of propagation channel, interference characterisation (especially for sparse multipath) and system-level testing in real-world environments presents new challenges as RF beamforming is essential to overcome the higher losses present at these frequencies. To support the growth at the network edge for Terabit transmission, new metrological support is required for coexistence with disruptive ultra-wide bandwidth short-range communication technology.

The near-field and far-field characterisation of large-scale multi-antenna systems presents additional challenges that stem from highly integrated antennas and RF circuitries, for example the potential imperfections within the substantially more complex hardware and effects due to nonlinearity, mutual coupling, and antenna-to-range interfaces. With the trend towards increasing the number of antenna elements, interactions between antennas and the propagation channel must be investigated to cater for the requirements of future mm-wave and THz mMIMO systems. Propagation channel models for sub-THz for in-city deployment are an active research topic and work has been reported at 90 and 200 GHz. The current state of the art for THz communication is a single-input and single-output (SISO) system at 237.5 GHz for transmitting data over 20 m at a data rate of 100 Gbit/s, showing that this technique has promise but the required technology is not widely available at present. There is currently no clear NR MIMO over-the-air (OTA) test methodology that can handle orthogonal frequency-division multiple access (OFDMA) scheduling in time, frequency and space at the same time for the evaluation of the realistic wireless performance of base stations and user equipment. Currently, all relevant verification is done by MIMO throughput comparison tests between OTA and cable connection (where feasible).

With the industrial exploitation and adoption of complex NR signals and large-scale multi-antenna technologies at different radio frequency bands in emerging wireless systems, several worldwide industries, research communities and standards bodies are now facing new measurement challenges for efficient and accurate multi-parameter multi-physics characterisation of diverse emerging wireless systems, especially for high-volume beam-reconfigurable product verification. In addition, public concern over potential perceived health risks from radiation exposure has led to stringent RF exposure compliance regulations in some countries. The time-burden in employing current RF exposure assessment methods applicable to specific absorption rate (SAR) and power density (PD) defined in European Committee for Electrotechnical Standardization (CENELEC), International Electrotechnical Commission (IEC), and Institute of Electrical and Electronics Engineers (IEEE) international standards will need to be addressed, especially, for high-volume beam reconfigurable product verifications. An internationally harmonised test standard to comply with absorbed PD limits as defined by the International Commission on Non-Ionizing Radiation Protection (ICNIRP) is lacking and will be required. However, current SAR and PD exposure assessment methods considered by IEC are time-consuming for beam reconfigurable product compliance testing.

Index

www.ingramcontent.com/pod-product-compliance
Lightning Source LLC
LaVergne TN
LVHW011536200926
848484LV00001B/270